EPM 87

Energy Pulse and Particle Beam Modification of Materials

Physical Research

Volume 8

EPM 87

Energy Pulse and Particle Beam
Modification of Materials

edited by
Klaus Hennig

EPM 87

Energy Pulse and Particle Beam Modification of Materials

International Conference held September 7—11, 1987
Dresden, G.D.R.

edited by
Klaus Hennig

With 470 Figures and 44 Tables

Akademie-Verlag Berlin
1988

Editor:

Prof. Dr. sc. nat. Klaus Hennig
Zentralinstitut für Kernforschung
Rossendorf
Akademie der Wissenschaften der DDR

Reproduction of the original authors' manuscript

ISBN 3-05-500540-6

Erschienen im Akademie-Verlag Berlin, DDR-1086 Berlin, Leipziger Str. 3—4

Lizenznummer: 202 · 100/414/88
Printed in the German Democratic Republic
Gesamtherstellung: VEB Druckerei „Thomas Müntzer", 5820 Bad Langensalza
Einband: Karl Salzbrunn
LSV: 1175
Bestellnummer: 763 859 7 (9132)

11000

The international conference EPM 87, the continuation of EPM 84, was devoted to

- implantation into silicon,
- implantation and annealing of compound semiconductors,
- implantation into metals,
- transient heat treatment of semiconductors,
- formation of silicides,
- ion beam assisted depostion and ion beam mixing,
- deposition, modification and structurization,
- silicon on insulator (SOI),
- diagnostics and ion microfocus beam,
- fundamental aspects.

About 220 scientists from 15 countries attended this conference.
The success of EPM 87 owed much to the cooperation of the International Advisory Committee. In agreement with the proposal of this committee the continuation with EPM 89 is planned.

We would like to express our sincere thanks to the members of the International and the Local Organizing Committee, to the invited speakers and the session chairmen and all participants for their active and successfull work.

Prof. Dr. G. Albrecht
chairman

Prof. Dr. K. Hennig
deputy chairman

Prof. Dr. E. Wieser
secretary

C O N T E N T S

1. Invited papers page

2. Implantation into silicon

3. Implantation and annealing of compound semiconductor

4. Implantation into metals

5. Transient heat treatment of semiconductors

6. Formation of silicides

7. Ion beam assisted deposition and ion beam mixing

8. Deposition, modification and structurization

9. Silicon on insulator (SOI)

10. Diagnostics and ion microfocus beam

11. Fundamentals

12. Miscellaneous

13. Author Index 592

EPM'87 - STATE OF THE ART

K.Hennig *)

1. Introduction

EPM'84 - Energy Pulse Modification of Semiconductors and Related Materials - concentrated on heat treatment of semiconductors. Since EPM'84 world-wide energy pulse and particle beams have been further used extensively for the modification and synthesis as well as for the characterization of surface and near-surface regions of different materials. The increasing interest in this country in the broader application of implantation and heat treatment techniques is taken into account by EPM'87 in changing the full title of the conference to: Energy Pulse and Particle Beam Modification of Materials.

As electronic and mechanical devices are miniaturized, the decreasing volume-to-surface ratio places greater demands on the reliability of surface properties. In the age of high technology products must maintain their surface characteristics. There is a continuing trend in materials science to develop methods for creating favourable surface properties.

Ion implantation together with special heat treatment allows very accurate surface modifications of semiconductors, metals and other materials, such as ceramics and polymers.

Implantation studies in semiconductors continue to be driven primarily by new requirements and developments in microelectronics due to the trend of decreasing device dimensions in silicon and the recent progress in layered materials. An intriguing example is the fabrication of a buried dielectric layer within silicon (SOI). However, the crystal quality of the overlying silicon layer has to be further improved. Other processes, such as silicide formation, ion beam assisted deposition (synthesis) and ion beam mixing play an important role as well.

New impacts are expected from the trend in microelectronics to three-dimensional packed circuits and many other areas, e.g. the new high-temperature superconductors.

The major use of energy pulse modification by ion and laser beams of metal surfaces is to control mainly the chemical and/or mechanical properties, e.g. the increase of wear and corrosion resistance.

Adequate methods for analyzing the modified surface have to be further developed.

The underlying processes are theoretically well understood by mathematical modelling using phenomenological descriptions.

---

*) Academy of Sciences of the GDR, Central Institute for Nuclear Research, 8051 Dresden, PF 19, GDR

2. Ion implantation

2.1. Implantation into semiconductors

There is no doubt ion implantation is now a well established technique in semiconductor technology. Modern very large scale integration (VLSI) devices and very high speed integrated circuits (VHSIC) can be produced only by ion implantation. The technology driver is the dynamic random access memory (DRAM),which currently peaks up to million bits of storage on a single silicon chip, with characteristic feature sizes below 1/um. Another trend is the rapid progress in GaAs and layered combinations of materials including superlattices and strained-layer superlattices, which give new electronic and optical properties. For the fabrication of high speed integrated circuits ion implantation into compound semiconductors and superlattices is essential.

VLSI applications of silicon have created interest in low energy ion implantation with $E < 10$ keV for producing very shallow doping profiles. Current studies combine the fabrication of shallow source/drain regions with silicide formation by implantation through metal layers using ion mixing effects. High energy implantation in the MeV region into silicon is another area of increasing interest with the aim to insulate active regions of circuits by the formation of buried layers.Due to simultaneous defect production special attention has to be paid to annealing procedures leading to device-quality silicon above the buried region.

Especially the production of retrograde wells for CMOS, the programming of read-only-memories (ROM) and the formation of buried p/n-grids for soft error reduction have forced the development of implanters and technology. At present implanters for the energy range up to 10MeV with currents up to 1mA are available (Nat. Electric Corp., General Ionex, Eaton, High Voltage Eng.) Further research deals with the formation of buried, highly doped layers and silicides and of buried damage layers to influence the local generation rate in power devices.This later direction needs energies in the 100 MeV region. In the case of ion beam induced epitaxial crystallization some advantages emerge by using MeV ions.

High dose implantation allows to synthesize chemical compounds, e.g. buried dielectric layers within silicon (SOI). At our institute activities in this field started already in 1979. At present, it is well established that implantation with energies > 150 keV and doses $>10^{17}$ cm^{-2} at a target temperature of about 500 ^{o}C followed by a heat treatment at least at 1200 ^{o}C for 5 h with Si_3N_4 and 1300 ^{o}C for 6 h with SiO_2 leads to optimized structures. That means the annealing gives appreciable improvements in the crystal quality of the overlying silicon layer with sharp interfaces in the case of oxygen. Due to the possibility to insulate the active regions of integrated circuits, both laterally and vertically they have improved properties, such as reduced single event upset, increased speed, less uncontrolled coupling (latch up), lower dynamic power consumption, larger packing density and possible new stacked structures exploiting, for example, two or more independent layers of devices, common gates or common device channels. Further elimination of disorder and dislocations within the overlying silicon layer and the production of extremely pure oxygen beams are necessary for successful SOI fabrication. The latest advanced oxygen implanter for 200 keV and 100 mA, built by Vacuum Generators Semicon Corp. in England and by Eaton Corp. in the United States, respond to this challenge. Some new results are concerned with the formation of these buried layers of a thickness < 200 nm and their stacking, the formation of oxynitrides and the formation of buried layers with implantation at relatively low energies (60 keV). Another example of high dose implantation is the formation of buried metallic conducting layers in silicon, for instance silicides. Further successfull methods for preparing SOI structures, for example strip heater recrystallization, will be also a subject of this conference.

The rapid development of micro-focus ion beams has led to beam diameters < 50 nm and current densities > 1 A/cm^2 with an energy in the region from 20 keV to 200 keV. Such a beam allows direct writing without proximity effects and direct doping of the semiconductor material.

2.2 Implantation into metals

The main advantage of ion implantation into metals is to control the chemical or mechanical properties of the near-surface region. Often commonly used coatings suffer from poor adherence or changes of the dimensions of the workpiece.

Ion implantation can essentially improve the surface properties of a metal or an alloy, increasing the hardness, the wear resistance, the corrosion or oxidation resistance, altering properties such as superconductivity, reflectivity, magnetism and others. In contrast to implantation into semiconductors, where a few parts per million of dopant concentration influence the electrical behaviour, implantation into metals requires implanted ion concentrations of several atomic percent. Nitrogen implantation into steel was one of the first implantation treatments successfully started at the Harwell Laboratories in England. At present the literature reports many cases of wear reduction by factors up to 10 for nitrogen implantation into steels and in special cases up to 100 for nitrogen implantation into chromium plating, itself applied for corrosion reduction.

The microscopic origin of wear resistance improvements is not completely understood. The following main suggestions have been made: - Damage in the surface due to implantation may arrest dislocation motion - Lattice expansion may seal microcracks - Surface compounds may serve as a lubricant in unlubricated applications.

A key aspect of this "implantation metallurgy" is the ability to form stable and metastable phases, i.e. conventional and non-conventional alloys. Typical implantation fluences are 10^{16} to 10^{18} ions/cm^2 and implantation energies from about 10 keV to 500 keV. The implantation process does not obey normal thermodynamic constraints such as the solubility limits or the Gibbs phase rule. Consequently unusual properties are expected. For instance, the high-dose implantation of non-metal ions can produce amorphous phases at room temperature (1.10^{17} B ions/cm^2 into Fe). Or refractory crystalline phases with very high melting points (> 2000 °C) and large microhardness (> 2000 $kgmm^{-2}$) can be produced as in the case of nitrogen ion implantation into aluminium which leads to the formation of hexagonal AlN.

Another aspect of ion implantation into metals is the effect of ion beam mixing of layered solid systems. In ion mixing experiments the amount of intermixing has two temperature regimes. At low temperatures the mixing is dominated by ballistic temperature insensitive processes which depend mainly on the energy and mass of the colliding atoms. In the high-temperature regime the displaced atoms have enough mobility and time to migrate and relax into a lower free energy configuration. The temperature behaviour can be calculated on the basis of a diffusion model, which includes the diffusion of irradiation-induced complexes. In this high-temperature regime distinct compounds are formed by this ion mixing process.

It should be mentioned that quasicrystalline phases possessing non-allowed crystalline symmetries, e.g. fivefold, may be produced by ion irradiation.

3. Transient heat treatment

The ion implanted semiconductor or metal surface region may be further modified by special heat treatments as a function of temperature and time.

In microelectronic technology rapid isothermal annealing, that is, annealing with pulse durations in the order of several seconds using for example an array of halogen lamps, is now at the beginning of industrial use. Besides annealing of implanted layers and silicide formation the fabrication of thin dielectric insulating layers on silicon such as SiO_2 or Si_3N_4 (or as buried layers) has been realized by rapid thermal annealing. One main motivation for the application of short time annealing arises from the possibility to control dopant redistribution in implanted profiles. During the last few years much work has been done to study the transient enhanced diffusion observed for very short annealing times, enhanced diffusion in comparison to steady state diffusion. A much deeper understanding of the diffusion mechanism correlated to the defect behaviour in the early stage of damage annealing has been

achieved.
In order to fulfil the increasing demands of VLSI technology on extremly shallow p-n junctions, a low sheet resistance of the source/drain regions and sure contacts with a low specific resistivity the so called SALICID technique has recently been developed. It combines different implantation steps, low temperature furnace annealing and rapid thermal annealing at high temperatures to form shallow p-n junctions and silicide contacts on the source and the drain as well as on the gate simultaneously. Implantation through a thin layer of refractory metals such as Ti or Mo ensures a shallow source/drain doping. The resulting ion mixing enables a homogeneous silicide formation at low temperatures between 500 and 600 °C and therefor avoids lateral silicide growth across the spacer oxide between the source/drain and the gate. A final high temperature short time annealing is used for activation of the source/drain dopants and in order to get the tetragonal silicide phase with a low resistivity and a good contact quality. In this way, for example, a sheet resistance in the order of $2\,\Omega_{\square}$ and a specific contact resistance of $10^{-6}\,\Omega\,cm^2$ was obtained by Kwang and Alvi with $TiSi_2$ contacts for p+ as well as n+ doping junction depth of about 150 nm.
The modification of metal surfaces by heat treatment is mainly used in connection with surface layers fabricated otherwise than by ion implantation. Lasers are used for hardening and/or alloying of the surface. Hard particles can be incorporated into the surface by melting. Laser pulse vapour deposition allows the production of diamond-like thin films.

4. Ion beam assisted deposition

Basic research in technological developement, devoted to all kind of thin film deposition, is of increasing interest all over the world. The motivations come from microelectronics, e.g. for the fabrication of superlattice structures, and from the possibility tho improve material properties by surface treatments. Already established methods, CVD or MBE, are further developed and new branches like rapid thermal processing or ion beam and laser assisted deposition are of fast growing importance. Ion beam assisted deposition and its application in material research will be considered more in detail in the following.
Energetic particle irradiation during film deposition can essentially modify the physical and chemical properties of the resulting thin film which is not achievable by conventional deposition.
Low energy ion bombardment takes place during almost all plasma-based film deposition processes but unfortunately in a quite uncontrolled way. Ion beam based deposition offers an independent procedure because the incident energetic particles with well defined energy and current densities change the dynamics of the growing film and may be incorporated forming a chemical compound. This may happen far away from

thermodynamical equilibrium conditions.

Ion beam assisted deposition (IBAD) is a very powerful new technique for producing adherent surface coatings with special properties. For many applications outside the semiconductor field surface layers modified by ion implantation are too thin to meet the requirements. For example, when the wear rate or the corrosion rate is very high the modified surface layer is consumed immediately.

Ion beam assisted deposition may offer a way out of these difficulties. Evaporated atoms and accelerated inert ions simultaneously arrive at the surface of the materials to be modified. To prepare thicker surface layers or coatings >1/um will be possible. In Japan, USA, the UK, Italy and the FRG the first facilities are in use for the preparation of cubic boron nitride, carbon titanium nitride or aluminium layers on steel surfaces. The wear resistance and the corrosion behaviour will be improved significantly.

Experimental set-ups combine the ion beam with evaporation or sputtering. Ion sources with low energies (up to 2keV), large areas and current densities up to 10^{17}ions/sec.cm^2 and more recently the electron-cyclotron-resonance source have been developed.

5. Diagnostics

Very well established experimental techniques for the detailed study of surface regions are electron microscopy, electron diffraction, Rutherford backscattering, secondary ion mass spectrometry, proton or ion induced X-ray emission, prompt nuclear reactions, ejected recoil detection, Auger electron spectrometry, spectroscopic ellipsometry and others.

The rapid development of micro-focus ion beams should be mentioned especially. The interaction of such a beam with the surface material causes secondary radiation: secondary electrons, secondary Auger electrons, X-rays, secondary ions which are used for element-mapping of the surface with very high resolution.

6. Fundamentals

If ions penetrate into solids they collide with the target atoms and electrons. For ion energies >10eV the binary collision approximation is used . One get a good description of implantation processes. In the case of sputtering molecular dynamics must be taken into account. In this concept of binary collisions elastic and inelastic processes are considered. The elastic part needs an interatomic potential suitable for cascade calculations, the so-called universal potential of Biersack and Ziegler. The inelastic processes mean that target electrons can be excited. The detailed theoretical procedure depends upon the energy of the incident ion. Therefore various phenomena are connected with the collision cascades: implantation of ions in the target, their reflection and transmission, recoil implantation and mixing sputtering, the ion energy deposition into the nuclear and electronic sub-systems, and defect creation. These processes are described by two alternative methods mainly on the assumption of an amorphous target: the analytical transport equation or Monte Carlo simulations.

Both methods have a number of advantages and disadvantages. The Monte Carlo simulations are very versatile and are widely used for practical purposes. The precision of their results depends strongly on the accuracy of the physical inputs. Recently the interatomic potentials and the elastic scattering formalism have been essentially improved. But the accuracy of the models for the inelastic processes used so far seems to be not quite satisfactory yet.

Photons interact via electrons with the material. For laser power densities $<10^6 W/cm^2$ solid phase transitions take place (hardening). In the region $<10^7 W/cm^2$ local melting is used for alloying.

7. New trends

The surface including the surface region of materials with a thickness from nm to μm continues to occupy the special interest of scientists, engineers and technologists. Demands for further accelerated development come from the VLSI and VHSIC programmes in microelectronics, from the wish to realize three-dimensional integration. The possibility to fabricate metastable phases is also of great interest. The whole diversity of layered structures with their astonishing physical parameters is another driving force in this respect. Last not least the present revolution in superconductivity will influence surface physics. A few publications have appeared concerning superconducting epitaxially grown films of YBaCuO on $SrTiO_3$, the multitarget sputtering of YBaCuO onto sapphire substrates, the electron-beam deposition of YBaCuO or LaSrCuO onto Al_2O_3 including a heat treatment in oxygen gas. Different features as the critical temperature, critical current, structure, texture, grain sizes and others are measured. Very exciting questions are raised, for instance, superconducting paths in the film, very high critical temperatures near the surface and critical current densities $>10^5 A/cm^2$ at 77K.

Acknowledgments

I would like to express my sincere thanks to Drs. E.Wieser, M.Voelskow, K.-H.Heinig, M.Posselt, F.Naehring, B.Rauschenbach, W.Skorupa at the Central Institute for Nuclear Research, H.-J.Dietze and S.Becker at the Central Institute for Isotope and Radiation Research for their helpful discussions and support during the preparation of this paper.

ACCELERATION OF MICROPARTICLES

Martin Setvak ①

1. Introduction.

Microparticles are of greater concern lately mostly as unwanted elements in vacuum technologies and space flights where they might cause serious damage. This paper deals with some problems of acceleration of conducting microparticles.

The presence of microparticles in ion implantation technology is a serious problem with economic impact. The presence of microparticles on silicon wafers obscures the underlying wafers from ion beam, thus causing defective integrated circuits especially on VLSI devices. The papers dealing with this problem are concentrated mostly on non-conducting particles created by wafer manipulation in the vacuum system of the implanter. The forces acting on a microparticle in an ion implanter, i.e. aerodynamic, electrostatic, gravitational, ion impact are described in the paper of D.H. Douglas-Hamilton and C. Taylor /1/. The particles sitting on a wafer during ion implantation are bombarded together with the wafer and charged and on some conditions can be eventually repelled by electrostatic forces. But any conducting particle which gets in touch with a high potential electrode in the vacuum can be very effectively charged by contact charging and in high field gradient can be accelerated and eventually hit and damage the target.

Ionized cluster beam deposition /ICB/ is a technology of thin film formation /2/. The term cluster in general means an aggregate of atoms or molecules loosely coupled whose size ranges from several atoms up to 10^5 or more atoms. The main interest is concentrated in crucible vaporized material clusters consisting of several thousand atoms and only some atoms are ionized by electron bombardment. On extracting the the clusters by several keV the energy/atom is in the order of 1 to 100 eV/atom. The clusters created by laser evaporation might be as big as one micrometer in size.

Space satellites are being bombarded during the space flight by micrometeorites which are conglomerates of matter whose maximum number is in the sizes between 0,1 to 10 micrometers and whose impact velocities are in the range of about 10 to 70 km/sec. The impacting micrometeorites create microcraters in surface layers of space satellites which cause degradation especially of optical and sensitive electronic surfaces placed on the outside of space vehicles. To find proper protection of the surface layers it is necessary to simulate the effects of microme-

① Technical University of Prague, Faculty of Nuclear Science and Physical Engineering - Brehova 7, 11519 Prague, Czechoslovakia

teorites in laboratory conditions.

Space satellites Vega and Giotto which were launched to Halley comet in 1986 were equipped among others with the equipment for studying dust particles /3/. The spectrometer PUMA was used for mass analysis of particles. The particles impacted on an electrode of pure silver and created a cloud of ions which were extracted and mass analysed. The detector was able to detect up to 100 particles/sec in mass range 10^{-16} to 10^{-10} g. Another analyser was DUCMA /Dust Particle Counter and Mass Analyser/ which was able to count up to 50000 particles/sec in the mass range $1{,}5.10^{-13}$ to 9.10^{-11}g. Laboratory simulation, i.e. acceleration of microparticles with comparable parameters is necessary for these experiments.

Several methods of accelerating microparticles were used like gas guns, shaped explosions, exploding wires, but the best results can be obtained by acceleration of charged microparticles in accelerators.

2. Particle charging

For acceleration purposes the particles can be charged positively or negatively, the speed of accelerated particle is proportional to the square root of electric field on the surface of the particle. A microparticle about one micrometer in size can hold on its surface about 10^5 electrons. The limiting value of field E on the surface of the particle is on negatively charged particles $E \doteq 10^9$ V/m, when a cold emission of electrons from the surface takes place. At positively charged particles the value is limited by ion evaporation and is about $E \doteq 10^{10}$ V/m. For this reason positively charged particles are mostly used for simulating experiments. Another factor limiting the charging experiments is the mechanical stress on particle and on the tip of the charging needle by a high electric field intensity on the surface. With $E = 4{,}5.10^{10}$ V/m the mechanical stress is about 0,9 tons/mm^2, which is many times larger than the bulk tensile strength of iron, but micron sized particles can withstand much greater strength than the bulk material. Many particles explode at high electric field/4/.

On Fig. 1. is a schematic representation of the charging and accelerating equipment which was built by the author of this paper at the University of Alberta in Edmonton, Canada/5, 6/. The charging equipment on Fig. 1a consists of a dust container Z which is partly filled with microparticles.For charging experiments carbonyl iron spheres 1 to 3µm in diameter were mostly used. Connecting a potential difference between the inner electrode A and the container Z the particles are charged by induction charging which provides the particles with a charge high enough to overcome the gravitation in the electric field of the central electrode. The particles are accelerated toward the inner electrode and

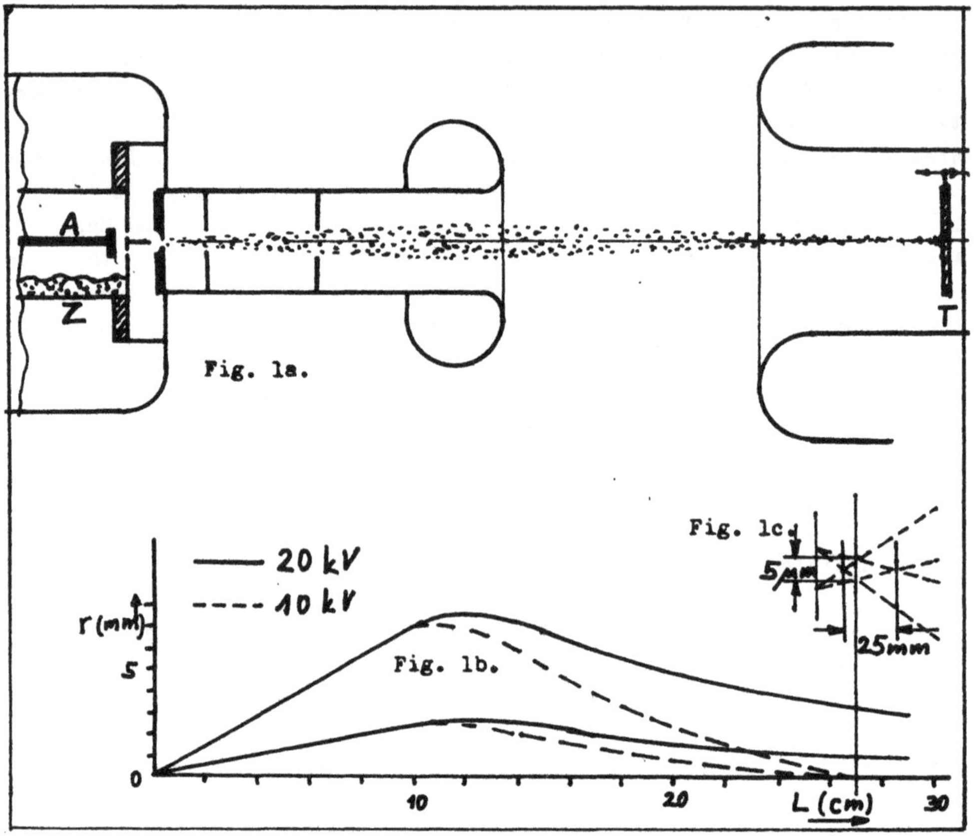

Fig. 1. - Schematic representation of the charging and accelerating equipment. Fig. 1a - Geometrical arrangement of the equipment
Fig. 1b - Computed trajectories for 10 and 20 kV extraction potential
Fig. 1c - Enlarged detail of the cross-over of the beam.

on contact with the electrode the conducting particles are immediately charged and repelled by electrostatic force. For a 2 µm sized particle and the distance between the central electrode and the level of particles of about one centimeter - the potential difference more than 10 kV is necessary to overcome the gravitation and excite the particles to a gas-like motion. The particles fill up the whole container, some of them oscillate between the front plate of the electrode and the wall in which small openings are located around the needle on the other side of the wall. Some particles penetrate the openings, get to the proximity of the needle-tip, which is on high potential of 10 to 20 kV toward the extraction hole 5 mm from the needle tip. The diameter of the extraction hole was 0,9 mm and the geometry needle-tip-hole defined the extracting angle of the particle beam. The needle-tip had to be adjusted very carefully under a microscope to the axis of symmetry of the extraction hole.

The needle was made of a tungsten wire 0,5 mm in diameter and the tip was tapered in a etching solution to a tip of about 5 µm in radius. After some time of operation the end of the needle was cowered with small craters which might mean that the particles bombarded the needle-tip with high speed. After contact with the needle-tip the microparticles are charged by contact charging in the high gradient of the needle-tip and get the maximum possible charge. The particles are extracted and further accelerated in a 210 kV potential difference gap lens.

On Fig. 1b are the computed trajectories for the given geometry of optics. The trajectories are given for 10 and 20 kV extraction potential and for two extraction angles/7/. Fig. 1c is a computed detail of the cross-over of the beam for 10 kV beam extraction. The computing was done for zero initial velocity of particles from the needle-tip. The measured cross-over of the beam was about 40 um. The contact potential influence was not taken into account because the charge gained in this way is negligible in comparison with the charge gained by contact charging. It is also assumed that the microparticle and the charging needle make contact for a sufficient length of time so that oscillations of charge between them ceased and electrostatic equilibrium has been established. If this would not be the case, i.e. the duration of contact is small compared to the the time constant of the charging process, then the microparticle could leave the charging needle at an instant at which it has either more or less charge than it would have if equilibrium had been reached.

The vacuum system of the equipment needs special attention. If the microparticle container would be evacuated very quickly,from atmospheric pressure, the particles would " explode ", i.e. the particles would be dragged out of the container and could damage rotary pumps. To prevent this, the evacuating of the system from atmospheric pressure to about 0,1 Pa had to be done very slowly- about half an hour. This effect and the measuring of the optics of the microparticle beam was investigated by trapping the accelerated particles in the thin layer of vacuum wax applied on glass under a heating lamp.

3. References

/1/ D.H. Douglas-Hamilton, C. Taylor: Particles and particle transport in ion implanters - Nucl.Instr.Meth. in Phys. Res. B6 /1985-/. 196-201.

/2/ I. Yamada: Ionized cluster beam deposition - Annual Report 1984 Kyoto University.

/3/ P. Lála: The probes to Halley comet - Vesmir 9/86 - Journal of the Czechoslovak Academy of Sciences - in Czech.

/4/ M. Setvak: The charging and acceleration of microparticles - Proc. of ISIAT'87 - Tokyo, Japan.

/5/ M. Setvak: A 250 keV injector for a microparticle accelerator-
The University of Alberta Report, Edmonton, Alberta, 1969.

/6/ M. Setvak: Some problems in particulate acceleration - Nucl.Instr. and Meth. in Phys. Res. B 21 /1987/ 378 - 80.

/7/ V. Sechovcov: PhD Thesis, TU Prague, 1970.

STUDY OF SILICON IMPLANTED BY HIGH DOSE OF IRON GROUP TRANSITION METALS

I.B.Khaibullin, V.Yu.Petukhov[1)]

I. Introduction

In recent years considerable interest has been shown in the investigation of the systems obtained by a high dose implantation of the different transition-group element ions into silicon where the concentration of implanted dopants becomes comparable with that of the substrate atoms.

The aim of this work is a comprehensive experimental investigation of the structural and phase transitions as well as the corresponding changes of surface texture, the optical, magnetic and electrical properties of silicon after the ion implantation by a high dose of Fe, Co, Ni ions and the following annealing at different regimes.

This problem is very actual because silicides of the transition-group elements are now considered to be one of the most promising materials for producing ohmic contacts, interconnections and resistive layers in VLSI production. Additional practical significance of the silicides of transition group elements is due to the fact that some of these compounds are ferromagnetic, and, therefore, may be used in electronics to produce various magnetic devices.

II. Experimental

Czochralski grown, ⟨111⟩ cut silicon wafers of p- and n-type resistivity (ρ = 10 Ohm·cm) were implanted along a random crystallographic direction with 40 keV Fe^+, Co^+ or Ni^+ ions. The maximum applied dose was $1.25 \times 10^{18} cm^{-2}$. The dose rate was varied from 0.6×10^{13} up to $40 \times 10^{13} cm^{-2} s^{-1}$. A sample heater stage allowed an implantation at elevated wafer temperatures up to 700 K. To determine the structure and identify the resulting phases of the polycrystalline silicides the RHEED technique and X-ray diffraction were used. The depth-dependent composition and thickness of the silicide layers were measured by RBS with 1.4 MeV He^+ ions. Some magnetic properties of the different phases were studied by the ferromagnetic resonance method (FMR), by the conversion electron Mössbauer spectroscopy (CEMS) and by the differential thermomagnetic analysis (DTMA). The study of the surface microrelief, optical

[1)] Kazan Phys.-Techn. Institute Academy of Sciences of the USSR
420029, Kazan, USSR

properties and swelling depending on the implantation condition was carried out by TEM, SEM , multibeam interferometry and optical spectroscopy in the visible and IR ranges. To investigate the thermally activated phase transitions and determine the temperature stability range of synthesized compounds, the samples were heated in a stream of dry nitrogen or helium gas ($T_a = 400 \div 1700$ K, $t_a = 5 - 60$ min).

III. Results and discussion

1. Swelling

The first result to be pointed out is the experimentally discovered effect of very high swelling which is estimated by the value of the linear expansion of a layer in the normal direction to the surface resulting in the formation of the step with the height "h" between the implanted and unimplanted regions. Under definite implantation conditions we observed the step which was equal to 800%, the typical value of the swelling for ion implanted silicon being only 7% [12].
It was experimentally established that the height of the step "h" is mainly determined by three components each of them reflecting a concrete physical process, namely: increase of the step due to the implanted ion accumulation, decrease of the step due to the ion etching and increase of the step due to the radiation pore formation. In Fig. 1 the dependences of h on the implantation dose, Φ , dose rate, j , pressure of residual gases in the accelerator chamber, P_r, and mass of bombarding ions, m_i, are presented. To explain the dependences obtained we used

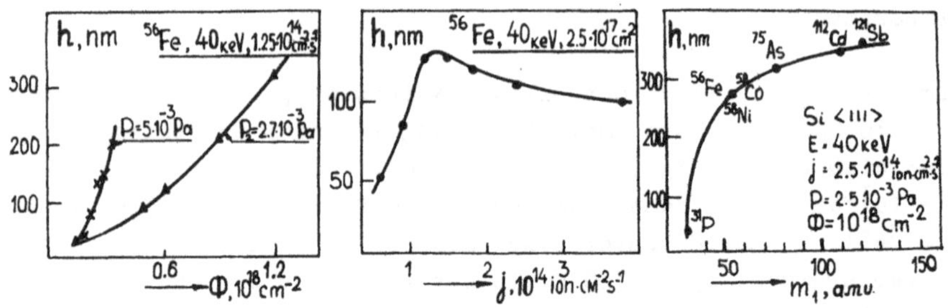

Fig. 1. Height of the step, h as a function of implantation dose, Φ at different residual pressure, P_r, dose rate, j , and bombarding ion mass, m_1.

the modified coagulation model of pore formation [6,7]. This model takes into account the generation, annihilation, diffusion of point defects and their successive coagulation into more complicated complexes containing up to "m" vacancies.

2. Changes in microrelief

Processes of pore formation, ion etching, ion synthesis during high dose implantation cause strong microrelief modification of the crystal surface: the initial plane and smooth surface of silicon is transformed into a rough one with the column texture. Such transformation of microrelief is illustrated by the surface micrographs obtained by electron microscopy at implantation dose increasing (see Fig.2). To explain the mechanism of appearence of a very specific microrelief, the following model of the process was suggested which is presented schematically in Fig.2. The micropores formed during ion implantation appear to be nucleation and precipitation centres of new phases (fine dispersed silicides as one can see below). Moreover, at high dose implantation (10^{17}- $10^{18}cm^{-2}$) we must take into account ion sputtering. As local grains of fine dispersed secondary phases, e.g. silicides, appear on the surface, the ion sputtering becomes very nonuniform and selective because of considerable difference in the sputtering rate of silicon and synthesized phases. As a result, the formation of the specific column-type microrelief with the diameter of columns 50 - 100 nm and the height 200 - 300 nm is observed.

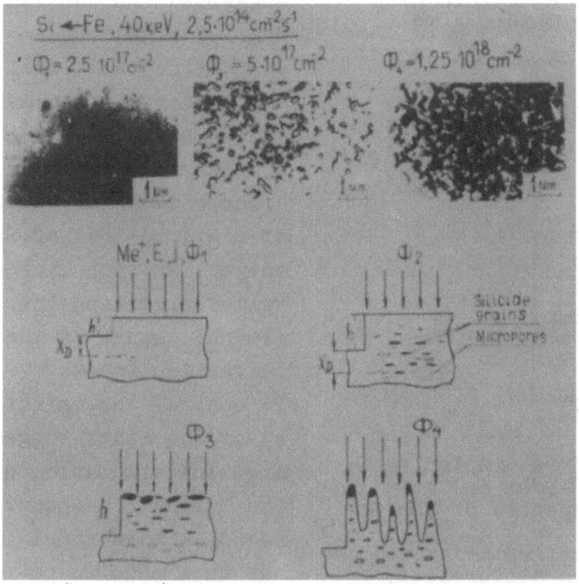

Fig.2. TEM micrographs of surface replica and a suggested model of microrelief transformation at dose increase

It should be noted that the absence of such microrelief on P^+-, As^+-, Sb^+- implanted silicon surface is also explained by this model. Although micropores are formed in this case, and swelling is observed, no secondary phases are formed due to a very high solubility of these elements in silicon. Thus, the inhomogeneity in the form of fine dispersed silicides is absent, and, as a result, there is no column texture.

3. Blackening effect

Strong changes in microrelief after ion implantation result in the decrease of Si reflectivity up to 5% in a visible and near-IR range (see

Fig.3a). Similar dependences are also observed for Co^+- and Ni^+- implanted silicon.

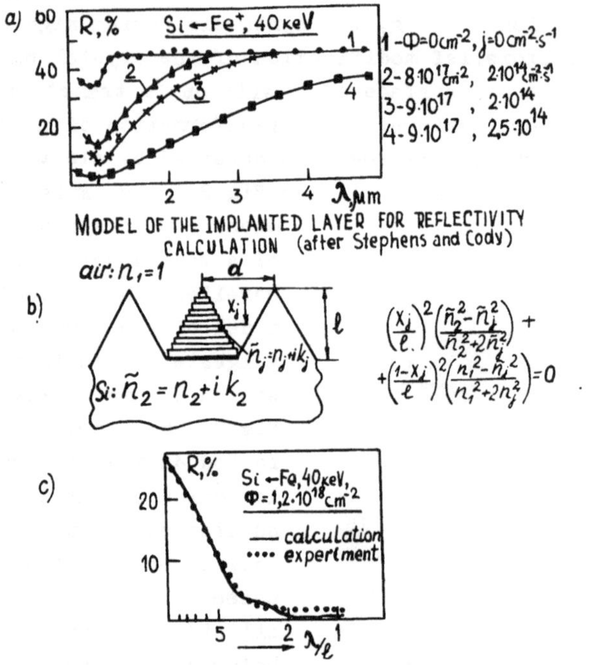

Fig.3. Optical properties of Fe^+-implanted Si : a) the effects of Fe^+-dose and dose rate on the reflectivity spectra of the implanted silicon: b) the model of the reflectivity calculation for column-type microrelief; c) comparison of the calculated and experimental reflectivity spectra

To explain the decrease in reflectivity of an irradiated sample which was named "blackening effect", the following model is suggested. As the result of high dose implantation, the transitional layer is formed on the surface where a refraction index increases from the value of n = 1 to that of Si, the column-type microrelief providing the gradual change of the refraction index.
To control the validity of the model suggested the simulation of the reflection spectra based on this model has been made in accordance with [10], the column structure being approximated by the cones (see Fig.3b). To determine the optical constants, the self-consistent field [13] or Maxwell-Garnett formulae were used [5]. The comparison of the calculated spectrum with an experimental one (see Fig.3c) has shown a good agreement. This testifies to the validity of our assumptions and the model as a whole.

4. Formation of thin magnetic film (TMF)

It is well known that the changes of the magnetic properties of silicon at low dose implantation consist in the generation and accumulation of the different kind of the radiation paramagnetic centres such as A-, E-, VV-centre etc, [11], in the case of 3d-metal implantation some peculiarities in the generation of VV-centres being observed [2].
At high dose Fe^+-implantation when the concentration of the implanted atoms can be comparable with that of host atoms of the material the formation of a thin ferromagnetic film with high Curie temperature in the near-surface layer was discovered. The intensive and relatively

broad absorption lines ($\Delta H = 50 - 100$ Oe) in the FMR spectra (Fig.4a) showed the formation of TMF, the angular dependence of FMR signal being in a good agreement with well-known Kittel formulae[3]. Using FMR spectra, the main magnetic characteristics of the formed TMF were determined, namely: the effective saturation magnetization M_{eff} and g-value (see table in Fig.4). The analysis of temperature dependence of M_{eff} revealed one Curie temperature $T_c = 580 \pm 10$ K for all ferromagnetic samples investigated. However, during next heating cycles we observed irreversible processes causing the increase of T_c up to 800 K, the saturation magnetization increasing too (see Fig.4b). To identify the phases with different T_c CEMS was applied. However, the CEMS spectrum consisted of a separate doublet with an isomeric shift $\delta E = 0.28 \pm 0.05$ mm·s^{-1} and the value of quadrupole splitting $\Delta E_Q = 0.5 \pm 0.05$ mm·s^{-1} and did not show any superfine structure which is inherent to magnetically ordered substances (see Fig.5a). In accordance with the data of Mangin and Marchal [4] such spectrum can be attributed to an amorphous alloy with composition approximately $Fe_{0.5}Si_{0.5}$ or stoichiometric compound FeSi, both of which are paramagnetic but not ferromagnetic. The superfine structure consisting of 6 lines we observed only placing the samples into the external magnetic field (see Fig.5a), the magnetic hyperfine field being $H_i = 200$ kOe. Such spectrum parameters are inherent for ferromagnetic iron compound, and the observed changes of the spectrum with external magnetic field application testify to superparamagnetism of a formed phase. This conclusion was confirmed by the static magnetization measurements using DTMA method (see Fig.5b). The behaviour of DTMA-dependences is typical for superparamagnets. Two minima at T = 580 and 800 K show the formation of two ferromagnetic phases, the second one with Curie temperature 800 K arising during the heating cycle.

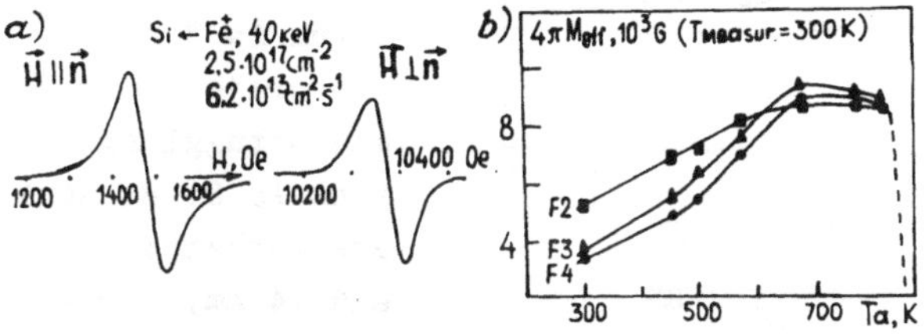

c) MAGNETIC CHARACTERISTICS OF Si←Fe$^+$ IN DEPENDENCE ON DOSE AND DOSE RATE

ION IMPLANTATION REGIME AND MAGNETIC PARAMETERS	Sample number						
	F1	F2	F3	F4	F5	F6	F7
Φ, $6{,}2\cdot10^{16}$ cm^{-2}	2	4	4	4	4	15	15
j, $6{,}2\cdot10^{12}$ cm^{-2}s^{-1}	10	10	20	30	50	10	50
$4\pi M_{eff}$, G	-	5050	3600	3000	-	8300	-
g-value	-	1.998	2.065	2.068	-	1.96	-

Fig.4. Magnetic properties of silicon implanted with high dose F$^+$-ions:
a) FMR spectra obtained at the different sample orientation in the magnetic field;
b) the effect of sample annealing on the effective saturation magnetization at 300 K for the samples listed in the table c);
c) the effects of Fe$^+$-dose rate and dose on the magnetic characteristics of the implanted silicon layer

Taking into account the experimental results, the following mechanism of the TMF formation at high dose Fe$^+$-ion implantation was suggested. As it

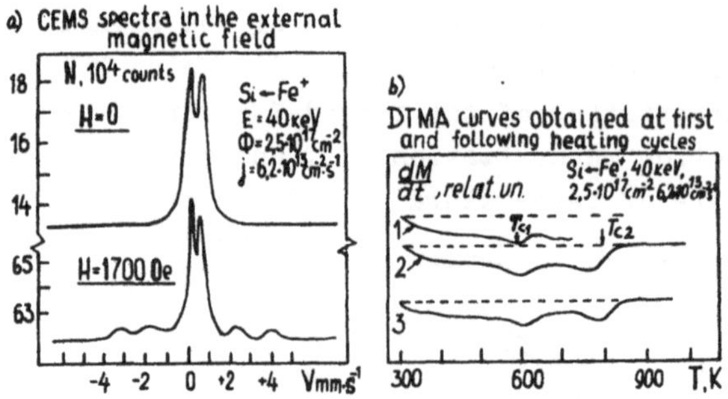

c) IDENTIFICATION OF SYNTHESIZED FERROMAGNETIC PHASES

Sample: Si←Fe$^+$, 40keV, Φ=2,5·10^{17}cm^{-2} j=6,2·10^{13}cm^{-2}·s^{-1}	MAGNETIC CHARACTERISTICS AND MEASUREMENT METHODS			
	H_i, 10^3 Oe CEMS	δ, mm·s^{-1} CEMS	T_c, K DTMA FMR	$4\pi M_{eff}$ 10^{3}G FMR
as implant.	200	0.28	580	6.5
570K, 30min	200	0.25	800 580	7.0
800K, 30min	non ferromagnetic			
Fe_3Si (I) (T.Shinjo et al)	305	0.1	803	12.4
Fe_3Si (II) (T.Shinjo et al)	195	0.35	803	12.4
$Fe_{0.7}Si_{0.3}$ (P.Mangin et al)	200	0.25	580	12.2

Fig.5. CEMS spectra (a), DTMA curves (b), and comparison of magnetic properties of synthesized and some known compounds (c)

has been mentioned above, numerous micropores in near-surface layer were formed. Since iron has a small solubility limit (maximum is 3 x 10^{16}cm^{-3}), and projected range and straggling of Fe^+-ions in Si at 40 keV are rather small (35 and 14 nm, respectively), the concentration of implanted metal is 4 or 5 orders higher than equilibrium solubility limit in the dose range investigated. In this case Fe atoms fastly migrate due to its high diffusivity, additionally stimulated by radiation and then precipitate on pore boundaries as effective sinks. Such precipitation under the ion mixing conditions stimulate a solid phase reaction resulting in the silicide formation. The comparison of measured magnetic characteristics of the samples with those of reference data [4,9] (see table in Fig.5) showed the presence of alloy $Fe_{0.7}Si_{0.3}$ in as implanted sample. During the following heating this phase was transformed into the stoichiometric compound α'-Fe_3Si.

Some differences in values of saturation magnetization of obtained ferromagnetic layers and reference data can be explained by two reasons: small thickness of the layer (d ≤ 20 nm) and island-like fine dispersed character of the film resulting in the arise of demagnetization effect.

5. Silicide synthesis

Although the ion synthesis is a nonequilibrium and complicated process, depending on a great number of parameters, three of them, as our experiments have shown, are basic. They are: the implantation dose, dose rate, and sample temperature. These parameters determine, firstly, minimum dose, Φ_{min} when synthesis of a new phase is observed; secondly, they effect the type of a synthesised compound, and, thereby, they determine the degree of the crystal lattice perfection of a synthesized phase.

Despite of an existing opinion that the molecular complexes of new compounds begin to form at the doses 10^{14}- $10^{15}cm^{-2}$ [1] in our experiments we observed the appearance of synthesized phases only at the doses exceeding $\Phi_{min} = 1.2 \times 10^{17}cm^{-2}$. It should be pointed out that this value Φ_{min} was determined taking into account the sputtering because at high dose implantation it limits the available concentration of the implanted atoms.

Dose rate influence on the ion synthesis is illustrated in Fig.6.where the electron diffraction patterns of the samples implanted with low (a) and high (b, c) dose rates are presented. Analysis of results enabled us to conclude that at high dose rates polycrystalline α'-Fe_3Si and ε-FeSi are synthesized whereas at low dose rates we have the amorphous alloys: $Fe_{0.7}Si_{0.3}$ and $Fe_{0.5}Si_{0.5}$.

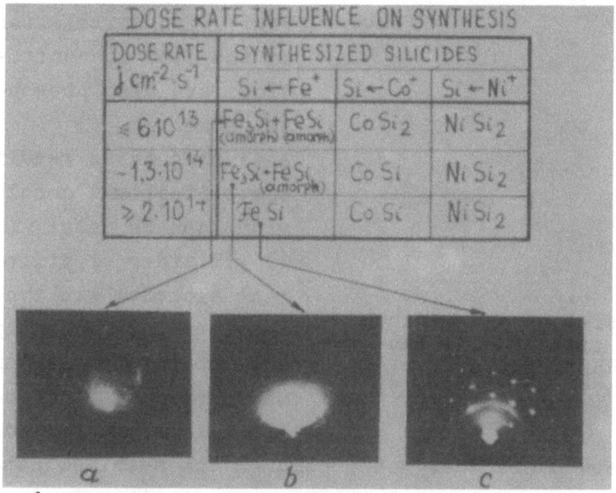

Fig.6. The effect of the dose rate on the ion synthesis of silicides

Similar results were obtained when we studied the ion synthesis vs sample temperature. Moreover, the correlation between the sample temperature and interval of thermal stability predetermines the type of a synthesized compound. In particular, it was established that at T > 630 K the phases Fe_3Si and $CoSi_2$ are not formed even at low dose rates since they are thermally unstable at these temperatures.

Identical results were obtained at ion synthesis of Co and Ni silicides. In ion synthesis the problem of a preferable nucleating phase on the metal-silicon boundary is of great scientific and technological significance. Experiments on simultaneous implantation of two elements (for example, Fe and Co or Fe and Ni) enabled us to clarify this complicated and yet unsolved question. Empiric rules for the initial nucleating phase suggested earliar by Walser and Bene [14] did not prove to be valid, at least, in our case. Comparison of this theory predictions with our experimental results shows discrepancy, particularly for Co-Si and Ni-Si systems. To study the basic items of ion synthesis, the experiments on simultaneous two-ion implantation of silicon with different dose rates were carry out. In this case, the implantation conditions (e.g. sample temperature, vacancy and complex concentrations etc.) are the same for both ions, except the implanted ion concentration.

Electron diffraction patterns of Si ← Fe^+ + Ni^+ and Si ← Fe^+ + Co^+ are presented in Fig.7. Identification showed that in the first case

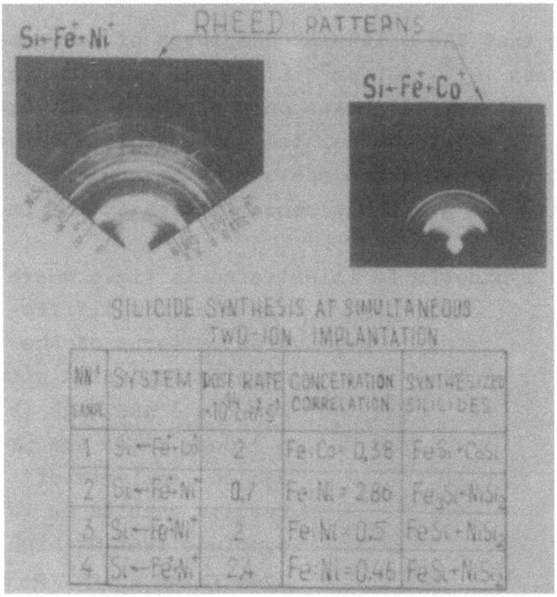

Fig.7. Concentration correlation and synthesized silicides at simultaneous two-ion implantation. RHEED patterns of samples No 4 and No 1, respectively, are presented

the silicides FeSi and $NiSi_2$ are formed while in the second - FeSi and CoSi.

Results of similar two-ion beam experiments for the other transition metal ions are presented in table in Fig.7. The correlation of the atom concentrations are also presented here.

Analysing these results three important conclusions can be made:

1) silicides of different elements are synthesized independently and separately, no ternary compounds and solid alloys are formed;

2) the type of a synthesized phase does not depend on the metal concentration (see Fig.7): e.g. in sample No 3 the Fe concentration is twice less than that of Ni; nevertheless, Fe-silicide is crystallized into a compound with high metal content, although the synthesis conditions are similar for both elements;

3) all synthesized silicides have the crystal structure belonging to cubic singony.

Comparing the lattice constants of the all known cubic silicides of Fe, Co, Ni with Si lattice constant (see Table 1) we can state that the lattices of Fe_3Si and $Fe_{11}Si_5$ among Fe-silicides, $NiSi_2$ among Ni-silicides and $CoSi_2$ among Co-silicides are closest to Si-lattice. These

Table 1. Lattice constants of silicon and cubic silicides of Fe-group elements

Silicide	Si	Fe_3Si	$Fe_{11}Si_5$	FeSi	CoSi	$CoSi_2$	$NiSi_2$	Ni_3Si
Lattice constant, Å	5.43	5.65	5.62	4.49	4.44	5.37	5.41	3.50

are the silicides we have observed experimentally at low dose rate implantation, except for $Fe_{11}Si_5$ silicide. As a result of a detailed analysis it was found that $Fe_{11}Si_5$ phase has a very small temperature interval of stability, and at T = 1020°C this compound decays, being not fixed even by quenching [8].

Thus, a dominating factor in the "choice" of a crystal structure of the

ion synthesized compounds is the degree of its similarity with a crystal structure of a substrate, i.e. Si in our case.
However, at ion synthesis with preliminary heating as well as at temperature increase of a sample at the high dose rate implantation the thermal stability of the formed phases starts to influence the ion synthesis, e.g. at sample heating at T_a > 630 K and at implantation with the dose rate $\dot{j}$ > 2 x $10^{14} cm^{-2} s^{-1}$ the formation of Fe_3Si and $CoSi_2$ phases is not observed, and only high temperature stable phases such as FeSi and CoSi, are formed.
One more important peculiarity of the ion synthesis is the influence of the dose rate on structure perfection of the synthesized phases. This effect enables to avoid additional treatments such as the high temperature or pulse annealings. We believe that $\dot{j}$ influence on the ion synthesized silicide crystallization is mainly due to thermal heating of a near-surface layer, though we do not exclude a stimulating effect of the other factors: ionization, defect generation etc.

6. Phase and stucture transformations at thermal treatments

Phase and structure transformations investigated by X-ray, RBS and magnetic methods in Fe-implanted silicon (with a high dose and a low dose rate) after subsequent isochronal thermal annealing are schematically shown in Fig.8. The following temperatures of the phase and structure

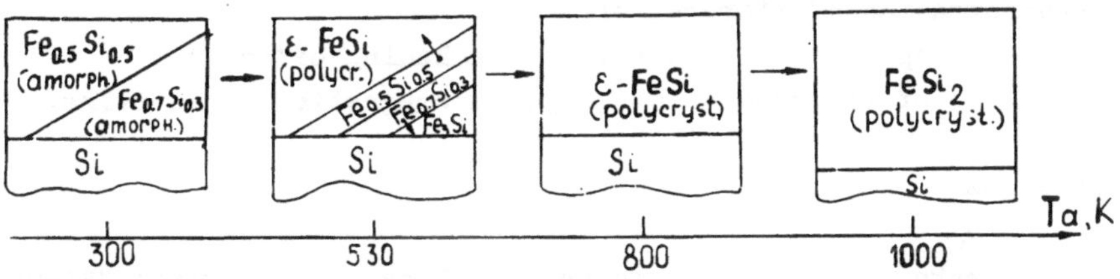

Fig.8. Phase and structure transformations in Fe^+-implanted silicon at the following heat treatments

transitions have been established.
<u>300 K</u> - as implanted surface layer consists of amorphous alloy $Fe_{0.5}Si_{0.5}$ and ferromagnetic alloy $Fe_{0.7}Si_{0.3}$. It is likely that they are amorphous ε-FeSi and α'-Fe_3Si compounds. At temperature increase the gradual crystallization of amorphous alloys takes place.
<u>600 K</u> - $Fe_{0.5}Si_{0.5}$ is crystallized into ε - FeSi; the ferromagnetic amorphous $Fe_{0.7}Si_{0.3}$ with Curie temperature T_c= 580 K is crystallized into ferromagnetic α'-Fe_3Si with T_c= 800 K.
<u>800 K</u> - α'-phase decays and transforms into a more stable ε-phase.
<u>1000 K</u> - ε-phase decays, the released Fe-atoms diffuse to a layer - substrate boundary, and $FeSi_2$ is formed, the thickness of $FeSi_2$ layer being 1.5 times increased.
Investigation of the sequence of phase and structure transitions showed

that in the samples implanted with high dose rates the phase transitions, taking place only at higher temperatures, are observed. For example, for both samples implanted with $\dot{j} = 1.3 \times 10^{14}$ and annealed at $T = 630$ K, the same phase transitions are observed at temperatures higher than 630 K. We can observe the similar behaviour also for both samples implanted with $\dot{j} = 2 \times 10^{14} cm^{-2}s^{-1}$ and annealed at $T = 800$ K.

Similarity in the sequence of phase and structure transitions which is observed at the subsequent T_a and $\dot{j}$ increase confirms our conclusion that a thermal factor is a dominating one at the dose rate increase in ion synthesis.

It should be pointed out that the phase and structure transitions can be stimulated also by the nonequilibrium thermal treatments (laser, RTA, electron-beam etc.) Some results of such experiments are presented in our reports (5.3 and 5.8) at this conference.

IV. Summary

At high dose silicon implantation by Fe-group ions the following basic effects take place:

- high swelling due to pore formation;
- silicide formation in a fine dispersed form;
- sputtering;
- transformation of microrelief;
- change in optical reflectivity (blackening effect);
- formation of ferromagnetic (superparamagnetic) thin films.

References

[1] Akimchenko, I.P., V.V.Krasnopevtsev: Proc. of the First USA-USSR Seminar on Ion Implantation. Albany, USA 1977.

[2] Antonenko, A.Kh. et al.: Fiz. Tehn. Poluprov. 13 (1979) 281-284.

[3] Frait, Z.: Phys. stat. sol. 2 (1962) 1417-1459.

[4] Mangin, Ph, G.Marchal: J.Appl.Phys. 49(1979) 1709-1711.

[5] Maxwell-Garnett, J.E.: Phil.Trans. 205 (1906) 237-288.

[6] Morozov, N.P. et al.: Phys.stat.sol. 37A (1976) 57-64.

[7] Petukhov, V.Yu.: Study of silicon implanted by high dose Fe-group transition metals, Diss., Kazan 1985 (in Russian).

[8] Samsonov, G.V. et al.: Silicides, Metallurgia, Moscow 1979 (Russian).

[9] Shinjo, T. et al.: J.Phys.Soc.Japan 18 (1963) 797-801.

[10] Stephens, R.B., G.D.Cody: Thin Solid Films 45 (1977) 19-29.

[11] Watkins, G.D.: Radiation damage in semiconductors, Dunod, Paris 1965.

[12] Whan, R.E., G.W.Arnold: Appl.Phys.Lett. 17 (1970) 378-380.

[13] Wood, D.M., N.W.Ashcroft: Phil.Mag. 35 (1977) 269-280.

[14] Walser, R.M., R.W.Bene: Appl.Phys.Lett. 28 (1976) 624-625.

MODIFICATIONS INDUCED BY PULSED LASER IRRADIATION OF METALLIC SURFACES IN PRESENCE OF A BREAKDOWN PLASMA IN THE AMBIENT GAS

I. Ursu[1)], I. N. Mihăilescu[1)], A. M. Prokhorov[2)], V. I. Konov[2)]

1. Materials (metals) processing with laser plasma

It is known (see for example /1/) that the laser breakdown (BD) of gases and vapours in front of solid (metal) samples can be achieved at moderate incident intensities. Once induced, BD plasma strongly absorbes most of the next incident laser intensity. Therefore, for a long time, preventing BD plasma formation was considered an important technological requirement-especially when a high degree of laser action localization was required.

It was only in 1970 that Yu. P. Raizer /2/ has introduced the term of laser plasmatron in order to describe the action of power laser radiation on solid (metal) targets by means of a BD plasma, induced and sustained in front of the solid surface under the action of laser radiation. As demonstrated since then laser plasmatrons exhibit specific advantages as against the classical plasmatrons among which we mention (i) no need of special elements for energy dissipation intro plasma; (ii) plasma can be produced at a large distance from the focussing element which enable both clean technological operations and plasma initiation in hardly accessible places; (iii) plasma can be initiated in small volumes ($\sim 1\ mm^3$) and hence one can accurately control the localization of the plasma-solid (sample) interaction; (iv) plasma can be displaced by moving the laser beam or, even simpler, by blowing gases in the discharge zone; (v) plasma can be heated up to higher (by $\gtrsim$ 20 000 K) than in other plasmatrons, a. s. o.

According to the operation made, laser plasmatrons can be classified as (i) steady state- when the gas BD plasma is sustained (even though not initiated) under the action of CW laser radiation; (ii) pulsed - which use pulsed gas BD plasma; and (iii) erosion - when (pulsed) plasma is induced into the vapours of the solid (mobil) sample.

Nowadays the CO_2 lasers are most largely used with laser plasmatrons. Indeed (i) CO_2 lasers - which are presently the most technological - developed laser systems - are reliable, simple in construction, highly efficient and can operate at high frequency repetition rate and with large output beams; (ii) BD plasma initiation is obtained with CO_2 laser radiation by the lowest BD threshold while (iii) plasma produces a minimum perturbation of the sample surface (e.g. by substance ablation). We note however that use of laser radiation of smaller wavelength allows for heating the plasma up to higher temperatures - due to the larger value of the critical electron density.

1) Central Institute of Physics Bucharest, Romania

2) Institute of General Physics Moscow, URSS

We review in this lecture some recent results obtained with pulsed CO_2 laser plasmatrons mainly in some mechanical and chemical processing applications.

2. Effective absorptivity

Most sintethycally, the efficiency of the energy coupling and transfer from laser radiation to solid (metallic) samples is characterized through the intermediary of the effective absorptivity (sometimes called as the energy coupling coefficient).

$$A_e = \frac{Q}{E_o} \quad (1)$$

[1)]

Here E_o is the incident laser energy and $Q = mc\Delta T$ is the energy absorbed into the sample m, c are the sample's mass and specific heat respectively while ΔT is the temperature excursion as an effect of laser irradiation.

In case of contact gas low-thereshold BD plasma action upon solid (metal) samples A_e was shown to depend /3-10/ to different extents on (i) the incident fluence (E_s) and intensity (I), (ii) the surface state, (iii) the radiation focussing geometry, (iv) the target's nature, (v) the ratio between the sample's and the focal spot's characteristic dimensions, (vi) the time-shape and the duration of the laser pulses, (vii) the ambient gas nature and pressure, (viii) the frequency repetition rate of the subsequent laser pulses directed to the same area on the sample's surface, (ix) the evolution of the "fire ball" - a domain of prolonged gas burning having a volume considerably exceeding the initial BD plasma volume, a. s. o. For selection of optimum irradiation parameters - e.g. in order to ensure the best energy coupling - a detailed analysis is required terefore in each particular case.

3. Metal processing under the action of gas BD pulsed plasma

Gas pulsed contact BD plasma provide for a two-fold-thermal and mechanical-processing source. With each laser pulse, BD plasma acting as an explosion source transfers to the target a pressure and a heat pulse. Different mecahnical processing operations (perforation, cutting, a. s. o) is then achieved by the combined effect of the two factors. In particular, the heating role in performing a certain operation can be enhanced by increasing the frequency repetition rate of the subsequent laser pulses or by preheating with external sources. We mainly refer in this lecture to plasmochemical processing applications of laser plasmatrons with emphasis on surface laser nitridation of transition metals (Ti, Zr, Hf) and steel.

Any laser activated chemical reaction between a solid (metallic) sample and an ambient gas proceeds in several steps : (1) the transport of gas molecules to surface;

[1)] We note that with CW or with high-frequency periodically-pulsed laser irradiation the definition $A_e = P_a/P_o$ is commonly used - where P_a and P_o stand respectively for the absorbed and respectively incided laser power.

(2) the adsorption of gas molecules on surface; (3) the dissociation of adsorbed molecules into atoms; (4) the gas atoms diffusion through the reaction products into the bulk material and (5) the chemical reaction. We note that in practical situations the steps (1) and (4) are the most limiting ones.

We point out the following advantages of using pulsed laser plasmatrons in order to induce and conduct surface chemical reactions as against the classical (pure thermal) methods : (i) the ionized gases exhibit an increased chemical activity; (ii) the advent of large electrical currents ($\sim 10\ A/cm^2$) crossing the plasma; (iii) the gas atoms diffusion (step 4) is facilitated due to the fact that the film forming as a result of chemical reaction grows with crackings - due to the mechanical action of BD plasma; and (iv) the development of surface instabilities causing turbulent motion into the surface molten layer resulting in faster saturation with gas. We also mention that besides the advantages listed under the first section of this paper the use of laser radiation of longer wavelength - i.e. of CO_2 laser radiation - allows for working at lower pressures of the ambient gas (of maximum ≈ 10 atm instead of ≈ 100 atm with glass : Nd laser sources)[1].

In some experiments /11-19/ we have used TEA-CO_2 laser sources which were able to generate pulses of a typical temporal shape-with a leading spike having a FWHM of $\simeq 150$ ns followed by an essentially lower intensity tail lasting for $\approx 2\ \mu s$ with a total energy $E_0 \simeq 2$ J (of which $\approx 30\%$ is to be found in the initial spike) succeeding at a frequency repetition rate of f = 100 Hz. The irradiation spot area was varied within the 1-9 mm^2 range.

We have used samples of high-purity Ti, Zr, Hf and steel of dimensions 30 x 30 x x 0.5 mm^3 which were carefully cleaned before laser irradiation. The irradiations were performed in a high-vacuum enclosure filled with a mixture of high-purity N_2 and O_2 up to a slightly superatmospheric pressure. The volume concentration of O_2 impurities in the irradiation chamber was controlled with the range $4 \times 10^{-7} - 5 \times 10^{-2}$. The samples were cased in three insulating needles and moved so slow as to ensure that every point to be exposed to the action of $N \simeq 1000$ subsequent laser pulses. Laser-treated zones were thus obtained with a length of the order of the target characteristic dimension and a width corresponding to the diameter of the focusing laser spot. This shape proved to the highly suitable for further post-irradiation investigations by electron-microscopy, X-ray diffraction and microhardness determinations.

As for results :

(i) We succeeded to get nitrified layers, hard and adherent by high-frequency periodically-pulsed CO_2 laser irradiation of Zr, Ti, Hf and steel. Typical processing velocities of $\simeq 0.6$ cm/min were applied.

---

[1] An alternative solution could be the shortening of the laser pulse duration which results however in a dramatic fall down of the effective absorptivity.

(ii) Plasma ignition by every laser pulse and the surface melting appeared to be the essential requirements for the formation of surface compounds layers by periodically-pulsed laser irradiation in a gas or under a liquid layer containing N_2.

(iii) The nitrified layers consisted of an outer film of nitride and a substrate metal layer saturated with dissolved N_2. By small f both layers have approximately equal thickness while when increasing f, the thickness of the outer nitride film keeps practically constant.

(iv) The layer thickness was increased and the final surface state of the sample was improved by increasing f (by f = 200 Hz a thickness of the nitrified layers of $\sim 100\ \mu m$ was obtained).

(v) The synthesis of the nitride (oxinitride) layers appeared to be determined by O_2 impurities concentration, C in the ambient gas, characterized by a critical value, C_o. ($C_o \simeq 10^{-1}$ per cent for Ti, $\simeq 10^{-2}$ per cent for Zr and only $\simeq 5.10^{-4}$ per cent for Hf).

Thus for $C > C_o$ the formed surface layer consists mainly of oxides, looks spongy, has a poor adherence and a rather modest microhardness.

The situation is improved for $C < C_o$ when the microhardness is increased and the aspect and the adherence of the surface layer are much better.

(vi) The utilization efficiency increased when performing irradiations under a layer of liquid N_2. However the interaction process was in this case also strongly influenced by the O_2 impurities dissolved into the liquid.

(vii) Large improvements without high purity gas requirements were obtained using a reducing atmosphere. The best results were obtained in case of irradiations performed in pure NH_3 at $p \simeq 0.1$ atm or N_2 of higher pressure containing small quantities of NH_3.

References

(1) Ursu I., Mihăilescu I.N., Prokhorov, A.M. and Konov, V.I. : Interaction of Laser Radiation with Metals, Editura Academiei R.S.România, Bucharest, 1987 (in Romanian)

(2) Yu.P.Raizer, Pis'ma v JETP 11 (1970) 195

(3) Ursu, I., Apostol, I., Dinescu, M., Drăgănescu V., Julea, Th., Morjan I., Moldovan M., Mihăilescu, I.N. and Konov, V.I., : Optics Comm. 35 (1980) 69; J.Phys 40 (1980) 765

(4) McKay, J., Bleach R., Nagel, D., Schriempf, J., Hale, R., Pond, C. and Manlief, S., : J.Appl.Phys. 50 (1979) 2321.

(5) Arzuov, M.I., Karasev, M.E., Konov, V.I., Kostyn, V.V., Metev, S.M., Silenok, A.S. and Chapliev N.I. : Kvant.Elektron. 5 (1978) 1567

(6) Maher, W. and Hall, R. : J.Appl-Phys 51 (1980) 1338

(7) Ageev, V.P., Barchukov A.I., Konov, V.I., Pujaev, S.B., Silenok A.S. and Chapliev, N.I. : Kvant.Elektron. 6 (1979) 78

(8) Maher W., and Hall, R. : J. Appl. Phys 49 (1978) 2254

(9) Ursu, I., Apostol, I., Bărbulescu, D., Mihăilescu, I. N., Moldovan, M., Prokhorov, A. M., Ageev, V. P., Gorbunov, A. A. and Konov, V. I. : Optics Comm. 39 (1981) 180.

(10) Ageev, V. P., Burdin, S. G., Goncharov, N. I., Konov, V. I., Mineev, I. M. and Chapliev, N. I. : Itoghi Nauki i Tekhniki, Radiotekhnika 31 (1983)

(11) Ursu, I., Mihăilescu, I. N., Prokhorov, A. M. and Konov, V. I. : Proc. 6-th EPS Gen. Conf "Trends in Physics" Eds. J. Janta and J. Pantoflicek, vol. 2 (1985) p. 442.

(12) Ursu, I., Mihăilescu, I. N., Prokhorov, A. M. and Konov, V. I. (in) Proc. Int. Winter School" Laser-Assisted Modification and Synthesis of Materials" (Gyuletchitza, Bulgaria, March 1985), Ed. S. M. Metev, University Press (Sofia, 1985), p. 191.

(13) Ursu, I., Mihăilescu, I. N., Prokhorov, A. M. and Konov, V. I., (in) Proc. Int. Conf. "Trends in Quantum Electronics", Springer-Verlag (Berlin, 1987), p. 175.

(14) Ursu, I., Mihăilescu, I. N., Prokhorov, A. M., and Konov, V. I. (in) Proc. E-MRS'86 Conf. Strasbourg, 1986, p. 223.

(15) Ursu I., Mihăilescu, I. N., Nistor, L. C., Popa, Al., Popescu M., Teodorescu, V. S., Nanu, L., Prokhorov, A. M., Konov, V. I., Tokarev, V. N. and Uglov S. A. : J. Phys D : Appl Phys. 18 (1985) 1693.

(16) Ursu, I., Mihăilescu, I. N., Prokhorov, A. M., Konov, V. I., Tokarev, V. N., and Uglov, S. A. : J. Phys. D : Appl. Phys 18 (1985) 2447.

(17) Ursu I., Mihăilescu, I. N., Nanu, L., Nistor, L. C., Popescu, M., Teodorescu, V. S., Prokhorov, A. M., Konov, V. I., Uglov, S. A. and Ralchenko, V. G : J. Phys. D : Appl. Phys 19 (1986) 1183.

(18) Ursu, I., Mihăilescu, I. N., Nistor, L. C., Teodorescu, V. S., Prokhorov, A. M. Konov, V. I. and Uglov, S. A. : Appl. Opt. 25 (1986) 2725

(19) Ursu, I., Mihăilescu, I. N., Nistor, L. C., Teodorescu V. S., Prokhorov, A. M., Konov, V. I. and Uglov, S. A. : J. Phys. D : Appl. Phys. 20 (1987) (in press).

IMPURITY EFFECTS AT ION BEAM INDUCED DISORDERING OF HIGHLY DOPED SILICON

A.V.Dvurechenskii[1], R.Grötzschel[2], and V.P.Popov[1]

The problem of crystalline to amorphous transition at ion implantation in semiconductors still attracts attention of scientists trying to understand the microscopy mechanism. Thermal spikes, displacement cascades, point defects accumulation are the different ways in explanation of amorphization of ion-bombarded semiconductors /1/. Impurity effects in crystalline to amorphous transition are expected to be important to clarify the mechanism of this transition. Therefore some attempts were undertaken to study the amorphization in heavily doped silicon /2, 6/. Italian scientists /2/ have found the impurity influence on amorphization dose of P^+-implanted silicon doped with boron at concentration of 10^{18} to 10^{20} cm^{-3}. Hirvonen and Eisen /6/ have made the detailed experiments to check this effect. They did not find the noticeable impurity effect on amorphization dose in ion-beam-bombarded silicon. We believe that this uncertain situation was connected with the problems in preparing of defect free heavily doped semiconductors. High impurity concentration leads to production of dislocation network, of precipitation and other extended defects, which may influence on amorphization of ion-implanted semiconductors /4/.

The laser or electron beam annealing is a new technique which allows to prepare defect free silicon layers. We have used this technique to obtain the crystalline heavily doped silicon layers after ion implantation of impurities. The impurity effects are expected when the impurity concentration is comparable with the concentration of displaced matrix atoms. Otherwise the defect-defect interaction may be dominant and impurity effects should be week. The concentration of displaced atoms depends on ion mass. For example, the relative concentration of displaced atoms in silicon is 10^{-3} for N^+, 10^{-2} for Si^+ and 10^{-1} for Xe at the middle range of ion energy (a few tens of keV). So, the impurity concentration should be close to (or exceed) the equilibrium solubility limit for usually used dopant elements of III and V groups.

P- and n-type, <100> cut Si wafers with resistivity of 10 to 20 Ohm.cm were implanted with B^+, P^+, As^+, and Sb^+. Samples with a double implantation (Si <B+As>, Si <B+P>, Si <B+Sb>) as the same single doses were prepared. The ion energy was varied from 35 to 200 keV and its value for double implantation provides the equal projected range of different ions. Subsequently the liquid phase epitaxy was performed by pulsed electron beam annealing. The used parameters (E=10 keV, τ =500 ns, energy density 2 J/cm^2) ensure a melting and impurity redistribution

[1] Institute of Semiconductor Physics, Novosibirsk, USSR
[2] Central Institute for Nuclear Research, Rossendorf, DDR

depth, respectively, of 0.5 μm.

The quality of recrystallization was checked by TEM and Rutherford backscattering spectroscopy (RBS) of axially channeled 1.5 MeV He^+ ions. According to the TEM results the recrystallized layers are free of extended defects. The minimum value χ_{min} of normalized backscattering yield in the aligned RBS spectra was ≈ 4% for the undoped Si after PEBA, 5% to 6% in the case of Si<B>, Si<As>, Si<P>, and 8 to 9% for Si<B+As>, 9 to 10% for Si<B+P> and 10% for Si<B+Sb> at average dopant concentration about 10^{21} cm^{-3} with maximum near the surface not exceeding $2x10^{21}$ cm^{-3}. The higher values of χ_{min} in doped samples indicate a certain lattice deformation and point defect production due to high impurity concentration.

Irradiation of highly doped silicon was made at room temperature with 80 (Ne^+), 30 (N^+), 500 and 7 (He^+), and 230 keV (Xe^+) ions. The bombardment was performed in increasing partial doses. After bombardment of each partial dose aligned RBS spectra have been taken with He^+ beam. As a measure for the accumulated damage we used the maximum value χ_{max} of the normalized scattering yield from the displaced fraction of Si atoms. The yield contribution from depth-dependent dechannelled part of the beam $\chi_d(E)$ has been calculated assuming a constant dechanneling cross section of the displaced Si atom only

$$\chi_{max} = (\chi(E) - \chi_d(E))/(1 - \chi_d(E))\Big|_{max}.$$

In the case of silicon crystals the crystalline to amorphous transition is complete at the amorphization dose φ_a, when aligned yield reaches the random yield level. This usually used criterion for amorphization is in good agreement with the results obtained by electron diffraction.

We have found very strong influence of the dopant material in highly doped silicon on defect accumulation rate and amorphization dose /4, 5/. For example, in the case of Ne^+ bombardment the amorphization (χ_{max} = 1) of the undoped Si was observed at the dose of $\varphi_a = 1.5x10^{15}$ cm^{-2}, which agrees well with the known values of amorphization dose /1/. Only insignificantly lower we found the amorphization dose ($\varphi_a = 1x10^{15}$ cm^{-2}) to be in the case of Si<As> (Fig. 1). For the heavily boron doped sample Si<B> the accumulation of defects is considerably retarded and amorphization is reached only at $\varphi_a = 3x10^{16}$ cm^{-2}. In contrary, the defect accumulation rate is enhanced in the case of Si<B+As>, where we have observed amorphization at the low dose of $\varphi_a = 4x10^{14}$ cm^{-2}. Similar effect of stimulation of defect accumulation in double doped silicon was observed in Si<B+P> and Si<B+Sb>.

We have studied the effect of retardation of amorphization in Si<B> with different concentration of impurity. The samples were prepared with 35 keV B^+ implantation into p-type Si with resistivity of 10 Ohm.cm. The implantation doses were $6.25x10^{15}$, $1.25x10^{16}$, $1.9x10^{16}$, $3.7x10^{16}$, and $6.25x10^{16}$ cm^{-2}. After PEBA χ_{min} were 3.4, 4.8, 5.4, 7.8, and 11.5% respectively for mentioned doses. The ion bombardment was performed with

80 keV Ne^+ by dose Φ = 1.5×10^{15} cm^{-2}, which is the amorphization dose of undoped silicon. The concentration of defects in Ne^+-bombarded Si<B> was found to decrease as the impurity content rises (Fig. 2). The minimum defect concentration takes place at the boron content of (5 to 6)$\times 10^{20}$ cm^{-3}. Further increase in the boron concentration leads to the week rising of the defect concentration in Ne^+-bombarded Si<B> apparently due to additional damage in highly doped Si (high value of χ_{min}).

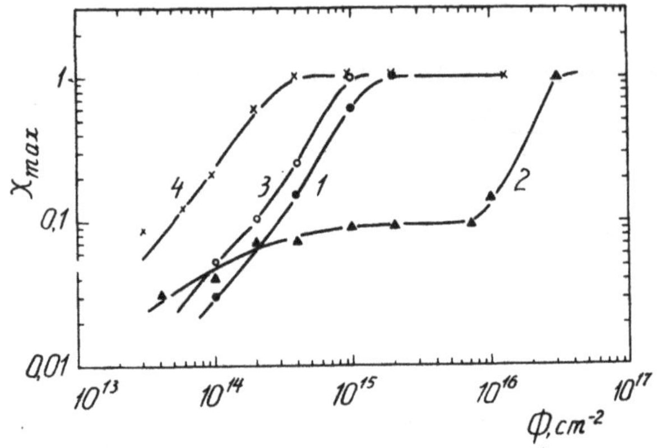

Fig. 1. Normalized backscattering yield of He^+ ions from displaced Si atoms versus the dose of 80 keV Ne^+ ions. 1-undoped Si; 2-Si<B> ; 3-Si<As>; 4-Si<B+As> /4/.

The similar impurity effects we have observed at heavy ion bombardment of Si with Xe^+. But the range of amorphization dose variation due to impurity effect was narrow as compared with Ne^+ bombardment (Fig. 3). Apparently it is connected with higher density of defects in displacement cascades for the heavy ions.

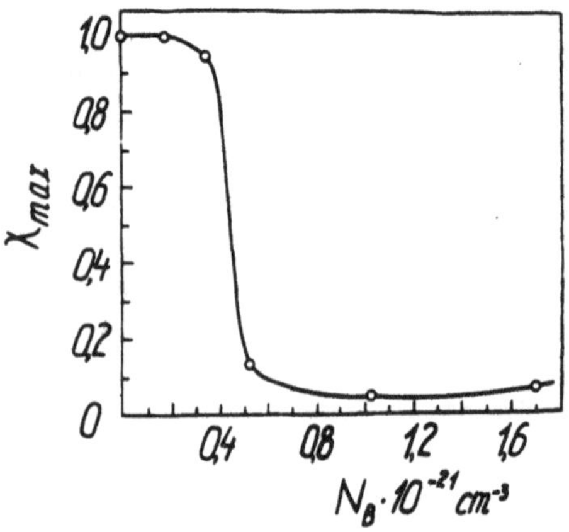

Fig. 2. Normalized backscattering yield as a function of the impurity concentration in Si<B> after 80 keV Ne^+ ions bombardment with dose 1.5×10^{15} cm^{-2}.

For very light He^+ ions we have found the impurity effect also. By contrast, there was no amorphization even at high doses up to 10^{17} cm^{-2}. For low dose bombardment the efficiency of defect production was higher in double doped Si as compare with undoped material. However at high doses the accumulation of defects was saturated. The level of saturation in double doped Si exceeded one in undoped material.

So we have found the strong impurity effects of retardation or acceleration of the defect accumulation which have not reported earlier. These effects can be understood at least qualitatively taking into consideration the influence of the different kind of dopants on the annihilation rate of the simple defects generated by bombarded ions.

It is known that boron, aluminium and also carbon atoms on substituti-

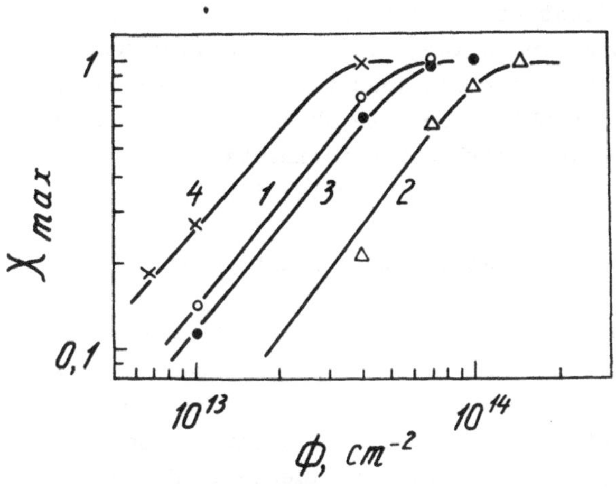

Fig. 3. Normalized backscattering yield versus the dose of 230 keV Xe^+ ions. 1 - undoped Si; 2 - Si<B>; 3 - Si<P>; 4 - Si<B+As>.

onal sites in silicon lattice are easily replaced by Si self-interstitials, ejecting them to interstitial positions /1/. This so-called Watkins mechanism can explain the increased defect annihilation rate at high boron concentration. The highly mobile self-interstitials, created in the impact cascades, are prevented from leaving the vacancy-rich zones by their site exchange with substitutional boron atoms. The resulting boron interstitials are much less mobile and remain in the impact cascade region where the probability for them to annihilate the migrating monovacancies is high. In this way they come back to substitutional sites and lower the divacancy formation rate (Fig. 4).

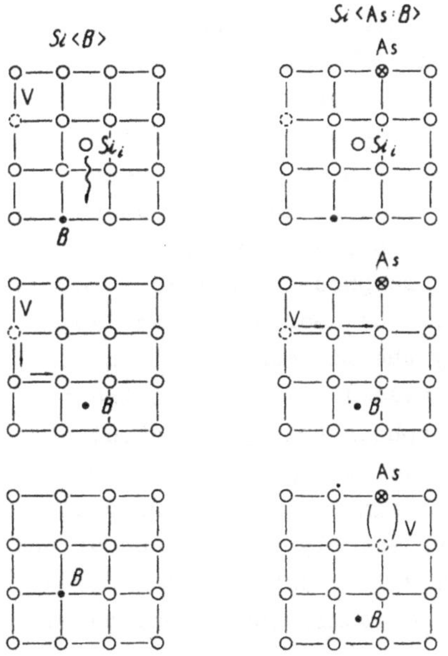

Fig. 4. The sketch of defect-impurity interaction in doped silicon.

For silicon doped with P, As, Sb this exchange process is not to be expected. It is known, that phosphorus atoms keep their substitutional sites even after high doses of neutron irradiation. The self-interstitials can escape from the impact zones and the remaining vacancies are forming both divacancies and vacancy-impurity pairs, leading to the defect accumulation rate similar to that of undoped silicon.

In the double doped Si both boron and impurity of V group are present at a high concentration level and like in the case of Si<B> here the substitutional boron atoms trade place with the generated self-interstitials as a result of the Watkins mechanism. But now the number of migrating monovacancies is essentially lower than for Si<B>, since the vacancies can be trapped effectively by the V group impurity to form practically immobile impurity-vacancy pairs. Therefore the effect of vacancy annihilation by the interstitial boron atoms is reduced and a large fraction of them remains on interstitial sites. This mechanism would lead to an encreased defect accumulation rate and a lower amorphization dose than for Si or Si<As>, Si<P>, Si<Sb>, which is in agreement with our experimental results.

Usually used parameter for the process of amorphization is the critical concentration of defects, which is above of 0.02 /1/. The bombardment with 500 keV He^+ ions produces mostly simple point defects (Frenkel pairs).

The relative concentration of defects in Si layer of 500 nm thickness as a function of dose of 500 keV He^+ ion bombardment is shown in Fig. 5. N_d is the sheet concentration of displaced Si atoms after He^+ irradiation, N_s is the sheet concentration of displaced Si atoms at the amorphization. The correction of the analized beam influence leads to dependence shown by dashed lines (Fig. 5). Irradiation with He^+ ions of the double doped Si allows us to reach the defect concentration about 10^{21} cm^{-3}. This value is approximately equal to impurity concentration. It means that the saturation of sinks for generated defect connected with depletion of impurity atoms.

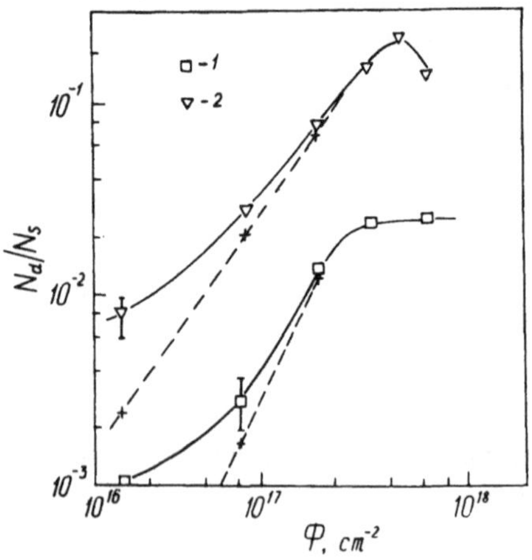

Fig. 5. Dose dependence of defect accumulation in undoped Si (1) and double doped Si<B+P> (2).

Although the concentration about 10^{21} cm^{-3} was reached, we did not get the amorphous regions. One may suggest that together with critical concentration there is another factor, which is important for the crystalline to amorphous transition. To produce the amorphous phase apparently the minimum size of nucleus is needed. The crystalline to amorphous transition should takes place, if the Gibbs energy of the damaged Si region of the critical size is higher than the Gibbs energy of the same region of amorphous material. This condition takes to define more exactly the concept of the critical concentration which is apparently valid just for one kind of defects, because the Gibbs energy depends on atomic configuration by defects.

REFERENCES

(1) V.V.Bolotov, A.V.Vasiljev, A.V.Dvurechenskii et al. A survey of semiconductor radiation technique, ed. L.S.Smirnov, MIR, Moscow, 1983.

(2) G.Dela Mea, A.V.Drigo, P.Mazzoldi et al. Appl.Phys.Lett., 16 (1970) 387-389.

(3) A.V.Dvurechenskii, L.P.Potapova, V.V.Kalinin. Phys.Stat.Sol.(a), 54 (1979) 457–462
(4) A.V.Dvurechenskii, R.Grötzschell, V.P.Popov. Phys.Lett.A, 116(1986) 399–402
(5) A.V.Dvurechenskii, V.P.Popov, B.P.Kashnikov, M.Gripentrog, Poverkhnost': Fizika, Khimia, Mekhanika, 9 (1986) 77–82
(6) J.K.Hirvonen, E.H.Eisen, Appl.Phys.Lett., 19 (1971) 14–16

DIRECTIONS OF DEVELOPMENT OF ION BEAM APPLICATIONS TO MODIFICATIONS OF TOOL PROPERTIES IN POLAND

Jan Kaczmarek[1/] , Witold Rosiński[2/]

Introduction.

Although some preliminary work on application of ion implantation to modification of surface properties of metals has started in Poland in late 70-ties the systematic research was undertaken in 1981. It was sponsored by the Institute of Fundamental Technological Research of Polish Academy of Sciences and was coordinated by its Department of Mechanical Systems. This research was conducted in two directions namely: the development of implantation methods and estimation of the advantages of ion implantation application. Many institutes were engaged in this work. The development of implantation methods was carried out by the Institute of Physics at Lublin University on their two implanters, by the Institute of Electronic Materials Technology at Warszawa, which has Balzers implanter type MPB-200 and by the Institute of Atomic Physics at Krakow,which used own production 60-keV implanter. Two main problems were investigated by Lublin and Krakow: one concerned with the development of new more efficient ion sources producing broad variety of atomic species, more intense ion beams and the other with the influence of implantation parameters on distribution and concentration of implanted species such as heating the implanted sample in course of implantation and after it. This last task was carried out chiefly by the Institute of Electronic Materials Technology. Some efforts were made here to increase the range of implanted species by installing in Balzers implanter an ion source designed by the Lublin Institute. Appreciable Al, Ti and V currents at the target were reached correspondigly 60 uA for Al, 40 uA for Ti and 10 uA for V.

1/ Institute of Fundamental Technological Research Polish Academy of Sciences Warszawa Poland

2/ Institute of Electronic Materials Technology-CEMAT Warszawa Poland

Results of Studies.

Since the work coordinating Department of Mechanical Systems is engaged in investigation of modification of surface properties of metals by various methods of surface treatment the studies of the influence of ion implantation on surface properties were naturally included in the programme of research. This was conducted in two main directions one concerning with the improvement of tools, the other of machine elements. In the first case wear of implanted tools was investigated in the other fatigue lifetime was mostly studied. Wear of ion implanted cemented-carbides of tungsten-titanium in cobalt cutting tools was throughly tested. Their wear was estimated from the length of the removed part of the cutting edge indicated by h in fig 1, and measured by suitably placed microscope. Measurements were made for three different speeds of the material.

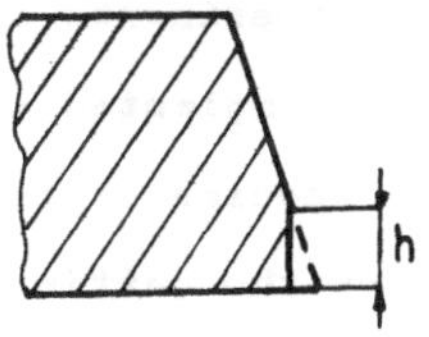

Fig.1. Shape of the tool edge after cutting

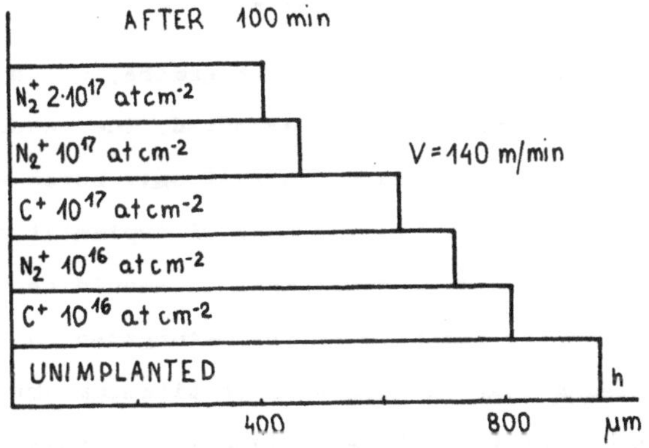

Fig.2. Wear h in um for different implantation condition

In the first part of the experiment carbon and nitrogen atoms were implanted with doses from 1.10^{16}at/cm^2 to 2.10^{17}at/cm^2 and energies 70 and 240 keV. The results of h measurements at speed of 140 m/min are shown in fig.2. The effectivness of carbon implantation in comparison to that obtained for nitrogen is small. There is also little advantage is using energies above 70 keV. Fig.3. indicates appreciable difference of slopes of wear rate for implanted and unimplanted tools and well over twice longer operation of implanted tool.

In the second part of the experiments the concept of implantation of suitably chosen pairs of atomic species in this case boron and nitrogen or nitrogen and titanium was examined.

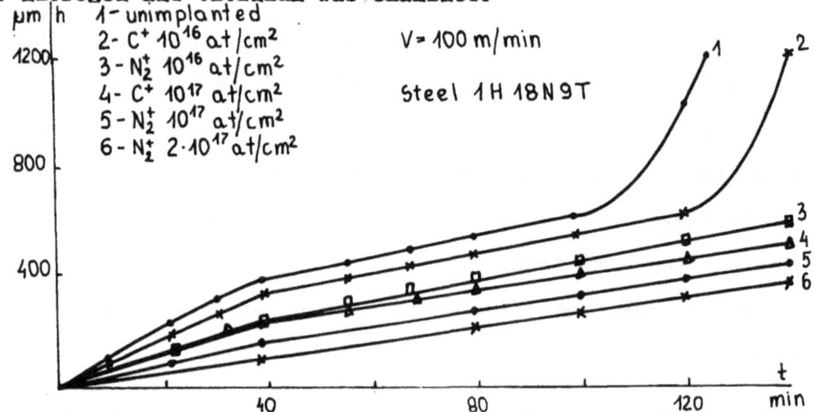

Fig. 3. Wear versus time of operation of the cutting tool

The formation of BN and TiN grains, which should improve surface hardness of the tool, was expected. The results of the experiments are shown in fig.4 and 5. Wear of only boron and titanium implanted tools is heavier than in the case of implantation of both species especially at high speed of operation. It should be noticed here the cutting tool was of the cemented titanium-tungsten carbides and it is a little surprising that the addition of titanium should improve cutting performance of the tool. Some explanation of this behaviour may be obtained from Auger eletrons spectroscopy observation made across the working surface

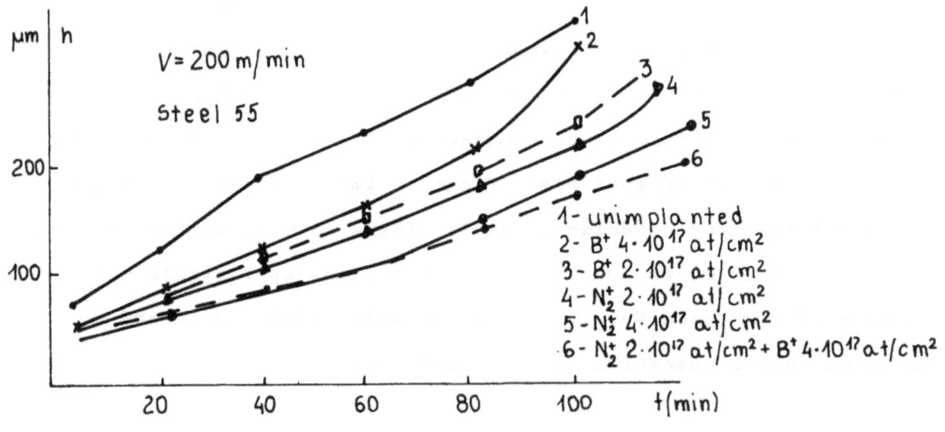

Fig. 4. Wear versus time of operation of the cutting tool

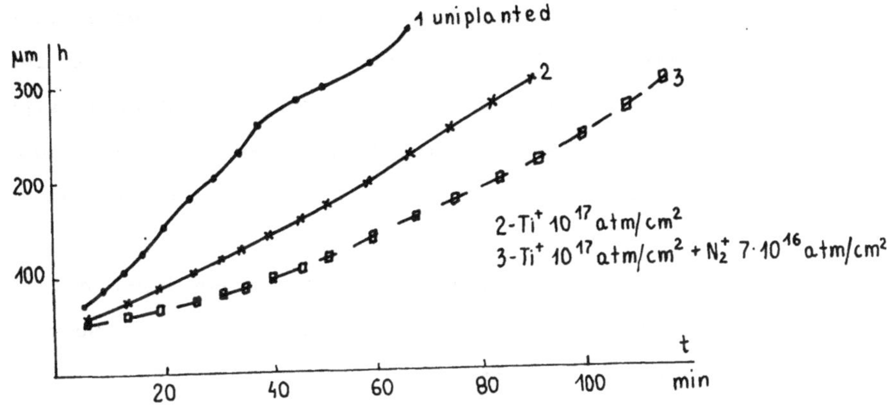

Fig. 5. Wear versus time of operation of the cutting tool

formed on the tip of the tool during cutting /h in fig.1/. In fig. 6 and 7 the relative intensities of signals coming from implanted atoms and atoms originally existing in the tool are shewn.

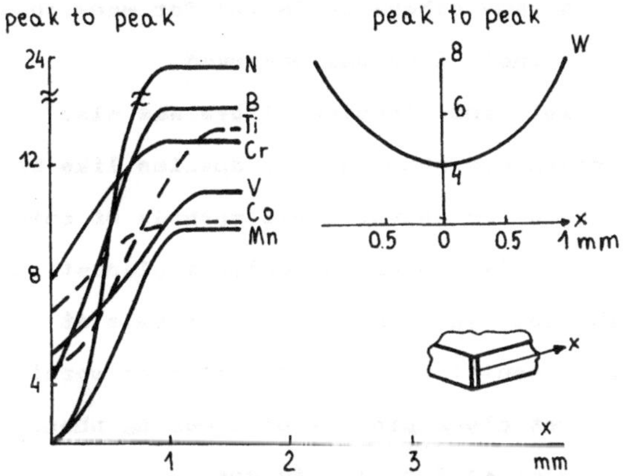

Fig.6. The segregation effects occuring at the tip of the tool

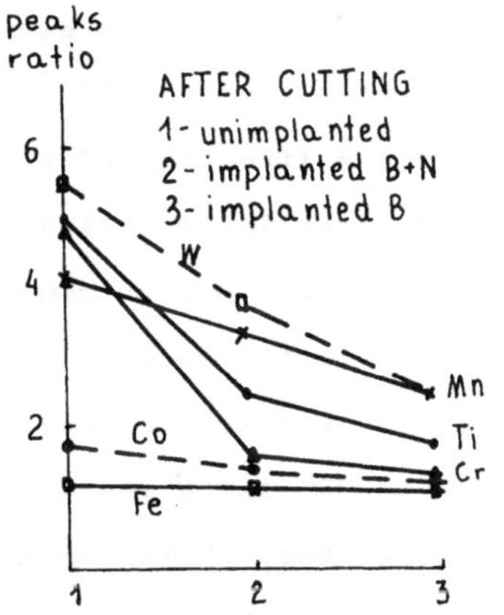

Fig.7. The influence of implanted species on segregation effects.

Fig. 7 clearly indicates the influence of the atomic species on the content of atoms in the surface layer of the tip. Over fourfold reduction of important for cutting process atoms of W, Ti occurs. In the insert of fig. 6 the distribution of tungsten across the surface formed during cutting process for unimplanted tool is shown. It is interesting to note the different degree of segregation for different types of implanted atoms. For unimpanted sample the reduction of tungsten amounts to about 5, for samples implanted with nitrogen and boron is about 3.75 and for

boron only about 2.5. The reason for such behaviour is not understood yet and will be investigated in detail, since it is also observed for other components of the matrix like Cr, V and Mn. The implanted nitrogen and boron were still observed at the depth of about 20 um at 10% of the top value due probably to the diffusion along crystallites boundaries. Concentration of boron was reduced only four times. Besides the above mentioned effects implantation of suitably chosen pairs of atoms allowed for quite appreciable increase of speed of operation, what is very important in workshop practise.

To confirm postive results of laboratory experiments large scale tests were arranged at the workshop, where over 400 tools, half of them unimplanted and half implanted, were tested in similar operating conditions It was shown that over 200% more operations were carried out with implanted tools. Similar results were found in case of many other tools like slot dies, gauging elements and many others implanted for many customers, two to fivefold increase of their life was observed.

Fatigue life-time of various steels and titanium alloys was also studied after implantation of nitrogen and other stomic species like boron, barium and carbon combined with other conventional methods of surface treatment used in metallurgy, applied before or after implantation. In case of steels very diverse behaviour was found in some steels it was up to 30% improvement in some, especially in martensitic or sorbitic structures, there was none. To get clear picture of occuring phenomena futher systematic studies are planned for near future.

In case of titanium alloys improvement due the implantation varied from 15% up to 40% and with additional treatment reached 50% Interesting fact is the independence of fatigue life-time from the increasing dose, starting from certain value in case of N and Ba atoms, but steadily growing with increasing dose of carbon. The implantation process made also the fatigue effect more uniform from sample to sample.

In conclusion the results of some studies concerning the methods of implantation should be mentioned. Since the temperature of the implanted layer depends upon the rate at which ions are delivered to it one should expect different distribution of implanted atoms resulting from diffusion effect. To confirm this the distributions of implanted nitrogen were measured by means of atomic reaction $^{14}N/d\alpha/^{12}C$ for samples of steel SW7M implanted with nitrogen ions rising the temperature of the sample to about $300^{0}C$ and for samples implanted at much lower rate of beam currents so as to keep the temperature of the sample below $70^{0}C$ and then heating it to temperature of $300^{0}C$ for about 1 to 2h. Fig. 8 and 9 show clearly the difference of diffusion process and a possibility of getting much deeper penetration at higher concentration level of implanted atoms by heating the sample by means of intense ion beam. Further tests are pending.

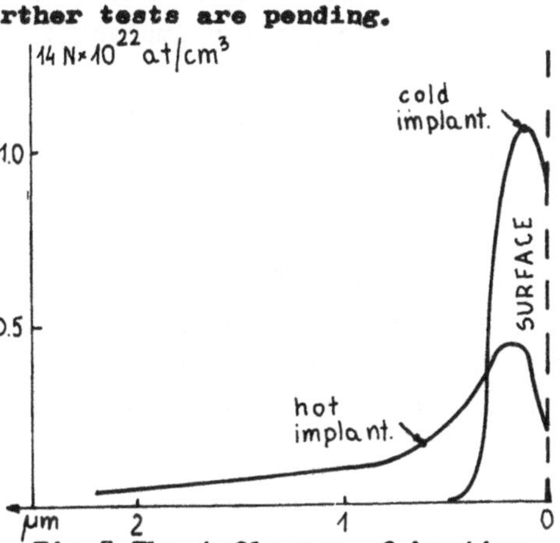

Fig.8. The influence of heating by implanted atoms on their distribution

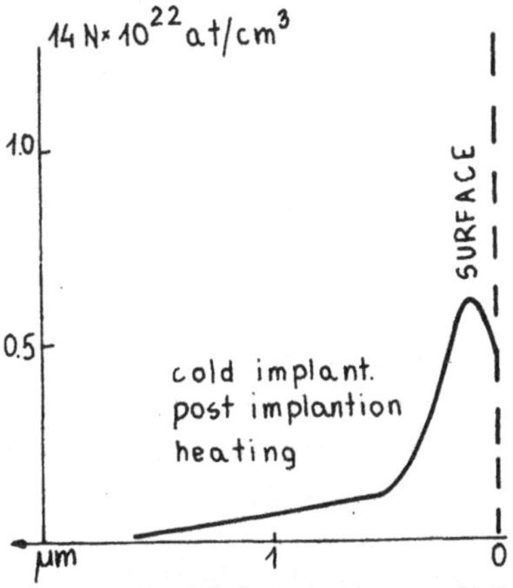

Fig.9. The influence of post implantation heating on the distribution of implanted atoms /cold implantation/

It was also found the heat treatment after implantation is lowering fatigue life-time probably due to reduction of maximum value of implanted nitrogen atoms /see fig. 9/. To get clear picture of phenomena affecting fatigue life time futher extensive research on statistical base is required.

COMPARATIVE INVESTIGATION OF PHASE FORMATION IN Fe, Al AND Ti AFTER NITROGEN ION IMPLANTATION

B.Rauschenbach[1)]

The effect of high fluence ion implantation on the near-surface composition and compound formation of metals and alloys has became the subject of many detailed investigations. This include increasing use of nitrogen ion implantation to form new materials with improved mechanical and chemical properties. Consequently, basic investigations are essential as to use this technique effectively.
The purpose of this paper is to present some results on the implantation induced phase formation of metals after nitrogen ion implantation (details see refs. /1/-/3/).
The metals Fe (bcc), Al (fcc) and Ti (hex.) were chosen as model substances for our fundamental investigations. The metals were implanted with nitrogen ions at temperatures between the temperature of liquid nitrogen (LNT) and 600 °C. The ion bombardment was performed at energies from 20 to 330 keV with doses from $1x10^{16}$ to $1.5x10^{18}$ N^+-ions/cm^2. The metal specimens were investigated using transmission electron microscopy, Auger electron spectroscopy, X-ray photoelectron spectroscopy, electron energy loss spectrometry and secondary ion mass spectrometry (details see ref. /4/).

Implantation in aluminium
Recently, we have demonstrated direct nitridation of evaporated aluminium layers by high-fluence nitrogen ion implantation /5/. The implantation always produced polycrystalline aluminium nitride. A thermal treatment up to 600 °C did not yield new crystalline phases. At low temperatures the growth of AlN precipitations takes place mainly coherently as a result of a high vacancy density. Larger AlN precipitates grow similarly to the Ostwald ripening process. But the implantation induced formation of AlN is depended on the implantation dose and the distance from the aluminium surface. In table the Al(2p), N(1s) and O(1s) binding energies, obtained from nitrogen implanted aluminium, are listed after different implantation and at various depths.

1) Central Institute for Nuclear Research Rossendorf
GDR-8051 Dresden, P.O. 19

Table: XPS-binding energies of aluminium (2p), nitrogen (1s) and oxygen (1s) after implantation with different implantation doses and at different depths

fluence (N^+-ions/cm^2)	depth	binding energy (eV) Al(2p)			N(1s)		O(1s)
5×10^{17}	surface	72.5	73.6	74.7	397.1		532.0
	R_p		73.8		396.5		
1×10^{18}	surface	72.6	74.2		396.7		531.6
	R_p	73.7			396.8	403.0	

The Al(2p)-binding energy 73.8...74.2 eV and the N(1s)-binding energy 396.5...396.8 eV can be attributed to AlN. The peaks corresponding to 73.6 eV, 73.7 eV and 397.1 eV can be taken as weakly bounded nitrogen atoms. The Al(2p)-binding energies higher than 74.2 eV characterize Al bound to oxygen. It can be seen from table that AlN is only formed in the near of R_p after implantation with $5x10^{17}$ N^+-ions/cm^2 and that AlN is also produced at the surface after implantation with $1x10^{18}$ N^+-ions/cm^2. A thick aluminium oxide layer was only detected for specimens implanted with fluences up to $5x10^{17}$ N^+-ions/cm^2. These XPS-results are in a good agreement with the results by Ohira and Iwaki /6/.
With the aid of the plasmon analysis using electron energy loss spectrometry the occurrence, as well as the amount of AlN was studied (Fig. 1) as function of nitrogen ion dose /7/.

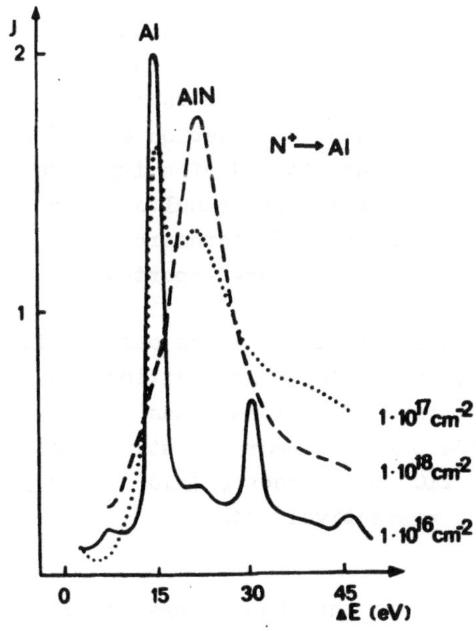

Fig.1: Relative electron energy loss intensity J of aluminium, implanted with different nitrogen doses

After implantation with low doses ($1x10^{16}N^+$-ions/cm^2), the high Al plasmon peak at about 15 eV and the high Al oxide peak at about 30 eV characterize the initial state of the AlN formation. After implantation with higher doses (e.g. $1x10^{18}$ N^+-ions/cm^2), the high AlN peak and the absence of the Al and Al oxide peaks indicate the complete AlN formation.

Implantation in titanium

Fig. 2 shows a bright and a dark field micrograph and an electron diffraction pattern of titanium implanted with $5x10^{17}$ N^+-ions/cm^2. These micro-

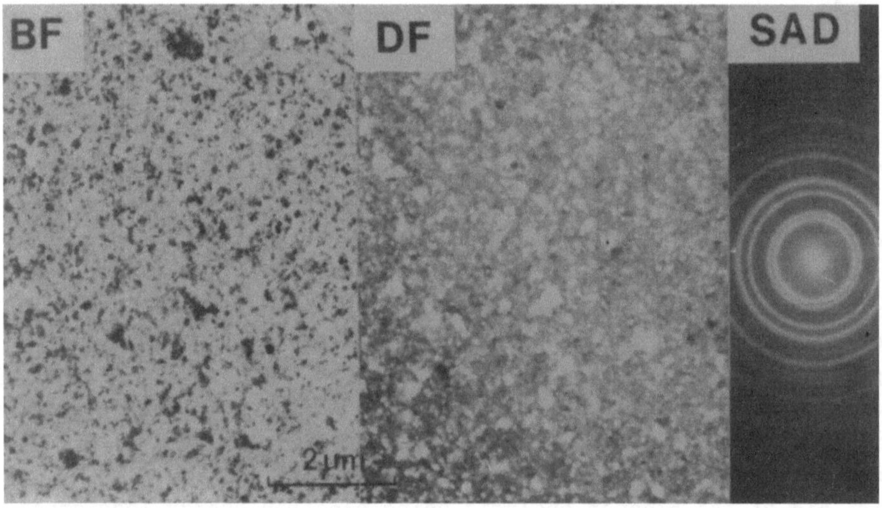

Fig.2: Bright field (BF), dark field (DF) micrographs and selected area diffraction pattern of titanium implanted with 50 keV N-ions to a fluence of $5x10^{17}$ N^+-ions/cm^2 at RT.

graphs of Ti demonstrate the formation of two different titanium nitride phases. The fcc phase δ-TiN_x can be identified in the total investigated dose range, where x is the stoichiometric factor which varied between 0.4 to 0.5 and 1.0 to 1.25, depending on the implantation dose /2/ /8/. The tetragonal ε-Ti_2N phase could be found after N^+-ion implantation in Ti at fluences $\geq$ 1 to $2.5x10^{17}$ N^+-ions/cm^2. The results of investigations of the nitrogen ion implantation-induced microstructure and its dependence on implantation dose and content were reported and discussed in comprehensive studies /2//4//8/.
Because of carbon adsorption during the implantation process and recoil implantation by nitrogen ion bombardment the Ti layer is additionally doped with carbon. Therefore the implantation of N^+-ions leads to formation of the cubic titanium carbonitride phase (isomorphic structure to TiN_x) and the titanium carbide phase TiC above doses $\geq 1x10^{17}$ to $2.5x10^{17}$ N^+-ions/cm^2 and $5x10^{17}$ N^+-ions/cm^2, respectively /4/.

Implantation in iron

The composition of the near surface region and the phase formation of iron after high-fluence nitrogen ion implantation were intensivly investigated (see e.g. refs. /1//4//9/-/12/).

An example of the depth profiling analysis is given in Fig. 3, an analysis of nitrogen ion implanted iron as function of the implantation dose. In accordance with results of AES-measurements by Singer /13/,

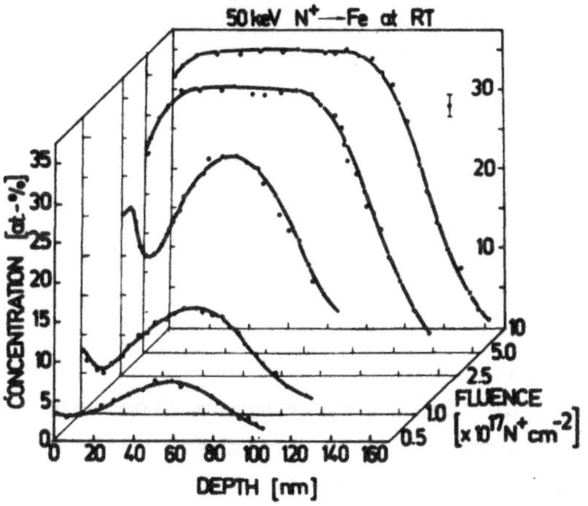

Fig.3: Concentration vs depth profiles of nitrogen ion implanted iron at 50 keV. Profiles were obtained after implantation with different fluences by Auger sputter depth profiling (details see ref. /4/).

the concentration distributions show distinctly non-Gaussian behaviour. At fluences between 2.5 and $5x10^{17}$ N^+/cm^2, the concentration of nitrogen at its peak depth begins to saturate (approximately 30...35 at.-%), and the nitrogen which continues to be retained in the implanted layer redistributes in a trapeziod-like profile. The peak at surface indicates a migration of nitrogen atoms to the surface. Chemical bond information obtained from AES- and XPS-spectra (see also this proceedings, /14/) helps account for the differences in the various profiles /4//15/.
The phase formation and phase transition processes of iron after nitrogen ion implantation are studied as function of the implantation dose, the annealing conditions and the temperature during implantation. But a coherent picture of the metallurgy of nitrogen implanted iron emerges from the compositional and microstructural studies exists not presently. In iron, the implanted nitrogen atoms form compounds or are mobile and move away from regions of high nitrogen concentration. A fraction of these atoms leave the iron target via the surface, while some get trapped at the metal/metal oxide interface and the other nitrogen atoms migrate in higher depths. These processes in combination with the variation of the nitrogen ion fluence and/or the temperature during implantation result in a manifold microstructure (morphology) of the implanted iron specimens. Examples are shown in the next figures (Fig.4).

The phases after nitrogen ion implantation were identified using diffraction contrast (morphology investigations), selected area diffraction (lattice structure analysis), and comparison of the measured and calibrated intensities with the calculated intensities of the diffraction reflections (details see ref. /4/). Results of the phase analysis after implantation at LNT and RT are summarized below.

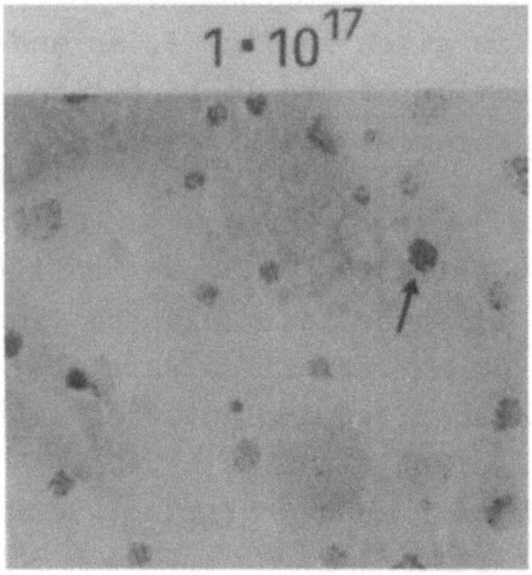

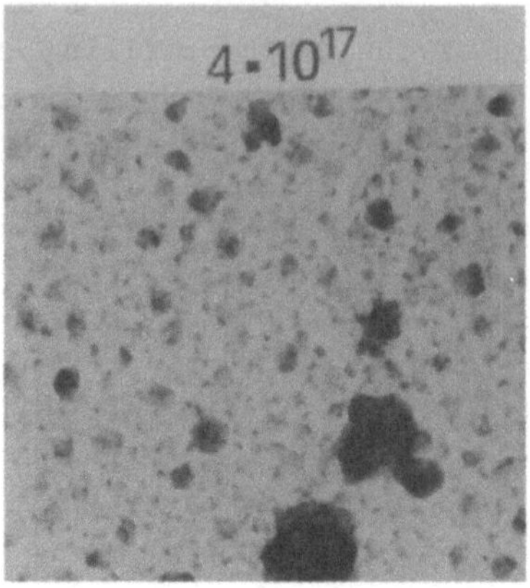

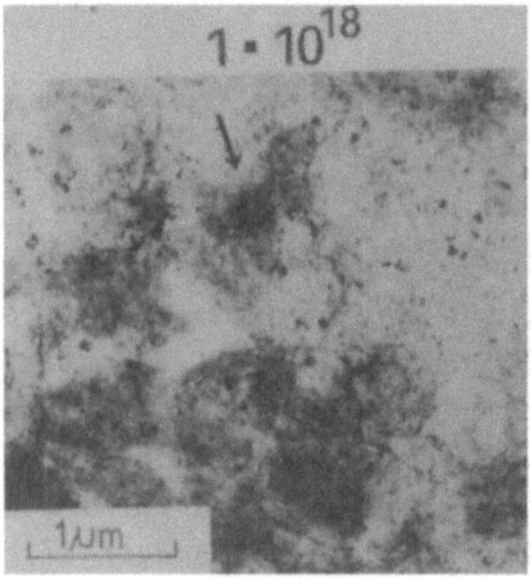

Fig.4:
Bright field transmission electron micrographs of Fe implanted with 50 keV N^+-ions to different doses at LNT.
The disc-like precipitates after implantation with $1x10^{17}$ N^+-ions/cm^2 indicate the α''-$Fe_{16}N_2$ phase (arrow). At higher fluences ($1x10^{18}$ N^+-ions/cm^2), ε-$Fe_2(C,N)_{1-x}$ precipitates (arrow) were found additionally.

Summary of implantation induced phases of iron after implantation at RT and LNT.

impl. temp.	impl.dose (10^{16} N^+/cm^2)	implantation induced phases and phase transition during vacuum-annealing		
LNT (-196 °C)	1...7.5	N-atom cluster		
	4...40	molecular β-N_2	dissolution →	
	> 5	α'-martensite	→	α''-$Fe_{16}N_2$
		ε-$Fe_2(C,N)_{1-x}$		
		α''-$Fe_{16}N_2$		
RT	≥ 1	γ-austenite	dissolution →	
	≥ 4	ε-$Fe_2(C,N)_{1-x}$	→	ζ-Fe_2N
		α'-martensite	→	α''-$Fe_{16}N_2$
		α''-$Fe_{16}N_2$	→	γ'-Fe_4N

In addition to the implanted nitrogen atoms, the surface contains carbon atoms absorbed from gas molecules in the vacuum champer. At the surface and at the near surface region carbon produces carbonitride compounds additionally. In higher depths, the carbon atoms are free or weakly bounded. Thus, distinct phases of nitrides carbonitrides and

and carbides exist in different depths in the implanted layer.

Acknowledgments
The author wishs to thank Prof. K.Hohmuth, Dr. V.Heera, Dr. K.Helming (ZfK Rossendorf), Dr. G.Kästner (IFE Halle), Prof. G.Leonhardt, Prof. K.Breuer (TU Karl-Marx-Stadt) and R.Dietsch (VEB Hochvakuum Dresden) for allowing various investigations and helpful discussions.

References

/1/ Rauschenbach,B.,E.Richter and K.Hohmuth: Nucl.Instr.Meth. 209/210 (1983) 249
/2/ Hohmuth,K.and B.Rauschenbach: Mater. Sci. Eng. 69 (1985) 489
/3/ Rauschenbach,B.: Nucl.Instr.Meth. B 18 (1986) 34
/4/ Rauschenbach,B.: Thesis B, and ZfK-report (1987), to be published
/5/ Rauschenbach,B., A.Kolitsch and E.Richter: Thin Solid Films 109 (1983) 37
/6/ Ohira,S. and M.Iwaki: Nucl.Instr.Meth. B 19/20 (1987) 162
/7/ Rauschenbach,B.: J.Less-Common Met. 117 (1986) 115
/8/ Rauschenbach,B.: J.Mater.Sci. 21 (1986) 395
/9/ Follstaedt,D.M.: Nucl.Instr.Meth. B 7/8 (1985) 1
/10/ Barnavon,T.,H.Jaffrezic, G.Marest, N.Moncoffre, J.Tousset and S.Fayeulle:Mater.Sci.Eng. 69 (1985) 531
/11/ Rauschenbach,B., A.Kolitsch and K.Hohmuth: phys.stat.sol. (a) 80 (1983) 211 and 471
/12/ Rauschenbach,B., K.Hohmuth and G.Kästner: Nucl.Instr.Meth. B 23 (1987) 323
/13/ Singer,I.L.: Vacuum 34 (1984) 853
/14/ Rauschenbach,B. and G.Leonhardt: this proceedings
/15/ Rauschenbach,B.:Nucl.Instr.Meth. B 15 (1986) 756

MECHANISMS OF STRUCTURE FORMATION IN LASER PULSE VAPOUR DEPOSITION

W. Pompe, G. Keßler, M. Klose, H.J. Scheibe[1)]
S. Becker, H.J. Dietze[2)], G. Seifert[3)]

1. Introduction

The vaporization of solid targets by laser irradiation with high power density has been applied as a new technique for thin film deposition. In the last years an increasing number of examples has been reported on the preparation of mono- or multi-layer structures of dielectrics, semiconductors, and metals by laser pulse vapour deposition (LPVD) /4, 5, 6, 10/. This term includes thermal evaporation and deposition with particle energies of about 0.5 eV as well as the explosive formation of a highly ionized plasma and subsequent film deposition with particle energies from 10 to 1 000 eV.

It is the aim of this paper to discuss the relevant physical mechanisms of structure formation in the evaporation and deposition process. LPVD can be subdivided into three partial processes:

- response of the target material to irradiation, resulting in evaporation and ionization,
- expansion of the plasma
- deposition including nucleation and film growth.

2. Response of the target to irradiation

The response of a solid target to the absorption of a power density j may be roughly estimated by means of a thermodynamic approach /9/. It predicts

1) Zentralinstitut für Festkörperphysik und Werkstofforschung, Helmholtzstr. 20, Dresden, 8027, GDR

2) Zentralinstitut für Isotopen- und Strahlenforschung, Permoserstr. 15, Leipzig, 7050, GDR

3) Technische Universität Dresden, Sektion Physik, Mommsenstr. 13, Dresden, 8027, GDR

- heating with eventual melting and microcracking
  for $j < j_o = (T_b/f(j)) \cdot (\lambda \varrho\, c/t_p)^{1/2}$ 4)
- evaporation and ionization
  for $j_o < j < j^* = \varrho \cdot q^{3/2}$
- detonation
  for $j^* < j$.

Evaporation may be accompanied by melting and surface cracking. Therefore, liquid droplets or solid particles may be emitted from the laser spot if no precautions are taken to avoid this.

Turbulences within the melt pool driven by pressure gradients $\frac{dp}{dx}$ in the plasma cloud could be one cause of droplet ejection. The flow velocity expected from this mechanism is

$$v \approx \frac{dp}{dx} \cdot \frac{t_p}{\varrho} \quad .$$

Droplets of radius r can be formed if the dynamic pressure $\varrho \cdot v^2/2$ of the turbulent flow outweights the pressure within the droplet, $2\gamma/r$. Considering that large droplets form easier than small ones, and the depth of the melt pool h is an upper limit of droplet size, we obtain

$$\left(\frac{dp}{dx}\right)^2 \cdot \frac{t_p^2}{\varrho} \cdot \frac{h}{4\gamma} < 1$$

as a condition for no droplet formation. It is seen that a low depth of the melt pool and a small pressure gradient will tend to avoid splashing. A shallow melt pool is formed with high power density in short pulses, and low gradients are obtained with homogeneously irradiated large spots. Expected trends were found with LPVD of W and Au, for instance.

The emission of solid particles is observed with brittle materials as ceramics, for instance. The microcracking is initiated by thermal stresses. It depends not only by the maximum temperature reached in irradiation but also on the pulse time: Short pulses make short cracks. Repeated pulses on the same spot may change the target surface in such a way that not only the absorption as well as the evaporation rate but also the droplet ejection rate change. Therefore the target should be

4) Meaning of the symbols: T_b = boiling temperature at normal pressure, ϱ = mass density, c = specific heat, λ = thermal conductivity, q = energy of vaporization, $f(j) \lesssim 1$, t_p = pulse duration, γ = surface tension.

moved such that the process becomes stationary.

The emission of debris from microcracking can be used for the preparation of microcomposites by LPVD. From ceramic targets of low porosity, large amounts of particles with sizes between 0.01 and 0.1 μm are emitted. Alternate deposition of particles from a brittle target and metal or carbon from another target forms a composite film. In this way, superhard boron nitride was embedded into a matrix /7, 8/. Obviously, stoichiometry and crystallographic structure are conserved if the transfer of material from the target to the substrate occurs in the shape of comparatively large particles.

As an additional complication of the LPVD process, the incoming radiation is partly or completely absorbed by the emerging plasma cloud, thereby heating the latter. At temperatures above 0.5 eV the vapour is sufficiently ionized to absorb the laser power to such an extent that the target is progressively shielded from irradiation /2/. Plasma clouds from 10^9 W/cm^2 pulses absorb the radiation completely. Fig. 1 shows theoretical curves for temperature and vapour pressure of a multiply ionized plasma from irradiation of a carbon target with a Nd laser pulse of $2 \cdot 10^{-8}$ s and 10^{10} W/cm^2 /14/. Theory and experiment agree sufficiently well. It is seen that in this range of power density the electron and ion temperatures do not much differ.

3. Expansion of the plasma cloud

Various plasma-chemical reactions proceed in the early stages of the plasma cloud. Aggregation, substitution and charge transfer reactions will occur as long as the density is high enough for collisions. At high temperature, dissociation and collision-induced fragmentation will be essential. All this goes on within the rapidly expanding plasma cloud, which means that it stops very soon due to cooling and rarification. What is left is a cold molecular beam. Its composition can be investigated by means of mass spectroscopy. Especially the laser ionization mass spectroscopy (LIMS) is favoured, because the experimental conditions of LIMS resemble those of the LPVD process, at least with respect to the power density, which reaches $10^8 \ldots 10^{10}$ W/cm^2. The LIMS technique suggests to apply external electric or magnetic fields to the LPVD process. In this way the particles could be selected before deposition, also they could be accelerated or recharged by field-induced collisions.

LIMS investigations reveal three ranges of power density characterized by different cluster size distributions in the molecular beam. Just above the evaporation threshold ($j_o \approx 10^6 - 5 \cdot 10^7$ W/cm^2) the composition of the cloud is mainly governed by the bonding state of the target. Therefore, atoms, non-stoichiometric fragment clustersor stoichiometric clusters can be observed.

With increasing power density, temperature and pressure of the plasma cloud increase also. Aggregation and fragmentation processes go on within the cloud until it has been cooled and rarified by adiabatic expansion. Dissociation of excited species may still occur in the cold beam.

Very large power densities may give rise to temperatures and pressures well above the critical point where no clusters survive. As this condition is usually reached with a very tiny spot, the detonation of the plasma cloud may be so rapid that there is no time for cluster formation.

Fig. 2 shows the C_n^+ cluster size distribution in the LIMS molecular beam plotted versus laser power density /3/. With increasing power density the average cluster size decreases. The size distribution shows an odd-even periodicity for $n < 6$ and a 4-cycle for $n > 6$. The observed periodicities coincide with those obtained from binding energy calculations for linear and cyclic C_n clusters (LCAO-X method) /13/. So-called magic numbers pertaining to clusters of high stability have been shown to exist. In agreement with the above discussion, the quenched-in cluster distribution seems to be determined by a limited number of favoured aggregation and fragmentation channels.

The intermediate stage of cluster formation and fragmentation in the plasma cloud can be used for the preparation of heteronuclear clusters from a multi-component target. The laser excitation offers also the possibility to form heteronuclear compounds of substances with high boiling temperatures. For the synthesis of the compounds, the different components can be distributed within the target as a mixture of various components as powders of metals, oxides, carbides etc. This offers a simple method of target preparation, especially in the case of high-melting compounds as carbides, silicides, and nitrides.

Fig. 3 shows results obtained with boron carbide for an example. Two kinds of targets were used, with the preparation of a B_4C target being more difficult than using pressed powder mixtures of BN and graphite. Both targets yield similar size distributions of BC_n clusters. The yield is higher with the BN/C mixture target.

The controlled formation of heteronuclear clusters is a prospective possibility of further applications of the LPVD technique /1/.

4. Nucleation and film growth on the substrate

The deposition process of the laser generated plasma on the substrate differs from the classical vapour deposition by a high instantaneous deposition rate R and by the presence of a significant percentage of particles with kinetic energies between 10 and 1 000 eV. As a consequence of the high deposition rate R, the classical growth modes of layer growth or island growth are modified. Fig. 4 shows schematically the relation between the various growth modes and the deposition rate R. In order to estimate which mode is dominating, one has to analyze the time scales of the various elementary processes:

- deposition at a given substrate point

$$\tau_R \sim (\tau_b/\tau_p)/a_o{}^2 R$$

a_o = average particle diameter, τ_p = pulse duration
τ_b = break between two pulses

- diffusion controlled aggregation

$$\tau_D \sim 1/D_{sub}\bar{Z}_1$$

D_{sub} = diffusion coefficient for the deposit on the substrate,
$\bar{Z}_1$ = average monomer density

$$\bar{Z}_1 = \begin{cases} \tau_{des} \cdot R \\ \tau_p \cdot R \end{cases} \quad \text{for } \tau_p \gtrless \tau_{des} = \frac{1}{\nu_o} \exp(E_{des}/kT)$$

The alternatives correspond to dominating desorption or deposition.

- island growth

$$\tau^{dep}_{sub}\Big] \sim \nu_o{}^{-1} \cdot \exp\,(E_{des} + \Delta E)/kT \approx \tau_{des}$$

- monolayer growth

$$\tau^{dep}_{sub}\Big\} \sim a_o^2 \cdot D_{dep}^{-1} \cdot \exp(-\Delta E'/kT) \approx a_o^2 \cdot D_{dep}^{-1}$$

D_{dep} = diffusion coefficient on the deposit surface

Within this frame, that growth mode with the shortest characteristic time is the dominating one. It is seen that with increasing rate the process shifts to a layer growth deposition mode, independent of the interaction between substrate and deposit. Therefore the LPVD technique offers unique advantages for the preparation of thin film structures such as multi-layer structures with layer thickness of only a few atomic distances as used in quantum well structures, for instance /6/.

The real structure of the films can be influenced by controlling the ratio of τ_R to the other characteristic times. The size distribution of independently growing clusters shifts with increasing deposition

rate to smaller clusters. Therefore, amorphous or microcrystalline films form preferably at lower temperatures /11, 12/.
Obviously the presence of particles with higher energy influences the film growth significantly. The critical substrate temperature for layer growth, for instance, decreases by about 100 K. This phenomenon could be connected with higher diffusivity of a large amount of interstitials implanted near the surface during the deposition process. A rough theoretical estimate shows that such an increase could be caused by the impingement of particles from the laser plasma with energies of about 20 ... 500 eV.

References

(1) Albrecht, G. and H.J. Dietze et al.: ZfI-Mitt. 134 (1987) 5
(2) Beyer, E.: Dissertation, Techn. Hochschule Darmstadt (1985)
(3) Dietze, H.-J. and S. Becker: ZfI-Mitt. 57 (1982) 1 - 187
(4) Dimitrow, D. and S. Metew et al.: J. Mat. Sci. Lett. 1 (1982) 33
(5) Dubowski, J.J. and D.F. Williams: Thin Solid Films 117 (1984) 289
(6) Gaponow, S.W.: Vestnik AN SSR 12 (1984) 3
(7) Keßler, G. and H.D. Bauer et al.: Thin Solid Films 147 (1987) L45-L50
(8) Keßler, G. and H.J. Scheibe et al.: WP C23C/282 2 133/30.10.85
(9) Pompe, W. and H.J. Scheibe et al.: "Trends in Quantum Electronics", Proceedings of the 2nd Conference, Bucarest, Sept. 1985; Springer (1986)
(10) Pompe, W. and H.J. Scheibe et al.: Wiss.Ber. ZFW Dresden 26 (1984) 87
(11) Pompe, W. and M. Bobeth et al.: Proceedings of EPM 87, Dresden, Sep. 1987
(12) Richter, A.: Proceedings of EPM 87, Dresden, Sep. 1987
(13) Seifert, G.: ZfI-Mitt. 134 (1987) 49
(14) Völlmar, S. and Th. Schlegel et al.: Proceedings of EPM 87, Dresden, Sep. 1987

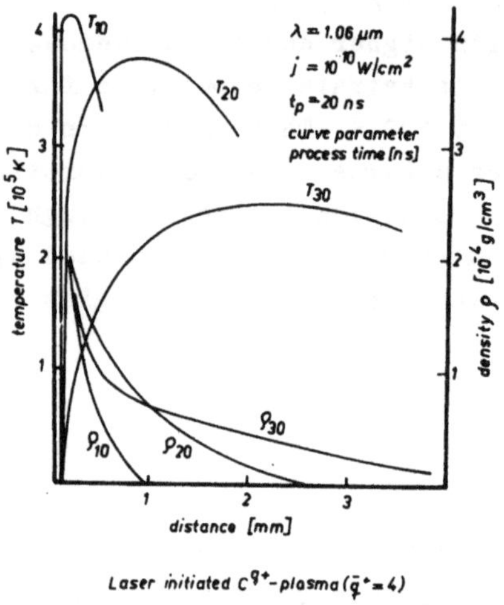

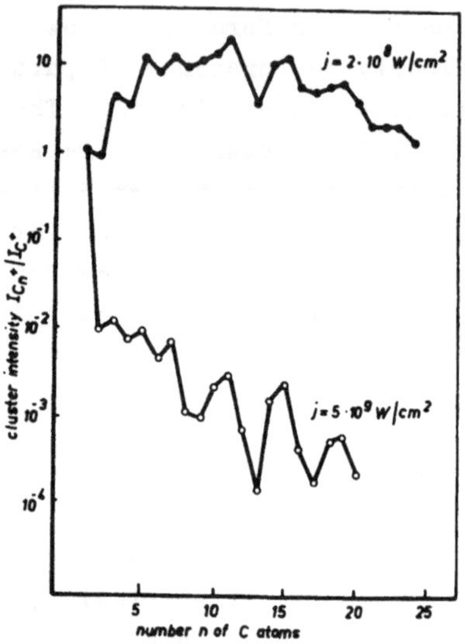

Fig.1 Temperature and mass density distribution of the LPVD plasma for different times

Fig.2 C_n^+ cluster size distribution in the LIMS beam

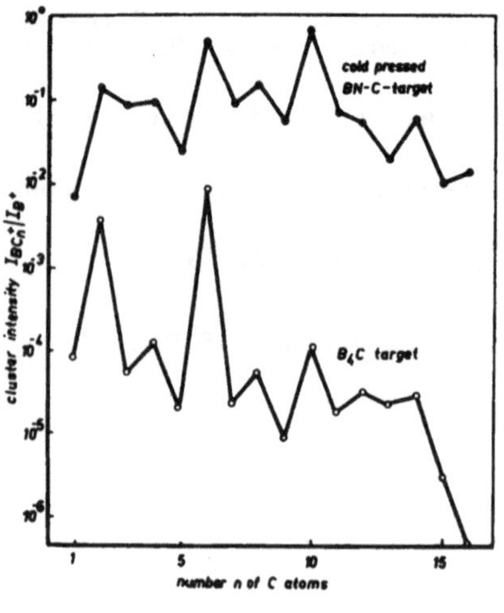

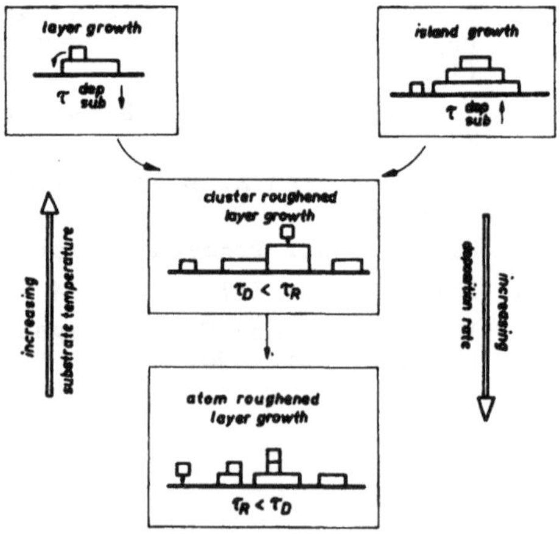

Fig.3 LIMS cluster distribution from different targets

Fig.4 Influence of temperature and deposition rate on the mode of film growth

FUNDAMENTAL ASPECTS OF THERMALLY INDUCED SOI LAYER FORMATION

K.-H. Heinig and H.-J. Mueller[1)]

1. Introduction

The thermally induced formation of silicon-on-insulator (SOI) layers, e.g. zone melting recrystallization (ZMR) and lateral solid phase epitaxy (L-SPE), is controlled by fundamental physical processes like nucleation, growth kinetics, heat conduction and so on. In general, very complex physical processes act at the same time. This makes it difficult to study the formation of SOI layers theoretically.

The method of computer simulation (CS) has been established to be an effective way for the improvement of the theoretical understanding of very complex physical processes. It has been applied succesfully to the description of phase transitions, nucleation, and normal grain growth (see e.g.[1]). The basic idea of these CS consists in a simulation of the kinetics of a large ensemble of microscopic particles, whereby the laws for the particle motion must be given by an appropriate model. In order to allow the numerical analysis of large ensembles which are representative for many particle systems, the models must be sufficiently simple, i.e. only crude approximations of the real interactions are considerd. This is why most of these simulations cannot be compared quantitatively with experiments. Nevertheless, these CS contribute to the qualitative understanding of the evolution of collective phenomena like phase transitions.

Contrary to the CS characterized above, we aim at the quantitative understanding of the complex physical processes of thermally induced SOI layer formation. Because these processes depend strongly on various external conditions, basic ingredients of our models are, for instance, temperature dependent rates for nucleation and growth. Thus, we rely rather on well established phenomenological laws than on a simulation on a microscopic scale.

2. Computer simulation of the lateral solid phase epitaxy of thin isolated Si layers

The solid phase transformation of thin amorphous Si films on insulating substrates into single-crystalline layers (SOI) is of growing interest for the production of high speed and ultra-large scale integrated circuits. Because of the low transformation temperature as low as about 600°C, it is of special interest for 3-dimensional integration. An unseeded solid phase transformation of isolated a-Si-layers results in polycrystalline structures which are formed by nucleation and growth of randomly oriented crystallites (primary crystallization PC). This material is of minor interest for circuit production. On the other hand, growth starting from seed openings through the insulating layer to the underlying bulk Si (L-SPE) is a promising technique for SOI-layer production. But L-SPE is limited by the competing process of PC, i.e. the

1) Zentralinstitut fuer Kernforschung Rossendorf, DDR-8051 Dresden, PSF 19, DDR

impingement and the resulting blocking of the L-SPE frontier by the growing poly-Si grains. The main characteristics of the L-SPE is the L-SPE overgrowth length (L-SPE length) from the seed to the region of blocking.

A semiempirical CS has been established for the investigation of the PC [2,3] and the competition between L-SPE and PC [4,5]. This simulation allows a detailed study of the topological and the kinetic features of the L-SPE. Fundamental theoretical considerations as well as applications to experimental conditions have been carried out.

The L-SPE length is determined by several external conditions. From the theoretical point of view, the influence of these conditions on the L-SPE process can be described by relations between three fundamental parameters. Using these parameters it is possible to investigate its behaviour for different external conditions.

The used method of CS combines a subdivision of space and time (similar to the method of finite elements) with a grand canonical Monte-Carlo method (which describes the random nucleation) and a special version of the cellular automata (which governs the growth of crystallites and gives the final grain boundary network).

Some of the results will be summarized:

(i) The competition between thermally activated L-SPE and PC results in a strong scaling behaviour of the topology as well as the kinetics. Thus, the ratio between the mean L-SPE length and the mean grain radius is about 1.3 in any case. For thermally activated processes, this value of the reduced L-SPE length represents an upper limit.

(ii) The probability distribution of the L-SPE length characterizes the roughness of the interface between the monocrystalline and the polycrystalline material and gives a measure for the likelihood that at a certain distance from the seed epitaxially grown material will be found. The probability distribution shows also the scaling behaviour mentioned above.

(iii) The enhancement of the L-SPE length and the shortening of the transformation time by doping, with respect to the undoped case, can be also calculated by our CS. They are in nice agreement with experimental results for B- and P-doped samples. Furthermore, from such experiments it is possible to extract information about the dopand induced variations of the rates for nucleation and growth of the crystalline phase.

(iv) The influence of an enhanced L-SPE growth rate with respect to the poly-Si growth rate, induced by focused ion beams, has been estimated too. A linear dependence of the achieved L-SPE length on the growth enhancement with a slope of 1.25 has been found.

3. Simulation of the melt instability of Si at light heating

Another fundamental aspect of SOI layer formation is the inhomogeneous melting when heating is performed by a radiant flux. Inhomogeneous melting by light heating has been observed in thin, isolated Si-layers [6,7] as well as at the surface of bulk Si [8,9]. Thus, inhomogeneous melting is apparently a quite general process.

The inhomogeneous melting arises naturally when a jump in reflectivity at the semiconductor-metal melting transition implies a negative feedback mechanism. For silicon, the reflectivity of the melt is 0.7 compared to 0.3 of the solid. As long as the temperature of silicon is below the melting temperature T_m, the temperature increases proportional to the incident light flux. Raising the light flux slightly from a steady state just below the melting point will create a small amount of liquid. Since the liquid reflects more light, less power will be absorbed. A further increase of the light flux increases the portion of the molten surface, but not the integral amount of absorbed power. Thus, for a wide range of light intensity, the surface will remain partly solid and partly liquid. Only if the absorbed power is sufficient to melt the surface completely, the temperature will continue to rise with increasing light flux.
Thus, we have a quite unusual situation in the inhomogeneously molten surface layer. Because the solid absorbes more energy than the liquid, there must be overheated solid in coexistence with undercooled liquid. The "slush" consisting of solid and liquid shows the tendency to form well-ordered structures like meanders and lamellae of liquid. The driving forces for this ordering process arise from the tendency to minimize the interfacial energy between the liquid and the solid regions and to form structures, which allow stationary heat flow pattern within the surface layer. The decrease of the interfacial energy is limited by the onset of nucleation in the superheated (supercooled) solid (liquid).
Recently, it has been found that there is a considerable impact of the melting instability of silicon on the formation of SOI layers. For instance, Biegelson [10] has pointed out that the presence of such a mobile, inhomogeneous melt region is one prerequisite for crystal orientation control in SOI layer production. If there are (100) precursor crystallites in the SOI layer, the formation of liquid and solid lamellae leads to (100) texture. It should be noted that non-radiant heating (e.g. with an electron beam) does not produce inhomogeneous melting and does not result in (100) texture.
Egami et al. [11] have observed that the liquid lamellae align with silicon pattern edges of a graphoepitaxy structure. Furthermore, in a lamp zone melting recrystallization process, Dutartre [12] found bursts of "explosive" melting at the leading edge of the molten zone. The "explosive" melting results in inhomogeneous melt regions consisting of molten and solid lamellae. This melting behaviour has a strong impact on the quality of the recrystallized films. And finally, recently Im et al. [13] have observed inhomogeneous melting at the rear edge of the molten zone in zone melting recrystallization experiments. This melt instability has a dramatic influence on the formation of low-angle grain boundaries and protrusions.
These examples prove that a deeper understanding of the inhomogeneous melt region is technological relevant for enhancing control of nucleation and growth in SOI systems.
A two-dimensional CS using the method of cellular automata (CA) has been established to simulate the melting kinetics during light heating [14,15]. The CA consists of 4096 lattice sites with periodical boundary conditions. The sites represents macroscopic blocks of many atoms. The values of the lattice sites are updated in a random sequence. 4096 alternations are called one Monte-Carlo step (mcs). Every site is surrounded by

eight neighbouring sites. Together with the central site, their actual states (liquid or solid) represent a configuration for which a special evolution code has been determined. The code governs the evolution of the configurations. It realizes a short range interaction between solid and liquid regions which is caused by the unusual thermal situation at the illuminated surface (see above). The short range interaction leads to a competition between the minimization of the liquid-solid interfacial energy and the limitation of superheating and supercooling by nucleation, melting and growth, respectively.

Our CS give pattern of the melt morphology, which are quite similar to that observed in experiments. Different experimental situations have been considered (e.g. pulsed and contineous irradiation, isolated thin layers and bulk material). For all cases, the variety of experimentally observed pattern have been obtained without a crucial change of the evolution code. This is a strong argument for the validity of our model.

Our model allows for the first time to study the pattern evolution theoretically. It has been found that after an initial stage, where a random solid-liquid pattern is formed, an ordering process can be observed. Various competing domains of lamellae are formed. One of them is selected by a process similar to the Ostwald ripening. The most important result is the observation of spontaneous symmetry breaking, although only an isotropic short-range interaction has been implied.

In contrast to existing theories, our approach requires no external source for symmetry breaking, but if there is any external anisotropy present, than the spontaneously formed "internal field" will align with the external one.

Furthermore, we can conclude that the formation of lamellae is a quite general process at light heating and, therefore, important for the further improvement of SOI-layer production.

4. CS of grain boundary confinement during zone melting recrystallization

One important method for the SOI layer formation is ZMR. The principal microscopic imperfections in ZMR SOI films are low-angle grain boundaries (or subboundaries). They form a network, which is roughly aligned to the scan direction. Several techniques to eliminate or localize subboundaries (subboundary confinement) have been proposed, but no theoretical investigation has been done systematically.

Therefore, a CS for the study of the subboundary formation and the possibilities of their confinement has been established [16,17]. It is based on a theoretical model of the subboundary formation [18], which has been generalized to include the calculation of subboundary confinement.

The model bases on assumptions mainly extracted from experimental results. Some of them are:

(i) The melt-solid growth interface consists of alternating (111) and $(11\bar{1})$ facets.

(ii) Subboundaries originate from the interior corners of the faceted growth front.

(iii) If a tip of a facet overtakes the melt isotherm, the tip melts and two additional facets are formed.

(iv) The growth rate of each facet is controlled by the nucleation of new monolayers.
Mathematically, the model is described by a set of coupled, nonlinear differential equations for the length of the facets. Especially, they describe also the motion of the interior corners of the solidification frontier, i.e. the subboundary network formation.

To simulate subboundary confinement, a variable undulation (with respect to amplitude, wavelength and phase) of the isotherm is included. Further parameters are the scan velocity as well as the tilt angle between the scan and the <100> in-plane crystal orientation.

Without temperature undulation, no subboundary confinement can be achieved. The undulation has been proven to be a necessary condition for stable subboundary confinement.

Using a temperature undulation we have found that, for the case of scan direction parallel to in-plane <100> orientation, strong confinement occurs only in a narrow range of scan velocities. This range is determined by the approximate equality between the natural distance of subboundaries and the wavelength of temperature undulation, and it should be mentioned that this method of confinement is very sensitive to parameter fluctuations, which are inevitable in experiment.

A more stable version of confinement has been found by our computer experiment: The scan direction should be tilted with respect to the <100> orientation. Then, confinement is achieved even when essential differences between the natural subboundary distance and the temperature undulation wavelength exist. Therefore, subboundary confinement depends not so sensitive on fluctuations of experimental parameters.

REFERENCES

[1] D.J.Srolovitz, M.R.Anderson, P.S.Sahni, and G.S.Grest, Acta metall.32(1984)795
[2] H.-J. Mueller, K.-H. Heinig, and K. Hennig, J. de Physique E-MRS XIV(1987)137
[3] H.-J. Mueller and K.-H. Heinig, Nucl. Instr.& Methods B22(1987)624
[4] H.-J. Mueller and K.-H. Heinig, J. de Physique E-MRS, to be published
[5] H.-J. Mueller and K.-H. Heinig, these proceedings
[6] M.A. Bosch and R.A. Lemons, Phys. Rev. Lett. 47(1981)1151
[7] J.S. Preston, H.M. van Driel, and J.E. Sipe, Phys. Rev. Lett. 58(1987)69
[8] K.-H.Heinig,in:Proc.Int.Conf.EPM84,Dresden,Ed.K.Hennig,ZfK-555,Dresden 1985(p.265)
[9] K.-H. Heinig and H.-D. Geiler, to be published
[10] D.K. Biegelsen, Proc. FED SOI/3D Workshop,March 19-21,1985,Shuzenji(Japan),p.31
[11] K. Egami, M. Kimura, T. Hamaguchi, and N. Endo, MRS Symp.Proc. 35(1985)669
[12] D. Dutartre, Appl. Phys. Lett. 48(1986)350
[13] J.S. Im, C.V. Thompson, and H. Tomita, preprint, Massachus.Inst.of Techn. (1987)
[14] K.-H. Heinig and H.-J. Mueller, J. de Physique E-MRS, to be published
[15] K.-H. Heinig and H.-J. Mueller, these proceedings
[16] K.-H.Heinig, H.-J.Mueller, G.Otto, and K.Hennig, J.de Physique E-MRS XIV(1987)185
[17] K.-H. Heinig and H.-J. Mueller, these proceedings
[18] L. Pfeiffer, S. Paine, G.H. Gilmer, Wim van Saarloos, and K.W. West, Phys. Rev. Lett. 54(1985)1944

PULSED ION IMPLANTATION OF SILICON WITH SELENIUM

G. Serfőző, R. Naujokaitis[1], I. Krafcsik, L. Dózsa[2], G. Battistig
P. Riedl, E. Klopfer, N.N. Gerasimenko[3] and J. Gyulai

ABSTRACT

In a repetitive mode pulsed ion implanter with magnetically insulated diode, selenium was used as dielectric source material. The electrical activity of multiple 250 kV selenium pulses was investigated by Deep Level Transient Spectroscopy /DLTS/.

Pulsed ion implantation using a Marx bank driven, magnetically insulated diode covered with suitable dielectric attracted attention as an alternative of laser annealing and simultaneous doping [1,2]. Experiments also showed certain advantages over laser irradiation, when liquid phase epitaxy of thin films was tried [3]. As a conclusion in [4], presented earlier at this conference it was stated that advantages of large surface treatment can only be of technological value, if improved machine concepts will appear, i.e. the machines, designed originally for plasma physics [1] will be modified to meet demands of semiconductor technology. The use of ion pulses as controlled depth heat treatment and eventually doping is worth as a research target, because the advent of 3-D integration will need controlled depth treatments and ions /pulses/ are certainly one of the contenders to bring solutions.

Other topics for these machines, as was first shown by Adler and Picraux [5], is the surface treatment of metals. Furthermore, the treatment of organics might also be a useful area.

As today's pulsed implanters do not have mass separation to select ion species, beam purity can be an important issue. The Adler-Picraux type machines has advantages at this point, but it is limited to metallic ions. On the other hand, the magnetically insulated diode needs a dielectric as a source material to be ionized.

There are not too many elemental insulators to explore as a source material in these latter type machines. In this paper we present preliminary results, when amorphous selenium was used on the diode of our repetitive mode pulsed ion implanter described earlier [6,7].

Electrical properties of implanted selenium in silicon was investigated at an early date [8] and with the advent of laser annealing studies, numerous papers appeared on the matter. Selenium turned out to be a dopant with relatively high electrical activity, especially, when laser treat-

Central Research Institute for Physics, P.O.Box 49, H-1525 Budapest
[1]Kaunas Polytechnic Institute, Kaunas, LSSR
[2]Inst.Technical Physics, P.O.Box 76, H-1325 Budapest
[3]Inst.Semiconductor Physics, Novosibirsk, USSR
Work partially supported by Hung.Acad.Sci. OTKA grant No.1794

ment was applied and it is associated with a deep level of about 120 meV. It was shown that this level is connected with defect structure of selenium in silicon. However, if high quality selenium doped layers were produced, certain applications seem to be imminent. These fact gave motivation to start work in this direction and preliminary results are now presented.

EXPERIMENTAL

Pulsed ion implantation was made on the repetitive-mode 250 kV machine of CRIP /KFKI/. This machine is still in a stage of development, but most parameters can be found in [7]. Different anode configurations were also tried and vacuum conditions $6x10^{-4}$ Pa/ seem now as limitation. During selenium implantation this problem was even a major issue. Repetition rate chosen in these experiments was about 0.25 Hz, the irradiated surface was about 20 cm^2. One to thousand, 250 kV pulses were repetitively applied to reach the desired dopant concentration. Because of duration and intensity of a single pulse no surface melting occurred, i.e. regrowth was in the solid state regime.

Helium ion backscattering was used to measure dose/pulse or ion retention. The yield was lower than the expected value based upon earlier experiments: for thousand pulses $3x10^{15}$ cm^{-2} selenium was found with a broad distribution from the surface. We believe that the low yield was caused by poor dielectric behavior of amorphous /glassy/ selenium.

Diodes were fabricated with Al evaporation on p-Si on wafers freshly etched /HNO_3+H_3PO_4 at 65°C/, rinsed 20 % HF refreshed and heat-treated for 15 minutes between 150-450°C in N_2. One wafer was annealed at 1000°C for ten minutes.

Before Al-evaporation, HF refreshment was used. To avoid later lithography steps, aluminium was evaporated through a mechanical mask with 0.5 mm openings. Samples were mounted onto TO frames, bonded ultrasonically. After this, I-V, C-V and DLTS measurements were made. The DLTS equipment, developed in Inst.Techn.Phys., was described elsewhere [9].

In this case /p-Si/, diode characteristics were not of high quality /Fig.1/. This is probably caused by the great number of defects after ion implantation. There is a compensated layer or even a pn-junction near the metal-semiconductor interface. This was the limiting factor in the depth scale for DLTS measurement.

An example of depth scale for a single defect line, E_v + 0.2 eV, produced by 100 Se-pulses in p-Si after heat treatment is shown in Fig.2. This electric level agrees well with the published data for divacancy /V_2/ complex [4]. It is obvious that this level reaches far beyond the R_p for 250 keV Se^+. Note also that in this experiment we were not able to detect the near-surface part of the level because of substrate damage due to high dose implantation.

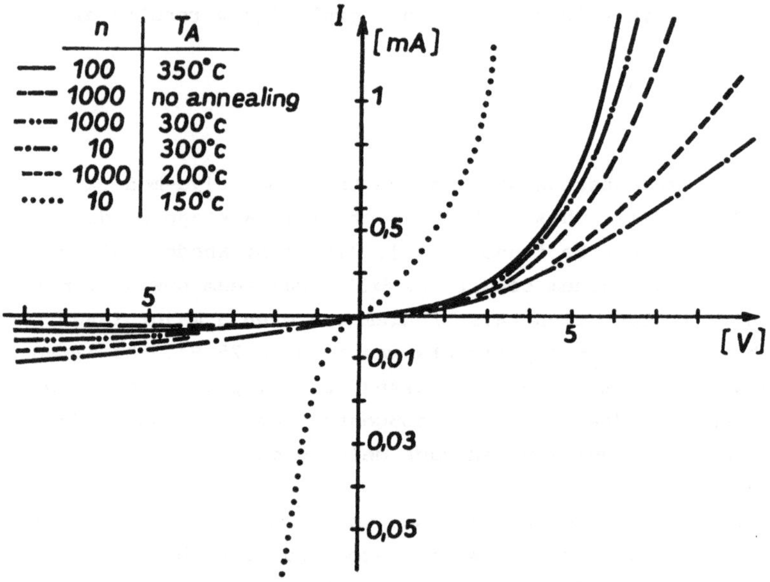

Fig.1. I-V characteristics of pulse implanted /250 kV; Se, n-times/ p-type Si Schottky diodes /15 min annealing at T_A; metallization: Al/

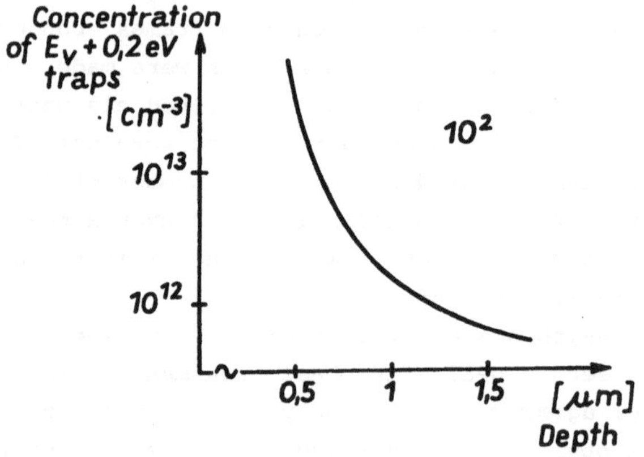

Fig.2. DLTS depth profiling for E_V + 0.2 eV peak of B-doped Si after hundred 250 kV Se ion pulses /15 min heat treatment at 350°C/

Figures 3 and 4 show spectra for heat treated 10^2 and 10^3 pulses taken at different depths. The strongest level reaching from a near-surface point to almost to one micron was an acceptor-like level 0.2 eV above E_v due to divacancies. At greater depths, two more acceptor type levels are visible /0.3 and 0.5 eV/. After measuring the capture cross section and annealing temperatures the exact identification of these defects can be realized probably, now the first one seems to be V-O-C /vacancy-oxygen-carbon/ complexe at this electric level [4].

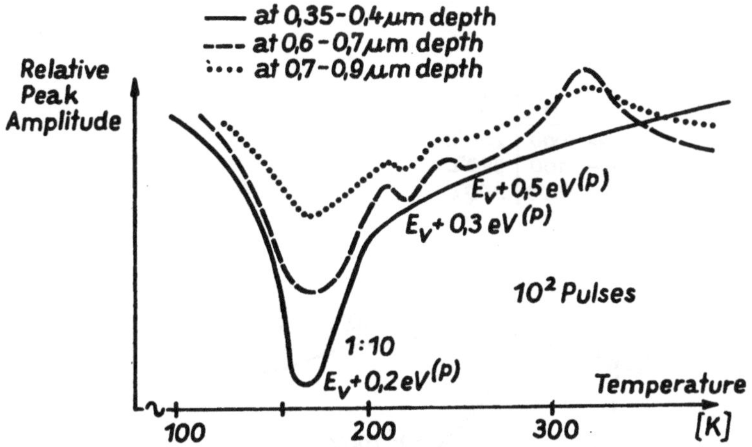

Fig.3. DLTS spectra of 1 ohmcm p-Si after hundred 250 kV Se ion pulses at different depths /DLTS frequency: 260 Hz, single pulse width: 200 μs/ 15 min heat treatment at 350°C

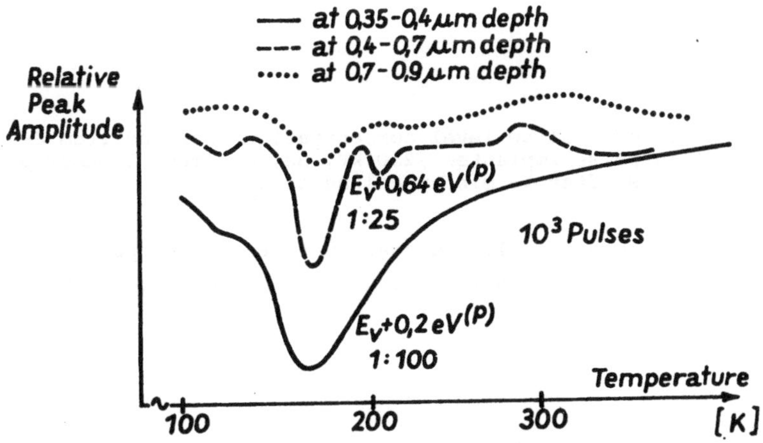

Fig.4. DLTS dpectra of 1 ohmcm p-Si after thousend 250 kV Se ion pulses at different depth /DLTS frequency: 260 Hz, single pulse width: 200 μs/. 15 min heat treatment at 300°C

For 10^3 pulses, the 0.2 eV level persists, but the two distinct levels at medium depths were not found, only the 0.64 eV one. Here, a donor-like level also showed up /0.4 eV/.

Defect profiling for different pulse numbers and heat treatments are shown in Fig.5. Shallow level distribution also was altered by ion implantation. Originally 10^{15} cm^{-3} level concentration has increased to above 10^{16} cm^{-3}]. This increased distribution reaches to greater depth /0.5-1.0 μm/. This concentration has even increased with heat treatment over 300°C. The nature of this defect is not understood.

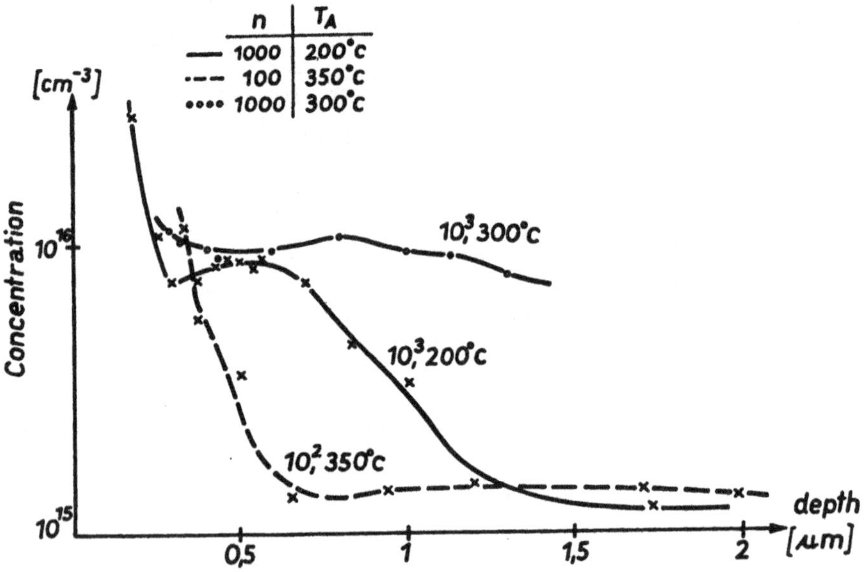

Fig.5. Depth profile of shallow level concentration /derived from C-V measurements/ in pulse implanted /250 kV, Se, n-times/ 1 ohmcm p-Si Schottky diodes after 15 min annealing at T_A

To draw a full picture including near-surface levels, possibly the level directly associated with selenium, another set of experiments is needed, which is underway.

REFERENCES

(1) Hodgson, R.T., J.E.E. Baglin, R. Pal, I.M. Neri and D.A. Hammer: Appl.Phys.Lett. 37 (1980) 187

(2) Gyulai, J., R. Fastow, M. Nastasi and J.W. Mayer: Proc. IUPAP Conf. Energy Pulse Modification of Semiconductors and Related Materials, Ed. K. Hennig, ZfK-555, 1985, p.95

(3) Gyulai, J., R. Fastow, K. Kavanagh, M.O. Thomson, C.I. Palmström, C.A. Hewett and J.W. Mayer: Laser Solid Interactions and Thermal Processing of Materials, Ed. I. Narayan, W.L. Brown and R.A Lemmons, Elsevier Publ.Co., New York 1983, p.455

(4) Mooney, P.M., L.J. Cheng, M. Sueli, Y.D. Gerson and J.W. Corbett: Phys.Rev. B15 (1977) 3836;
Oehrlein, G.S.: Dissertation, 1981. State Univ.of New York at Albany, Dept.of Physics, p.37

(5) Adler, R.I. and S.T. Picraux: Proc.5th Internat.Conf.Ion Implantation Equipment and Techniques, E.D. J.F. Ziegler and R.L. Brown, North Holland, Amsterdam 1985, p.123

(6) Krafcsik, I., L. Királyhidi, P. Riedl and M. Fried: Phys.Stat.Sol. a94 (1986) 855

(7) Krafcsik, I., L. Királyhidi, P. Riedl, M. Fried and J. Gyulai: Nucl. Instr.Meth. B21 (1987) 604

(8) Gyulai, J., O. Meyer, R.D. Pashley and J.W. Meyer: Rad.Effects 7 (1971) 17

(9) Ferenczi, G, P. Horváth, F. Tóth and J. Boda: U.S. Patent No.4437060, 1984;
Phys.Stat.Sol. a94 (1986) K119

RAPID THERMAL PROCESSING FOR VLSI APPLICATION

E. Wieser[1], D.Gerisch[2], and V.Kahlert[2]

1. Introduction

The dominating doping process in modern microelectronic technologies is ion implantation connected with subsequent high temperature annealing. Thereby, the necessity to control carefully the dopant redistribution during annealing at approaching submicrometer dimensions and the trend to single wafer treatment favours rapid thermal processing (RTP) against conventional furnace tempering. If the substitution of furnace tempering by RTP is discussed not only annealing of the implanted structures has to be considered. All high temperature steps following implantation will cause dopant redistribution. In this connection high temperature processes incorporated in the preparation of metallization schemes are of special interest.

Therefore, in this contribution the redistribution and electrical activation of dopants implanted in single crystalline silicon as well as annealing of implanted polysilicon, SILOX reflow and silicide formation as typical steps during the preparation of interconnects and contacts are considered. Concerning dopant activation and redistribution experimental results are discussed with respect to modelling of these processes because process simulation is a very important tool in VLSI technology development. For the optimum choice of the temperature-time cycle of RTP the leakage current of the p-n junctions and structural defects created in the wafer are very important parameters and will be discussed. The applicability of RTP in a complex fabrication technology for LSI devices is demonstrated.

As RTP thermal treatment at maximum temperatures between 1000 and 1200 °C with a dwell time in the order of seconds is considered. Typical temperature rise times are in the same order of magnitude. In this contribution the time t = 1s describes a tempering where the irradiation is finished after approaching T_{max}. The experiments were carried out using a set of halogen lamps.

2. Electrical activation of dopants and transient enhanced diffusion

Due to the temperature dependence of the solubility of boron in Si annealing at T_{max} between 1000 and 1200 °C results in a remarkably higher

1 - Central Institute for Nuclear Research Rossendorf, AdW der DDR, PSF 19, DDR-8051 Dresden
2 - Center for Microelectronic Research Dresden, Kombinat Carl Zeiss Jena, PSF 34, DDR-8080 Dresden

electrical activation compared to conventional furnace annealing at 900 to 1000 °C if the maximum concentration of the implanted profile exceeds 10^{20} cm^{-3}. Fig. 1 compares the equilibrium solubility of boron /1/ with the maximum carrier concentration measured after RTP of 1 x 10^{16} Bcm^{-2} implanted at 30 keV /5/. It is shown that non-equilibrium states with a supersaturation of carriers are formed by RTP. Therefore, the gain of activation by RTP compared to furnace tempering is still higher than estimated from the temperature dependent solubility.

For room temperature implantation of boron the silicon becomes not amorphized by fluences of technical interest. Therefore, the electrical activation proceeds slowly compared to the recrystallization of an amorphous layer. In /3/ the time evolution of the electrical activation during RTP has been investigated. The time constant for activation has been found to be 25 s at 1000 °C and 2 s at 1100 °C. This finding is contrary to our results. We observed complete electrical activation in the temperature range between 1000 and 1100 °C already at t = 1 s if the maximum boron concentration was below the solubility limit (1000 °C - 1 x 10^{15} B cm^{-2}, 1100°C - 3x10^{15} B cm^{-2}, E=30 keV). This means complete activation takes place already during the temperature rise time.

If the implanted layer is amorphized dopant activation proceeds very rapidly during the epitaxial regrowth ($<$ 1 s). We observed for B implantation in preamorphized Si complete activation for 3 x 10^{15} B cm^{-2} ($c_{max} \approx 3 \times 10^{20}$ B/cm^3) already after annealing at 900 °C/1 s /8/. In this case the maximum carrier concentration is a factor 4 above the equilibrium solubility at 900 °C.

After passing of the recrystallization front the formation of electrically inactive clusters or precipitates takes place. Tsai et al. /10/ proposed a model describing very well the deactivation of As in Si for furnace tempering. We have found for 10^{16} As/cm^2 a deactivation of 50 % already after very short annealing of 1 s at 1050 °C. According to the Tsai model only 20 % deactivation are expected. This difference between experiment and model vanishes at t $\gtrsim$10 s.

In order to optimize the temperature-time cycle of RTP an accurate knowledge of the dependence of the diffusion coefficient on temperature and dopant concentration is necessary. If tempering treatments with durations in the order of several seconds are considered also the time becomes an important parameter.

For boron implanted into Si a remarkable transient enhanced diffusion compared to conventional long time annealing has been observed in several investigations. This effect can be described by multiplication of the steady state diffusion coefficient, e.g. the expression used in SUPREM II with an average enhancement factor $\bar{f}$:

$$D = \bar{f} \cdot D_o(1 + \beta p/n_i)(1 + \beta)^{-1} \exp(-E_a/kT) \qquad (1)$$

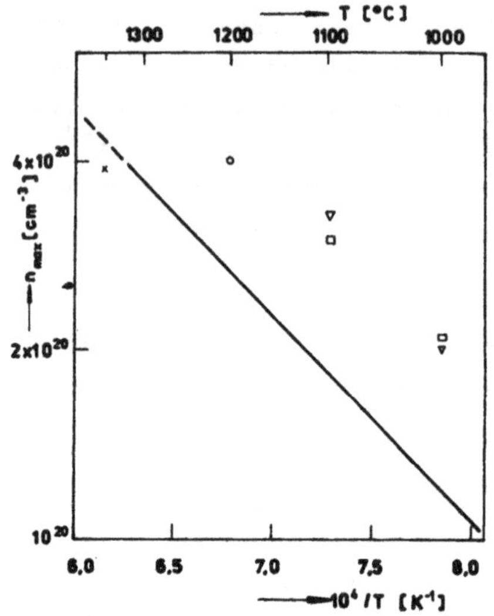

Fig.1: Dependence of the carrier concentration on the RTP parameters (t: ▽ - 20s, □ - 5s, ○ - 1s) full line equilibrium solubility /1/

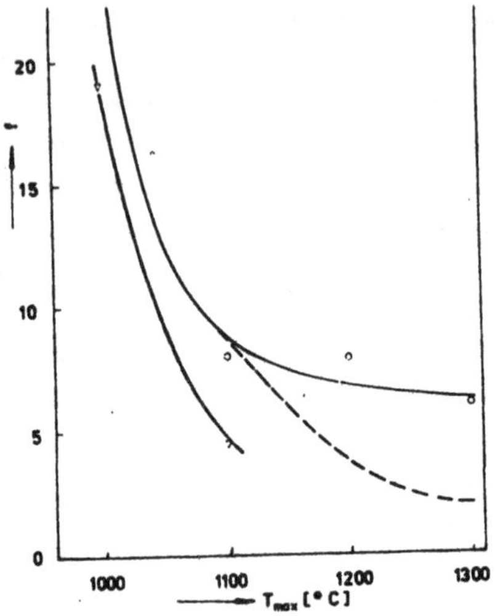

Fig.2: Diffusion enhancement f in dependence on the temperature (t: ▽ - 20s, ○ - 5s), broken line behaviour according to /4/

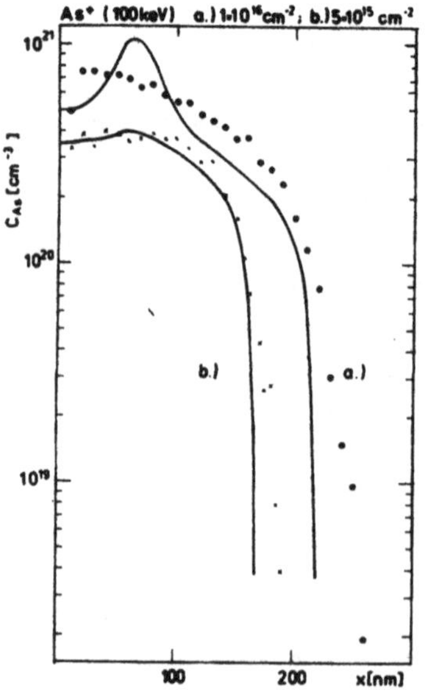

Fig.3: As depth profiles
a) 1 x 10^{16} cm^{-2}, 1100 °C/10s,
b) 5 x 10^{15} cm^{-2}, 1180 °C/1s,
full line - computer simulation

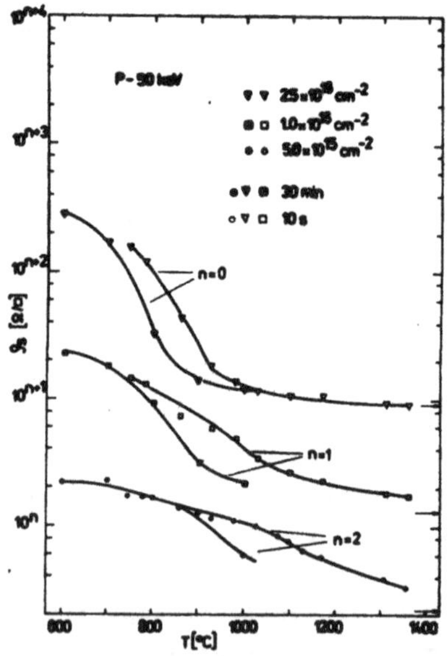

Fig.4: Sheet resistance of P implanted poly-Si (d=350 nm) in dependence on the annealing temperature

(D_o = 3.33 . $10^9 \mu m^2$/min, β = 3, E_a = 3.43 eV, p = hole concentration, n_i = intrinsic carrier concentration). $\bar{f}$ is chosen in such a way that the computer simulation fits the experimental profile /3, 5/. Therefore, $\bar{f}$ represents the effective diffusion enhancement for a given temperature-time cycle. Fig. 2 demonstrates the decrease of $\bar{f}$ with increasing annealing time and temperature. Fair /4/ proposed the following expression in order describe the temperature and time dependence of f:

$$f(t,T) = 1 + 10^{-6}\exp(2eV/kT)\exp(-t/4.4s) \qquad (2)$$

The behaviour of theoretical $\bar{f}$ values calculated by integration of (2) over corresponding temperature-time cycles is also shown in Fig. 2. The agreement with the experiment is good within the temperature range of 1000 to 1100 °C. The saturation behaviour of $\bar{f}$ at higher temperatures is not obtained with the Fair model. From a large number of computer simulations we come to the conclusion that a diffusion coefficient obtained by multiplication of equation (1) and (2) works very well for annealing at 1000 to 1100 °C and 5 to 30 s if the maximum concentration of the implanted profile is between 5 x 10^{20} and 10^{21} B/cm^3.

The existence of a transient enhanced diffusion during RTP of As implanted Si is considered controversely. Only a weak diffusion enhancement should be expected due to the relatively perfect lattice structure resulting from the epitaxial regrowth of the amorphized layer. In /10/ also a diffusion model for As is proposed. The diffusion coefficient has the same structure as equation (1). It takes into account diffusion by neutral and singly charged vacancies. The parameters for As are D_o = 22.9 cm^2/s, β = 100, and E_a = 4.1 eV. The electrically inactive As is assumed as immobile. The diffusion enhancement can be taken into account in the same way as for B by multiplication with a constant enhancement factor $\bar{f}$ or with the time dependent expression proposed by Fair /4/ (analogous to (2)). Typical enhancement factors are $\bar{f}$ = 5 for 1100 °C/5s and $\bar{f}$ = 3 for 1100 °C/20s. The decrease with increasing t shows the transient behaviour.

In /7/ a diffusion coefficient is proposed which describes a transient enhanced diffusion without an additional factor considering transient point defect supersaturation. It has the form

$$D = [0.066\exp(-3.44\ eV/kT) + 12\exp(-4.05\ eV/kT)n/n_i + 125\exp(-4.6\ eV/kT)(n/n_i)^2]\ cm^2s^{-1} \qquad (3)$$

(n - electron concentration, n_i - intrinsic carrier concentration)

The third term in equation (3) takes into account the diffusion by two-fold charged vacancies. Due to its n^2 dependence this term decreases rapidly with increasing deactivation in the very early stage of annealing. A further decrease is caused by the decreasing carrier concentration connected with the increasing diffusional broadening of the profile.

Therefore, it results in a higher effective diffusion coefficient only for a short time. Fig. 3 shows the computer fit of two experimental distributions. In this simulation the Tsai model was used, but the diffusion coefficient (1) was replaced by (3). The fit is quite good if we consider the limited experimental resolution.

3. Formation of interconnects and contacts

In MOS metallization schemes the combination of a refractory-metal silicide with a polysilicon layer to a polycide structure is now widely used. A high temperature treatment is necessary for annealing of the implanted poly-Si and for silicide formation.

A nearly homogeneous distribution of the implanted dopants in the poly-Si layer is desired in order to ensure high quality contacts. Due to grain boundary diffusion this can be achieved by annealing times in the order of 10s using temperatures >1000 °C for As and B and >900 °C for P (poly-Si thickness 200-300 nm).

Fig. 4 compares the temperature dependence of the sheet resistance of P-implanted poly-Si (d=350 nm) for furnace annealing and RTP. The resistance obtained by RTP at 1100 °C corresponds to the result of a 1000 °C furnace tempering. For the B-implanted polycide structures investigated in /11/ a decrease of the poly-Si resistance from 3300µΩcm to 580µΩcm was found increasing the annealing temperature from 1000 to 1100 °C.

Fig. 5 shows the dependence on the RTP temperature (t = 15s) of the resistivity of a $MoSi_2$ layer deposited by cosputtering on poly-Si. A resistivity below 100µΩcm, comparable to the results of furnace annealing in pure N_2, has been measured after RTP at 1050 to 1100 °C in air.

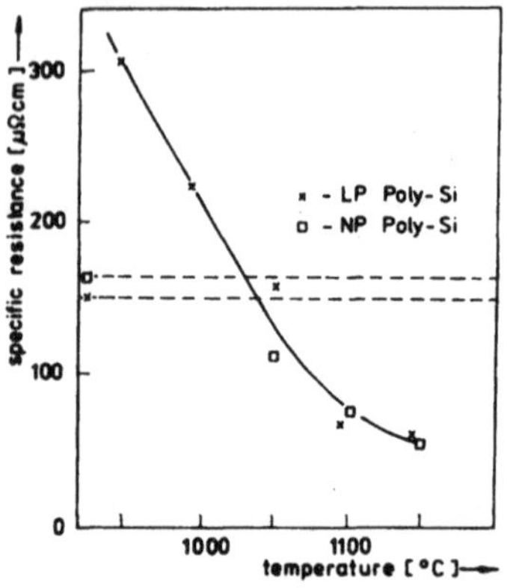

Fig.5 Dependence of the resistance of $MoSi_2$ on the RTP temperature (t = 15s)

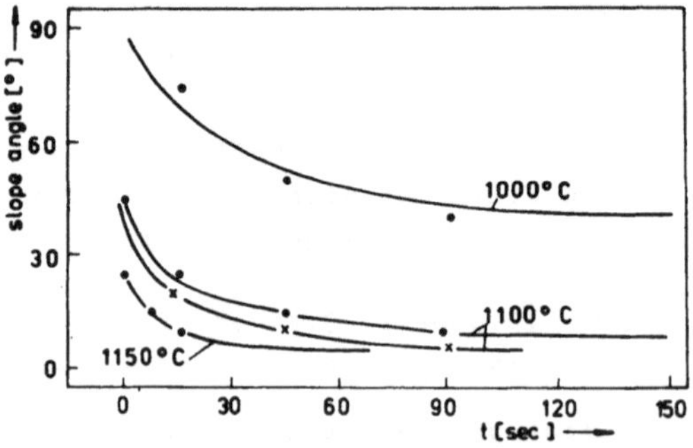

Fig.6 Slope angle of PSG(20 vol % PH_3) and BPSG (15/15 vol % B_2H_6/PH_3)on the RTP parameters (- PSG, - BPSG)

The mechanical stability of polycide structures is more critical in the case of RTP compared to conventional tempering /9/. Considering the influence of dopants As is the most disadvantageous one. Methods to improve the stability are the use of non-stoichiometric, silicon rich silicides (e.g. $MoSi_{2.75}$) or the in situ deposition of a very clean Si film between silicide and poly-Si.

Another important process is the reflow of deposited insulating layers like silicate glasses doped with P or B and P (PSG, BPSG). Fig.6 shows the dependence of the slope angle of PSG and BPSG layers over a vertical step on the RTP conditions. For an annealing temperature of 1100 °C a slope angle below 30 degree is easy to obtain while at 1000°C about 50 degree are the minimum.

4. Results on devices

A very important parameter for the choice of optimum RTP conditions is the reverse bias current I_r of the p-n junction. I_r is strongly influenced by residual defects within the depletion layer of the junction. For As implanted Si defect free annealing by RTP is only possible for $T_{max} \geq 1200$ °C /2/. Therefore, it is necessary to shift the junction depth x_j beneath the defect band in the depth of the original interface amorphous/crystalline (about 150 nm for As, 80 keV) by diffusion.

Table 1: Dependence of I_r /nA/cm^2/ on the RTP parameters (diode area 0.116 mm^2)

	1000 °C/20s	1050 °C/20s	1100 °C/20s
As: 6 x 10^{15}cm^{-2} (80 keV)	0.32 ± 0.06	0.37 ± 0.06	0.49 ± 0.08
B: 3 x 10^{15}cm^{-2} (40 keV)	0.31 ± 0.05	0.34 ± 0.09	0.45 ± 0.08
BF_2: 3 x 10^{15}cm^{-2} (160 keV)	---	2.8 ± 0.20	1.7 ± 0.15

This is achieved for As implanted layers already by annealing at 1000 to 1050 °C. Table 1 shows that also for B implanted p^+ structures excellent I_r values are obtained by such a RTP treatment. BF_2 implantation up to the same depth results in a remarkably higher leakage current. This can be explained by F trapped on defects within the depletion region /6/. In this case I_r can be reduced by a lower implantation energy (e.g. 90 keV), by furnace postannealing at 850 °C,or according to /6/ by preannealing at 600 °C/30 min.

The increase of I_r at 1100 °C found for As and B in Table 1 becomes steeper at 1150 °C. It can be explained by defects formed in the bulk Si due to thermal stress. By Lang topography an increasing extent of dislocations starting from the wafer edge could be detected.

Table 2: Comparison of 16k s RAM parameters for RTP and standard technology

		standard technol.	RTP (1100 °C/10s)
S/D junction depth	n^+:	500 nm	480 nm
	p^+:	650 nm	500 nm
S/D sheet resistance	n^+:	23 $\Omega_\square$	23 $\Omega_\square$
	p^+:	74 $\Omega_\square$	73 $\Omega_\square$
resistance of poly-	n^+:	50 $\Omega_\square$	51 $\Omega_\square$
Si interconnects	p^+:	124 $\Omega_\square$	82 $\Omega_\square$
slope angle (SILOX 1)	:	30°	40°

The applicability of RTP in a complex fabrication technology has been tested on example of a CMOS 16k s RAM and a n SGT 64k d RAM. Source/drain annealing and SILOX reflow tempering were substituted by a single RTP treatment. The efficiency was the same as for the standard technology. Table 2 compares important parameters of the 16k s RAM. Remarkable is the lower resistance of the poly-Si interconnects after RTP. Also the access time is slightly lower although the effective channel length was larger due to the small lateral diffusion (e.g. 64k d RAM: gate length 2.2µm, effective channel length 2.1 µm). For the 64k d RAM the junction depth has been reduced from 550 to 260 nm by RTP.

References:

/1/ Armigliato, A., Nobili,D., Ostoja,P., Servidori,M., and Solmi,S., in: Semiconductor Silicon 1977, Ed. H.R.Huff and E.Sirtl, Electrochem.Soc.Inc., Princeton 1977, p. 638

/2/ Baither, D., Kögler,R., Panknin,D., and Wieser,E., phys.stat.sol. (a) 94 (1986) 767

/3/ Cowern,N.B.E., et al., MRS Symp. Proc. 52 (1986) 65

/4/ Fair, R.B., MRS Symp. Proc. 35 (1985) 381

/5/ Guimaraes, S., Kögler,R., Landi,E., Solmi,S., and Wieser,E., phys.stat.sol. (a) 93 (1986) 543

/6/ Hill, C., Nucl.Instr.Meth. B 19/20 (1987) 348

/7/ Hoyt, J.L. and Gibbons,J.F., MRS Symp.Proc. 52 (1986) 15

/8/ Kögler, R., et al., in this proceedings

/9/ Portius, R., et al., phys.stat.sol. (a) 100 (1987) 199

/10/Tsai, M.Y., et al., J.Appl.Phys. 51 (1980) 3230

/11/Wieser,E., et al., in this proceedings

SILICON ON INSULATOR STRUCTURES FOR VLSI FORMED BY ION BEAM SYNTHESIS

P.L.F. Hemment[1] and K.J. Reeson[1]

Abstract: The formation of SIMOX and SIMNI substrates for microelectronics is described. Recent device results, highlighting the relative merits of the two technologies, are reported, together with speculation about future developments.

1. INTRODUCTION

Ion beams are playing an increasingly important role in many areas of materials science and technology. the availability of particle accelerators for materials studies during the early 1960's led to the successful development of Ion Implantation as a means of achieving chemical doping of semiconductors. This has had a major impact on the development of microelectronics. Initially, ion implantation was seen only to be a novel, nonequilibrium method for the introduction of impurity atoms into a solid. Subsequent work, during the late 1960's and early 1970's, established the real strengths of the technology, which are precision in the positioning of impurity atom profiles (both lateral and vertical) and excellent control and reproducibility, these being inherent features of all ion beam processes. The principal benefits which have resulted are greatly improved yields of working circuits and reduced unit costs.

In this paper the use of synthesised silicon on insulator structures for application to VLSI circuits will be described. Interest in the generic technology of "Ion Beam Synthesis" (IBS) has been stimulated by the desire to develop a viable SOI technology and recent research has highlighted the versatility of the technique. Examples of novel dielectric structures and MOS devices will be given.

2. HISTORICAL BACKGROUND

The synthesis of SiO_2 by the implantation of high doses of O^+ ions into silicon was first reported by Watanabe and Tooi(1) in 1966. In the same year Pavlov et al(2) used N^+ ions to study nitride formation and this was followed by much activity during the next few years, which included studies of buried layers formed by implantation of 2 MeV O^+, N^+ and C^+ ions(3). Much later, similar (MeV) experiments were repeated by Hubler et al(4) to form multilayer optical elements. At Harwell, Freeman and co-workers(5)(6) used ions of 40 keV energy to study the

----

[1]Department of Electronic and Electrical Engineering, University of Surrey, Guildford, Surrey GU2 5XH, England

synthesis of surface oxides, nitrides and carbides. In 1973, Dexter et al(7) formed buried nitride layers by implanting 150 keV ions and, after annealing at 1200°C, successfully deposited a layer of epitaxial silicon. In common with many of the early experiments, compound synthesis was confirmed by infra-red transmission measurements. During the mid-1970's, further work was reported(8)(9) and, in 1977, it was first proposed by Badawi and Anand(10) that buried oxide layers might be suitable for microelectronics applications. In the following year Izumi and co-workers(11) reported the first successful fabrication of MOS transistors in an epitaxial layer grown on the silicon overlayer above the buried oxide. This group proposed the acronym "SIMOX", with other workers later proposing "SIMNI" and "SIMCA" for the N/Si and C/Si analogies, respectively.

Many groups have successfully fabricated small geometry CMOS circuits on SIMOX substrates and the technology has now reached a sufficient maturity that recipes are available for the preparation of high quality substrates(12), which are able to support circuits exhibiting all the benefits predicted for SOI. In contrast, the adverse physical properties of Si_3N_4 which are discussed in Section 4, have outweighed the strengths of SOI/SIMNI technology and have discouraged the general application of these substrates. At the present time, despite the limited supply of material, SIMOX is the front running SOI technology, holding great promise as the substrate for future generations of VLSI circuits and, arguably, bipolar and BIMOS circuits, also.

3. FORMATION OF BURIED LAYERS OF OXIDE AND NITRIDE

The synthesis of dielectric layers in silicon has been reviewed by a number of workers(13)(14)(15). Here we will concentrate upon recent experimental results and draw attention to the differences between oxide and nitride formation.

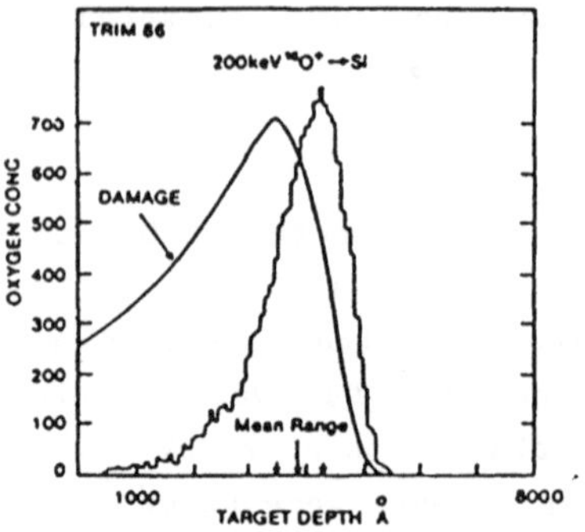

Figure 1. Computed damage and atomic depth profiles for 200 keV O^+ into silicon(16).

The instantaneous depth profile of implanted ions may be adequately described by the projected range and moments of the distribution(16). Figure 1 shows damage and atomic profiles for 200 keV O^+ in silicon. The former is representative of the distribution of point defects and lies closer to the surface. In this case the oxygen (nitrogen) profile is clearly buried beneath a surace layer of silicon. Thus the synthesis of a buried layer may be viewed as an "internal" chemical process, in which there are two moving interfaces (corresponding to the upper and lower wings of the profile). The movement of these interfaces may be expected to be different, due to the asymmetry of the profiles and the absence of registration of the two distributions. Additionally, the manner in which the excess oxygen (or nitrogen) is incorporated into the (damaged) silicon matrix will influence their movement. Diffusion and segregation of the impurity will control the evolution of the buried layer, particularly during the final stage of the implantation.

Several models of the evolution of the oxygen profile in SIMOX material have been developed. Those due to Maydell-Ondrusz(17) and Jager(18), based upon a finite element simulation and a diffusional equation respectively, provide reasonable fits to experimental values, although discrepancies in the fit do exist.

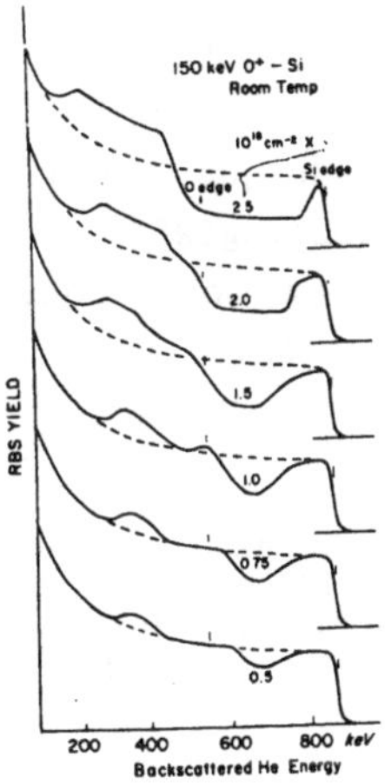

Figure 2. RBS spectra of 1.5 MeV $He^{=}$ ions from (100) silicon samples implanted at room temperature with 150 keV O^{+} ions, over the dose range $0.5x10^{18}$ to $2.5x 10^{18}$ cm^{-2} (from Namavar et al.(22)).

It is well known that Si_3N_4 is an excellent diffusion barrier and the low self-diffusion of nitrogen in Si_3N_4 causes a build up of excess nitrogen in the synthesised compound. Experiments by Meekison et al.(19) and Chang et al.(20) show that at high doses a porous layer forms, which is presumed to consist of nitrogen gas bubbles. These introduce stress, which during subsequent thermal processing can cause delamination of SIMNI material. In contrast, oxygen is able to redistribute (as O_2) during implantation, to give rise to the flat topped oxygen profiles reported by many groups(21) (Figure 2)(22). Figure 3 shows a schematic of SiO_2 formation at 500°C with the oxide beginning to form at the critical dose (ϕ_C = 1.4 x 10^{18} cm^{-2})(23). With increasing dose, the excess oxygen diffuses to the upper interface, which grows preferentially. Excess oxygen in the silicon overlayer is able to precipitate out into the second phase (SiO_2). Evidence for segregation exists for both the O/Si(24) and N/Si(25) systems. Figure 4

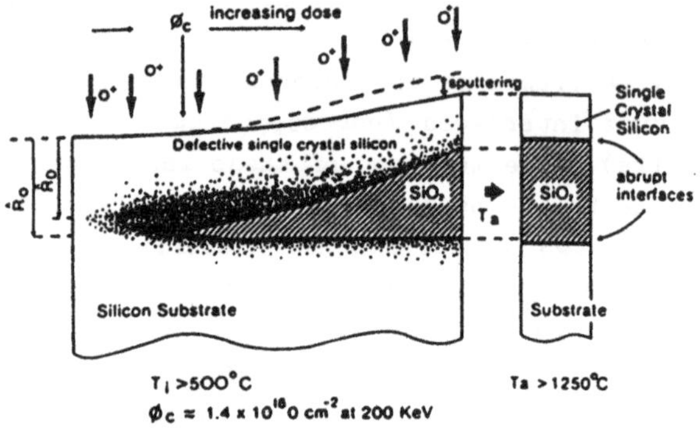

Figure 3. Evolution of SiO_2 layer. Hatched region is SiO_2, centred initially on peak of the oxygen distribution. Annealing causes segregation of oxygen to the buried layer(58).

shows a micrograph of an ordered "superlattice", of cubic symmetry, consisting of 20 Å oxide precipitates with a periodicity of 50 Å(26). Nitogen rich islands in the overlayer above a buried nitride have been reported by Nesbit(27). The presence of these structures in "as implanted" material seems to be important if a high quality silicon overlayer is to be formed during subsequent annealing(28). Their presence is critically dependent upon the processing conditions, in particular good temperature stability during implantation.

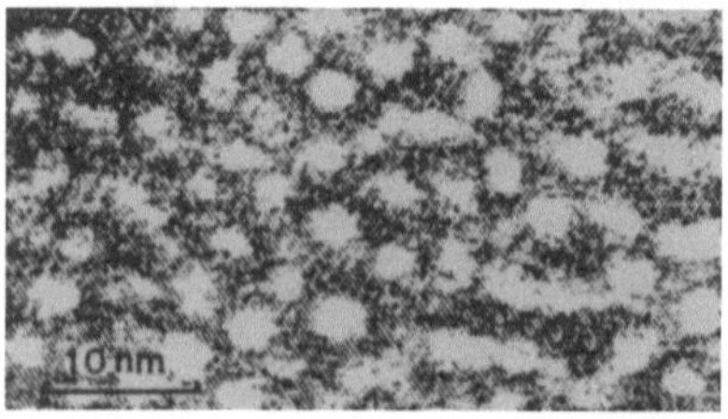

Figure 4. Cross section TEM micrograph, showing periodic array of amorphous zones. O^+ implanted silicon, no anneal (van Ommen and Viegers(26)).

Redistribution of oxygen (nitrogen) in the silicon overlayer may be achieved during a post-implantation anneal. However the controlling processes appear to be different in the two systems. Oxygen is found to redistribute by a process of diffusion controlled precipitate growth, in a manner which is similar to the precipitation of oxygen in bulk CZ silicon(30). During high temperature (>1250°C) and long time anneals, the oxygen is able to segregate onto the buried oxide layer, whilst defects, with the exception of threading dislocations, are annihilated (31). Figure 5, due to Marsh(32), shows the evolution of a typical SIMOX structure with increasing temperature. The oxide layer is amorphous and continuous but usually contains silicon islands(28) ("X"), the presence of which depends upon the processing conditions. Recently van Ommen(28) has reported that much improved SIMOX material, with $<10^5$ dislocations/cm², is formed during high temperature annealing of substrates which contain the oxide superlattice in the "as implanted" overlayer. He proposes that, as silicon interstitials are ejected during precipitate growth, these defects must play an important role in the evolution of the final microstructure. Such ideas are consistent with new descriptions of the mechanism for oxide growth in bulk silicon.

In contrast, buried nitride layers prefer to recrystallize as polycrystalline α-Si_3N_4(19)(37) during high temperature annealing. In SIMNI substrates the recrystallization occurs in two steps, firstly the nucleation and growth of a spherulitic structure consisting of α-Si_3N_4 grains and secondly, the cellular structure grows into the silicon overlayer by consuming excess nitrogen and at the same time traps silicon on the cell walls(20). The nitride layer is found to be a leaky dielectric(34), due to conduction along grain boundaries, and attempts to inhibit recrystallization(35) and to re-amorphize(36) the layer have only been partially successful. The presence in the silicon overlayer of excess nitrogen and anomalous electrical activity above the solid solubility level ($\approx 10^{15}$ cm^{-2}), has been reported.

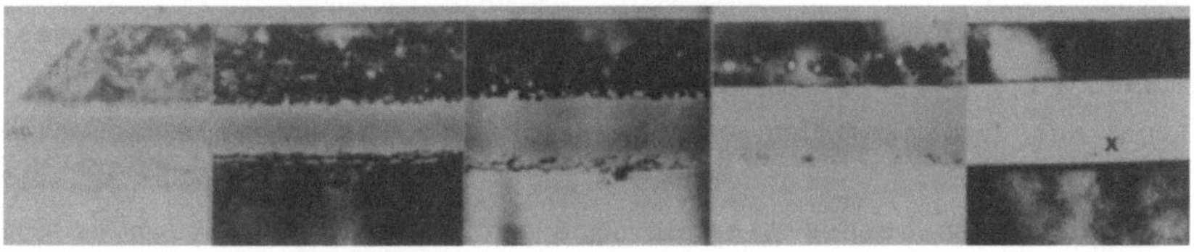

(a) (b) (c) (d) (e)

Figure 5. Cross sectional TEM micrographs of SIMOX samples annealed for two hours at a) no anneal, b) 1150°C, c) 1200°C, d) 1250°C and e) 1405°C for 30 min (Marsh(32)).

4. SIMOX AND SIMNI TECHNOLOGIES

The preparation of substrates is a two stage process consisting of (i) implantation and (ii) high temperature anneal with a third stage (epitaxial growth) being optional, depending upon the device application and implantation energy used in stage (i). For CMOS circuits, the SOI substrates require a silicon overlayer of thickness 0.1 to 0.3µm above a buried dielectric layer of thickness about 0.5µm, sufficient to provide a high breakdown voltage, low leakage and low capacitance coupling between the overlayer and substrate. These structures may be formed by implantating >10^{18} ions cm^{-2} at energies typically 150 to 200 keV. Alternatively, a lower implantation energy may be used, say 80 keV(38) when the silicon overlayer is thickened by epitaxy to the optimum value for the particular CMOS or bipolar device. In order to maintain the crystallinity of the silicon overlayer during implantation, it is necessary to hold the wafer at an elevated temperature, typically 500°C(39).

5. DEVICE RESULTS

The pioneering experiments from which SOI/SIMOX technology developed were carried out by Izumi et al.(11)(1978). The substrates were annealed at 1150°C for two hours and fabricated MOS devices exhibited a leakage current of 1×10^{-12}A at 10V. Nineteen stage CMOS ring oscillators, with 5µm design rules, were fabricated and a propagation delay of 960pS/stage and power dissipation of 406mW were recorded. Subsequently, in 1983, Zimmer and Vogt(40) reported the operation of 1µm PMOS and 2µm NMOS transistors in SIMNI substrates, with the devices fabricated in the silicon overlayer without epitaxy. The substrates were annealed at 1200°C for 5 hours to achieve square nitrogen depth profiles (segregation). Mobilities, subthreshold behaviour and leakage currents were nearly the same as in bulk CMOS devices. Ring oscillators showed a delay time of 230pS. Many groups have now demonstrated the device worthiness of SIMOX substrates(41)(13)(42). An example of one of the more complex circuits is a 4k SRAM with a minimum geometry of 2.5µm, reported by Chen et al.(43). Recessed LOCOS was used to provide lateral isolation. Mobility values of 90% of bulk silicon and leakage currents less than 0.1pA/µm were reported. Circuit performance was recorded as delay times of 490nS for ring oscillators, whilst an access time of 55nS and power dissipation of 85mW were measured for the SRAM.

New device applications for SIMOX substrates are actively being researched. Recently MacIver and Jain(44) have reported the fabrication of low voltage power switches (j-MOS) in a 2.2µm thick n-type epitaxial film deposited at 1150°C. The electron mobility and minority carrier lifetime were measured and found to be 910 cm²/Vs and 3µm, respectively. The integration of CMOS logic and COMFET devices (lateral conductivity modulated FET) has been reported by Colinge and Chiang(45). In this case, the SIMOX substrates were annealed at 1125°C in a N_2/1%O_2 ambient and the devices were fabricated directly in the 230Å silicon overlayer. Techniques for realizing hybrid bulk/SOI circuits have been explored by Matsumoto et al.(38) and they report nMOS/bulk and pMOS/SOI structures which have been implemented on the same chip. The bulk devices were fabricated in epitaxial silicon, which was selectively grown in areas where both the overlayer and the buried oxide had been removed. Gate lengths were 2µm and the subthreshold slope and carrier mobilities were almost the same as those of bulk control devices. Leakage currents of 0.002pA/µm and 0.03pA/µm were measured for the n-and p-channel transistors, respectively.

The properties of the electric field shielding (EFS) layer, which can be formed in SIMOX substrates, have been exploited by Nakashuma et al.(47) for the fabrication of high voltage CMOS devices. A source-drain breakdown voltage of 180V was reported and at least a factor of two reduction in the chip size and power dissipation was achieved. More novel applications of SIMOX include vertical npn transistors(59) and the formation of local (1μm) SOI areas by the implantation through windows in deposited metal (W)(48) and Si_3N_4(49) masks. The latter process is the basis for a self aligned oxygen-implanted SOI technology (SALOX-SOI), in which the active channel region is connected to the bulk substrate.

The devices referred to above have been fabricated in substrates annealed at modest temperatures, typically, 1150°C, when only partial redistribution of the implanted oxygen occurs. These substrates consisted of a thin oxygen denuded surface layer (1000-1500Å), with good transport properties and lifetime, located above a thicker (2000Å), heavily disorded region, adjacent to the buried oxide (see Figure 5). Nevertheless, the reported device performances have been good and the circuits exhibited the predicted benefits of SOI (freedom from latch-up, low leakage, improved speed and simplicity of processing). As a consequence, commitments have been made for the long term exploitation of SIMOX technology and companies, notably Eaton-Nova (50) in the USA, have undertaken the development of high current particle accelerators, dedicated to the production of substrates at commercially attractive rates.

The introduction of higher anneal temperatures (1250°C to 1405°C), during the last two years, has enabled the complete segregation of the implanted oxygen to be achieved(31) (cf. SIMNI material annealed at 1200°C), leading to the realization of near ideal, planar SOI structures with atomically abrupt interfaces (see Figure 5). Mao et al.(51) have reported the improvement in carrier transport properties with increasing temperature up to 1275°C, (see Figure 6). Davis et al. have fabricated small geometry (1μm gates) CMOS transistors in material annealed at

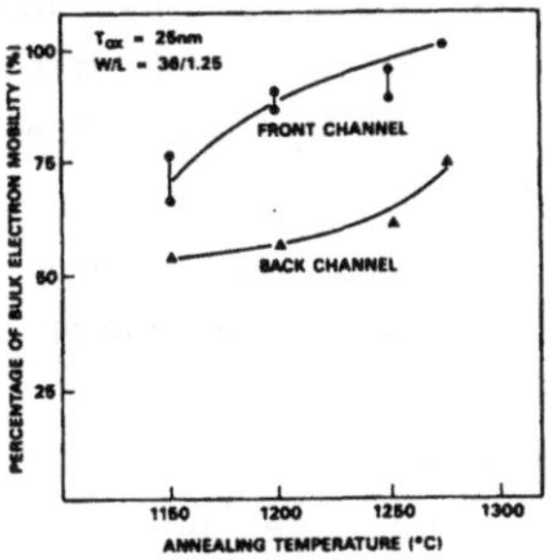

Figure 6. Normalized front and back channel electron mobilities of carriers in the overlayer of a SIMOX substrate as a function of post implant anneal temperature (Mao et al.(51)).

1300°C. In the latter work, low field inversion mobilities of 580 cm^2/Vs for electrons and 205 cm^2/Vs for holes were obtained for the n- and p-channel transistors with optimum implantation doses of boron. These values are within a few percent of those obtained in bulk transistors with similar doping levels. Drain junction leakages with the transistors off and $|V_D| = 1V$ were typically below 5×10^{14} A/μm for both transistor types. In most cases the subthreshold slope was less than 100mV/decade. The output characteristics are shown in Figure 7. The "kink" effect(51)(13), seen in substrates annealed at lower temperatures, is absent, due to the full depletion of the thin, high quality silicon overlayer, which inhibits the formation of a neutral region. Additionally, anomalous lateral diffusion of the source/drain depants, which is a problem in SOS materials, did not occur, due to the absence of major crystallographic defects (other than threading dislocations).

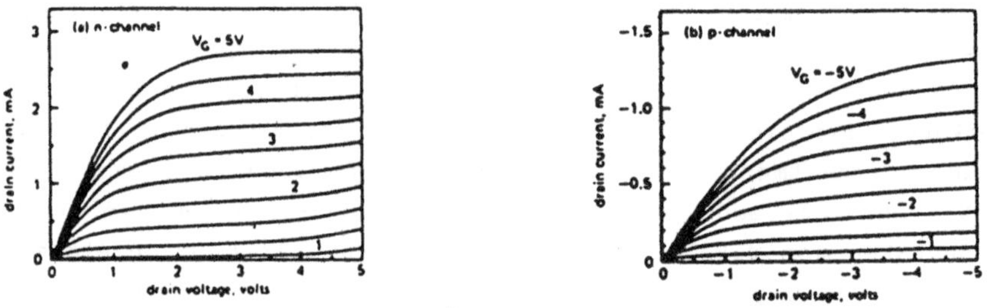

Figure 7. Output characteristics for short channel CMOS transistors. Gate length = 1μm, channel width = 8μm. (a) n-channel, (b) p-channel (Davis et al.(41)).

Experiments to fabricate devices on SIMNI material are relatively few, despite the early successes reported by Zimmer and Voit(40), in 1983. However, vertical difused npn transistors with base widths of about 1μm have been reported(52). The substrates were formed by implanting a dose of 1×10^{18} N^+ cm^{-2} at 180 keV and annealing at 1150°C for 2 hours. Epitaxy was used to achieve a 5μm thick silicon overlayer. Leakage currents of 5×10^{-9} Å/mm were a factor of two higher than in bulk silicon. Emitter current gains of up to 100 and emitter-collector breakdown of 35V were similar to bulk silicon devices. The good quality of the epitaxial layers over a buried nitride has been confirmed by Skorupa et al.(53), who fabricated 500μm MOS capacitors and recorded generation lifetimes of the minority carriers in the range 20-200μS.

6. THE FUTURE ROLE FOR ION BEAM SYNTHESIS

Recent studies of the formation and use of SIMOX and SIMNI materials have led to a number of exciting new developments. These relate to new applications for the generic technology of ion beam synthesis of compounds and the recognition that improved device performance, and possibly new devices, may be achieved with SOI materials.

6.1 Materials

Lateral isolation of MOS device islands by the selective implantation of O^+ ions, in those regions where a thick field oxide is required, has been proposed by Ratnam and Salam(33). The presence of damage and excess oxygen is then utilised to accelerate the thermal oxidation of the defined areas. These authors report the implantation of doses of up to 1.5×10^{18} cm^{-2} at an energy of 150 keV and, after thermal oxidation, they measure an oxide thickness of 1µm, which is a factor of two thicker than in bulk silicon. The "bird's beak" effect is reduced by a factor of five compared to a LOCOS oxide. The implanted oxide exhibits a much higher threshold voltage than the LOCOS oxide and lower leakage. Control of the "bird's beak" has also been reported by Berruyer and Bruel(29), who implanted a dose of 10^{17} N^+ cm^{-2}. Selected area, high dose implantation of 185 keV nitrogen has been proposed by Kenyan(55) as a means of forming high resistivity islands in the vicinity of the source-drain regions in 3D (stacked) polysilicon joint-gate SOI/MOSFETS. The implantation is used to convert regions of the polysilicon gate (0.001Ωcm) into a nitrogen rich, high resistance (>500Ωcm) layer. Stacked devices were successfully fabricated but lack of process optimisation inhibited good performance.

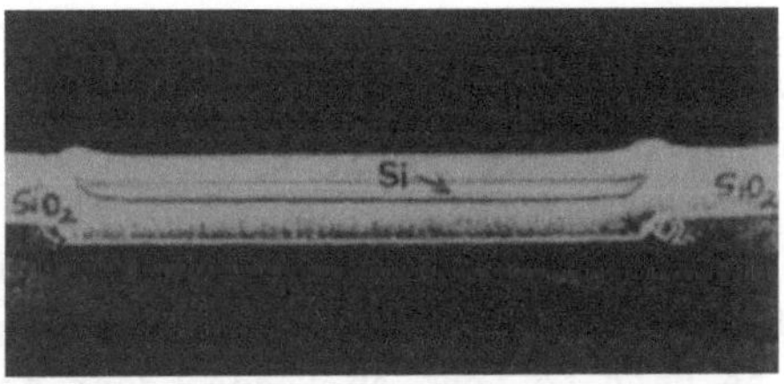

Figure 8. Cross section SEM micrograph of a sample containing a silicon device island, which is totally isolated from the substrate by a synthesised layer of SiO_2 (Davis et al.(54)).

Total dielectric isolation (both vertical and horizontal) of device islands by the synthesis of a continuous but non-planar layer of SiO_2, has been domonstrated by Davis et al.(54) as an extension of SOI/SIMOX technology. The benefits of this technique are simplicity of processing and, as LOCOS and Mesa etching are not required, fewer compromises need be made in order to optimise circuit performance. Figure 8 shows a cross sectional SEM micrograph of a silicon island. The technology involves implantation of high doses of O^+ ions through a patterned masking layer, in which windows are opened to define the dimensions and location of the islands. The mask thickness is such as to absorb about

70% of the O^+ ion energy and the shape of the mask edges ensure that the oxide layer is continuous. Thus the SOI and isolating field oxide regions are formed at the same time and in such a manner that the dielectric completely surrounds the device island.

The formation of oxynitrides has been studied by a number of groups in order to form either thin gate dielectrics with improved integrity and dielectric properties(55), or SOI structures with improved radiation tolerance and the combined strengths of SIMOX and SIMNI technologies (56). The benefits of implanting equal doses of oxygen and nitrogen as N^+ have been studied by Reeson et al.(56), who found that, after annealing, the structures had abrupt interfacial regions with an overlayer of high quality single crystal silicon.

6.2 Devices

A novel MOS devices, which uses the buried oxide in SIMOX material as a second gate, in order to achieve volume inversion of the complete silicon overlayer, has been proposed by Balestra et al.(57). The volume inversion transistor (VI-MOS) has a normal SOI configuration but is operated with simultaneous bias on both gates, the normal gate (G1) and the secondary gate (G2), the gate biases being scaled to allow for the different oxide thicknesses, in order to induce inversion at both interfaces. By selecting an appropriate thickness and doping level for the overlayer, it is possible to drive the device into strong inversion throughout the volume of the overlayer. These workers report that, as current is carried by minority carriers, which are not constrained to an interface, the following benefits accrue: higher current, reduced sensitivity to interface properties, higher carrier mobilities and higher transconductance and effective mobility.

Measurements show a factor of three increase in the current and a subthreshold slope of 29mV/decade, to be compared with a theoretical minimum value of 64mV/decade for bulk silicon devices. Balestra et al. suggest that double SIMOX layers(58) formed by two implantations of different energies could be used to form an independent second gate that is isolated from the substrate.

Other new devices and circuit configurations can be expected in the future as increasing effort is now being directed to the study of SOI technologies.

7. CONCLUDING REMARKS

The benefits of SOI/SIMOX substrates for small geometry circuits have been confirmed in many laboratories and worldwide activity now is aimed at the exploitation of SIMOX technology. In contrast SOI/SIMNI substrates have produced disappointing results, due to their poor mechanical stability and the tendency of Si_3N_4 to recrystallize during processing. However, studies of both systems are leading to a better understanding of the formation of these SOI materials and, as a result, new and exciting ideas for further exploitation of ion beam synthesis now exist.

The increasing availability of high current implantation equipment is improving the supply of substrates and this will permit more detailed assessment of the technology in terms of yield and unit cost. The eventual inclusion of ion beam synthesis, as a processing stage in the fabrication of future generations of high performance circuits, is confidently predicted, as it enables the benefits of SOI materials to be combined with the strengths of silicon and ion beam technologies.

ACKNOWLEDGEMENTS

The authors thank their many colleagues who have contributed to this paper by providing details of their work and reprints.

REFERENCES

1. M. Watanabe and A. Tooi, Japan J. Appl. Phys. 5, 737, 1966.
2. P.V. Pavlov, E.I. Zorin, D.I. Telel'baum and Yu S. Popov, Soviet Phys. Doklady 10, 786, 1966.
3. G.H. Schwuttke and K. Brack, Acta. Crystallography A25 (Pt3):S43, 1969, 8th Int. Conf. of Int. Union of Crystallography, AIP, Buffalo, Stony Brook and Upton, New York, USA, 1969.
4. G.K. Hubler, P.R. Malmberg, C.N. Waddell, W.G. Spitzer and J.E. Fredrichson, Rad. Effects 60, 34, 1982.
5. J.H. Freeman, G.A. Gard. D.J. Mazey, J.H. Stephen and F.B. Whiting, European Conf. on Ion Implantation, Reading UK, Peregrinus, Hitchin p.74, 1970.
6. J.H. Freeman, Proc. Proc. Int. Conf. "Applications of Ion Beams to Semiconductor Technology", Grenoble (Editions Ophrys), p.75, 1967.
7. R.J. Dexter, S.B. Watelski and S.T. Picraux, Appl. Phys. Letts. 23, 445, 1973.
8. J. Dylewski and M.C. Joshi, Thin Solid Films 35, 227, 1976.
9. S.S. Gill and I.H. Wilson, Thin Solid Films 55, 435-448, 1978.
10. M.H. Badawi and K.V. Anand, J.Phys. D 10, 1931, 1977.
11. K. Izumi, M. Doken and H. Ariyoshi, Electron. Lett. 14, 593, 1978.
12. P.L.F. Hemment, Mat. Res. Soc. Symposium, Proc. 33, 41, 1984.
13. M.E. Burnham and S.R. Wilson, SPIE85, Los Angeles, 1984.
14. P.L.F. Hemment, Mat. Res. Soc. Symp. Proc. 53, 207, 1986.
15. I.H. Wilson, Ion Beam Modification of Insulators, Eds. P. Mazzoldi and G. Arnold (Elsevier Sci. Publishers BV 1987).
16. J.F. Ziegler, J. Biersack and U. Littmark, The Stopping and Ranges of Ions in Solids, Ed. J.F. Ziegler (Permagon 1985).
17. E.A. Maydell-Ondrusz and I.H. Wilson, Thin Solid Films 114, 357, 1984.
18. H.U. Jager, Nucl. Inst. Meths. B15, 1-6, 748, 1986.
19. C.D. Meekison, G.R. Booker, K.J. Reeson, P.L.F. Hemment, R.J. Chater, J.A. Kilner and J.R. Davis, Vacuum 36, 11/12, 925, 1986.
20. P.H. Chang, C. Slawinski, B.Y. Mao and H.W. Lam, J. Appl. Phys. 61, 1, 166, 1987.
21. R.J. Chater, J.A. Kilner, P.L.F. Hemment, K.J. Reeson and J.R. Davis, Nucl. Inst. and Meths. B19/20, 290, 1987.
22. F. Namavar, J.I. Budrick, F.H. Sanchey and H.C. Hayden, Mat. Res. Soc. Symp. Proc., 53, 233, 1986.
23. P.L.F. Hemment, K.J. Reeson, J.A. Kilner, R.J. Chater, C. Marsh, G.R. Booker, J.R. Davis and G.K. Celler, Nucl. Inst. and Meths. B21, 129, 1987.
24. C. Jaussaud, J. Stoemenos, J. Margail, M. Duprey, B. Blanchard and M. Bruel, Appl. Phys. Letts. 46, 1064, 1985.

25. P.L.F. Hemment, R.F. Peart, M.F. Yao, K.G. Stephens, R.J. Chater, J.A. Kilner, D. Meekison, G.R. Booker and R.P. Arrowsmith, Appl. Phys. Letts. 46, 10, 925, 1985.
26. A.H. van Ommen and M.P.A. Viegers, Microscopy of Semiconductor Materials, Oxford, 1987.
27. L. Nesbit, S. Stiffler, G. Slusser and H. Venton, Mat. Res. Soc. Symp. Proc. 53, 373, 1986.
28. A.H. Van Ommen and M.P.A. Viegers, Inter. Conf. Insulating Films on Semiconductors, Leuven, April 1987.
29. P. Berruyer and M. Bruel, Appl. Phys. Letts. 50, 2, 89, 1987.
30. R.C. Newman, M.J. Binns, W.P. Brown, F.M. Livingstone, S. Mersoloras, R.J. Stewart and J.G. Wilkies, Physica 116B, 264, 1983.
31. G.K. Celler, P.L.F. Hemment, K.W. West and J.M. Gibson, Appl. Phys. Letts. 48, 532, 1986.
32. C. Marsh, private communication.
33. P. Ratman and C.A.T. Salama.
34. R. Kwor, private communication.
35. K.J. Reeson, private communication.
36. W. Skorupa, K. Wollschlager, U. Kreissig and R. Grotzschel, Nucl. Inst. and Meths. B19/20, Pt. I, 285, 1987.
37. J. Belz, E.H. te Kaat, G. Zimmer and H. Vogt, Nucl. Inst. Meths. B19/20, Pt. I, 279, 1987.
38. S. Matsumoto, T. Ohno and K. Izumi, submitted to Electron. Letts.
39. C.G. Tuppen, M.R. Taylor, P.L.F. Hemment and R.P. Arrowsmith, Thin Solid Films 131, 233, 1985.
40. G. Zimmer and H. Vogt, IEEE Trans. ED-30, 1515, 1983.
41. J.R. Davis, K.J. Reeson, P.L.F. Hemment and C.D. Marsh, IEEE EDL-8, 7, 291,1987.
42. D.J. Foster, Electron. Letts. 19, 17, 684, 1983.
43. C.E. Chen, T.G.W. Blake, L.R. Hite, S.D.S. Malki, B.Y. Mao and H.W. Lam, IEDM Tech. Digest IEEE, New York 1984, p.702.
44. MacIver and Jain, IEEE ED-33, 12, 1953, 1986.
45. J.P. Colinge and S.Y. Chiang, IEEE ED-33, 1, 126, 1986.
46. As reference 38.
47. S. Nakashuma, Y. Maida and M. Akiya, IEEE ED-33, 1, 126, 1986.
48. T.I. Kamins, P.J. Marcoux, J.L. Moll and L.M. Roylance, J. Appl. Phys. 60, 1, 423, 1986.
49. J.C. Tzeng, W. Baeng, C. Ting and B. Siu, IEEE ED-33, 11, 1841, 1986.
50. J.P. Ruffell, D.H. Douglas-Hamilton, R.E. Kairn and K. Izumi, Inter. Conf. Ion Implantation Tech., Berkley, 1986.
51. B.Y. Mao, M. Malloubian, C.E. Chen, R. Sundaresan and C. Slawinski, IEEE EDL-8, 7, 306, 1987.
51. S.L. Partridge.
52. H. Munzel, G. Albert and H. Strach, IEEE EDL-5, 7, 283, 1984.
53. W. Skorupa, U. Kreissig and H. Oertel, Vacuum 36, 11/12, 933, 1986.
54. J.R. Davis, A.K. Robinson, K.J. Reeson and P.L.F. Hemment, Appl. Phys. Letts., accepted for publication, 1987.
55. P. Kenyon, IEEE ED-33, 11, 1841, 1986.
56. K.J. Reeson, P.L.F. Hemment, J.A. Kilner, R.J. Chater, C.D. Meekison, C. Marsh, G.R. Booker and J.R. Davis, Vacuum, 36, 891, 1986.
57. F. Balestra, S. Cristoloveanu, M. Benachir, J. Brini and T. Elewa, IEEE, Electron. Device Lett., in press, 1987.
58. P.L.F. Hemment, K.J. Reeson, J.A. Kilner, R.J. Chater, C. Marsh, G.R. Booker, J.R. Davis and G.K. Celler, Nucl. Instr. and Meths. B21, 129, 1987.
59. E.W. Greeneich and R.H. Reuss, IEEE EDL-5, 91, 1984.

EFFECTS OF IMPURITY-DEFECT INTERACTIONS ON THE LATTICE SITE OCCUPATION OF IMPURITIES IN ION IMPLANTED METALS

A.TUROS[1)]

INTRODUCTION

The prediction of solid solubility in alloys in one of the most important problems in metallurgy. Based on the thermodynamical considerations several attempts to solve this problem have been made in last decades (6). However, the question arises if such methods, even appreciably succesful in the case of equilibrium alloys, are applicable at all for ion implanted systems. Ion implantation is a nonequilibrium process which leads to the formation of alloys of unique metallurgical properties. The surface alloys can be formed which are outside the limits of equilibrium phase diagrams (1). These alloys are metastable.
Moreover, other strange properties have been observed: impurity atoms which apparently fit well in the host lattice can occupy distorted substitutional positions i.e. these atoms are slightly displaced (~ 0.02 nm) from lattice sites, some largely oversized atoms are found to be located in very small interstitial sites and finally, the recent experiments of the Karlsruhe group (8) have shown that the fraction of atoms occupying substitutional sites depends strongly on the temperature. Any comprehensive theory if applied to ion implanted systems must be able to explain these phenomena.

The majority of results reviewed here (4) was obtained by the use of ion channeling (2). The implantation and analysis have been performed in situ in the temperature range from 5K to 293K.

The annealing behavior of irradiated metals is governed by distinct recovery stages which appear at temperatures where different defects become mobile (8). Two of them are of special interest for the study of defect - impurity interactions. Self - interstitials (SIA) perform long - range migration in the stage I whereas the vacancies can migrate in the stage III. An attractive interaction between point defects and impurities leads to the formation of defect - impurity complexes. Such an association produces displacements of impurity atoms from their initial lattice position (7). As lattice defects are always created by ion implantation itself, these mechanisms are expected to play an important role.

1) Institute for Nuclear Studies,
ul. Hoża 69, 00-681 Warsaw, Poland

The defect - impurity interaction potential consists of two parts: the long range interaction is due to the strain field produced by the atomic size mismatch and the short range one which resulted from the electronic density rearrangement when a defect occupies a neighbouring position to the impurity atom. Consequently, the trapping probability can be scaled with the parameter $\Delta H = \Delta H_{sol} + \Delta H_{size}$, where ΔH_{sol} is the heat of solution and ΔH_{size} is the size mismatch energy as given by Miedema (5).

LATTICE SITE OCCUPATION OF IMPLANTED ATOMS

Impurity atoms with positive heats of solution are nonsoluble in metals. Detailed studies on the substitutional fraction f_s performed by in - situ implantation and channeling analysis revealed that the temperature is the key parameter for the lattice site occupancy. Some of the results are shown in Fig.1.

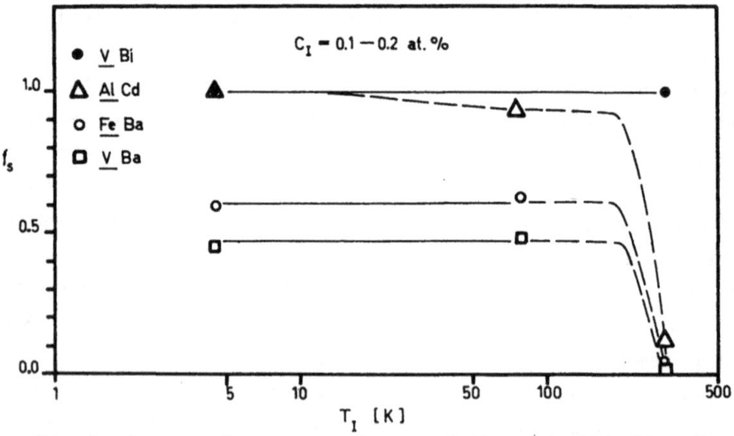

Fig.1. Temperature dependence of the substitutional fraction for different ion implanted systems

There are system like VBi where f_s of the impurity is equal 1.0 independently of the lattice temperature. For other systems like VBa, FeBa or AlCd f_s is 1.0 or less and is independent of the lattice temperature for temperatures below 77 K. For these systems a pronounced decrease of f_s is noted after implantation at 293 K.

These results can be understood by noting the correlations between the recovery stages and the temperatures at which the change of the substitutionality has been observed. Since stage I sets in below 77 K and stage III lies between 200 K and 300 K it can be concluded that in this case the interaction of impurity atoms with mobile vacancies leads to the decrease of f_s via vacancy capturing and subsequent displacement of the impurity.

The correlation between f_s and the corresponding ΔH - value for V and Al - based alloys produced by low dose (0.1 - 0.2 at%) ion implantation is shown in Fig.2. f_s is seen to decrease with increasing ΔH. The decrease is much more pronounced for systems implanted at 293 K. Some other effects have also been observed. For AlGa system, which is an equilibrium one, it was found that Ga is located on a distorted substitutional site after implantation at 293 K. Several other systems such as VCe, VXe, FeAu and FeBi exhibit the same behaviour. This displacement is probably caused by a vacancy on a neighbouring lattice site. The location of impurities on perfect or distorted substitutional sites depends on a delicate balance between vacancy trapping by impurities and by other competing traps on the one hand and by the increasing density of competing traps and the dissolution of impurity - vacancy clusters in overlapping cascades on the other hand. These effects can lead to an anomalous increase of the substitutional component. At $T_I < 100$ K, where vacancies are not mobile f_s is always larger than at 293 K. Nevertheless, the f_s values are smaller than 1.0 for systems exhibiting a large positive ΔH_{sol} value. The non-substitutional component for these systems can be attributed to vacancy impurity interaction in the cooling phase of the collision cascade (within $\sim 10^{-11}$ s).

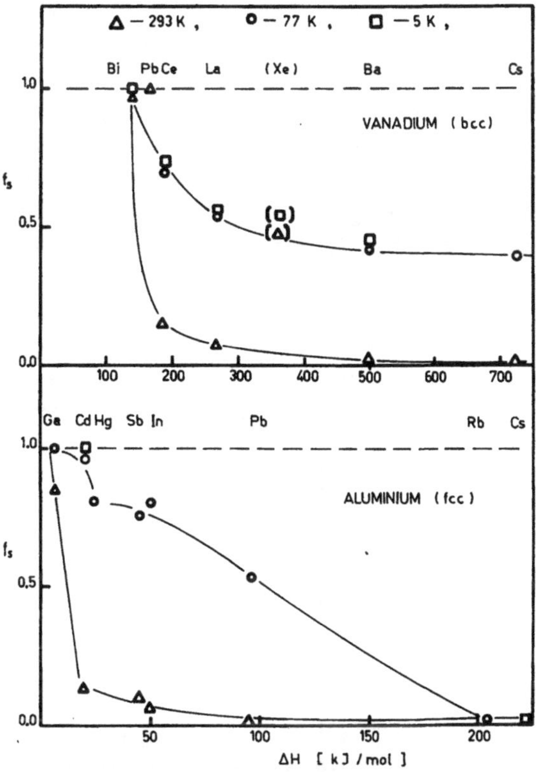

Fig.2. Substitutional fraction of different V - and Al - based systems as a function of ΔH for different implantation temperature.

Elements with negative heats of solution are known to form intermetallic phases and reveal rather low solid solubility. They usually are undersized with respect to the host atoms. Such atoms are known to interact with selfinterstitial atoms forming mixed dumbbells. In contrast to systems with positive heats of solutions where ΔH_{sol} increases with increasing ΔH_{size} this correlation does not exist for systems with negative ΔH_{sol}. Thus it is possible to select systems with different ΔH_{sol} and ΔH_{size} in order to study the influence of these parameters separately (3).

Systems with moderate ΔH_{sol} values (between - 100 kJ/mol and 0.0 kJ/mol) and small ΔH_{size} form supersaturated solid solutions independently of the implantation temperature. The nonsubstitutional component increases for systems with large negative ΔH_{sol} values and moderate ΔH_{size}. For these systems the effect of SIAs mobility was clearly seen: f_s was appreciably higher after implantation at 5 K than at 77 k. Fig. 3 shows the influence of ΔH_{size} exceeding 40 kJ/mol f_s was zero. Since the implantation has been performed at temperatures where vacancies are immobile the decrease of f_s is due to the formation of a variety of SIA - impurity complexes.

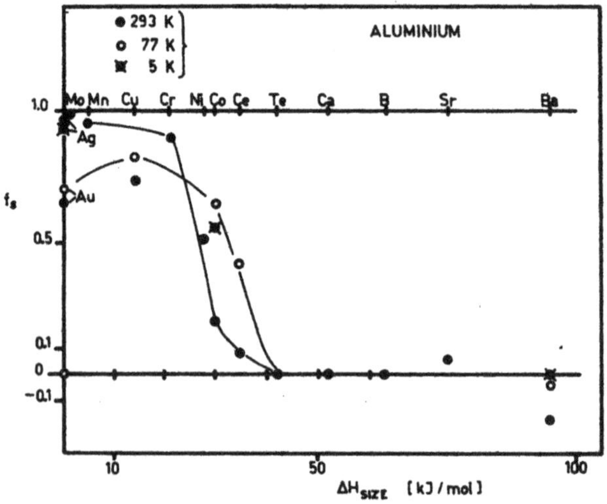

Fig. 3. Substitutional fraction of various elements with ΔH_{sol} values between - 150 and - 10 kJ/mol as a function of ΔH_{size}.

CONCLUSIONS

The empirical models which are based on thermodynamical equilibrium are not successful in predicting the substitutionality of non-soluble elements implanted in metals. The basic mechanism for lattice occupation of implants in metals is the spontaneous recombination of impurity atoms with lattice vacancies within the cascade and subsequent point defect trapping either within the cascade or in the delayed regime. The important trapping of defects also occurs at temperatures where different point defects become mobile. These processes lead to the displacements of impurity atoms from lattice site thus decreasing f_s. They can be correlated with the magnitude of ΔH_{sol} and ΔH_{size}. Also the sign of ΔH_{sol} plays a role. For systems with positive ΔH_{sol} the trapped defect are vacancies whereas for those with negative ΔH_{sol} SIAs trapping dominates.

REFERENCES

(1) Borders J.A.: Ann. Rev. Mat. Sci. 9(1979)313.

(2) Feldman L.C., J.W.Mayer and S.T.Picraux: Material Analysis by Ion Channeling, Academic Press, N.Y. 1982.

(3) Meyer O., A.Azzam, M.K.Kloska and A.Turos: Mat.Sci.Forum, 15-18(1987)757.

(4) Meyer O. and A.Turos: Mat.Sci.Rep. (in press).

(5) Miedema A.R., P.F.de Chatel and F.R.de Boer: Physica 100B(1980)1.

(6) Singh V.A. and A.Zunger: Phys.Rev. B25(1982)907.

(7) Turos A.: phys. stat. sol. (a), 94(1986)809.

(8) Turos A., A.Azzam, M.K.Kloska and O.Meyer: Nucl.Instr. and Meth. B19/20(1987)123.

THE PROCESS OF COMPOUNDS FORMATION BY ION IMPLANTATION

Masaya IWAKI, RIKEN, Wako +)

Introduction

Ion implantation as an impurity doping process in developing semiconductor devices is an indispensable technology in the LSI manufacturing process. One of the most salient features of ion implantation based on the constituents of ion implantation equipment is free selections of both ions and substrates. Yet another merit emphasized in the industrial use as surface treatments is that impurities can be introduced in substrates at room temperature.

In the early 1970s the initial application of ion implanatation for modifying the nonelectric properties of materials such as the mechanical or chemical properties of materials led to unexpected large effects, but these studies have been limited in the field of basic researches[1]. In recent years, the demand for new materials gradually grew in materials science and engineering. Futhermore, with the progress of ion beam technology, efforts gained momentum to apply ion implantation to metals, ceramics, and polymers[2].

Metal ion implantation in metals is a useful technology for the formation of new metastable alloys without thermal equilibrium. The process is carried out by using special features of ion implantation, one of which is the fact the density of doping particles can exceed the solid solubility limit. Moreover, this technique is useful in the formation of compounds near the metal surface layers by using light ions such as boron, carbon, nitrogen and oxygen. Normally, these compounds consisting of host atoms and implanted elements are produced at the interior due to direct implantation. Sometimes they are also formed near the surface by knock-on doping of surface contaminations.

In this report, the formation of compounds such as nitride and oxide near the surface and in the interior induced by ion implantation is introduced, and the formation processes are discussed.

High fluence implantation

Implanted ion distributions actually measured of specimens subjected to a low fluence implantation showed a fair agreement with distributions calculated according to LSS theory, which are indicated as the Gaussian distribution. However, it is important to note that since ion implantation is part of the ion bombardment technology, it will be accompanied by a chipping effect by incident ions themselves; sputtering will take place. Sometimes the knock-on doping of contaminations during ion implantation also takes plase. These phenomena were insignificant when the fluence of ion bombardment is low.

Sputtering by implanted ions themselves, which inevitably accompanies ion implantation, is in no way negligible with high fluence implan-

---

+) Saitama, 351-01 Japan

tation. Taking into account the sputtering that occurs during ion implantation, the implanted ion distribution is approximately represented by the integral fomula of a Gaussian distribution[3]. Fig. 1(a) shows the typical depth distribution of titanium implanted in iron at 50 keV as a function of fluence, calculated from the modified equation. It is clear that the surface concentration of implanted ions is very low at the initial stage of ion implantation and gradually increases as the fluence increases. Finally, its concentration is saturated at the higher fluence. On the other hand, the depth at the maximum concentration can be obtained by subtracting a half of a total sputtering thickness to an average projected range. If sputtered at double the average projected range or more, the maximum concentration appears on the surface.

When the sputtering thickness or sputtering ratio is small, the form of distribution for a low fluence is Gaussian, ignoring sputtering, as shown in Fig. 1(b). For a high fluence the maximum concetration of implanted species may exceed that of host atoms.

Nitrogen implantation in aluminum

The easy formation of nitrides at the metal surface layers by ion implantation is significant for the fabrication of electronic devices and the improvement of wear and corrosion resistance. Some results suggest that nitrogen implantation in iron results in the formation of various iron nitrides[4]. The nitrogen implantation into aluminum sheets results in the formation of aluminum nitride (AlN)[5].

Fig. 2 shows the distribution of atomic concentration of nitrogen implanted aluminum measured AES combined with Ar sputtering[5]. The fluences for (a), (b) and (c) are $5x10^{16}$, $1x10^{17}$ and $1x10^{18}$ ions/cm^2, respectively. The nitrogen depth profiles differ significantly from each other with the implanted fluences; they show nearly Gaussian-like shape at the low fluence and become rectangular-like at the high fluence. The binding energy spectra of Al_{2p} and N_{1s} measured by means of XPS supported the formation of AlN in nitrogen implanted layers except near the surface.

Direct evidence of AlN layer formation was observed by XRD, TEM and TED studies. The XRD patterns of nitrigen-implanted poly-crystalline aluminum surface showed weak diffraction from a (100) direction of AlN for the low fluence, a hcp AlN crystal structure with a strong prefered (002) orientation for the intermediate fluence, and the enhancement of the (002) orientation for the highest fluence. The TED patterns taken from the nitrogen implanted surface layers of polycrystalline and single crystal aluminum indicated the formation of polycrystalline and single crystal AlN, respectively.

Fig. 2 indicates two undesirable parts in the profiles; one is a hump of nitrogen concentration in the interior and the other is a oxygen migration from the surface which forms a diffusion-like distribution

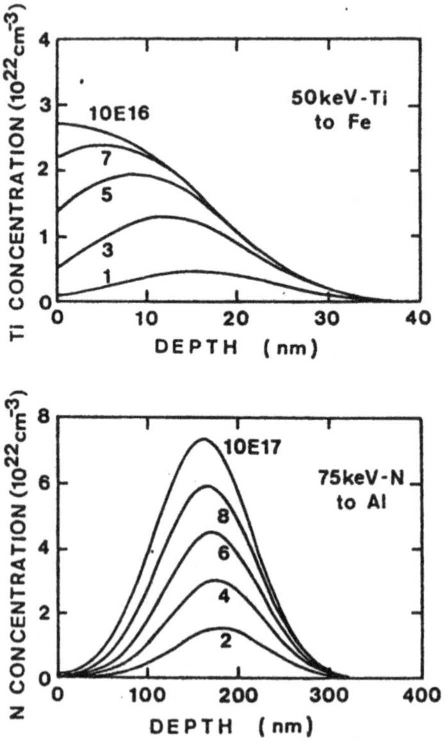

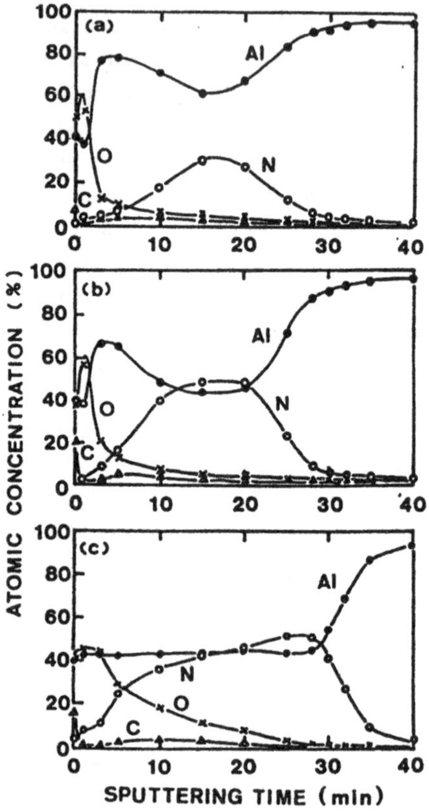

Fig. 1(left) Depth distributions of titanium (a) and nitrogen (b) implanted irons, calculated by the modified range equation.

Fig. 2(right) Depth distributions of atomic concentrations in nitrogen implanted aluminum measured by means of AES.

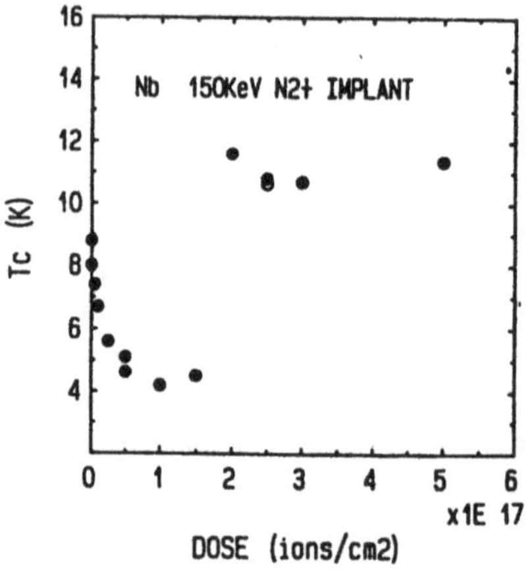

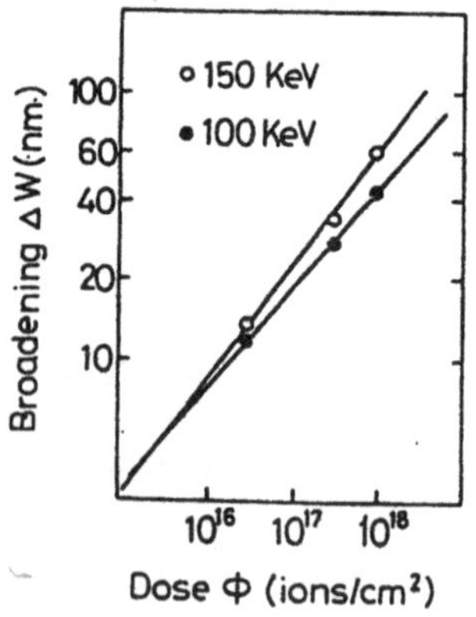

Fig. 3(left) Tc as a function of fluence for nitrogen implanted niobium thin films.

Fig. 4(right) Broadening width ΔW as a function of fluence ϕ obtained by Fig. 2.

near the surface. These phenomena can be also observed for oxygen implantation into aluminum[6].

Nitrogen implantation in niobium

The ion implantation technique is known to be useful as a method to control a superconducting transition temperature of resistance (Tc) by radiation damage produced during ion implantation or chemical effects of implanted ions. Some researchers reported Tc of nitrogen implanted niobium, and both increase and decrease in Tc took place due to the implantation conditions[7].

Fig. 3 shows the typical Tc of nitrogen implanted niobium thin films as a function of fluence[8]. The pristine niobium filmes (150 nm) deposited on a quartz glass has Tc of about 9K. The Tc of nitrogen implanted specimens gradually decrease in proportion to increase in fluence up to $1x10^{17}$ ions/cm^2. Beyond the fluence of $2x10^{17}$ ions/cm^2, the Tc drastically increase. The observations of XRD indicate the diminish of diffraction peak from niobium polycrystalline at the lower fluence, and the appearance of weak diffraction from nitrides at the higher fluence.

The decrease in Tc may be caused by increase in lattice distortion according to radiation damages. The new compounds formation, which will reduce the radiation damage and lattice distortion will lead the Tc to increase. The internal friction measurements showed that the internal stress increased at the low fluence and decreased at the high fluence[9], corresponding to the increase in lattice defects and the formation of new compounds, respectively. From these results, the depth distribution of nitrogen implanted in niobium is considered to be similar to that for nitrogen implantation into aluminum.

Self ion beam induced crystallization

From the surface analyses of nitrogen implanted aluminum layers, the formation of the crystalline AlN layer can be considered to occur in bulk aluminum sheets as follows: At the initial stage of the nitrogen implantation in aluminum, implanted nitrogen combined to aluminum and existed as AlN precipitates in the aluminum metal. With progress of nitrogen fluence, these AlN precipitates become larger than remaining areas of aluminum metal, and the stoichiometric ratio of AlN is reached at the depth of maximum nitrogen concentration. This high fluence implantaiton induces the compression in the implanted layers if nochange in volume occurs. The compression may be the driving force for crystallization of AlN, i.e. the following excess nitrogen implantation induces the radiation damage, both the radiation damage and compression cause the excess nitrogen to migrate toward the surface, and consequently the compression is relieved. Nitrogen migrated toward the surface combines with aluminum to form AlN.

This phenomenon of crystallization with high fluence nitrogen implantation into aluminum will occur by the incorporated effects of chemical bonding of implanted species and host atoms, radiation damage, and the induction and reduction of compression. It can be called self ion-beam induced crystallization (SIBIC).

Mechanism of oxygen migration in aluminum

Fig. 1 indicates that oxygen atoms migrate toward the deeper regions of aluminum as more nitrogen ions are implanted aluminum. In order to examine the process of oxygen migration of aluminum during nitrogen implantation, we evaluated the amount of the broadening by AES depth profiles in Fig. 1. Fig. 4 shows the relationship between the oxygen broadening width ΔW and nitrogen fluences ϕ. ΔW increases with O as $\Delta W \propto \phi^{0.43}$ at 150 keV and $\Delta W \propto \phi^{0.37}$ at 100 keV are nearly equal to the value of 0.5 predicted for the case of cascade mixing. Oxygen migration by cascade mixing may be caused as the following process is repeated: nitrogen implantation causes a part of oxygen atoms in Al_2O_3 at the surface to impinge into aluminum and/or to sputter, and subsequently the surface without sufficient oxygen atoms combines with oxygen in the residual gas to form Al_2O_3. Therefore, nitrogen implantation in aluminum is always carried out through Al_2O_3, and this process is similar to ion beam mixing.

Method of reduction of nitrogen hump and oxygen migration

As mentioned above, the wurtzite structure of poly- or single crystal AlN can be formed by mass-analyzed nitrogen molecule ion implantation into aluminum at room temperature without any thermal annealing. However, the depth profile has a hump of nitrogen concentration at the deeper region than the average projected range, and it also shows the diffusion-like distribution of oxygen near the surface. In order to diminish the hump of nitrogen, the concept of SIBIC suggests that the heavy radiation damage should be induced into the humped region of implanted layers. If a non-mass-analyzed nitrogen ion beam including N^+

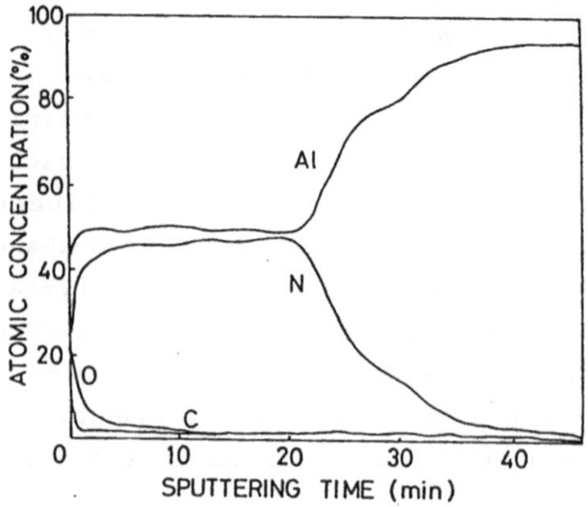

Fig. 5 Depth distribution of atomic concentration for non-mass analyzed nitrogen implantation of an atomic fluence of 10^{18} ions/cm^2.

and N_2^+ was implanting, N_2^+ ion beams form the AlN layers and N^+ ion beam induces the radiation damage beyond the AlN layers produced by N_2^+ ion beam, because the ratio of N_2^+ ions and N^+ ions is normally 3:2 whose atomic ratio is 3:1. Oxygen maigration may diminish due to feeding nitrogen gas on the surface or using a UHV system, because it is difficult that cascade mixing occur in these conditions.

Fig. 5 shows the distribution of the atomic concentration for nitrogen implanted aluminum measured by AES using non-mass-analyzed ion beam[10]. The nitrogen depth profile has a rectangular-like distribution, and the atomic concentration of nitrogen is almost the same as that of aluminum in all implanted layers. Comparing it with Fig. 1, we can not see a hump of nitrogen concentration. The result supports the concept of SIBIC. The migration of oxygen atoms in aluminum cannot be also observed. The phenomenon can be considered to occur as follows: The pressure of the target chamber is originally $\sim 10^{-4}$ Pa which is almost the same as the implantation condition of Fig. 1 and 1.3×10^{-3} Pa during non-mass-analyzed nitrogen implantation, so that the residual gas is regarded to be fulled with nitrogen atmosphere. Therefore, the probability of oxygen entrapment from gas is reduced, and consequently oxygen migration into aluminum by cascade mixing will hardly happen.

Summary

A high fluence implantation causes the formation of crystalline compounds in metal surface layers due to SIBIC process. In this treatment, we must take care of the invasion of undesirable atoms near the surface layers, because cascade mixing of contaminations occur. Anyway, according to this SIBIC model, bombardment of energetic ions can induce crystallization in a mother crystalline target, producing a new phase transformation. A high fluence of accelarated ions may give rise to a phase transition by radiation-induced nucleation and crystal growth.

References

1) Applications of ion beams to metals, ed. S.T. Picraux et al., Plenum Press (1973).
2) Ion Implantation, ed. J.K. Hirvonen, Academic Press (1980)
3) M. Iwaki et al., Trans. ISIJ, 24 (1984) 581.
4) M. Iwaki et al., Mater. Sci. Eng., 69 (1985) 211.
5) S. Ohira et al., Nucl. Instrum. Methods, B19/20 (1985) 162.
6) S. Ohira et al., Proc. Ion Beams in Materials Science, Hosei Univ. Press (to be published).
7) K. Gamo et al., Jpn. J. Appl. Phys., 16 (1977) 1853.
8) S. Tanda et al., Proc. ISIAT'85 (1985) 495.
9) T.R. Bulmer et al., Mater. Sci. Eng., 69 (1985) 429.
10) S. Ohira et al., Nucl. Instrum. Methods, B21 (1985) 588.

MICROSTRUCTURAL AND TRIBOLOGICAL STUDY OF IRON AND STEELS ION IMPLANTED WITH NITROGEN

G. MAREST[*], S. FAYEULLE[**], N. MONCOFFRE[*], J. TOUSSET[*]

Abstract

Iron and various steels have been implanted with nitrogen ions under different conditions (fluence, temperature, current density). The implanted layers have been characterized with CEMS, TEM, GXRD and NRA techniques. Dry friction tests on a 100C6 steel have been performed. A correlation between the wear behaviour and the microstructure of the implanted layer is presented.

1. INTRODUCTION

For many years numerous laboratories have been interested by the improvement in the mechanical properties of high fluence nitrogen implanted steels (1,20,4,5). In order to determine the best implantation conditions and to understand the basic phenomena involved in the alteration of tribological properties both the influence of the implantation parameters (energy, fluence, temperature, current density) and of the nature of steels (chemical composition, structure) have been studied (16,17,21,7,12).

In this paper we present some new results concerning the microstructural characterization of implanted layers with different techniques and the correlated wear resistance improvements.

2. EXPERIMENTAL TECHNIQUES

Polished iron and steel samples were implanted with 40 keV $^{15}N^+$ ions using the isotope separator of the Institut de Physique Nucléaire of Lyon. The implantation doses were 10^{17} or 2.10^{17} ions.cm^{-2}. The implantation temperature was varied from 20 to 200°C by heating the sample holder. The current density was normally maintained at 6 µA.cm^{-2} but also was raised to 18 µA.cm^{-2} for comparison.

Studies of the chemical phases formed by implantation were carried out using Conversion Electron Mössbauer Spectroscopy (CEMS), Transmission Electron Microscopy (TEM) and Glancing Incidence X-Ray Diffraction (GXRD). CEMS and GXRD techniques are non-destructive and allow a depth analysis of the order of magnitude of the implanted layer (~ 100 nm). By varying the incident angle of the CuK_α X-radiation between 0.5° and 1.2° the GXRD technique probes depths ranging from 20 to 90 nm. Nitrogen profiles were determined using the $^{15}N(p,\alpha\gamma)^{12}C$ nuclear reaction.

3. RESULTS

It is now well known that a saturation of the retained nitrogen concentration not due to sputtering is always obtained when the implantation fluence increases at a given energy (6). In our experimental conditions this saturation occurs for a fluence of 2 to 3 x 10^{17} ions.cm^{-2} and corresponds to around 30-35 nitrogen atoms per 100 iron atoms. The formation of nitrides is then favoured but for lower fluences only compounds poor in nitrogen will be formed. In **figure 1** TEM micrographs of pure iron and XC06 (AISI 1006) steel implanted at 20°C

(*) *Institut de Physique Nucléaire (and IN2P3), Université Claude Bernard Lyon-1 43, Bd du 11 Novembre 1918 - 69622 Villeurbanne Cedex (France)*

(**) *CNRS/UA-447, Ecole Centrale de Lyon, B.P. 163 - 69130 Ecully Cedex (France)*

with 1 and 2 x 10^{17} $N^{+}.cm^{-2}$ respectively show the formation of fine dispersed $\varepsilon Fe_{2+x}N$ precipitates (**figure 1a**) which grow with fluence. For fluence of 2 x 10^{17} $ions.cm^{-2}$ the size of these precipitates is about 25 nm (**figure 1b**). For both fluences α'-(N) martensite is also detected by electron diffraction.

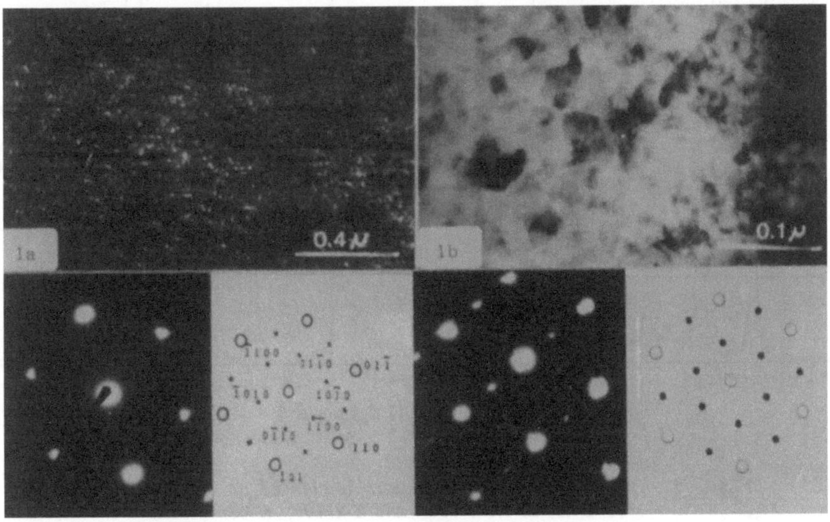

__Figure 1__ - TEM micrographs of implanted samples :

(a) Pure iron implanted with 10^{17} $N.cm^{-2}$ (dark field micrograph on $10\bar{1}0$ nitride spot)

(b) 1006 steel implanted with 2 x 10^{17} $N.cm^{-2}$ (bright field micrograph)

Corresponding electron diffraction :
BCC iron. Zone axis z = $\bar{1}11$
ε nitride $Fe_{2+x}N$. Zone axis z = 0001

In **figure 2** CEMS spectra display both the influence of the fluence and of the temperature during implantations of pure iron. For a 10^{17} $N^{+}.cm^{-2}$ fluence the sextuplet due to α-Fe predominates and only 20% of the spectra correspond to ε-$Fe_{2+x}(C,N)$ carbonitrides (three fields around 250, 210 and 115 kOe). For 2 x 10^{17} $N^{+}.cm^{-2}$ fluence two doublets representative of Fe_2N (isomer shift : IS = 0.39 $mm.s^{-1}$, quadrupole splitting QS = 0.29 $mm.s^{-1}$) and εFe_{2+x} (C,N) (IS = 0.23 $mm.s^{-1}$, QS = 0.92 $mm.s^{-1}$) appear but Mössbauer spectra corresponding to implantations performed at 20°C and at 150°C obviously differ. It has been shown, in particular by Moncoffre et al. (8) and Rauschenbach (18) that when the temperature during implantation is higher than 20°C the nitrogen profiles change rather rapidly. Nitrogen migrates towards the surface thus forming a superficial peak. Due to carbon pollution and different nitrogen concentration the phase composition evolves. This effect of implantation temperature is general as confirmed by CEMS spectra of a Z200C13 (AISI D3) steel implanted with 2 x 10^{17} $N^{+}.cm^{-2}$ at 100 and 200°C (**figure 3**). The main features are a large decrease in the amounts of α-(N) martensite (24, 19, 9%) and nitrides (5, 6, 3%) when the implantation temperature increases (20, 100 and 200°C) whereas the carbonitride formation increases (30, 29, 39%). Between 20 and 100°C the evolution of the spectra is not drastic due to the stabilizing effect of chromium on the nitrogen profile. However above 100°C the destabilization of martensite and Fe_2N phases allows the diffusion of nitrogen and consequently the increase of carbonitrides. We have to point out that carbonitrides are not more abundant in Z200C13 (2 at.% C) than in XC06 (0.06 at.% C) (7).

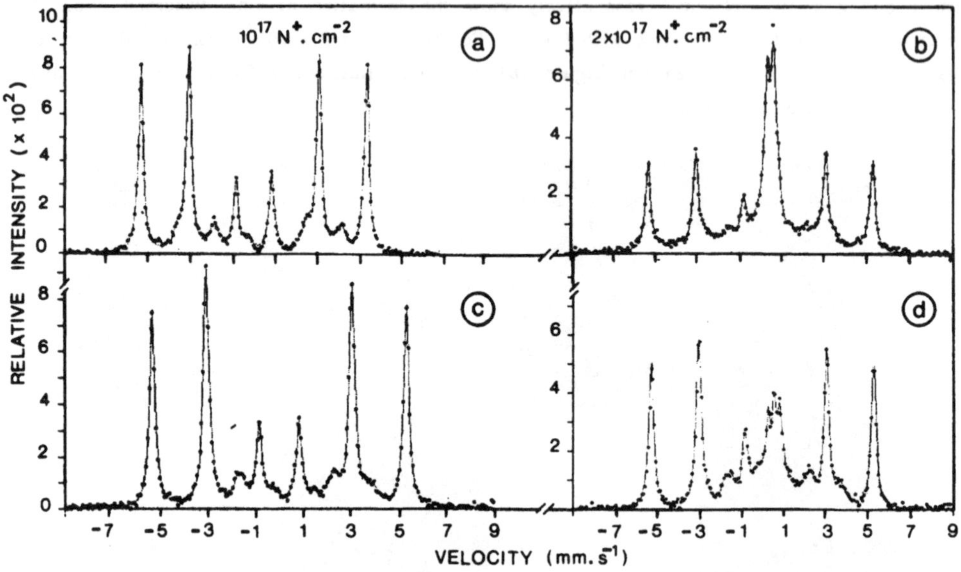

Figure 2 - CEMS spectra of pure iron implanted at 20°C **(a, b)** and 150°C **(c, d)** with 10^{17} and 2×10^{17} $N^+.cm^{-2}$.

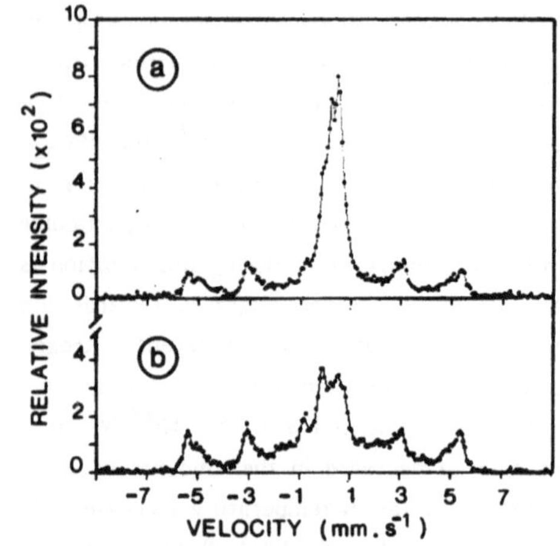

Figure 3 - CEMS spectra of a Z200C13 steel implanted at 100°C **(a)** and 200°C **(b)** with 2×10^{17} $N^+.cm^{-2}$.

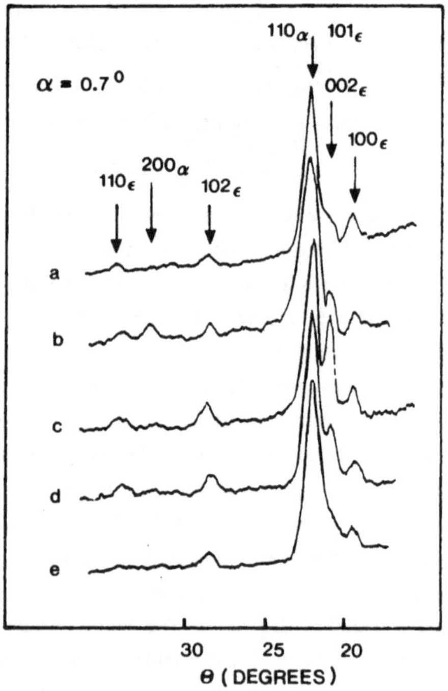

Figure 4 - XRD spectra taken at 0.7° of a 100C6 steel implanted with 10^{17} $N^+.cm^{-2}$ at 20°C **(a)**, 100°C **(b)**, 150°C **(c)**, 200°C **(d)** and **(e)** with 2×10^{17} $N^+.cm^{-2}$ at 20°C.

Implanted samples have been analyzed by GXRD under various incident angles (α = 0.5°, 0.6°, 0.7°, 0.9° and 1.2°) corresponding to X-ray penetration depths of about 20, 30, 40, 60 and 80 nm respectively. For example **figure 4** represents diffraction spectra obtained at 0.7° for a 100C6 (AISI 52100) steel implanted with 10^{17} ions.cm^{-2} at 20, 100, 150 and 200°C and 2.10^{17} ions.cm^{-2} at 20°C. In each case the formation of $\varepsilon Fe_{2+x}N$ nitrides (or Fe_{2+x} (C,N) carbonitrides) is identified by the presence of (100), (002), (101), (102) and (110) reflexions. The most intense peak around d = 0.204 nm corresponds to the (101) ε-reflexion or to the superposition of (110) and (101) reflexions belonging to martensite with the (101) belonging to nitrides. The position of this peak allows to deduce the relative amount of these two compounds. It has been concluded that for 100 and 150°C the outermost layers (~ 40 nm) contain mainly nitrides or carbonitrides whereas deeper martensite predominates for example in implantation with 2×10^{17} N^+.cm^{-2} (2).

A very important feature is the variation of the intensity of the (002) peak. This peak, almost non-existent at 20°C, becomes very intense at 100 and 150°C and decreases at 200°C. This phenomenum can be explained by the existence of a texture created by implantation and which evolves with temperature (9).

The current density of the ion beam is principally considered as influencing the sample temperature (15,13). In the following study the temperature of the sample was maintained at 150°C and implantations in pure iron and XC06 were performed with 10^{17} N^+.cm^{-2} and 6 or 18 μA.cm^{-2} current densities. The nitrogen distributions thus obtained using the $^{15}N(p,\gamma)^{12}C$ nuclear reaction are displayed in **figure 5**. When implanting at 6 μA.cm^{-2} (that is in 45 min) the distribution maximum is shifted from about 40 nm (at 20°C) to 20 nm whereas for 18 μA.cm^{-2} only a fraction of nitrogen reaches the surface. As the temperature of the sample was kept constant the only varying parameter is here the implantation time.

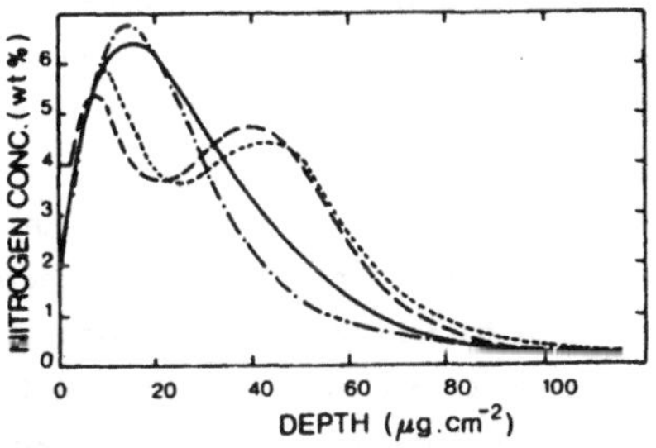

Figure 5 - Nitrogen profiles obtained for pure iron and XC06 implanted at 6 and 18 μA.cm^{-2} current densities i (φ = 10^{17} ions.cm^{-2}, T = 150°C).

Iron	*i = 6 μA.cm^{-2}*	*XC06*	*i = 6 μA.cm^{-2}*
Iron	*i = 18 μA.cm^{-2}*	*XC06*	*i = 18 μA.cm^{-2}*

From the tribological point of view implantation conditions can play a very important role. Fayeulle (2) has shown that the very high friction coefficient (0.8) of 100C6 in dry conditions of friction is only weakly affected by nitrogen implantation (f = 0.76 for 10^{17} ions.cm^{-2}). Meanwhile the weight loss as a function of the friction time depends on the implantation conditions and only implantations performed at 100 and mainly at 150°C are favourable **(figure 6)**. This has been confirmed for a XC06 steel implanted in the same conditions and for the 100C6 steel implanted with 2×10^{17} ions.cm^{-2} at various temperatures (2).

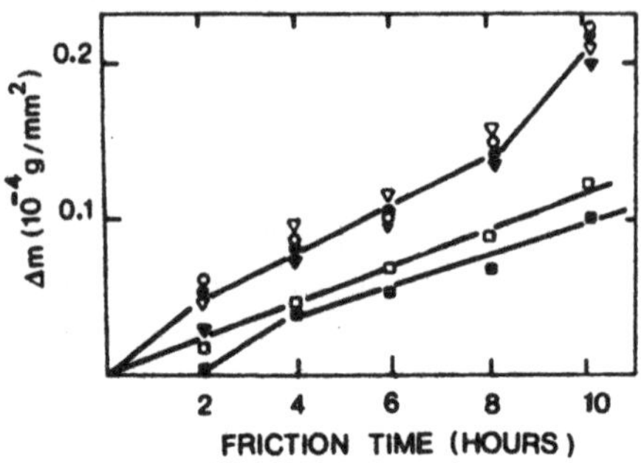

Figure 6 - Weight losses as a function of the friction time of a 100C6 steel unimplanted ( ) and implanted with 10^{17} $N^+.cm^{-2}$ at 20°C ( ), 100°C ( ), 150°C ( ), 200°C ( ) and with 2 x 10^{17} $N^+.cm^{-2}$ at 20°C ( ).

4. DISCUSSION

The amount of carbonitrides formed following implantation is not very different in iron, XC06 and Z200C13 steels and so the bulk carbon content is not the fundamental and determining factor in the carbonitride formation. However it is necessary to compare steels having the same crystalline structure because of the alloying element influence (10).

In variance with results of Rauschenbach et al. (10) no austenite has been detected after implantation at 20°C. This difference could be explained by the fact that their samples are thin deposited layers of iron and that the temperature can not be controlled.

When implanting at high temperature the amount of carbonitrides increases due to several simultaneous phenomena : nitrogen migrates towards the surface (8), carbon contamination increases (10) and diffuses inside the matrix (18). A very interesting observation is that implantation at 20°C leads to the formation of nitrides or carbonitrides with a c-axis preferentially oriented parallel to the surface plane. This texture effect could be due to strong compressive strains induced by implantation and also to the low c/a ratio of ε-nitrides (1.60) favouring this orientation (2). When implantations are performed at 100 or 150°C the implanted layer consists in nitrides with their c-axis preferentially perpendicular to the surface plane. This orientation change could be due both to a strain release at 150°C and to the superficial position of carbonitride phases at this temperature.

It has been shown that implantations performed at 100 and 150°C give the best wear resistance improvement. During wear tests X-ray diffraction spectra have been systematically performed after 2, 4 and 10 hours of friction (2). α-Fe_2O_3 oxide is immediately formed (except for samples implanted at 100 and 150°C) and even after 10 h of friction the surface under the oxide layer remains unchanged. In our friction conditions the weight losses of unimplanted or implanted steel at 20°C occur by the mild oxidational wear mechanism (14). Two fundamental properties explaining this wear behaviour, hardness and oxidizability of the surface, can be affected by implantation. A layer containing only nitrides is hardened and superficial strains are known to have a great importance on friction properties (3,19). At the same time the presence of nitrides affects the oxidation behaviour of the surface. However these two effects are not sufficient to explain the results since an implantation at 20°C has no influence on wear resistance. Implantation at 100°C produces two additional effects : better homogeneity of the superficial layer (higher hardness and lower oxidation rate) and new orientation of nitrides with (0001) planes parallel to the surface. These planes are the densest in the hcp structure and thus they have the lowest surface energy and consequently the lowest reactivity with oxygen.

These results emphasize that the understanding of friction mechanisms requires a good microstructural characterization of the implanted layers.

REFERENCES

(1) Dos Santos,C.A., M.Behar, and I.J.R.Baumvol: J. of Phys. D-Appl. Phys. **17**(1984) 551-562

(2) Fayeulle,S.: Doct. ès-Sciences Physiques Thesis, Lyon (1987)

(3) Ho,J.W., C.Noyan, J.B.Cohen, V.D.Khanna,, and Z.Eliezer: Wear **84**(1983) 183-202

(4) Hohmuth,K., E.Richter, B.Rauschenbach, and C.Blochwitz: Mat. Sci. and Eng. **69**(1985) 191-201

(5) Hubler,G.K., and F.A. Smidt: Nucl. Instr. Meth. in Phys. Res. **B/7-8**(1985) 151-157

(6) Marest,G., C.Skoutarides, T.Barnavon, J.Tousset, S.Fayeulle, and M.Robelet: Nucl. Instr. Meth. in Phys. Res. **209/210**(1983) 259-265

(7) Moncoffre,N., G.Marest, S.Hiadsi, and J.Tousset: Nucl. Instr.Meth. in Phys. Res. **B 15**(1986) 620-624

(8) Moncoffre,N., G.Hollinger, H.Jaffrezic, G.Marest, and J.Tousset: Nucl. Instr. Meth. in Phys. Res. **B 7-8**(1985) 177-183

(9) Moncoffre,N.: Mat.Sci. and Eng. **90**(1987) 99-109

(10) Moncoffre,N., H.Jaffrezic, G.Marest, J.Tousset, F.C.Zawislak, S.Fayeulle,, and J.Brissot: to be published in "MRS 1987 Spring Meeting Proceedings"

(11) Moncoffre,N., M.Brunel, P.Deydier, and J.Tousset: Surf. and Interf. Analysis **9**(1986) 139-143

(12) Principi,G., S.Lo Russo, and C.Tosello: in "Industrial Applications of the Mössbauer Effect", Eds G.J.Long, and J.G.Stevens (Plenum Press, 1987) 267-294

(13) Principi,G., S.Lo Russo, and C.Tosello: J. Phys. F:Mat. Phys. **15**(1985) L207-L210

(14) Quinn,T.F.J.: in "Microscopy Aspects of Adhesion and Lubrication", Ed. J.M.Georges, (Elsevier Publ., 1982) 579-597

(15) Ramous,E., G.Principi, L.Giordano, S.Lo Russo, and C.Tosello: Thin Solid Films **102**(1983) 97-106

(16) Rauschenbach,B., A.Kolitsch, and K.Hohmuth: Phys. Stat. Sol.(a) **80**(1983) 471-482

(17) Rauschenbach,B., A.Kolitsch: Phys. Stat. Sol.(a) **80**(1983) 211-222

(18) Rauschenbach,B.: Nucl. Instr.Meth.in Phys. Res. **B 15**(1986) 756-759

(19) Rigney,D.A., and J.P.Hirth: Wear **53**(1979) 345-354

(20) Singer,I.L.: Appl. of Surf.Science **18**(1984) 28-62

(21) Singer,I.L.: Vacuum **34**(1984) 853-859

ELECTRON BEAM AND UV-LASER MICROSTRUCTURE FORMATION BY DECOMPOSITION OF SOLID METAL RESINATE FILMS

V.P.Ageev, V.I.Konov, A.I.Maslakov, V.M.Mel'nikov, and L.V.Velikov[1)]

INTRODUCTION

At the present time chemical methods for thin film deposition due to a thermally induced decay of organometallic compounds above heated surfaces have been widely extended into many fields of science and technology /1/. Being successful in producing a variety of high-purity materials, as well as some their unique modifications these methods need further development to form planar film structures of required topology. The main achievements in this direction have been so far related to using direct laser writing due to thermo- or photostimulated local deposition of materials from gas phase or liquids onto substrates /2/. Recently, another approach to a surface patterning by metal features has begun developing /3-8/ based on laser produced decomposition of solid metal resinate films.

In this paper new experimental results are presented concerning the action on resinate films of 308-nm pulsed radiation of an excimer XeCl - laser and of high energy (15-30 KeV) electrons. The following two basic regimes for film processing are considered below:

(1) - intensive thermal decomposition of resinate films into volatile products and deposit produced by laser pulses, and

(2) - the photon- and electron lithography.

The former is a modification of the traditional way for resinate film treatment, some aspects of which have also been studied previously /3-7/ by using VIS and IR radiation of cw Ar^+- and CO_2-lasers. Here, the main attention will be paid to the dynamics of material losses at multiple-pulse irradiation, and to energetical characteristics of the process.

The latter originates from photon and electron induced changes in film solubility in some organic oils /8,9/. We shall consider lithographic characteristics of resinate films and trends of their alterations under different irradiation conditions.

Experimental investigation has been carried out using as an example Fe(III)-abietate-based micron-thick films applied to silicon substrates. The structural formula for an organic ion-radical of the resinate molecule: $(C_{19}H_{29}COO)_3Fe$ is presented in Fig.1.

---

1) General Physics Institute of the USSR Academy of Sciences, 117942 Moscow, Vavilov str.38, USSR

Film decomposition at multiple-pulsed laser irradiation

First pay attention to some details of equilibrium decay of the compound used, especially as many peculiarities of the film processing by intensive pulsed UV radiation are predetermined by them. Fig.2 shows thermogravigrams of decay of Fe-abietate (in a powder form) during its heating in the air at the rate of temperature scanning of 5 °C per minute. According to the curves of differential thermal analysis (DTA) and thermogravimetric ones (TG and DTG) resinate decomposition accompanied by mass losses starts from the temperatures T $\simeq$ 100 °C, goes through a few characteristic stages and finishes at T=530 °C by an appearance of iron oxides (Fe_2O_3, mainly). The most intensively the mass losses proceed during a pronounced endothermic step at the temperatures near to T=380 °C (see DTA curve). After that the decomposition rate decreases practically to negligible values at T $\simeq$ 420 °C (see the DTG curve) indicating the appearance of an intermediate stable product, the latter being decomposed only at further heating in exothermic reaction, probably of burning of carbon adsorbed in oxygen.

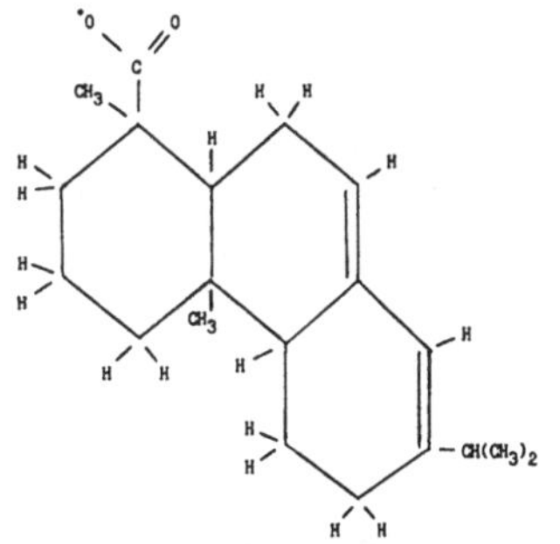

Fig.1. Structural formula of organic ion-radical of Fe(III)-resinate: $(C_{19}H_{29}COO)_3Fe$

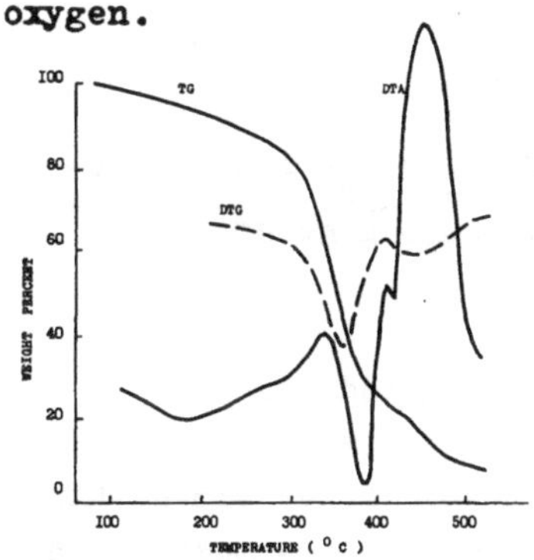

Fig.2. Thermogravigrams for decay of Fe(III)-abietate (in a powder form) at heating in the air.

In experiments on laser radiation processing of resinate films, decomposition proceeded in the form of laser produced ablation of material from the film surface. The measurements carried out under various irradiation conditions have shown the process to be first of all of a pronounced threshold character: it could start provided the pulsed laser fluence E (energy per unit area) has exceeded some critical value of $E_{th} \simeq 25$ mJ/cm^2(for films thicknesses X_0=1 μ).

Further, on exceeding the threshold E_{th}, the laser produced film etching appeared to be controlled not only by fluence, but also by a number of successive laser shots, N. Fig.3 shows some typical curves of etching expressed in terms of normalized current thicknesses X/X_0 as a function of dose D=E·N correspondent to various fluences. It is well seen from the curves presented that in a general case the intensive decomposition is preceeded by some activation stage, even though $E > E_{th}$, but material removal starts only after the dose would reach the necessary level of $D=D_a$ being conditioned in its turn by laser fluence. For the fluence-dependent behaviour of the activation dose see Fig.4. Note, the activation dose does not depend on fluence ranging from $E=E_{th}$ to $E \simeq 50$ mJ/cm^2 and is as high as $D_a \simeq 4$J/cm^2. At fluences increasing above the upper boundary of the mentioned interval the dose D_a decreases sharply until the etching starts from the first laser pulse at fluences $E > E_{th}^1 \simeq 80$ mJ/cm^2, where E_{th}^1 is the etching threshold for a single shot event determined by equality $D_a(E)=E$.

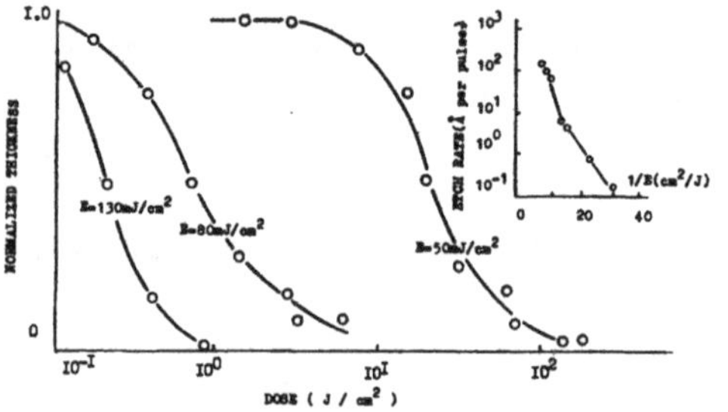

Fig.3. Normalized current thicknesses of films X/X_o (X_o=1 μ) vs dose of exposure D at different fluences E.

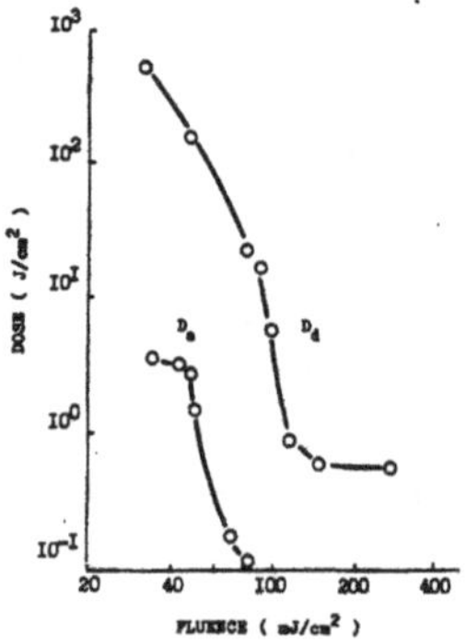

Fig.4. The fluence-dependent behaviour of the activation dose for laser produced films etching, D_a, and of dose D_d of full decomposition of films with thicknesses X_o=1 μ.

Having started the process exhibits at comparatively low fluences some acceleration at first. Then the etch rate decreases gradually as the films become a few times thinner than the initial ones. According to VIS/UV transmission measurements when using similar resinate films on quartz substrates, preliminary furnace-annealed in the air at different temperatures, the observed decrease of the etching rates may be attributed to decoupling of laser radiation with the resinates decomposed somewhat to a stage of intermediate stable product formation.

An analysis of average rates V of etching performed using lgV vs (1/E)-plots (see inset in Fig.3) shows the process to obey an activation-type scaling law equivalent to the Arrhenius one: $V \propto \exp(-E_a/E)$ with fluence-dependent parameters of E_a ($E_a' = 0.73$ J/cm^2 at $E > E_{th}^1$ and $E_a'' = 0.23$ J/cm^2 at $E_{th} < E < E_{th}^1$). On the one hand, this cofirms the etching process to be entirely of thermal nature. In this case an extremely sharp decrease in the values of dose D_d required to produce full decomposition of the film at fluence increasing becomes clear, as shown in Fig.4. Note that the following saturation of the decomposing dose by its minimum value of $D_{dd} \simeq 0.6$ J/cm^2 at fluences being above some characteristic value of E = 120-150 mJ/cm^2 does not either contradict to the thesis of thermal nature of etching process. In our opinion, such a behaviour of $D_d(E)$ points to realization in our experiments of the regime of developed decomposition (in analogy with the developed laser produced evaporation of a solid target /10/) characterized by the values of laser energy input into material being much higher than that required for film heating up to the temperature interval of resinate decay. This is also confirmed by the proximity of the value of the developed decomposition dose D_{dd} to the higher value of efficient activation energy per unit area E_a'.

On the other hand, one should note that a less effective activation density of energy E_a'' corresponds to the etching of the films produced by lower-energy pulses. Besides, a relation appears to exist between the activation parameters and the values of characteristic threshold fluences, such that $E_{th}^1 / E_{th} \simeq E_a^1 / E_a''$. Both of these facts make us to conclude that preliminary exposure of the films during the activation stage changes the optical properties of the films in such a way that the efficiency of the radiation energy input per unit volume of material increases. Therefore, laser pulses give rise to the material heating up to higher temperatures enabling the etching process to develop at fluences below a single shot threshold E_{th}^1. It is interesting to mention that the effect discussed seems also to be responsible for the observed self-acceleration of the etching process at fluences $E < E_{th}^1$ (see Fig.3).

What kind of the internal processes gives rise to a laser-to-film coupling improvement is not established. But the experimental results obtained let us speak about their nature just now. So, a nontrivial fluence-dependent behaviour of the activation dose, especially the appearance of a plateau in the $D_a(E)$-curve at comparatively low laser intensities present an evidence that these processes are rather photolic than thermal. Therefore, one may conclude that we deal here with photostimulated thermodecomposition of resinate films. "Pure" thermal etching seems to take place only at the fluences $E > E_{th}^1$.

UV laser and electron lithography

Preliminary series of experiments have shown that the films exposed to UV radiation begin to exhibit resistivity with respect to their dissolution in some organic solvents. Thus, films of thicknesses $X_o \simeq K^{-1}$ (where K is the absorption coefficient) can be classified as a photoresist. In experiments with films of thicknesses $X_o \simeq 1\,\mu$ ($KX_o \simeq 1.4$ for the wavelength used) the effect of solubility changing was not strong enough, so that the exposed films were developed in a weak nonpolar solvent CCl_4. Fig.5 shows some typical characteristic curves for resinate resists, taken as dose-dependent thicknesses of the exposed and developed films, X_d, normalized to those of the films, X(D), before dissolution. For a comparison, where it is necessary, current film thicknesses during irradiation are also presented there.

The behaviour of the characteristic curves is seen to be influenced to a great extent by laser fluence, or more definitely, by the fact whether material ablation takes place or not. So, in the case of exposure at fluences E E_{th} the characteristic curve rises up progressively with the dose increasing until it saturates by the upper limiting value of $X_d/X_o = 1$ (see the curve for $E = 5mJ/cm^2$ in Fig.5); that is typical of negative photoresists.The behaviour of the characteristic curves correspondent to irradiation at fluences above the etching threshold E_{th} appears to be more complicated. Initially, as under low energetic exposure, there also exists a slow increase of the $X_d/X_{(D)}$-value (here $X(D) = X_o$) at the rising dose. This tendency fails however when we continue with laser shots: at doses above D_a, sharp increasing of the characteristic curve is observed, while the values of X_d/X_o decrease. Note that the contrast of the ablating films may be much higher as compared to the previous case of the nonetched exposure. Another difference is that ablation of the resinate films changes also the resist sign (i.e. it becomes a positive one) at higher doses, as observed in a subsequent decreasing of the $X_d/X(D)$-values.

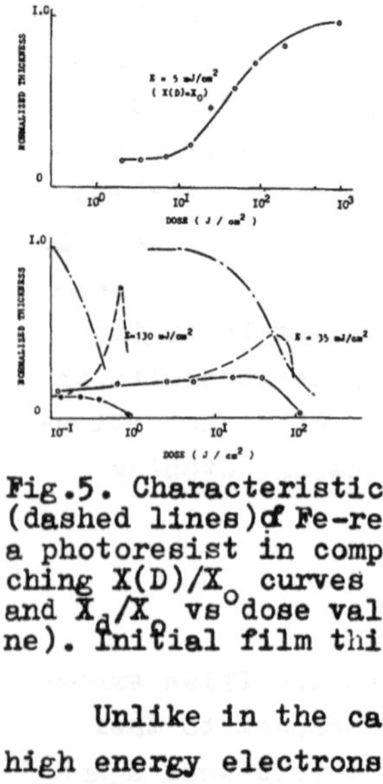

Fig.5. Characteristic $X_d/X(D)$ curves (dashed lines) of Fe-resinate film as a photoresist in comparison with etching $X(D)/X_o$ curves (doted line) and X_d/X_o vs dose values (solid line). Initial film thickness $X_o = 1\,\mu$.

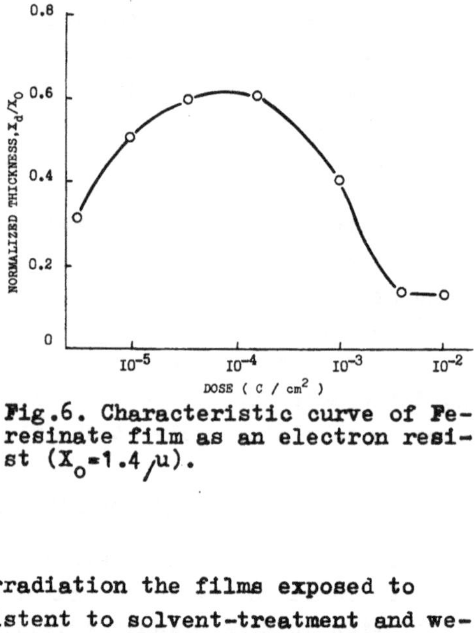

Fig.6. Characteristic curve of Fe-resinate film as an electron resist ($X_o = 1.4\,\mu$).

Unlike in the case of UV laser irradiation the films exposed to high energy electrons became more resistent to solvent-treatment and were developed in a solution of methyl-ethyl-keton: isopropyl alkohol (1:3). The corresponding characteristic curve of the resinate films classified as an electron resist is shown in Fig.6. Here the current values of X_d are normalized to the initial film thickness X_o, since any appreciable changes have not been observed in the film thicknesses (less than 15%) during exposures at doses ranging from 10^{-6} to $10^{-2}C/cm^2$. Besides, it is important to point out that according to evaluations the electron exposure did not increase the film temperature above the thermal decomposition threshold. As it is seen, the curve exhibits the properties analogous to some extent to those of the ablating photoresist. Namely, at comparatively low values of dose $D \lesssim 10^{-4} C/cm^2$ the films represent a negative resist. Then, at dose rising, the resist's sign conversion also takes place. Note also, the positive resist stage is not perfect here, as the characteristic curve becomes stable at high doses $D \sim 10^{-2} C/cm^2$ on the final level of $X_d/X_o = 0.15$.

The appearance of both photon- and electron negative resists' properties should be attributed to the exposure induced polymerization process. Moreover, there exists evidence that these processes proceed mainly by the same channel in both cases (but of course at different rates). So, according to preliminary IR spectral transmission measurements by testing the films exposed to electrons or to photons, the irradiation leads to cutting, first of all, of methyl radicals, destroying primarily the isopropyl groups of a resinate molecule (film absorption weakens within the CH_3-vibrational bands near the frequencies of $1180 cm^{-1}$ and $1360-1380 cm^{-1}$). Additionally, in the case of laser action some doubled C=C cyclic bonds are broken. Thus, a possibility appears of cross-linking between monomer resinate molecules presumeably due to radical mechanisms. It is clear, an amount of these cross-linkages will increase with dose rising, thus affecting film solubility.

The above qualitative picture, dealing with a situation where the system of resinate molecules responds to any individual photon or electron absorbed, can not be obviously applied here to fluence-dependent phenomena. In particular, it cannot explain the effect of the contrast arising, which accompanies the laser-produced film etching. We suppose that this effect reflecting additional changes in film solubility is thermally induced by its origin, but stronger resistance to solvent development is connected here with a substantial reduction of the organic radical mass.

The mechanisms of forming positive resist properties in both cases of electron and laser exposure is not well-understood now, and will be the subject of our nearest investigations. Most probably, they are associated with the pores appearance within the films (as it sometimes happens during exposure of positive photoresist /11/), that improves solvent penetration into the film, as well as increases tremendously the surface of liquid-resinate interaction. The base for such a proposal is originated from the facts that resinate molecules are destroyed drastically not only by laser pulse heating, but also by electron beam at doses as high as $D > 10^{-3} C/cm^2$, as it follows from the results of the analysis of films by the methods of transmission IR- and X-ray photoelectron spectroscopy. So, for an example, in IR absorption spectra vibrational bands of the cyclic rings disappear, i.e. the distruction touches the very foundation of the resinate molecule. In addition to this, note that according to the results of ESCA some fraction of metal atoms (about 10% at dose $D \simeq 10^{-3} C/cm^2$) appeared to be in a free state.

To conclude the discussion of some aspects of resinate films modification due to electron exposure let us mention one more unexpected to some extent fact concerned with analysis of chemical composition of the exposed films by using layer-by-layer sputtering of materials by 2KV-argon ions. It appears that ion etching gives rise to a substantial increase of the metal atom percentage: the ratio of metal-to-oxygen concentration becomes 5:1 at the etched film thickness $X \simeq 500 Å$ instead of the initial one 1:6. It is interesting that in this case metal atoms are mainly non-bonded with oxygen ones. The mechanisms of this film enrichment by metal is not clear. As a hypothesis, on the one hand, it may be connected with variations in sputtering coefficients of the elements from different energy states /12/. If this is true, then ion-etching would be used as an additional film processing to obtain the metallized structures. But on the other hand, we do not either exclude that the enrichment proceeds during the electron exposure and rises from the real moving of metal ions into the direction of the substrate due to self-generated static electric fields within the film volume.

Fig.7. SEM of pattern in Fe-resinate film obtained by UV photon lithography using excimer XeCl laser and contact mask.

Fig.8. SEM of strip obtained by electron lithography in Fe-resinate film

Basing on photon and electron resist properties of films discussed, the experiments have been performed on surface structure formation. Some results of these preliminary tests are illustrated in Figs.7,8. Figure 7 shows a pattern in the Fe-resinate film obtained using multiple-pulsed irradiation of the surface through a chrom-on-quartz mask contacted to the substrate. Fig.8 presents SEM of the strip written by a scanning electron beam. Note, the strip is characterized by a high height-to-width aspect ratio (about 10), and by the absence of its broadening near the substrate, which is not typical of all negative electron resists.

Conclusion

The investigations performed have proved that the metal resinate films can be successfully applied to surface microstructuring using pulsed UV laser radiation and electron beam. The intrinsic properties of metal resinates make it possible to write inorganic material features directly by pulsed radiation induced thermodecomposition of the organometallic precursor, as well as to form metal resinate-based surface structures using UV photon and electron lithography in films. In the frame of inorganic material deposition procedure one may consider these structures to be an intermediate product which will be subsequently subjected to (maybe complex) development, such as thermal annealing in various gases, etching by ions, chemically active plasma, and others. Along with this, the structures may also be used independently of various purposes at an appropriate fit of organometallic materials.

(1) Razuvaev G.A. (Ed.): Application of Organometallic Compounds for Obtaining Inorganic Coatings and Materials, Nauka, Moscow 1986 (in Russian).

(2) See for example Bäuerle D. (Ed.): Laser Processing and Diagnostics. Springer Ser. Chem. Phys., vol.39, Springer, Berlin, Heidelberg 1984.

(3) Jan R.Y., S.D.Allen: SPIE 459, 71 (1984).

(4) Fisanik G.J., M.E.Gross, J.B.Hopkins, M.D.Fennel, K.J.Schnoes., and A.Katzir: J.Appl.Phys. 57, 1139 (1985).

(5) Fisanik G.J., J.B.Hopkins, M.E.Gross, M.D.Fennel, and K.J.Schnoes: Appl.Phys.Lett. 46, 1184 (1985).

(6) Gross M.E., G.J.Fisanik, P.K.Gallagher, K.J.Schnoes , and M.D.Fennel, Appl.Phys.Lett. 47, 923 (1985).

(7) Ageev V.P., S.D.Allen, V.I.Konov, V.M.Mel'nikov, and H.-P.Preiswerk: Preprint No143 of Gen.Phys.Inst. of the USSR Acad.Sci., Moscow 1986.

(8) Ageev V.P., I.V.Baranenkov, K.A.Valiev, L.V.Velikov, V.I.Konov, A.I.Maslakov, V.M.Mel'nikov, and A.M.Prokhorov: Dokl.Acad. Nauk SSSR 294, 1358 (1987).

(9) Graihead H.J., and L.M.Schiavone: Appl.Phys.Lett. 48, 1748 (1985).

(10) Anisimov S.I., Ya.A.Imas, G.A.Romanov, and Yu.V.Hodyko: Action of High-Power Radiation on Metals, Nauka, Moscow 1970 (in Russian).

(11) Bogdanov A.L., K.A.Valiev, L.V.Velikov, D.Yu.Zaroslov: J.Vac.Sci. Technol. B5(1), 391 (1987).

(12) Kelly R.: Surface Sci. 100, 85 (1980).

APPLICATIONS OF MICROFOCUSSED ION BEAMS

F.G. Rüdenauer *)

Microfocussed ion beams have found four main applications in materials research:

(a) microanalysis (spatially 1- 2- and 3- dimensional compositional analysis of solids by secondary ion mass spectrometry);

(b) microimplantation (patterned implantation of ions in solids for modification of material properties);

(c) micromachining (utilizing ion induced sputtering for controlled pattened removal of material from solid surfaces);

(d) microdeposition (utilizing ion induced chemical reactions for controlled patterned deposition of material onto solid surfaces).

1. Spatially Multidimensional Microanalysis

Spatially 3 - dimensional elemental distribition analysis in solids by means of the scanning ion microprobe has been devised in the author's lab (1),(2) and is continuously being developed into a valuable microanalytical technique (3). A focussed ion beam isscanning in a 2 - dimensional raster across the surface. Secondary ions are synchronously detected by a mass spectrometer and ion count numbers for each scanned pixel are stored in a computer. Pixel counts for each scanned frame can be arranged as a 2-dimensional ion image; the collection of all 2D - frames represents the 3D - elemental distribution in a surface - near volume of the solid because due to the sputter removal of surface layer during scanning, each 2D - frame represents the elemetal distribution at successively larger sputtering depths. Analogous 3D - acquisition techniques have been developed for ion microscopes (4),(5),(6). Computer graphic techniques have been developed to comprehensively display 3D elemental distributions. **Fig.1.** shows the reconstructed distribution of Al on an integrated - circuit section. The sputtered analytical volume is intersected by three planes and the secondary ion intensity in these planes is displayed in intensity modulation. It can be seen that Al (from contact pads) extends straight into depth without any significant changes in lateral distribution. Such changes are not expected since the thickness of the Al - pads is larger than the total sputtering depth (0.3 µm). Previously known modes of SIMS analysis, i.e. point analysis, depth profiling, line scan analysis and imaging can be considered special cases of 3D analysis. It is obvious, that on a given 3D data set all these analytical modes can be performed retrospectively by proper selection of stored

*) Austrian Res. Ctr. Seibersdorf, A-2444 Seibersdorf, Austria

data. Artifacts, causing the reconstructed 3D distribution to deviate from the true distribution have been identified (surface topography, matrix effect, differential sputtering, redeposition, etc.) and discussed (7),(8),(9). Complete correction of artifacts is impossible at the moment.

2. Microimplantation and Micromodification

Microimplantation, in particular "maskless" ion implantation for production of spatially patterned implantation zones in solids, can be performed by electrostatically deflected microfocussed beams of a particular ion species. Systems have been designed for fabrication of microelectronic devices by direct - write maskless implantation of multiple ion species. A recent review of the technique has been given by Melngailis (10). Another application is production of microimplant - elemental concentration standards which can be used for sensitivity calibration of microanalytical instruments. Here, multiconcentration and multielement standards (11) have been produced in the shape of multiple 100 x 100 μ implant zones at 200 μ central distance (11).

3. Micromachining and Microdeposition

Micromachining makes use of the sputtering effect to "write" topographic patterns into a solid substrate. This technique has found application in repair of dark defects in lithographic masks (12),(13). The complementary effect, patterned deposition of material by writing ion beams (for repair of clear defects) is also under development (14),(15). In this process, a focussed ion beam is bombarding a surface in an atmosphere of a metal - containing gas ("precursor gas"). The gas adsorbed at the surface is locally cracked by the action of the ion beam. Metal - containing radicals remain permanently adsorbed, volatile reaction products escape into vacuum. Continuous bombardment and gas adsorption may result in local buildup of a metal - containing layer with sufficient adsorption for visible, UV- and X - radiation. Ion beam induced deposition of Al, Cr, W and Au has been reported (10), however no physical model for the deposition process has been given. We have developed a simplified process model (16) describing the process in equilibrium as simultaneous precursor adsorption, ion induced transformation (fractionation) of adsorbed precursor molecules and into permanently adsorbed metal-radicals, ion - induced sputtering of layer material. It is based on the following assumptions:

(a) precursor molecules from the gas phase may adsorb on already fractionated molecule sites only but not on undissociated precursor sites. A sticking probability of 1 is assumed.

(b) adsorbed precursor molecules may only be transformed into permanently adsorbed metal - containing radicals but not sputtered as a whole from the surface.

(c) transformed metal - radicals only may be sputtered.

Consequently, the following parameters are sufficient to describe the process: fractionation cross section σ [cm^{-2}], sputtering yield Y, gas adsorption rate ν [$cm^{-2}.sec^{-1}$] and ion current density j [$cm^{-2}.sec^{-1}$].

A short calculation shows (15) that the equilibrium deposition rate dD/dt [layer molecules/cm^2/sec] is given by

$$(dD/dt)/\nu = \frac{1 - (Y/\sigma n_o).(j\sigma n_o/\nu)}{1 + 1/(j\sigma n_o/\nu)} \qquad [1]$$

Obviously, a threshold exists for the ratio ν/j :

$$\nu/j > Y \qquad [2]$$

Above this gas flux / ion flux ratio net deposition occurs, below threshold the net effect is sputtering of the deposit (or substrate). At a given gas flux ν the deposition speed in (1) in dependence on ion flux j goes through a maximum. The optimum deposition rate can be calculated as

$$(dD/dt)_{opt} = \nu.[1-1/(1+R)]/[1+(Y/\sigma n_o).(1+R)] \qquad [3a]$$

and occurs at an optimum current density

$$j_{opt} = \nu/[Y.(1+R)] \qquad [3b]$$

with $R = [1+(\sigma n_o/Y)]^{1/2}$ and no the number of adsorption sites / cm^2.

Fig.2 shows normalized deposition rate in dependence on normalized j/ν - ratio for different sputter yield / cross section ratios. The deposition maximum and the transition from deposition to sputtering regime can be seen. **Fig.3** shows calculated optimized deposition rate and corresponding optimized current density in dependence on sputter yield / cross section ratio. The model as such does not predict numerical values for the physical parameters Y and σn_0 but allows to predict deposition rates for any value of process parameters j, ν provided the physical parameters are known. For the case of Al - deposition from trimethyl - aluminum (TMA) under Ar - bombardement Y and σno have been determined by fit of equ. (1) to experimentally measured deposition rates. Fig.4 shows measured values for deposition ratios RD (deposited molecules / ion) in dependence on TMA - pressure (=gas flux) at constant ion current density for Ar ion energies of 3.5, 6.0 and 10.0 keV. The dashed lines show optimum - fits of our model to the data. Numerical values of process parameters derived from the optimum fit curves are given in Tab.1.

<u>Tab.1</u> Fractionation cross-section σno and sputter yield Y for Al - deposition form TMA under Ar - ion bombardement. ΔRD/RD...mean fitting error.

E [keV]	σn_0	Y	ΔRD/RD [%]
3.5	1.32	7.4	11.4
6.0	1.22	6.5	3.8
10.0	1.01	5.9	4.3

The fit is reasonable with exception of the near - threshold region of the 10 keV data. It may be necessary to modify the model by including a sputtering yield dependence on TMA - coverage of the deposit during deposition. Such a dependence seems reasonable since near threshold the lower TMA - coverage changes the chemical environment and binding energy and consequently the sputtering yield of adsorbed metal radicals. Fig.5 shows an Auger - depth profile of an "Al" - layer deposited at 10 keV Ar - energy and 5.E-5 torr TMA - pressure. It can be seen that the layer contains relatively large impurities of C and O in accordance with similar published data (17). The depth profile however seems to show that C is enriched at the interface. The large O - concentration probably arises due to exposition to air after layer deposition, since the background pressure of the vacuum system was in the 10^{-9} torr region. Another possible explanation for the C - enrichment lies in the sputter - fractionation of the measured deposit during Auger - depth profiling. This, because of the low C - sputtering yield, would tend to gradually enrich C in the analyzed thin deposit layer (150 nm).

This work was supported in part by the Austrian Fonds zur Förderung der wissenschaftlichen Forschung, Proj. No S43/04.

References

(1) Rüdenauer, F. G., and W. Steiger, Mikrochim. Acta [Wien], Suppl.II (1981) 375.
(2) Rüdenauer, F. G., in "Secondary Ion Mass Spectrometry; SIMS III", A. Benninghoven, J. Giber, J. Laszlo, M. Riedel and H. W. Werner, Eds., Springer Verlag, Berlin (1982), p. 222.
(3) Fathers, D.J. and J.R. Wolstenholme, in "Proc. 11th ICXOM", J.D. Brown and R. Packwood, Eds., Univ. W. Ontario, London (1987), p.222.
(4) Patkin, A.J. and G.H. Morrison, Anal. Chem. 54 (1982) 2.
(5) Bryan, W.S., Woodward and R.W. Linton, J. Vac. Sci. Technol. A3 (1985) 2102.
(6) Brown, J.D., in "Scanning Electron Microscopy 1987", in print.
(7) Rüdenauer, F.G. and W. Steiger, in "Proc. 11th ICXOM", J.D. Brown and R. Packwood, Eds., Univ. W. Ontario, London (1987) p.211.
(8) Levi - Setti, R., J. Chabala and Y.L. Wang, Scanning Microscopy, in print.
(9) F.G.Rüdenauer, Surf. Interf. Anal. 6 (1984) 132.
(10) Melngailis, J., J. Vac. Sci. Technol. B5 (1987) 469.
(11) Rüdenauer, F.G., in "Proc. SIMS VI" (1987) in print.
(12) Cleaver, J.R.A., H. Ahmed, P.J. Heard, P.D.Prewett, G.J. Dunn and H. Kaufmann, Microel. Eng. 3 (1985) 153.
(13) Müller, K.P., U. Weigmann and H. Burghause, Microel. Eng., 5 (1986) 481.
(14) Shedd,G.M., A.D. Dubner, H. Lezec and J. Melngailis, Appl. Phys. Lett.49 (1986) 1584.
(15) Yamamoto, M., M. Sato, H. Kyogoku, K. Aita, Y. Nakagawa, A. Yasaka, R. Takasawa and O. Hattori, SPIE 632 (1986) 97.
(16) Rüdenauer,F.G., W. Steiger, D. Schrottmayer, M. Kammerhofer and G. Stangl, in "Proc.Sem. Grundlagen und technologie Mikroelektron. Bauelemente", Inst. f. Allg. Elektrotechn. u. Elektronik, TU Vienna, Ed., Vienna (1987) p. 64.
(17) Gamo, K., N. Takagawa, N. Samoto, R. Shimizu and S. Namba, Jap. J. Appl. Phys. 23, (1984) L293.

Fig.1 Al - distribution on IC - section; scan field 250 x 250 µm, sputterdepth 0.3 µm. Primary beam 10 keV O_2^+.

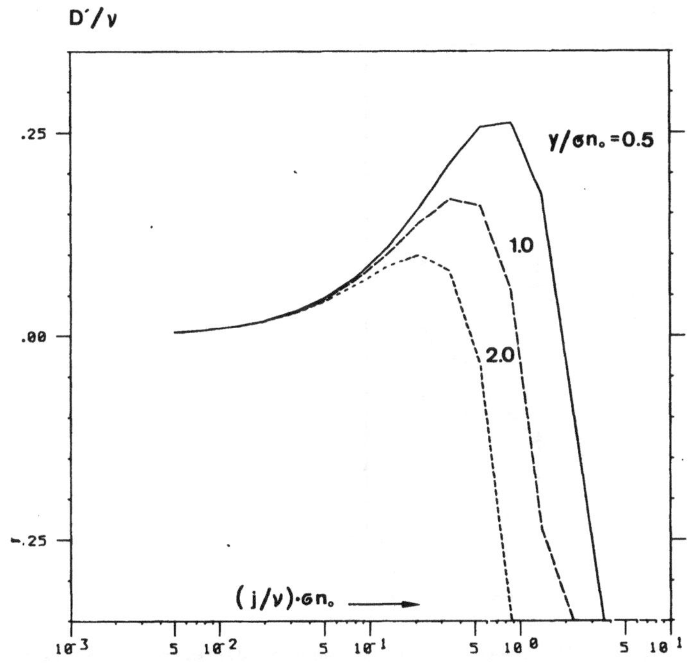

Fig.2 Ion - induced metal deposition; dependence of normalized deposition rate on normalized ion / precursor flux ratio with $Y/\sigma n_0$ as parameter.

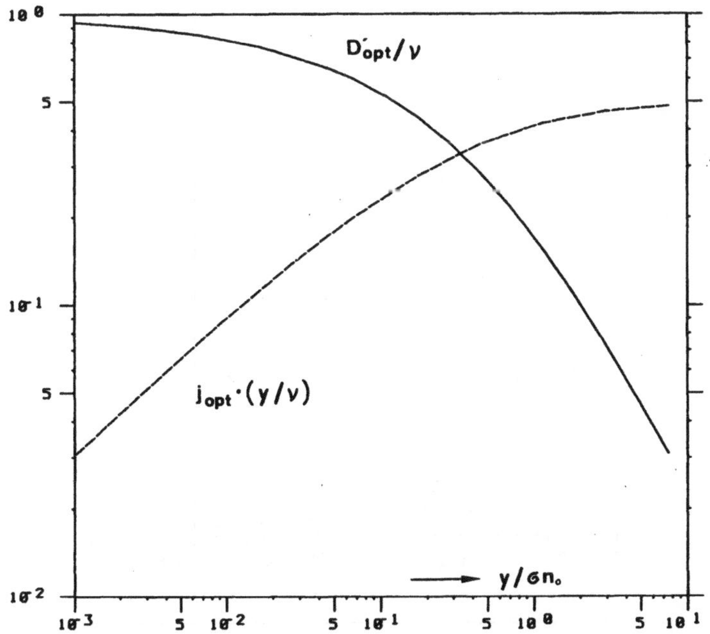

Fig.3 Ion induced metal deposition; dependence of normalized optimum deposition rate and optimized current density on sputter yield / cross section ratio.

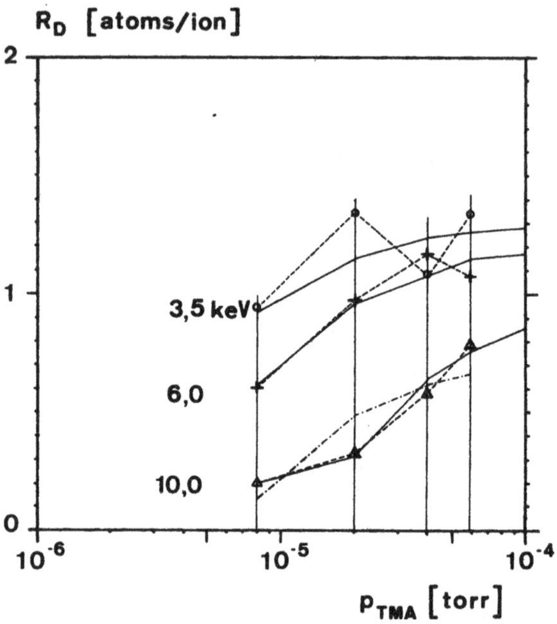

Fig.4 Deposition of Al from TMA under Ar - bombardment at different ion energies; dependence of deposition ratio [molecules / ion] on TMA - pressure. o + Δ ... experimental points, _____ optimum model fit; primary current density 1 μA / cm².

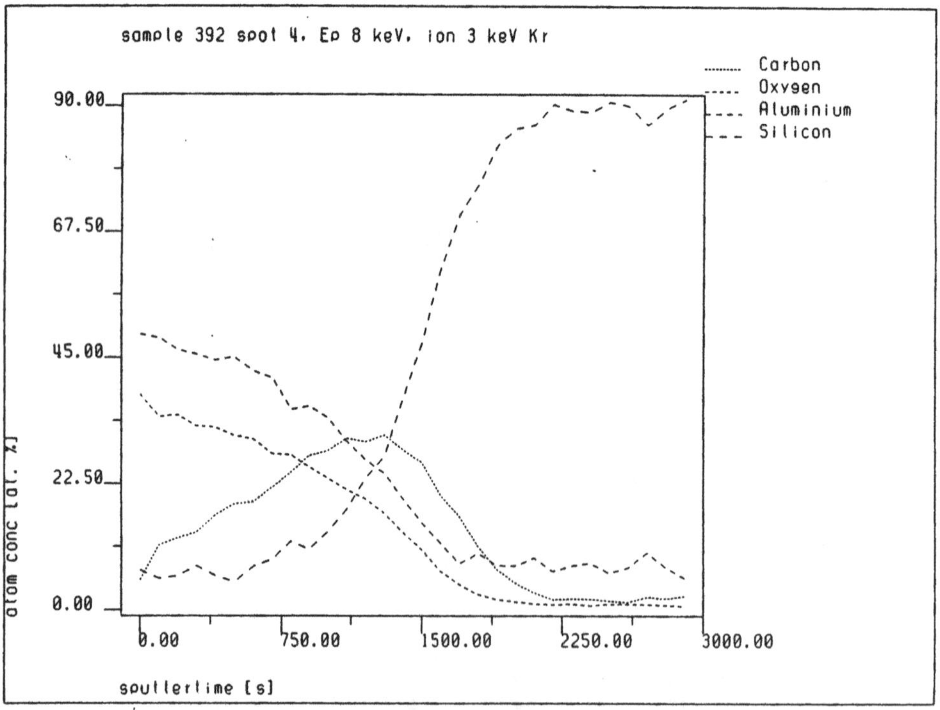

Fig.5 Auger depth profile of Al - layer deposited from TMA by 10 keV Ar - ions onto Si Total layer thickness 0.15 μm; sputter beam: 3 keV Xe.

STRUCTURE OF CRYSTAL FILMS PREPARED BY VAPOR DEPOSITION AND SIMULTANEOUS ION BOMBARDMENT

F. Fujimoto[1], M. Satou[2]

Properties of crystal structure of TiN, MoN and BN films prepared by the dynamic mixing method are reported. Films prepared by evaporation and simultaneous ion beam bombardment have preferential orientation which depends on the preparation conditions, such as ion beam direction, deposition rate of energy from the ion beam and temperature of the substrate. The present method has potential application in production of various types of films with controlled crystal structure and properties.

1. Introduction

Film formation is one of the most important techniques not only in today's electronics, but also in surface modification. The authors have developed a method of coating film formation using ion beam and vapor deposition (Dynamic mixing or IVD method)[1-9]; nitride films of TiN, AlN, BN and MoN were prepared.

This method has various advantages such as the strong adhesion between substrate and film, formations of films thicker than usual implantation method, films consisting of elements which are not included in the substrate, and film preparation at low temperature, etc. In most cases, films prepared by the present method consist of crystallines with a definit prefered orientaion.

In the present report, the film formation method is briefly introduced and crystal structures of films, as well as the orientation dependence on preparation conditions are reported.

2. Appratus[1,2,4,9]

We used various types of machines for the dynamic mixing method. A schematic diagram of one of them is shown in Fig. 1. The ion beam from a backet-type ion source was accelerated in the range 5-40kV. The maximum current was 100mA or higher and the beam diameter at the sample holder was 5cm. Another element was deposited onto the holder from an evaporator with an electron gun. The composition of the prepared film was controlled by measuring the ion current and deposition rate, using a Faraday cup and a thickness monitor, respectively. The temperature of the specimen holder was controlled by water cooling and electric heater.

3. Results and Discussions

a) TiN Films[6,9]

Coating films of titanium nitride were prepared on various substrates by evaporation of titanium and simultaneous bombardment

---

[1]) College of General Education, University of Tokyo, Komaba, Meguro, Tokyo 153, Japan.

[2]) Government Industrial Research Institute, Osaka, Ikeda, Osaka 563, Japan.

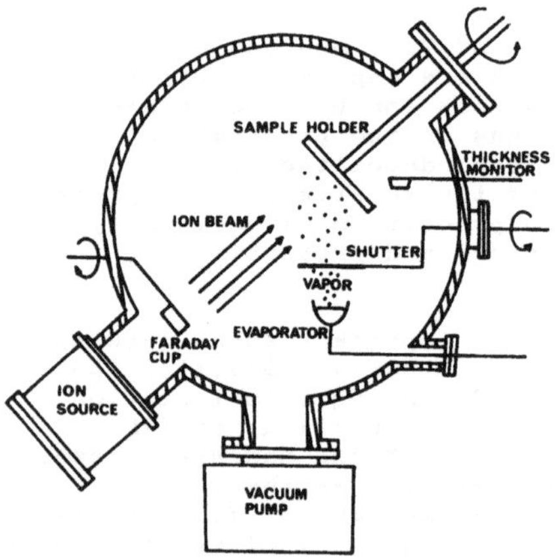

Fig. 1. Schematic diagram of the dynamic mixing machine

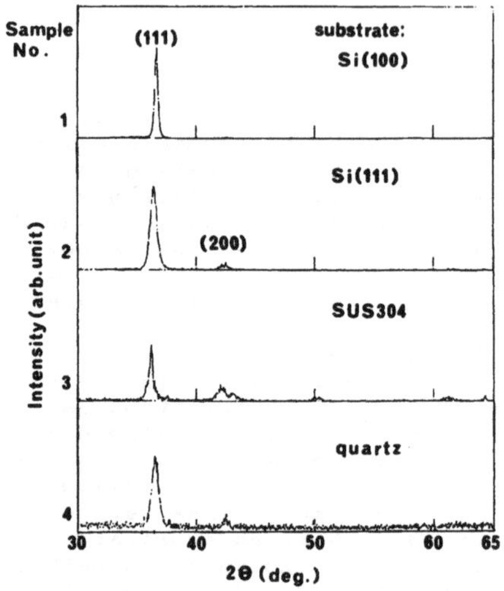

Fig. 2. X-ray diffraction of Cu-Kα X-rays for samples No.1-4.

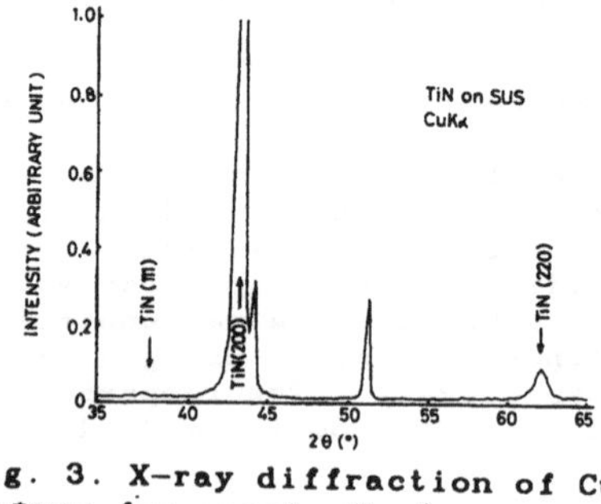

Fig. 3. X-ray diffraction of Cu-Kα X-rays for sample No.5.

of nitrogen ions accelerated at 30kV. The beam direction was normal to the substrate surface. The numerical ratio of atomic to molecular ions, obtained by a separate measurement using a magnet, was about unity. The basic pressure of the chamber was $10^{-6}-10^{-7}$ Torr and the pressure during the film preparation was of the order of 10^{-5} Torr. The difference is due to leakage of nitrogen from the ion source. The temperature during the film formation was about 300°C or lower. The preparation conditions and substrates are shown in Table 1.

Table 1. Preparation condition of TiN coating films

Sample No.	Sub-strate	Deposition rate of Ti (Å/s)	Current density of ion beam (mA/cm^2)	Film thick-ness (Å)
1	Si(100)	6.8	0.1	7000
2	Si(iii)	8.7	0.3	10000
3	SUS304	8.7	0.3	10000
4	quartz	8.7	0.3	10000
5	SUS304	0.6	0.061	3000

The structure of the films was studied by the θ-2θ method of Cu-K_α X-ray diffraction for the surfaces. Diffraction patterns observed from samples No.1-4 are shown in Fig. 2, where the (111) reflections from TiN crystal with B1 structure (NaCl type) are very strong. This result reveals that the [111] axes orient preferentially normal to the film surface.

Fig. 3 shows a diffraction pattern from sample No.5, where the (200) reflection from TiN crystal is very strong. This result represents that the surface normal of the sample No.5 is the [100] axes and is different from that of samples No.1-4.

The difference in preparation conditions between samples No.1-4 and 5 is the deposition rates of two elements. For the former samples the rate Ti/N is 2.6-6.1 which is much greater than that for the latter one, 0.9. Despite the different rate, the crystal structure was cubic. This situation arises from the properties of titanium that it easily absorbs nitrogen of leakage gas from the ion source and titanium nitride has the cubic structure in the wide range of nitrogen concentration shown in the phase diagram.

These results indicate that the structure of the films is independent of the crystal structure of the substrates, but depends on the preparation conditions. The essential difference for the two cases is the energy deposition from the nitrogen ion beam per titanium atom. The energy difference must cause the orientation difference of the crystal growth.

In order to study the mechanism of the crystal growth in the dynamic mixing method, the cross-sectional crystal structure and morphology of a TiN film deposited on a Si(100) substrate are observed by a transmission electron microscope. Fig. 4 shows the micrograph, where the amorphous intermixed layer with 400Å thickness is clearly observed between the silicon crystal substrate and the TiN film. The thickness of intermixed layer agrees well with that obtained by SIMS for sample No.5[6].

Fig. 5(a) represents the electron diffraction pattern of the whole area of the cross section. Fig. 5(b), (c) and (d) show

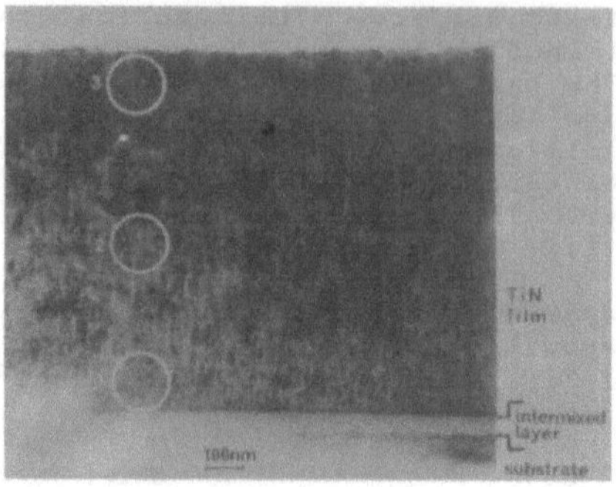

Fig. 4. The cross-sectional profile of TiN film on Si(100) by TEM. Circles indicate the area which gave the diffraction patterns shown in Fig. 5(b), (c) and (d).

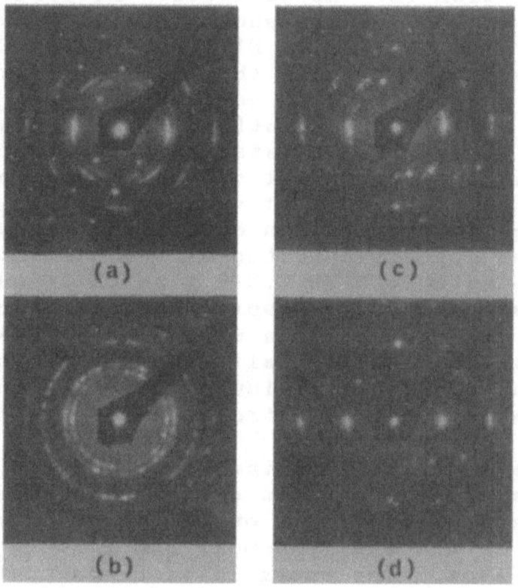

Fig. 5. (a) The diffraction pattern from the sectional profile of TiN film, (b) From the bottom part (1), (c) From the middle part (2) and (d) From the upper part (3) in Fig. 4.

those from selected areas indicated by 1, 2 and 3 in Fig. 4, respectively. As seen in the series of Fig. 5, crystal nuclei with all orientations are formed during the primary stage of film formation. Then the crystal structure varies from isotropical orientation to preferential. Moreover, we can see lines running parallel to the incident ion beam in the TiN film in Fig. 4. This is a column structure with preferential orientation.

b) MoN Films[7]

Molybdenum nitride with B1 stracture (B1-MoN) is a superconductive material whose transition temperature has been predicted from an energy band calculation to be 29.4K. However, B1-MoN does not exist in the equilibrium phase diagram of the Mo-N system.

The present authors and collaborators have tried to produce B1-MoN by the dynamic mixing method. The deposition rate of molybdenum was fixed at 1Å/s and the accelerating voltage, current density of nitrogen ion beam and substrate temperature were changed. Silicon wafers with their surface of (100) plane were used as substrates. The preparation condition is shown in Table 2, where the composition ratio, N/Mo, was measured by the Rutherford backscattering spectroscopy (RBS) of 1.5 MeV helium ion beam.

Table 2. Preparation condition of B1-MoN coating films

Sample No.	Subst. temp. (°C)	Mo dep. rate (Å/s)	N Current ($\mu A/cm^2$)	Acc. volt (kV)	N/Mo	Pref. orient.	Trans. temp. (K)
1	RT	1	60	20	1.0	[110]	5.8
2	RT	1	82	20	1.4	[110]	5.3
3	450	1	81	5	1.4	[100]	4.9
4	500	1	60	5	1.0	[100]	5.0

In Fig. 6, a reflection diffraction pattern (RHEED) from sample No.1 which was prepared at room temperature is shown. This pattern indicates that the film is formed by crystallites with cubic structure and that their [110] axes orient preferentially normal to the substrate surface, though the surface of silicon substrate is the (100) plane. The reflection microscopic image reveals a bumpy surface. More detailed surface information can be seen in a scanning electron microscopic image (SEM) shown in Fig. 7. This image shows that the film has a column structure of diameter 400-2000Å. This is very similar to the case of TiN films.

Samples No.3 and 4 were made by 5keV nitrogen ion beam at substrate temperatures 450 and 500°C. The X-ray diffraction studies showed that all films prepared at high substrate temperature have a cubic structure with the [100] axes normal to the substrate surface, differing from the samples prepared at room temperature. Fig. 8 shows an X-ray diffraction pattern from sample No.3 measured by the θ-2θ method for the surface where only the (200) reflection is dominant. We can say that the films were formed by the epitaxial growth on silicon substrates.

The superconducting transition temperature of films with the B1 structure prepared by the present method was between 4.5 and

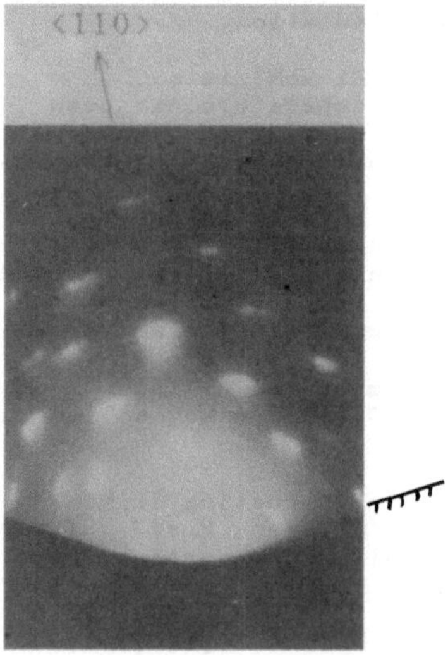

Fig. 7. RHEED pattern from sample No.1 of MoN.

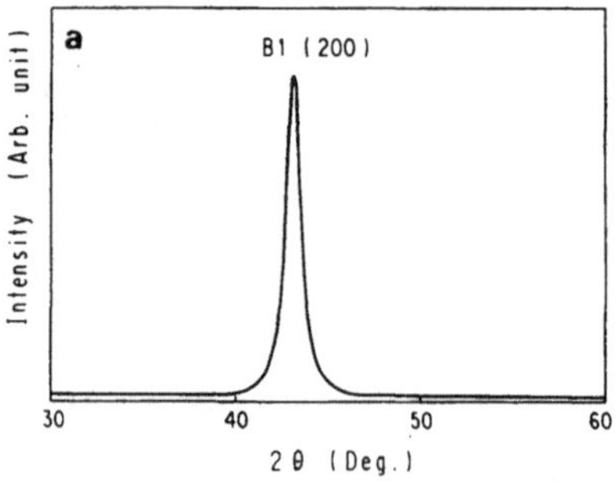

Fig. 8. X-ray diffraction of Cu-Kα X-rays from samle No.3 of MoN.

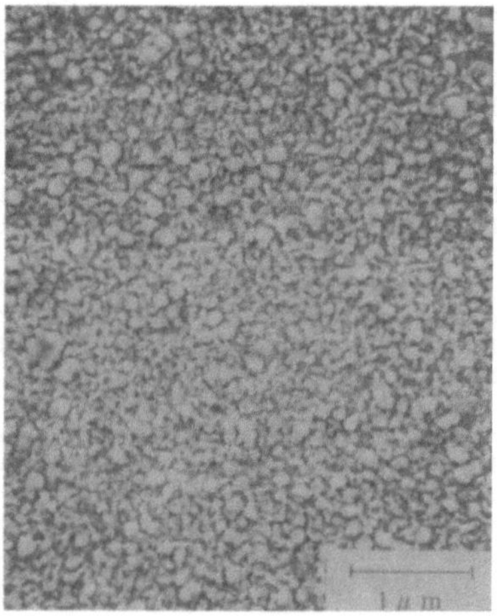

Fig. 6. SEM image of the surface of sample No.1 of MoN.

5.8K. These values are much lowere than expected. As a reason for a low transition temperature , we may consider crystal imperfection due to the ion bombardment.

c) BN Films[3,5,7]

It is well known that stoichiometric boron nitride has three kinds of structure, namely cubic (c-BN), wurtzite (w-BN) and hexagonal (layered, h-BN). The c-BN has almost the same hardness as diamond and is very stable thermally. Usually c-BN is produced under a condition of 1500°C and 5GPa. Concerning the formation of c-BN films by using ion beam deposition, Weissmantel has reported evidence for its presence[10].

The film formation of boron nitride was carried out by the present method. The accelerating voltage of the nitrogen ions was changed in the range of 2-40kV and the current density was 1-8 Å/s. Silicon wafers with their surface of (100) plane and cleaved rocksalts were used as substrates and were kept at room temperature. The film thickness was 0.1-1μm.

The composition ratio, B/N, in prepared films was measured by X-ray photoelectron spectroscopy (XPS) and RBS of 2.0 MeV proton beam for films prepared on thin aluminum foils. The B/N ratio for the boron bound with nitrogen measured by XPS spectra is almost unity, despite the change of B/N ratio. This fact indicates that all of the nitrogen atoms in films bind with boron ones while the excess boron exists as metallic boron.

The structure of films was studied by X-ray diffraction of Cu-K_α, the infrared absorption spectra for the samples prepared on silicon wafers and the transmission electron microscope at 150 kV for those on cleaved rocksalts which were floated off on water. These results reveal that films consist of mixed structures of h-, w-, and c-BN. Electron micrographs and their diffraction patterns are shown in Fig. 9(a), (b), (c) and (d) which are obtained from samples prepared at 30-35 kV and B/N = 1.1, 0.95, 0.5 and 0.5, respectively. The structure of the sample shown in Fig. 9(a) consists of small crystallites with structure close to c-BN. The sample of Fig. 9(b) has amorphous structure with atomic distance of h-BN. Structures seen in Fig. 9(c) and (d) are w- and c-BN, respectively. These structures are always observed in all films independent of the preparation condition and the composition.

Films with the w-BN structure have always preferential orientation of its c-axis normal to the film surface as seen in Fig. 9(c). For samples with the c-BN structure shown in Fig. 9(d), the surface normal is parallel to the [100] axes of c-BN. The grain size in Fig. 9(c) and (d) is 100-3000Å. Sometimes, big grains of c-BN of a few microns were observed. Films formed by grains of w- and c-BN structures are considered to have a column structure.

Vickers hardness was measured on films with thickness of about 1μm. The result shows that the films are 3-5 times harder than those of silicon (about 900Hv) in spite of the difference of the composition ratio B/N.

4. Conclusions

Properties of crystal structure of TiN, MoN and BN films

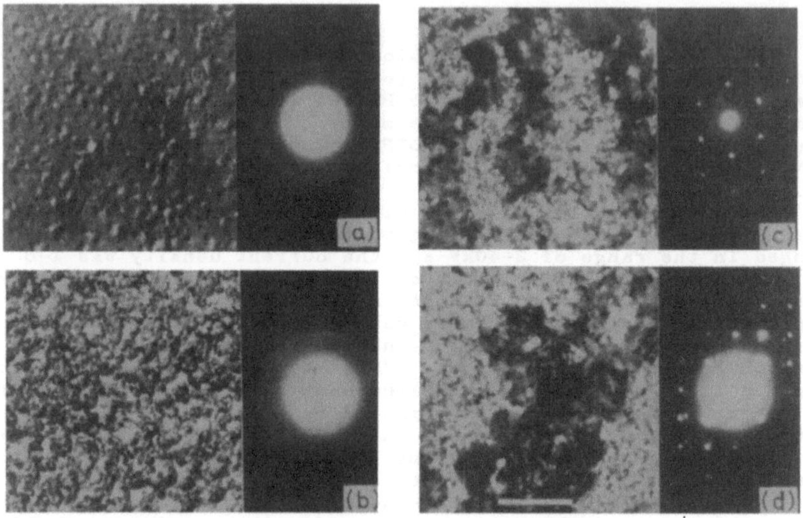

Fig. 9. Electron micrographs and their diffraction patterns. composion ratios are (a) B/N=1.1, (b) B/N=0.95, (c) B/N=0.5 and (d) B/N=0.5. The scale marker shown in the lowest micrograph =2μm.

prepared by the dynamic mixing (or IVD) method are reported. Films produced at temperatures lower than 300°C have an amorphous intermixed layer which serves as an isolating zone for crystalline intercorrelation between the film and substrate. This intermixed layer plays an important role of strong adhesion of film with substrate. Produced films had preferential orientations which depend on preparation conditions such as the ion beam direction, deposition rate of energy from the ion beam and the temperature of substrate. The present method has potential for wider application to produce various types of films with controlled crystal structure and properties.

References

1) M. Satou, F. Fukui and F. Fujimoto, Proc. Int. Workshop by Professional Groups on Ion-based Techniques for Film Formation (Ionics Corp. Ltd., Tokyo) p.349.
2) M. Satou, K. Matsuda and F. Fujimoto, Proc. 6th Symp. Ion Sources and Ion-Assisted Technology, Tokyo (1982) p.425.
3) M. Satou and F. Fujimoto, Jpn. J. Appl. Phys. 22 (1983) L171.
4) Y. Andoh, Y. Suzuki, K. Matsuda, M. Satou and F. Fujimoto, Nucl. Instr. and Meth. B6 (1985) 111.
5) M. Satou, K. Yamaguchi, Y. Andoh, Y. Suzuki, K. Matsuda and F. Fujimoto, Nucl. Instr. and Meth. B7/8 (1985) 910.
6) M. Satou, Y. Andoh, K. Ogata, Y. Suzuki, K. Matsuda and F. Fujimoto, Jpn. J. Appl. Phys. 24 (1985) 656.
7) Y. Andoh, K. Ogata, Y. Suzuki, E. Kamijo, M. Satou and F. Fujimoto, Nucl. Instr. and Meth. B19/20 (1987) 787.
8) F. Fujimoto, Y. Nakane, M. Satou, F. Komori, K. Ogata and Y. Andoh, Nucl. Instr. and Meth. B19/20 (1987) 791.
9) M. Kiuchi, M. Tomita, K. Fujii, M. Satou and R. Shimizu, Jpn. J. Appl. Phys. 26 (1987) L938.
10) C. Weissmantel, K. Bewilogua, D. Dietrich, H.-J. Erler, H.-J. Himerberg, S. Klose, W. Nowick and G. Reisse, Thin Solid Films 72 (1980) 19.
C. Weissmantel, J. Vac. Technol. 18 (1981) 179.

MULTILAYER STRUCTURE CREATION AND CRYSTAL SURFACE MODIFICATION BY LASER-PRODUCED PLASMAS

S.V.Gaponov and M.D.Strikovsky[1)]

1. Laser-produced plasma technique (LPPT) /1-3/

The LPPT uses a conventional ion-beam processing scheme where a source of particles and a crystal, the surface of which is bombarded, are in the vacuum volume (Fig. 1). The LPPT is distinguished by that a particle flux is produced by a short (10^{-8} - 10^{-7} s) powerful laser pulse focused on the target surface. The particle flux is produced as a microburst in which a small evaporated portion of matter is ionized in an intense ($q \sim 10^9$ W cm^{-2}) laser beam absorbing the beam energy almost entirely and transforming to a dense plasma bunch with a temperature about 10 eV. With expanding into the vacuum normal to the target surface, the plasma is recombinated and accelerated producing a beam of erosion particles bombarding the crystal surface.

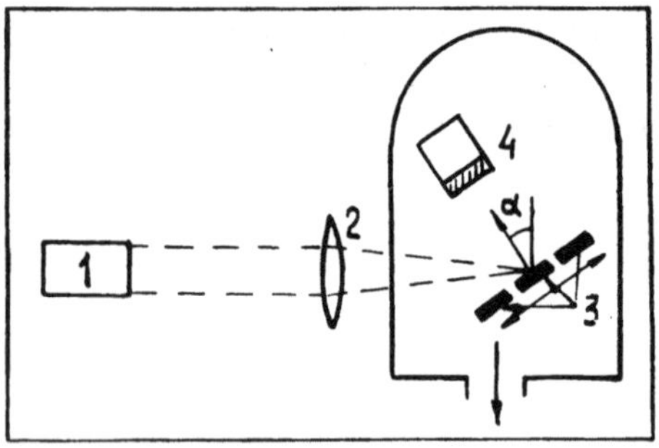

Fig. 1. LPPT scheme: 1 - laser, 2 - focusing lens, 3 - accessory targets, 4 - substrate with heater

Three specific features of the LPPT can be pointed out. The first one is the thickness h of the layer deposited on the surface with one laser pulse. The plasma production on the target weakens the dependence of the amount H of evaporated matter on the laser radiation intensity: $H \sim q^{1/2}$. Also, the plasma temperature increases with q and the angular distribution diagram broadens (Fig. 2a). As a consequence, the film thickness becomes stabilized in a rather wide range of q; the typical value is $h \sim 0.2 - 0.5$ nm.

1) Institute of Applied Physics, Academy of Sciences of the USSR, 46 Uljanov Street, 603600 Gorky, USSR

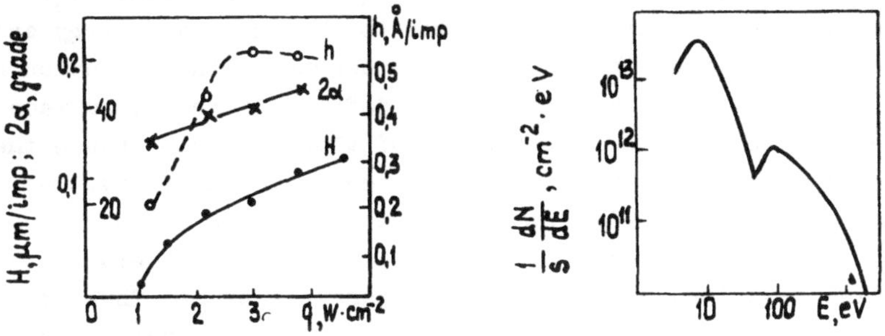

Fig. 2. Plasma production parameter vs laser radiation density (see paper) and typical energy spectrum of plasma flux particles

The second feature of the LPPT is related to the matter arriving rate at the crystal surface. The particle flux density is $j \sim 10^{20}$ $cm^{-2} \cdot s^{-1}$ which is not achievable by conventional techniques. When these particles are condensed, the adatoms interact with each other more often than with the sinks on the surface. This leads to the production of continuous films beginning with a monomolecular thickness.

The third feature is the energy spectrum of particles. The typical spectrum (Fig. 2b) covers the range up to 5 to 1000 eV including the value of threshold energy ($\sim$ 20 eV) needed for the lattice point defect production. The action of particles with such a broad spectrum onto the crystal surface is a complex process accompanied by implantation of fastest ions, partial sputtering, production of nonequilibrium lattice point defects and condensation of particles onto the excited surface. Normally, the result of the particle action depends on the ratio of low-energy particles to defect-producing particles. In practice, a film is produced if the condensation is not specially prevented.

2. Laser deposition of films and multilayer structures /4-6/

The burst-like production of erosion laser-produced plasma eliminates any limitations on the composition and thermophysical characteristics of the target material and provides a nonfractional (steheometric) deposition of multicomponent films. Of particular interest is the obtaining of superthin single-crystal semiconductor films useful for producing a "superlattice". The potentialities of the LPPT are illustrated in Table 1. The high rate of condensation, $V \sim 10^4 - 10^5$ nm s^{-1}, at which individual atoms become critical nuclei, allows to obtain superthin continuous films. The initial temperature of oriented film growth (T_e) with laser deposition is lower than that with the conventional thermal deposition. There are two reasons for decreased T_e. The first is an additional heating effect due to the internal energy evolved by condensed matter on the surface. The second is the formation of a network of nonequilibrium vacancy-type defects in the subsurface layer of the substrate; the electric fields of these defects induce the epitaxial growth with the adatoms having restricted mobility. The latter is strong enough to provide the film orientation on the surfaces of single crystals covered by thin (1 - 3 nm) amorphous sublayers if the plasma contains fast ions passing through these sublayers. Thus, the LPPT has allowed to obtain continuous superthin single-crystal films of many semiconductor (PbTe, PbSe, InSb, CdTe) and dielectric materials.

Table 1

Continuous single-crystal films	Thickness 1 - 100 nm, $V \sim 10^5$ nm s^{-1} InSb (T_e = 150°C), PbTe (T_e < -160°C), CdTe (T_e = 100°C), GaAs
Superlattices with single-crystal layers	InSb (5 nm) - CdTe (2 nm) Bi (5 nm) - CdTe (2 nm)
Superlattices with amorphous barrier layers	InSb (5 nm) - GaAs (1.5 nm) InSb (5 nm) - C (2.5 nm)
Multilayer X-ray mirrors	Ni (1.85 nm) - C (4.3 nm) λ = 4.47 nm, R = 28%, $\lambda/\Delta\lambda$ = 60

The stable thickness of layers deposited with one pulse allows to use a program for the film deposition of precisely specified thickness and to create multilayer periodic structures. The structures of superthin semiconductor films (superlattices) seem to be promising for electronics and laser engineering. Applying the LPPT, superlattices of new pairs of semiconductors and superlattices with amorphous barrier layers were obtained, which are a completely new type. The absorption spectrum of a quantum superlattice InSb (5 nm) - CdTe (2 nm) has minibands and the static I-V characteristic contains a region of negative resistance (Fig. 3).

The second type of multilayer structures obtained by the LPPT is normal-incidence mirrors for low X-rays (λ = 1 - 30 nm) /7, 8/. Here, the problem concerning the obtaining of smooth superthin films of precisely pre-set thickness is successfully solved. The parameters of these multilayer mirrors (see Table 1) are of the world-standard level.

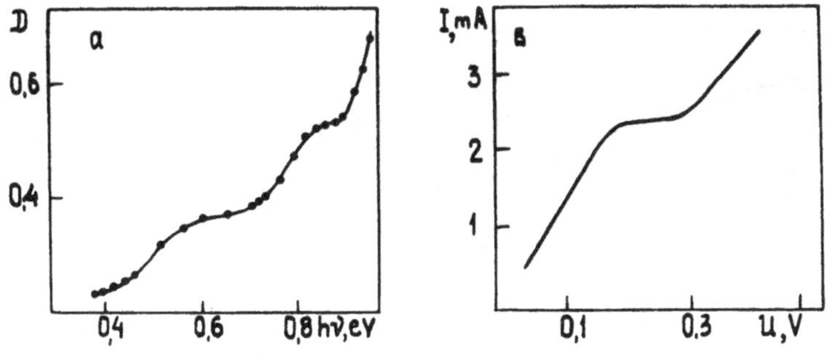

Fig. 3. Spectrum dependence of optical absorption (a) and static I-V characteristic (b) of a ten-period superlattice InSb (5 nm) - CdTe (2 nm)

3. Modification of crystal surface by laser-produced plasmas /9-11/

Particles with the energy 10^2 - 10^3 eV in laser-produced plasma allow to use a scheme of thin-layer structure creation differing from the film deposition one. This is due to the intense fast-ion generation of nonequilibrium point defects (vacancies) which are able to diffuse to a depth of $L \sim 0.1 - 1\ \mu m$ during their lifetime. In the excited surface layer of thickness $\sim L$, processes whose rate is determined by the collision frequency with migrating vacancies (diffusion of impurities, formation and destruction of electrically-active centres and of defect complexes) can be sharply increased in activity at low temperatures. The crystal surface is modified by the variation of the electrophysical characteristics of the subsurface layer. The nonthermal induction of the processes by laser-induced plasma features a higher flux density (three to four orders of magnitude) than that of conventional sources of low-energy ions. As a result, vacancies are produced in such a large number that reactions which are nonlinear with respect to the vacancy density can dominate.

One way of surface modification by the LPPT is production of doping layers of Si, wherein the flux of erosion plasma of donors (P, As and Sb) is a source of impurities while providing its induced diffusion. The doping depth x is governed by the irradiation time t ($x \sim \ln D_o t$) in the range of 10 to 300nm. The carrier density near the surface is determined by the donor type and ranges from 10^{19} to $10^{20} cm^{-3}$. The nonequilibrium diffusion coefficient D_o is increased in proportion to the flux density j and reaches $\sim 10^{-7}\ cm^2\ s^{-1}$ being nondependent of the surface temperature (Fig. 4). The submicron layers of Si obtained in such a manner are considered to be attractive in solar cells engineering.

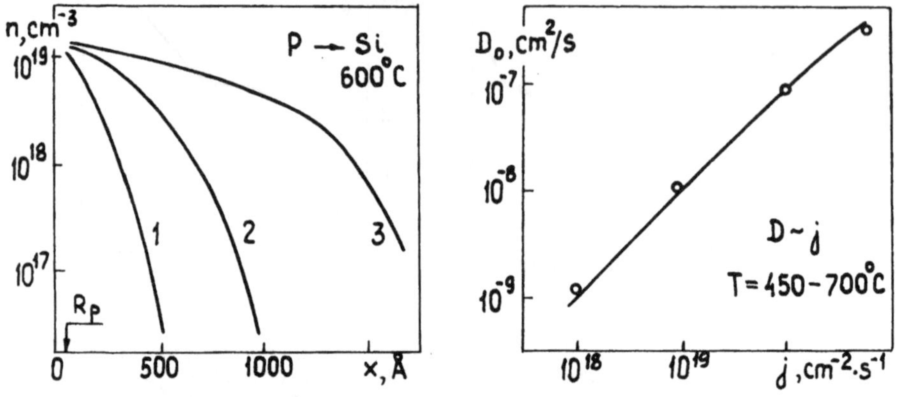

Fig. 4. Dynamics of doping layer formation (a) and phosphorous flux density dependence of diffusion coefficient when exposing Si

When GaAs crystals are exposed to laser-produced plasma, the formation of a carrier-depleted subsurface layer is a dominating process (Fig. 5). The effect is attributed to the capture of free carriers to the deep levels of accumulating complexes of defects, the rate of the process increasing quadratically with the particle flux density. The effect is strongly pronounced for intense flux of laser-produced plasma. The conductivity is compensated irrespective of the type of the bombarding particles but depends on the type of the doping impurity in the original GaAs. This way of surface modification gives isolating regions on the surface of GaAs as deep as 0.5 μm since the specific conductivity of a depleted layer is more than 10^7 $\Omega\cdot$cm.

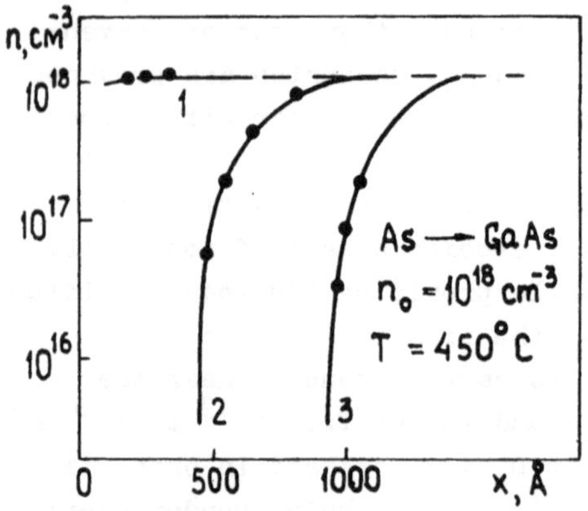

Fig. 5. Dynamics of carrier-depleted layer formation when exposing GaAs to plasma As

The examples above show that the flux of erosion laser-produced plasma and the crystals interaction is a rather complex process. The investigation of this process leaves plenty of room for the creation of thin-layer structures by the LPPT.

References

(1) Akhsakhalyan, A.D., Ya.A.Bityurin, S.V.Gaponov et al: Zhurnal teoreticheskoj fiziki 52 (1982) 1584-1596
(2) Bityurin, Yu.A., S.V.Gaponov et al: Elektronnaja promyshlennost N 5 (1981) 110-115
(3) Gaponov, S.V., A.A.Gudkov et al: 52 (1982) 1843
(4) Gaponov, S.V. and N.N.Salashchenko: Elektronnaja promyshlennost N 1 (1976) 11-20
(5) Gaponov, S.V., B.M.Luskin and N.N.Salashchenko: Fizika i tekhnika poluprovodnikov 14 (1980) 1468
(6) Gaponov, S.V., B.M.Luskin and N.N.Salashchenko: 39 (1981) 301
(7) Gaponov, S.V., S.A.Gusev et al: Sol.St.Comm. 38 (1981) 7
(8) Gaponov, S.V., S.A.Gusev et al: Zhurnal teoreticheskoj fiziki 56 (1986) 891
(9) Gaponov, S.V., M.D.Strikovsky and M.A.Kalyagin: Izvestija vuzov - fizika 49 (1985) 758
(10) Bityurin, Yu.A., S.V.Gaponov, E.B.Kljuenkov and M.D.Strikovsky: Fizika i tekhnika poluprovodnikov 18 (1984) 1729
(11) Gaponov, S.V., M.D.Strikovsky and M.A.Kalyagin: Preprint No. 134, Physical Institute, Acad. Sci. USSR, Moscow 1986

RANGE PROFILE CALCULATIONS FOR THE HIGH ENERGY IMPLANTATION OF PHOSPHORUS INTO SILICON

M. Posselt, W. Skorupa[1)]

1. Introduction

The main difficulty in the theoretical prediction of high energy implantation profiles, especially for the implantation of phosphorous into silicon, is caused by the fact that for MeV energies the existing formulae for the electronic stopping cross-section S_e are not very precise. In a previous work we have reviewed some essential models for S_e (7). We have also compared the experimental data with the theoretical range profiles which have been calculated using different models for S_e. We have found that in the energy range from 0.99 MeV to 4 MeV the experimental values are in satisfactory agreement with the theoretical range profiles which have been determined employing the original Brandt-Kitagawa approach (3).Recently, new range data have been published (6)(8)(9)(10)(11)(12). In the present work we compare the available experimental results for the MeV implantation of phosphorous into silicon with the theoretical calculations. The choice of the appropriate electronic stopping model for precise range profile predictions is investigated. The discussion shows that some of our previous conclusions (7) must be modified.

2. Comparison of experimental range data with theoretical predictions

Table 1 gives an overview over the available experimental data. They have been obtained by three different methods: (i) SIMS, (ii) spreading resistance profiling (SRP), and (iii) pulsed CV-measurements. The SIMS data of the as-implanted samples are most appropriate for the comparison with the theoretical range profile calculations. The other techniques use annealing steps which may lead to a redistribution of the implanted species. But this effect is generally not very important: In the case of short time annealing the redistribution of the dopants is negligible. Further, we can assume that furnace annealing does not influence the depth of the profile peaks. However, furnace annealing may yield broader range distributions.
To consider the choice of an appropriate electronic stopping formula for range profile predictions we study three models(7): (i) the equipartition between the Lindhard-Scharff and the Bethe-Bloch formulae (LSBB), (ii) the original Brandt-Kitagawa approach (BK), and (iii) the modified Brandt-Kitagawa model of Ziegler, Biersack and Littmark (ZBL). At first we want to get an overview over the range values. Fig. 1 shows the mean projected ranges $R_p(E)$ for the three models. The analytical method of Biersack (1) has been employed to calculate R_p. In the LSBB model and in the ZBL model there are the adjustable parameters C_k and C_Λ, respectively (2)(3). In fig. 1 curves for different values of these parameters are shown .The experimental range data are also shown in this figure. For implantation energies above 2 MeV only few experiments have been performed.

---

1) Academy of Science of the GDR, Central Institute for Nuclear Research, Rossendorf, Postfach 19, DDR-8051 Dresden, GDR

In some cases the scattering of the range data measured by different authors is considerable. In particular, for the 4 MeV implantation the data of Ingram et.al.(5) show important deviations from the other experimental results. Unfortunately, these data have been used in the discussion of our previous paper (7). Note that many experimental range values concern the depth R_m of the profile peak instead of R_p. But due to the scattering of the experimental results this difference is not visible in fig. 1. In the entire energy range considered in fig. 1 the best agreement between the experimental and the theoretical ranges is achieved for the LSBB curve with C_k=1.2. The BK curve yields reasonable ranges below energies of about 1.2 MeV. The ZBL model seems to be applicable with different values of C_Λ ($1.0 \geq C_\Lambda \geq 0.8$) for different energy intervals.
In the following we discuss the comparison of available experimental profiles with the theoretical predictions. Taking into account the scattering of the experimental results published by different authors we want to limit our considerations to the two cases in which two or more experimental range profiles have been found in the literature. All range profile calculations have been carried out using the TRIM Monte Carlo simulation program (2)(13). In fig. 2 the cases of 0.99 MeV and 1.0 MeV implantation are shown. The experimental values have been obtained by the normalization of the data from three different authors. The agreement of the experimental data with the theoretical range profiles drawn in fig. 2 is rather good. The BK and the LSBB (C_k=1.2) profiles are very similar. The ZBL (C_Λ=1.0) profile is narrower than the other range distributions.
Fig. 3 shows the curves for the 2 MeV implantation. For the LSBB model (C_k=1.2) and the ZBL model (C_Λ=0.9) a satisfactory agreement between the theoretical profiles and the experimental data is achieved. The SRP profiles in figs. 2 and 3 reveal that the activations of the dopants in the corresponding samples have been different.

Table 1

energy/MeV	range prof.	range param.	method (annealing)	refs.
0.1,0.2,0.4	X	X	SIMS(--)	(11)
0.8,1.2,2.0	-	X	SIMS(--)	(6)
			SIMS(furnace)	(6)
			SRP(furnace)	(6)
1.2,4.0,6.0	X	-	SRP(furnace)	(5)
0.99	X	-	SIMS(--)	(7)
			CV(flash-lamp)	(7)
1.0,2.0,3.0	X	X	SRP(furnace, lamp)	(4)
1.0,1.3,1.5, 1.6,2.0	X	-	SRP(furnace)	(8)
1.0	X	-	SRP(furnace)	(10)
2.0	X	-	SIMS(furnace)	(9)
1.0,2.0,3.0, 4.0,5.0,6.0, 7.0	-	X	SIMS(--)	(12)

Fig. 1

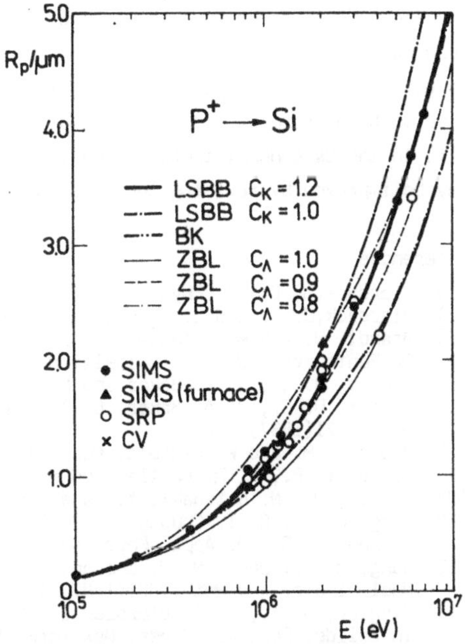

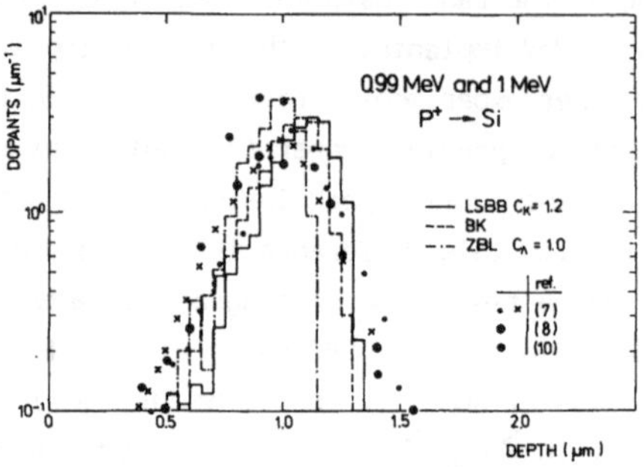

Fig. 2

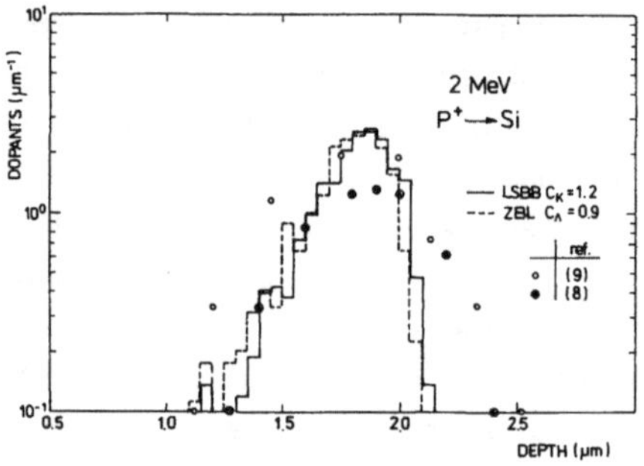

Fig. 3

3. Conclusions

Three models for the electronic stopping cross-sections have been investigated: (i) A rather satisfactory description of the most experimental range data can be given by the theoretical calculations which employ the LSBB electronic stopping model with the parameter C_k=1.2. (ii) Below implantation energies of about 1.2 MeV the BK model leads to a good agreement between the theory and the experiments. (iii) Changing the parameter C_Λ from 1.0 at lower energies to 0.8 at higher energies also the ZBL model gives reasonable results.

To get reliable range profile predictions for the MeV implantation further investigations have to be performed: (i) More and sufficiently precise experimental data are necessary. (ii) In the BK concept there are some rough approximations. We suppose that this model may be improved by revising these approximations.

References

(1) Biersack, J. P.: Z. Phys. A305(1982) 95
(2) Biersack, J. P. and L. G. Haggmark: Nucl. Instr. and Meth. 174(1980) 257
(3) Brandt, W. and M. Kitagawa: Phys. Rev. B25(1982) 5631
(4) De Pontcharra, J. , P. Spinelli and M. Bruel: Mat. Res. Soc. Symp. Proc. 45(1985) 115
(5) Ingram, D. C., J. A. Baker and D. A. Walsh: Nucl. Instr. and Meth. B7/8(1985) 361
(6) Ingram, D. C., J. A. Baker, D. A. Walsh and E. M. Strathman: Nucl. Instr. and Meth. B21(1987) 460
(7) Posselt, M. and W. Skorupa: Nucl. Instr. and Meth. B21(1987) 8
(8) Spinelli, P., A. M. Cartier and M. Bruel: Nucl. Instr. and Meth. B21(1987) 450
(9) Tamura, M., N. Natsuaki, Y. Wada and E. Mitani: Nucl. Instr. and Meth. B21(1987) 438
(10) Wang, L. K., C. T. Chuang and G. P. Li: Solid-State Electronics 29(1986) 739
(11) Wilson, R. G.: J. Appl. Phys. 60(1987) 2797
(12) Wong, H., E. Deng, N. W. Cheung, P. K. Chu, E. M. Strathman and M. D. Strathman: Nucl. Instr. and Meth. B21(1987) 447
(13) Ziegler, J. F., J. P. Biersack and U. Littmark: The Stopping and Range of Ions in Solids, Pergamon Press, New York 1985

DEFECT FLUX EFFECTS DURING ION IMPLANTATION

K.Holldack, H.Kerkow+)

Introduction

Dose rate effects of damage accumulation during ion bombardment have already been observed in the seventies /1-3/, but there is a lack of systematical experiments, because often machine parameters play the dominant role. In order to get reproducible initial conditions for annealing and thus, for the electrical activation of the inserted dopands it seems still nessecary to collect more informations about such dynamical events occuring already during implantations at room temperature /4,5/. This paper deals with flux effects of damage accumulation at different target temperatures and for Ne- and Ar- bombardment.

Experimental

The dose rate effects of damage accumulation were measured with the RBS-method in connection with the channeling technique. For each implantation temperature different doses were chosen to guarantee the linear dependency of the RBS-yield on the defect density. For a given temperature the ion dose had been kept constant and only the beam current was varied. The total number of defects was calculated from the damage peak in the well known manner described in /6/ and then normalized with the corresponding dose to get the average damage contribution per incident ion.

All implantations were carried out with a 350 kV-accelerator. The ⟨111⟩-oriented silicon wafers were irradiated with 60 keV Ne- and 120 keV Ar-ions into random direction with the help of a scanned ion beam of 2 mm diameter. There was no dependency on the scanning frequency.

The RBS-analysis was performed with a 300 keV-proton beam registrating the energy spectra of the backscattered protons electrostatically.

+) Humboldt-University Berlin, Dep. of Phys., Invalidenstr.11o, 1040,GDR

Results

As visible in fig.1 and 2, a dose rate effect could not only be measured for Ne-bombardment (see fig.1 and ref./5/), but also for Ar-implantation (fig.2).

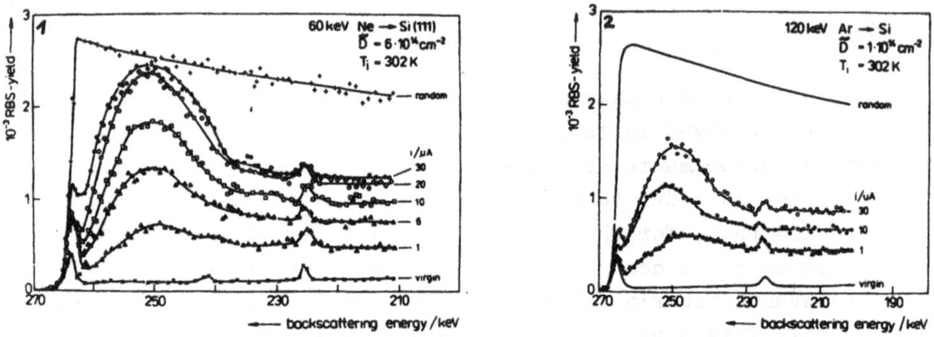

fig.1 and 2: RBS-spectra after Ne- (1) and Ar- (2) implantation with different beam currents.

In the figures the beam current was given instead of the dose rate. The current density is given by $j=30 \times i\ cm^{-2}$ (see /5/) under the beam conditions used. Trying to extract the static effects of damage accumulation and to obtain the pure dynamical flux effect the total amount of damage at given flux was normalized with the corresponding dose. The real additional damage increase per incident ion $N-N_o/\tilde{D}$ was plotted versus the current, where $N_o/\tilde{D}$ represents the total damage per ion for the extrapolation $j \to 0$, N the number of displaced atoms per unit area and D the ion dose. As shown in fig.3 the damage increase with increasing current density sensitively depends on the implantation temperature.

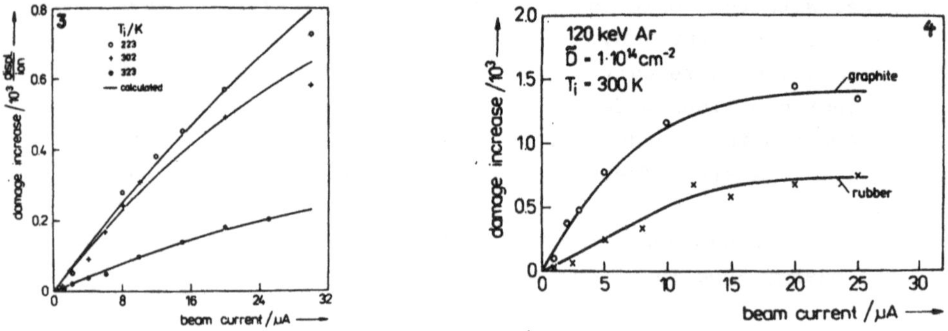

fig.3 and 4: Damage increase per incident ion as a function of beam current due to Ne- (3) and Ar- (4) bombardment

For the heavier Ar-ions one is forced to take higher energy in order to get about the same range values. In this case the flux effect is disturbed by the competing enhanced in-situ-annealing due to the higher beam power. This could be shown by comparing the damage increase under different heat contact conditions between the samples and the target holder (see fig.4). In one case the specimen were sticked with a colloidal graphite solution onto the aluminium holder and in the other case they were pressed onto a special rubber, which defines a weaker heat contact.

For the interpretation of the damage increase with increasing beam current density a dynamical overlapping model was proposed (see ref./5/). It was assumed that there is a thermally stable region around the ion track which is surrounded by a time dependent point defect enriched region where recrystallization and clusterung proceed. If this region is overlapped by subsequent collision events during it's relaxation time, the in-situ-annealing is disturbed and the mobile point defects will be partly converted into stable defect configurations by dynamical gettering effects. Assuming that the damage increase due to the flux effect is proportional to the first order overlapping probability according to /5/ one can calculate it from (see the solid lines in fig.3):

$$N-N_0/D = \alpha (1-\exp(-\beta j)$$

with the abbreviation $\beta = (A_0 + \pi D\tau)\frac{\tau}{e}$, where D is the diffusity of the point defects, τ their lifetime and α the gettering efficiency. An independent determination of D and τ was possible assuming that molecule implantation will correspond to the case of two simultaneously incident ions and it could be estimated $D \approx 10^{-4} cm^2 s^{-1}$ and $\tau \approx 7 \cdot 10^{-7} s$. These values characterize the properties of point defects and we concluded that highly mobile Si-interstitials govern the in-situ-annealing at about room temperature.

References

/1/ Eisen,F.H.,Welch,B.: Europ.Conf. on Ion Implantation,Reading, (Mass.), Stevenage 1970, p. 227

/2/ Picraux,S.T.,Vook,F.L.: Rad.Eff. 11,(1971) 179

/3/ Crowder,B.L.: Proc. US-Japan.Seminar on Ion Implantation in Semiconductors,Kyoto 1971, p.63

/4/ Holland,O.W.,Fathy,D.,Narayan,J. and Oen,O.S.: Rad.Eff. 90,(1985) 127

/5/ Holldack,K.,Kerkow,H.: phys.stat.sol.(a) 98,(1986) 527

/6/ Westmoreland,J.E.,Mayer,J.W.,Welch,B.,Eisen,F.H.: Rad.Eff. 6,(1970) 175

IMPURITY REDISTRIBUTION IN SILICON DUE TO ION BOMBARDMENT

K.Holldack, H.Kerkow+)

Introduction

As it is well known, particle irradiation is able to enhance the diffusity of foreign atoms such as boron and phosphorus in silicon not only for the case of proton bombardment but also due to postbombardment of predoped crystals with heavier ions /1,2/. Recent investigations could show that the point defect kinetics which seems to govern the process is sensitively affected by sink formation during the irradiation /3/. In order to switch a suitable balance between the diffusity enhancement and the competing processes one has to satisfy a need of systematical experimental investigations.

Experimental

Czochralsky grown ⟨111⟩-oriented crystalline silicon wafers previously passivated by a 20 nm thick oxide layer were firstly predoped at room temperature with boron (50 keV, $1x10^{15}cm^{-2}$) or phosphorus (120 keV, $1x10^{15}cm^{-2}$) and then postimplanted with different ion species (He...Xe) at 750°C, a temperature at which the enhancement was found to become maximal /2,4/. The incident energy of the different ion species was chosen in the way that nuclear energy is only deposited within the predoped layer. That means, the energy values were fitted after range tables /5/ to accomplish the requirement R_{pd}(He...Xe)=R_p(B,P), where R_{pd} and R_p are the mean projected damage ranges and the projected ranges, respectively. In this case also the condition $\Delta R_{pd}=\Delta R_p \approx 80nm$ is given (compare ref./3/).
All implantations were carried out with a 350 keV implantation equipment and a special target holder based on a graphite heater.
After the performance the resulting profiles were measured by SIMS. During the implantations one part of the samples had been screened by a mask in order to get a profile for comparison which had been in the same way thermally treated like the irradiated parts.

+)Humboldt-University Berlin,Dep.of Phys., Invalidenstr.110,1040, GDR

Results

As fig.1 shows, the postbombardment of the predoped layers leads to an enhanced outdiffusion of boron and to an extended tail in the diffusion profiles. For a fixed dose and current density the extension of the tails firstly increases with the nuclear deposited energy of the ions (defect generation rate) up to Ne-bombardment and then shrinks again for ions heavier than silicon. Taking a fixed extension l of the profiles at a definite concentration (10^{17} cm^{-3}) one can approximately obtain an excess diffusity D_{irr} by the expression

$$D_{irr} = (l^2 - l_{th}^2)/t ,$$

where l_{th} and t are the corresponding extension of the unbombarded profile and the irradiation time, respectively. As represented in fig.2a the excess diffusity depends on the ion species and the current density. The same experiments made for the predoping with phosphorus had yielded an equivalent dependency on the ion species (fig.2b).

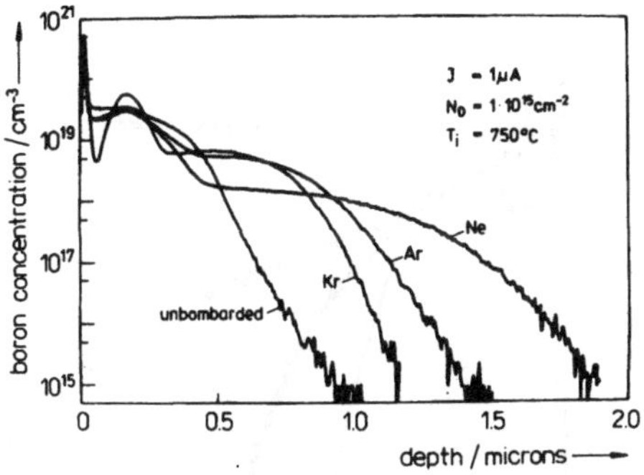

fig.1 Boron profiles after postbombardment at 750°C (SIMS)

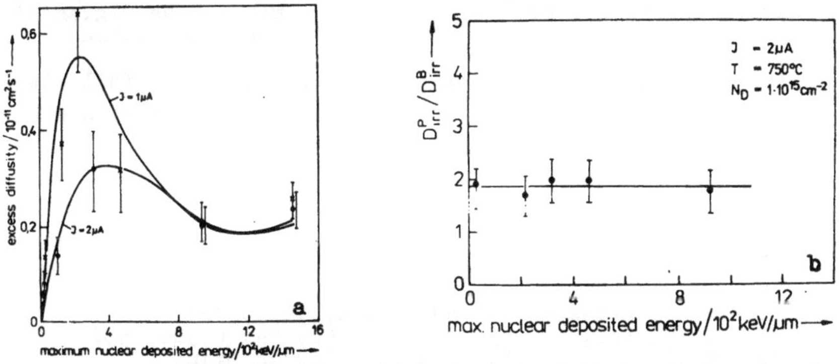

fig.2 Dependency of the excess diffusity on the ion species (nuclear deposited energy of the postbombarding ions) for boron (a) and in relation to boron for phosphorus (b)

In another recent work we could show the connection between extended residual defects and the shrinkage of the profiles by comparison of SIMS and increasing defect density in TEM-images /6/.
As well as dynamical sinks also static defects restrict the enhanced diffusion as visible from the experiments represented in fig.3. In this case a part of the sample was preamorphized by Ar-bombardment (5×10^{15} cm^{-2}, 100keV) and annealed for 30 min at 800°C so that a thermally stable damaged layer of about 100 nm thickness remains.

After this treatment the samples were postbombarded in the described way with Ne (80 keV, $1x10^{15}cm^{-2}$) and again partly masked to extract the thermal effect. The result in fig.3 could show that there is no enhanced diffusion if dynamical damage is generated within a layer of high statical sink density.

The results could be explained by radiation-enhanced diffusion of the impurities due to a dynamical excess of point defects which may act as diffusion vehicles. The number of dynamical mobile point defects will be partly reduced due to sink formation and dynamical gettering due to the overlapping of the defect enriched regions around the individual ion tracks. The dynamical formation of extrinsic stable defects /6/ and a dip in the profiles at the damage range occuring very pronounced during Ne-bombardment at 750°C /7/ could be explained in terms of the inverse Kirkendall effect /8/ and lead to the assumption that self-interstitials play the dominant role. Because there is a full electrical activity in the tail regions /6/, an interstitialcy diffusion mechanism seems probable for both boron and phosphorus.

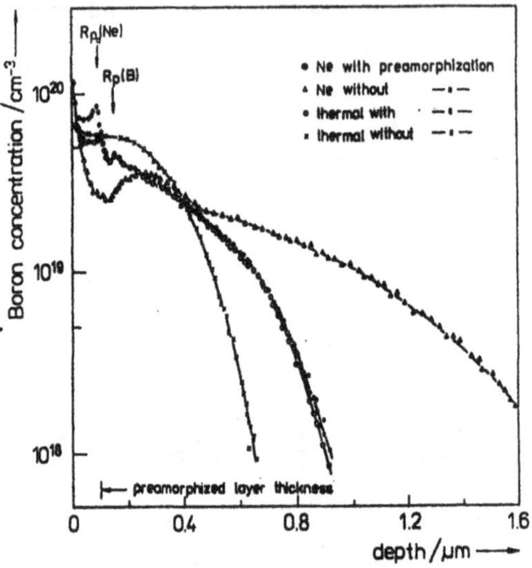

fig.3 Enhanced diffusion of B under different conditions of the crystal perfection within the region of defect generation

References

/1/ Tsuchimoto,T.,Tokuyama,T.: Rad.Eff. 6,(1970) 121
/2/ Holldack,K.,Kerkow,H.: phys.stat.sol.(a) 94, (1986) 357
/3/ Holldack,K.,Kerkow,H.: phys.stat.sol.(a) in print
/4/ Holldack,K.: Thesis,Humboldt-University Berlin 1987
/5/ Burenkov,A.F.,Komarov,F.F.,Kumachov,M.A.,Temkin,M.M.: Prostranstvennye raspedeleniya energii videlennoi v kaskadje atomnykh stolknovennii v tverdykh telakh, Energoatomizdat, Moskva 1985
/6/ Holldack,K.,Kerkow,H.,Reiche,M.: to be published
/7/ Holldack,K.,Kerkow,H.: to be published
/8/ Baruch,P.: Inst.Phys.Conf.Ser.Nr.31,(1977) 120

6 MeV Ni HIGH DOSE IMPLANTATION INTO SILICON

J. K. N. Lindner and E. te Kaat+)

Structural changes of silicon crystals after 6 MeV Ni high dose implantation up to $1.3 * 10^{18}$ Ni/cm² at 450 K are studied by optical reflectivity measurements and TEM analysis. Interaction between damage and implanted Ni atoms occurs. Amorphization of silicon is suppressed at depths of high Ni concentrations. Epitaxial $NiSi_2$ preferentially precipitates at dislocations. Layers with high densities of epitaxial $NiSi_2$ crystallites are formed.

Introduction

Buried metallic layers can be formed by high dose implantation of metals into silicon. Recently, metal implantations with ion energies lower than 200 keV have been performed /1,2,3,4/. In these studies problems due to surface sputtering arise. Therefore, in order to study silicide formation, implantations in the MeV energy region are advantageous. Moreover, the relatively large distance between the maxima of nuclear stopping power and impurity atom concentration permits the separation of influences of the energy deposited in nuclear collisions and influences of the implanted impurity atoms on structural changes of the target.

Implantation

6 MeV $^{58}Ni^{2+}$- ions were implanted from a 7°-off-normal direction into (111) - oriented high purity silicon crystals at the Dynamitron-Tandem-Laboratory at Bochum (FRG). Current densities were 0.3 µA/cm² for doses up to $5 * 10^{14}$ Ni/cm², 7 µA/cm² for doses up $4 * 10^{17}$ Ni/cm², and 12 µA/cm² for doses up to $1.3 * 10^{18}$ Ni/cm². Implantation temperatures, measured at the substrate holder, were kept constant at 300 and 450 K, respectively.

Results and Discussion

The depth of maximum damage R_d as well as the depth of maximum Ni concentration R_{max} have been determined by optical reflectivity depth profiling in combination with small angle bevelling /5,6/. For irradiations at 300 K we find $R_d = 3.0 \pm 0.1$ µm and $R_{max} = 3.3 \pm 0.15$ µm, in good agreement to ranges calculated by a TRIM /7/ simulation ($R_d^{TRIM} = 2.82$ µm, $R_{max}^{TRIM} = 3.17$ µm).
Depth profiles of the optical reflectivity change $\Delta R/R_c$ (R_c: reflectivity of undamaged crystalline Si) for 450 K - samples are shown in fig. 1. Additionally

+)Institute of Physics, University of Dortmund, POB 500500,
D-4600 Dortmund 50, FRG

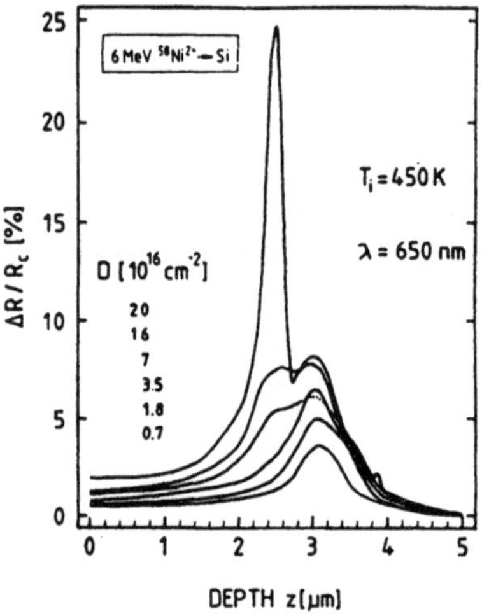

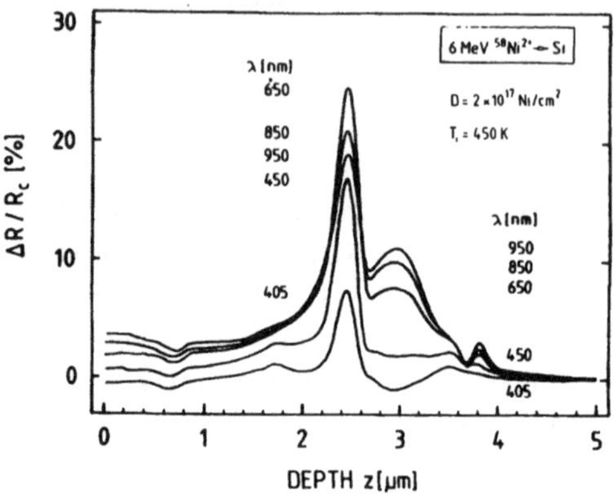

Fig. 2 Relative reflectivity change vs. depth at different wavelengths λ for a 450 K - sample at 2 * 10^{17} Ni/cm².

Fig. 1 Depth profiles of the relative reflectivity change of 450 K - samples.

to the maximum at about 3 µm, at doses from 3.5 * 10^{16} Ni/cm² to 2 * 10^{17} Ni/cm² a second maximum develops at z = 2.45 µm. The wavelength dependence of the reflectivity /5/ for a 2 * 10^{17} Ni/cm² sample (fig. 2) at z = 2.45 µm is typical for amorphous silicon. However, the monotonous increase of $\Delta R/R_C$ with λ at z = 3 µm indicates a different structural state at this depth.

Transmission electron microscopical (TEM) investigations on 450 K - samples reveal dislocations already at a dose of 1 * 10^{17} Ni/cm², similar as for 2 MeV Si - self irradiations at these implantation parameters /8/, and silicide precipitates (fig. 3). They are identified as epitaxial $NiSi_2$ crystallites by means of their (200) - spots in TEM diffraction patterns.
At a dose of 2 * 10^{17} Ni/cm² $NiSi_2$ precipitates are observed between z = 1.8 µm and z = 3.6 µm, with maximum concentration at z = 3.0 µm. They are preferentially found at dislocations. In regions of low Ni concentration and low dislocation densities, small (40 Å) $NiSi_2$ crystallites are grouped in loops and lines. Dislocations are therefore thought to act as sinks for diffusing Ni atoms, leading to high Ni concentrations at depths of high dislocation density, i.e. near R_d. From z = 2.3 µm to z = 2.6 µm, isolated amorphous volumina (Ø ~ 500 - 4000 Å) occur, which form a thin continuous amorphous layer at z = 2.45 µm (compare fig. 2). The layer contains some small (140 Å) $NiSi_2$ crystallites with orientations near that of the silicon substrate.

At doses above 2 * 10^{17} Ni/cm² the amorphous zone near z = 2.45 µm disappears and layers with high densities of epitaxial $NiSi_2$ crystallites are formed (fig. 4). A detailed description of precipitates will be given elsewere /9/. A dose

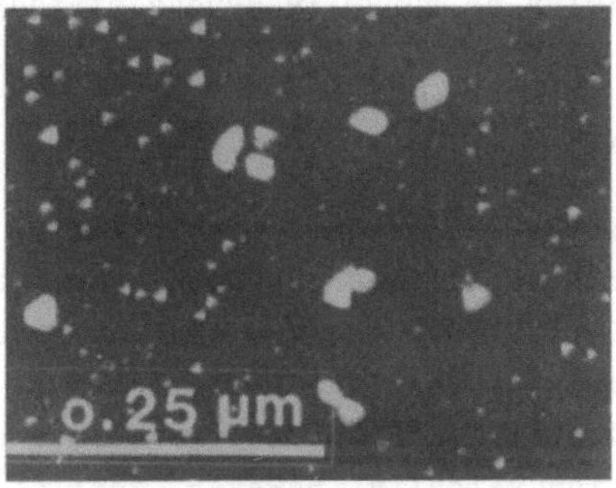

Fig. 3 (200)-TEM dark-field micrograph showing epitaxial $NiSi_2$ precipitates at a dose of $1 * 10^{17}$ Ni/cm^2.

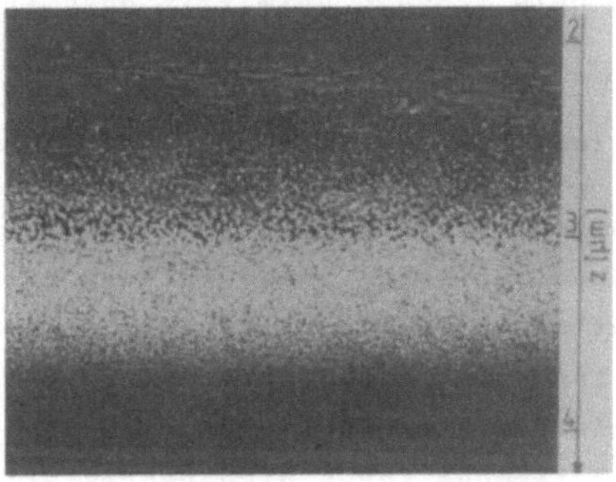

Fig. 4 (200)-XTEM dark-field image of the $NiSi_2$ - rich layer at a dose of $3 * 10^{17}$ Ni/cm^2.

increase to $1.3 * 10^{18}$ Ni/cm^2 results in a layer of adjoining $NiSi_2$ crystallites, while amorphous layers cannot be observed.

At 450 K the critical dose for amorphization by 2 MeV Si ions is $5 * 10^{17}$ Si/cm^2 /8/. Although the critical dose should be considerably lower for 6 MeV Ni ions due to their stronger nuclear stopping power, amorphous layers cannot be observed even at doses as high as $1.3 * 10^{18}$ Ni/cm^2. This result indicates, that implanted Ni suppresses amorphization of the silicon matrix. Moreover, a suppression of amorphization at depths of high Ni concentration explains, why an amorphous layer at a depth far above R_d can be formed and disappears at higher doses. The suppression of amorphization by Ni is probably due to the metallic nature of Si-Ni bonds, which may be responsible for the low activation energy for the crystallization of $NiSi_2$ /10/.

(1) A. E. White, K. T. Short, R. C. Dynes, J. P. Garno, and J. M. Gibson, Appl. Phys. Lett. 50 (2),(1987),p.95.
(2) G. J. Campisi, H. B. Dietrich, M. Delfino, and D. K. Sadana, Mater. Res. Soc. Proc. Vol. 54,(1986),p. 747.
(3) V. Yu. Petukhov, I. B. Khaibullin, and M. M. Zaripov, Phys. Chem. Mech. Surfaces, Vol. 4 (2),(1986),p. 522.
(4) F. H. Sanchez, F. Namavar, J. I. Budnick, A. Fasihudin, and H. C. Hayden, in: Beam-Solid Interactions and Phase Transformations, Boston, MA, USA, 2-4 Dec. 1985 (Pittsburgh, PA, USA: Mater.Res. Soc. 1986),p. 439.
(5) K. F. Heidemann; Phil. Mag. B, Vol.44 No. 4, p. 465 (1981).
(6) For applications of this method for MeV implantations see : N. Hecking and E. te Kaat, these proceedings, and references herein.
(7) J. F. Ziegler, J. P. Biersack, and U. Littmark, The Stopping and Range of Ions in Solids, in: The Stopping and Range of Ions in Matter, Vol. 1, ed., J. F. Ziegler (Pergamon Press, New York, 1985).
(8) J. Belz, K. F. Heidemann, H. F. Kappert, and E. te Kaat, phys. stat. sol. (a) 76 K81 (1983).
(9) to be published at MRS fall meeting, Boston 1987.
(10) M. Mäenpää, L. S. Hung, M.-A. Nicolet, D. K. Sadana, and S. S. Lau, Thin Solid Films, 87 (1982) 277.

ION BEAM INDUCED EPITAXIAL CRYSTALLIZATION OF DOPED AMORPHOUS SILICON LAYERS

M. Voelskow[1], W.Skorupa[1], and J.Matthäi[1]

Ion beam induced epitaxial crystallization has become an interesting new method for solid phase epitaxy of amorphous silicon at temperatures below that of conventional furnace variant ($\sim$550 °C). Because the recrystallization rate is proportional to the energy of the ions deposited into nuclear processes /1, 2/, the use of low implantation energies should be advantageous for a high recrystallization velocity. However, most of the previous work was concerned with high energies (> 300 keV). Otherwise, only self ion amorphized layers were studied systematically /1, 2/.

In this contribution we report on a special mode of IBIEC at a relatively low energy of 165 keV applied to doped amorphous silicon layers. The amorphous silicon layers were produced by implantation of arsenic ions at an energy of 100 keV with a dose of 1 x 10^{15}cm^{-2} into monocrystalline silicon wafers (<100>, p-type). IBIEC was carried out with 165 keV N^+(330 keV, N_2^+) ions at a target temperature of 400 °C. Target heating was realized by clamping the samples on a stainless steel holder and irradiating the central 10 x 10 mm^2 area with a beam current density of 8 μAcm^{-2}.

In Fig. 1a RBS spectra of an amorphous silicon layer with an initial thickness of 140 nm are shown for different regrowth steps by IBIEC.

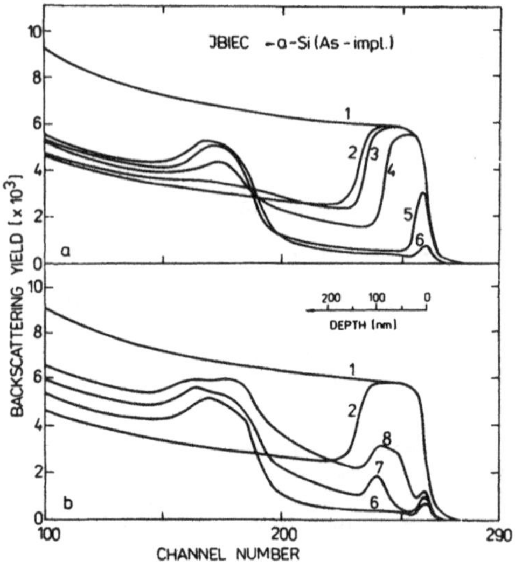

Figure 1: RBS spectra (1.2 MeV, $^4He^+$) of amorphous silicon layers (100 keV, $^{75}As^+$, 1 x 10^{15}cm^{-2}, RT) on monocrystalline silicon after different steps of IBIEC (165 keV, $^{14}N^+$, 400 °C): random (1), aligned (2-8); 2- as amorphized, 3-1 x 10^{16} cm^{-2}, 4-3 x 10^{16}cm^{-2}, 5-5 x 10^{16} cm^{-2}, 6-7 x 10^{16} cm^{-2}, 7-9 x 10^{16} cm^{-2}, 8-12 x 10^{16} cm^{-2}

1 - Central Institute for Nuclear Research, Rossendorf
PSF 19, DDR-8051 Dresden, GDR

The regrowth process begins at doses of about 1 x 10^{15} cm^{-2}, and after implanting a dose of 7 x 10^{16} cm^{-2} full recrystallization is reached. At the depth corresponding to the projected range of the nitrogen ions (~ 380 nm) a defect peak grows due to the high nitrogen concentration in this region.

Surprisingly, further irradiation up to a dose of 1.2 x 10^{17} cm^{-2} leads to the evolution of a second defect peak in the region of the regrown amorphous layer as visible in Fig. 1b. The depth of this peak is about 100 nm.

In Fig. 2 the amount of arsenic atoms on lattice sites evaluated from the channeling spectra is shown (upper curve). It can be seen that up to the dose necessary for full recrystallization, this value increases reaching about 80 %, but in the dose range corresponding to the appearance of the second defect peak it is decreasing again.

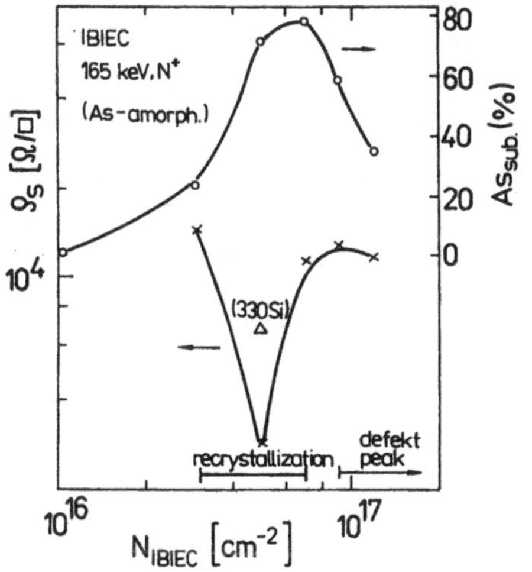

Fig. 2: Dependence of the concentration of substitutionally arsenic atoms (upper curve) and of the sheet resistivity (lower curve) on the IBIEC dose

To investigate in more detail this defect behaviour, some samples were prepared for cross sectional electron microscopy (XTEM). It was shown that in the as implanted case an amorphous surface layer extends to a depth of aproximately 140 nm. In the initial stage of the irradiation the amorphous/crystalline interface begins to move toward the surface leaving a region with low defect concentration. The implantation of a dose of 7 x 10^{16} cm^{-2} leads to full recrystallization of the amorphous layer, although it contains now a considerable amount of small dislocation loops. The highest defect concentration is located near the original amorphous/crystalline interface. The implantation of higher doses results in a strong increase of the defect concentration in the arsenic implanted surface layer. This is in good agreement with the appearance of the second defect peak in the RBS spectra.

Because of this defect behaviour, which is unusual compared to conventional annealing, one can expect that the dopant activation should be influenced by the IBIEC process too.

In Fig. 2 (lower curve) the dependence of the sheet resistivity on the IBIEC dose is shown. In the initial stage of the IBIEC process the resistivity decreases drastically what gives rise to the assumption that more and more arsenic atoms are built in on electrically active lattice sites in the recrystallized part of the layer. But by a further increase of the implantation dose the sheet resistivity begins to increase again, although the recrystallization front has not yet reached the surface.

This unexpected behaviour can be explained in terms of point defect accumulation in the doped layer during the movement of the crystallization front. These point defects will be created especially at the projected range of the annealing ions. Due to the high target temperatures they are expected to migrate during IBIEC. A part of those moving to the surface may be trapped by arsenic atoms thereby forming electrically inactive point defect clusters or secondary defects like loops.

The question arises whether the second defect peak is a general problem of the IBIEC process or is connected with the presence of impurity atoms. To circumvent the influence of arsenic in our experiment self-ion amorphized layers were regrown by IBIEC in the same manner as the arsenic amorphized ones. The corresponding RBS spectra are shown in Fig.3.

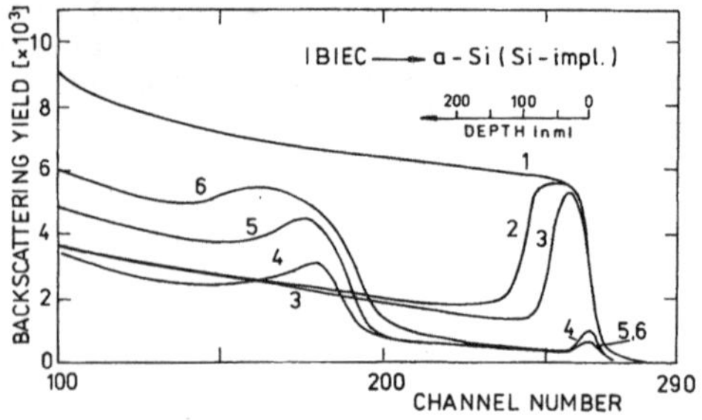

Fig. 3: RBS spectra (1.2MeV, $^4He^+$) of amorphous silicon layers (40 keV, $^{28}Si^+$, 3×10^{15} cm^{-2}, RT) on monocrystalline silicon after different steps of IBIEC (165 keV, $^{14}N^+$, 400°C): random (1), aligned (2-6); 2-as amorphized, 3-1 x 10^{16} cm^{-2}, 4-3 x 10^{16} cm^{-2}, 5-4 x 10^{16} cm^{-2}, 6-12x 10^{16} cm^{-2}

The initial thickness of the amorphized layer ($^{28}Si^+$, 40 keV, 3×10^{15} cm^{-2}, RT) amounted only to 100 nm leading to a lower dose of 4×10^{16} cm^{-2} sufficient for full recrystallization. Also, the defect peak at the range of the annealing ions is visible. But although irradiation was continued up to the same dose as in the case of the As amorphized layer, no second defect peak was found in this experiment. That means, the presence of arsenic within the amorphous layer in the first experiment (Fig. 1) is of crucial importance for the evolution of the second defect peak described above.

References
/1/ R.G.Elliman, J.S.Williams, D.M.Maher and W.L.Brown, MRS Fall Meeting, Boston 1986

/2/ J.S.Williams, W.L.Brown, R.G.Elliman, R.V.Knoell, D.M.Maher and T.E.Seidel, MRS Spring Meeting, San Francisco 1986

HIGH ENERGY IMPLANTATION AND ANNEALING OF PHOSPHORUS IN SILICON

W. Skorupa, E. Wieser, R. Grötzschel, M. Posselt (1); A. Armigliato, A. Garulli (2);
A. Bèyer, W. Markgraf (3)

The use of high energy ion beams to modify the properties of solids has become of increasing importance in recent years. Also, high energy implantation of semiconductors is an interesting new field of research. A lot of results concerning 1 MeV phosphorus implantation up to doses of 7.5 · 10^{14} cm^{-2} were reported in Ref. 1. Here, new experiments on radiation damage evolution and electrical activation after implantation of higher doses up to 6 · 10^{15} cm^{-2} will be presented. 1 MeV implantation was realized by accelerating P^{3+} ions with 330 keV. The target temperature was kept at ~ 300 K. Annealing was carried out by flash lamps (FL) at 1350 °C, 20 ms, air, by halogen lamps (RTA) at 1100 °C, 10 s, air and by furnace (FA) at 900 °C, 30 min, N_2.
Further experimental details are given in Ref. 1.

After implantation and annealing Gaussian shaped profiles with negative skewness were found as reported extensively in Refs. 1, 2. It was also shown that despite sufficient energy deposition into nuclear processes by implanting a dose of 7.5 ·10^{14} cm^{-2} no amorphization was possible /1/. A further increase of the implantation dose led to amorphization in the range of 2-4 ·10^{15} cm^{-2} as deduced by RBS. The exact dose is strongly dependent on target temperature fluctuations during implantation. In Fig. 1 RBS spectra are shown after implanting doses of 2 · 10^{15} cm^{-2} and 6 · 10^{15} cm^{-2} leading to a buried damaged but not amorphized layer (spectrum a) and a buried amorphous layer (sp. d), respectively. The damage maximum lies at about 0.85 µm which coincides with the maximum of energy deposition into nuclear processes /1/. The damaged layer as shown by XTEM is visible in Fig. 2a. The amorphous layer has a thickness of about 0.75 µm as evaluated by RBS (Fig. 1, sp. d) as well as XTEM (Fig. 2d). After annealing by the methods mentioned above aligned spectra were found showing only some de-

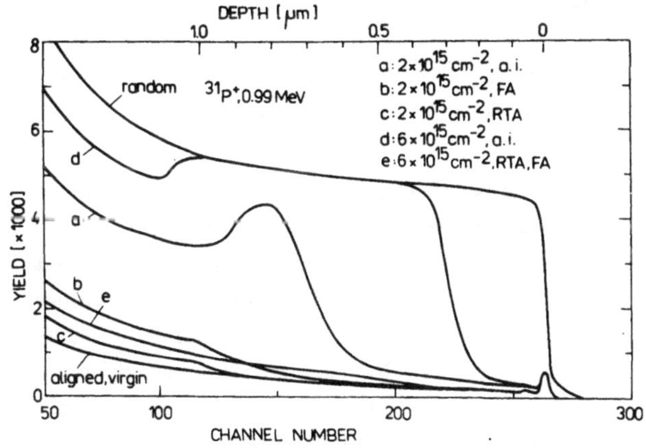

Fig. 1

RBS spectra after 0.99 MeV phosphorus implantation into silicon and annealing

---

(1) Central Institute for Nuclear Research Rossendorf, Academy of Sciences of the GDR, PF 19, DDR-8051 Dresden
(2) CNR-Istituto LAMEL, Bologna, Italy
(3) Dept. of Physics of Electron Devices, Techn. Univ. Karl-Marx-Stadt, GDR

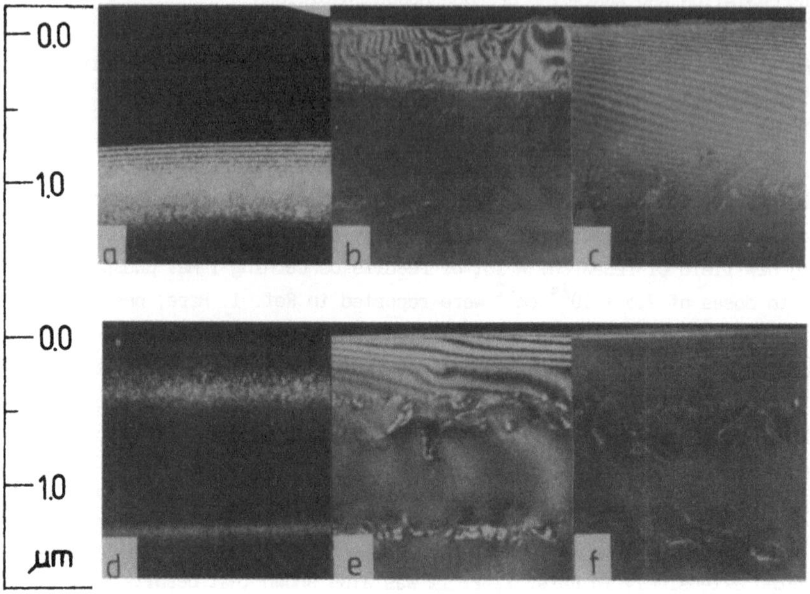

Fig. 2

XTEM micrographs after 0.99 MeV phosphorus implantation into silicon: $2 \cdot 10^{15}$ cm^{-2} (a-c), $6 \cdot 10^{15}$ cm^{-2} (d-f); as impl. (a,d), FA (b,e), RTA (c,f)

channeling (Fig. 1). By XTEM (Fig. 2b) two bands of secondary defects (dislocations, loops) were found after FA of the $2 \cdot 10^{15}$ cm^{-2} implant. For RTA only a single band is visible (Fig. 2c). By annealing of the buried amorphous layer in each case two defect bands develop (Fig. 2e,f). Generally, if two bands are present than they are located at the transitions between the as-implanted damaged or amorphized layer and the undisturbed crystal. The single band in Fig. 2c coincides with the depth range of the as-implanted damage region (Fig. 2a).

The results of van der Pauw-measurements are shown in Fig. 3 for FL which gives the highest electrical activity at negligible impurity redistribution. The dashed line represents an activity of 100 %. Most of the experimental values are in the range of 80 - 100 %. The Hall mobility behaviour points to impurity scattering indicating well annealed material. For the low dose range

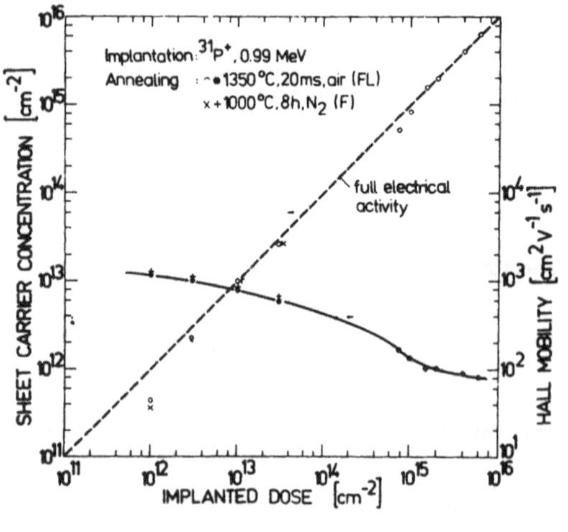

Fig. 3

Sheet carrier concentration and Hall mobility in dependence on the implanted dose after 0.99 MeV phosphorus implantation also the values reached after furnace annealing at 1000 °C, 8 h, N_2 are included. Only a little difference was found. Such an anneal step was used to simulate the influence of technological processes after CMOS-well formation.

REFERENCES

/1/ Skorupa, W., E. Wieser, R. Grötzschel, M. Posselt, H. Bücke, A. Armigliato, A. Garulli, A. Beyer and W. Markgraf: Nucl. Instr. Meth. B19/20 (1987) 335

/2/ Posselt, M. and W. Skorupa, Nucl. Instr. Meth. B21 (1987) 8

DEPTH PROFILES OF RADIATION DAMAGE AFTER HIGH ENERGY SELF-IRRADIATION OF SILICON FROM OPTICAL REFLECTIVITY MEASUREMENTS ON BEVELLED SAMPLES

N. Hecking and E. te Kaat +)

Si wafers are irradiated with high energy Si-ions in the range of 1 to 20 MeV at a target temperature of 210 K. Depth distributions of damage are determined by optical reflectivity measurements after small angle bevelling of the irradiated sample. Experimental results are compared to Monte Carlo simulations at corresponding energies. The dose dependence of damage is reproduced by model calculations.

1. Introduction

Difficulties arise when high energy ion beam modifications of materials are analysed by standard methods. The RBS accessible depth is restricted to a few μm. Sputtering, used for both AES and SIMS depth profiling, includes the difficult problem of sputter time to depth conversion.

Highly resolved optical reflectivity measurements after small angle bevelling have been successfully applied to determine radiation damage profiles /1,2,3/ as well as ion ranges /4,5/ after high energy ion irradiation of silicon. In this paper radiation damage after self-irradiation of silicon in the energy range of 1-20 MeV is studied. Measurements are compared to energy deposition profiles from Monte Carlo simulations /6/.

2. Experiment

Si wafers of (111) orientation at 2-20 MeV and (100) at 1 MeV are irradiated 7° off the channel axis with silicon ions at the Dynamitron Tandem Accelerator (Ruhr-University Bochum, F.R.G.). Irradiations are performed in the dose range 1×10^{13} cm^{-2} to 3×10^{14} cm^{-2} at a constant target temperature of 210 K. The dose values, determined from implanted area and integrated ion current, are accurate within 5 % . To reveal the damage depth profiles the irradiated samples are bevelled to angles of 1 to 2.5° depending on ion energy. Bevel angles are determined with an accuracy of 0.01°.

A vidicon detector system (OMA 2, PAR) attached to an optical microscope records the radiation induced optical reflectivity change ΔR across the bevelled surface in reference to the value R_c of the un-

---

+)Institute of Physics, University of Dortmund, POB 500 500, D-4600 Dortmund 50, F.R.G.

disturbed crystalline substrate. Details have been reported by Heidemann /1/. The normalized reflectivity change

$$s = (R - R_c) / (R_{ia} - R_c) = \Delta R / \Delta R_{max}$$

is used as a measure of optically detected lattice damage /2,3,7/, where R_{ia} is the reflectivity of amorphous silicon.

3. Results and discussion

Fig. 1 shows high energy damage depth profiles. The dose dependence of damage at profile maximum is presented in fig. 2.

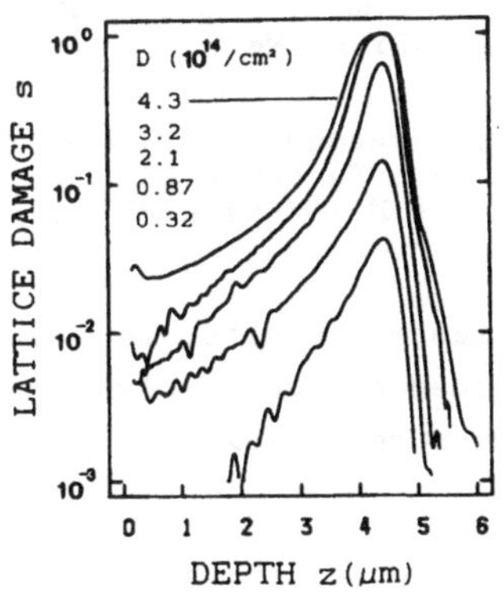

Fig. 1 Optically detected damage (λ =650 nm) after 8 MeV Si -> Si implantation at T=210 K

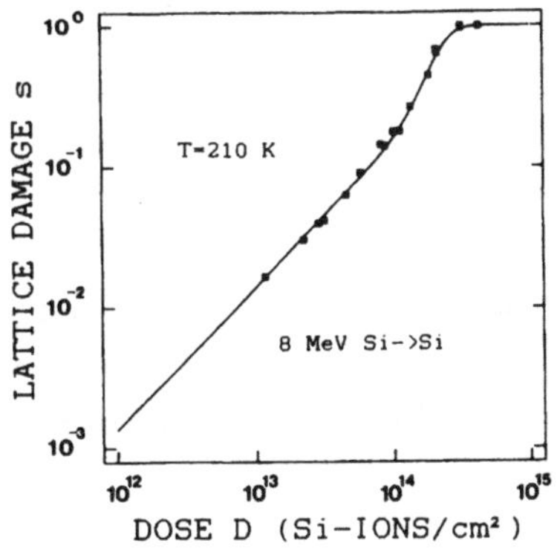

Fig. 2 Dose dependence of optically detected damage at the profile maximum (solid line is calculated /2/)

The sigmoidal shape of damage curves has been observed at this temperature for all energies investigated and is also found for low temperature irradiations of silicon with other ion species (B,N,Ni..).
As previously reported /2,3/ the linear increase of damage s in the low dose region is due to the simultaneous generation of point defects (production parameter P_p) and amorphous regions (production parameter P_a). The stimulated growth of amorphous regions leads to an increasing damage efficiency until saturation is reached by amorphization.
Model calculations /2/ have been carried out to describe the dose dependence at different energies (example in fig. 2). The slope at low dose, which is the sum of generation parameters P_p and P_a is shown in fig. 3. At energies from 2 to 20 MeV damage generation decreases about 10 % . This corresponds to decreasing simulated values of energy deposited in nuclear collisions at profile maximum. The lower damage efficiency of 1 MeV irradiation may be due to the (100) target material /8/.

Fig. 4 shows the measured depths of maximum damage R_d in comparison to depths of maximum nuclear stopping as calculated by Monte Carlo simulation code TRIM /6/.

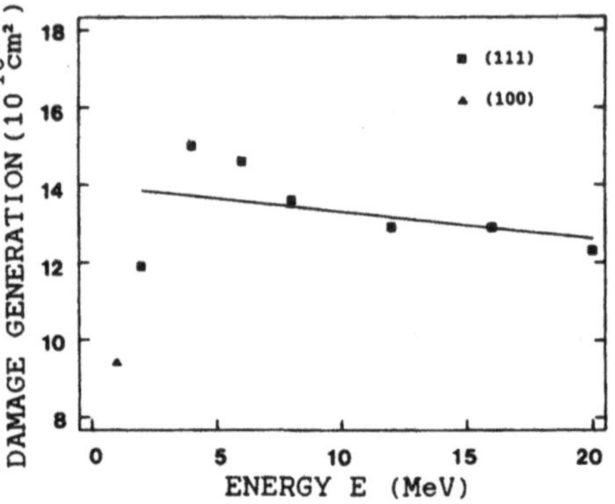

Fig. 3 Energy dependence of damage generation (straight line: fit from 2 to 20 MeV)

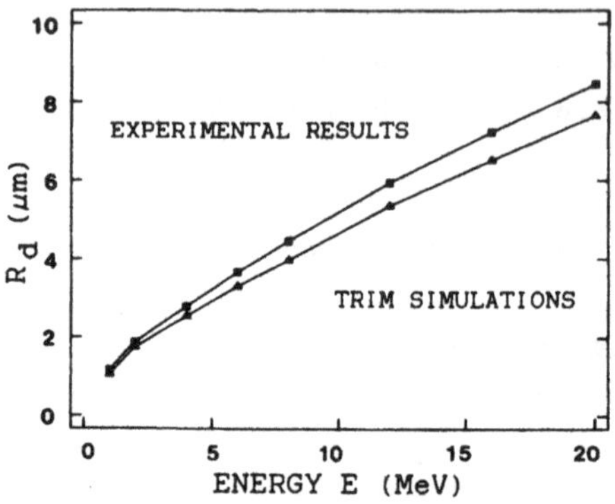

Fig. 4 Energy dependence of depth of maximum damage R_d compared to corresponding TRIM /6/ simulation results

TRIM values are significantly smaller than experimental results. In the whole investigated energy range, simulated values can be simply corrected by $R_d = 1.11 * R_{d,\,TRIM}$. A detailed discussion is to be published.

References

/1/ K. F. Heidemann, Phil. Mag. B 44 (1981) p. 465.

/2/ N. Hecking, K. F. Heidemann and E. te Kaat, Nucl. Instr. and Meth. B 15 (1986) p. 760.

/3/ J. K. N. Lindner, N. Hecking and E. te Kaat, Nucl. Instr. and Meth. B 26 (1987) p. 551.

/4/ U. Bussmann, N. Hecking, K. F. Heidemann and E. te Kaat, Nucl. Instr. and Meth. B 15 (1986) p. 105.

/5/ K. F. Heidemann, M. Grüner and E. te Kaat, Rad. Eff. 82 (1984) p. 103.

/6/ J. F. Ziegler, J. P. Biersack and U. Littmark, The Stopping and Range of Ions in Solids, in: The Stopping and Range of Ions in Matter, vol. 1, ed., J. F. Ziegler (Pergamon Press, New York, (1985).

/7/ K. F. Heidemann, H. F. Kappert, in: Defects and Radiation Effects in Semiconductors (Institute of Physics, London, (1979) Conf. Ser. 46, p. 492.

/8/ I. F. Bubb, D. J. Chivers, J. R. Liu, A. P. Pogany, K. T. Short, H. K. Wagenfeld and J. S. Williams, Nucl. Instr. and Meth. B 2 (1984) p. 761

JONOTRON - A HIGH INTENSITY ION PULSE PRODUCING MACHINE FOR MATERIAL PROCESSING

J. Langner, K. Czaus, E. Górski, M. Gryziński, A. Horodeński,
J. Piekoszewski, C. Pochrybniak, Z. Werner, M. Wójtowicz +)

It was shown that in a high-current, low-pressure discharge between penetrable, cylindrical, multi-rod electrodes of RPI-type devices an effective generation of powerful ion beams takes place [1,6]. Operation principles of those machines have been presented in detail at the EPM'84 conference [2]. The generation of kiloamp beams constitutes a basis for a new implantation technique - Pulse Implantation Doping (PID) [3]. In the PID process an ion stream transports an ion dose necessary to dope a semiconductor, and simultaneously it supplies an energy sufficient to melt a surface layer of the solid. The technique has been successfully used to produce, e.g. p-n silicon junctions for photovoltaic cells [4].

Recently, a new generation of the high-intensity ion-pulse devices for pulse implantation (the so-called JONOTRONs) has been developed. In particular, the JONOTRON'85 laboratory device - see Fig.1. - and the JONOTRON'86 laboratory-scale technological processing device - see Fig.2 - have been designed.

Fig.1. JONOTRON'85.

Fig.2. JONOTRON'86.

+)Institute for Nuclear Studies, 05-400 Swierk-Otwock, Poland

The machines presented in the paper constitute relatively compact systems with completely screened high-voltage modules, automatic control of discharge parameters and semi-automatic change of samples. A rotary holder makes it possible to locate the samples at various distances away from the ion source and from the beam propagation axis. The JONONTRONs are equipped with a set of calorimeters for beam energy-density measurements; an energy dose can be displayed on a screen of a microcomputer.

Technical tests and series of solid-surface modification experiments have been carried out; some of them are presented in the other paper of this conference [5,7].

Basic parameters of the JONOTRON devices are as follows:

total ion current	≲100 kA
number of particles	$10^{17} \div 10^{18}$
total pulse energy	≲ 1 kJ
mean ion energy	≲ 30 kV
ion pulse duration	0.5 ÷ 1.0 μs
diameter of the beam	1 ÷ 15 cm

In the both JONOTRONs, it is possible to vary the energy/mass dose by a change of the gas discharge preconditions and by a change of the electrode-sample distance. Ion-beam energy density distributions at different operation modes are shown in Figs. 3 and 4; an exhibited area corresponds to 15*15 cm².

A reliability of the ion beam producing devices and a reproducibility of the ion beam parameters play a decisive role in the technological applications of jonotron-type machines. The beam reproducibility is illustrated in Fig. 5.

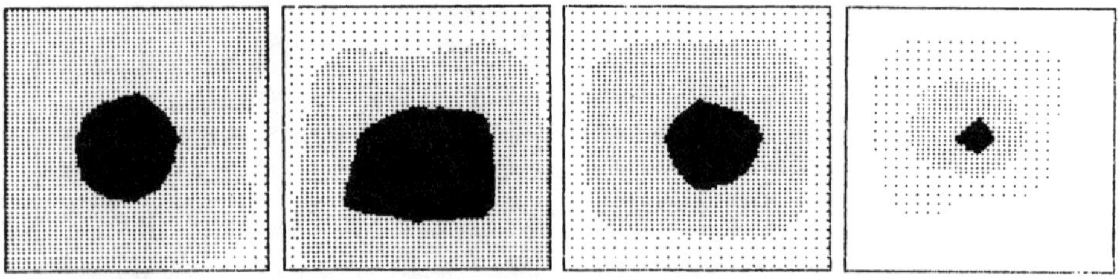

Fig. 3. Ion beam energy density distributions vs initial gas conditions for BF3 operation (from left to right): delay time 210 μs, the maximum (central) energy 5.5 J/cm²; 215 μs, 4.1 J/cm²; 220 μs, 3.8 J/cm²; 225 μs, 2.7 J/cm². Source-target distance 25 cm; isotherms correspond to 90, 70 and 50% of the maximum value.

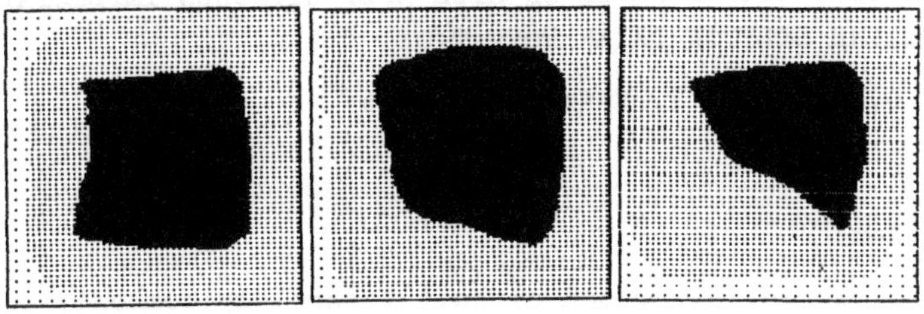

Fig. 4. Ion-beam energy density distributions vs source-target distance for H2 operation (from left to right): distance 20 cm, the maximum (central) energy 2.5 J/cm²; 30 cm, 1.8 J/cm²; 40 cm, 1.6 J/cm². Delay time 120 µs; isotherms as in Fig. 3.

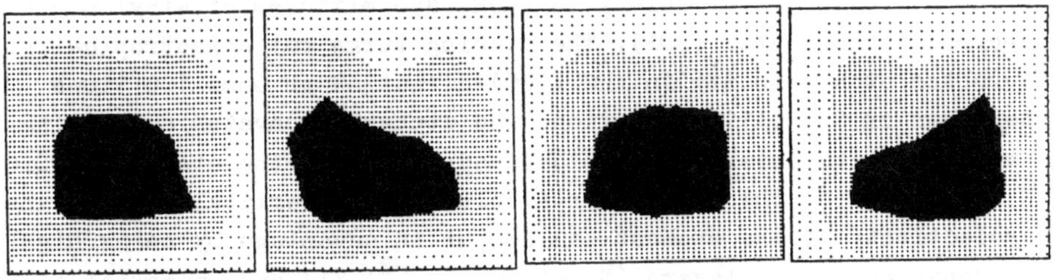

Fig. 5. Repeatibility of the ion beam energy density distribution for BF3 operation. Source-target distance 25 cm, delay time 215 µs, the maximum (central), energy densities are (four subsequent pulses, from left to right): 4.3, 4.6, 4.1 and 4.6 J/cm². Isotherms as in Fig. 3.

The obtained parameters, e.g. deposited-energy density range from 1.0 to 6.0 J/cm², an energy deposition area of the order of 100 cm², long-distance energy/mass transport ability, and a good reproducibility of ion beams, make the JONOTRONs interesting from the point of view of different technological programs on solid surface modification. By now, they can provide a relatively effective, economic, low-scale production of the photovoltaic cells.

REFERENCES

(1) Gryziński, M., et al.: Proc. 10th Eur. Conf. Contr. Fusion Plasma Phys., vol.II, p.22, Moscow, 1981.

(2) Langner, J., et al.: Proc. Conf. Energy Pulse Modification of Semiconductors and Related Materials, vol. 1, Dresden, 1984.

(3) Piekoszewski, J., et al.:J.Physique 43 (1982) 1353.

(4) Piekoszewski, J., et al.: Proc. Conf. Energy Pulse Modification of Semiconductors and Related Materials, vol. 1, Dresden, 1984.

(5) Piekoszewski, J., et al.: this conference.

(6) Skladnik-Sadowska, E., et al.: J. Physique 43 (1982) 715.

(7) Werner, Z., et al.: this conference.

A NOVEL ETCH STOP SYSTEM BASED ON HIGH DOSE NITROGEN IMPLANTATION FOR THIN CRYSTALLINE SILICON LAYER FORMATION

B. Schmidt, U. Kreißig, H. Münzer, E. Rost +)

1. Introduction

The highly-developed integrated-circuit technology commonly employs planar process only. Semiconductor sensor fabrication, however, often requires three-dimensional structuring of the substrate. For this purpose, anisotropic selective etching is well suited, playing an important role in the recent advance of semiconductor sensor technology. Selective anisotropic etchants for monocrystalline silicon are generally alkaline hydroxide-water solutions (KOH) and mixtures of ethylenediamine-pyrocatechol-water (EDP). The etching turns out anisotropically, because the etch velocity in the (100)-direction is some 100 times faster than in the (111)-direction. The etchant excavates an opening in the silicon, with lateral limits defined mainly by the etch mask (SiO_2, Si_3N_4 etc.) and by the (111)-crystal planes. For vertical structuring, the selectivity of the etchant can be put to advantage by using highly-boron-doped epitaxial layers ($N_B \sim 1 \cdot 10^{20}$ cm^{-3}) /1/ and epitaxial layers with oposite doping in the case of electrochemical etching /2/. Insulating etch stop layers, such as oxides and nitrides are at present used only for polycrystalline silicon or nonsemiconducting layers /3,4/. The aim of the present work was the study on the etch characteristics (etch stop behavior) of buried high dose implanted Si_xN_y-layers using alkaline-water solutions as an anisotropical and selective etchant.

2. Experimental

The buried Si_xN_y-layers were formed by N^+-ion implantation in p-type, (100)-oriented silicon wafers. The implantation was carried out at E = 330 keV in the dose range D = $(1 \cdot 10^{16} \ldots 1.2 \cdot 10^{18})$cm^{-2}. The target temperature induced by ion beam heating during implantation was around 500 °C. After implantation the samples were thermal annealed in dry N_2 at T = 1000 °C and t = 2 h. At these implantation and annealing conditions we can assume an amorphous Si_xN_y-layer in the mean depth of about 600 nm with the layer thickness of about 300 nm /5/. Undoped epitaxial silicon layers with thicknesses between 0.5 and 5.0 µm were deposited on the crystalline silicon top layer above the buried nitride layer. The anisotropic and selective etching (40 % KOH at T = 80 °C) was chosen through an etch mask window from the front side (epi-layer) or from the back side of the wafer. The etch studies were focused on the investigation of the etch rate in dependence on the implanted nitrogen dose and on the formation of thin monocrystalline silicon membranes.

+) Zentralinstitut für Kernforschung, DDR-8051 Dresden, PF 19

3. Results

The dose dependence of the etch rate for buried Si_xN_y-layers in 40 % KOH at T = 80 °C is shown in Fig.1. In the figure the etch rate in the (100)-direction and the etch rate of CVD-Si_3N_4 are indicated. The implanted nitrogen dose for stoichiometric Si_3N_4-formation was assumed at $D = 1.2 \cdot 10^{18}$ cm^{-2} /5/. Up to N^+-doses about $D = 1 \cdot 10^{17}$ cm^{-2} no reduction of the etch rate was observed. The etch rate than decreases by a factor of about $6 \cdot 10^3$ increasing the implanted N^+-dose up to the stoichiometric one. From Fig.1 one can conclude that after implantation of one third of the stoichiometric dose ($D = 4 \cdot 10^{17}$ cm^{-2}) the etch rate decreases by two orders of magnitude compared to nonimplanted (100)-Si. Such an etch rate lowering is quite enough for an etch stop in many cases of three-dimensional structure formation. For usually applied alkaline solutions (EDP, KOH) the etch stop is caused by heavy boron doping in epitaxial layers. A reduction of the etch rate can be observed for boron concentrations higher than $1 \cdot 10^{19}$ cm^{-3} /6/. Comparing the etch rates of low doped silicon ($N_B = 1 \cdot 10^{18}$ cm^{-3}) to high doped silicon ($N_B = 1.5 \cdot 10^{20} cm^{-3}$) KOH shows a ratio of 100:1 for 10 % KOH and 10:1 for 60 % KOH. A higher ratio of 500:1 was observed for EDP (from /6/).

The observed results show, that increasing the implanted N^+-dose up to the stoichiometric one it is possible to increase the etch ratio $R_{(100)}/R_{etch\ stop}$ by one order of magnitude compared to the highest known ratio in high boron doped epitaxial layers. Fig.2 shows the comparison between the reduction of the etch rates for high boron doped layers in EDP-solution /1/ and for implanted buried Si_xN_y-layers in KOH-solution. We can conclude, that the reduction gradient of the etch rate in both cases is the same. But one can get a more efficient etch stop for implanted stoichiometric Si_3N_4-layers, espacially in the case of anisotropical and selective etching in KOH compared to high boron doped layers of the same thickness.

Etched membranes were fabricated using buried Si_3N_4-layers as an etch stop. A good flatness for small etched areas until 135 x 135 μm^2 was observed. Etched areas until 10 x 10 mm^2 were wavy, but mechanical very stable.

References

/1/ Csepregi, L. et al.: Forschungsbericht BMFT-FB-T 84-209 (1984)

/2/ Hirata, M. et al.: Proc. 3rd Int. Conf. Transducer '85, Boston (1985) 287

/3/ Müller, J.E. et al.: Proc. 2nd Symp. Measurement in Industry and Science, Suhl (1984) 179

/4/ Choi, I.H., K.D. Wise: Proc. 3rd Int. Conf. Transducer '85, Boston (1985) 132

/5/ Scorupa, W. et al.: Vacuum 36 (1986) 933

/6/ Seidel, H., L. Csepregi: Sensors and Actuators 4 (1983) 455

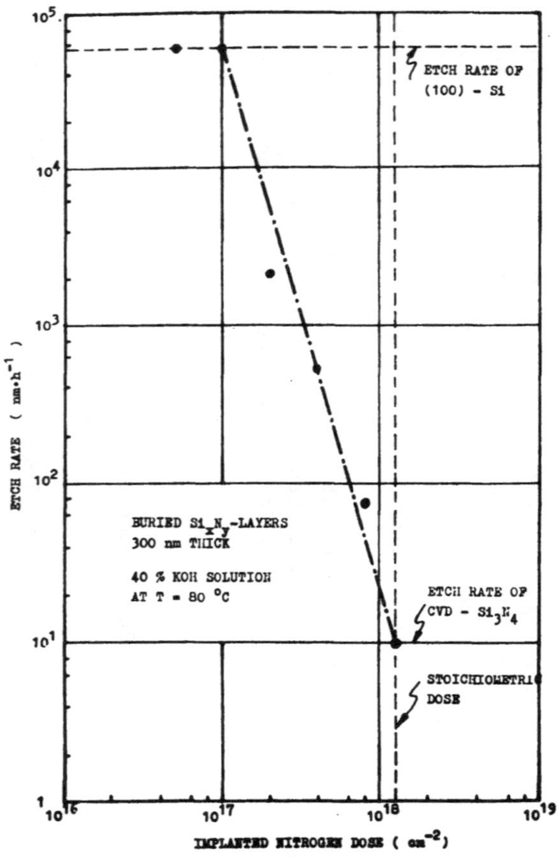

Fig.1: Dose dependence of the etch rate for buried Si_xN_y-layers in 40% KOH at T = 80 °C

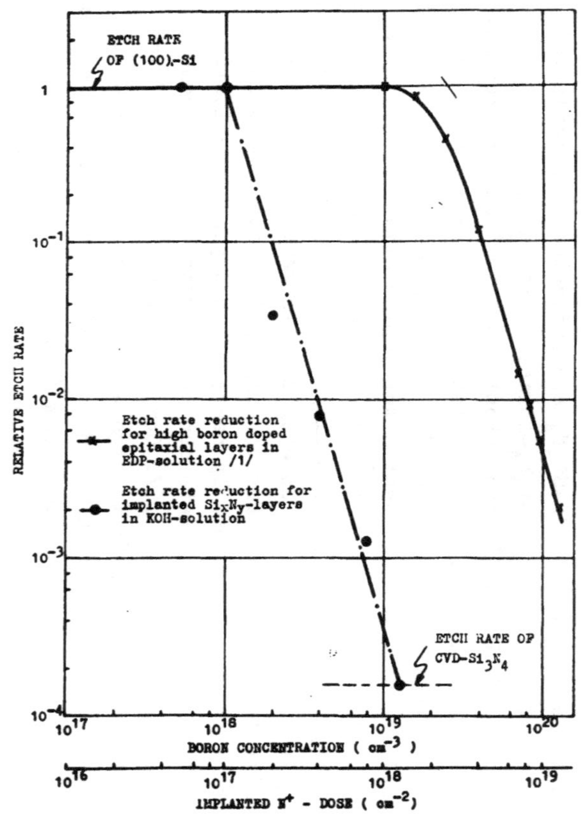

Fig.2: Comparison between the reduction of the etch rates for high boron doped layers in EDP and for implanted buried Si_xN_y-layers in KOH

EFFECT OF HIGH ENERGY HYDROGEN BEAM INTENSITY ON DEFECT FORMATION IN SILICON

R.F.Konopleva, I.V.Nazarkin, V.A.Evseev, B.A.Borisov, V.A.Chekanov*)

Temperature and irradiation intensity are known [6] to affect substantially the nature of radiation defect formation in semiconductors. In the case of low-energy electron and gamma-ray bombardment producing primarily defect - impurity complexes, one observes a monotonic variation of the defect formation efficiency with irradiation intensity [10]. To explain this behaviour, models were used which are based on concepts involving the variation of the complex formation barrier height with the charge state of the complex components [6]. However in the case of high-energy electron irradiation (10 MeV $\leqslant$ E $\leqslant$ 1200 MeV) [5] one observed an essentially nonmonotonic dependence of the defect complex formation rate on irradiation intensity.

It is known [2,3] that high-energy particles produce, apart from point defects, also complex structural damage in the form of defect clusters or disordered regions (DR). Under these conditions, the presence of DRs under sufficiently strong compensation ($K \geqslant 0.3$) results in the creation of a large scale potential relief associated with a spatially inhomogeneous distributions of defect complexes and DRs [1,7]. Hence, in semiconductors with DRs the filling of the defect levels by electrons and, accordingly, their charge state will be governed by two factors, namely, the large scale potential relief created by the DRs and the particle beam intensity resulting in defect recharging via the trapping of nonequilibrium carriers. The effect of the second factor on the efficiency of defect formation in silicon with DRs was considered in ref. [4], where the influence of the potential relief created by the DRs on the charge restructuring of the defects was not adequately analysed. To account for the results of an experimental study of the defect formation efficiency in silicon under different levels of 1000 MeV proton irradiation, a phenomenological model is proposed here, by which the variation of the volume - averaged carrier concentration is due to fluctuations of the DR-induced large scale potential relief. The samples used were n- and p-Si with initial concentrations from $2 \cdot 10^{12}$ to $2 \cdot 10^{15}$ cm^{-3}. 1000 MeV proton irradiation was performed at 300K on the LNPI synchrocyclotron [1]. One measured the dose dependences of the carrier concentration and lifetime for various proton irradiation intensities. Fig. 1 presents the initial rate of carrier removal vs. proton irradiation intensity. The quantity $S = -\left[\frac{dn(p)}{d\Phi}\right]_{\Phi\to 0}$ is seen to

[1] Beam parameters: pulse current 1 μA, pulse duration 10 ms, repetition frequency $\sim$ 40 Hz.

*) Leningrad Nuclear Physics Institute
Gatchina, Leningrad 188350, USSR

vary nonmonotonically with increasing irradiation intensity, a maximum being reached at 10^{10} $cm^{-2}s^{-1}$. The dependence of S on the type and initial concentration of the impurity was studied.

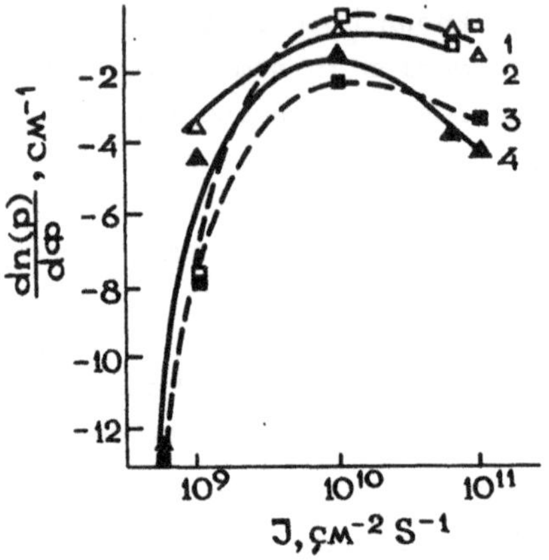

Fig.1. Initial carrier removal rate in Si vs 1000 MeV proton irradiation intensity: 1-p-Si (P_0 = $5.6\cdot10^{12}cm^{-3}$); 2 - n-Si (n_0=$2.5\cdot10^{12}cm^{-3}$); 3 - p-Si (P_0=$1.4\cdot10^{14}cm^{-3}$); n-Si (n_0 = $5.2\cdot10^{13}$ cm^{-3}).

Presented in fig.2 is the dependence of the lifetime degradation coefficient $K_\tau = \Delta(1/\tau)/\Phi$ on the initial dopant concentration for various irradiation intensities. It is obvious that the variation of K_τ as a function of concentration and intensity follows a power law: $K_\tau \sim \bar{J}^{\alpha}[n_0(P_0)]^{\gamma_{\tau n(p)}}$, where $\alpha = 0.45\pm0.05$, $\gamma_{\tau n(p)} = 0.4\pm0.05$. It is known that the dependence $K_\tau = f(n_0, P_0)$ in the case of one deep level in the gap should have a slope of ±1 [8]. The deviation from unity implies the presence of several deep levels in the gap. The slope of $K_\tau = f(n_0)$ of ±30 obtained by us suggests the radiation defects originating from 1000 MeV protons to be dominated by DRs. Hence, the defect formation under irradiation by high energy particles should be governed by both the nature of DRs and the character of the associated large scale potential relief.

The amplitude of the DR potential relief was calculated from the Poisson equation

$$\frac{d^2\Psi}{dr^2} + \frac{2}{r}\frac{d\Psi}{dr} = -\frac{4\pi e}{\varepsilon kT} Q(r) \qquad (1)$$

where $\Psi(r)$ is the potential distribution function, e is the electronic charge, K is the Boltzmann constant, Q(r) is the charge density distribution inside a DR. The volume-averaged free concentration can be derived from the expression

$$\bar{n} = \frac{3}{4\pi r_2^2}\int_0^{r_2} \exp(\Psi-\varphi)\, 4\pi r^2 dr \quad , \qquad (2)$$

where $\Psi(r)$ is the DR potential distribution function, φ is the Fermi potential, r_2 is the size of the spherical cell occupied by one DR. The initial rate of free carrier removal was found from the expression

$$S = -(dn/d\Phi)_{\Phi\to0} \qquad (3)$$

A calculation of S by eqs. (1)-(3) carried out for an n-Si sample with $n_0 = 2.3\cdot10^{13}$ cm^{-3} and for I = 10^9 $cm^{-2}s^{-1}$ showed a satisfactory agreement with the experimental values of the initial carrier removal rate.

In the calculations it was assumed that the negative charge of the DR core is compensated by the positive charge of an envelope consisting of vacancy-impurity atom complexes which in our case are E-centres [6] .

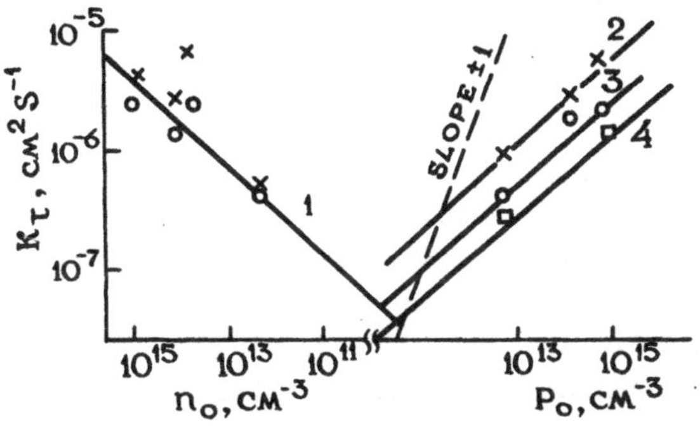

Fig.2. K_τ vs initial dopant concentration in n- and p-Si irradiated with 1000 MeV protons at different flux intensities (cm^{-2}s^{-1}): 1 - $1.8 \cdot 10^{11}$; 2 - $8 \cdot 10^{10}$; 3 - - $2 \cdot 10^{9}$; 4 - $2 \cdot 10^{10}$.

The actual type of the stable predominant defect complex depends on the irradiation intensity [6]. Therefore the composition of the DR envelope will change (at a fixed size of its core) with increasing intensity. Since different complexes are characterized by different depth of the levels in the energy gap, the size of the space charge region (r_2) will likewise be different. In the case of n-Si, the various complex formation rates vary with increasing irradiation intensity in such a way that the fraction of divacancies in the total spectrum of defects grows noticeably [4]. Since a positively charged divacancy occupies a substantially deeper level in the gap (E_v+0.27 eV) than the E-center (E_c-0.45 eV) does, and, hence, provides a less effective screening of the DR core charge, the magnitude of r_2 should increase with growing irradiation intensity. This will result in a reduced spread of fluctuations of the potential relief created by the DRs and, accordingly, in a decrease of the carrier removal rate. At still higher irradiation intensities the probability of the divacancy annihilation with interstitial atoms increases [9] . This results in an increase of the contribution of E-centers to the defect spectrum and a corresponding growth of S which is observed experimentally. By comparing the calculated and experimental values of S one could evaluate the DR size r_2 which depends nonmonotonically on intensity: r_2 = 90 nm for I = 10^{9}cm^{-2}s^{-1}, r_2 = 143 nm for I = = 10^{10}cm^{-2}s^{-1}, and r_2 = 80 nm for I = 10^{11}cm^{-2}s^{-1}. For the same values of intensity the maximum amplitude of the potential relief was found to be $\Psi^{(1)}$= 0.25 eV, $\Psi^{(2)}$= 0.14 eV, $\Psi^{(3)}$= 0.26 eV. The nonmonotonic behaviour of S as a function of irradiation intensity should apparently be observed in all cases where the flux density varies over a sufficiently wide range. In the case of low energy bombardment the nonmonotonic course of S=f(I) is due to the direct trapping of carriers by the defect complexes whose type varies with growing intensity. Under the conditions of high energy irradiation where the volume-averaged free carrier concentration in the semiconductor is governed by the structure of the potential relief produced by DRs the nonmonotonic variation of the DR size r_2 results from S(I).

The above scheme of a successive change of stable predominant defect complexes with growing irradiation intensity in the presence of DRs is universal and apparently valid for any view of high energy ionizing radiation.

References

1) Konopleva R.F.: Preprint PTI-670 (1980).
2) Konopleva R.F., V.L.Litvinov and N.A.Ukhin: Radiation Damage of Semiconductors by High Energy Particles, Atomizdat, Moscow 1971.
3) Konopleva R.F. and V.N.Ostroumov: Interaction of High Energy Charged Particles with Ge and Si, Atomizdat, Moscow 1975.
4) Kozhevnikov V.P. and V.V.Mikhnovich: Zh.Tekh.Fiz. 53(1983)1361-1367
5) Lugakov P.F. and V.V.Lukyanitsa: Fiz.Tekh.Polupr. 18(1984)345-348.
6) Physical Processes in Irradiated Semiconductors, ed. L.S.Smirnov, Nauka, Novosibirsk, 1977.
7) Shklovski B.I. and A.L.Efros: Electronic Properties of Disordered Systems, Nauka, Leningrad 1979.
8) Shockley W. and W.I.Read: Phys.Rev. 87 (1952) 835.
9) Vinetsky V.A. and G.A.Kholodar: Radiation Physics of Semiconductors Naukova Dumka, Kiev 1979.
10) Vitovskii N.A., T.V.Mashovets and O.V.Oganesyan: Fiz.Tekh.Polupr. 12 (1978) 2143.

RAPID THERMAL ANNEALING OF ION IMPLANTED GaAs BY MEANS OF A GRAPHITE STRIP HEATER

T.Bachmann, J.Kleemann, W.Wesch and G.Götz[1)]

1. Introduction

Selenium ion implantation has generated a great deal of interest for the fabrication of n-conducting layers in semiinsulating GaAs [1,2]. The annealing of the damaged layers by unconventional methods like halogen lamps [2] or graphite strips [1] shows a lot of advantages in contrast to furnace annealing. The short time annealing is therefore a favoured method for electrical activation of the dopands. To characterize annealed layers we used Rutherford Backscattering Spectrometry (RBS) and electrical measurements.

2. Experimental

Semiinsulating Cr-doped GaAs was implanted in a nonchanneling direction with $5*10^{12}$ to $5*10^{14}$ Se^+ cm^{-2}. The implants were performed at a fixed ion energy of 300 keV and several implantation temperatures (80, 300 and 473 K). Some pieces of the samples implanted at 80 and 300 K were encapsulated with 120 nm SiO_2 by a sputtering process on both sides before annealing. Annealing was carried out with the graphite strip heater in a vacuum chamber (10 Pa). The temperature regimes used are characterized by $A/T_a/t$ and $B/1100^oC/t$ (T_a - annealing temperature, t - annealing time). A and B correspond to heating rates of 900 Ks^{-1} and 120 Ks^{-1}, respectively. Assuming that the sample and heater temperatures are equal we applied a photodiode for the temperature measurement. The calibration with a known melting point using Planck's law allows an accuracy better than 20 K.

After annealing the cap was removed by etching in HF-acid and RBS-measurements were performed with 1.4 MeV He^+-ions at a backscattering angle of 170°. Recrystallized samples were submitted to a photolithographical process and following Hall-measurements to determine sheet resistivity, mobility and carrier concentration.

---

1) Friedrich-Schiller-Universität Jena, Sektion Physik, Max-Wien-Platz 1, Jena, DDR 6900

3. Results and discussion

(a) RBS-measurements

For the characterization of crystal damage the normalized minimum yield [3] was used as an average value of disorder at the peak maximum.

The degree of damage in GaAs for a given ion fluence is strongly dependent on the implantation temperature. Low dose implantations performed at 300 and 473 K lead to weakly damaged layers characterized by low χ_{min} in the range of 6 to 9%. A strongly damaged layer is generated in the case of $5*10^{13}$ Se^+ cm^{-2} by implantation at 300 K.

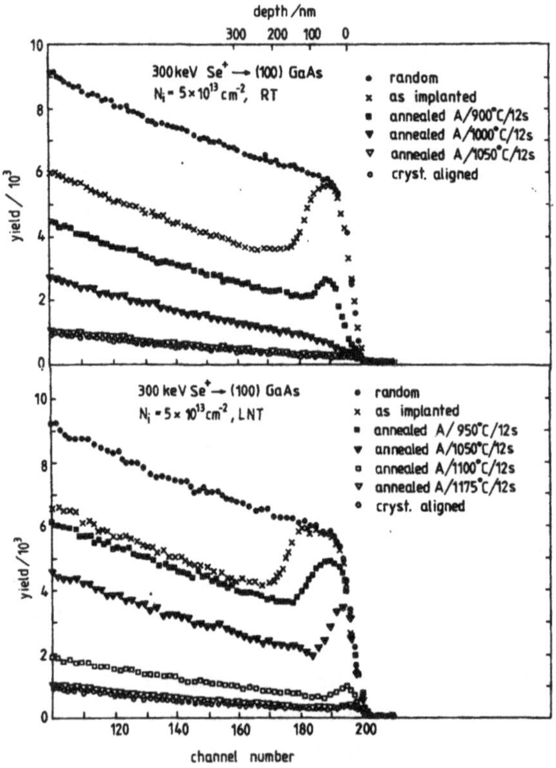

Fig.1. RBS-aligned spectra for Se^+ implanted GaAs partly annealed

The annealing behaviour shows strong dependence on the implantation damage. Fig.1 demonstrates typical spectra for a dose of $5*10^{13}$ Se^+ cm^{-2} implanted at 80 and 300 K, respectively. To recrystallize the amorphous layer (implantation at 80 K) higher annealing temperatures or longer times are necessary. One can observe that for implantation at 300 K (nonamorphous layer) the damage is reduced from the surface and the depth. The residual disorder peak corresponds to the maximum of radiation damage during the ion implantation. Amorphous layers which are extended to the surface show a successive defect reduction from the bulk material. Fig.2 gives an overview about all temperature-time-cycles applied. For annealing times longer than 18 seconds at 1100°C only SiO_2 capped samples were used to prevent the degradation of the surface during annealing. It must be mentioned that for all samples the recrystallization process occurs at low temperatures. In situ-reflection-measurements [4] have shown a recrystallization velocity of about 200 nms^{-1} for temperature risetimes of 900 Ks^{-1}. That means this stage is finished after nearly 1 second for

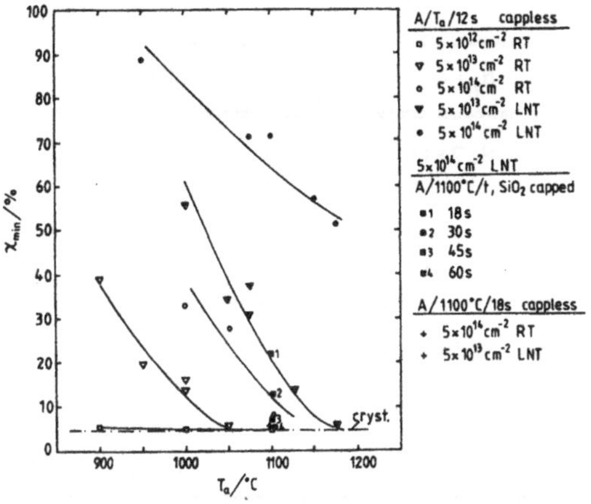

Fig.2. Overview about the RBS-measurements as a function of the annealing regime

the thickest amorphous layer (220 nm, see also fig.1) and the observed disorder reduction in the RBS-aligned spectra is connected with defect conversion. The quality of the recrystallized layers decreases with increasing thickness of the implanted amorphous layer, therefore thicker layers need longer annealing cycles (or higher temperatures) for reaching the χ_{min}-value of the virgin crystal [5,6].

(b) Electrical measurements

Implantation at 80 and 300 K

Investigations about the electrical activation of the implanted selenium were carried out for a fixed annealing temperature of 1100°C and a varied annealing time. As mentioned the visible degradation of the surface limits the annealing time for samples without a cap and yields p-conducting layers independent of the implanted dose. Typical sheet resistivities are in the range of 2 to 40 kΩ/□. Encapsulated samples show no or weak electrical activity at annealing times smaller than 30 seconds, higher times create n-conducting layers with a mobility of 200 to 400 $cm^2V^{-1}s^{-1}$ and sheet resistivities of 1 to 2 kΩ/□. The corresponding sheet carrier concentration of $1*10^{13}$ cm^{-2} could be measured on both sides of the sample. An explanation of these results could be the interaction of the SiO_2-cap with the substrate. Ga outdiffusion during annealing is a wellknown process [7] and the creation of Si_{Ga}-V_{Ga}-complexes which act as hole traps could be proved with photoluminescence [3]. As a consequence there would be a net electron concentration contrary to capless annealing producing p-conducting layers due to possible degradation or a changed impurity or chromium redistribution, respectively.

Implantation at 473 K

The table gives an overview about the electrical parameters reached for annealed hot implanted samples. A maximum activity of 24% for $5*10^{13}$ cm^{-2} and 5% for $5*10^{14}$ cm^{-2} implants is achieved for the capless annealing. The regimes A and B yield no remarkable differences with regard to the electrical properties. In comparison to the implantations at the low temperatures we think these results are due to lower implantation damage.

Dose	Regime	ρ_s [$\Omega/\square$]	N_s [cm^{-2}]	μ_s [$cm^2V^{-1}s^{-1}$]
$5*10^{13}$	A/1100/10s	600	$8.7*10^{12}$	1180
	B/1100/12s	$>10^6$		
	B/1100/14s	470	$8.7*10^{12}$	1530
	B/1100/17s	440	$1.2*10^{13}$	1170
$5*10^{14}$	A/1100/10s	250	$2.6*10^{13}$	1075
	B/1100/12s	$>10^6$		
	B/1100/14s	275	$2.5*10^{13}$	1180
	B/1100/17s	190	$2.2*10^{13}$	1480

[1] Patel, K.K., R.Bensalem, M.A.Shahid, B.J.Sealy: Nucl.Instr. and Meth. 1985, B7/8, 418-422
[2] Sealy, B.J., M.A.Shahid, M.Anjum, S.S.Gill, J.H.Marsh: Nucl. Instr. and Meth. 1985, B7/8, 423-428
[3] Chu, W.K., J.W.Mayer, M.A.Nicolet: Backscattering Spectrometry, Academic Press, New York, San Francisco, London, 1978
[4] Herold, J., to be published
[5] Grimaldi, M.G., B.M.Paine, M.A.Nicolet, D.K.Sadana: J.Appl.Phys. 52 (6), June 1981
[6] Williams, J.S., Laser Processing of Semiconductors eds J.M.Poate, J.W.Mayer, Academic Press, New York 1982, p.285
[7] Bell, E.C., A.E.Glaccum, P.L.F.Hemment: Rad.Eff. 1974, 22, 253-258
[8] Berger, L., private communication

SILICON AND CHROMIUM DISTRIBUTION IN IMPLANTED AND SHORT TIME ANNEALED GaAs

D.Panknin[1], C.E.Richter[2], M.Gericke[2] and Ya.V.Fattachov[3]

1. Introduction

Ion implantation connected with rapid thermal annealing (RTA) is a preferred method also for doping of GaAs. For a good electrical activation it is necessary to carry out the annealing with temperatures and times slightly below the dissociation threshold of the GaAs surface /2,4/. In this contribution we report the results of detailed studies of Si and Cr distribution in Si-implanted semi-insulating Cr doped GaAs in the as-implanted state and after capless RTA.

2. Experimental

Si is implanted in the dose range between 5 . 10^{12} and 1 . 10^{16} cm^{-2} at an energy of 200 keV into semi-insulating Cr-compensated (100) GaAs. The implantation temperature was about 30 and 100 °C, respectively.

Annealing was performed in N_2 atmosphere without an encapsulant by flash lamp irradiation with pulse durations between 50 and 500 ms at maximum temperatures above 1000 °C or using a set of tungsten halogen lamps at temperatures between 900 and 1000 °C with dwell times longer 0.5 s.

The depth distribution of Si and Cr was measured by SIMS using Cs^+ and O^+ ions, respectively.

3. Results

Fig. 1 shows for two implantation temperatures the depth distribution of Si and Cr in the as-implanted state. The profiles are both remarkably broadened compared to a Gaussian distribution calculated with R_p and ΔR_p according to the LSS theory /3/. From the observed dependence on the implantation temperature we conclude that the broadening is due to radiation enhanced diffusion already at these low temperatures /4/. Annealing at about 1070 °C, 50 ms as well as at 920 °C, 2 s leads to no remarkable additional dopant redistribution.

The depth distribution of Cr in the as-implanted state is characterized

1 - Central Institute of Nuclear Research of Academy of Science of the GDR, Rossendorf, PSF 19, DDR-8051 Dresden

2 - VEB WF Berlin, GDR

3 - Kazan Physico-Technical Institute, Kazan, USSR

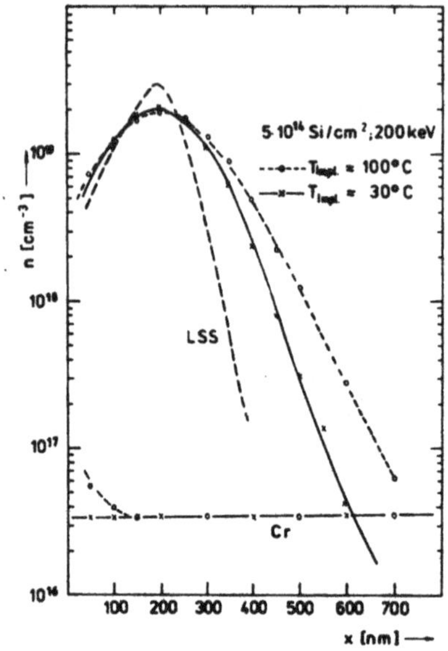

Fig. 1: Si and Cr distribution in the as-implanted state

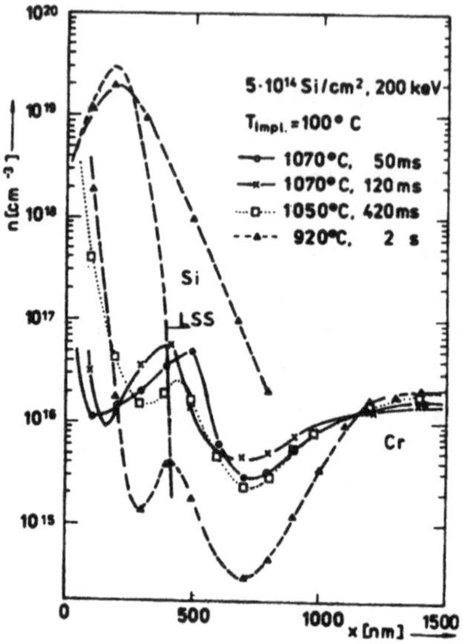

Fig. 2: Cr distribution for 100 °C implantation after capless annealing

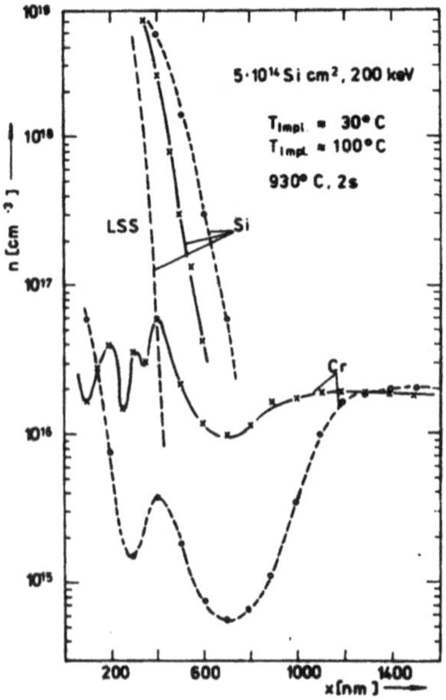

Fig. 3: Cr distribution after implantation at different temperatures and capless annealing

by a weakly increasing concentration at the surface region only for the 100 °C implantation. Whereas after the annealing treatment a remarkable segregation of Cr at the surface is observed (Fig. 2). Increasing annealing time causes an increase of the surface pile-up. Consequently the surface near region shows a Cr depletion. Within the depleted layer an enhanced Cr concentration is found in the depth between R_p and $R_p + \Delta R_p$ of the Si profile. For longer times ($\geq$ 400 ms) at adequate lower temperatures we observed a decrease of this Cr peak and the peak is also shifted to the surface up to a depth corresponding to the penetration depth of the LSS profile.

Fig. 3 shows the Cr distribution obtained after implantation at different temperatures (30 °C and 100 °C) and capless annealing (930 °C, 2 s).

For comparison the penetration depth of the Si profile is also represented. Compared to the 100 °C implantation the room temperature implanted sample shows a very complicated Cr distribution with a second Cr peak in the depth of R_p of the Si profile.

Considering the residual defects after RTA we have found for all implantation and capless annealing conditions an one-to-one depth correlation between the position of the dislocation bands and the Cr peaks. The distribution of the defects and their diameter is dependent on the implantation and the annealing conditions. The detailed results of TEM and XTEM are shown in /1/.

4. Discussion

At capless annealing conditions, which are necessary for a good electrical activation, some lost of As from the surface is not to avoid, as proved by RBS /4/. Thereby, As vacancies are found causing the Cr segregation to the surface. Consequently, near the surface region Cr depletion arises. In consequence of the concentration gradient Cr diffuses from the depth to the surface. Within the depletion region Cr will be trapped on dislocations formed during annealing. As consequence of the Cr diffusion the trapping begins from the bulk and for increasing time the trapped Cr decorates the whole dislocation band. The weak shift of the Cr peak in Fig. 2 is caused by the shift of the maximum defect density. Annealing at lower temperatures and adequate longer times (920 °C, 2 s) leads to a lower density of larger loops. This defect state is obviously not so effectiv in trapping of Cr.

Contrary to the 100 °C implantation the 30 °C implantation produces an amorphous layer /4/. After annealing we observed by XTEM two dislocation bands: the usual band in the depth of the transition region to the undisturbed crystal and an another band at the interface of the amorphous surface layer. The Cr peaks in Fig. 3 are also observed in this depth.

From the one-to-one depth correlation of the Cr peaks and the dislocation bands for different implantation and annealing conditions we conclude that the main process of the formation of the Cr peaks is trapping of Cr on dislocations and not the formation of Cr-defect or Cr-impurity-complexes /5,6/. Cr redistribution in the Si implanted region has not been found for doses $<10^{13}$ Si/cm^2 or for annealing modes where no dislocations are formed (face-to-face or 0.5 s - RTA with Si_3N_4 cap layer).

5. References

/1/ Baither, D. et al.: This conference
/2/ Kuzuhara, M. et al.: MRS 23 (1984) 651
/3/ Lee, D.H. and R.M. Malbon: Appl.Phys.Letters 30 (1977) 327
/4/ Panknin, D. et al.: Nucl.Instr.and Methods B19/20 (1987) 492
/5/ Sadana, D.K. et al.: J.appl.Phys. 53 (1982) 6413
/6/ Wilson, R.G. et al.: J.appl.Phys. 54 (1983) 6868

INVESTIGATION OF LOW-TEMPERATURE RECRYSTALLIZATION OF ION IMPLANTED AMORPHOUS GaAs LAYERS

J.Herold, W.Wesch, G.Götz[1], and H.Bartsch[2]

1. Introduction

n^+-doped layers in GaAs offer many important applications in semiconductor technology, particular for fabricating high-frequency devices and integrated circuits. The high-dose implantation necessary mostly produces amorphous layers. Such layers have to be heated to anneal damage and to realise electrical activation of the dopants. In GaAs different annealing stages can be identified. The recrystallization occurs in the temperature range between about 450K and 700K. As a result layers with a large amount of extended defects, especially twins, are created, which prevent electrical activation of the dopants. Temperatures higher than 1200K are mostly neseccary to reduce the defects and to make the layers electrically active.

Contrary to silicon during the low-temperature recrystallization of GaAs a rough interface is built up [1]. As a consequence twins are created due to wrong capture of atomic groups at <111>-oriented facets [3]. To investigate the evolution of the interface roughness during the recrystallization process, time-resolved reflectivity (TRR) measurements in combination with TEM and RBS investigations were carried out.

2. Experimental

Chromium doped semi-insulating <100> GaAs substrates were implanted with $5*10^{14}$ cm^{-2} P^+ ions of energy 85 keV at the temperature of 80K. With the chosen parameters amorphous surface layers of a thickness of about 147 nm were obtained. The recrystallization process was induced by heating the sample with a 10 W cw-argon laser. A He-Ne-laser test beam of the wavelength $\lambda = 0.6328\,\mu m$ was used for the registration of the reflectivity. The recrystallized samples furthermore were investigated by RBS and TEM measurements.

---

1) Friedrich-Schiller-Universität Jena, Sektion Physik, Max-Wien-Platz 1 Jena, DDR 6900

2) Institut für Festkörperphysik und Elektronenmikroskopie der AdW der DDR, Halle, DDR 4050

3. Results and discussion

The TRR method is based on the differences between the complex refractive indices $\tilde{n} = n - ik$ for amorphous ($\tilde{n}_a$) and crystalline ($\tilde{n}_c$) material, the dependence of the reflectivity on the temperature of recrystallized GaAs $\frac{\Delta R'}{\Delta T}$, and the oscillation of the reflectivity as a consequence of the movement of the amorphous/crystalline interface. As a supposition for the theoretical simulation of the epitaxial process in a first step the optical parameters were determined to $\tilde{n}_a = 4.24-1.09i$, $\tilde{n}_c = 3.89-0.22i$ (at $\lambda = 0.6328\,\mu m$) and $\frac{\Delta R'}{\Delta T} = 0.0055\%/K$. The calculated dependence of the reflectivity on the thickness d_a of an amorphous layer is shown in Fig.1 for a planar (a) and a non-planar (b,c) interface.

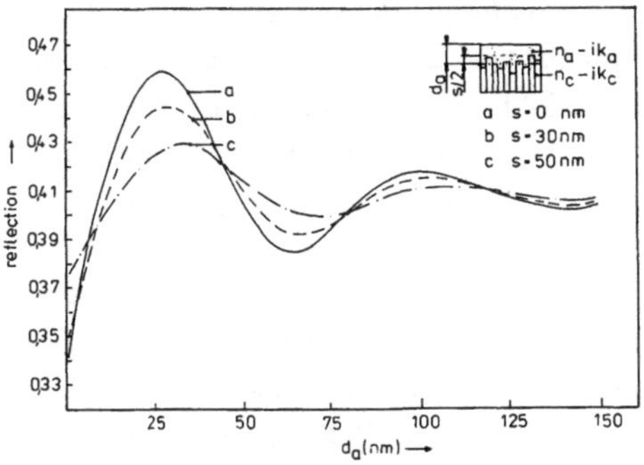

Fig.1. Calculated reflection as a function of the thickness of the amorphous layer for a system consisting of an amorphous layer with a rough interface on a crystalline substrate

The reflectivity for the non-planar case was calculated assuming, that the interface consists of steps with step heights randomly distributed around the thickness of the amorphous layer in the limits of $(d_a - \frac{s}{2})$, where s characterizes the extent of the nonplanarity. One can see that the amplitude of the reflectivity curve decreases with increasing maximum step height s.

The reflectivity measured during recrystallization of the P^+ implanted sample at 505K is depicted in Fig.2 as a function of the time. In the time interval between t_1 and t_2 (beginning and end of the heating) distinct oscillations of the reflectivity are observed. By comparing the experimental curve with calculated ones the connection

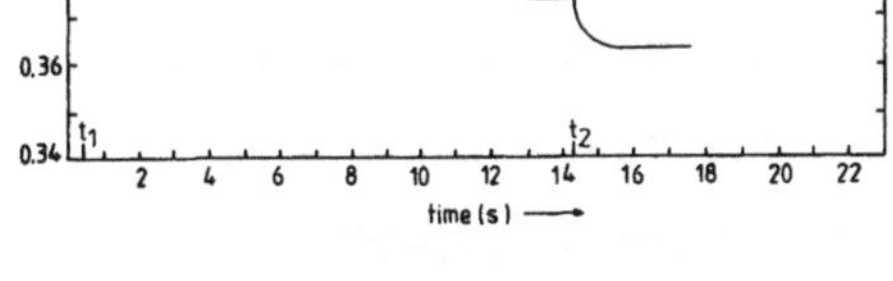

Fig.2. Reflection as a function of the annealing time for recrystallisation of P^+-implanted GaAs at $T_A = 505K$

Table 1. Roughness of a moving interface for different recrystallization times t_A and interface positions d_a determined by different experimental techniques

		TRR	TEM cross-section	RBS
t_A (s)	d_a (nm)	s (nm)	z (nm)	z_{ir} (nm)
0	147	0±10	10±5	10±10
3.5	118	-	-	21±10
4.2	105	36±3	30±5	43±10
6.0	75	48±3	43±5	-
7.3	40	74±3	-	-

between the thickness of the residual amorphous layer and the time necessary for the realization of the respective annealing stage can be determined with the procedure described in [2]. Furthermore, in this way the comparison of the measured reflectivity-values with the theoretical curves calculated for different extents of the nonplanarity yields the nonplanarity for a given annealing stage. The results achieved for different times and thicknesses d_a are shown in Table 1. With increasing annealing time and decreasing thickness of the residual amorphous layer the nonplanarity s increases. The increase of the roughness with the recrystallization time is also illustrated in the TEM cross section pictures in Fig.3. Within the error limits the values of the roughness determined by TRR, TEM as well as by RBS (Tab. 1) are in good agreement.

By different methods it was shown that during recrystallization of amorphous GaAs layers a rough interface is produced connected with the creation of extended defects. Because these defects strongly influence the electrical activation extent and concentration of such defects must be minimised. It was illustrated that especially the TRR method is a relatively simple and fast experimental technique for investigating the dynamics of epitaxial recrystallization process in GaAs.

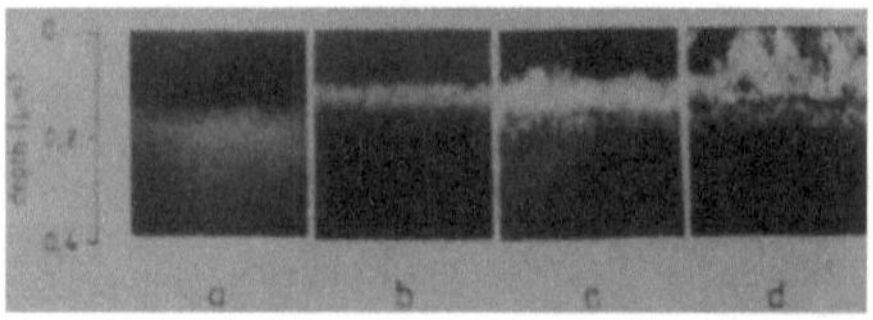

Fig.3. TEM cross-section pictures of a P^+-implanted GaAs layer (a) and after recrystallisation at T = 505K for 4.2 s (b), 6 s (c), and 7.3 s (d) annealing time

[1] Bhattacharya, R.S., A.K.Rai, P.P.Pronko, J.Narayan, S.C.Ling and S.R.Wilson, J.Phys.Chem.Solids, 44 (1983) 61-69

[2] Olson, G.L., S.A.Kokorowsky, J.A.Roth, R.S.Turley and L.D.Hess, Proc. MRS Symposium Boston 1981

[3] Sadana, D.K., Nucl.Instr. and Meth. B 7/8 (1985) 375-385

INVESTIGATION OF THE TEMPERATURE DEPENDENCE OF THE DECHANNELING IN WEAKLY DAMAGED GaAs

A.Jordanov, W.Wesch, K.Gärtner[1)]

1. Introduction

In GaAs under certain implantation conditions (light ions, sufficiently low dose or high temperature for heavier ions) weakly damaged layers are produced which are characterized by an increased near edge optical absorption depending exponentially on the photon energy $h\nu$. Both the average absorption coefficient $\bar{K}$ the characteristic tailing energy E_1 increase with the nuclear deposited energy density $\bar{G}_n$ and reach saturation values ($\bar{K} = 4*10^4$ cm^{-1}, $E_1 = 0.52$ eV) at $\bar{G}_n = 3*10^{20}$ keV cm^{-3} for room temperature implantation [1,2]. Rutherford backscattering investigation of such layers show a small increase of the minimum yield after ion implantation.

In a previous paper [3] it was shown that the temperature dependence of the dechanneling of light ions in crystals with displaced atoms is very sensitive to the position of the displaced atoms in the lattice cell. Preliminary investigations of the temperature dependence of the dechanneling of ions in N^+ implanted GaAs have shown high concentrations of slightly displaced lattice atoms [4]. In the present paper we have investigated the dependence of concentration and position of the displaced atoms in N^+ implanted GaAs on the ion fluence.

2. Method

<100> n-GaAs wafers were implanted with N^+ ions of energy 280 keV at room temperature. The N^+ fluence was varied between $7*10^{13}$ cm^{-2} and $1*10^{16}$ cm^{-2}. During all implantations the current density was kept constant at 0.5 µA cm^{-2}.

The defect production was investigated by means of the RBS channeling technique using 1.4 MeV He^+ ions. The samples were mounted on a special goniometer which allows measurements in the temperature region between 300 K and 80 K. From the difference of the minimum yields of implanted and virgin GaAs χ_{min} measured at 300 K and 100 K the distance r_a of displacement of the lattice atoms and the relative concentration n_{pd} of displaced atoms were calculated using the procedure described in [3],

1) Friedrich-Schiller-Universität Jena, Sektion Physik, Max-Wien-Platz 1 Jena, DDR 6900

where the defects are assumed to consist of a part of lattice atoms all displaced by the same value r_a from the string of atoms.

In order to obtain a sufficient statistical accuracy of the investigations (statistical error of the minimum yield less than 3%) for the measurement geometry used here a total He^+ fluence of $5*10^{13}$ cm^{-2} (beam diameter 1 mm) is necessary for recording a spectrum. However, to get realistic information about the implantation induced damage the contribution to the minimum yield due to the defects produced by the He^+ beam itself must be eliminated. Therefore each backscattering measurement was performed step by step using He^+ fluences of $6.25*10^{12}$ cm^{-2}. The results show that the minimum yield for a fixed depth aproximately depends lineary on the He^+ fluence. A linear extrapolation to the zero point of the He^+ fluence for all depths provides the corrected minimum yield.

3. Results and discussion

In Fig.1 the depth distribution of the relative point defect concentration $n_{pd}(z)$ is depicted for different N^+ fluences D.

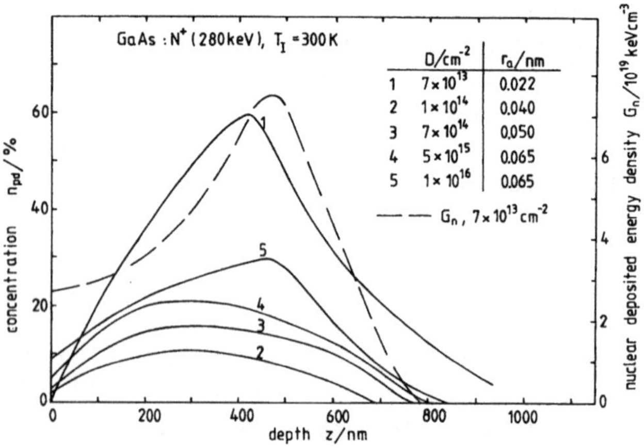

Fig.1. Depth dependence of the relative point defect concentration $n_{pd}(z)$ in <100> GaAs implanted with different N^+ fluences at room temperature and profile of the nuclear deposited energy density G_n for $D = 7*10^{13}$ cm^{-2}.

For $D = 7*10^{13}$ cm^{-2} a negative temperature dependence of the difference of the minimum yield $\Delta\chi_{min}$ is measured, which was theoretically predicted in [3] for slightly displaced lattice atoms, and experimentally verified in [4] for GaAs for the first time. The analysis of the spectra shows that 60% of the lattice atoms in the maximum of the distribution are displaced to $r_a \propto 0.02$ nm from their atomic strings (curve 1). As can be seen the defect distribution approximately corresponds to the energy density deposited into nuclear processes, which was calculated with a TRIM program [5]. With increasing N^+ fluence the calculated displacement distance r_a increases, but the concentration of point defects becomes remarkably lower. For $D = 1*10^{14}$ cm^{-2} about 10% of the lattice atoms are displaced by $r_a \approx 0.04$ nm (curve 2). This considerable change of $n_{pd}(z)$ and r_a in the fluence range $7*10^{13}$ $cm^{-2} \leq D \leq 1*10^{14}$ cm^{-2} indicates a transformation into a different defect structure. Obviously low ion fluences (in our case $D < 1*10^{14}$ cm^{-2}) cause a slight distortion of the lattice with small displacement of the lattice atoms and a high percentage of displaced atoms. Large values of r_a (i.e. atoms on interstitial sites) can be definitely excluded in this case by the negative temperature dependence of $\Delta\chi_{min}$ measured [3].

The results for $D>10^{14}$ cm^{-2} may be understood by the generation of interstitials (or amorphous clusters) in the distorted lattice. In this case there exist displacements with both small and large values of r_a which is not taken into account in the quantitative analysis performed. Therefore the results obtained for r_a and n_{pd} represent only mean values. The continuous increase of r_a until $D = 5*10^{15}$ cm^{-2} is a consequence of the increasing number of interstitials (or amorphous clusters) because their dechanneling contribution dominates more and more the contribution from the atoms weakly displaced by the lattice distorsion.

For $D>5*10^{15}$ cm^{-2} we find a constant value of r_a, i.e. the dechanneling contribution due to weakly displaced lattice atoms can be neglected. Additional calculations with randomly distributed r_a did not provide the measured temperature dependence of $\Delta\chi_{min}$. This indicates that the displaced atoms are more probably interstitials than amorphous clusters.

In curve 1 n_{pd} is the part of weakly displaced atoms but in the other cases n_{pd} approximately represents the number of interstitials because they provide the main contribution to the dechanneling. For $D>10^{14}$ cm^{-2} the concentration n_{pd} increases with increasing N^+ fluence because of the increasing number of interstitials.

[1] Wendler, E., W.Wesch and G.Götz: phys.stat.sol. (a) 93 (1986) 207-212.

[2] Wesch, W., K.Gärtner, E.Wendler and G.Götz: Rad.Eff. 97 (1986) 313-319.

[3] Gärtner, K., K.Hehl and G.Schlotzhauer: Nucl. Instr. and Meth. B4 (1984) 55-62.

[4] Wesch, W., K.Gärtner, E.Wendler and G.Götz: Nucl. Instr. and Meth. B15 (1986) 431-434.

[5] Biersack, J.P. and L.G.Haggmark: Nucl. Instr. and Meth. 174 (1980) 257-269.

SHORT TIME THRESHOLD FOR ELECTRICAL ACTIVATION OF IMPLANTED AND ANNEALED GaAs

D. Panknin[1], E. Wieser[1], Ya. V. Fattachov[2] and I. B. Khaibullin[2]

1. Introduction

For a good electrical activation it is necessary to carry out the annealing at temperatures and times slightly below the dissociation limit of the GaAs surface /3,4/. For very short annealing times ($<$100 ms) the possible maximum temperature is only weaklydependent on the annealing time /4/.
This contribution presents results concerning the short time threshold for which electrical activation of the non-amphoteric dopants Se, S, Te and the amphoteric dopants Si, Ge, Sn can still be achieved by capless annealing using temperatures in the range of about 1100°C. The temperature dependence of the electrical activation for annealing times between 50 ms and 2 s is also discussed.

2. Experimental

Semi-insulating Cr compensated (100) GaAs was implanted using the parameters given in Tab. 1. In the last column of Tab. 1 the defect state of the as implanted sample is characterized as determined by RBS.

Tab. 1

species	$T_{impl.}$ /°C/	dose /cm^{-2}/	energy /keV/	RBS characterization of the as-implanted state
Si	100	$5 \cdot 10^{14}$	200	weakly damaged
	30	$5 \cdot 10^{14}$	200	amorphous
Ge	100	$5 \cdot 10^{14}$	200	damaged
Sn	100	$5 \cdot 10^{14}$	300	damaged
Se	100	$5 \cdot 10^{14}$	300	damaged
	200	$5 \cdot 10^{14}$	300	weakly damaged
S	200	$1 \cdot 10^{15}$	300	weakly damaged
Te	300	$1 \cdot 10^{14}$	300	weakly damaged

Annealing was performed in N_2 atmosphere without cap layer by flash lamp irradiation with pulse durations between 30 and 500 ms at maximum

1- Central Institute of Nuclear Research, Academy of Sciences of the GDR, Rossendorf, PSF 19, DDR-8051 Dresden

2- Kazan Physico-Technical Institute, Kazan, USSR

temperatures above 1000°C or using halogen lamps at temperatures between 900 and 1000°C with dwell times longer 0.5 s. Temperature and time pairs were used which are slightly below the dissociation threshold of the GaAs surface shown in Fig. 1.

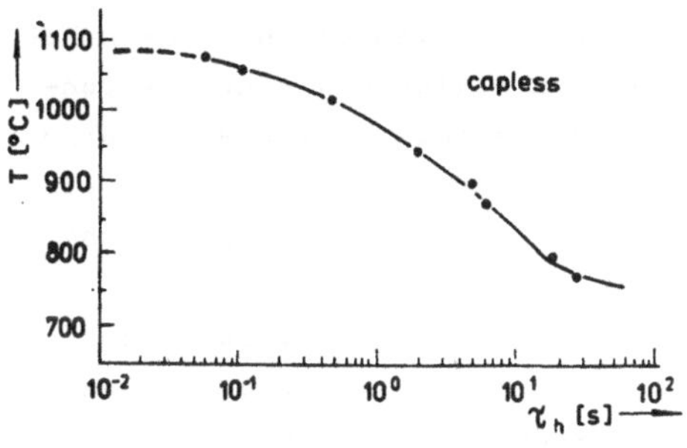

Fig. 1 Dependence of the optimum annealing temperature on the dwell time at T_{max} for capless annealing.

3. Results

In Fig. 2 the sheet carrier concentration is shown in dependence on the dwell time at the maximum temperature for capless annealing of the non-amphoteric dopants. Fig. 3 shows the same dependence for the amphoteric species.

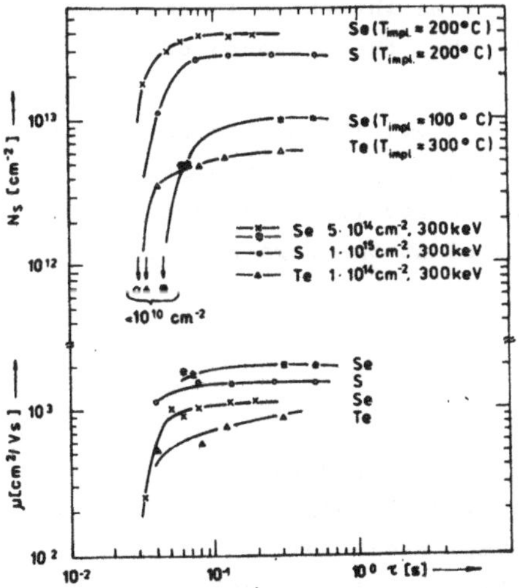

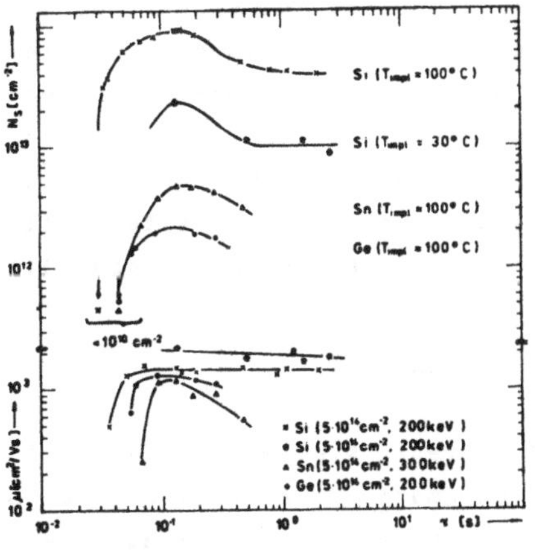

Fig. 2 Sheet carrier concentration and mobility of the non-amphoteric species (Se, S, Te) in dependence on the annealing time at the corresponding maximum temperature according to Fig. 1.

Fig. 3 Sheet carrier concentration and mobility of the amphoteric species (Si, Ge, Sn) in dependence on the annealing time at the corresponding maximum temperature according to Fig. 1.

A time threshold τ_{th} for the electrical activation is observed at about 30 ms if the implantation was carried out at elevated temperatures. Whereas after implantation at room temperature τ_{th} was found at about 45 ms. The time dependence of the sheet carrier concentration of the non-amphoteric species is characterized by a transition region and a nearly constant value at longer times, representing the maximum of the achievable carrier concentration for capless rapid thermal annealing in N_2 atmosphere. The time dependence of the mobility is very similar to that of the activation.

For the amphoteric species the time dependence of the sheet carrier concentration is characterized by a maximum at about 150 ms and a nearly constant value at times $>$ 1 s. The time dependence of the mobility is characterized by a transition region and, esspecially for the heavier elements, a maximum at about 100 ms.

4. Discussion

In the literature are given different values for the activation energy of the implant activation for different kinds of dopants (Si: 0.51 eV /1/, Se: 1.2 eV /2/). We have determined comparable values from our experiments for typical RTA conditions.

The time threshold of the electrical activation is independent on the kind of species, but it depends on the damage state caused by the implantation. From this we conclude that the time threshold is an attribute of the lattice structure. It might be that the presence of As vacancies plays an important role.

The maximum of the carrier concentration at about 150 ms for amphoteric dopants can be explained by the assumption of a Si concentration on Ga sites higher than the equilibrium value found for longer annealing times. Obviously, the Ga sites will be occupied at the beginning of the annealing with a higher probability.

5. References

/1/ Barrett, N.J. et al.: J. appl. Phys. 56 (1984), 3503

/2/ Hiramoto, T. et al.: Jap. J. appl. Phys. 24 (1985), L 193

/3/ Kuzuhara, M. et al.: Mat. Res. Soc. Symp. Proc. Vol. 23 (1984), 651

/4/ Panknin, D. et al.: Nucl. Instr. and Methods B19/20 (1987), 492

ELECTRICAL PROPERTIES OF RAPID THERMAL ANNEALED IMPLANTED GaAs

J. G r ň o [1/]

1. Introduction

Electrical activation of implanted impurities is depended on impurity occupation of substitutional lattice sites in implanted semiconductor substrates. The electrical activation in implanted samples is currently determinated from the ratio sheet free carrier concentration and the dose of the implanted impurities. In this way determined electrical activation is called apparent activation. There can be considerable differences between the actual and apparent electrical activation in the meaning of occupated substitutional lattice sites. The reduction of free carrier concentration in the implanted electrically active GaAs layer can be caused by the existence of depleted regions near the semiconductor surface. There are also various other traps in the implanted layers causing the reduction of the free carrier concentration as well. Otherwise these traps are generally known like the residual defects of the accptor type in the n-type layers. The both factors mentioned above are mainly shown at low dose implantation.

2. Surface depletion in GaAs

It has been shown that when 0.01-0.1 monolayer of different metals or oxygen cover the surface of GaAs and other compounds as well, the Fermi level at the surface becomes pinned within a small range of energies, and that the pinning energy is roughly independet of the contaminating metal or oxide (1). For GaAs this pinning position is about 0.7 eV bellow the conduction band minimum. Several models and empirical rules have been proposed to explain pinning. One of the best appreciated theory is Spitzer´s unified defect model. Spitzer et al. (2) have suggested that pinning is due to surface states induced by the interaction of either metal or oxygen atoms with the GaAs surface and that the surface states are associated with native defects e. g. arsenic and gallium vacancies, produced by interaction. The captured free carriers in surface states do not contribute to electrical conductivity.

[1/] Microelectronics Department, Slovak Technical University, Bratislava, Czechoslovakia

For calculation of the reduction of the free carriers we used the Taylor's work (3) where the implanted MESFET model is analyzed. In Fig. 1 a dependences of apparent electrical activations of Si^+ ions implanted into GaAs as function of implanted doses for different energies and for surface barrier of 0.7 V are shown.

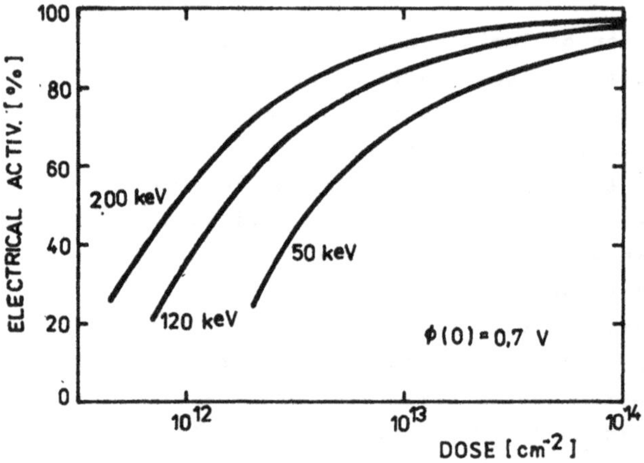

Fig. 1 Calculated electrical activation /apparent/ of Si^+ ions implanted into GaAs as a function of implanted doses

3. Residual defects

In generally, the residual defects behaving as traps results in the free carrier concentration in the GaAs electrical active layers. These defects can be connected with technology of the samples. The total concentration of the residual defects and the value of the actual electrical activation can be determinated from the linear dependence of the measured sheet free carrier concentration on the dose. A set of the linear dependences of sheet carrier concentration as functions of implantation doses for different implantation energies of Si^+ ions implanted into Cr doped semi-insulating /100/ GaAs are in Fig. 2. Annealing of the samples was performed by a method of RTA with halogen lamps in vacuum. No capsulation of the samples was used. All samples were annealed at uniform regime with temperature of 975^0 C and annealing time of 5 s. The measured values of the sheet free carrier concentrations N_s for a given energy were approximated to a line of the form $N_s = N_d \cdot A - N_{rd}$, where N_d is implantation dose, A is electrical activation and N_{rd} is concentration of the residual defects of an acceptor type. The values for electrical activation lies between 55 and 70% and the concentration of residual defects is in a range from 1.9 to 1.5×10^{12} cm^{-2} in a dependence on implantation energies.

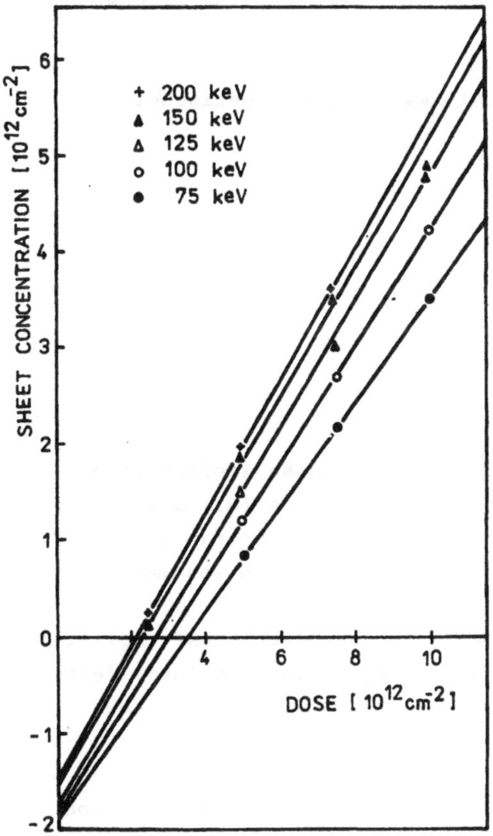

Fig. 2

Linear dependences of sheet carrier concentration as function of implanted doses for different energies of Si ions implanted into GaAs

(1) Van Laar, J., A. Huijser and T. L. Van Rooy: J. Vac. Sci. Technol. 14 (1977) 89

(2) Spicer, W. E., I. Lindau, P. Skeath and C. J. Su: J. Vac. Sci. Technol. 17 (1980) 1019

(3) Taylor, G. W., H. M. Darley, R. S. Frye and P. K. Chatterjee: IEE Trans. El. Dev., ED-26 (1979) 172

FORMATION OF AMORPHOUS GaP LAYERS BY ION IMPLANTATION AT LOW TEMPERATURE

J.Krynicki[1)], A.Kozanecki[2)], A.Olszewski[1)], R.Groetzschel[3)]

INTRODUCTION

Overlaping of defect clusters during high dose implantation results in the formation of continuous amorphous (α) layer in the crystal. Thus a certain level of the deposited energy is needed to generate α - layer. This critical value depend on ion mass and substrate temperature [1].

In this communication the formation of α-layer in GaP implanted with light (N^+) and heavy (Cd^+) ions respectively is investigated.

EXPERIMENTAL PROCEDURE

Sulphur doped ($5*10^{17} cm^{-3}$) GaP samples with <111> orientation have been used for the studies. Low temperature implantations were carried out using liquid nitrogen cooled stage allowing the sample temperature during bombardement to be held below 130 K.

The samples were implanted with 150 keV N^+ ions ($D = 5*10^{13}$, $2*10^{14}$, $6*10^{14}$, $2*10^{15}$, $6*10^{15}$, $2*10^{16}$ cm^{-2}) and with Cd^+ ions ($D = 6*10^{12}$, $2*10^{13}$, $5*10^{13}$, $1.5*10^{14}$, $5*10^{14}$, $1.5*10^{15}$ cm^{-2}).

Room temperature implantation has also been performed with doses $1.8*10^{16}$ cm^{-2} of N^+ ions and $5*10^{14}$ and $1.5*10^{15}$ cm^{-2} of Cd^+ ions.

Rutherford backscattering (RBS) technique has been applied to study damage in the implanted layers. RBS and channeling

---

1) Institute of Nuclear Chemistry and Technology, Warsaw, Poland.

2) Institute of Physics, Polish Academy of Sciences, Warsaw, Poland.

3) Zentralinstitut fuer Kernforschung, Rossendorf, GDR.

measurements were performed using 1.7 MeV $^4He^+$ ions at the Rossendorf Van-de-Graaff generator. Backscattered particles were detected at 170° angle to the incoming beam.

RESULTS AND DISCUSSION

Fig.1 presents deposited energy distribution calculated for 150 keV Cd^+ as well as 150 keV N^+ ions implantations. Calculations were done for all implanted doses on the basis of the procedure described by Winterbon [2].

Damage distribution in the implanted layers has been found from the RBS spectras. The following standard formula for estimation of damage density has been used.

$$\mathrm{Damage}(x) = \frac{\chi_{ali}(x) - \chi_{dech}(x)}{\chi_{ran}(x) - \chi_{dech}(x)} * 100\%$$

where: $\chi_{ali}(x)$ - counts in the aligned spectra at the depth x.
$\chi_{ran}(x)$ - counts in the random spectra at the depth x.
$\chi_{dech}(x)$ - counts in the dechanneling fitline.

Damage calculated from the experimental data is shown in Fig.2. It can be estimated that threshold dose for amorphization is about $4*10^{14}$ cm^{-2} for N^+ implantation and $4*10^{13}$ cm^{-2} for Cd^+ ions. It can also be seen that at room temperature even the dose as high as $2*10^{16}$ cm^{-2} of 150 keV N^+ ions is too low to form α-layer, while for Cd^+ implantation the temperature have no influence on the depth of amorphous layer.

Figure 3 illustrates the dose dependence of the depth of amorphous layer for N and Cd implantation.

Analysis of the experimental and theoretical data allows to estimate threshold damage density (TDD) for amorphization. The TDD value seems to be the same for both ions, approximately of $8*10^{20}$ keV/cm^3 for 130 K implantation. The depths of the α-layers have been calculated using this TDD value (broken horizontal line in Fig.1) and compared with the values extracted from the experimental data (Fig.3). It is seen that for Cd^+ implantation the agreement between experiment and theory is fairly good, while for N implantation the experimentally found depths are much lower than calculated theoretically. This remarkable discrepancy can be explained by self - annealing processes during implantation. Therefore, to move α/crystalline interface deeper into the crystal the deposited energy density at the depth of the interface must many times exceed the estimated value of $8*10^{20}$ keV/cm^3.

CONCLUSIONS

Threshold damage density for amorphization of GaP is the same for N^+ and Cd^+ low temperature implantation and is about $8*10^{20}$ keV/cm^3.

For low temperature implantation the threshold dose for amorphization is $\approx 4*10^{13}$ cm^{-2} for 150 keV Cd^+ ions, and about $4*10^{14}$ cm^{-2} for 150 keV N^+ ions.

Calculated amorphous layer depth D_a for Cd^+ ion implantation agrees well with D_a value found from RBS measurements. Oppositely, calculated D_a for N^+ implantation exceeds considerably the experimental D_a value. It indicates that dynamical annealing of damage during implantation is a substantial competing effect to the damage accumulation particularly at low density of deposited energy.

REFERENCES

1. S. Prussin, D.I. Margolose, R.N. Tauber, J. Appl. Phys. 57, 180, 1985.
2. K.B. Winterbon Ion Implantation Range and Energy Deposition Distribution, vol.2 Low Energy, Plenum Press 1975, p.17-20.

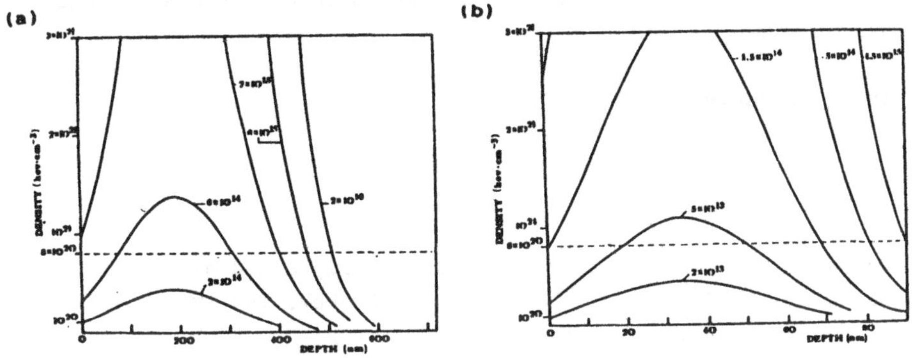

Fig.1 Calculated damage density in GaP for 150 keV implantation for N^+ (a) and Cd^+ (b) ions

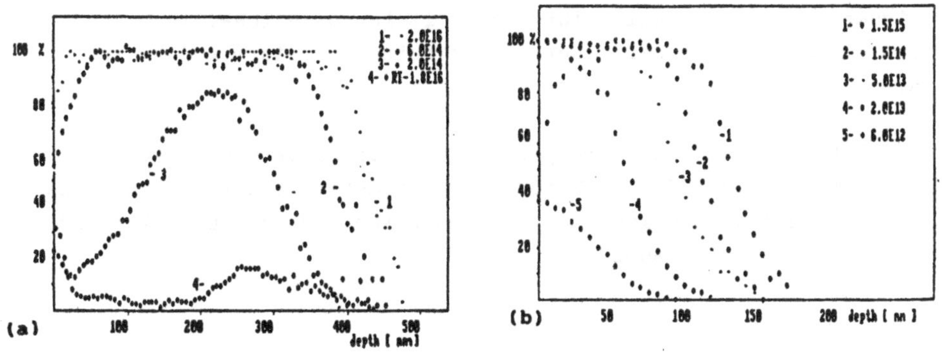

Fig.2 Damage density distribution for 150 keV N^+(a) and Cd^+(b) ion implantation obtained from RBS measurements (100% corresponds to full amorphization).

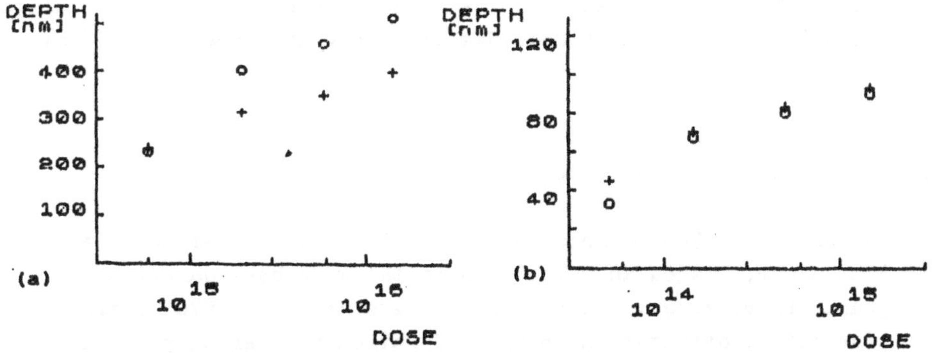

Fig.3 Amorphous layer depth for N^+(a) and Cd^+(b) implantation as a function of implanted dose (+ - experimental, o - calculated for the critical value of $8*10^{20}$ keV/cm^3).

RESIDUAL DEFECTS IN IMPLANTED GaAs AFTER RAPID THERMAL ANNEALING WITH INCOHERENT LIGHT

D. Baither, H. Bartsch[1] and D. Panknin[2]

EXPERIMENTAL

Semi-insulating Cr-compensated (100) GaAs was implanted with Si at doses between 5×10^{12} and 10^{15} Si cm^{-2} at 200 keV. The implantation temperatures were about 30 (RT) and 100 °C, resp.
For these investigations capless rapid thermal annealing was carried out at temperatures between 760 and 1050 °C for 0.5 and 25 s, using a set of tungsten halogen lamps.
The thermal treatment of GaAs without a cap layer is limited due to the dissociation of arsenic; that means it is not allowed to exceed a critical value of the energy deposition which is determined by the temperature and the pulse duration.

TEM and XTEM observations were performed by means of a high voltage electron microscope (HVEM) JEM 1000. To the defect imaging mainly the weak beam method was applied to yield narrower contrast profiles and consequently a higher resolution.

RESULTS

Using the HVEM, defects were resolved only for doses higher than 10^{13} Si cm^{-2}.
For that reason the dependence of the defect structure on the annealing parameters is represented in more detail mainly at a dose of 10^{15} Si cm^{-2}.

Implantation at elevated temperature (100 °C)

After annealing at 760 °C for 5 s defect clusters of diameters less than 5 nm occur, imaged as white dots in the weak beam dark-field micrograph. Increasing the annealing temperature up to 870 °C yields a similar defect structure at a pulse duration of even only 0.6 s. For longer pulse durations the defects grow and it is possible to resolve them as prismatic dislocation loops . Their diameters

1) Institute of Solid State Physics and Electron Microscopy, Halle
2) Central Institute for Nuclear Research, Rossendorf

amount to about 10 to 20 nm after annealing for 1.6 s, and to 20 to 40 nm for 9 s. Further increasing the time up to 25 s does not change the loop diameter. The defect density seems to be constant within the considered interval of annealing time.

Higher annealing temperatures cause the acceleration of the defect formation. At 920 ^{o}C loops arise with diameters between 20 and 40 nm al-

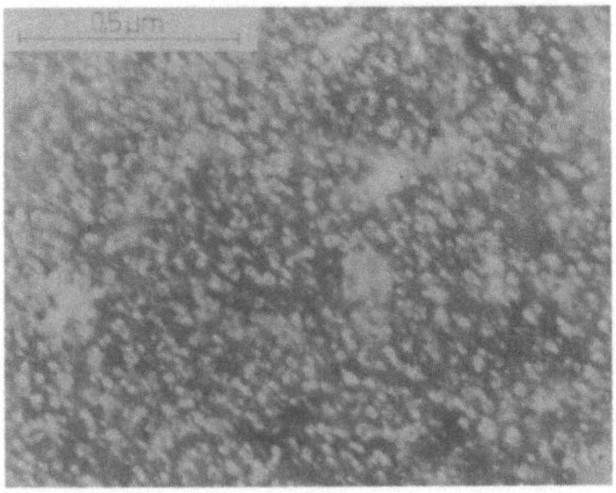

Fig. 1: Si implantation at elevated temperature (100 ^{o}C) $D = 1.10^{15}$ cm^{-2}, T_a= 870 ^{o}C, t_a= 1.6 s

ready after 1.7 s (Fig. 1). Annealing at 1000 ^{o}C allows a maximum pulse duration of about 0.4 s. This time is not sufficient to create elongated defects; in the micrographs only white dots of diameters less than 5 nm are resolvable.

Fig. 2: Si implantation at elevated temperature (100 ^{o}C) $D = 1.10^{14}$ cm^{-2} T_a= 880 ^{o}C, t_a= 1.6 s

Lower implantation doses (5 x 10^{14}, 10^{14} Si cm^{-2}) are associated with diminished defect densities, compared with the dose of 1 x 10^{15} cm^{-2}, whereas the defect sizes are nearly unchanged under the same annealing conditions (Fig. 2).

Implantation at room temperature (30 ^{o}C)

Unlike the implantation at elevated temperatures, at room temperatures an amorphous layer,about 200 nm thick,is created at the surface.

After annealing at 900 ^{o}C for 1.2 s loops arise, which are smaller than those after high-temperature implantation (Fig. 3a). The opposite appears after annealing for 2.2 s. Large loops of 20 and 50 nm in diameter and with Burgers vectors of the type a/2 <110> occur, which interact forming networks (Fig. 3b).

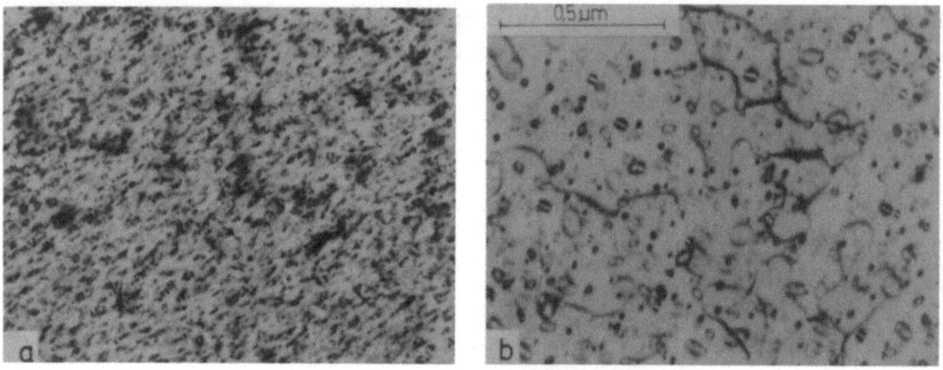

Fig. 3: Si implantation at room temperature (30 °C)
a) $D = 1.10^{15}$ cm^{-2}, $T_a = 900$ °C, $t_a = 1.2$ s
b) $D = 5.10^{14}$ cm^{-2}, $T_a = 900$ °C, $t_a = 2.2$ s

Differences between high and room-temperature implantations become most evident in the depth distribution of defects, observed by XTEM. For high-temperature implantation defects exist over the depth range of 80 to 500 nm, with a maximum of concentration at about 250 to 350 nm in depth (Fig. 4a).

After room temperature implantation two discrete layers of loops are observed in cross section (Fig. 4b). One layer is formed again at 300 nm in depth. The upper layer containing large loops and straight dislocation lines is centered at about 150 nm in depth, which corresponds to the former position of the amorphous-/crystalline interface.

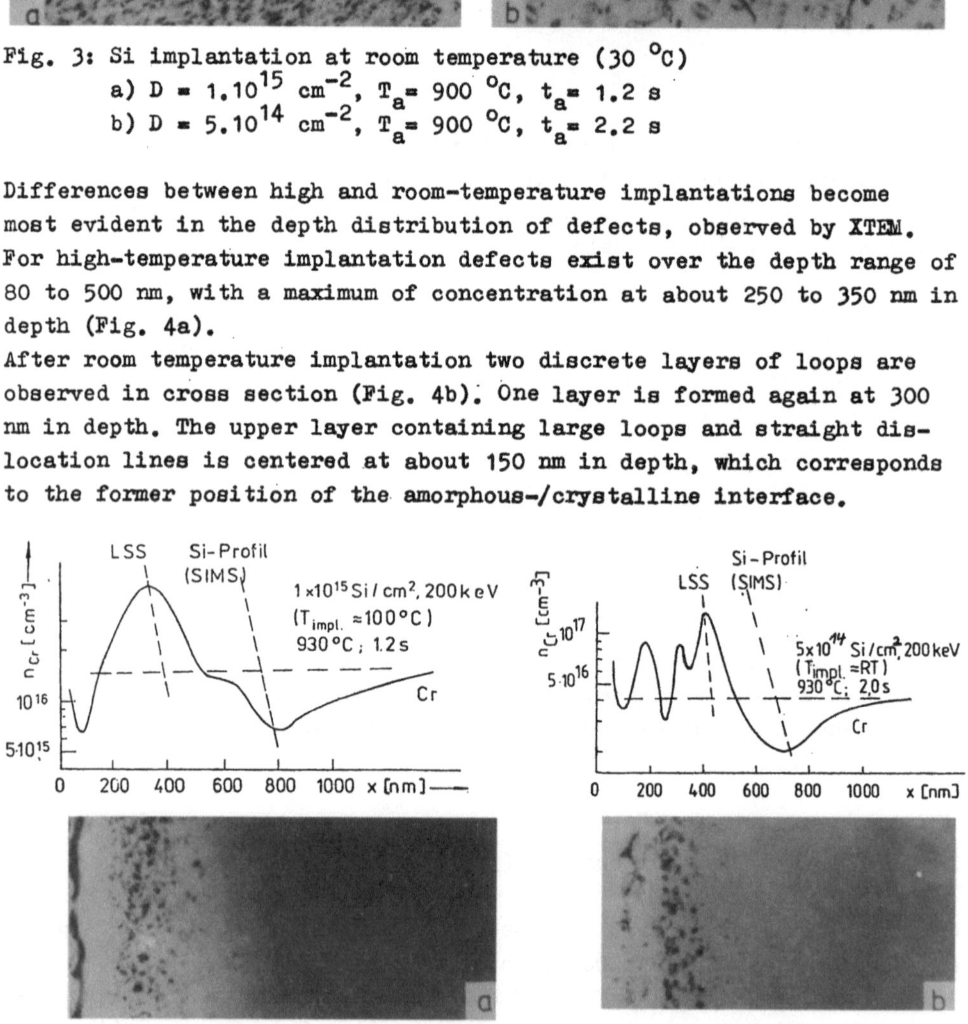

Fig. 4: Depth distribution of defects after Si implantation
a) at 100 °C, b) 30 °C

Besides the cross section micrographs in Fig. 4 the Cr depth distributions are shown, which are in good agreement with the defect profiles, indicating that Cr is trapped at the dislocations. Detailed investigations of the Cr distribution are discussed in /1/.

References:

/1/ D. Panknin et al., this volume

HALOGEN LAMP ANNEALING OF Au AND AuGe CONTACTS TO GaAs

P. Thiele+)

Introduction

An important contact material to GaAs and other compound semiconductors is gold, but the interaction between gold and III-V semiconductors is very strong if conventional annealing technique is used.

One possibility to improve the contact properties is to shorten the annealing times. Laser annealing [1-3] and electron beam annealing [4,5] have been used recently with some success, but process reliability and effectivity have been low. In particular with regard to process reliability, better results can be expected from halogen lamp annealing. In this investigation a rapid thermal annealing (RTA) system with halogen lamps was employed to produce ohmic contacts on n-type GaAs. The results of halogen lamp annealing were compared with the results of thermal furnace annealing (TFA).

Experimental

Except for annealing, the contacts were made by conventional evaporation technique. For halogen lamp annealing times up to 2 seconds were used and for furnace annealing 10 minutes. The Au-GaAs samples were analysed by optical microscope, SEM and SIMS. In addition, the electrical contact resistivity, the sheet resistivity of the goldlayer, the etching time, the roughness of the contact surface and the roughness of the interface Au-GaAs were measured.

Results

Figure 1 and 2 show that the halogen lamp process produces specific contact resistances, which are lower (Au-GaAs) or in the same order (AuGe-GaAs) compared with conventional furnace annealing.

In contrast to these results, at the contact resistance minimum, the interaction between Au and GaAs is very small in halogen lamp

+)Department of Physics
Humboldt-University, 1040 Berlin, GDR

annealing (Figure 3). The sheet resistance of AuGa layers is higher than that of pure gold layers. Therefore the sheet resistance measurements of annealed Au layers on semiinsulated GaAs confirm the SIMS measurements (Figure 4).

The advantages of halogen lamp contact annealing compared with thermal furnace annealing are

- improvement of the surface morphology and edge roughness of gold contacts
- lower roughness of the Au-GaAs interface
- strongly reduced diffusion from Ga into Au
- improved chemical etching of the goldlayer
- equal or better specific contact resistance
- improved wire-bonding-process.

The results demonstrate that halogen lamp annealed contacts on GaAs have properties, which are superior to those produced by a conventional furnace annealing process.

References

1 A. H. Oraby et al., Appl. Phys. Lett. 38 (1981) 562

2 O. Aina et al., J. Appl. Phys. 52 (1981) 6997

3 P. Sircar et al., phys. stat. sol. (a) 85 (1984) 649

4 C. P. Lee et al., Appl. Phys. Lett. 39 (1981) 556

5 P. Thiele, EPM 84, Sept. 1984

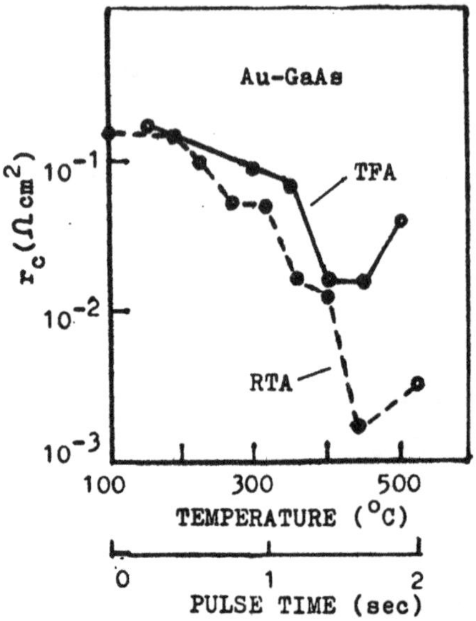

Fig. 1. Comparision of the furnace annealing(TFA) and halogen lamp annealing (RTA) of Au-GaAs contacts (temperature and pulse time versus specific contact resistance r_c)

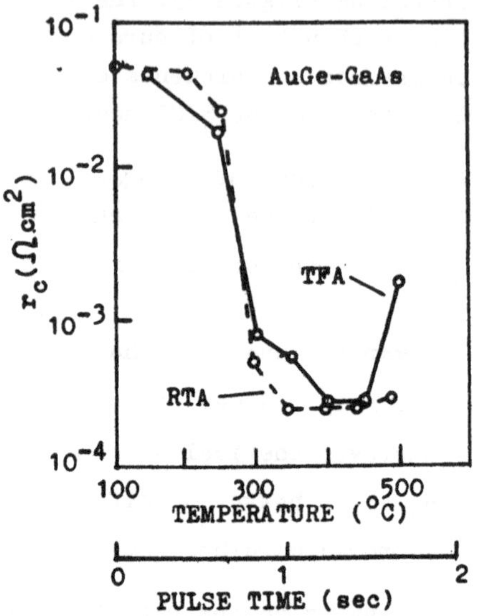

Fig. 2. Specific contact resistance r_c of AuGe-GaAs contacts plotted against temperature (TFA) and pulse time(RTA)

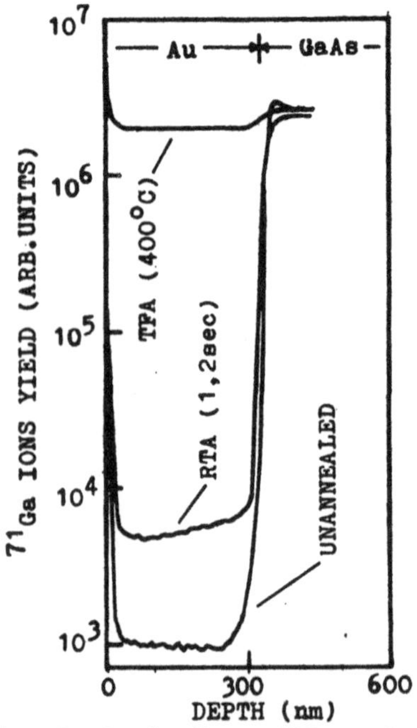

Fig. 3. Ga depth profiles of furnace annealed (TFA) and halogen lamp annealed (RTA) contacts (SIMS measurements)

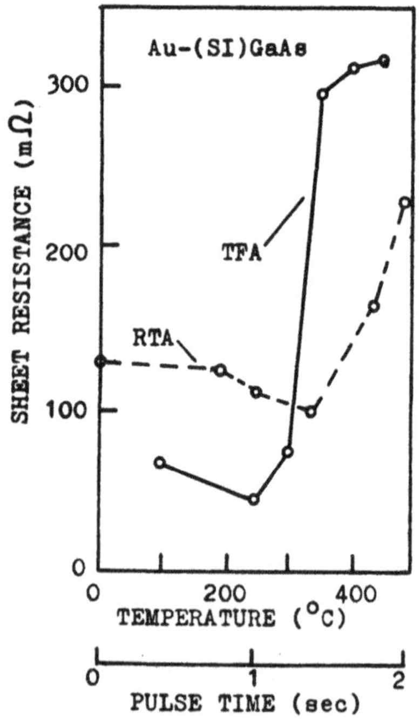

Fig. 4. Sheet resistance of Au on (SI)GaAs as function of annealing temperature (TFA) and pulse time (RTA)

FORMATION AND STRUCTURE OF IMPLANTATION-INDUCED AMORPHOUS ALLOYS

V. Heera and B. Rauschenbach [1])

1. INTRODUCTION

In a previous paper (1) we proposed a simple model for the amorphization of iron layers by boron implantation, which should also apply to all metalloid implantation into transition metals. The main assumption of this model is the formation of small amorphous clusters around the stopped boron ions kept stable by covalent boron-iron bonds. From this model follows that not the energy transferred by atomic collisions provide the amorphization threshold in metals, but the content of implanted glass-forming (i.e. disorder stabilizing) material. This seems to be the key to understand the large differences in ion doses needed for amorphization of semiconductors or metals.

Ion irradiation experiments of preimplanted metals can serve to prove this assumption. In this paper we present the results of ion irradiation of recrystallized iron-boron layers prepared by high dose boron implantation into iron (2). If the suggestion formulated above is correct the amorphous state of boron preimplanted material should be reached under ion irradiation at doses much less than those necessary for the primary amorphization. In addition, a variation in ion masses should provide information about the relevance of spike effects in the amorphization process. Two target temperatures, liquid nitrogen temperature (LNT) and room temperature (RT) were chosen to investigate the influence of defect relaxation.

Besides the formation process we also investigate the structure of implantation induced amorphous metal-metalloid alloys by means of electron diffraction (3). Because of the substantial differences in the formation process of implantation induced and rapidly quenched amorphous alloys it can be assumed that there are differences in their structure, too. By comparison of radial distribution functionc it should be examined whether such differences exist.

2. AMORPHIZATION OF BORON PREIMPLANTED IRON

Iron targets were amorphized by high dose boron implantation (50 keV, $1 \ldots 5 * 10^{17}$ ions/cm^2, RT) and then recrystallized by annealing (1.5 h at 450°C). Subsequent irradiation of these polycrystalline specimens by He^{+}- (40 keV), B^{+}- (30 keV) and Ne^{+}-ions (60 keV) leads to amorphization both at RT and LNT. The phase transformation was studied by high voltage electron microscopy (HVEM) and transmission high energy electron diffraction (THEED). The obtained amorphization threshold doses and the corresponding values of displacements per atom (DPA) are summarized in table 1.

[1]) Zentralinstitut für Kernforschung Rossendorf, PF 19, DDR-8051 Dresden, GDR

Table 1

ion	LNT		RT	
	D_A (ions/cm^2)	DPA	D_A (ions/cm^2)	DPA
He^+	2.5×10^{15}	0.074	3.5×10^{15}	0.102
B^+	2.0×10^{14}	0.061	3.5×10^{14}	0.107
Ne^+	7.5×10^{13}	0.072	1.0×10^{14}	0.096

From these results the following conclusions can be drawn:

(i) The boron concentration provides the amorphization threshold during boron implantation into iron. Whereas the amorphization dose for pure iron amounts to about 3×10^{16} B^+-ions/cm^2, the amorphization of boron preimplanted and recrystallized specimen already takes place at a dose of about 3.5×10^{14} B^+-ions/cm^2. This confirms the dominant role of the disorder stabilizing covalent metalloid-metal bonds for the amorphous state of metals. Only when a sufficient amount of metalloid preexists in the metal the amorphization process is governed by energy deposition during irradiation.

(ii) The crystalline to amorphous transition in the iron-boron layers occurs under ion irradiation with all ion species used in our experiments. Even the light He^+-ions are able to produce the amorphous state. There is, however , a strong ion mass dependence of the doses needed for the amorphization: the higher the ion masses, the smaller the amorphization doses.

(iii) There exists a uniform critical number of displacements per atom of 0.07 at LNT and 0.10 at RT irradiation, respectively, which results in amorphization independent of ion species and energies utilized for the irradiation of the iron-boron targets. In this sense, the mass dependence of the amorphization dose can be completely explained by the differences in the nuclear stopping powers. No thermodynamic or hydrodynamic effects in high density cascades (spikes) are required to understand the amorphization process.

(iv) The critical number of DPA obtained in our experiments agrees very well with the results of ion irradiation of Ni_3B single crystals at LNT (4) and of Fe_3B precipitates in annealed $Fe_{75}B_{25}$ alloy at RT (5). This is noteworthy as far as these irradiation experiments were performed with ion species and energies very different from ours and thus corroborate conclusion (iii).

(v) The differences in the critical DPA's between LNT and RT experiments illustrate the influence of thermally activated defect relaxation on the amorphization process. In contrast to the results of Thome et al. (4) we did not find any mass dependence of the critical DPA at RT.

3. AMORPHOUS STRUCTURE OF METALLOID IMPLANTED METALS

The amorphous structure of metalloid (B, P) implanted (10^{17} ions/cm^2, 50 keV) metal (Fe, Ni) films was investigated by HVEM and THEED. Information on the atomic structure of the amorphous layers was obtained from the elastically difracted electron density. On the basis of the scattering curves, the total interference function and the pair correlation function were determined (3). Peak positions of the pair correlation function, their ratios and the calculated coordination number of the first atomic sphere are given in table 2.

Table 2

system		Fe/B	Fe/P	Ni/P	Ni/B
peak position	r_1	0.25	0.25	0.245	0.25
/nm	r_2	0.425	0.42	0.425	0.415
	r_3	0.485	0.5	0.48	0.477
	r_4	0.64	0.66	0.65	0.64
ratio	r_2/r_1	1.7	1.68	1.73	1.66
	r_3/r_1	1.94	2.0	1.96	1.9
	r_4/r_1	2.56	2.64	2.65	2.56
coordination number	Z_1	12.5	12.3	12.4	12.6

The peak positions of implantation induced amorphous alloys lie rather well between the data obtained from rapidly quenched alloys and melts, respectively. Furthermore, the correlation length for the topological short-range order is shorter for implantation induced amorphous alloys compared with the amorphous alloys formed by rapid quenching and longer compared with melts. Thus, it can be assumed that metalloid implantation into metals leads to a more "disordered" amorphous state than rapid quenching of metallic melts.

REFERENCES

(1) Heera, V. and B. Rauschenbach: Rad. Eff. 91 (1985) 71

(2) Rauschenbach, B. and V. Heera: phys. stat. sol. (a) 100 (1987) 423

(3) Rauschenbach, B., G. Otto, K. Hohmuth and V. Heera: J. Phys. F in press

(4) Thome, L., J. C. Pivin, F. Pons and A. Benyagoub: Nucl. Instr. Meth. B 19/20 (1987) 563

(5) Johnston, W. G., A. Mogro-Campero and J. L. Walter: Mater. Sci. Engng. 55 (1982) 121

STRUCTURAL AND PHASE TRANSFORMATION IN PULSED ION IMPLANTATION OF BCC METALS

S.P.Bugaev*, Yu.R.Kolobov**, A.D.Korotaev**, A.O.Savchenko*

Structure evolution and molybdenum and vanadium surface layers phase composition changes were studied by electronic microscopy, electronography and augerspectroscopy methods.

Copper ion pulsed implantation ($E = 100$ keV, $f = 10^{17} cm^{-2}$, $T = 0.2$ ms) into recrystallized molybdenum results in formation of metastable ultradispersion structure with microregions less than 3-5 nm in a thin superficial layer. In this case electronogramms have the form of continuous hoops corresponding to BCC-lattice of initial molybdenum. The diffraction from monocrystalline underlayer and disperse structure were observed at the same time in very thick places of molybdenum foil still transparent for electrons ($t \leq 130$ nm). It witnesses that the thickness of superficial layer with changed structure does not exceed the above mentioned size. It was established by augerspectroscopy method that besides copper atoms the carbon ones also penetrate into the surface layer during implantation. The carbon concentration profile rather differs from the one of copper atoms. The carbon maintenance is practically constant from the surface to the depth of 25 nm, then it decreases and carbon maintenance reaches initial level on the depth of 50 nm. At that time copper concentration maximum is on the depth of 20 nm and decreases to zero towards 45 nm.

The annealing of implanted specimens at 1273 K for an hour results in coarsening of disperse structure and ordering of implanted copper atoms in molybdenum lattice which are recognized by appearing of superstructural reflexes on electronogramms. The revealed structural changes of superficial layers are due to losing resistance of crystal lattice and appearing of ω -phase displacements by admixture of the penetration.

Series of experiments were performed to find out the influence of implantation conditions on the molybdenum surface layer structure. The pulse frequency and thermal conductivity from the target were changed and the layer of carbon film absorbed on the model surface was artificially increased. This task of investigation is connected with the fact that the influence of carbon, penetrating into the surface layer of the target as a result of absorbed carbon film mixing, remains unclear.

* Institute of High Current Electronics of the USSR Academy of Sciences, Siberian Division, Tomsk
** Siberian Physical-Technical Institute, Tomsk

It was established that as a result of self-ion implantation with pulse frequency of 10 Hz to the dose of $10^{17} cm^{-2}$, lattice sizes uniformly increase to 0.339 nm (8 %) and diffusive crosses, settled down at matrix reflexes, appear on the electronogramms. Besides, the hoops of small disperse carbidum of MoC_x-type appear on electronogramms.

Ultradisperse carbidum MoC_x forms in surface layer during carbon ion implantation with pulse frequency of 10 Hz. The molybdenum lattice size increases to 6% and diffusive crosses appear in the area of matrix reflex. The size of the surface layer lattice increases to 4% and the formation of disperse carbidum of Mo_2C, MoC -types with the size of 4.0-7.0 nm is noted when pulse frequency increases to 50 Hz.

Ultradisperse structure forms in a surface layer, as above mentioned, in molybdenum implantation by copper ions with pulse frequency from 1 to 10 Hz in condition of good thermal conductivity. Lattice size increases to 2% in this case. Besides the molybdenum hoops the MoC carbidum ones are on the electronogramms.

The continuous turning of molybdenum lattice flats to 60° angle in a surface molybdenum layer takes place when ion dose increases to $10^{18} cm^{-2}$ with simultaneous increasing of a lattice size of the model surface reaching 10-11%.

Decreasing of thermal conductivity from a target in copper ion implantation with pulse frequency of 10 Hz leads to dispersion and appearance of diffusive cross-pieces on the electronogramms situation of which does not coincide with correct crystallographyc direction. The appearance of such effects on electronogramms is possible on the initial stage of formation of FCC-liberations in BCC-matrix.

The experiment shows that disperse of molybdenum surface layer takes place in the process of copper ion implantation in artificial increased layer of carbidums on the target surface inspite of thermal conductivity decreasing. The presence of a strong lattice distortion reaching 30°/µm is characteristic for all implantation cases. This effects in vanadium implantation by chromium leads to turning of near-surface volume as a whole regarding to the fundamental material on the angles to the several dosens degrees.

Considerable soiling of models surface in the process of copper ion implantation and implantation of inner-oxyded strongly deformated molybdenum models which have a dispersed structure, lead to the surface layer amorphysation.

So, the following order of structural transformation in pulsed ion implantation of BCC-metals may be supposed. The formation of numerous defects have radiation origin and there are defects due to appearance of high strains in implanted layer in the initial process stage. Dislocative high density charges which produce distortion of matrix lattice are formed as a result of fixation of dislocation by admixtures penetration. Later on the lattice relazes with formation of high-angles borders when extreme value of distortion is reached, that is, the process of dispergation takes place. The complete amorphyzation of a surface layer is the extreme case of such a process.

STRUCTURAL MODIFICATIONS OF NITROGEN IMPLANTED HIGH-SPEED STEEL

J.Jagielski[x], G.Gawlik[x], A.Podgórski[x], and W.Zieliński[xx]

Introduction

In the literature on the nitrogen implanted steels the effect of ion dose on mechanical properties such as microhardness and resistance to wear has been established [1,4]. Up to a dose of $5x10^{16}$at.cm^{-2}, no change in the wear rate has been observed. The significant effect of a dose on the mechanical properties has been revealed at the dose range of $5x10^{16}$ - $5x10^{17}$at.cm^{-2}. With increasing doses over $5x10^{17}$at.cm^{-2} no further change or even increase in the wear rate has been found. The formation of nitrided phases as well as structural modifications such as vanishing of primary polycrystalline structure, distraction of the precipitates, homogenization of the implanted layer have been induced by the nitrogen implantation of the steel [2,3].

The purpose of the present work was to examine the effect of doses, ranging as was mentioned above, on the structural modifications of nitrogen implanted high-speed steel.

Experimental procedure

High-speed steel of composition: 0.88% C, 4.2% Cr, 6.5% W, 5% Mo, 1.9% V preliminary hardened and tempered up to 64 HRC, has been investigated. All the samples used in the experiment were prepared from one rod of material. Prior to the implantation process the samples were electrolytically polished. The samples were 50 keV nitrogen implanted with doses: $1x10^{16}$at.cm^{-2}, $1x10^{17}$at.cm^{-2} and $0.8x10^{18}$at.cm^{-2}. After implantation the samples were thinned by etching the non-implanted side. The implanted side of the sample was protected by varnish coating. TEM observations were performed using Philips EM 300 microscope with 100 kV accelerating voltage.

Results and discussion

The typical TEM pictures as well as the changes in diffraction patterns of the samples unimplanted and implanted with doses: $1x10^{16}$at.cm^{-2}, $1x10^{17}$at.cm^{-2} and $0.8x10^{18}$at.cm^{-2}, are demonstrated in Figures 1-4, res-

x/ Institute for Electronics Materials Technology, Wólczyńska 133, 01-919 Warsaw, Poland
xx/ Institute of Materials Science and Engineering, Warsaw University of Technology, Narbutta 85, 02-524 Warsaw, Poland

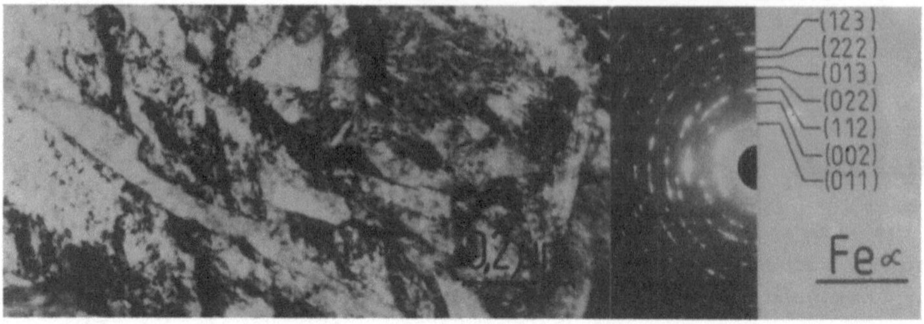

Fig. 1. Unimplanted steel

Fig. 2. N-implanted steel. Ion energy: 50 keV. Dose: $1x10^{16}$at.cm^{-2}

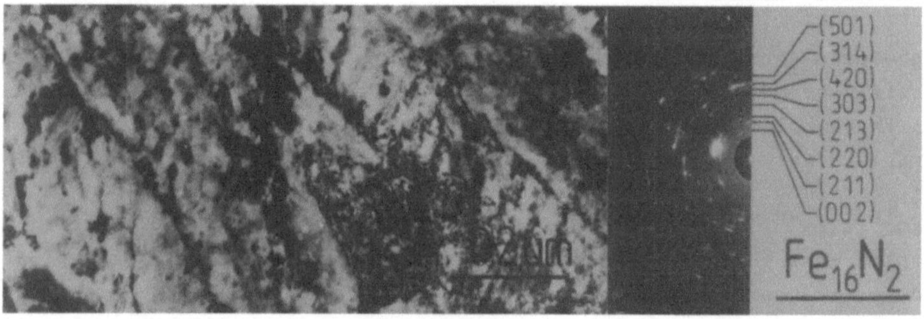

Fig. 3. N-implanted steel. Ion energy: 50 keV. Dose: $1x10^{17}$at.cm^{-2}

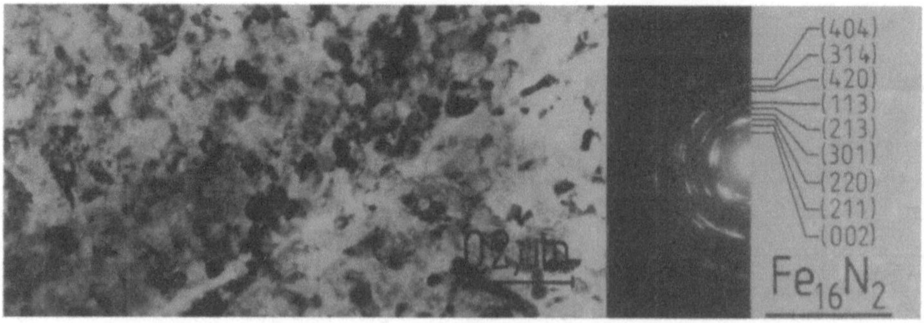

Fig. 4. N-implanted steel. Ion energy: 50 keV. Dose: $8x10^{17}$at.cm^{-2}

pectively. No visible modifications of the structure were observed for the samples implanted with the dose $1x10^{16}$at.cm^{-2} (Fig. 2). With increasing doses to $1x10^{17}$at.cm^{-2} the significant structural changes have been revealed (Fig. 3).The study of diffraction patterns has allowed the identification of the detected precipitates as $Fe_{16}N_2$. The dimensions of the precipitates varied from 10 to 30 nm. The original structure of the samples implanted with the dose $0.8x10^{18}$at.cm^{-2} was almost completely destroyed (compare Fig. 1 and Fig. 4). The amount of the $Fe_{16}N_2$ precipitates has increased. The dimensions of these precipitates were found to be ranging from 30 to 40 nm. The reported structural changes have been compared to the level of the radiation damage induced by the ion implantation. The number of atoms displaced from the lattice site by a single ion dpa was determined using the model presented previously by Torrens, Norgett and Robinson [5]. The effect of dose on the structural modifications and the extent of the radiation damage have been shown in Table 1.

Table 1

Doses at.cm^{-2}	Radiation damage dpa	Nitride precipitates	Structure
$1x10^{16}$	5	invisible	undamaged
$1x10^{17}$	50	visible	partly damaged
$0.8x10^{18}$	400	visible	destroyed

The data presented in Table 1 have revealed, that the level of the radiation damage seems to achieve the saturation at the whole range of the applied doses.
We concluded, that the observed structural modifications, attaining almost complete destruction of the original structure at the dose $0.8x10^{18}$at.cm^{-2}, could be ascribed to the ion beam mixing.

References

(1) Dimigen,H., K.Kobs, R.Leutenecker, H.Ryssel and P.Eichinger: Mater. Sci. and Eng. 69 1985 181-190
(2) Fayeulle,S., O.Treheux and C.Esnouf: Appl. Surface Science 25 1986 288-304
(3) Gawlik,G., J.Jagielski and A.Podgórski: Phys. Status Solidi A, 95 1986 453-458
(4) Hartley,N.E.W.: Wear 34 1975 427-438
(5) Norgett,M.J., M.T.Robinson and I.M.Torrens: Nucl. Eng. and Design 33 1975 50-54

WEAR PROPERTIES OF HIGH-SPEED STEEL IMPLANTED WITH NITROGEN AT VARIOUS THERMAL CONDITIONS

G.Gawlik[x], J.Jagielski[x] and A.Podgórski[x]

Introduction

The nitrogen ion implantation has became an accepted method for improving resistance to wear in a wide range of steels [2, 6]. Some authors[1, 4] have proved, that the distribution profiles of the nitrogen implanted into steel have been strongly affected by thermal conditions over the implantation process. It was recently found [3, 5], that the tribo-enhanced diffusion has played a significant role in the wear process. The differences in the nitrogen concentration profile can substantially alter the wear behaviour.

The main aim of the presented work was to examine the role of thermal conditions in the resistance to wear of N_2^+ implanted high-speed steel.

Experimental procedure

High-speed steel of composition:0.88% C, 4.2% Cr, 6.5% W, 5% Mo, 1.9% V, preliminary hardened and tempered up to 64 HRC, implanted with N_2^+ ions has been investigated. The shape and configuration of the specimens used for the wear test are presented in Figure 1.

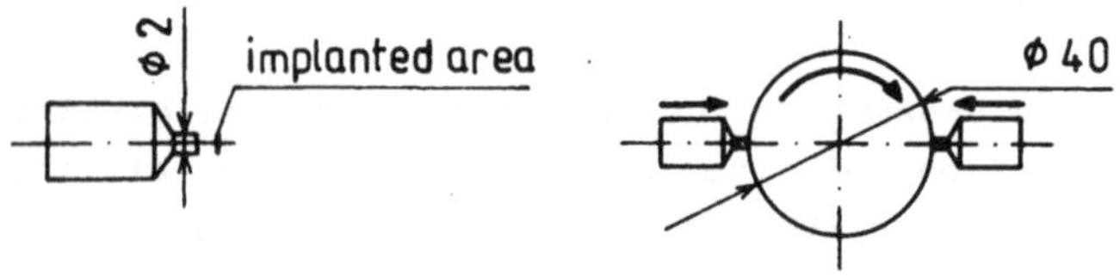

Fig.1.The shape and configuration of the samples used for the wear tests

Prior to implantation process the samples were preliminary worn to provide the required contact area. The data concerning implantation process such as: energy (E), beam current density (J), dose (D) and sample's temperature during the implantation (T) are listed in Table 1. The implanter used was Balzers MPB 202 RP. Postimplantation annealing at 100°C for 6h was applied to the samples of series D.

Tribological tests at the loads varied from 48 up to 240 MPa were carried out to determine the relative wear rates. The sliding velocity was

x Institute for Electronics Materials Technology, Wólczyńska 133, 01-919 Warsaw, Poland

Table 1.

Series	E [keV]	J [μAcm^{-2}]	D [$x10^{17}cm^{-2}$]	T [°C]
A	100	7	2	appr. 300
B	100	7	1	appr. 300
	70	2	1	70
C	70	2	2	70
D	70	2	2	70

0.2 ms^{-1}. Paraffine oil was used as a lubricant. The wear rate was determined by monitoring the axial shift of the sample during the test. Nuclear reaction analysis was applied for nitrogen depth profile determination.

Results and discussion

The distribution profiles of the nitrogen implanted into samples at different thermal conditions are demonstrated in Figure 2.

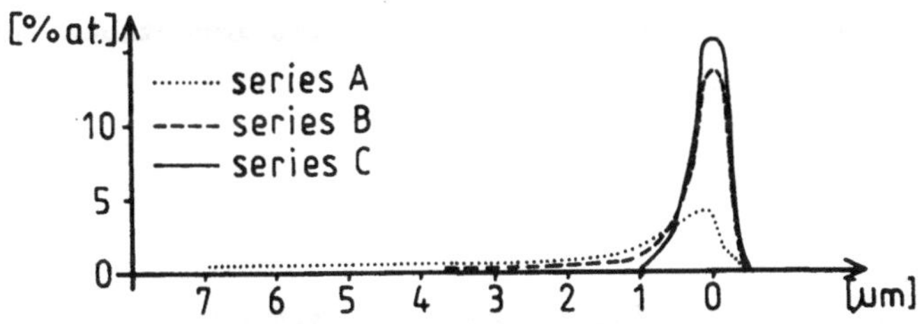

Fig.2.The distribution profiles of nitrogen implanted into samples at different thermal conditions (see Table 1).

With increasing temperature during the implantation up to appr. 300°C a significant diffusion deep into the sample has been revealed. No effect of the postimplantation annealing on the concentration profile was observed.The effect of loads used in the tribo-test on the relative wear rates is presented in Figure 3.

The samples of C series showed the lowest wear rates at the loads up to 72 MPa. The implantation at a low temperature has led to the high nitrogen concentration at the surface (Fig. 2). With increasing loads over 180 MPa, the samples implanted at appr. 300°C (series A) revealed relatively lower wear rates. At such temperature, the nitrogen concentration at the surface has been reduced and a diffusion tail has appeared (Fig. 2). The samples of B series showed the lowest wear rates at the

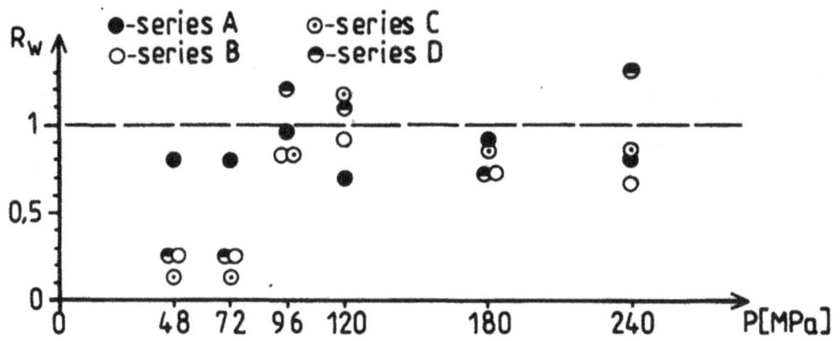

Fig.3. The effect of loads used in the tribo-tests (P) on the relative wear rates (R_w). The samples of A - D series were implanted at thermal conditions listed in Table 1.

whole range of the applied loads. As was mentioned above Table 1 , those samples were implanted in two steps at the different temperatures. Thermal conditions of the experiment have assured the high nitrogen concentration at the surface as well as the diffusion tail broadening the depth profile (Fig. 2). The beneficial effect of the postimplantation annealing series D on the wear properties was not confirmed.

We concluded, that the thermal conditions during the implantation can considerably alter the tribological behaviour due to the significant changes in the nitrogen concentration profiles. The importance of the applied loads was also proved.

Acknowledgments

The authors would like to thank Dr. A.Turos, Mr. A.Dygo and Mr. R.Pietrak for performing the nuclear reaction measurements.

References

(1) Barnavon, Th., H.Jaffrezic, G.Marest, N.Monocoffre, J.Tousset and S.Fayeulle: Mater. Science and Eng. 69 (1985) 531-537
(2) Dimigen, H., K.Kobs, R.Leutenecker, H.Ryssel and P.Eichinger: Mater. Science and Eng. 69 (1985) 181-190
(3) Feller, H.G., R.Klinger and W.Benecke: Mater. Science and Eng. 69 (1985) 173-180
(4) Gawlik, G., J.Jagielski and A.Podgórski: Phys. Status Solidi A, 95 (1986) 453-458
(5) Korycki, J., A.Dygo, R.Pietrak, A.Turos, G.Gawlik and J.Jagielski: Nucl. Instr. and Methods in Phys. Res. B19/20 (1987) 177-179
(6) Wittkower, A. and J.K.Hirvonen: Nucl. Instr. and Methods in Phys. Res. B6 (1985) 78-87

TRIBOLOGICAL PROPERTIES OF ION IMPLANTED Fe-Cr STEELS

Jun Sasaki [1], Masaya Iwaki [2]

INTRODUCTION

Ion implantation is a useful technique for improvement of wear and corrosion resistances of Fe-based alloys. (Sioshansi 1985, Iwaki 1987) Steels used for plastic injection moulds or rolling mills have a variety of Cr content. In this report, surface mechanical properties of N_2^+-, O_2^+- and Ti^+-implanted Fe-Cr alloys with different Cr content are investigated in terms of hardness and friction coefficients during reciprocal sliding motions. SEM, AES and XPS analyses were carried out to investigate surface micro-characteristics.

EXPERIMENTAL

Substrates used were Fe-Cr alloy plates prepared with four different Cr content of 5, 10, 15 and 20 wt.%. A surface of each specimen was mechanically mirror-polished with diamond particles down to 0.5micron. N_2^+-, O_2^+- and Ti^+-ion implantations into Fe-Cr alloys were performed at near room temperature with the following conditions; (1) 50keV N_2^+-implantation with $5x10^{16}$ N_2^+/cm^2 (2) 100keV O_2^+-implantation with $5x10^{16}$ O_2^+/cm^2 (3) 150keV Ti^+-implantation with $1x10^{17}$ Ti^+/cm^2. A part of each sample was masked by Al sheets in order to clarify the implantation effect on tribological properties.

The Vickers hardness of implanted specimens was measured at the load of 2gf using a micro-Vickers hardness tester. Friction measurements were carried out in atmospheric room temperature using a Bowden-Leben type friction tester. An unimplanted AISI1045 pin was reciprocally slided on the specimen between implanted region and unimplanted region at velocity of 0.04mm/sec without lubrication. The applied force was 0.245 or 0.49N. Surface micro-characteristics were estimated by means of AES, SEM and XPS.

RESULTS AND DISCUSSION

(NITROGEN AND OXYGEN)

Figure 1 shows the micro-Vickers hardness of both unimplanted Fe-Cr alloys and N_2^+- or O_2^+-implanted specimens as a function of Cr content. The control specimens are harder at higher Cr content. Both N_2^+- and O_2^+-implantations caused the hardness to increase in comparison with that of the controled specimens. Its increase in hardness takes place at all specimens with various Cr content. The effect of ion implantation on surface hardness is similar in both ion species.

Figure 2 shows the friction coefficients of the N_2^+- or O_2^+-implanted specimens as a function of Cr content. The friction coefficients of the control surfaces indicated by open triangles in the figure are little dependent on the Cr content. Friction coefficients of N_2^+-implanted surfaces shown by open circles in the figure are lower than those of unimplanted surfaces, and slightly depend on Cr content. While the friction coefficients of O_2^+-implanted specimen shown by open square in the figure also

[1]) Electronics & Control Systems R & D Lab., Nippon Steel Co.,
1618 Ida, Nakahara-ku, Kawasaki 211, Japan
[2]) RIKEN, 2-1 Hirosawa, Wako, Saitama 351, Japan

decrease and depend on the Cr content; the higher the Cr content is, the lower the friction coefficient is.

In general, the friction coefficient becomes lower as the surface is hardened. In the present experiments, the control specimen does not indicate this tendency the friction coefficients of control specimens with various Cr content is similar to each other, but their hardness is dependent on the Cr content. From the comparison between the implanted and the unimplanted specimens with the same hardness, decrease in friction coefficient induced by ion implantation seems to be enhanced by decrease in adhesion rather than increase in hardness. These results show that the friction coefficient of the implanted surfaces are affected not only by the hardness but also by the surface adhesion.

In order to investigate these tribological properties with micro-characteristics of the implanted layers, surface micro analyses were carried out. X-ray photoelectron spectroscopy (XPS) results show that the implanted nitrogen combines with Cr to form Cr_2N in the interior. Formation of FeN_x could not be detected. Some oxides, such as iron oxides and chromium oxides exist only at thin surface layers. While, in the O_2^+-implanted specimen, surface oxide layers were enhanced in comparison with those in control specimen, and implanted oxygen combined with Cr to form Cr_2O_3 in the interior. These results suggest that the formed compounds such as Cr_2N and Cr_2O_3 may cause the hardness to increase and the friction to decrease. SEM observation of the wear tracks on the N_2^+-implanted surfaces shows the decrease of ploughing.

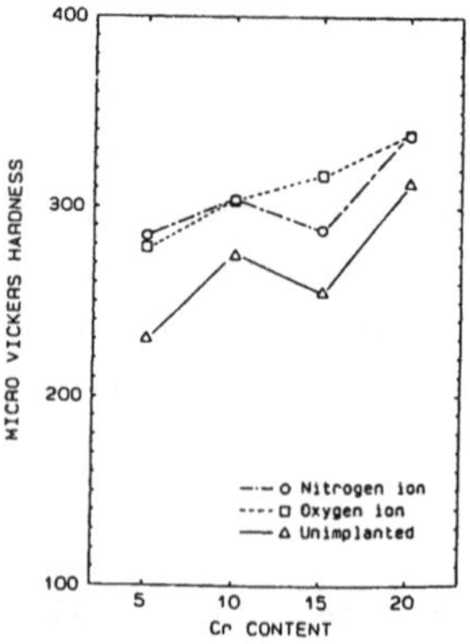

(Figure 1) Micro-Vickers hardness of N_2^+- (○), O_2^+-implanted (□) and control (Δ) specimens at the load of 2gf as a function of Cr content

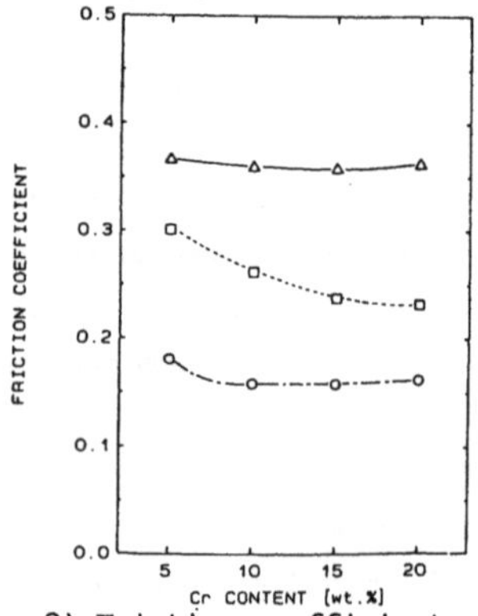

(Figure 2) Friction coefficient of N_2^+- (○), O_2^+-implanted (□) and control (Δ) specimens as a function of Cr content

·(TITANIUM)

Ti^+-implantation also results in the change of friction coefficients depending on the dose and the Cr content. Figure 3 shows the friction coefficients as a function of Cr content. The friction coefficient of implanted surfaces has a tendency to decrease as the Cr content becomes lower.

Relative change in micro-Vickers hardness for Ti^+-implanted specimens was small. Hence the remarkable decrease in friction coefficient after Ti^+-implantation is not caused by increase in hardness.

SEM observation of the wear tracks shows apparent suppression of ploughing during the sliding motions on the Ti^+-implanted surfaces even after the reciprocal sliding test. It signifies that Ti^+-implantation causes the decrease of adhesion between a carbon steel pin and the implanted surface.

A concentration profile the Ti^+-implanted 20Cr-Fe alloy surface layers obtained by AES is shown in figure 4. As typically shown iron-based alloys, diffusion like invasion of C from the surface is observed. Auger peak signals related to the invaded C appear in the form of carbide. One may expect the formation of TiC, because the peak signals related to Ti in the carburized layer are different from those in uncarburized region. Results from XPS analyses of this specimen showed, however, that TiC compound was never formed. These results suggest that Ti^+-implantation in Fe-Cr alloys forms Fe-Cr-Ti-C metastable alloy layer, which results in decrease in adhesion and decrease the friction coefficient of the surfaces. Decrease in the friction coefficient is more evident when the Cr content is lower.

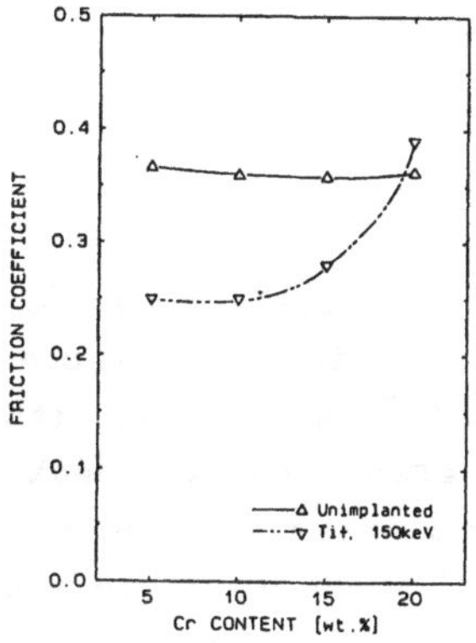

(Figure 3) Friction coefficients of Ti^+-implanted specimens as a function of Cr content

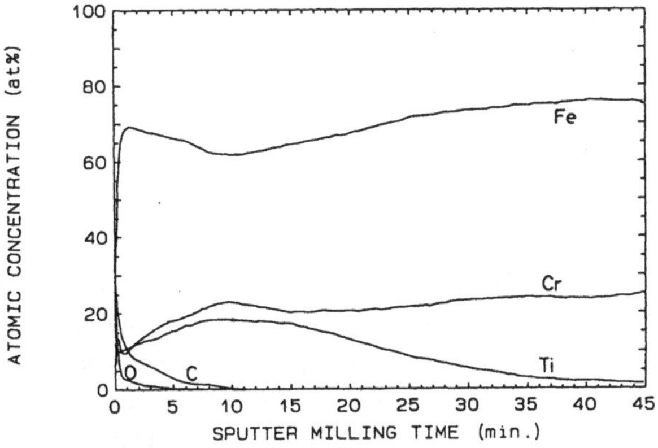

(Figure 4) A concentration frofile of the Ti^+-implanted 20Cr-Fe specimen obtained by AES with the dose of $1x10^{17}$ Ti^+/cm^2 at energy of 150keV

CONCLUSION

Tribological properties of N_2^+-, O_2^+- and Ti^+-implanted Fe-Cr alloys with various Cr content were investigated utilizing the surface micro-analyses. Micro-Vickers hardness of N_2^+- or O_2^+-implanted specimens are both increased. Decrease in the friction coefficient was observed for both specimens. This decrease was more evident for the nitrogen implanted specimens. Although increase in hardness caused by Ti^+-implantation was small, the friction coefficient of the specimens was decreased.

It is concluded that for the improvement of the tribological properties of Fe-Cr alloy nitrogen and oxygen implantations are appropriate for higher Cr content and titanium implantation is suitable for lower Cr content.

References

(1) Iwaki, M.: Materials Sci. & Eng. 90 (1987) 263-271
(2) Sioshansi, P.: Materials Sci. & Eng. 69 (1985) 161-166

MICROSTRUCTURE OF TI-IMPLANTED FE

Y. FUKUI[1)], Y. HIROSE[1)] and M. IWAKI[2)]

1. INTRODUCTION

Ti implantation changes tribological[1,2] and electrochemical[3] properties of iron and steel surfaces, which may be caused by the change in microscopic properties[2,3] of surface layers implanted with Ti. In this report, the structure and hardness of Ti-implanted layers is studied as a function of fluences of Ti ions, and the relation of each properties is discussed.

2. Experimental

The iron specimens with purity above 99.9% recrystallised at 600 °C in a vacuum were implanted with Ti^+ ions on conditions indicated in table 1. AES analyses were performed in a UHV chamber (<1X10^{-8} Pa) equiped with a PHI model of 610 microprobe using 5 keV primary electrons. AES peak to peak depth profiles of Fe, C, O and Ti of the implanted iron were recorded during ion milling with a rastered 4 keV Ar^+ ion beam at an angle of 45 degrees from a normal to the surface. TEM observations of specimens made by electro-polishing from an unimplanted side were carried out using a JEOL 200FX microscope at 160 kV.

Table 1 Conditions of Ti ion implantation

Ion energy ------------	150keV
Fluence ---------------	1~30X10^{16} ioms/cm^2
Current density -------	0.4~0.8 μA/cm ^{2}
Substrate temperature -	Room temperature
Vacuum pressure -------	~2X10^{-6} Torr

3. Result

AES sputter depth profiles of the irons implanted with 3X10^{16} and 10X10^{16} ions/cm^2 are shown in Fig. 1. At the fluences of 3X10^{16} ions/cm^2, C and

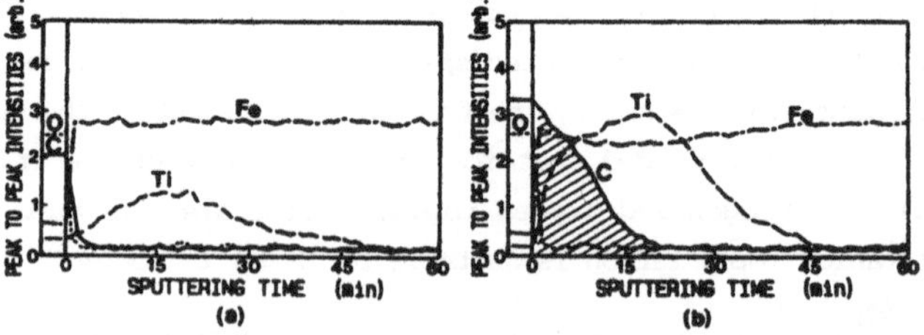

Fig. 1 Auger sputter depth profiles of iron implanted with Ti: (a) 3X10^{16} ions/cm^2; (b) 10X10^{16} ions/cm^2.

------

1) Hanshin Research and Development Labolatories, Nisshin Steel Co. LTD., 5, Ishizunishi-machi, Sakai, Osaka, 592, Japan.
2) The Institute of Physical and Chemical Research (RIKEN), Hirosawa 2-1, Wako, Saitama, 351-01, Japan.

O atoms do not incorporate into the iron surfaces. At the fluence of $10X10^{16}$ ions/cm^2, C atoms incorporate into the iron surfaces. Contrary to C-migration, O atoms do not incorporate into the surface.

Fig. 2 shows depth profiles magnifying the depth with pristine and implanted specimens. There is an oxidized and contaminated surface layer which was originally formed before implantation. The comparison of depth profiles of C and O atoms in the pristine iron with in the specimen implanted with $3X10^{16}$ ions/cm^2 indicates that the originally oxidized and contaminated layer is a little sputtered with a fluence of $3X10^{16}$ ions/cm^2. As shown in Fig. 2(c) after Ti implantation with $10X10^{16}$ ions/cm^2 the oxidized layer is thinner than those in Fig. 2(a) and (b). At the depth indicated by the arrow the peak to peak intensity of C(KLL) is minimum. Auger KLL spectra of C depicted together with depth profiles in Fig. 2(c) are different between 0 and 1.2 min sputtering time; Auger KLL spectra of C at the shorter sputtering time than the arrow show that C atoms are those in a contamination formed in air. Consequently, it is considered that the position indicated by the arrow is just on the surface produced by Ti implantation in a vacuum. The originally oxidized and contaminated layer were completely sputtered off by bombardment of Ti ions at the high fluence. These results show that incorporation of C atoms does not occuer when the surface oxide layers exist. It is proposed that the incorporation of C atoms starts after the originally oxidized and contaminated layer is sputtered off.

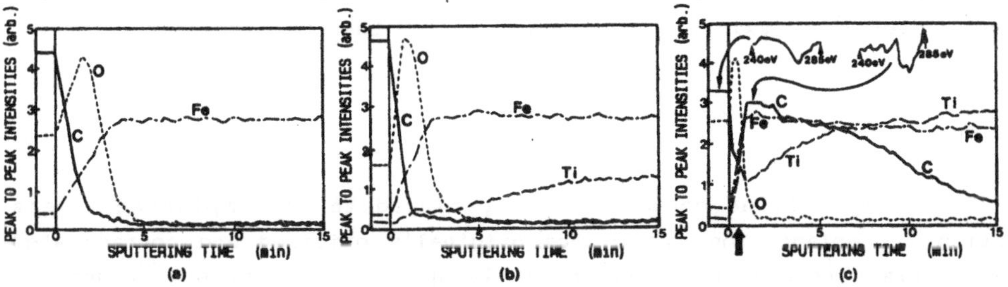

Fig. 2 Auger sputter depth profiles at the iron surface implanted with Ti: (a) unimplant;(b) $3X10^{16}$ ions/cm^2; (c) $10X10^{16}$ ions/cm^2.

Transmission electron diffraction(TED) images from Ti-implanted iron with fluences from $1X10^{16}$ ions/cm^2 to $10X10^{16}$ ions/cm^2 are shown in Fig. 3. In $1X10^{16}$ and $3X10^{16}$ ions/cm^2 weak diffraction spots and rings unrelated to Fe nor Ti are observed in addition to diffraction spots of Fe. The spots and rings indicate that crystalline compounds are produced in the implanted layer. In TED images of $6X10^{16}$ and $10X10^{16}$ ions/cm^2 there exist broad diffuse amorphous rings. From the AES analyses and TED images it is clear that the Fe-Ti-C amorphous layer produced when the incorporation of C atom occures. It is concluded that the high fluence Ti-implantation removed the originally oxidized and contaminated layer to incorporate C atoms into the iron surface and consequently it results in formation of the amorphous layer.

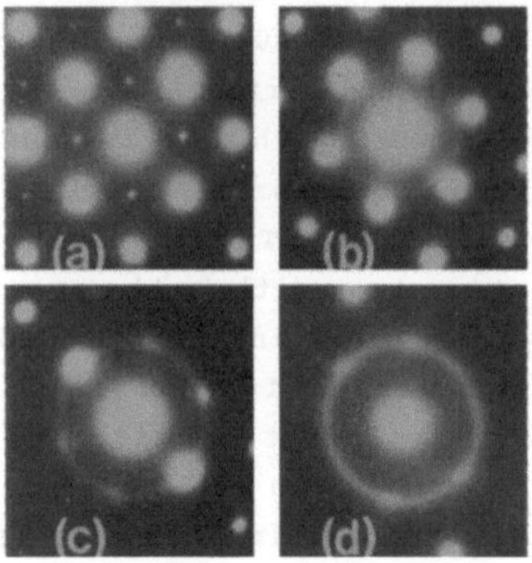

Fig. 3 Electron diffraction patterns from the iron implantedwith Ti: (a) 1×10^{16} ions/cm^2; (b) 3×10^{16} ions/cm^2; (c) 6×10^{16} ions/cm^2;(d) 10×10^{16} ions/cm^2.

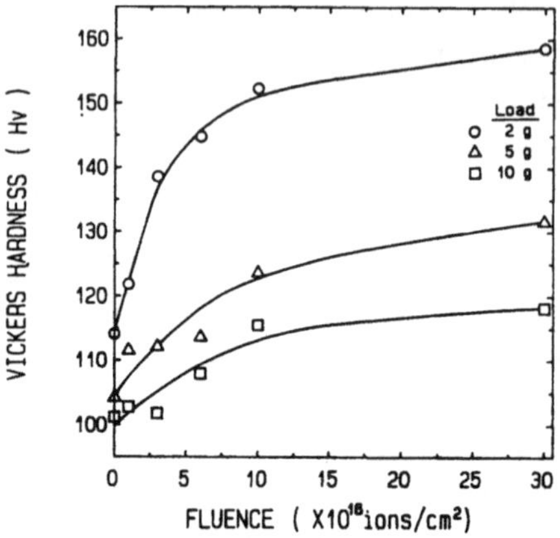

Fig. 4 Relationships between hardness of the implanted iron surfaces and Ti fluences.

Fig. 4 shows the relationships between fluences of Ti ions and surface hardnesses. Ti implantation results in increase in hardness, that is clear from increases in the hardness as the load is lowered at the same fluences. The hardness increases in proportion to the fluence up to 6×10^{16} ions/cm^2 but saturates beyond the it. It indicates that the amorphous layer induced by the incorporation of C atoms does not contribute to surface hardnesses; the amorphous layer has lower strength than Fe-Ti layer which exists in the deeper region than the amorphous layer.

5. Conclusion

Fe-Ti crystalline compounds are produced in the implanted layers during Ti implantation while the originally oxidized and contaminated layer formed before implantation little sputtered off by bombardment of Ti ions. A Fe-Ti-C amorphous alloy layer is produced resulting from incorporation of C atoms after the originally oxidized and contaminated layer is sputtered off. The surface harndness increases rapidly at low fluences but saturates after the amorphous layer of low strength is produced.

References

(1) D. M. Follstaedt: Nucl. Instrum. Methods B10/11(1983)549-555
(2) I. L. Singer: Appl. Surf. Sci. 18(1984)28-62
(3) M. Iwaki, K. Yabe, M. Suzuki and O. Nishimura: Nucl. Instrum. Methods B19/20(1987)150-153

DOSE AND TEMPERATURE DEPENDENCE OF ARSENIC REDISTRIBUTION IN ION IMPLANTED NICKEL

V.P.Goltsev,V.V.Khodasevich,V.V.Uglov[1)],G. Götz[2)],K.Berndt and V.Bayer[3)]

Ion implantation of metals causes considerable changes in their physical properties.They are most pronounced in the case of ion bombardment of chemically active elements resulting in structure and phase transformations and accumulation of a great amount of radiation-induced defects.The knowledge of the redistribution of the impurity depending on the dose and annealing temperature will add to better understanding of the nature and mechanisms of these transformations.

The present study deals with As redistribution in thin-film and massiv Ni samples.The samples were bombarded with As^+ iones of 60 keV and 10^{16}... 10^{17} ion/cm^2 doses.The redistribution of As was analysed using the Rutherford back scattering (RBS),X-ray microanalysis (XMA) and transmission electron microscopy (TEM).

The calculations of the concentration profiles (CP) of As atom distribution obtained from RBS spectra have shown the deflection of CP from the theoretical one with increassing irradiation dose (Fig.1).

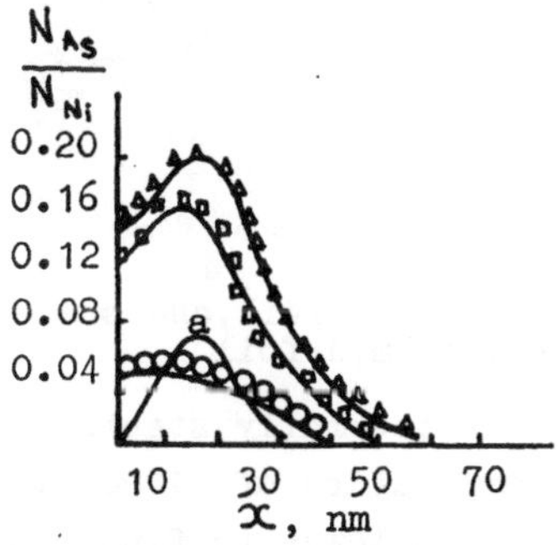

Fig.1 Concentration profiles for As^+ ion implanted into Ni.Doses : o-10^{16}; □-$6\cdot10^{16}$; Δ-10^{17}ion/cm^2; a-theoretical profile for D=10^{16} ion/cm^2

It has been found that the doped As atoms penetrate 40-60 nm deep wich is 1.5 to 2 fold as large as the calculated values [1] . The changes in the As CP are shown to be related to two simultaneous processes.One the one hand,it is surface sputtering resulting the changes in the profile left side.The variation of the right part are caused by the radiation-induced diffusion of the impurity taking place in the field of mechanical stress gradient. As the irradiation dose increases, As is accumulated in the thin-film

1) Byelorussian State University,4 Lenin Ave.,Minsk,220080,USSR
2) Friedrich Schiller University,Section of Physics,Jena,GDR
3) M.Luther University,Section of Physics,Halle,GDR

samples. In the dose range $10^{16}...6\cdot10^{16}$ ion/cm^2, the layer concentration (Nt) of As increases by $3,7\cdot10^{16}$ at/cm^2, while at $D \geqslant 8\cdot10^{16}$ (Nt) of As achieves saturation and amounts to $5\cdot10^{16}$ at/cm^2. The As concentration at a depth corresponding to R_p constituted $1,6\cdot10^{22}$ cm^{-3} (18 at.%) for a dose 10^{17} ion/cm^2. The mean As concentration for the same dose in the bombarded layer was $1,0\cdot10^{22}$ cm^{-3} (11 at.%).

As behavior is similar in the massive Ni samples. Fig.2 presents the intensity spectra of the characteristic X-ray AsL_α radiation obtained due to excitation with the electrons accelerated to 10 keV. The XMA spectra show that for doses $10^{16}...6\cdot10^{16}$ ion/cm^2, As is homogeneously distributed over the surface. At doses of $8\cdot10^{16}$ ion/cm^2, one can see deposition of As [2] . Such a As redistribution with the dose results in the phase β-Ni_5As_2 formed at $D \geqslant 10^{16}$ ion/cm^2, as follows from the TEM studies of thin-film samples (Fig.3 a,c). The new hexagonal closely packed phase

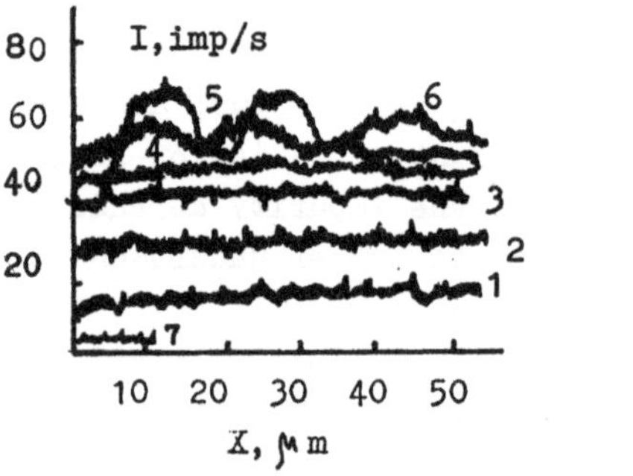

Fig.2 Characteristic AsL_α X-ray radiation spectra for Ni samples bombarded by As^+ ions. Doses: 1-10^{16}, 2-$2\cdot10^{16}$, 3-$4\cdot10^{16}$, 4-$6\cdot10^{16}$, 5-$8\cdot10^{16}$, 6-10^{17} ion/cm^2, 7-phone

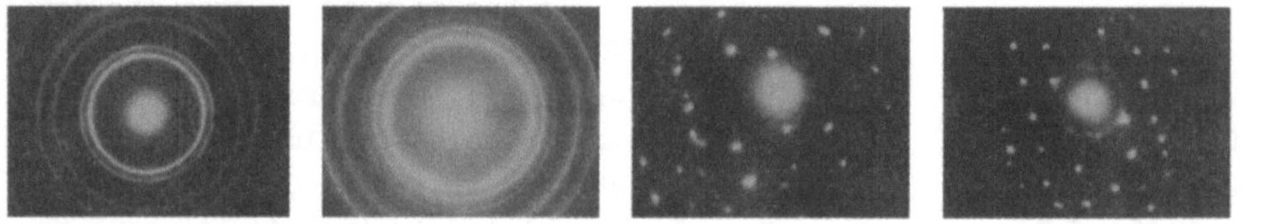

Fig.3 Electronograms of initial Ni samples : a- film, b- foil, and samples bombarded by As^+ ions with dose $4\cdot10^{16}$ ion/cm^2: c-film, d-foil

with the structure of nickel arsenide β-Ni_5As_2 is thermally stable exhibiting no decomposition up to the annealing temperature of 1000°C. The electron microscopy studies of massive samples allowes to find out that for the initial radiation doses $10^{16}...$ $4\cdot10^{16}$ ion/cm^2 the hexagonal closely packed phase is coherently related to the matrix (Fig.3d); at the doses above $4\cdot10^{16}$ ion/cm^2, the loss of coherence is observed due to overlapping of the stress fields of individual depositions as well as the radiation-induced defects and their complexes.

The As behaviour in ion-implanted Ni was studied at annealing temperatures T_{an}=20...1000°C. The XMA studies of massive samples exhibited 3-fold decrease of the AsL_α characteristic X-ray radiation (Fig.4). An extra As redistribution is supposed to occur at such annealing temperatures however, the mechanism of this redistribution is unclear. The behaviour of experimental As concentration profiles for massive Ni samples obtained from RBS data show transformation of As CP at T_{an}=20...800°C

(Fig.5). As T_{an} grows, the CP chifts to the surfase and becomes narro-

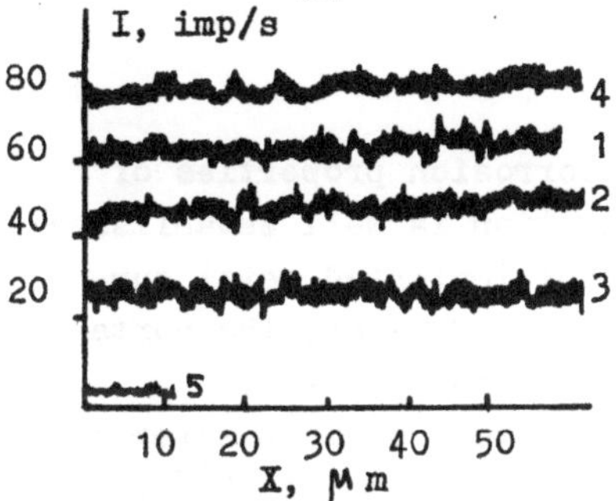

Fig.4 Characteristic AsL_α X-ray radiation spectra for Ni samples bombarded by As^+ ions with dose 10^{17} ion/cm^2 and for the samples annealed at: 1-400°C, 2-600°C, 3-1000°C, 4-20°C, 5- phone

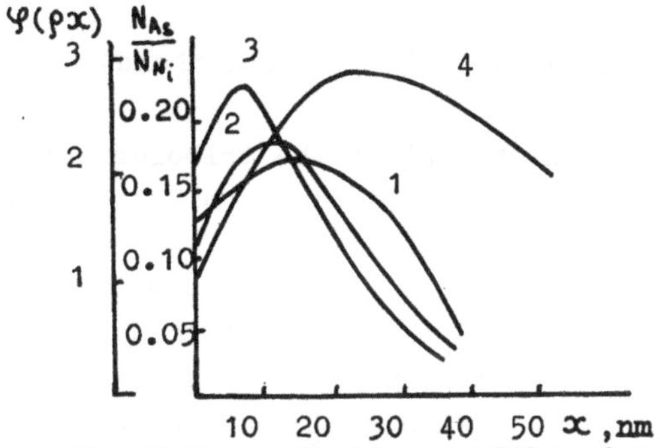

Fig.5 Concentration profiles for As ion implanted into Ni (D = 10^{17} ion/cm^2) at different T_{an}: 1-20°C, 2-600°C, 3-800°C, 4- function φ (ρx)

wer. For T_{an}=800°C, the relative As concentration showes 1,3-fold increase at a maximum point. In this case, (Nt) of As does not vary with the increasing of T_{an} constitutes $4{,}9\cdot10^{16}$ at/cm^2. Such an As redistribution causes deviation from the distribution function, $\varphi(\rho x)$, the primary characteristic radiation throughout the sample depth. $\varphi(\rho x)$ was calculated using the approach proposed by Filiber who derived the analytical expression for $\varphi(\rho x)$:

$$\varphi(\rho x)=4\exp(-\sigma\rho x)-(4-\varphi_0)\exp\left[-\sigma(1+1/h)\rho x\right],$$

where φ_0 is the function $\varphi(\rho x)$ in a surface layer; h= 1,2 A/Z^2 where A and Z are the atomic number and weight; σ is the Lenard absorptivity of an electron beam; ρ is the absorber density. The calculation results are presented in Fig.5. One can see that the function $\varphi(\rho x)$ is shifted to the right relatively the CP. Such a deviation is evidently due to the decrease in the intensity of AsL_α characteristic X-ray radiation (Fig.4).

Thus, it has been found out experimentally that the introduced As atoms distribute over the depths 1,5 to 2 times greater than the calculated values. The increase of T_{an} causes displacement of As concentration profiles to the surface and its narrowing. The As concentration around the CP maximum increases.

REFERENCES

(1) Goltsev V.P., Komarov F.F., Getz G., Scwabe F., Drako V.M., Khodasevich V.V., Uglov V.V. : In Proc.XY All-Union Conf.Physics Inter.Charged Particles Crystals.Moscov, MGU.(1986)139-141

(2) Khodasevich V.V., Uglov V.V., Baier W., Goltsev M.V., Drako V.M. :Coll Abstr.XYI All-Union Conf.Charged Particles Crystals.Moscow, MGU. (1986)117

INFLUENCE OF TEMPERATURE ON NITROGEN IMPLANTED IRON

B.Rauschenbach[1)] and G.Leonhardt[2)]

The improvement of the tribological and anticorrosion properties of metals and alloys due to nitrogen ion implantation is well established. It is to be expected that the properties of the implanted metal surfaces will be affected by the implant itself, the fluence, the contamination and the temperature during the implantation process.
In recent papers, we have studied the microstructure of iron after implantation with nitrogen ions in the dose range between 1×10^{16} and 1×10^{18} N^+-ions/cm^2 at room temperature /1/ and at the temperature of liquid nitrogen (LNT) /2/. We have also reported about the distribution and redistribution of the implanted nitrogen as function of the implantation temperature /3/. The purpose of this paper is to illustrate the influence of temperature during nitrogen ion implantation on the microstructure.
Fine-grained iron layers were deposited in a high vacuum. The polycrystalline iron layers were implanted with nitrogen ions with energies from 30 to 60 keV in the dose range of 1×10^{16} to 1×10^{18} N^+-ions/cm^2 at temperatures between -196 °C and +500 °C. The crystallographic structure before and after implantation were investigated by high-voltage electron microscopy (HVEM) and transmission high-energy electron diffraction (THEED). In the papers /1//2/ were reported about the conditions of preparation of the iron specimens and the transmission electron microscopical phase analysis. All XPS measurements of the nitrogen implanted iron surface were performed on a VG Scientific photoelectron spectrometer with AlK_α (1486.6 eV) radiation. An ultrahigh vacuum of 10^{-9} Torr was maintained throughout the measurement. The C(1s) line (284.6 eV) was used for calibration purposes and the Ar^+-ion beam was used for depth profiling.
The implantation of nitrogen ions at different temperatures leads to significant changes in the morphology of the formed phases. Fig. 1 shows transmission electron micrographs of unimplanted iron and iron implanted at different temperatures. The morphology is significantly changed. These various morphologies are caused by the existence of implantation-induced iron nitride and iron carbonitride phases.

[1)] Zentralinstitut für Kernforschung Rossendorf
8051 Dresden, PF 19
[2)] Technische Universität Karl-Marx-Stadt
Sektion Chemie und Werkstoffwissenschaften
901 Karl-Marx-Stadt, PF 964

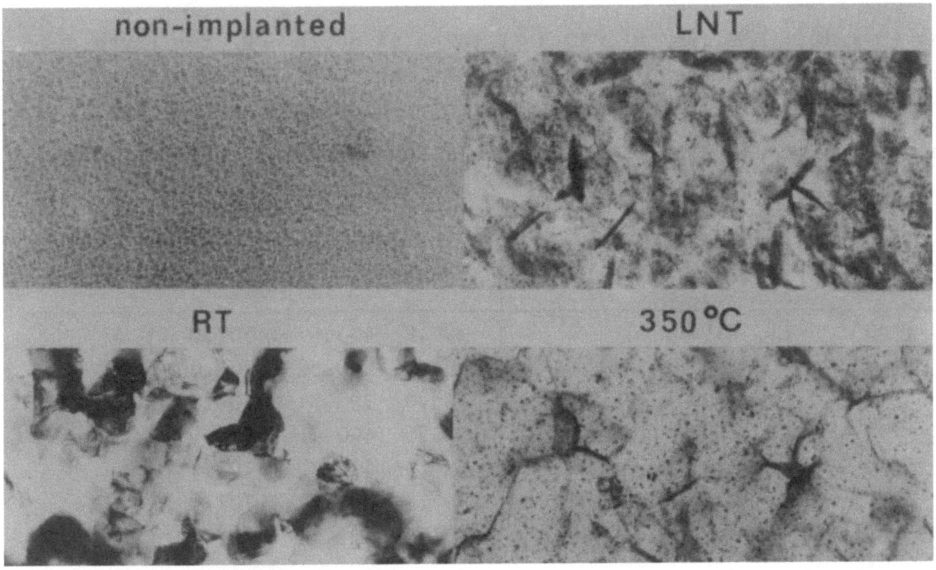

Fig.1: HVEM-micrographs of iron implanted with 5 x 10^{17} N^+-ions/cm^2 at different temperatures

Recently, we have reported on the analysis of iron nitride phases after nitrogen ion implantation in the fluence range from 1 x 10^{16} to 1 x 10^{18} ions/cm^2 at room temperature /1/. Occurrence of the iron nitride phases and solid solutions γ-austenite, α'-martensite, ε-Fe_2N_{1-x} and α''-$Fe_{16}N_2$ was shown to be fluence dependent. We have also reported on the formation of nitrogen atom clusters, a molecular solid nitrogen phase, α'-martensite, α''-$Fe_{16}N_2$ and ε-$Fe_2(C,N)_{1-x}$ after implantation at LNT /2/. In a good agreement with the results by Moncoffre et al. /4/, a strongly favour ε-iron carbonitride formation could be detected for nitrogen ion implantation between 150...250 °C /3/.

It is possible to detect the presence or absence of various binding states by XPS. Fig. 2 shows the N(1s) spectra of implanted iron speci-

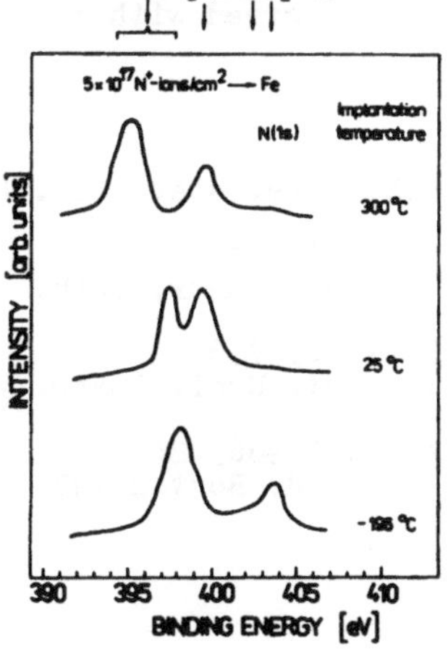

Fig.2:
N(1s)-XPS spectra of iron after implantation with 5 x 10^{17} N^+/cm^2 at different temperatures

mens (5×10^{17} N^+-ions/cm^2, 50 keV) taken at depth $\approx R_p$ for three different implantation temperatures. In the table, the N(1s) and Fe(2p) binding energies are included for iron specimens at various depths and

Table: Binding energies of nitrogen (1s) and iron (2p) after nitrogen ion implantation in iron ($^{+)}$ annealing time)

fluence (N^+-ions/cm^2)	temperature (°C)	depth (nm)	binding energy (eV) N(1s)		Fe(2p)	
5×10^{17}	-196		398.2	403.5	707.0	
	25		397.6	399.5	707.0	711.1
	300		395.0	399.5	706.8	711.3
	300/1h$^{+)}$		398.5	400.0	706.9	711.0
1×10^{18}	25	~5	396.9	400.1	707.1	710.8
	25	~10	396.9	399.1	707.6	710.9
	25	~50	397.4	399.1	707.9	

implantation temperatures. It can be seen that the N(1s) spectra shows two values of nitrogen binding energies corresponding to 398.2 eV and 403.5 after implantation at LNT, 397.6 eV and 399.5 eV after implantation at 25 °C and 395.0 eV and 399.5 eV after implantation at 300 °C. The energies between 394 eV and 398 eV can be attributed to nitrides /5//6/. It can be observed a shift of the nitride binding energies towards the lower binding energy scale with increased implantation temperature. Thus, the nitride formation is increased with increasing of the implantation dose. Furthermore, the spectra of the N(1s) level of the annealed specimens do indicate noticeable shift in the binding energy, which is a good indication for decomposition of nitride or carbonitride during annealing. The peaks corresponding to 398...400 eV can be taken as weakly bonded nitrogen or free nitrogen gas. It can be concluded that the fraction of the free or weakly bonded nitrogen is decreased with increasing of the implantation temperature. The observed N(1s) peak at 403.5 eV after implantation at LNT agreed with the binding energy of solid nitrogen.

References

/1/ Rauschenbach,B., Kolitsch,A. and Hohmuth,K.:phys. stat. sol. (a) 80 (1983) 211 and 471
/2/ Rauschenbach,B.: Nucl. Instr. Meth. B 18 (1986) 34
Rauschenbach,B., K.Hohmuth and G.Kästner: Nucl. Instr. Meth. B 23 (1987) 323
/3/ Rauschenbach,B.: Nucl. Instr. Meth. B 15 (1986) 756
/4/ Moncoffre,N., G.Marest, S.Hiadsi and J.Tousset: Nucl. Instr. Meth. B 15 (1986) 620
/5/ Zemek,J. and J.Kral: phys. stat. sol. (a) 94 (1986) 845
/6/ Bodart,F., G.Terwagne and M.Piette: MRS-Proceed. Boston 1986 (to be published)

CONVERSION ELECTRON MÖSSBAUER SPECTROSCOPIC STUDIES OF IRON IMPLANTED WITH BORON, CARBON AND PHOSPHORUS

H. Reuther [1], E. Richter [1], M. Fujinami [2], Y. Ujihira [2]

Ion implantation has been established as a method of surface improvement of metals (5). Particularly, it was demonstrated that an amorphous layer can be formed if large doses of certain metalloid ions were implanted (1,2) and that this amorphous layer seems to cause the surface improvement.
Such amorphous layers are produced if iron as target is implanted with high doses of boron, carbon, and phosphorus.
Boron implantation causes an increase of corrosion resistance and fatigue lifetime, carbon implantation shows a desirable effect in reducing wear, and phosphorus implantation influences the corrosion behaviour too.
An elucidation of the mechanism responsible for the amorphization and the structure of the amorphous layer is of considerable interest. The implanted layer is produced at the near surface region (some 10 to 100 nm). Iron-57 conversion electron Mössbauer spectroscopy (CEMS) is one of the most suitable techniques for the study of thin surface layers (4). The Mössbauer parameters depend sensitively on the nearest-neighbour surroundings of iron atoms, so they can provide much information on the compositions and atomic arrangements of amorphous iron-metalloid alloys.
Iron samples were implanted with high doses of boron, carbon, and phosphorus (10^{18} ions/cm^2, energy 50 keV). In all samples amorphous layers are formed. To study the crystallization effects the samples were annealed in a vacuum of 10^{-1} Pa and better at various temperatures.
The Mössbauer measurements were carried out at room temperature. The CEM spectrum of boron implanted iron consists of a broad doublet and the six-line-pattern of alpha-iron, with broad lines too. The very spreaded hyperfine field distribution indicates that the Fe-B alloy produced by boron implantation is more disordered than that prepared by conventional quenching techniques. This conclusion could be drawn because the hyperfine field distribution envelopes both the magnetic field values of the crystalline iron borides as known in the case of conventionally produced amorphous Fe-B alloys and ranges which could not be reproduced by superposition of the values

[1] Central Institute of Nuclear Research Rossendorf, Post Box 19, Dresden, 8051, GDR

[2] Engineering Research Institute, Faculty of Engineering, The University of Tokyo, Bunkyo-ku, Tokyo, Japan

of the crystalline compounds. Annealing at 400 °C results in the precipitation of FeB which is converted into Fe_2B at 500 °C.
Carbon implantation in iron results in a Mössbauer spectrum consisting of three sextets with internal magnetic fields strongly correlated with those of crystalline Fe_5C_2 and Fe_2C. It suggests that the amorphous iron-carbon alloy in the implanted layer has certain preferences for the local atomic configuration and composition similar to these crystalline compounds. Annealing at 300 °C and 400 °C let crystallize the amorphous layer and Fe_5C_2 is formed. After annealing at 500 °C it is transformed into Fe_3C.
The CEM spectrum of phosphorus implanted iron (Fig.1) is very similar to that of boron implanted iron. The doublet indicates an amorphous portion representing neighbourhood properties like Fe_2P which is paramagnetic. Besides the alpha-iron no other ferromagnetic component could be found. This is in contrast to conventionally produced amorphous iron-phosphorus alloys which are ferromagnetic (2). Annealing the sample at 300 °C and more results in the precipitation of Fe_2P. At 500 °C it vanishes and Fe_3P is formed. The transformation is all at once without any coexistence between the two phosphides. Temperatures of 700 °C and more let the phosphorus migrate out of the investigated region.

References

(1) Ali, A., W.A.Grant and V.P.J.Grundy: Phil. Mag. B37(1978)353
(2) Eibschütz, M., M.E.Lines, H.S.Chen and T.Masumoto: J. Phys. F: Metal Phys. 14(1984)505
(3) Hohmuth, K., B.Rauschenbach, A.Kolitsch and E.Richter: Nucl. Instrum. Methods 209/210(1983)249
(4) Reuther, H.: phys. stat. sol. (a) 93(1986)201
(5) Smidt, F.A. and B.D.Sartwell: Nucl. Instrum. Methods Phys. Res. B6(1985)70

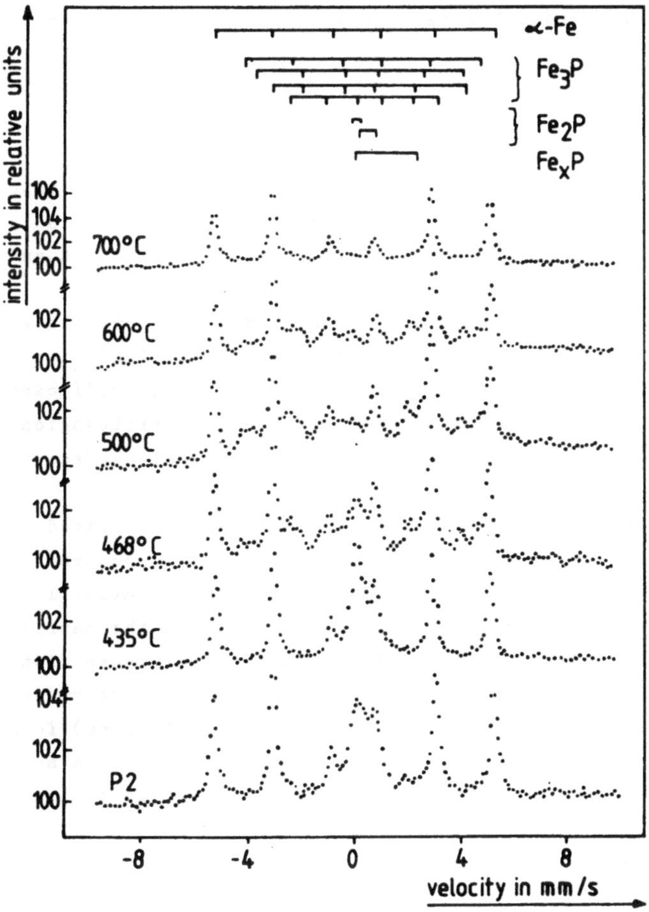

<u>Fig.1:</u> CEM spectra of iron implanted with 10^{18} cm^{-2} phosphorus ions at 50 keV and annealed at different temperatures

WEAR BEHAVIOUR OF ION IMPLANTED STEELS AND HARD METALS

Gerhard Dienel, Andreas Kolitsch, Edgar Richter and Uwe Scholz *)

1. Introduction

It has been established that the implantation of nitrogen into steel and hard metals improve their mechanical properties especially wear, friction and fatigue (1-3). Some tests on industrial tools and machine parts shown a significant increasing of their lifetime.
Generally, three different ways are possible to improve the wear resistance and the surface hardness of steels by ion implantation:

- Surface layers are prepared with a lower friction coefficient. If the friction decreases the surface temperature in a tribological system will be lower and the oxidation rate of the steel surface will be reduced.
- Metastable crystalline or amorphous phases are stabilized or the phase transformation will be diminished (e.g. austenite/martensite).
- Hardening of surface by interstitials and precipitates. In the surface layer of steel nitrided by nitrogen implantation the matrix coexists with supersaturated solutions of interstitial nitrogen wich segregate to dislocations to form Cottrell atmospheres and a very fine dispersion of nitride and crabonitride precipitates (1-10 nm) formed during implantation. These very finely dispersed precitpitates and the segregated interstitials pin the dislocations and impede their motion in the surface region[4,7].

2. Results and discussion

In the following some results are presented for change of wear behaviour and surface hardness of the steel typs 38CrMo21.14 and 67SiCr5 and the tungsten carbide hard metal typ HG-110 after implantation of nitrogen with different doses near room temperature. The chemical composition of the steels are given in table 1.
The hardness was measured using a Vickers indenter and for the wear experiments a "pin-on-disk"-arrangement was used. A fixed steelness bearing ball with 3 mm diameter served as pin with a definite load. The pin-ball slided on the disk with about 1 ms^{-1}. The volume abrasion was investigated in relation to the sliding way of the bearing ball on the

*) Academy of sciences of the GDR
Central institute for nuclear research Rossendorf
Postfach 19, Dresden, DDR-8051

rotating disk without lubrication. The load of the pin was 0.08 N in all experiments. All samples were implanted with nitrogen without mass separation. The implantation energy was 50 keV and the N_2^+/N^+-rate about 60 : 40.

Table 1

Chemical composition (wt-%) of the investigated steel types

steel typ	C	Si	Mn	Cr	Mo	V	P_{max}	S_{max}
67SiCr5	0.67	1.30	0.50	0.50	./.	./.	0.030	0.030
38CrMoV24.14	0.38	0.85	0.50	5.25	1.30	0.38	0.030	0.030

Figure 1 showes the results for the steel 38CrMoV24.14. With increasing dose in the range from $1x10^{17}$ to $8x10^{17} N\ cm^{-2}$ the wear rate was reduced significantly because of formation of various nitride and carbonitride precipitates during implantation (5,6). Doses higher than $8x10^{17} N\ cm^{-2}$ lead to increasing of the wear rate. The measured microhardness rises with increasing doses in the same way (figure 2). The steel 67SiCr5 showes no significantly effects for wear and hardness after nitrogen implantation.

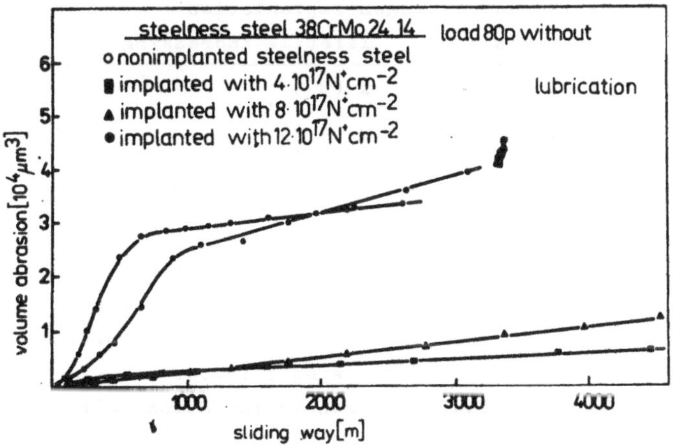

Figure 1
Volume abrasion versus sliding way for the steel typ 38CrMoV24.14 after Implantation of different doses of nitrogen ions

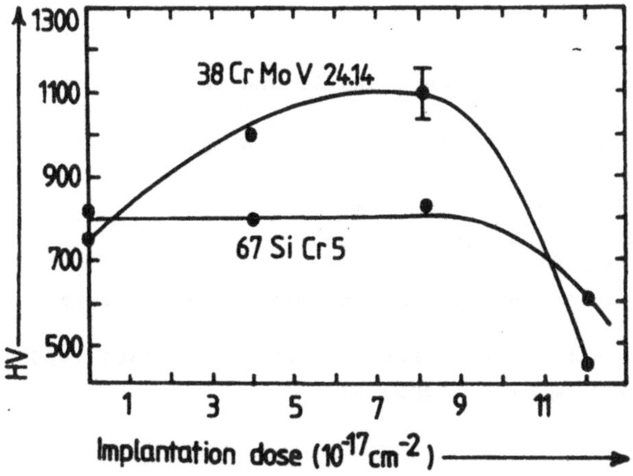

Figure 2
Microhardness versus implanted nitrogen dose for the steel types 67SiCr5 and 38CrMoV24.14. Implantation energy 50 keV

At higher implantation dose probably a part of the free nitrogen can precipitate in small bubbles. For doses > $8x10^{17}$ N/cm² the nitrogen pressure in the bubbles is getting so high that the skin can burst (blistering). Simultaneously growth the nitride and carbonitride precipitates abd produce also very high stresses in the surface region which probably support the process of blistering and forming microcracks. Consequently the surface roughness increases and therefore the wear resistance and the surface hardness decreases with increasing implantation dose.
In the other hand a significant reduction of wear will be found for tungsten carbide hard metal after nitrogen implantation in the same dose range like for steels. Comparing the different metalloide ions (boron, carbon and nitrogen) in view of the influence on the wear behaviour of tungsten carbide the nitrogen implantation give the best results (2).
In some tests on different industrial steel and tungsten carbide hard metal tools were demonstrated the improvement of their lifetime by nitrogen implantation. In table 2 some examples are given for multiplying the lifetime of production tools.

Table 2

Examples for practical applications of nitrogen implantation

Application	materials	lifetime increasing
Punching of aluminium	tool steel	x (3-4)
Stamping of soft alloys	tool steel	x (3-10)
Drilling of plastics	hard metal	x (2-4)
Cutting of plastics	tool steel	x (3-10)

Conclusion

In agreement with results of many other groups obtained for other steel types it was shown that the wear behaviour and the surface hardness of the steel 38CrMo24.14 and tungsten carbide hard metals will be improved by nitrogen implantation. The steel 67SiCr5 gives no effects.
Tests on industrial tools have shown that nitrogen implantation multiply their lifetime by factors between 2 and 10.

References

(1) Baumvol, I.J.R.: Ion implantation science and technology, Academic press,New York 1984

(2) Dienel, G., Kolitsch, A. and E. Richter: Proc. 11.th int. Plansee-Seminar, Reutte (Austria), Vol.1, 1985, 757

(3) Hirvonen, J.K.: MRS Proceedings 27 (1984) 621

(4) Iwaki, M., Yoshida, K., Sakudo, N. and S. Satou: Nucl. Instr. Meth. B6 (1985) 51

(5) Rauschenbach, B., Kolitsch, A. and K. Hohmuth: phys. stat. sol (a) 80 (1983) 471

(6) Rauschenbach,B. and K. Hohmuth: Cryst. Res. & Techn. 19 (1984) 1425

(7) Sioshansi, P.: Thin solid films 118 (1984) 61

HEAT PRODUCTION DURING ION IMPLANTATION INTO METALS AND ALLOYS

Roland Küchler and Edgar Richter *)

Heat input and maintenance of a low substrate temperature represented one of the most difficult problems for the high dose in implantation into metals and alloys. A typical implanter with a current beam density of 30 μAcm^{-2} and an acceleration voltage of 100 kV produce a heat load of about 3 Wcm^{-2} on the implanted surface area. Radiation cooling in vacuum is below about 400 °C totally inadequate for this heat input and the aspired low workpiece temperature. If the materials temperature exceeds 200 - 400 °C the beneficial effects of ion implantation anneal at this temperatures soon. The improvements in wear resistance or corrosion will be very small and not long lasting. The implanted nitrogen partially diffuses to the surface and goes out (3). Grabowski (2) discussed three possible regimes of cooling by heat conduction: - heat diffusion limited, - heat capacity limited and - steady state conduction to a heat sink. For the most implantation doses and workpiece sizes considered heat capacity alone is insufficient to cool the target effectivily. The beam current density must maintained at an uneconomically low level if the workpieces not will be overheated. Conduction to a heat sink will be essential. For practical application two systems are usefull for workpieces cooling during implantation: - elastomers with a high conductivity (4) or - eutectic metal alloys with a low melting point (1). Working with an eutectical metal bath gives both good cooling as well as for some workpieces a simple arrangement for holding during implantation.

For such cooling mode it is very important to know the temperature gradient between the cooled base of the workpieces and their parts standing out from the eutectic metal bath into the ion beam area. The rise of the equilibrium temperature depends upon the thermal conductivity of the target materials and the target geometry. For three selected geometrical configurations wich are mathematically simple to describe the equilibrium temperature on the tip were calculated in the following.
The temperature distribution satisfies the LAPLACE equation. The temperature in therefore assumed to be stationary. At first we consider a rectangular flate plate wich is illustrated in figure 1. The ion beam treates the plate on the edge a and the other edges satisfy the following bondary conditions:

$$\left.\frac{\partial T}{\partial x}\right|_0 = 0 \quad \text{and} \quad \left.\frac{\partial T}{\partial x}\right|_d = 0$$

---

*) Academy of sciences of the GDR
Central institute for nuclear research Rossendorf
Box 19, Dresden, DDR-8051

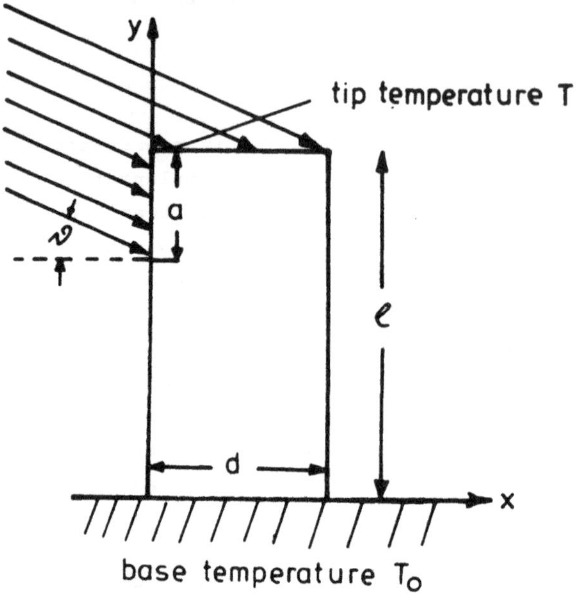

Figure 1
Schematic drawing of a rectangular flate plate in the ion beam cooled on the workpiece base

The equlibrium temperature T on the tip of the flat plate is given for $a/l = 0.1$ and $d/l = 1$ by

$$T = T_0 + \frac{\cos\vartheta \cdot U \cdot j\, \ell}{K}\left[\frac{\ell}{d}\left(\frac{a}{\ell} - \frac{1}{2}\left(\frac{a}{\ell}\right)^2\right) + \frac{d}{\ell}\left(\frac{1}{3} - \frac{\ell}{5}\right)^{-3\frac{a}{d}} + \tan\vartheta\right] \tag{1}$$

with K - thermal conductivity, j - current density, U - acceleration voltage, l - object length, d - object thickness and ϑ - angle between the normal vektor of the plate and the incidence of the beam.
The final formula includes geometrical parameters only and gives the temperature at the tip of the plate ($x = 0$, $y = 0$). Assuming the following parameters: $U.j = 2\ W/cm^2$, $\vartheta = 0$, $l = 4$ cm, $K_{steel} = 0.5$ W/cm K and $d/l = 0.5$ we find for $a/l = 0.25$ 8 K as difference between tip temperature and base temperature and for $a/l = 0.5$ we find 15 K. The rising of the tip temperature during implantation is very small only.
For a circular sector prism the tip temperature T is given by

$$T = T_0 + \frac{U \cdot j \cdot \cos\vartheta \cdot R}{K} \cdot \frac{1}{\varphi_0} \tag{2}$$

with φ_0 - tip angle, R - edge length and the boundary condition for the unirradiated edge is $\left.\frac{\partial T}{\partial \varphi}\right|_{\varphi_0 2} = 0$. The circular arc acts as heat sink with a temperature T_0.
If we assume that $U.j = 2\ W/cm^2$, $\vartheta = 0$, $K_{steel} = 0.5$ W/cm $\cdot$ K, R = 4 cm and $\varphi_0 = \pi/4$ we get a difference between base temperature T_0 and tip temperature T at about 20 K.

Third, we consider a rectangular triangle. Either side adjoining the hypotenuse will irradiated the other one acts as head sink with a temperature T_0. In this case we find for the tip temperature the equation

$$T = T_0 + \frac{U \cdot j \cdot \cos\vartheta \cdot \ell}{K} \cdot 1.35 \qquad (3)$$

Assuming the same parameters like in the other examples we get for the tip temperature about 22 K.
All calculations for the three different forms of tool types show that the difference between the cooled tool base in the eutectic metal bath and the irradiated by the ion beam tool tip is very small only. We assume that the little higher tip temperature have not a significant influence on the properties of the implanted tool. It is very important, however to have a good heat sinking by an intensive cooling system for the tools during implantation.

References

(1) Denholm, A. S. and A. B. Wittkower: Nucl. Instr. Meth. B6 (1985) 89 - 93
(2) Grabowski, K. S. and R. A. Kant
in: Ion implantation: equipment and techniques,
Springer-Verlag, New York 1983, 364
(3) Moncoffre, N., G. Hollinger, H. Jaffrezic, G. Marest and J. Tousset: LYCEN-Report No. 8445, July 1984
(4) Smidt, F. A. and B. D. Sartwell: Nucl. Instr. Meth. B6 (1985) 70 - 77

COMPOSITION DEPENDENCE OF MAGNETIZATION OF ION IMPLANTED COBALT LAYERS

A.Unkroth, G.Götz[1)], U.Heiber[2)], W.Andrä[3)]

1. Introduction

It is well known that alloying a metalloid to a transition metal (TM) the magnetic moment per TM atom decreases. Different models have been developed to describe the dependence of the magnetic moment per TM atom on the composition. The models seem to fit some systems with reasonable accuracy but fail in some cases [1,2]. This situation makes it necessary to work out systematical trends of experimental data. Ion implantation is an especially useful method for alloying because of the possibility of alloying any desired element up to concentrations even over that of the thermodynamic solid solubility.

2. Experimental

Cobalt films with a thickness of (30...100) nm were sputtered or evaporated onto oxidized silicon. P^+ or Si^+ ions with doses in the range of $8*10^{15}$ cm^{-2} - $2*10^{17}$ cm^{-2} have been implanted at room temperature or at liquid nitrogen temperature in order to produce $Co_{1-x}P_x$ and $Co_{1-x}Si_x$ with $0 < x \leq 0.4$ respectively. Dual implantation with different ion energies has been carried out to achieve a nearly homogeneous concentration over the whole depth of the Co-layer. The magnetic moment of the layer volume M_sV has been measured with a torque magnetometer [3]. Rutherford Backscattering Spectrometry (RBS) has been used for determination of both the number of Co-atoms per area $(N\ d)_{Co}$ and the metalloid content in the layer. Fig.1 shows typical RBS-spectra before and after implantation of various doses of phosphorus ions. The areal density of Co-atoms $(N\ d)_{Co}$ has been determined from the total yield Y^{Co}_{tot} of ions backscattered from Co-atoms [4]. Fig.1 shows that $(N\ d)_{Co}$ decreases with increasing phosphorus dose. The relative phosphorus concentration c_p in the Co-layer was calculated by means of Eq.(1), where $Y^{Co}_{o,Co}$ and $Y^{Co(1-x)P(x)}_{o,Co}$ are the cobalt yields of the unimplanted and implanted layer extrapolated to the surface

1) Friedrich-Schiller-Universität Jena, Sektion Physik, Max-Wien-Platz 1 Jena, DDR 6900
2) Kombinat Mikroelektronik "Karl Marx", Rudolf-Str. 47, Erfurt, DDR 5010
3) Physikalisch-Technisches Institut der AdW der DDR, Helmholtzweg 4, Jena, DDR 6900

$$\frac{Y_{o,Co}^{Co_{1-x}P_x}}{Y_{o,Co}^{Co}} = (1-c_p)\frac{[\varepsilon_o]_{Co}^{Co}}{[\varepsilon_o]_{Co}^{Co_{1-x}P_x}} . \qquad (1)$$

$[\varepsilon_o]_{Co}^{Co_{1-x}P_x}$ and $[\varepsilon_o]_{Co}^{Co}$ are the stopping cross section factors in surface energy approximation for scattering processes with Co atoms within the $Co_{1-x}P_x$ and Co layers, respectively [4].

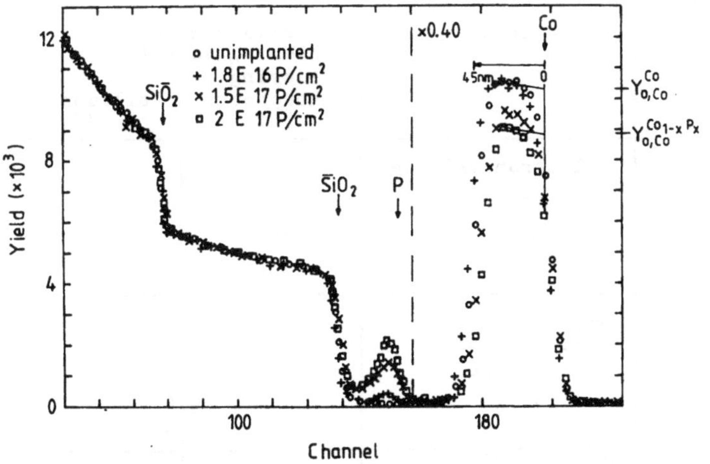

Fig.1. Backscattering spectra of cobalt layers on oxidized silicon before and after implantation of various doses of P ions at room temperature

3. Results and discussion

The magnetic moment per Co-atom n_B has been calculated by means of formula (2), whereas A means the sample area,

$$n_B = \frac{M_s V}{A(N d)_{Co}} \qquad (2)$$

Fig.2 and Fig.3 show the composition dependence of n_B of $Co_{1-x}P_x$ and $Co_{1-x}Si_x$, respectively. Within the experimental errors n_B decreases linearly with increasing metalloid content. The slope of the lines is similar for $Co_{1-x}P_x$ and $Co_{1-x}Si_x$ but it differs from results achieved on electrodeposited Co-P alloys [5] which are shown in Fig.2 for comparison. Known models for the description of the composition dependence of the magnetic moment per Co-atom are the rigid band model [1], the valence bond model [2]and the assumption that an nonmagnetic compound like Co_2P or Co_2Si is formed. An essential parameter for the distinction between these models is the slope of the lines or the metalloid concentration at which n_B goes to zero (see Table1). In the case of our measurements the scattering of the experimental points is too large in order to find out the correct model. Further investigation including annealing, TEM and X-ray diffraction results would help to bring light in the matter.

Table 1. Models of the composition dependence of the magnetic moment per Co-atom n_B in cobalt-metalloid alloys

model	main criterion	metalloid concentration at which n_B goes to zero by alloying phosphorus [at%]	silicon [at%]
rigid band model	electron valence		
	z=5	25.5	-
	z=4	30	30
	z=3	36.3	36.3
valence bond model	number of neighbouring atoms (structure)		
	z=12	29.4	
	z=9	35.7	
nonmagnetic compound	stoichiometry		
	Co_2P, Co_2Si	33.3	33.3
	Co_3Si		25

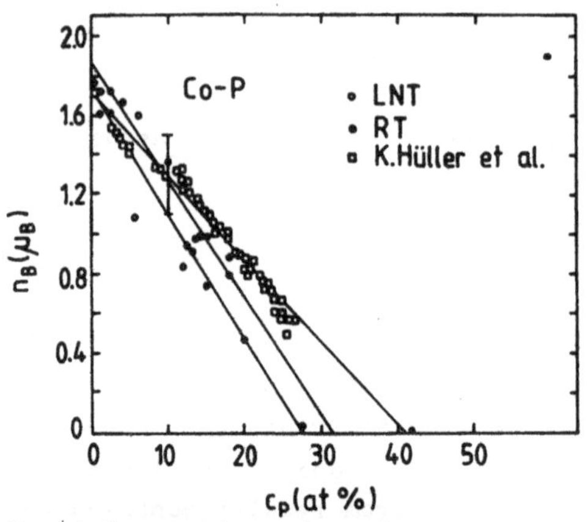

Fig.2.Composition dependence of the magnetic moment per Co-atom after implantation of P ions at RT (•) and LNT (o). For comparison results measured at electrodeposited Co-P alloys by K. Hüller et al.[5] (□) are depicted.

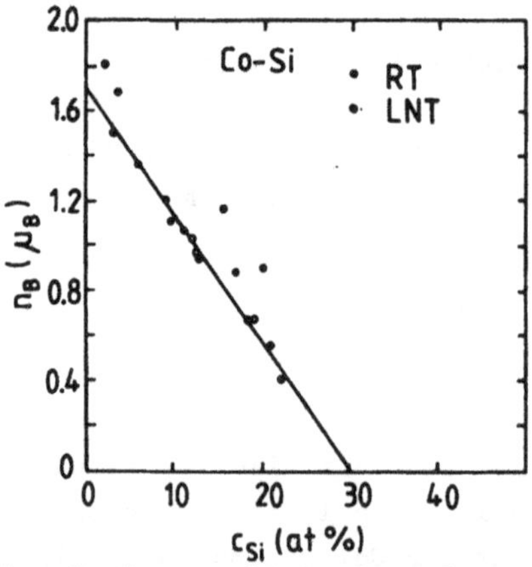

Fig.3. Composition dependence of the magnetic moment per Co-atom after implantation of Si ions at room temperature (•) and liquid nitrogen temperature (o).

[1] Malozemoff,A.P., A.R.Williams and V.L.Moruzzi: Phys.Rev. B29 (1984) 1620-1632
[2] Corb,B.W., R.C.O'Handley and N.J.Grant: Phys.Rev. B27 (1983) 636-641
[3] Griffiths,J.H.E.: Physica 17 (1951) 253-258
[4] Götz,G., K.Gärtner: High Energy Ion Beam Analysis of Solids, Akademie-Verlag, in press
[5] Hüller,K., G.Dietz, R.Hausmann and K.Kölpin: J.Magn.Magn.Mat. 53 (1985) 103-110

AMORPHIZATION PROCESSES STUDIED BY HIGH DOSE ION IMPLANTATION INTO METALS

G. Linker, A. Seidel, B. Strehlau [+)]

Results of amorphization studies by high dose ion implantation in different ion-target systems are summarized. For all systems studied the amorphization proceeds in a similar way, namely in three steps: accumulation of impurities on interstitial or substitutional sites in the crystalline phase, transition into the amorphous state of the bulk of the material and total amorphization in a saturation region. In the accumulation step an increase of strain is observed which may be the driving force for amorphization.

Introduction

Ion implantation and ion irradiation quite frequently are used in amorphization studies since these techniques have some unique advantages. While irradiation may be applied in self-amorphizing materials, i.e., systems which do not require additional stabilizers for amorphization, ion implantation must be used e.g., in most elemental metals which will not transform into the amorphous state without the introduction of suitable stabilizers. In comparison to various other amorphization techniques the benefit of implantation in binary alloys is that the amorphization process can be observed quasi continuously as a function of the concentration of one of the constituents. The results of such experiments may resolve the intrinsic reason for amorphization and can be compared with theoretical models.

We have studied the amorphization processes in different ion-target systems. Suitable glass formers like B,P,S or Mn have been implanted into targets of bcc Nb and Mo or fcc Ni and Al. Some of the results have been already published (1-4). In this contribution the common features of the crystalline to amorphous transformation in the different systems will be summarized and underlined with some recent exemplary data.

Experimental and Analysis

The implantations were performed homogeneously into evaporated crystalline thin films or single crystal surfaces at liquid nitrogen temperature. The amorphization processes in the thin films were studied by X-ray diffraction at room temperature employing a Guinier thin film diffractometer, and by in situ channelling-backscattering experiments in single crystals. The bulk of the data has been deduced from the X-ray measurements. Lattice parameter and homogeneous strain values could be evaluated from the line positions following a procedure described by Feder and Berry (5). The amorphous fraction and static displacements of the host lattice atoms were determined by comparison of integral line intensities before and after each implantation using a procedure described in detail elsewhere (6). Material properties like resistivity or superconductivity (transition temperature) reflecting the amorphization process were also measured.

Results and Discussion

The amorphization process is illustrated in Fig. 1 where for the example of B implantation into Mo, X-ray spectra are shown before and after implantation of different B concentrations. It can be seen that for 5 at% B the crystalline structure is preserved however with an appreciable intensity loss. At 15 at% B almost all of the crystalline structure has disappeared but a sharp peak on top of a broad diffraction maximum

---

[+)] Kernforschungszentrum Karlsruhe, Institut für Nukleare Festkörperphysik
P.O.B. 3640, D-7500 Karlsruhe, FRG

indicates that still crystalline regions are present in the material. Finally, at 19 at%B the amorphization is completed. At low impurity concentrations where the structure is still crystalline large lattice parameter changes occur. It depends on the type of incorporation (interstitial or substitutional) and on the atomic radii of the impurity and target atoms whether an increase or a decrease is observed. Interstitial incorporation on, e.g., octahedral sites (7) like observed for B implantations always leads to an increase of the lattice constant. For substitutional incorporations the atomic radii determine the behaviour of the lattice parameter. Mn in Al, e.g., is incorporated almost 100% substitutionally as has been shown by channelling measurements (8). Since the atomic radius of Mn is smaller than that of Al, the lattice parameter decreases. This effect is demonstrated in Fig. 2. The decrease, however, is observed only to an average Mn concentration of about 7 at%. This is the critical concentration for the onset of amorphization. For higher average concentrations more and more material is rendered amorphous and the lattice constant values are determined in the crystalline rest regions with a lower Mn concentration than the critical one. Therefore, after the onset of amorphization no further decrease of the lattice parameter or even an increase over the minimum value occurs depending on the distribution of the impurities and the crystalline regions. Since the amorphization is accompanied by a volume change presumably also additional effects like intrinsic stress influence the lattice parameter in the mixed amorphous-crystalline state.

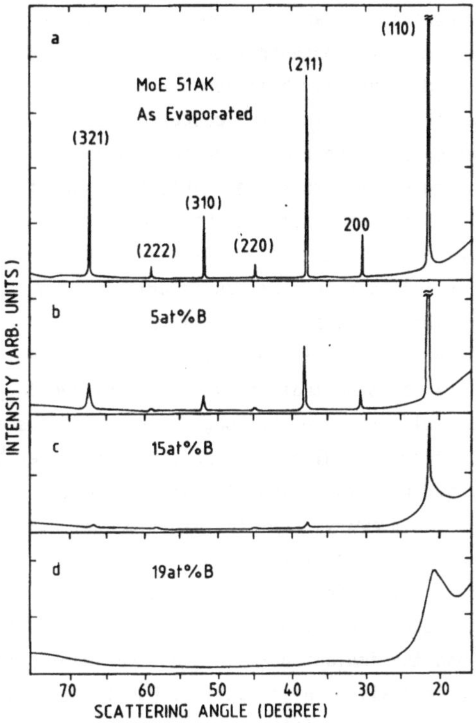

Fig. 1. X-ray spectra of a Mo film before (a) and after implantation of different B concentrations (b-d) illustrating the amorphization process.

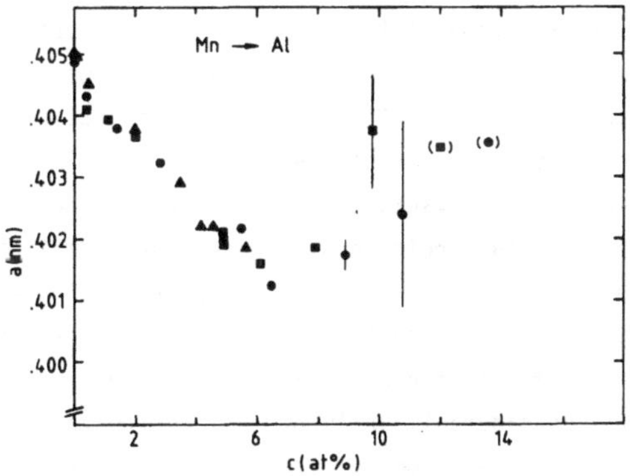

Fig. 2. Lattice constant values a of different Al films vs. implanted Mn concentration c.

Since in the systems we have studied the equilibrium solubility of the implanted impurities is very low (<1 at%) the forced interstitial or substitutional incorporation leads to structural distortions and to accumulation of strain. We describe the structural distortions by small average static displacements of the host lattice atoms. Their average values increase in the crystalline phase coinciding with the increase or decrease of the lattice parameter. As an example, we show in Fig. 3 an amorphization curve, i.e., a plot of the amorphous fraction vs. impurity concentration, and the homogeneous strain vs. impurity concentration for B implanted into Mo. The amorphization curve reflects the three stages of the process, namely formation of a supersaturated interstitial solid solution with a negligible fraction of amorphized material, then transition into the amorphous phase of the bulk of the material within a rather narrow impurity concentration range, and finally a saturation range where total amorphization is achieved. It can be seen that strain is increasing in the solid solution as long as the amorphous fraction is negligible. Then, strain is released with the onset of amorphization.

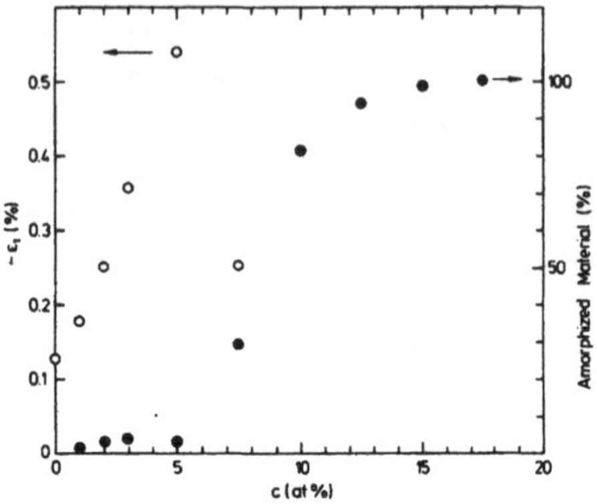

Fig. 3. Amorphization curve (full circles) and homogeneous strain values ε_1 vs. B concentration in a Mo film. The strain values decrease drastically upon amorphization.

In summary, from the experimental results of the different systems we have studied so far it emerges that amorphization by ion implantation can be described by an accumulation model of amorphization where effects like quenching of dense collision cascades are of minor importance. In the accumulation phase the implanted impurities are incorporated interstitially or substitutionally into the crystalline host lattice. Due to the size mismatch of the constituent atoms also strain is accumulated in the matrix. At a critical strain value the matrix will undergo a phase transformation leading to strain release. This representation is in accordance with an amorphization model of Egami and Waseda (9) which assumes strain as a necessary condition, i.e., as a driving force for amorphization. In addition, however, for the stabilization of the amorphous phase, locally also a critical concentration of suitable (10) glass forming impurities is necessary. This is in agreement with the analysis of Cohen et al. (11) who deduce critical concentrations and minimum volumes of amorphous regions from amorphization curves applying statistical models. Since impurities and strain in implantation studies are accumulated simultaneously it is difficult to untangle their mutual role in the phase transformation process. In our interpretation the role of the impurities is twofold, namely: strain generation by forced incorporation into the host lattice, and then stabilization of an amorphous phase upon transformation. Other mechanisms of strain release than phase transformations like impurity-defect interactions may considerably influence the amorphization process. Therefore, if implantation is performed at room temperature where vacancies are mobile, the shape of the amorphization curves will be modified (4).

References

1) G. Linker, Nucl. Instr. Meth. 209/210 (1983), 969
2) G. Linker, Materials Science Eng. 69 (1985), 105
3) G. Linker, Sol. State Commun. 57 (1986), 773
4) G. Linker, Nucl. Instr. Meth. B 19/20 (1987),526
5) R. Feder. B.S. Berry, J. Appl. Cryst. 3 (1970), 372
6) G. Linker, Nucl. Instr. Meth. 182/183 (1981), 501
7) K. Beck, B. Feuser, K. Kopitzki, G. Krauss, G. Mertler, Nucl.Instr.Meth. 209/210 (1983), 969
8) A. Seidel, G. Linker, unpublished data
9) T. Egami, Y. Waseda, J. Non-Cryst. Solids 64 (1984), 113
10)B.Rauschenbach, K. Hohmuth, Physica Status Solidi A 72 (1982), 667
11)C.Cohen,A.Benyagoub,H.Bernas,J.Chaumont,L.Thome,M.Berti,A.V.Drigo,Phys.Rev.B31(1985),5

IMPLANTATION EFFECTS ON MICROCRACK INITIATION IN FATIGUED METALS

by C.Blochwitz [1]) and B.Rauschenbach [2])

I.INTRODUCTION

Ion implantation (IP) has been widely used to improve the fatigue life of metals immediately. Moreover, the modification of shallow surface layers should be a sensible approach to study the role of the free surface in the fatigue process at all.

In fcc metals, strain localization in persistent slip bands (PSB) is the typical prestage of microcrack initiation (m.c.i.). The density of PSB surface traces decreases considerably with increasing implantation dose /1/. Consequently, m.c.i. in the notch-peak relief of PSBs should be reduced by IP. But, if the plasticity of the near-surface layer is suppressed by IP to too great extent, the material can change to a grain boundary mode of m.c.i./2/, leading often to a fast, brittle-like crack propagation. In order to optimize the implantation regime, the influence of IP on the different m.c.i. mechanisms has to be investigated in greater detail.

First experimental results concerning the interaction between PSBs and the implanted layer and the occurence of different m.c.i. types in boron and in nitrogen implanted nickel are discussed in the present work.

II. EXPERIMENTAL

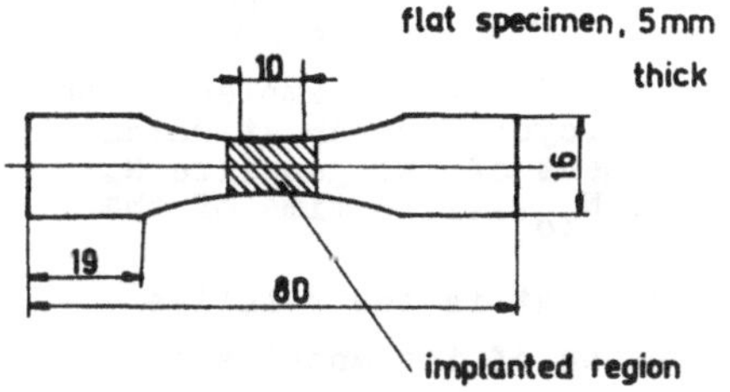

Fig.1 Specimen shape

The ion bombardment was performed at room temperature with the dose 5 x 10^{17} cm^{-2} at 50 keV energy. Cyclic deformation of nickel polycrystals (99.9 % purity) was done in symmetric push-pull at room temperature and the constant plastic strain amplitude ε_{ap}. Some specimens were fatigued until different cycles N_I at first and then implanted. Concentration profiles of implanted atoms were determined by Auger electron spectroscopy (AES).

---

[1]) Sektion Physik der Pädagogischen Hochschule"K.F.W. Wander", Wigardstraße 17, DDR - 8060 Dresden (G.D.R.)

[2]) Zentralinstitut für Kernforschung Rossendorf, Postfach 19, DDR - 8051 Dresden (G.D.R.)

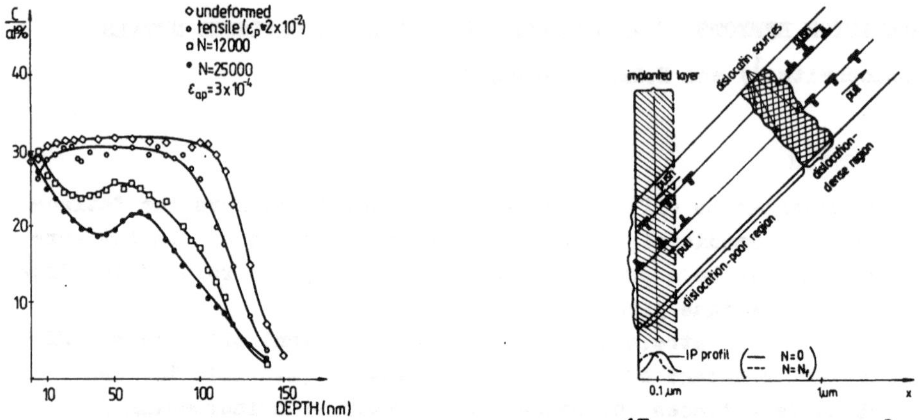

Fig.2 Change of concentration profiles (5×10^{17} nitrogen ions/cm^2) during mechanical deformation and model of dislocation motion within PSBs near the surface of implanted specimens.

Note that after tensile deformation a striking change of the concentration profile can not be observed.

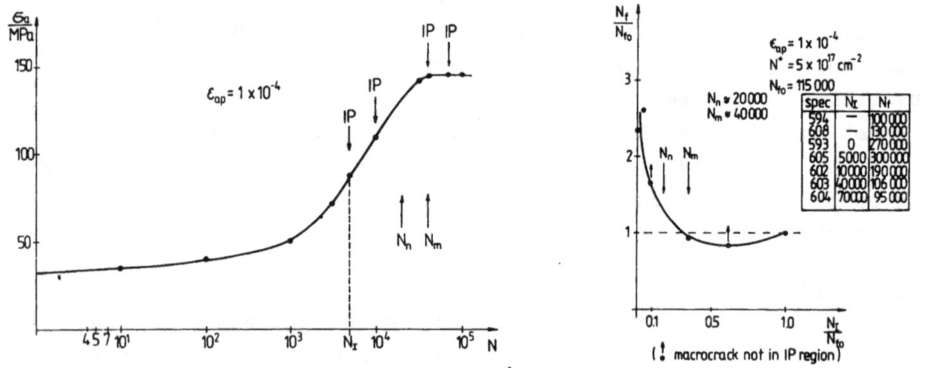

Fig.3 Cyclic hardening curve (left) showing the different implantation cycle numbers N_I (σ_a...stress amplitude, N_n...cycle number at which the first PSBs can be observed) and improvement of fatigue life N_f after IP at different fatigue stages (right), N_{fo}...life time of the unimplanted material.

In the Table, the PSB surface densities (that is the relative number of grains with visible PSB surface traces of implanted specimens in relation to unimplanted specimens) can be seen. From the total number of microcracks of different types, investigated at an area of about 3 mm^2, the microcrack density was calculated. The absolute number of microcracks which have been observed in the grain boundaries (A-type, cf./3/) or in the PSB relief (B-type) after fatigue fracture is listed in parentheses.

III. CONCLUSIONS

1. The penetration of PSBs through the strengthened IP layer is inhibited, but PSBs carrying high plastic strains exist in the grains below the IP layer (cf./4/). During the fatigue process, however, the PSBs are able to penetrate the IP layer successively, especially at large grains /1/ and at higher amplitudes, note e.g. the PSB density

in the field of high local strains near growing macrocracks (Fig.4).

2. If PSBs penetrate the IP layer successively, the shift of implanted ions to the free surface (Fig.2) might be caused by dislocations emerging from the dislocation walls and moving through the IP zone to the free surface. The high cumulative plastic strain within PSBs enables such a transport phenomenon also at room temperature fatigue.

3. The propagation of microcracks can not be prevented by IP, as soon as the PSBs have penetrated the surface layer (cf. the vanishing improvement of life time at $N_I > N_n$ in Fig.3).

4. Under the conditions of the present work, the PSB and the microcrack densities are remarkably reduced when IP was performed before fatigue ($N_I = 0$), but grain boundary m.c.i. was observed also in few cases. This agrees well with the maximum degree of life time improvement of boron implanted nickel at the dose 5×10^{17} cm^{-2} /1/.

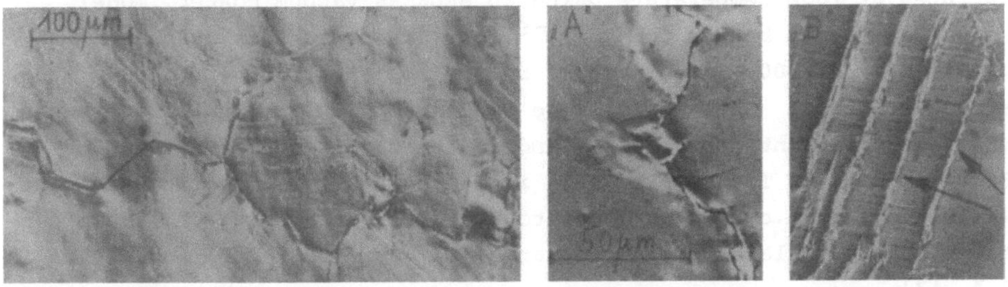

Fig.4 Examples for a macrocrack, growing through the IP region (spec. 553), and for the microcrack types A and B.

spec.	ion	N_I/N_{fo}	$N_f/10^3$	PSB surface density	microcrack density per mm^2 and absolute number of A/B type microcracks	
					unimplanted	implanted
						(A / B)
604	N^+	0.62	95	1.00		17 (8 / 41)
603	"	0.35	106	0.96		12 (10 / 22)
602	"	0.10	190	0.80		4 (7 / 0)
593	"	0	270	0.40	13	1 (2 / 0)
553	B^+	0	120	0.58	4	1 (3 / 1)

REFERENCES:

/1/ K.Hohmuth, E.Richter, B.Rauschenbach, C.Blochwitz, Mat.Sci.and Engng. 69 (1985) 191-201

/2/ R.G.Vardiman, in "The applications of ion plating and implantation to materials", Ed. R.F.Hochman, ASM 1986

/3/ G.Dörr, C.Blochwitz, Cryst.Res.Technol. 22 (1987) 113-121

/4/ B.Rauschenbach,C.Blochwitz,phys.stat.sol.(a),94 (1986), 827-831

STRUCTURAL MODIFICATIONS OF Ni EVAPORATED THIN FILMS IRRADIATED WITH P^+ IONS

L.C.Nistor, V.Teodorescu, G.Dincă, R.Medianu*

Ion implantation is emerging as a powerful tool for the production of amorphous metals and alloys. Previous investigations [1,2] have demonstrated that ion implantation of P^+ at high doses ($\geq 10^{17}$ions.cm^{-2}) can be used to produce non-crystalline surface regions in Ni, metastable at room temperature.

In this paper we have investigated the structural changes induced by energetic ion implantation at medium doses ($\sim 10^{16}$ions.cm^{-2}) and low fluence ($\leq$ 2μA.cm^{-2}) of P^+ ions in vacuum evaporated Ni films.

EXPERIMENTAL

Ni thin films were evaporated on freshly cleaved KCl or mica substrates at room temperature using an BALZERS vacuum coater, model BA 510 in a vacuum better than 5×10^{-6} torr. The thickness of the films approximately of 600 Å was measured before implantation by Tolansky interference method. The samples were implanted with P^+ ions delivered by a VEZUVIU-2B implanter at 120 keV and in the dose range of 5×10^{15}-2.6×10^{16}ions.cm^{-2}. The samples were implanted at room temperature and at low fluence ($\leq$2μA.cm^{-2}), in order to avoid target heating. In these conditions, the calculated projected range, $\bar{R}_p$, for 120 keV P^+ ions in Ni is 527 Å [3].

Transmission electron microscopy (TEM) and selected area electron diffraction (SAED) were performed using a JEM-200-CX electron microscope operating at 200 kV. Samples for transmission studies were prepared after implantation by dissolving the KCl support in distiled water, or by cleaving the mica substrate to very thin foils (tenth of nm).

RESULTS AND DISCUSSION

Uniform polycrystalline Ni thin films have been obtained by evaporation on both KCl and mica substrates. Figure 1 a gives the TEM and SAED images Ni thin film deposited on mica. Both layers are similar in morphology, with crystallites dimensions of 10-20 nm. The SAED image shows polycrystalline Ni and single crystal diffraction spots corresponding to mica substrate.

After implantation at relatively low doses (5×10^{15} ions.cm^{-2}) the amorphisation of Ni thin film has occured, figure 1 b on both thin films. The morphology of the layer has significantly changes after implantation, but still retains a grainy appearance.

* Central Institute of Physics, P.O.Box MG-6, Bucharest-Măgurele, R-76900, Romania.

It is interesting that the amorphisation effect occured in evaporated Ni thin films at least one order of magnitude lower doses of P^+ implanted ions as in thinned foils of bulk Ni [1,4]. At doses of $5x10^{15}ions.cm^{-2}$ and 120KeV, the irradiation damage in the sample is high enough to produce a large number of defects. Usually the crystalline metallic lattice recovers easily, so that no metastable phases could occur, as in the case of semiconducting materials [5] . The presence of P atoms as impurities can stabilized the disorder. Also it must regard that the energy of P^+ ions was high enough as many of them to cross the Ni layer and to reach the substrate. Atoms from the substrate produced by radiolysis [7] entering the Ni layer can also act as impurities stabilizing the irradiation defects introduced by implantation. Other impurities which could hinder the recovering process are the C atoms, appearing at low doses, as recoil implantation in the contamined layer formed during the long implantation cycles [6] .

Another striking aspect observed at these low doses was the appearance of crystalline islands in the amorphous layer, without any subsequent annealing treatment. Figure 2 a, c, d gives the TEM and SAED images of such crystalline zones. The crystalline island (Figure 2a) have an ellipsoid shape with dimensions ranging between 0.3-1 μm. At higher magnifications (Figure 2c) details in the micrograph show a superposition of the crystalline Bragg contrast to the grainy morphology of the amorphous layer, as the crystalline island had grown inside the Ni film or more probable at its interface to the KCl substrate. The SAED pattern (Figure 2d) show that these regions are monocrystalline. They are formed of Ni_2P, a stable hexagonal phosphide. At doses exceeding $10^{16}ions.cm^{-2}$ a complete polycrystalline Ni_2P layer is formed (Figure 2b). The layer is formed by joining the crystalline Ni_2P islands and consuming the amorphous film.

The mobility of P atoms must be high enough especially at the amorphous film - substrate interface to allow the formation of a compact crystallized Ni_2P layer. According to the displacement spike mechanism one suppose that the displacement atoms created by an incident ion tend to rearrange themselves into a new structure, without the occurence of actived jumps. To estimate wheather this mechanism could play a role in the growth of crystalline Ni_2P film , we note that a crystallized Ni_2P region formed at a dose of $5x10^{15}ions.cm^{-2}$ will have been subjected to 10^{11} displacements . Since the Ni_2P crystallites contain roughly $5x10^9$ atoms, results that the displacement effects occur on a sufficient scale to make them important to this process.

REFERENCES

[1] A.Ali, W.A.Grant, P.Y.Grundy, Phyl.Mag. 37 (1978) 353.
[2] W.A.Grant, Nucl.Instr.Meth., 182/183 (1981) 809.
[3] G.Dincă, Ph.D.Thesis, Bucureşti (1987).
[4] Z.Y.Al Tamimi, W.A.Grant, P.Y.Grundy, Vacuum 34 (1984) 861.
[5] C.Jech, R.Kelly, J.Phys.Chem. Solids 30 (1969) 465.
[6] J.A.Knapp, D.M.Follstaedt, S.T.Picraux, Proc.Symp. on Surface Modifications of Materials by Ion Implantation,Cambridge ,Mass.,USA (1979).
[7] D.J.Elliot, P.D.Townsent, Phil.Mag. 23 (1971) 249.

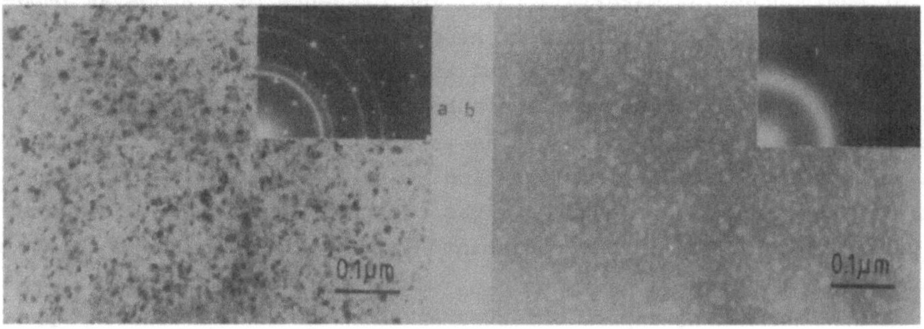

Figure 1. TEM and SAED images of Ni thin films just after evaporation (a) and after implantation with 5×10^{15} P^+ ions.cm^{-2} (b) .

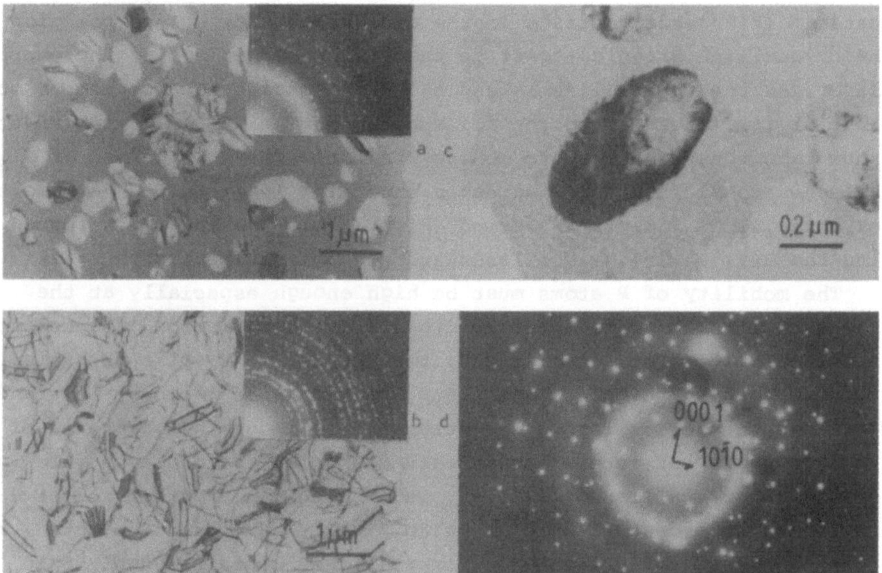

Figure 2. TEM and SAED images of Ni thin film deposited on KCl and implanted with 5×10^{15} P^+ ions.cm^{-2}, at 120 kev : (a) and (c) crystalline islands of Ni_2P in the amorphous layer; (d) <1210> zone axes for Ni_2P, and with 10^{16} P^+ ions.cm^{-2}, at 120 KeV; (b) compact crystallized Ni_2P layer is formed.

ANNEALING OF LASER INDUCED DEFECTS IN SILICON

W. Andrä[1)], I. Bryłowska[2)] and G. Götz[1)]

The phenomena governing the annealing of shallow implanted layers in silicon by laser processing have been a subject of intensive scientific investigation for about ten years. Ultrafast thermal treatment subsequential to ion implantation for the removal of the lattice damage and for doping activation generates high dopant concentration in excess of conventional solubility limits and without dopant redistribution. However, laser annealing in spite of its advantages, induces electrically active defects in silicon [1]. These defects are known to anneal at temperatures between 500 and 700°C in a conventional post-annealing step [2]. In this work we study the properties of p/n-junctions and electrical active defects in the space charge region induced by the laser treatement.

Circular dots with 1 mm^2 were prepared by LOCOS technique on n-type (phosphorous doped) <100> Czochralski silicon of 1-5 Ωcm resistivity. The following BF_2 implantation were made through a 40 nm SiO_2 layer. The implantation dose was $5*10^{15}$ cm^{-2} at 155 keV producing an amorphous layer of 0.185 μm measured by Rutherford backscattering spectrometry (RBS). After removal of the SiO_2 cover layer the structures were laser annealed using a Q-switched Nd:glass laser (pulse duration 40 ns) of energy density 3 J/cm^2 and wavelength 1.06 μm. A couple of samples were furnace post-annealed in dry nitrogen for 20 min within the temperature range 400-800°C. Preparation was completed with a contact metallization. The junction depth was determined to be about 0.3 μm by bevel method combined with a decoration of the junction. That means the junction depth is somewhat deeper than the edge of the amorphized layer after ion implantation.. Junction depths of 0.3 μm and higher are connected with minimum leakage currents.

After laser annealing minimum leakage current densities of our structures are measured to be about $1-5*10^{-7}$ A/cm^2. A furnace post-annealing treatement at 600°C for 20 min in dry nitrogen improves the junction properties. The minimum leakage densities decrease for about two orders of magnitude to about 10^{-9} A/cm^2.

To get more informations about the origin of the high leakage currents after laser annealing deep level transient spectroscopy (DLTS) was used. A typical DLTS spectrum of a laser-annealed sample is shown in

---

1) Friedrich-Schiller-Universität Jena, Sektion Physik, Max-Wien-Platz 1 Jena, DDR 6900
2) Uniwersitet Marii Curie-Sklodowskiej, Institut Fisiki, pl. Marii Curie-Sklodowskiej 1, 20-031 Lublin, Poland

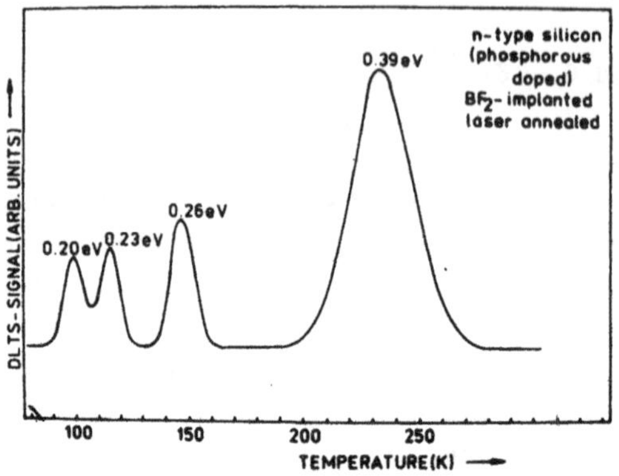

Fig.1. DLTS spectra recorded at emission rate 20.5 s^{-1} and pulse duration 50 µs for Q-switched Nd:glass laser (1.06 µm, 3 J/cm^2) annealed n-type Si sample. The curve was measured at a reverse bias of 1 V.

Fig.1. Four electron traps E (0.20eV), E (0.23eV), E (0.26eV), and E (0.39eV) are observed at an emission rate of 25.5 s^{-1}. The energy values have been obtained from measurements of the shift in DLTS peak temperature as a function of reference time constant.
Other authors [1,3] didn't measure exact this four levels. However, it is well known [4] that melt depth, melt cooling rate and velocity of solidification, and also resistivity of the bulk material and species of dopants are the essential parameters in defect formation. Generally the structural and chemical indentities of the defects are not revealable by DLTS measurements. Three of the four levels shown in Fig.1 are possibly corresponding to defect levels measured by other authors. The E (0.23eV) level is close to the E (0.24eV) donor level measured by Fan et al. [5] in phosphorous-doped laser annealed silicon. The E (0.26eV) level corresponds to the E(0.26 eV) level of Wang [4] which is a quenched-in defect after laser annealing. L.C. Kimerling et al. [6] described a donor center at E (0.3-0.4eV) induced in low resitivity n-type silicon after laser treatment.

All levels are measured in laser annealed silicon although the melt depth (0.3 µm) is sufficient deeper than the amorphous thickness (0.185 µm) to remove all implantation defects and to save a perfect liquid phase epitaxy. Similar spectra, not shown here, have been obtained subsequential to a post-annealing step at 400°C for 20 min in dry nitrogen.

In the DLTS spectra of laser annealed samples post-annealed at 800°C not shown here no electron traps are plotted. Only one hole trap E (0.07eV) very close to the band edge is found.

The defects observed subsequential to ultrafast thermal treatment were annealed at 600°C. This property corresponds to the improvement of the junction characteristic by a furnace post-annealing step at this temperature. Furthermore from literature it is known that defects induced by laser treatment anneal at this temperatures.

[1] Benton, I.L., L.C.Kimerling, G.L.Miller, D.A.H.Robinson, and G.K.Celler, MRS Conf. "Laser-Solid Interactions and Laser Processing", Boston, 1978
[2] Mesli, A., J.C.Muller, and P.Siffert, European Materials Research Society Conference Proc., Strasbourg, France C5, 1983, p.281
[3] Pensl, G., M.Schulz, P.Stolz, N.M.Johnson, J.F.Gibbons, and J.Hoyt, MRS Conf. "Energy Beam-Solid Interactions and Transient Thermal Processing", Boston, 1983
[4] Adekoya, W.O., J.C.Muller, and P.Siffert, Appl.Phys.Lett. 49, 1429 (1986)
[5] Fan, Z.K., V.Q.Ho, and T.Sugano, Appl.Phys.Lett. 40, 418 (1982)
[6] Kimerling, L.C., and J.L.Benton in "Laser and Electron Beam Processing of Materials", ed. by C.W.White and P.S.Perey (Academic, New York, 1980) p.385

ORIGIN OF THE DEFECTS WITH DEEP LEVELS IN PULSED ANNEALED SILICON

V.P.Popov, S.S.Shaimeev, A.V.Dvurechenskii, L.N.Safronov[1)]

The interface compensated i-layer is observed between ion doping layer and base material after pulse liquid phase annealing /3, 4, 8/.

We confirmed these results by C-V measurements of base carrier concentration on sharp asymmetric p^+-n, n^+-p junctions or Shottki diodes (Fig. 1, 2). For example, the space charge area was greater by 0.3 μm for n-Si (with ρ =4.5 Ohm.cm) Shottki diodes after melting the base material by electron beam (E=2.0 J/cm^2, τ =0.5 μs). The depth of the partially compensated area arrived ~2 μm for n^+-p and p^+-n junction. At this time, as shown by calculation, the depth of the melting layer arrived only ~0.6 μm for this treatment /1/. A question came concerning the origin of the defects up. So long as the defects formed on initial (unimplanted) samples, it is possible, that the defects may be formed in solid phase as the result of thermal quenching. The deepest part of the defects may keep for implanted layers for depth, that is greater than the melting layer thickness.

We studied DLTS-spectra of the Si layers after pulse annealing of the virgin and implanted samples for determination of the defect formation mechanism. Carrier concentration was typically ~10^{15} cm^{-3}. A p-n structure was gotten by B^+ and As^+ ion implantation at high ($3\cdot10^{14}$-10^{16} cm^{-2}) and very high (> 10^{16} cm^{-2}) doses. Pulse annealing was carried out by electron (as stated above) and ruby laser (E=1.5 J/cm^2, τ =20 ns) beam. The lower defect concentration was observed outside the space charge area at zero bias voltage (U=0 V) in the case of greater dose of As^+ ions ((2 to 5)x10^{16} cm^{-2}) and electron beam annealing. Total concentration of the defects was less than 10^{13} cm^{-2} in the lower part of the forbidden gap at depths x > 0.7 μm. At the same time a decrease in hole concentration (Fig. 1) evidences on the existence of compensated defects in upper part of the forbidden gap near by n^+-p junction (x_{n-p}= 0.25 μm). The deep levels in upper half were observed actually by DLTS-measurements on the p^+-n diodes and Shottki diodes.

The change in DLTS-spectra in virgin n-Si with increase in pulse energy and with transition from solid phase annealing to the liquid phase epitaxy is reflected in Fig. 3. We determined the temperature and phase state of heating layers by structure measurements with RBS and electron difraction. These was gotten on the surface part of the same samples with control amorphous layer (a-Si) formed by ion implantation of 100 keV As^+ ions. The temperature of the monocrystalline surface is determined from data /1/ in points of phase transition from a-Si to polycrystalline and monocrystalline Si. A maximum defect concentration N_D~ 10^{14}

[1)] Institute of Semiconductor Physics, Academy of Sciences of the USSR, 630090 Novosibirsk, USSR

cm^{-3} is formed at energy density, corresponding to the point of melting of the monocrystalline surface (spectrum 3). This density is characterized by the maximum recrystallization velocity and maximum thermal gradients in solid state.

In the spectra we observed three main peaks A, B and C (Fig. 3) with energy levels A: E_c-0.18 eV, B: E_c-0.33 eV, C: E_c-0.42 eV (accuracy is ±0.02 eV). The B-level dominates as a rule. An average defects concentration arrives $\sim 3 \cdot 10^{13}$ cm^{-3} in 1 μm layer. The origin of the defects with B-level is connected with thermal quenching /3, 4/, but their generation is observed also at neutron /9/ and high energy electron /2,5,7/ irradiation. In our experiments on thermal quenqhing we observed the increase in this defect concentration at doping the crystal with high diffusing Ni atoms, which effectively interact with free vacancies and lower their concentration in volume. By this reasons it seems to be likely, that B-type defects content the self interstitials.

The A-level is connected with ordinary A-centre but as distinguished from /4/ we observed the defect with level E_c-0.19 eV in about the same position in spectra at time windows Δt=4 ms. The cross section of C-level corresponds to the E-centre, which was registrated repeatedly after scanning electron or laser beam annealing. We hold the opinion, that the defect with C-level is formed in the result of cooling of subsurface layer, with high vacancy concentration. The broad strip of the defect levels predominated between B and C-levels over others (spectrum 3). The appearence of this strip correlates with increase and relaxation of thermoelastic stresses beyond melting layer /4/.

At the study of the deep levels in p^+-n junction fabricated by B^+ implantation (E=35 keV, Φ =3x10^{14} to 2x10^{16} cm^{-2}) and following laser or electron treatment we observed the new D-level E_c-0.49 eV with cross section σ =(1 to 10)x10^{-16} cm^{-2} (Fig. 3, spectrum 5). The appearence of D-level is controlled by implantation conditions. If the temperature of the samples exided 50 °C at the implantation, the defects with this level were not observed. Such behavior evidences the origin of these defects as the result of the deep penetrating and keeping the part of the defects beyong melting layer. To the opinion of author /6/ the defect with the same parameters as D-centre is the self interstitial in neutral charge state, which annealed at the temperature $T_a \geq 100$ °C.

Thus, the energy density dependence of the defect concentration evidences, that the main reason of the defect origin is the relaxation of thermoelastic stresses at pulse annealing.

REFERENCES

(1) Balandin V.Yu, Dvurechenskii A.V., Aleksandrov L.N.:Phys.Stat.Sol. (a) 93(1986) K105-K109.
(2) Baranov A.I., Vasiljev A.V., Komolova N.I.: Fiz.Tekhn.Poluprov. 18 (1984) 2177-2188.
(3) Dvurechenskii A.V., Kachurin G.A., Nidaev E.V., Smirnov L.S.:Pulsed Annealing of Semiconductor Materials, Nauka, Moscow (1982).
(4) Kimerling L.C., Benton J.L. In: Aip Proc. on Laser and Electron Beam

Processing of Materials, Acad.Press., New York (1980).
(5) Krynicki J., Bourgoin J.C., Vassal G.: Rev.Phys.Appl. 14 (1979) 481–484.
(6) Lefevre H.: Appl.Phys.A 29 (1982) 105–111.
(7) Mesli A., Müller J.C., Siffert P.: J. de Physique, 44(1983) 281–284.
(8) Popov V.P., Dvurechenskii A.V., Kashnikov B.P., Popov A.I.: Phys. Stat.Sol.(a) 94(1986) 569–572.
(9) Vasiljev A.V., Kopshick I.A., Smagulova C.A., Shaimeev S.S.: Fiz. Tekhn.Poluprov. 17 (1983) 1155–1157.

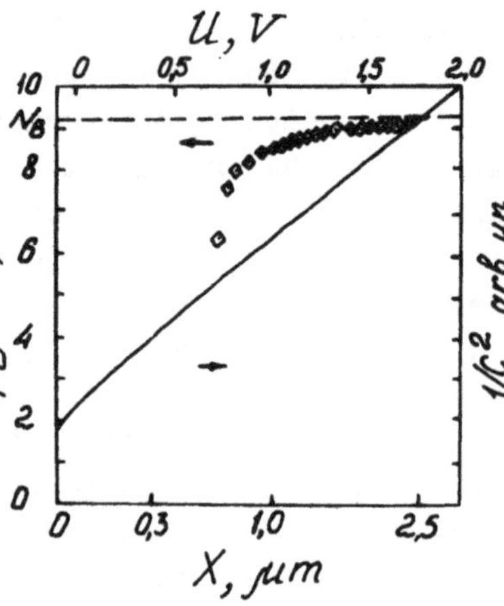

Fig. 1. Charge carriers concentration profile in p-Si (ρ = 7.5 Ohm.cm) beyond n$^+$-layer, provided by As$^+$ implantation and electron pulse annealing. Dotted line-initial concentration; solid-dependence 1/C^2 (U).

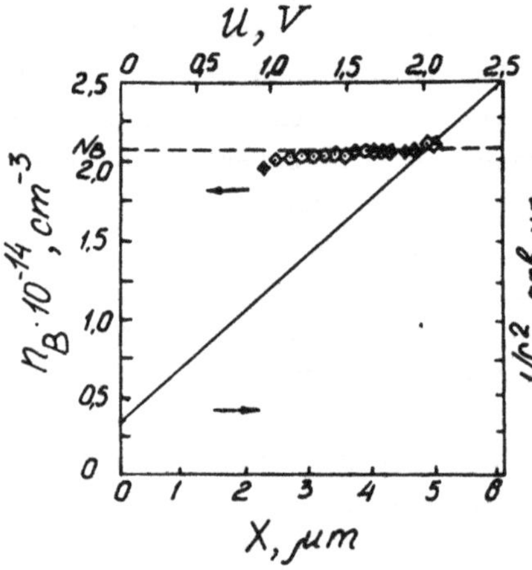

Fig. 2. As in Fig. 1, but in n-Si (ρ = 20 Ohm.cm), implanted B$^+$.

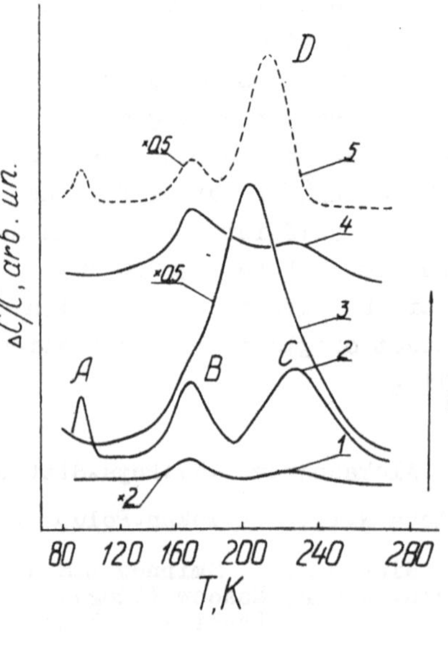

Fig. 3. A change of the DLTS-spectra of initial n-Si (ρ = 4.5 Ohm.cm) with increase of the surface temperature T_s, K: <1400 (1), ≃1400 (2), ≃1700 (3), >1700 (4); p$^+$-n junction (5).

ANNEALING OF SILICON BY INCOHERENT LIGHT PULSES OF 50 ms - 10 s DURATIONS

Ya.V.Fattachov, R.M.Bayazitov, I.B.Khaibullin, E.M.Misyuryov, M.A.Mchitarov+)

The insufficiently studied range of pulse light annealing with the duration of hundreds milliseconds, as well as seconds, was investigated. Pulse annealing of implanted and virgin silicon by flash lamps (PAFL), operated in stroboscopic regime, in air and nitrogen atmosphere, as well in vacuum, was carried. PAFL duration ranged from 30 ms to 15 s. A maximum temperature of annealing is the melting point of monocristalline silicon. Flash power depends on the duration of connected state of lamps (Δt) to circuit during one half of period alternating current. For durations $\Delta t = 2 \div 5$ ms the power is accordingly 10 - 90 kW, and the threshold of melting of silicon samples (with thickness 400 μm) reduces from 10 s to 80 ms (Fig.1). The heating rate of semiconductor samples ranges from 10^2 to 10^4 $K \cdot s^{-1}$.

Monocrystalline silicon of (100) and (111) orientation was implanted by B^+, P^+, As^+, Si^+ ions with the energies 5-80 keV and with the dose 10^{12} - 10^{17} cm^{-2}. Some samples were annealed using the Q-switched ruby laser (pulse duration is 25 $\div$ 70 ns) (NLA). Thermal annealing (TA) was carried out in a quartz furnace in a flow of dry nitrogen (T= 800 °C, t= = 30 min).

The dependence of electrical activation of implanted species and the degree of restoration of crystalline structure of silicon on PAFL power and duration was investigated (Fig.2). The character of electrical activation dependence is the same for boron implantation. For all the powers maximum values of N_s and ρ_s are the same (Fig.3).

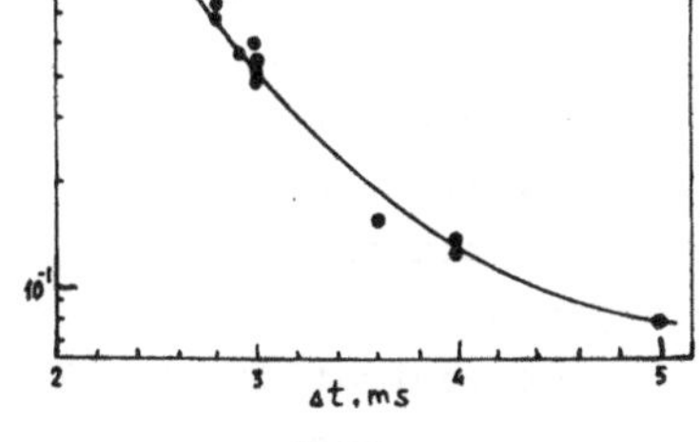

Fig.1

The distribution of concentrations and mobilities of carriers for silicon implanted by B^+, P^+, As^+ ions was investigated. Maximum concentrations of carriers for all layers are equal to equilibrium solubility limits, and the mobilities are equal to diffusion values. This testifies to the highe quality of PAFL.

A comparison of the results of three types of annealing of silicon implanted by 40 keV P^+ ions: PAFL, TA, and NLA shows that PAFL gives the higher level of electri-

+) Kazan Fysical-Technical Institute, Kazan, USSR

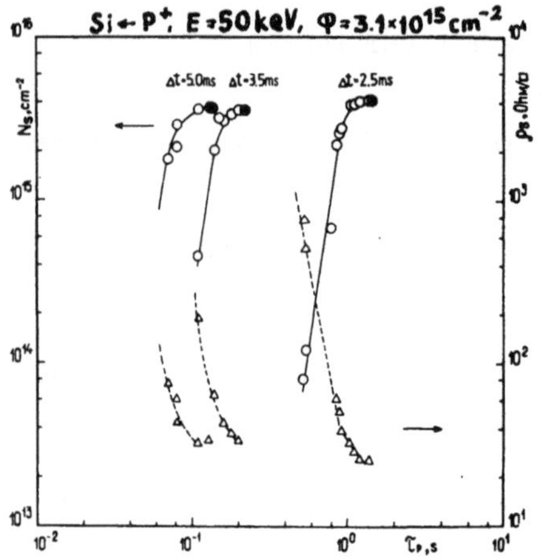

← Fig.2

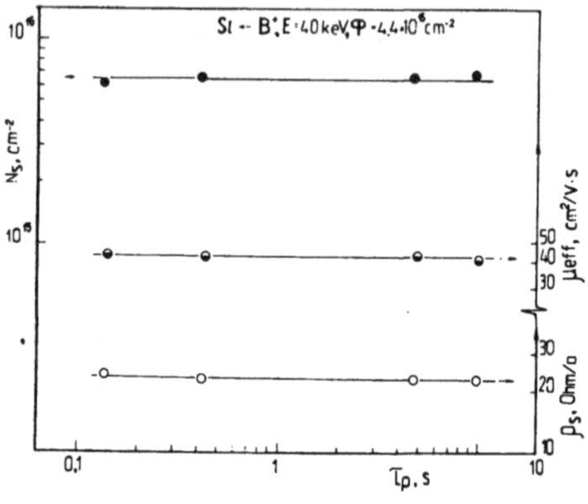

Fig. 3

cal activation than NLA in the dose range from 10^{12} to 10^{14} cm^{-2}, and higher level than TA in the whole investigated range (10^{12} - $10^{17} cm^{-2}$), Fig.4.

It is known, that during TA the degradation of life time of minority carriers takes place (20-fold decrease in our case). After PAFL only 2-fold decrease of life time in bulk material is observed. These results in a good agreement with numerically calculated ones. As we were interested in the study of volume life time of minority carriers, special chemical technology of reduction and stabilisation of surface recombination (before the measurements) was used.

Using RHEED high degree of restoration of crystalline structure after PAFL with different powers was established.

The numerically calculations of temperature fields in semiconductors during PAFL taking into account periodical character of light intensity were made. The experimentally measured threshold of silicon melting is in good agreement with the calculated values. Alternative component of light intensity does not make visible contribution to the sample temperature, because cooling between the two flashes of lamps is negligible.

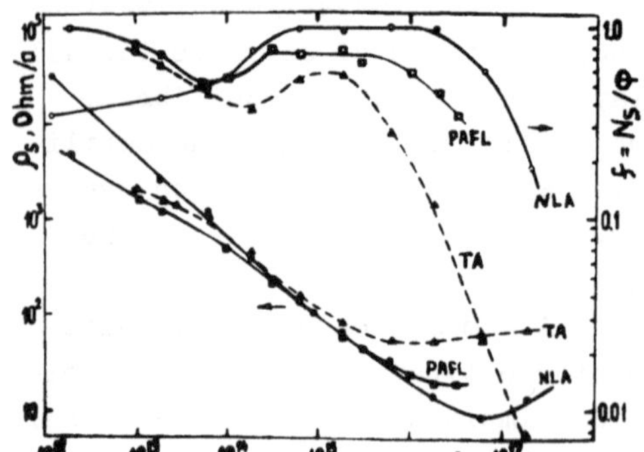

Fig. 4

The influence of PAFL regimes on the quantity and size of local molten segments (1) in silicon was investigated. With the growth of exposition at constant power (and with the growth of sizes of segments) the concentration of molten segments remains nearly constant. The higher the powers of PAFL the higher is the concentration of these regions. This means, that the generation of local molten segments is not only the results of semiconductor defects, but also the result of generation and/or reconstruction of melting nuclei by light pulse.

The exact geometrical shape of molten segments at the surface of implanted silicon gradually deteriorates with the increasing of ion dose. The crystalline structure of silicon was checked by RHEED and RBS technique (Fig.5 for 40 keV P^+ implantation). TEM-micrograf of molten and nonmolten silicon of (100) orientation is shown in Fig. 6. As it was expected, dislocations in local molten segment are absent.

We thank Dr. R.Groetschel for RBS-measurements, and Dr. A.A.Sitnikova for TEM-investigations.

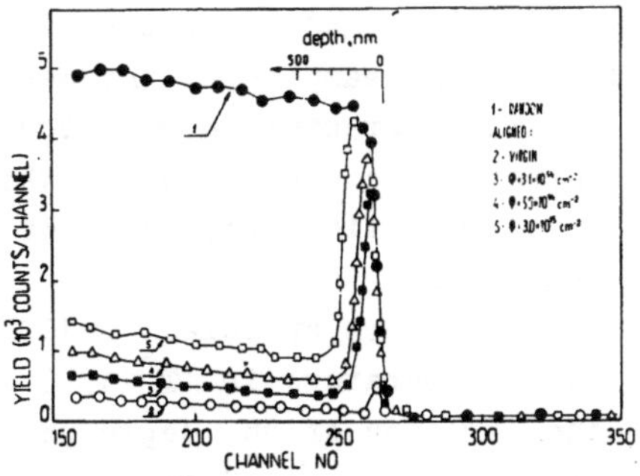

Fig. 5

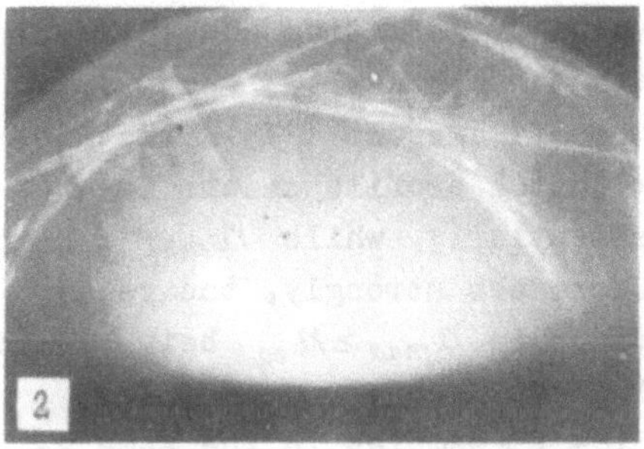

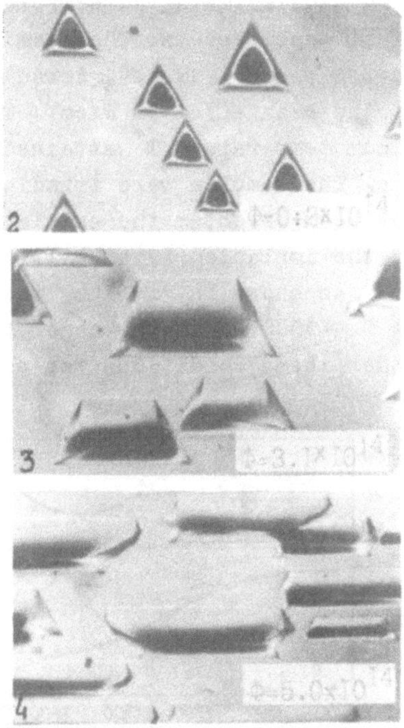

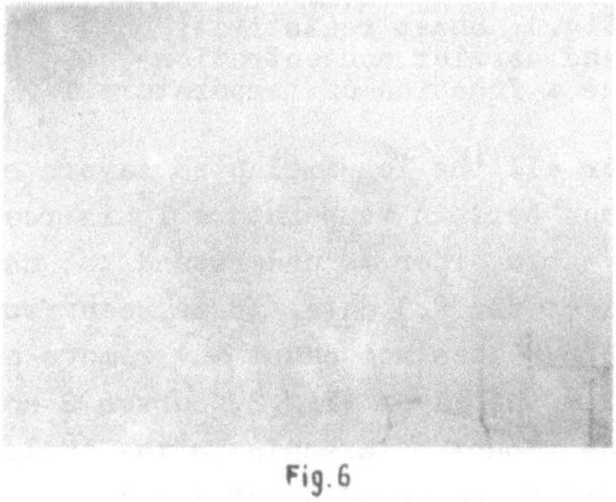

Fig.6

(1) K.-H.Heinig. in: Proc. of the Conference on Energy Pulse Modification of Semiconductors and related Materials. Dresden, GDR, ed. by K. Hennig, 1984, p. 265.

CONTINUOUS ELECTRON BEAM ANNEALING OF ION-IMPLANTED SILICON

V.S.Budishevsky[1], R.Grötzschel[2], V.A.Kagadei[1],
N.I.Lebedeva[1], D.I.Proskurovsky[1], E.B.Yankelevich[1]

Properties of silicon layers implanted with As^+ (100 keV, 5×10^{15} cm^{-2}), P^+ (40 keV, 1×10^{16} cm^{-2}), and Sb^+ ions (60 keV, $2,2 \times 10^{15}$ cm^{-2}) and annealed by means of an continuous electron beam [1] were studied using Rutherford backscattering plus channeling and electrical measurements. The beam parameters were as follows : the electron energy was 5 ÷ 10 keV, irradiation time - 10 ÷ 20 s, and power density of the beam - 7 ÷ 30 W/cm^2. The heating time of the samples at the maximum temperature T^* attained during the electron irradiation was about 5 s. The samples were irradiated from the side of the substrate, so that the beam affected the sample only thermally. Irradiation from the side of the implanted layer will be discussed separately.

As shown in Fig.1(b) the activation of Sb is most efficient (~100%) at $T^* \simeq 1000$ K. In the case of Si <As^+> this temperature is $T^* \simeq 1375$ K; under these conditions the peak carrier concentration of the depth profile ($n_{max} \simeq 7 \times 10^{20}$ cm^{-3}) exceeds the maximum equilibrium value of As active concentration ($n_{eq} \simeq 3,2 \times 10^{20}$ cm^{-3} [2]). When T^* exceeds 1375 K the distribution profile widens inessentially, while n_{max} decreases strongly, the relationship $n_{max} \gtrsim n_{eq}$ being valid. The dependence $n_s(T^*)$ for Si <P^+> samples is the same as for Si <As^+> samples with the maximum at $T^* \simeq 1400$ K. Thus, for all the ion-implanted layers studied the reverse annealing occurs when certain temperature T^* is reached.

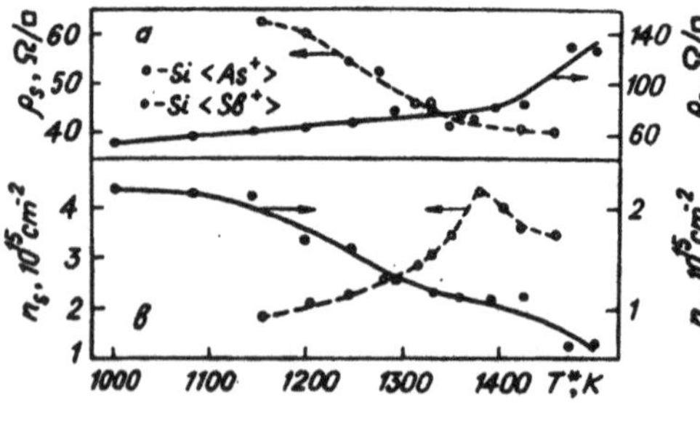

Fig.1. Sheet resistivity ρ_s (a) and carrier concentration n_s (b) as a function of temperature T^*.

In order to understand the nature of reverse annealing let us consider the RBS data. As is seen from this data the sample heating below 1375 K does not cause the remove of any quantity of implanted arsenic from the layer (Fig.2, curves 2 and 3). At temperatures $T^* > 1375$ K the amount of arsenic in the implanted layer decreases. Thus, at T^* = 1450 K (curve 4) about 1×10^{15} cm^{-2} of As atoms, i.e. ~20 % of the implanted dose, sublimate. From the fact that $n_{max} \gtrsim n_{eq}$ it follows that

1) Institute of High Current Electronics of the USSR Academy of Sciences, Siberian Division, Tomsk
2) Central Institute of Nuclear Research, Academy of Sciences of the GDR, Rossendorf

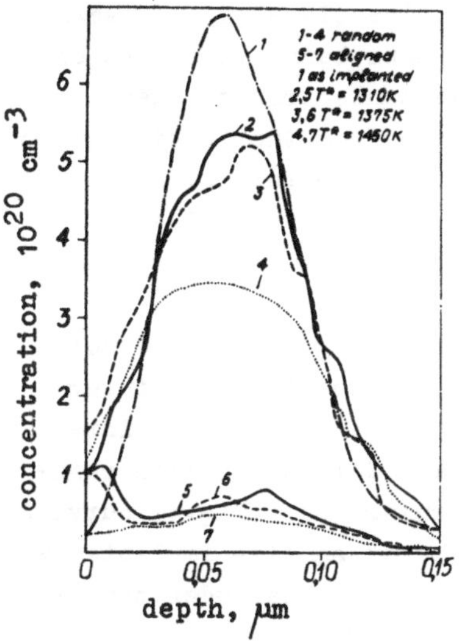

Fig.2. Depth distribution of As in Si after electron beam annealing.

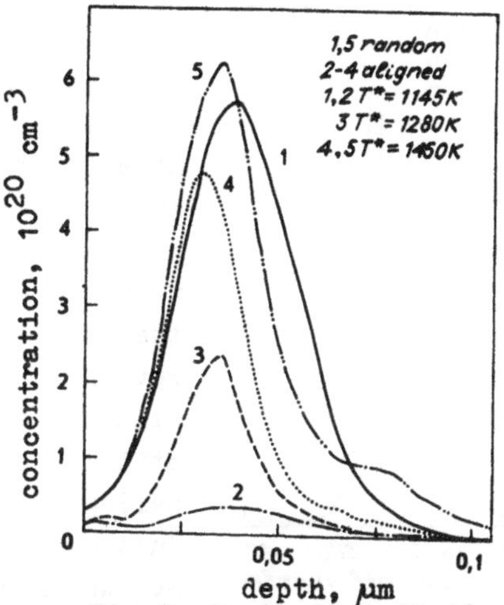

Fig.3. Depth distribution of Sb in Si after electron beam annealing. Depth profiles for as implanted samples (random and aligned) coincide with curve 1.

only those As atoms sublimate which are in excess with respect to the equilibrium active concentration. The fraction of As atoms in the interstices is practically independent of T^*, while for Si $<Sb^+>$ samples the amount of Sb in interstices continuously increases with the temperature increase (Fig.3). The sublimation of Sb was not observed within the whole temperature range investigated.

The RBS spectra presented in Fig.4 were measured for two samples (III) - Si $<As^+>$ which were irradiated at $T^* \simeq 1400$ K simultaneously, but from different sides. When the sample was irradiated from the implanted layer side, no appreciable sublimation of arsenic proceeded (cf. curves 1 and 3) and the number of As atoms in the $<111>$ channel increased essentially (cf. curves 2 and 4). Nevertheless, the electrophysical parameters of the layers (ρ_s and n_s) as well as the fraction of arsenic embedded into the regular lattice sites are practically the same for both samples. The explanation for the weaker As sublimation off a sample irradiated from the layer side is not clear at present. This, however, does not prevent us from drawing the principal conclusion from the latter experiment : only those As atoms sublimate that have displaced into interstices during the annealing process.

As the estimates show, at T^* = 1450 K about 70 % of the implanted As^+ ions are in lattice sites. About 68 % of the implanted ions are electrically active (see Fig.1), that means that practically all the ions remained in substitutional sites are electrically active. The value $\chi_{min}(Si)$ = 0,031 indicates that crystalline structure of the layer is perfect. All the above said prevents one from relating the observed deactivation of As^+ ions at $T^* > 1375$ K to the formation of any non-active defect-impurity complexes when the impurity remains

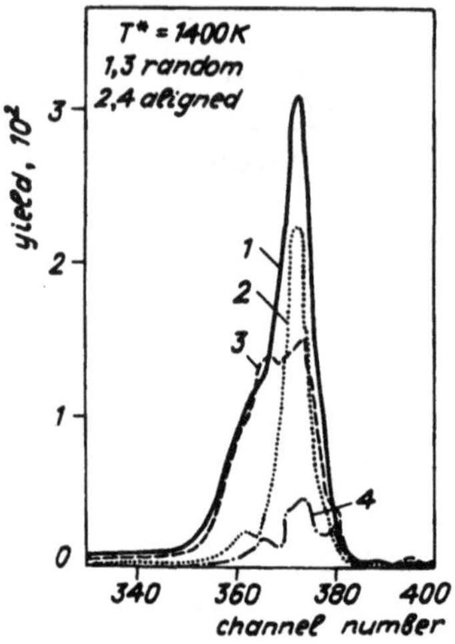

Fig.4. RBS spectra of Si <As>.
1,2 - irradiation from the side of the implanted layer.
3,4 - irradiation from the side of substrate.

in the regular lattice sites [2]. It seems to be so that As^+ ions being in excess of equilibrium active concentration displace into the interstices from the lattice sites at sufficiently high temperature. As is shown above, just this portion of arsenic can sublimate when the sample is irradiated from the side of the substrate. This is the reason why channeling spectra do not reveal As atoms displaced into the interstices. If sublimation is prevented interstitial As atoms are detectable with channeling without angular scan (Fig.4, curve 2), just as in the case of samples implanted with Sb^+ which does not sublimate under conditions of our experiment.

Thus, one can state that the reverse annealing observed in our experiments is caused by deactivation of P^+, As^+ and Sb^+ ions in silicon occuring due to the displacement of impurity ions from the lattice sites into the interstices at high temperature heating during a few second after the solid-phase recrystallization of an ion-implanted layer.

In conclusion it should be noted that, according to the data of several works (for example [3]), heat treatment of samples during a few seconds (at $T \simeq 1400$ K) after liquid-phase epitaxial recrystallization at nanosecond electron annealing leads to As deactivation via formation of coherent and semicoherent precipitates. This difference from our results seems to be due to the fact that solid-phase epitaxial recrystallization results in the formation of a layer qualitatively different from that being formed at liquid-phase epitaxial recrystallization. Therefore even identical subsequent heat treatment may cause different processes in this kind of layers.

REFERENCES

[1] Yu.E.Kreindel, N.I.Lebedeva, V.Ya.Martens, G.A.Mesyats and D.I.Proskurovsky : Pis'ma v Zhurn. Tekn. Fiz. 8, (1982) 1465-1468

[2] A.Lietoila, J.F.Gibbons, T.W.Sigmon : Appl. Phys. Lett. 36(9) (1980) 765-768

[3] V.P.Popov, A.V.Dvurechenskii, B.P.Kashnikov and A.I.Popov : Phys. Stat. Sol. (a) 94 (1986) 569-572

IN SITU ANNEALING OF ION-IMPLANTED SILICON IN A HIGH-VOLTAGE ELECTRON MICROSCOPE (HVEM)

D. Hoehl[1)]

Introduction

A high-temperature heating stage, specially constructed for our HVEM /1/, enables also in situ annealing experiments to be carried out. These in situ experiments permit the direct observation of the annealing of the implantation damage. According as the implantation parameters are, we observe recrystallization processes, the formation of dislocations and their interaction.

Experimental

Using this annealing method implies some special experimental problems. Before the annealing results are presented, these problems will be pointed out and a short description will be given of the heating stage. Fig. 1 shows the schematic cross-section view of the heating inset for the top-entry 20°x-y-goniometer stage.

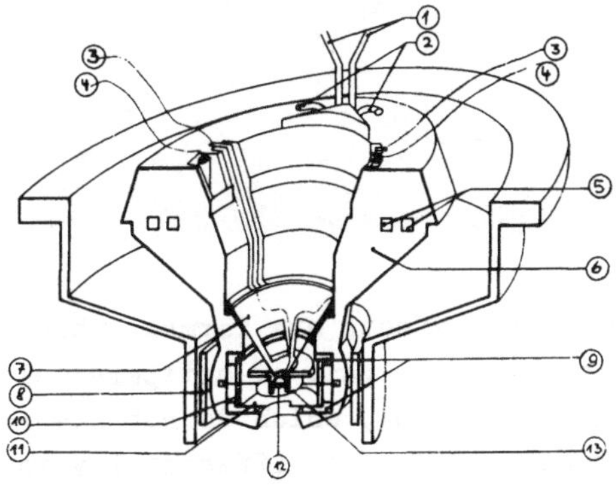

Fig. 1 (1) - tilting drive fork, (2) - inlet/outlet pipes for cooling-water, (3) - contacts for the thermocouple, (4) - connecting wires to power supply, (5) - cooling-water channel, (6) - tilting cone, (7) - specimen holder with thermocouple, (8) - gimbal of cardanic suspension (with two pairs of agate balls), (9) - ceramics, (10) - Mo cylinder fixing the ring cathode (13),(11) - potential aperture, (12) - specimen, (13) - ring cathode.

Because of a nonreproducible temperature difference between the specimen and the specimen holder, the specimen temperature is measured directly by a pyrometer.

For the annealing experiments very thin specimens are used, as they have to be transparent to the imaging electrons. Thus, to avoid an adulteration of the annealing behaviour the substrate below the implanted layer should remain sufficiently thick and undamaged (the investigation of these necessarily thick specimens can be performed only in the HVEM).

In Si the imaging electron beam of the HVEM creates radiation damage. Under normal investigation conditions, thus to avoid an influence on the annealing behaviour, during heating the electron beam is switched off. (But it turned out that at an annealing temperature $T > 1000°C$, also under illumination, no adulteration is visible).

1) Institute of Solid State Physics and Electron Microscopy, Academy of Sciences of the GDR, Weinberg 2, Halle, DDR-4050

Results

Fig. 2 shows the formation of annealing defects and their interaction during the heat treatment. From cross-section investigations /2/ we know that in the as-implanted state a damaged layer exists (500... 600 nm thick, up to the surface), consisting of small amorphous clusters in the crystalline matrix - Fig. 2a.

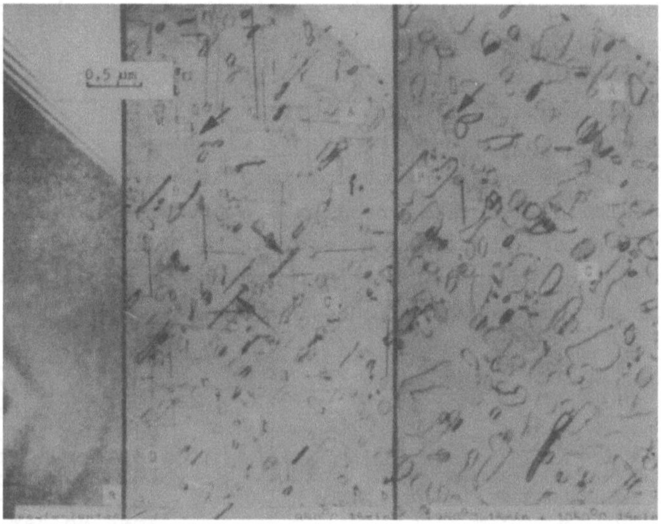

Fig. 2

In situ annealing of ion-implanted Si(100) - B^+, 150 keV, $10^{15} cm^{-2}$ - in a high-voltage electron microscope

After annealing at 950°C for 15 min rod-like defects are formed, which begin to transform into dislocation dipoles - see pair of antiparallel arrows and (B) in Fig. 2b. Crossing dipoles (at D), which do not influence each other, prove that the defects lie in two different layers. An additional heat treatment at 1050°C for 15 min - Fig. 2c - shows that the loops and dipoles have grown (A, C). If neighbouring dipols grow, they can touch one another, forming big loops, by annihilation of dislocation segments (B). Because of the small thermal mass of the heating stage also rapid-thermal-annealing-like (RTA) conditions /3/ can be realized. Fig. 3 shows a typical heating and cooling profile.

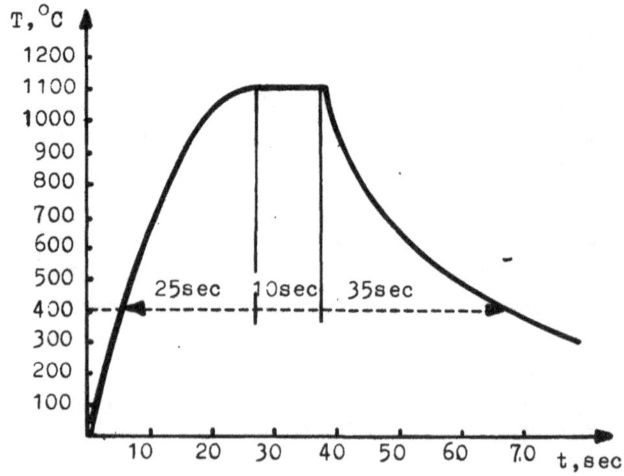

Fig. 3

time vs. temperature profile for in situ annealing under RTA-like conditions

In Fig. 4 a conventional annealing sequence is compared with such a RTA-like annealing (using a temperature/time profile as in Fig. 3). Meandering dislocation lines, forming large closed shear loops /4/ result from annealing at 900°C for 15 min - Fig. 4a (dashed line marks the same loop configuration).

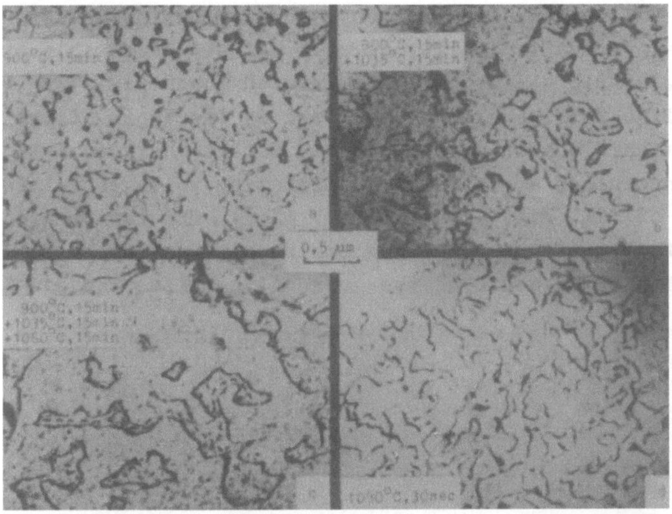

Fig. 4
In situ annealing of ion-implanted Si(100)
- BF_2, 50 keV, $5x10^{15}cm^{-2}$ -
in a high-voltage electron microscope

Further annealing - Figs. 4b, c - causes the dislocation segments of the shear loops to glide. Neighbouring dislocation segments of opposite sign attract each other and will be annihilated, resulting in a decrease in the dislocation density. Fig. 4d shows that a RTA-like annealing at 1090°C for 30 sec is comparable to the annealing state at 900°C for 15 min (Fig. 4a).

The results evidence that this annealing method is an additional tool -under the special experimental conditions- to investigate the annealing mechanism of ion-implanted Si. It enables the direct observation of the annealing process.

References

/1/ Hoehl, D., Kästner, G.: Proc. 8th Europ. Congr. on Electron Microscopy, Budapest 1984, p. 469

/2/ Bartsch, H., Hoehl, D., Werner, P.: see do. Congr. (10.12)

/3/ Wilson, S.R.: Semiconductor Silicon 1986, Pennington N.J. The Electrochem. Soc. 1986

/4/ Mader, S., Michel, A.E.: phys. stat. sol. (a) 33, 793 (1976)

CW LASER-BEAM INDUCED MICRODEFECTS IN SILICON DETECTED BY SEM AND TEM

H. Johansen[1], H. Bartsch[1], and B. Lämmel[2]

It is well-known that cw laser beam annealing of ion-implanted semiconductors results in a material which indeed is free from extended defects and which shows nearly 100 % electrical activation of the incorporated dopant atoms. But such layers frequently contain point defect complexes leading to significant minority carrier effects detectable by SEM-EBIC (Sheng et al.) (EBIC - Electron Beam Induced Conductivity). Depending on the specific foreign atom species and on the depth of the spatial distribution within the substrate one can detect characteristic sharp EBIC contrast inversions at selected accelerating voltages V_c of the primary electrons of SEM, Fig. 1 (V_c = 15 kV - b)). The latter are distinguished by the fact that the annealing stripe contrast appears "embedded" in the as implanted state of the surrounding, i.e. the dif-

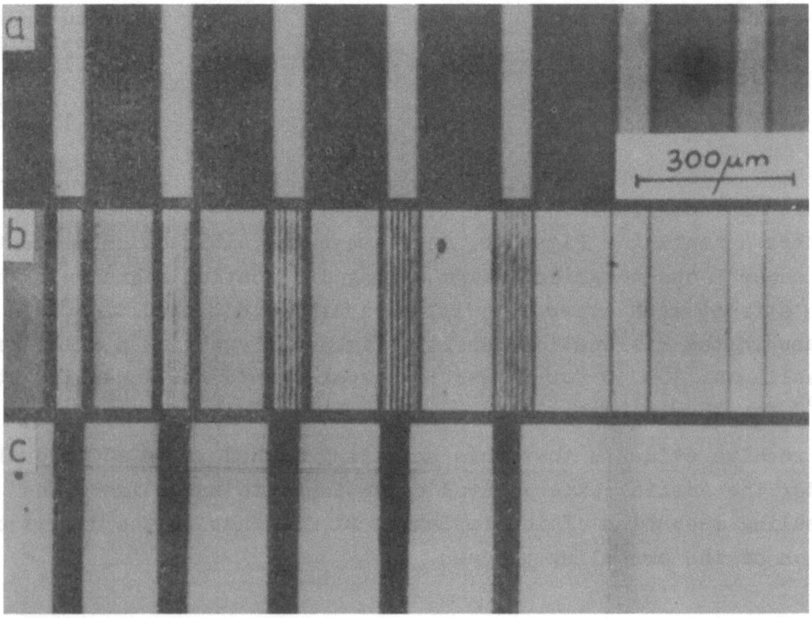

Fig. 1 EBIC micrograph series of cw argon ion laser beam annealing in stripes for different accelerating voltages, boron implantation $D_i=2x10^{15}$ B^+/cm^2, E_i=60 kV; T_i=LNT a) 9 kV, b) 15 kV, c) 20 kV P_{cw}=3.2 W (on the left) to 2.0 W (on the right), ΔP_{cw}=0.2 W.

ference between EBIC signals of the p^+n junction and the amorphous regions vanishes independent of the annealing power P_{cw}. For $V > V_c$, the strong activity of recombination sites originates in pronounced dark EBIC contrast features (Fig. 1c, V = 20 kV; not valid for the small values of P_{cw} = 2.2 W and 2.0 W on the right).

[1]AdW der DDR, Institut für Festkörperphysik und Elektronenmikroskopie Halle (Saale)

[2]Ingenieurhochschule Mittweida

Starting from such fundamental EBIC surface contrast studies, after appropriate bevel preparation one finds that the ratio of the depth of laser-induced quenched-in defects of $x_d = 6.5\,\mu m$ and the produced junction depth of $x_j = 0.3\,\mu m$ by cw laser beam annealing at $P_{cw} = 3.2$ W amounts to $R > 20$, Fig. 2. There are large volumes of information in the literature about thermally induced quenched-in defects in silicon (i.e. using ESR (Lee et al.) or DLTS (Gerson et al.)), but now our EBIC bevel face investigations of such matérials for the first time yield a direct imaging method to measure the real spatial extent of the quenched-in

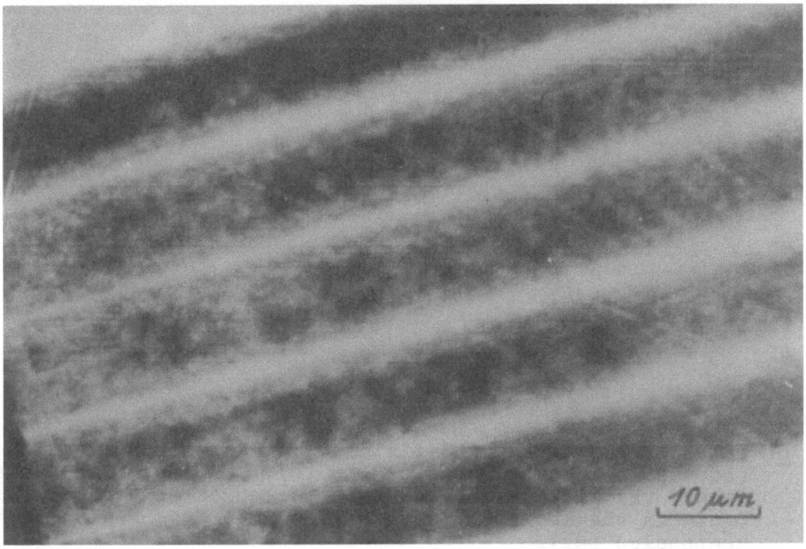

Fig. 2 Evidence of the depth distribution of cw argon ion-laser induced quenched-in defects in boron-implanted silicon, Schottky-barrier bevel face imaging by EBIC, $E_p = 9$ kV; $P_{cw} = 3.2$ W; $t_{Au} = 15$ nm; $\alpha = 0.3^o$

defect distribution into the depth of the substrate below the p^+n junction. The measurement was possible by carefully optimizing the respective Schottky barrier onto the bevel face by a gold evaporation layer, yielding an optimum value of 15 nm thickness on n-type silicon. For the same aim one can use electron-beam induced p^+ inversion layers on bevel faces having sufficiently high charge collection efficiency together with the underlying n-type substrate (Johansen et al.).
The TEM micrograph of the laser-irradiated stripes shows an excellent crystallinity of the recrystallized stripe (diffraction pattern, on the left in Fig. 3), whereas the electron diffraction pattern taken of regions out of the stripe shows amorphous rings in addition to the substrate reflexes (on the right in Fig. 3). During annealing dislocation loops have formed lying in a buried layer ranging from 90 to 250 nm below the surface (determined by cross-sectional TEM) corresponding to the depth range of the initially buried amorphous layer after implantation.

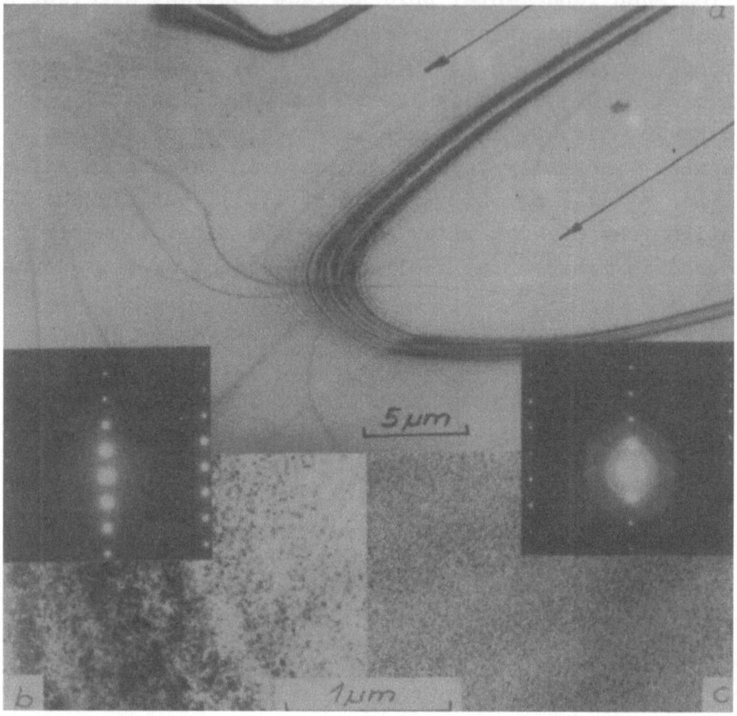

Fig. 3 TEM micrographs (bright-field) of a laser-beam annealed stripe (boundaries marked by arrows in (a)) and its amorphous surroundings. (b) dislocation loops in the annealed stripe, (c) defects in the amorphous/crystalline interface region of the unannealed part.

Heat transition into the surroundings causes the recrystallization to start also in the interface region of the neighborhood of the stripe. In this interface region defects are created, which are observed in more extended contrast than in the as-implanted state (Fig. 3c).

References

(1) Sheng, N.H., M. Mizuta and J.L. Merz: J. Appl. Phys. 55 (1984) 3072 - 3082

(2) Lee, Y.H., R.L. Kleinhenz and J.W. Corbett: Appl. Phys. Lett. 31 (1977) 142

(3) Gerson, J.D., L.J. Cheng and J.W. Corbett, J. Appl. Phys. 48 (1977) 4821

(4) Johansen, H., W. Rechner and J. Heydenreich, Scanning (1988) in press

RAPID THERMAL ANNEALING AND PROPERTIES OF B AND P IMPLANTED SILICON

J. Hlávka[1], J. Grňo[2], J. Dragula[3], M. Líbezný[1]

Rapid thermal annealing (RTA) is widely used for crystal defect annealing and adjusting the electrical activity of implanted layers in semiconductors. The connection of electrical activity of implanted layers with crystal defects is well established. Excess carrier recombination is considered as a very sensitive tool for defect detection because crystal defects act usually as recombination centres. That was the reason why the correlation of electrical activity of annealed layers with excess carrier recombination in these layers was investigated. We expected that such a study gives detail information on annealing mechanism of RTA. Another method for crystal defect detection is RBS method but the sensitivity to defect concentration is low. So, the much more sensitive excess carrier recombination method was applied. The second aim of the study was to compare results of RTA with TA.

It was found that an implanted layer contains implanted impurities and crystal defects which are responsible for a space charge region (1). Detail information on the defect distribution in implanted layers can be obtained in such a way that thin layers of the implanted region are removed step by step (e.g. by anodic oxidation stripping technique) and recombination properties of the sample after each step are measured. In such a way, so called recombination profile is obtained from which the total depth of damaged region and the recombination centre distribution can be appreciated (2). Excess carrier recombination was measured by the use of the Photomagnetoelectric effect method (3) which gives values proportional to the effective lifetime in a measured sample. The measured sample consists of an implanted region and the substrate. The thickness of the substrate is much greater than the thickness of the implanted layer, and, therefore, the implanted region can be con-

---

[1] J.E. Purkyně University, Brno, [2] Slovak Technical University, Bratislava, [3] Tesla Piešťany, Piešťany, Czechoslovakia.

Table 1

B, energy 80 keV, dose 1. 10^{14} cm^{-2}, N- Si

R_p= 0,32 μm, R_p= 0,08 μm

$T/^{o}C$	t/s	τ/τ_{st}	τ_b/τ_{bst}	d_1/μm	d_2/μm	R_s/ohm□	$\mu/cm^2V^{-1}s^{-1}$
0	0	0,05	1	0,3	1,0	-	-
1000	1800	1	1,5	0,5	0,6	-	-
1050	5	0,4	0,5	0,4	1,0	700	90
1050	10	0,6	0,5	0,5	0,9	600	94
1050	15	0,6	0,4	0,5	0,7	540	90
1050	20	0,7	0,4	0,5	0,7	640	85
1050	25	0,7	0,4	0,5	0,7	580	85
1015	10	0,4	0,5	0,4	0,9	650	88
1050	10	0,6	0,5	0,5	0,9	530	92
1095	10	0,5	0,4	0,5	0,7	500	95
1135	10	0,7	0,4	0,5	0,6	500	95

Table 2

P, energy 120 keV, dose 1.10^{14} cm^{-2}, P-Si

R_p= 0,15 μm, R_p= 0,04 μm

$T/^{o}C$	t/s	τ/τ_{st}	τ_b/τ_{bst}	d_1/μm	d_2/μm	R_s/ohm□	$\mu/cm^2V^{-1}s^{-1}$
0	0	0,07	1 st	0,6	1,0	-	-
1005	5	0,7	0,5	0,5	0,7	364	222
1005	15	0,8	0,5	0,5	0,7	368	240
1005	25	0,9	0,4	0,5	0,6	370	220
800	10	0,5	0,5	0,6	0,8	1156	370
895	10	0,7	0,4	-	0,7	430	240
950	10	0,7	0,5	0,6	0,7	369	226
1005	10	0,8	0,4	0,5	0,6	368	240
1055	10	1 st	0,4	0,5	0,6	365	210

sidered as a surface recombination approximately.

Silicon samples implanted with B and P ions were annealed at various temperatures and for various time intervals as indicated in the tables. Samples were annealed in vacuum (2.10^{-2} Pa) and the annealing apparatus exploits radiant flux of halogen lamps. In the experiment, sheet resistance, surface mobility, and effective lifetime were measured. Sheet resistance was measured by the van der Pauw method. Surface mobility was evaluated on the basis of the Hall effect measurement. Effective lifetime was measured by the use of the Photomagnetoelectric effect method.

Experimental results of temperature and time dependences of properties of annealed samples are summarized in the Tables 1.,2. The following quantities are presented in the Tables: relative effective lifetime (the ratio of effective lifetime of a given sample to the effective lifetime of a standard sample as indicated in the table by "st"), the thickness of the layer in which the majority of recombination centres is distributed d_1, the total thickness of the damaged region d_2, sheet resistance, and surface mobility.

On the basis of our results the following conclusions were made. The set of measurements of P implanted n-type Si samples (150 keV, 1.10^{14} cm^{-2}) could not be included in the present paper due to its limited extent.

The recombination in annealed (RTA) implanted layers decreases as the annealing temperature, or time, increasis. Recombination in layers annealed by RTA is slightly higher than the one in thermally annealed layers. The recombination of the substrate is increased due to RTA. The total thickness of the damaged region decreases as the annealing temperature, or time, increases. There is no substantial difference between recombination in layers annealed by RTA and TA. Recombination measurements do not provide detail information on annealing mechanism as was expected.

1 J. Hlávka and K. Kouba, Radiat. Eff. 40, 193 (1979).

2 J. Hlávka, phys. stat. sol. a 94, 353 (1986).

3 J. Hlávka, Rev. Sci. Instrum. 54, 1386 (1983).

THE EFFECT OF OXYGEN ON ELECTRICAL ACTIVATION AND DIFFUSION OF As IN Si BY R T A

R. Koegler[1] and E. Wieser[1]

High dose (~ 7×10^{15} cm^{-2}) As implantation into p-type Si is an usual method to produse n^+-areas in VLSI device processing.

As implantation often is carried out through an oxide cap. But an implantation of 1×10^{16} cm^{-2} through a cap of only 30 nm reduces the sheet carrier concentration by about 40 % after RTA (T_{max} = 1000 °C, t_h = 1s) in comparision to implantation in bare Si.

This is mainly caused by recoil implantation of oxygen into the silicon substrate and the subsequent formation of As-O or As-Si-O clusters /2, 5/.

We have investigated this effect by implantation of arsenic and oxygen in bare silicon. Beside the electrical activation also the influence of oxygen on the diffusion behaviour of arsenic is considered.

The samples investigated consisted of (100) Si wafers (5-10 Ωcm) implanted with 100 keV As ions to doses of 1×10^{15} cm^{-2} and 1×10^{16} cm^{-2}. Some of the wafers were implanted through a 30 nm thermal grown SiO_2 cap. Another part of the samples was implanted additional with 30 keV O ions to doses of 7×10^{15} cm^{-2} and 1.5×10^{16} cm^{-2}.

Rapid thermal annealing using a halogen lamp equipment was carried out in air. The RTA cycle was characterized by T_{max} = 900 to 1000 °C and dwell times at T_{max} between 1 s and 30 s.

All temperature time curves were recorded by a pyrometer. In order to avoid differences of temperature the different kinds of samples were annealed together.

The analytical techniques employed include use of RBS, SIMS and sheet carrier concentration measurements using the Van der Pauw method.

Figs. 1 and 2 As depth profiles after the same RTA treatment (T_{max} = 1000°C, t_h = 25 s) for implantation in bare silicon and through a 30 nm silicon dioxide cap, respectively. Experimential As depth distributions were compared with computer simulations using a theoretical model for diffusion and electrical activation of As in Si proposed by Tsai et al. /6/. The model takes into account the formation of electrically inactive, non mobile arsenic clusters at arsenic concentrations above the solubility limit.

Transient enhancement of diffusion during RTA is taken into consideration by the use of the diffusion coefficient given by Hoyt and Gibbons.

$$D = D_o + D_- (n/n_i) + D_= (n/n_i)^2 \quad /1/$$

---

1 - Central Institute of Nuclear Research, Rossendorf, GDR
Academy of Sciences of GDR, PF 19, DDR-8051 Dresden

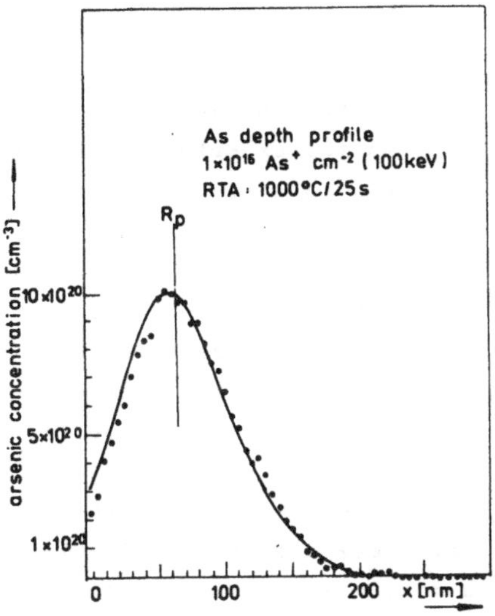

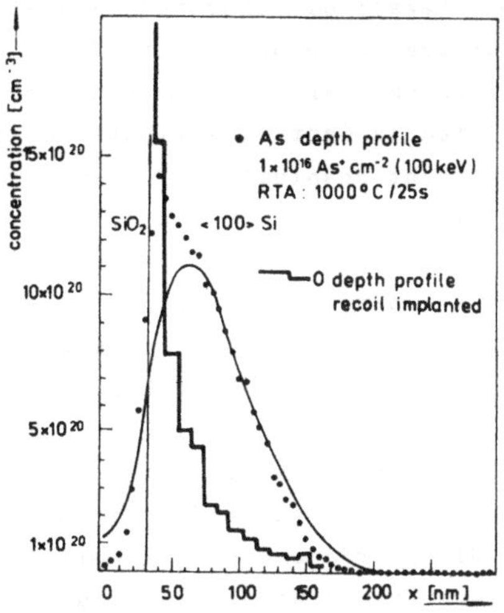

Fig. 1: As depth profile (RBS) for implantation in bare Si, result of simulation (see text) included (line)

Fig. 2: As depth profile (RBS) for implantation through SiO_2 cap, TRIM calculated O recoil implanted profile after /4/

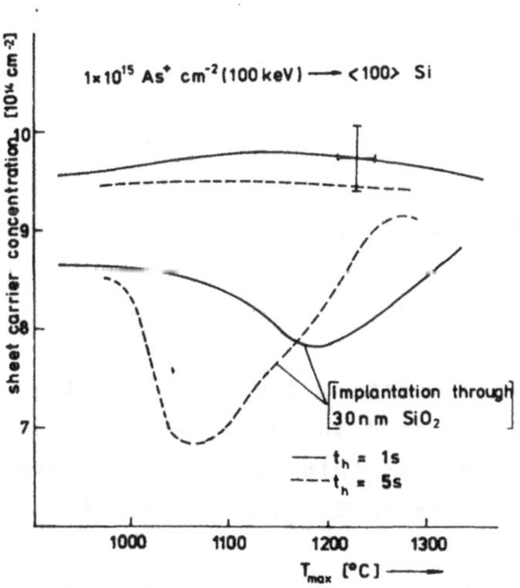

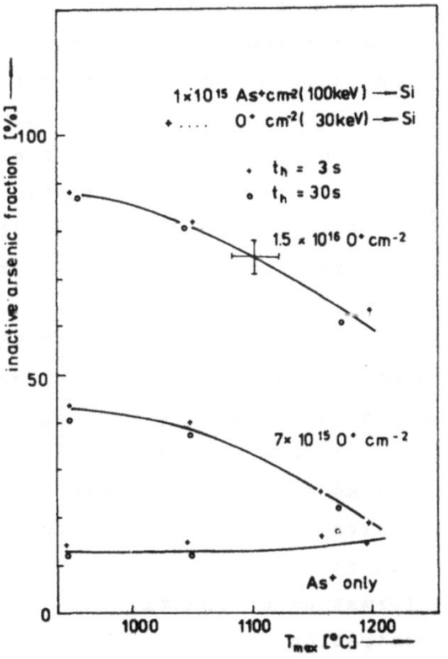

Fig. 3: Sheet carrier concentration in dependence on RTA conditions

Fig. 4: Inactive As fraction in dependence on RTA conditions

This expression corresponds to As diffusion with neutral, single and double negatively charched vacancies.

Surface and SiO_2-Si interface, respectively, are modeled to act as a diffusion barrier because no arsenic lost has been observed. The simulated As depth distributions are faulted with the experimental depth measurement resolution of 20 nm. The model is able to account for the electrical activation and the As depth profile in the case of implantation of arsenic in bare silicon (see Fig. 1). But the model fails in the case of implantation through the SiO_2 cap. The measured As concentration at the SiO_2-Si interface is always higher than the model predicts. A comparison of samples with and without cap shows also shallower penetration of As in samples with cap. Moreover, the electrical activation is smaller in this case.

The cause of the effect is the high oxygen concentration near the SiO_2-Si interface due to recoil implantation. There are As-O or As-Si-O clusters formed which are assumed to be non mobile and electrically inactive /3/.

However the clustering or precipitation of As at concentrations above the solubility limit has the same effect (decreased activation and diminished diffusivity). Therefore, we investigated also concentrations of As well below the solubility limit (1×10^{15} As cm^{-2}).

The sheet carrier concentration for 1×10^{15} As cm^{-2} is shown in Fig.3 in dependence on the RTA conditions.

For the samples with implantation in bare silicon there is nearly full electrical activation reached. But the samples implanted through a SiO_2 cap show always a lower electrical activation and a revers annealing behaviour. This behaviour can be explained by the diffusion of As from the region of highest concentration around R_p to the region near the SiO_2-Si interface where the concentration of recoiled oxygen is very high. Because of the formation of As-O or As-Si-O clusters the arsenic is trapped there.

At temperatures above 1100 °C these clusters are partly resolved and the sheet carrier concentration increases again.

The discussed effects are more clearly demonstrated considering the samples where the full As depth profile is placed in an oxygen rich region caused by an additional oxygen implantation. For this case Fig. 4 shows the inactive fraction of arsenic in dependence on T_{max}. From this figure we conclude:

- the formation of inactive As-O clusters is proportional to the oxygen concentration,
- the As-O clusters need temperatures above 1100 °C to resolve,
- the decay of the As-O clusters is independent on t_h in the time scale of RTA.

The As depth profiles after RTA with and without an additional O implantation give informations about the mobility of As-O clusters. The SIMS analysis does not resolve any difference between the As depth profiles before and after a RTA treatment of T_{max} = 1050 °C, t_h = 20 s in the case of O implantation (1.5×10^{16} cm^{-2}). The activation of arsenic is 20 % in this case.

After an additional RTA treatment (T_{max} = 1075 °C, t_h = 15 s) the activation is increased to 35 % and the As depth profile is now moved by about 7 nm, but, in the case without O implantation the diffusion shifts the profile by about 30 nm (see Fig. 5). This leads to the conclusion that the As-O clusters are less or non mobile.

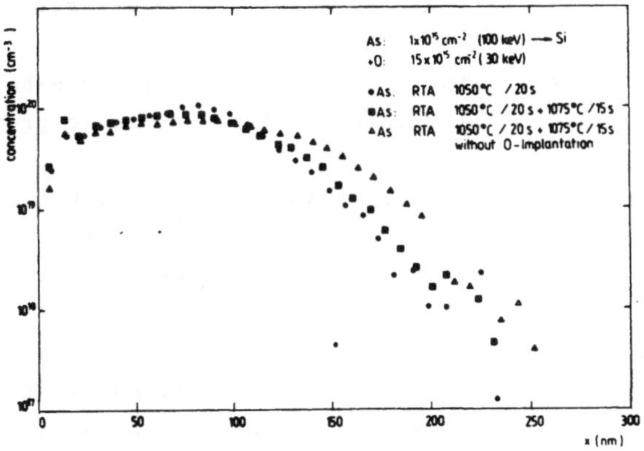

Fig. 5: SIMS measured As depth profiles

References

/1/ Hoyt, J.L. and J.F. Gibbons
Mat.Res.Soc.Symp.Proc.Vol. 52(1986)p 15

/2/ Izumi, T., T. Matsumori, T. Hirao, Y. Yaegashi, G. Huse and K. Inoue
Nucl.Instr. and Meth. 182/183 (1981) 483

/3/ Koegler, R., E. Wieser, M. Voelskow and G. Otto
Nucl.Instr. and Meth. B 19/20(1987) 516

/4/ Posselt, M. private communication

/5/ Sadana, D.K., N.R. Wu, J. Washburn, M. Current, A. Morgan, D. Reed and M. Maenpaa,
Nucl.Instr. and Meth. 209/210 (1983) 743

/6/ Tsai, M.Y., F.F. Morehead, J.E.E. Baglin and A.E. Michel
J. Appl. Phys. 51(1980) 3230

INFLUENCE OF ELECTRON-BEAM AND HALOGEN LAMP ANNEALING ON THE ELECTRICAL PROPERTIES OF MOS STRUCTURES

N. Sieber[1], D. Proskurovsky[2]

Over the last several years, short time annealing has been developed as an alternative to conventional furnace annealing. Most investigations were made using heating systems based on radiation sources such as tungsten halogen lamps with pulse duration of about 10s. An electron-beam heater can also be used for thermal treatment of large surfaces in vacuum /1/.

The present paper is concerned with the influence of electron-beam (EBA) as well as halogen lamp annealing (HLA) on the electronic properties of the Si/SiO_2 interface.

The used material was n-type, (100) oriented Si. SiO_2 (thickness 100 nm) was grown thermally at 1000 °C. On the edge of the wafers the SiO_2 was ring-shaped removed in order to enable a charge flow from the substrate to the sample holder during the electron-beam irradiation. The samples were subjected to the HLA and EBA for 10s using temperatures in the range from 800 °C to 1300 °C. EBA was performed by means of an electron-beam heater with a large area plasma cathode /1/ at a vacuum of $\sim 10^{-2}$ Pa. The annealing ambient for the HLA was nitrogen. The temperature-time dependence for HLA and EBA is shown in Fig. 1.

After Al deposition MOS capacitors were fabricated. A pastmetallization annealing (PMA) was performed in forming gas at 400 °C for 20 min. The samples were investigated by high-frequency and quasistatic CV, by TVS and pulsed CV methods.

The dependence of the effective charge density (N_{eff}) and of the interface state density (N_{st}) on the annealing temperature are qualitative similar (Fig. 2). These values reach a maximum at approximately 900 °C and then decrease with increasing temperature. In the temperature range of 800 °C – 1100 °C the irradiation from back side causes a lower increase of N_{st} and N_{eff} then the electron irradiation from the front side. The increase of N_{st} and N_{eff} can be accounted for by the superposition of effecting mechanisms such as atomic reconstruction of the interface and non-equilibrium point defect generation. In the case of irradiation from SiO_2 side we have to take into account the damage introduced by ionization during the electron exposure. For the electron energy of 7 keV utilized here the beam penetrates the oxide and causes

---

1 - Central Institute for Nuclear Research Rossendorf, Academy of Sciences of GDR, PSF 19, DDR-8051 Dresden

2 - Institute of High Current Electronics of the USSR Academy of Sciences, Tomsk, pr. Academichesky, 4

a buildup of positive charge in oxide and creates surface states.

The front side irradiation influences also the carrier profiles in silicon determined from pulsed CV-measurement (Fig. 3). For the unirradiated reference sample a well-known weak accumulation of donor like centers is obtained. This carrier profile remains unaffected by irradiation from the back side. In the case of front side irradiation an apparent carrier profile can be due to the influence of deep traps in the silicon on the pulsed CV measurement. A decrease of the apparent carrier concentration with growing annealing temperature may be associated with an enhanced annealing of damage during the heating cycle.

The effect of HLA is in a contrast to what is observed in the case of EBA (Fig. 4). The values of N_{st} and N_{eff} increase continuously with increasing temperature. In this case a clear correlation between N_{eff} and N_{st} is observed also. This correlation can be explained by a similar physical origin of N_{eff} and N_{st}. The difference in the results for HLA and EBA can be explained by the following effects:

(i) The shorter cooling times in the case of HLA (Fig. 1) cause a "freezing" of defects created in the high temperature heating phase and consequently an increase of N_{st} and N_{eff}.

(ii) The concentration of mobile ions (N_{Na}) estimated by a TVS measurement at 400 °C is higher for HLA (Fig. 5). In the case of EBA the values for N_{Na} are at the detection limit of the TVS technique and are smaller then those found for the reference sample ($N_{Na(ref)} = 3 \cdot 10^9\ cm^{-2}$).

These results indicate that the effective interface charge N_{eff}, the interface state density N_{st} and sodium concentration N_{Na} are dependent on annealing temperature as well as on the annealing method.

The effective interface charge N_{eff} and sodium concentration N_{Na} are remarkably smaller for EBA.

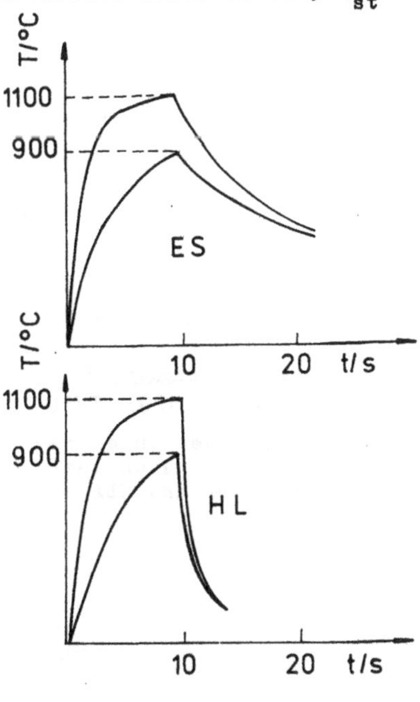

Fig. 1: Temperature-time dependence for HLA and EBA

References

/1/ Kreindel Yu.E. et al.: Proc. of the Conf.on Energy Pulse Modification of Semiconductors and Related Materials, Dresden GDR, 25-28 Sept.(1984)226

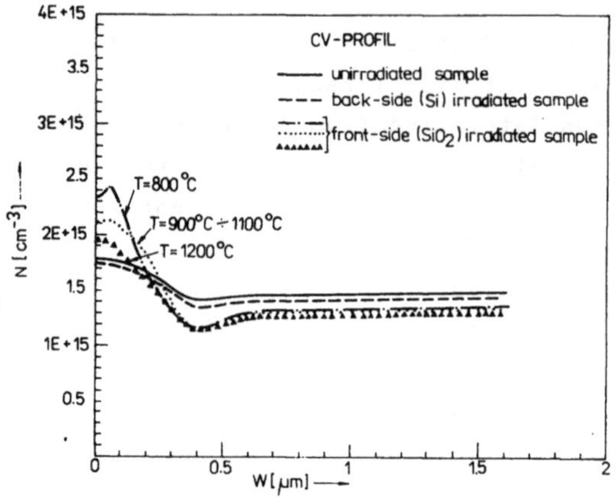

Fig. 3: Measured carrier profiles in Si of the samples irradiated by electrons from back and front side

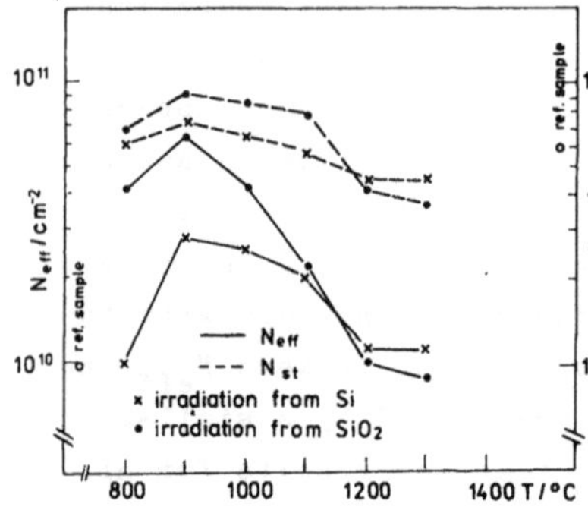

Fig. 2: Dependence of interface state density and effective charge density on EBA temperature

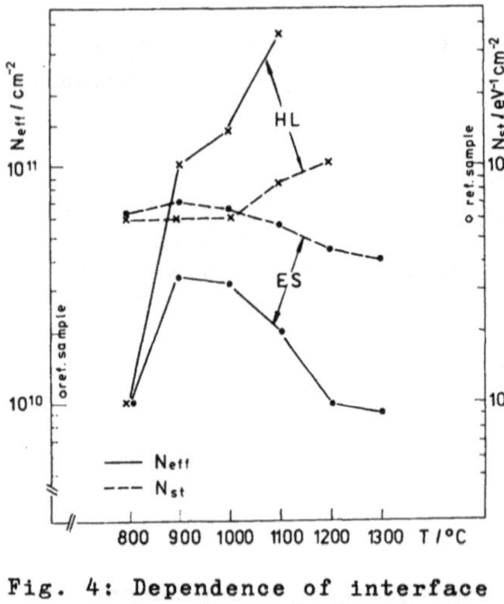

Fig. 4: Dependence of interface state density and of effective charge density on temperature for EBA and HLA

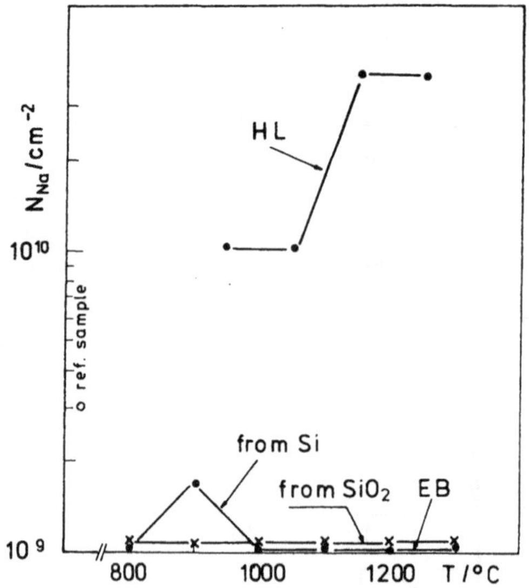

Fig. 5: Dependence of the concentration of mobile ions on temperature for EBA and HLA

ELECTRICAL ACTIVATION OF BORON IMPLANTED SILICON DURING THE EARLY STAGE OF RAPID THERMAL ANNEALING

R.Kögler[1], E.Wieser[1], A.Armigliato[2], E.Landi[2], and S.Solmi[2]

A boron implanted silicon layer is normally not amorphized. Therefore, electrical activation by annealing proceeds relatively slowly and is a complex process (e.g. defect annealing, precipitation above the solubility limit).

Cowern et al. /1/ determined a characteristic time constant for boron activation of τ= 25 s at 1000 °C and τ= 2 s at 1100 °C, respectively, from the analysis of rapid thermal annealing (RTA) experiments. Seidel and Mc Rae /2/ proposed a model for the electrical activation of boron on the base of low temperature furnace annealing (600 to 900 °C). An activation energy of 5 eV has been found. The preexponential factor of the time constant for activation ($\tau = \tau_0 \exp(E_a/kT)$) is assumed as increasing with increasing defect density. This means it varies within the implantation profile. Moreover, a significant time dependence of τ_0 should be expectet for RTA conditions due to the strong change of the defect state during the early stage of annealing.

The present investigation aims to contribute more information to the dependence of the boron activation on dose, temperature, and time during the early stage of annealing. Concerning the solubility effects at high doses also implantation into preamorphized silicon is considered.

(100) silicon wafers are implanted at room temperature with B at an energy of 30 keV in the dose range of 5×10^{14} to 1×10^{16} cm^{-2}. Preamorphized silicon substrates were produced by a three step silicon implantation (E = 50 keV, 5×10^{14} cm^{-2}; E = 100 keV, $1{,}5 \times 10^{15}$ cm^{-2}; E = 180 keV, 3×10^{15} cm^{-2}). Annealing is carried out in air using a set of halogen lamps at maximum temperatures T_{max} between 900 and 1100 °C with dwell times t from 1 to 35 s. The temperature rise time was at about 10 s. The dwell time t = 1 s characterizes a temperature - time cycle where the irradiation is finished at appraoching T_{max}. Van der Pauw measurements and stepwise sheet removal by anodic oxidation were used in order to determine the sheet carrier concentration and carrier profiles. For selected samples the distribution of residual defects after annealing was observed by cross section TEM.

The analysis of the temperature and time dependence of the activation process shows remarkable differences for maximum temperatures below and above 1000 °C.

1 - Central Institute of Nuclear Research Rossendorf, Academy of Sciences of GDR, DDR-8051 Dresden, PSF 19
2 - C.N.R. Istituto LAMEL, Via de Castagnoli 1, I-40126 Bologna

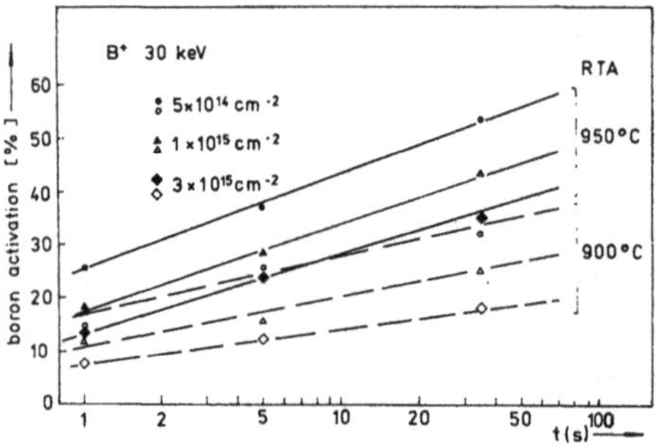

Fig.1: Temperature and time dependence of the activation (T = 950 °C, C C_{Sol})

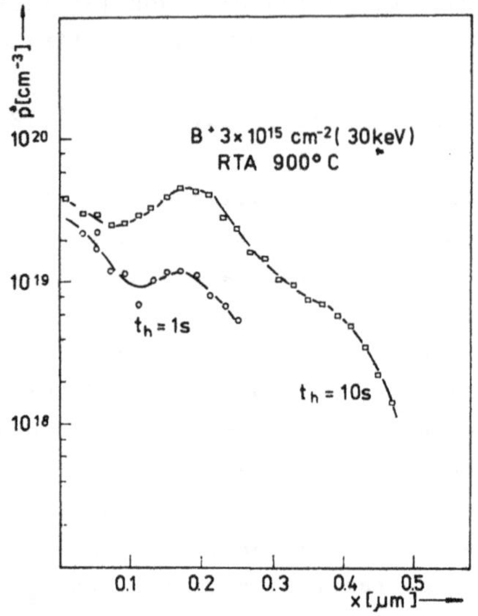

Fig.2: Time development of the carrier profile at 900 °C

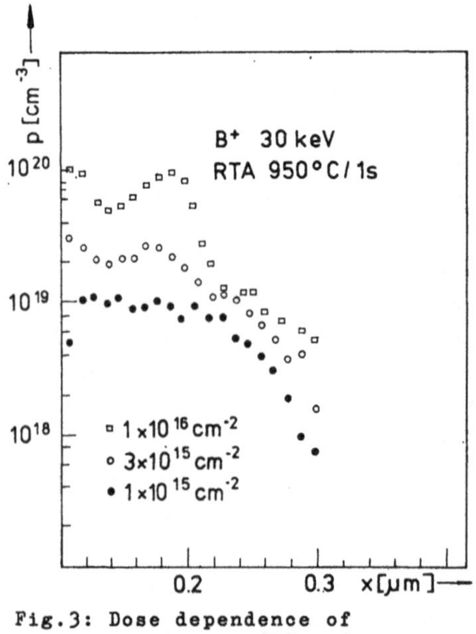

Fig.3: Dose dependence of carrier profiles for annealing at 950 °C/1s

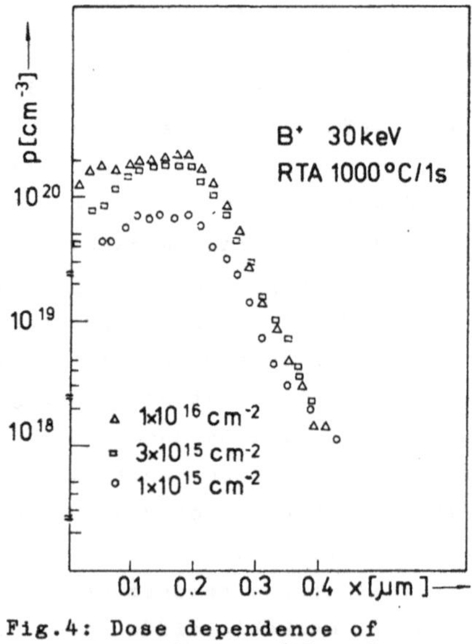

Fig.4: Dose dependence of carrier profiles for annealing at 1000 °C/1s

Fig. 1 shows the time dependence of the electrical activation at 900 and 950 °C for doses where the solubility effects are still unimportant. The maximum boron concentration c_{max} for the dose of 1×10^{15} cm^{-2} is below the solubility limit of about 1.2×10^{20} cm^{-3} determined for RTA at 900 °C in this work. For the fluence of 3×10^{15} cm^{-2} the profile maximum corresponds to 2.5×10^{20} cm^{-3}. Fig. 2 presents an example for the time development of the charge carrier profile. The following conclusions can be drawn from Fig. 1 and 2:

Complete activation is not possible in times characteristic for RTA using $T_{max} \leq 950$ °C. The low activation is due to the low activation velocity at these temperatures and not to the solubility limit. The carrier concentration increases over the whole profile depth with increasing t and the maximum concentration is well below the solubility as proved by Fig. 2.

The observed temperature und time dependence of the activation according to Fig. 1 is in reasonable agreement with the activation energy of 5 eV proposed in /2/. The increase of activation with decreasing dose at given values of T_{max}, t confirms the growth of τ_0 with increasing defect density according to /2/.

For T_{max} = 1000 °C an activation of 90 % was measured already after annealing with t = 1s for the dose of 1×10^{15} cm^{-2} (no solubility effect). Taking into account the temperature rise time and the small effect of $T_{max} \leq 950$ °C we conclude that complete activation is obtained at 1000 °C in $t < 5s$ for $c_{max} < c_{sol}$ in disagreement to /1/.

The strong increase in activation going from 950 to 1000 °C can not be explained by the activation energy of 5 eV. A possible explanation is a strong decrease of the preexponential factor τ_0 due to remarkable defect annealing during the temperature rise between 950 and 1000 °C.

Carrier profiles measured after annealing below 1000 °C show a characteristic minimum near R_p (somewhat shiftet to the surface) as demonstrated in Fig. 2. The dose dependence of corresponding carrier profiles is presented in Fig. 3. The minimum near R_p arises if the maximum dopant concentration exceeds the boron solubility at the used T_{max}. On the other hand is the maximum carrier concentration in these cases still well below the solubility limit.

The occurrence of the minimum in the carrier distribution is correlated with a band of residual defects around R_p arising at the same dose. These defects may be precipitates ore small lattice defects connected with the beginning precipitation.

From the oberved dose dependence we conclude that the minimum is caused by beginning precipitation of electrically inactive boron complexes, superimposed to the still incomplete process of electrical activation. Therefore, two conclusions about the precipitation are possible:

i) Precipitation is driven by the concentration of boron on interstitial sites instead of substitutional boron.

ii) Due to compensating defects the measured carrier concentration is lower than the concentration of subtitutional boron determining the precipitation process.

For doses $>3 \times 10^{15}$ cm^{-2} the temperature dependence of the boron solubility becomes dominating for the activation at $T_{max} = 1000$ °C. Thereby, the maximum charge carrier concentrations p_{max} obtained by RTA are significantly higher than the equilibrium solubility:

1000 °C:	equil c_{sol}	p_{max}(RTA, 1s)	p_{max}(RTA,10s)
/10^{20} cm^{-3}/	1.2	2.0	1.7

This effect is much more pronounced for implantation into preamorphized silicon. Due to the epitaxial recrystallization of the amorphous layer electrical activation proceeds very fast already at 900 °C. Complete electrical activation has been obtained by RTA at 900 °C/1s up to 3×10^{15} B/cm^{-2}. The measured maximum charge carrier concentration of 2.7×10^{20} cm^{-3} in this case corresponds to a supersaturation by a factor 4. At still higher supersaturation electrically inactive boron complexes are already formed during annealing at 900 °C/1s. For 1×10^{16} B/cm^2 only 60 % activation are obtained. This corresponds to a supersaturation by a factor of about 8.

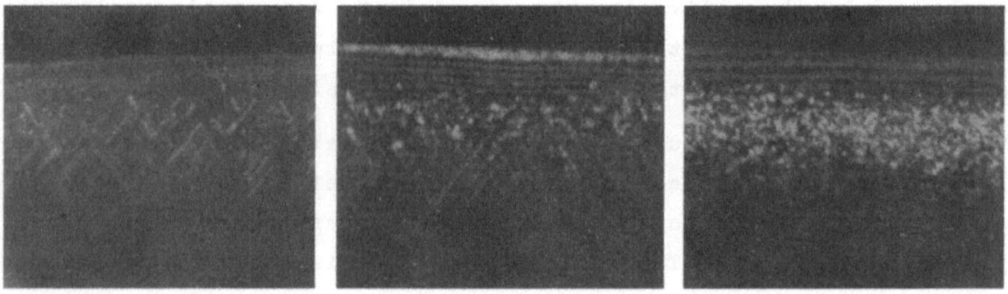

Fig. 5 a b c

cross section TEM after RTA 900 °C/1s

a) 1×10^{15} B^+ cm^{-2}, b) $3 \times 10^{15} B^+$ cm^{-2}, c) 1×10^{16} B^+ cm^{-2}

References:

/1/ Cowern, N.E.B., K.J. Yallup, D.J.Godfrey, D.G.Hasko, R.A.Mc.Mahon, H.Ahmed, W.M.Stobbs and D.S. Mc Phail
Mat.Res.Soc.Symp.Proc. Vol. 52 (1986) 65

/2/ Seidel, T.E. and A.U. MacRae
Proc. 1st.Int.Conf. on Ion Implantation in Semiconductors
Thousand Oaks, CA Gordon and Breach, New York, 1970

NONEQUILIBRIUM IMPURITY SEGREGATION IN PHOSPHORUS IMPLANTED POLYCRYSTALLINE SILICON SUBJECTED TO TRANSIENT HEATING

V.A.Samujlov, V.F.Stelmakh[1)]

The process of segregation in ion implanted polycrystalline silicon may be nonequilibrium under pulse annealing, when time duration of radiation is comparable with diffusion redictribution time of impurity atoms by average grain size distances. The nonequilibrium segregation model in polycrystalline silicon films doped by ion implantation was proposed. This model includes the following processes: diffusion redistribution of implanted impurity atoms inside the grains; capture of impurity atoms into electric passive states on grain boundaries; thermal ejection of impurity atoms into grains with their electric activation.

The theoretical task of electrically active impurity time-dependent reduction through the impurity atom diffusion from the grains to their boundaries was previously solved /1/. But the result didn't agree with the equilibrium impurity segregation theory /2/ because of the assumption that the boundaries were the infinite drains /1/.

Our system of equations has the following form. The diffusivity equation for the electrically active impurity concentration N(r,t) inside the grain with radius R is:

$$\frac{\partial N(r,t)}{\partial t} = D\,\frac{1}{r^2}\frac{\partial}{\partial r}\left[r^2\,\frac{\partial N(r,t)}{\partial r}\right]. \qquad (1)$$

The initial conditions for N(r,t) and $N_{GB}(t)$ - electrically inactive impurity atom number on the grain boundary have the form:

$$N(r,0) = n_o\,, \quad 0 \leq r < R\,, \qquad (2)$$

$$N(R,0) = N_{GB}(0) = 0\,. \qquad (3)$$

The boundary condition in linear form is:

$$\frac{dN_{GB}(t)}{dt} = \gamma\,\frac{4}{3}\pi R^3 N(R,t) - \beta\,N_{GB}(t)\,, \qquad (4)$$

where γ - the impurity atom capture rate in electrically inactive states on the grain boundary, β - the impurity atom thermal ejection rate in electrically active states in the grain. The full number of impurity atoms inside the grain and on the grain boundary is:

$$N_o = \frac{4}{3}\pi R^3 n_o = N_G(t) + N_{GB}(t)\,, \qquad (5)$$

where

$$N_G(t) = \int_0^R N(r,t)\,4\pi r^2 dr\,. \qquad (6)$$

The equations (1) - (6) admit the analitical solution in the form:

1)Physics Department of the Byelorussian State University, Lenin Avenue 4, 220080, Minsk, USSR

$$N_G(t) = N_o \left\{1 - \frac{\gamma}{\gamma+\beta}\left[1-\exp(-6\varphi \frac{\gamma+\beta}{\gamma} t)\right]\right\} \quad , \quad (7)$$

where $\varphi^{-1} = R^2/D$ - the average impurity atom diffusion time inside the grain. But there are two undetermined parameters γ and β there. The equation (7) must correspond to the equilibrium segregation theory equation /2/:

$$N_G(t)\Big|_{t\to\infty} = N_o\left[1 + \frac{AQ_s}{N_{si}} \exp\left(\frac{Q_o}{kT_A}\right)\right]^{-1} \quad , \quad (8)$$

where $A = \exp(-S/k)$, S is the difference between impurity atoms oscillation entropy on the boundaries and inside the grains, k is Boltzmann's constant, Q_s and N_{si} are density atom states on the boundary and total density on the boundary and inside the grain respectivelly; $Q_s/N_{si} = 1.62 \cdot 10^{-7}/L$, where L is the average grain size (cm), Q_o is the segregation heat, T_A is the annealing temperature.

So, according to (8):

$$\frac{\gamma}{\beta} = \frac{AQ_s}{N_{si}} \exp\left(\frac{Q_o}{kT_A}\right) \quad , \quad (9)$$

Thus, the analitical equation for changing the electrical active impurity concentration was received. It depends on time duration, tempetature and thermodinamic and structuralproperties of polycrystalline silicon.

Sheet resistivity and electron concentration, total and electric active impurity concentration profiles depending on the incoherent light pulse annealing temperature and its time duration have been investigated for the experimental proof of the nonequilibrium segregation in phosphorus implanted polycrystalline silicon. The polycrystalline films with thickness of 0.40-0.45 μm were deposited by low pressure decomposition of SiH_4 onto monocrystalline (111) oriented, boron doped silicon wafers. Accoding electron microscopy data the grain size was about 0.15 μm. The implantation of 30 keV P - ions was carried out up to dose $1.3 \cdot 10^{16}$ cm^{-2}. The experimental samples were heated by intense light of halogen lamps for time duration from 6-40 s. Light power density was chosen to vary samples induced temperature in the range 800-1000°C.

From Fig.1 one can see that the sheet resistivity and electric impurity concentration dependences on pulse time duration at 800°C are nonmonotonic. The superposition of the electrical activation and the nonequilibrium impurity segregation in phosphorus doped polycrystalline silicon is the reasons of these nonmonotonic dependences. The estimation based on the equations (7), (9) is shown by the broken curve, which is good approximation of the experimental time - sheet electron concentration dependence under $t > 10$s. For the calculation we use the parameters: $D \simeq 1 \cdot 10^{-12}$ cm^2/s /3/; $Q_o \simeq 10$ Kcal, $A = 2.46$ /2/. To prove the segregation nature of sheet resistivity and sheet electron concentration decreasing under $t > 10$s, $T \simeq 800^o$C we have investigated the total and electric active impurity (electron) concentration profiles (Fig.2). One can see that if the total phosphorus concentratuion profile doesn't vary under T=800°C, the electron concentration profile varies nonmonotonically. This experimental fact proves the nonequilibrium impurity segregation mechanism in the annealing temperature range ~800°C.

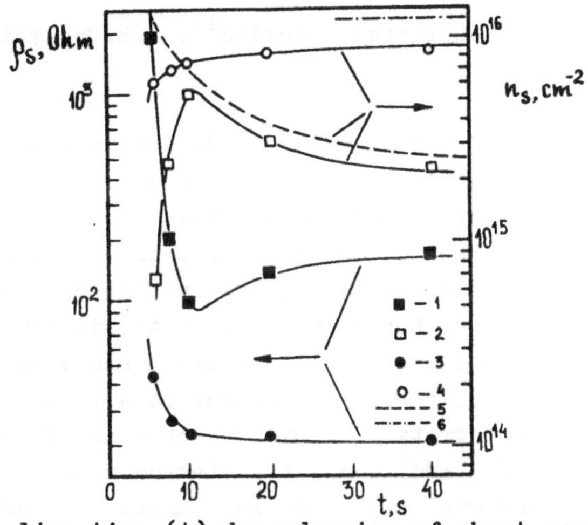

Fig.1. The annealing time (t) dependencies of sheet resistivity (1,3) and sheet electron concentration n_s (2,4) in polysilicon layer-monocrystalline silicon substrate structures. The annealing temperature: 1,2 - 800°C, 3,4 - 1000°C. 5 - the theoretical calculation based on the equations (6), (8). 6 - the implanted dose $1.3\cdot10^{16}$ cm^{-2}

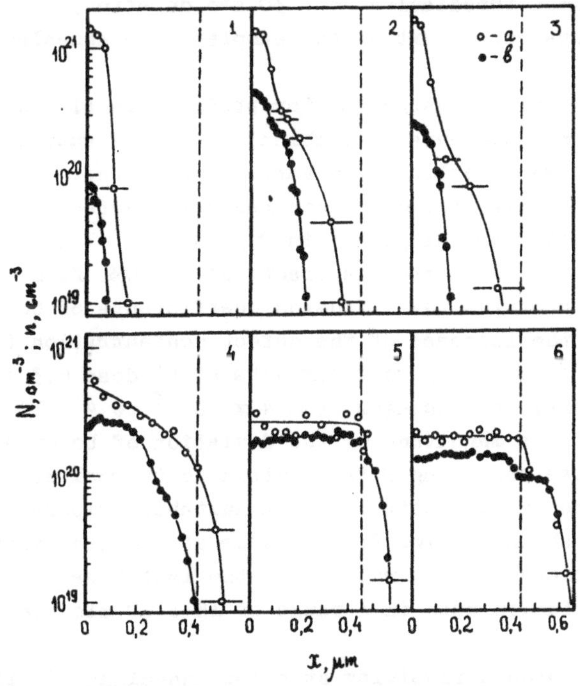

Fig.2. The phosphorus N (a) and electron n (b) profiles in polysilicon layer-monocrystalline silicon substrate after short time annealing. The annealing temperature: 1,2,3 - 800°C, 4,5,6 - 1000°C. The exposure time: 1,4 - 6s; 2,5 - 10s; 3,6 - 40s. The broken line is the boundary between poly-Si and mono-Si

/1/ Kadyrakunov K.B., et.al.:Phys.Stat.Sol (a) 75 (1983) 483-488
/2/ Mandurah M.M, et.al.: J.Appl.Phys. 51 (1980) 5755-5763
/3/ Borisenko Y.E., et.al.:Proc.Conf.Energy Pulse Modif.of Semicond.and Rel.Mater, GDR, 1984, 331-336

TIME DEPENDENCE OF BORON DIFFUSIVITY DURING PULSE ANNEALING OF ION-IMPLANTED SILICON

V.F.Stelmakh[1), Yu.R.Suprun-Belevich[1), A.R.Chelyadinskii[1)

Observed anomalous diffusion phenomena of ion implanted impurities in silicon caused by radiation defects must depend on thermal treatment duration in consequence of defects annealing.

In order to determine the character of the influence of radiation defects on the dopant diffusion in silicon in time of annealing the diffusion redistribution of boron implanted into silicon has been investigated. Different exposures (in a range 5 sec ÷ 60 min) of pulse annealing and conventional thermal treatment in a furnace were performed. The defect concentration in silicon crystals implanted with B^+ ions was varied by the additional irradiation with Si^+ ions. 45 keV B^+ and 100 keV Si^+ ions were implanted into silicon wafers (n-type, ρ_0 = 0,3 Ohm x cm) with (III) orientation. The degree of the radiation damage of ion implanted layers was investigated by an X-ray diffraction method by measuring the lattice period (4). Pulse annealing was carried out by halogen lamps irradiation (5 ÷ 30 sec duration) at 1050°C in a settled temperature rate. Part of the samples were annealed in a furnace.

Fig. 1a shows typical charge carrier profiles in silicon obtained from Hall effect measurements in combination with a succesive layer removing after pulse annealing with various exposures. Charge carrier profiles in silicon with the different radiation defects concentration after pulse annealing are displayed in Fig 1b. The electrical activity of implanted boron in all cases is practically 100%. From the experimental diffusion profiles the diffusion coefficients of boron were calculated. With the increase of the defect concentration from 5×10^{18} to $2{,}5 \times 10^{20}$ cm^{-3} according to X-ray data at B^+ dose $6{,}3 \times 10^{13}$ cm^{-2} the diffusion coefficient changes from 4×10^{-13} to $4 \times 10^{-12} cm^2 s^{-1}$. The alteration of the magnitude of the deviation of boron diffusivity determined at pulse annealing D from intrinsic thermal diffusion constant in a furnace in bulk material at a suitable temperature D_i versus exposure time is shown in Fig. 2, the value of D characterizing the diffusion redistribution of boron during the last 5 sec of pulse annealing. Thermal diffusion data in a range of 15 ÷ 60 min are shown in the same figure too.

Ion implanted boron diffusivity at pulse annealing of silicon samples which prior to pulse treatment were thermally annealed in a furnace was investigated for comparison. The obtained in this way boron diffu-

1) Byelorussian State University, Lenin avenue 4, 220080, Minsk, USSR

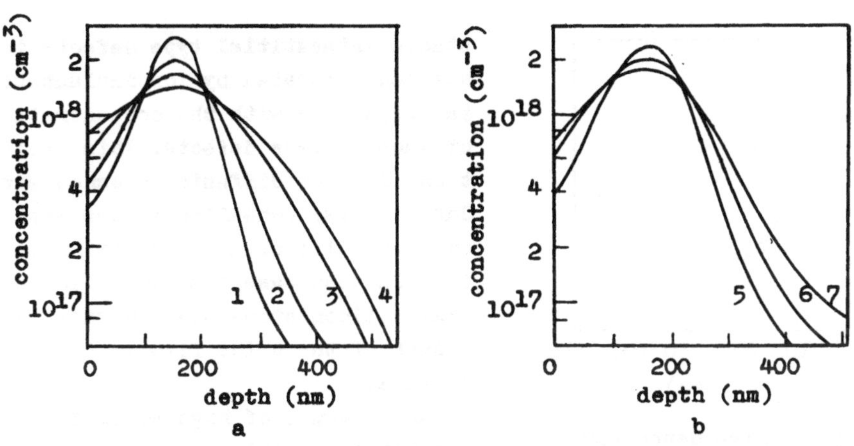

Fig. 1. Charge carrier profiles in B^+-ion implanted silicon after pulse annealing with various exposures (a) and with different additional defect concentrations (b). B^+ dose - 6,3 x 10^{13} cm^{-2}. Exposure time: 1 - 7 sec; 2 - 10 sec; 3 - 20 sec; 4 - 30 sec; 5,6,7 - 8 sec. Doses of the additional Si^+-ion irradiation: 6 - 6 x 10^{13} cm^{-2}; 7 - 6 x 10^{14} cm^{-2},

sivity does not depend on pulse duration and coincides with the intrinsic boron diffusion constant D_i.

We can see from Fig. 2 that the boron diffusion coefficient in ion implanted silicon during pulse annealing for 7 ÷ 15 sec exposure is essentially greater than the intrinsic boron diffusivity in bulk silicon D_i. With the increase of the exposure of pulse annealing from 7 to 30 sec D decreases by a factor ~20 and drops down below D_i. With the increase of the annealing exposure to dozens of minutes (thermal annealing in a furnace) D approaches to D_i.

The accelerated boron diffusion in silicon during pulse annealing was observed by some other authors (2, 6) also. They explain this phenomenon by the fact that after ion implantation the considerable part of dopant atoms occupy interstitial positions and are very mobile. In initial stage of the annealing process they can rapidly move via interstitially mechanism until they are captured by vacancies. Then dopants become substitutional and relatively immobile. However from this consideration does not follow the obtained in present experiments increase of dopant diffusivity involved by the increase of the defect concentration in ion implanted layers (in samples additionally irradiated with Si^+ ions). In our opinion high value of diffusion coefficient D of implanted boron in silicon during pulse annealing and the increase of D with the increase of the density of radiation defects are caused by the fact that in the first stage of the annealing process the dopant diffusion mainly via interstitials is assisted by succesive replacement of boron atoms from the lattice sites with silicon interstitial atoms (Watkins substitution (5)) released during dissociation of stable interstitial type complexes. According to (1) the concentration of the

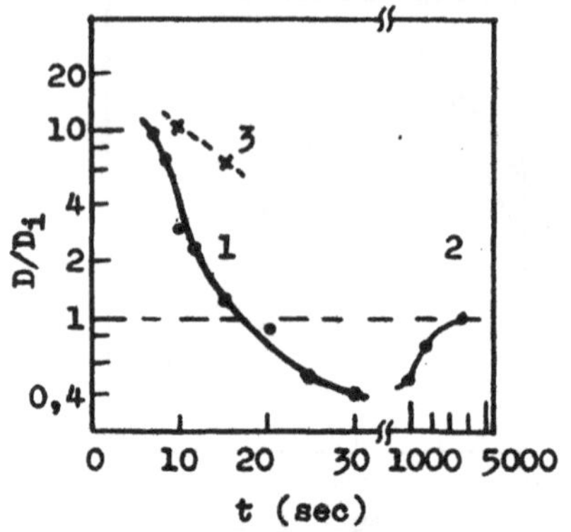

Fig. 2. Time dependence of D/D_i. B^+ dose - 6,3 x 10^{13} cm^{-2}. 1 - pulse annealing; 2 - annealing in a furnace. 3 - the samples were additionally irradiated with Si^+ ions, dose - 6 x 10^{14} cm^{-2}.

stable interstitial type defects in silicon generated by ion implantation is comparable with the concentration of vacancy type defects. This explanation of boron diffusion enhancement during pulse annealing agrees with data on boron diffusion in oxidizing medium (3). In that case the excess interstitial silicon atoms are generated during growth of the oxide film on the silicon surface.

As a result of high mobility of the interstitial silicon atoms their concentration decreases rapidly in course of annealing involving the decrease of D. The diffusion coefficient D becomes not equel to D_i but takes anomalous low value. Low magnitude of the diffusion coefficient is caused by the radiation defects also since in the case of the preliminary annealing of radiation defects prior to pulse annealing no anomalous phenomenon is observed. The retardation of boron diffusion (at the annealing exposures greater then 15 sec) can be explained by the capture of B atoms by excess vacancies. As it was established in (4) the diffusion coefficient has minimum value if the impurity concentration is comparable with the vacancy type defect concentration. The increase of the vacancy defect concentration involves an increase of D, but even in the limiting case of amorphous silicon D remaines lower than D_i (4).

After the heating exposure of about 60 min the radiation defects are annealed and D approaches to D_i (Fig. 2, curve 2).

Hence, interstitial and vacancy type radiation defects involve the redistribution of moving dopant atoms between vacancy and interstitially diffusion mechanisms and determine the observed anomalous boron diffusivity during pulse annealing.

References

(1) Berezhnov, N.I., V.F.Stelmakh and A.R.Chelyadinskii: Phys. stat. sol. (a) 78 (1983) K121 - K125

(2) Hodgson, R.T., V.Deline, S.M.Mader, F.F.Morehead and J.Gelpey In: Energy Beam - Solid Interact. and Transient Therm. Processing, North-Holland, New York (1984) 254 - 257

(3) Matsumoto, S., Yu.Ishikawa and T.Niimi In: Ext. Abstrs 15 Conf. Solid State Devices and Mater., Tokyo (1983) 19 -22

(4) Stelmakh, V.F., Yu.R.Suprun-Belevich, V.D.Tkachev and A.R.Chelyadinskii: Phys. stat. sol. (a) 89 (1985) K45 - K49

(5) Watkins, G.D.: Radiation Damage in Semiconductors, Dunod, Paris 1965

(6) Wilson, S.R., W.M.Paulson, R.B.Gregory, A.H.Hamdi and F.D.McDaniel: J.Appl.Phys. 12 (1984) 4162 - 4170

COMPARATIVE ANALYSIS BY THE X-RAY DIFFRACTION METHOD OF PULSE AND THERMAL ANNEALED ION-IMPLANTED SILICON

N.I.Berezhnov[1)], A.R.Chelyadinskii[1)], Yu.R.Suprun-Belevich[1)], N.L.Prokhorenko[2)], A.K.Polonin[2)], N.T.Kvasov[2)]

Present development of microelectronics is characterized by the replacement of a traditional thermal treatment of semiconductor wafers by the pulse annealing. Short time heating processing using halogen lamps with the exposure of the order of seconds has the significant advantages over the conventional furnace annealing. However, insufficient amount of the experimental data on kinetics of pulse annealing keeps back the development of theoretical notion about this process.

In present investigation the character of the annealing and the structural defect reconstruction during traditional furnace treatment and during pulse heating of dislocation-free p-type silicon implanted with P^+ and Si^+ ions has been studied.

Lattice period change obtained from X-ray measurements was used as a parameter characterizing integral degree of radiation damage (5). The investigations were carried out by the X-ray double crystal diffractometer using $CuK\alpha_1$ irradiation in the fourth order (III) reflection. The isochronal 15 min annealing of the irradiated samples was carried out in evacuated quartz tubes. Temperature deviations did not exceed $\pm$ 2°C. Pulse annealing was carried out in the air by halogen lamps irradiation with pulse duration from 0 to 12 sec with discrete step 0,3 sec, the temperature being in a range 20 - 1200°C. More detail annealing technique is described in (4). The sample temperature during annealing was monitored by pyrometer with an accuracy $\pm$ 3°C. The degree of the dopant electrical activity was obtained using Hall effect measurements in Van der Paw configuration.

Fig. 1 shows heating kinetics of the silicon crystal and electrical activation process of the impurity. Lattice period reconstruction during isochronal 15 min annealing and pulse heating is shown in Fig.2. The lattice period reconstruction during the isochronal annealing of intrinsic radiation defects in silicon implanted with Si^+ ions takes place in two stages (Fig. 2, curve a): 1 - 80 ÷ 260°C; 2 - 380 ÷ 600°C. In the P^+-implanted samples the second stage of annealing is shifted to the high temperature region: 460 ÷ 700°C (Fig. 2, curve b). Two stages of lattice period reconstruction in P^+-implanted silicon take place during pulse annealing also (Fig. 2, curve c). But stage 2 of pulse annealing is prolonged some more to high temperature region (to

1) Byelorussian State University, Lenin Avenue 4, 220080, Minsk,USSR
2) Minsk Radioengineering Institute, P.Browka 6, 220069, Minsk, USSR

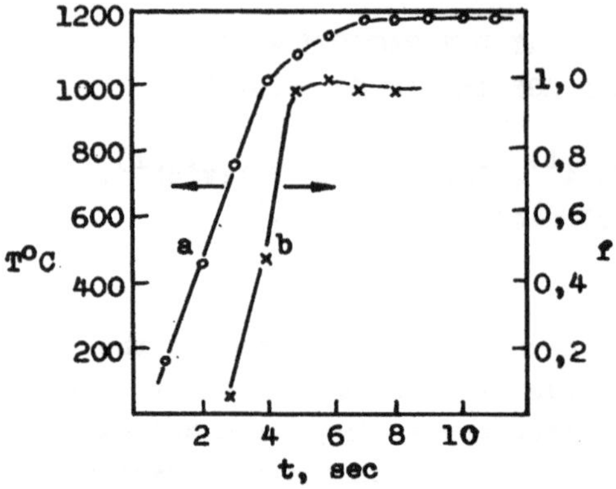

Fig. 1. Temperature of the silicon wafer T (a) and coefficient of the dopant electrical activity f (b) versus time exposure of pulse annealing t.

1000°C). It is important to note that the degree of the radiation damage of silicon after pulse annealing is higher then after furnace isochronal annealing. It should be pointed out that both thermal and pulse heating of P^+-implanted silicon have similar peculiarity: there is some increase of lattice period in a temperature region 450 ÷ 500°C for furnace annealing and within 700 ÷ 800°C for pulse annealing.

Stage 1 of the lattice period reconstruction is related to the annealing of double vacancies as the main one among the defects which have the same annealing temperature (1). Stage 2 of thermal annealing as it follows from available EPR studies (2,3) is determined by the annealing of multivacancy complexes, for example Si-P1, generated during thermal treatment as a result of the defect structural reconstruction. In P^+-implanted silicon the phosphorus atoms are included into these complexes increasing their annealing temperature (Fig. 2, curve b).

During pulse treatment stage 1 is delaied to 450°C and the annealing of the multivacancy complexes (stage 2) is completed only at 1000°C (Fig. 2, curve c) which can be caused by short duration of pulse heating.

The inverse annealing regions in curves b, c (Fig. 2) of the lattice period reconstruction of P^+-implanted crystals are stipulated, in our opinion, by the ejection of phosphorus atoms from the lattice nodes with self-interstitials which appear during annealing of stable interstitial type defects at about 560°C (1). Interaction of interstitial silicon atoms with elements of Group III was described in (6). It is not excluded that the same substitution process can take place for the elements of Group V (phosphorus).

Electrical activation of the impurity is started at pulse duration about 3 sec and is finished at the end of 5 sec exposure. The degree of the dopant alectrical activity is a linear function of the time exposure of pulse annealing. The begining of the activation process coincides with the stage 2 of the annealing (the dissociation of the multivacant complexes including phosphorus atoms) (Fig. 2, curve c). We can expect some decrease of the degree of dopant activity in the inverse annealing region at the temperatures preceding the dissociation of multivacant complexes. With further increase of the annealing exposure the degree of the electrical activity of phosphorus atoms increases due to their annihilation with vacancies generated during the dissociation of multivacant complexes.

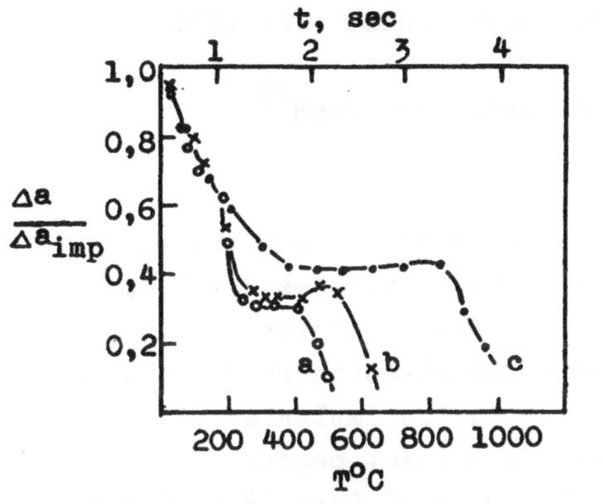

Fig. 2. Lattice period $\Delta a/\Delta a_{imp}$ reconstruction of Si:B crystals (ρ_0 = 10 Ohmxcm) irradiated with Si^+ (a) and P^+ (b,c) ions (200 keV, 4 x $10^{14}cm^{-2}$) versus temperature T. a, b - isochronal furnace annealing; c - pulse annealing. t - pulse exposure corresponding to the sample temperature T.

Thus, the results of X-ray and Hall effect measurements indicate that electrically neutral phosphorus atoms are included into the structure of multivacant complexes and are replaced into crystal lattice nodes during dissociation of these complexes.

Hence, both thermal and pulse treatment have similar stages of the defect annealing which are slightly shifted to higher temperatures during pulse heating in consequence of short duration of the process in which the defect reconstruction and dissociation occurs.

Qualitative coincidence of the curves of lattice period reconstruction for pulse photon heating and thermal treatment confirms that thermal effects have the fundamental influence on annealing process during pulse heating.

References

(1) Berezhnov, N.I., V.F.Stelmakh and A.R.Chelyadinskii: Phys. stat. sol. (a) 78 (1983) K121 - K125

(2) Botvin, V.A., Yu.V.Gorelkinskii, V.O.Sigle and M.A.Chubisov: Fiz. Tekh. Poluprov. 6 (1972) 1683 - 1687

(3) Brover, K.L., F.L.Vook and J.A.Borders: Appl. Phys. Lett. 15 (1969) 208 - 211

(4) Naumenko, N.N., N.L.Prokhorenko, V.V.Shimanovich, N.T.Kvasov and A.K.Polonin: Radiotekhnika i elektronika 15 (1986) 113 - 118

(5) Stelmakh, V.F., V.D. Tkachev and A.R.Chelyadinskii: Fiz. Tverd. Tela 20 (1978) 2196 - 2200

(6) Watkins, G.D.: Radiation Damage in Semiconductors, Dunod, Paris 1965

INTERNAL MECHANICAL STRESS AND ELECTRICAL ACTIVATION OF IMPURITY IN ION-IMPLANTED Si DURING PULSE ANNEALING

N.L.Prokhorenko[1), Yu.R.Suprun-Belevich[2)

A specific feature of pulse annealing of semiconductor wafers is the high level of thermoelastic mechanical lattice stresses right up to crystal destruction. The results of the experimental investigations of the influence of the dynamical thermoelastic stresses on the electrical activation and redistribution of impurity in ion implanted silicon during pulse photon annealing are given in present paper.

P-type silicon crystals were implanted with P^+ ions ($E = 30$ keV, $F = 4{,}7 \times 10^{13} \div 3{,}1 \times 10^{15}$ cm^{-2}). Pulse annealing of the implanted Si wafers was carried out with halogen lamps irradiation in the special heating installation (1). Annealing exposure was varied in the range 1 - 10 sec with a discrete step 1 sec. The sample temperature during annealing was monitored by pyrometer with an accuracy $\pm 3^oC$ (Fig. 1).

Dynamical strains of the implanted silicon during pulse annealing were measured in a real time scale (4). The registration of the measured in sity dynamical strain was carried out by camera. The original control setup was used to determine the dynamical sagging in the wide range of values. Holographic interferometer to measure low dynamical sagging to 20 mu and the method of spatial misfocusing of the object picture to measure the dynamical sagging within 20 - 1000 mu were used in the same equipment. The sign (direction) of the dynamical strain was established by the preliminary experiments to be determined by the sign (direction) of the static strain of the implanted silicon wafer. The initial silicon wafers with the different values of simmetrical sagging of one sign were selected for the investigations. During pulse treatment the sign of the dynamical strain did not depend upon what side of the wafer was irradiated with light.

The degree of the electrical activiy (Fig. 2) and profiles (Fig. 3) of the impurity were obtained from Hall effect measurements in combination with a successive controlled anodic stripping.

Fig. 1 presents the magnitude of the dynamical sagging of the implanted silicon wafer versus annealing exposure for the different values of initial static sagging. According to Fig. 1 the character of the dynamical strain during pulse heating depends on the magnitude of the initial sagging of the wafer. The magnitude of the dynamical sagging increases with the increase of the initial sagging of the wafer. As it has been shown by the special experiments the surface temperature

---

1) Minsk Radioengineering Institute, P.Browka 6, 220069, Minsk, USSR
2) Byelorussian State University, Lenin Avenue 4, 220080, Minsk, USSR

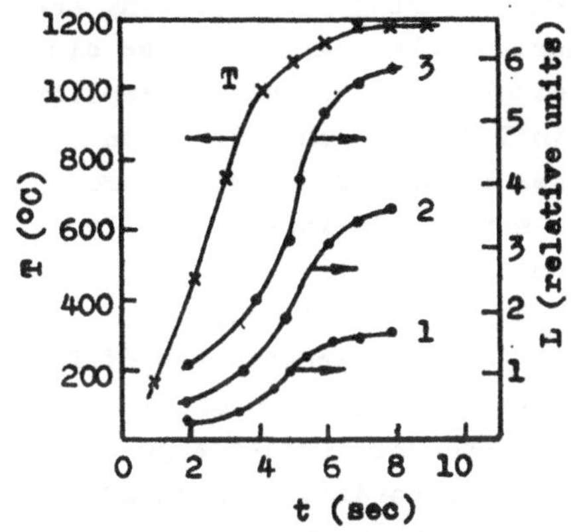

Fig. 1. Temperature T and dynamical sagging L of the silicon wafer versus pulse annealing time t. The initial static sagging: 1 - 5 mu; 2 - 20 mu; 3 - 40 mu.

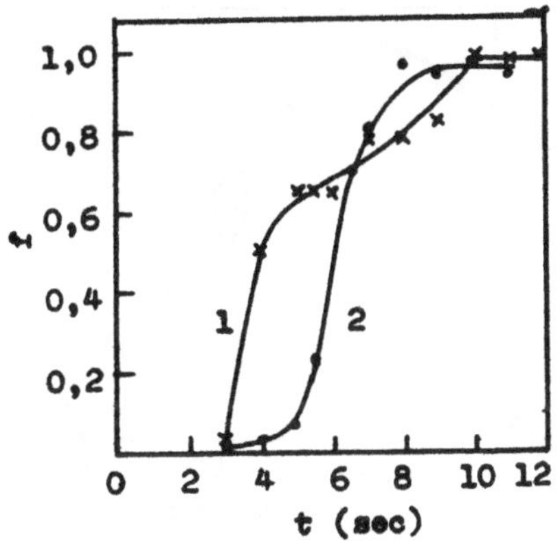

Fig. 2. Degree of the dopant electrical activity f versus annealing exposure t. 1 - the initial static sagging 40 mu; 2 - control sample without sagging.

gradient has the fundamental influence on the dynamical strain. The investigations of the distribution of thermal fields on the surface of the heated silicon wafer (1) indicate that during pulse annealing of small samples (~0,8 x 1 cm size) no surface temperature gradient exists. The small silicon samples were used as a control units because of the absence of the dynamical strain during pulse annealing.

The first stage of the dopant electrical activation in samples with a large values of the dynamical thermal strain proceeds at a shorter time exposure and lower temperatures respectively than in the control samples. Thus, the dynamical strain intensify the electrical activation process in the implanted crystal.

The results of the investigation of the influence of the dynamical stress on the dopant diffusion during pulse treatment are displayed in Fig. 3. Different values of the dynamical strain were determined by selection of initial silicon wafers with the definite static sagging. Phsphorus diffusion coefficient obtained from experimental diffusion profiles increases from 2×10^{-12} to $3,5 \times 10^{-12}$ cm^2/s with the increase of the initial static sagging from 0 to 40 microns and with the corresponding increase of the dynamical strain. Curve 4 in Fig. 3 conforms to the sample with opposed sign of the strain.

The observed behaviour of the implanted impurity under the influence of the dynamical stress can be explained in the following assumptions. The dynamical stress generated during pulse heating accelerates the dissociation of the defect complexes reducing the temperature of the dopant electrical activation. Disposed in silicon regular lattice sites phosphorus atoms involve the decrease of the lattice period of the implanted layer of silicon crystal due to differences in covalent radii between Si and P atoms and cause the elastic discrepancy stress. Thermal dynamical pressing stresses generated in the surface layer of

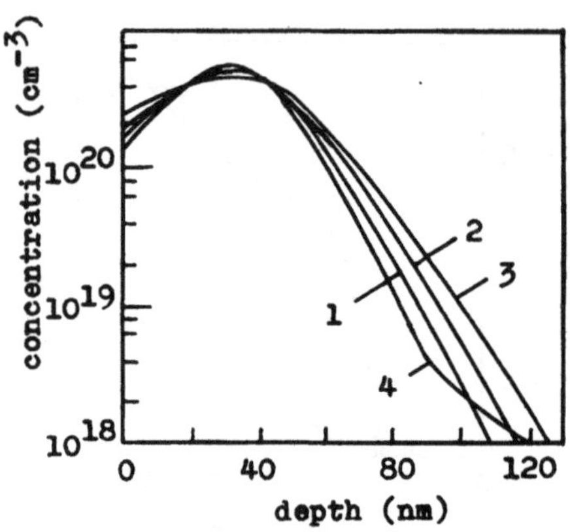

Fig. 3. Diffusion profiles of implanted P in silicon (dose - 3,1 x 10^{15} cm^{-2}) after pulse annealing 8 sec. Initial static sagging: 1 - 0 (control sample); 2 - 20 mu; 3 - 40 mu; 4 - -15 mu (inverse direction of the strain).

the wafer during pulse heating are summarized algebraic with the elastic discrepancy stress and amplify its influence on the diffusion (3). Besides that, part of phosphorus atoms moving via interstitially diffusion mechanism might increase under the action of mechanical stress (2) accelerating the diffusion also. Thermal dynamical stress with the opposed sign (direction) (according to sign of the initial static sagging of silicon wafer) influences inversely on dopant diffusion leading to appearance of the flex point on the diffusion profile (Fig. 3, curve 4).

Hence, the electrical activation and diffusion of ion implanted impurity is strongly affected by the dynamical thermal stress in silicon crystals produced during pulse heating and determined by the initial static sagging of silicon wafers. The dynamical stress may cause the observed inaccuracy of the electrical parameters of implanted silicon wafers after pulse annealing and in some cases can involve the oscillation behaviour of these parameters during annealing.

References

(1) Naumenko, N.N., N.L.Prokhorenko, V.D.Shimanovich, N.T.Kvasov and A.K.Polonin: Radiotekhnika i elektronika 15 (1986) 113 -118
(2) Panteleyev, V.A. and V.A.Muravyev: Fiz. tverd. tela 19 (1977) 682 - 686
(3) Shaw, D. (Ed.): Atomic Diffusion in Semiconductors, Plenum Press, London and New York 1973
(4) Vest, C.: Holographic Interferometry, Chichester Brisbane, New York 1978.

THE 13 MeV ELECTRON PULSE INDUCED MODIFICATION OF DYNAMIC PARAMETERS OF SILICON DIODES, THYRYSTORS AND TRANSISTORS

P.P.Pańta +)

1. INTRODUCTION:

Radiation-induced effects in electronic systems using semiconductors, mainly silicon, devices have become of increasing interest. Traditional approach focused mainly on radiation tolerance of semiconductor devices , however , recently an ionizing radiation can be applied for radiation modification of some parameters. An illustrative example is radiation-induced formation of recombination centers in silicon, which can improve dynamic parameters of diodes and thyrystors. The switch high power thyristors and fast recovery power diodes displays rising importance both for special applications (frequency inverters, uninterruptible power supply) and also according to electrical energy saving stimulated by energy crisis. Pulsed electron beams begin to compete with with traditional high-temperature diffusion of gold or platinum atoms in silicon lattice. In bipolar transistors radiation-induced reduction of diffusion lenght of carriers decreases the current gain. Thus, it is worth to compare classical results of research of radiation defects in silicon monocrystals with defects investigated in thin epitaxial layers of semiconductor devices in industrial scale.

2 SOME AUXILIARY CALCULATIONS

The value of cross-section of atomic displacements in irradiated silicon determines initial concentration of vacancy-instertitial pairs.
Primary and secondary (cluster) displacement cross-sections calculated for electron energy from 1 to 15 MeV were performed using method suggested by Lomer and Michell, [1]. Obtained results are listed in Table 1.

+) Institute of Nuclear Chemistry and Technology
Warsaw, Dorodna 16, Poland

Table 1. Calculated cross-sections for atomic displacement in silicon by fast electrons.

ENERGY [MeV]	CROSS-SECTION, [barn] primary	secondary
1	52	60
2	58	71
5	61	110
10	62	133
13	63	143

It is expected and confirmed experimentally by many authors, that isolated point defects formation rate is proportional to cross-section of atomic displacents. At room temperature vacancies recombine significantly. In the opinion of van Lint [2] the concentration of radiation-induced defect remainig after self-annealing at room temperature is around 10%.

3. EXPERIMENTAL

Polish production silicon power thyristors and diodes structures, low frequency bipolar transistors, high-voltage switching diode structures and some integrated circuits have been investigated using the electron linac LAE-13/9, described elsewhere, [3]. The electron scanned beam technique was applied for reduction of initial level of minority carrier lifetime (dynamic parameters) and shortening of diffusion length (current gain). The irradiations were carried out at room temperature with 5.5 μs pulses at the initial accelerator electron energy from 10 to 13 MeV. The dosage was calibrated with the quasi-adiabatic water calorimeter and checked routinely by chemical dosimeters. The silicon p-n-p-n power thyristors and p-n-n* diodes structures have a high concentration of 10^{14} cm^{-3}. The low frequency transistors and high-voltage switching diodes have a thin n-base region of several μm.

4. RESULTS and DISCUSSION

A typical radiation-induced tailoring of turn off time , t_q , in power thyristors can be fitted by simple exponential approximation:

$$t_q = 42.734 \exp(- D/D_{1/2} \ln(2))$$

where t_q in μsec ; D-dose in e^-/cm^2 ; $D_{1/2}$ - dose that will reduce given value of t_q in half.

However, for larger doses this simple exponent turns gradually into a composite curve. Similar behavior exhibit power diodes.

Bipolar transistors reveal hiperbolic relationship between current gain and dose as follows:

$$h_{21E} = 1/(0.01 + D^{1.26745})$$

h_{21E} is expressed in percent of initial value;
D- dose in e^-/cm^2; 0.01 and 1.26745 are constans applicable for given type of transistor.

Irradiated integrated circuits exhibit very low radiation tolerance. Long-term stability investigation of radiation-induced recombination centers in silicon devices have been performed , by isothermal annealing (furnace). Migration energy of centers was determined as 1.96 eV. Obtained results are slightly higher than the older data from 1.3 to 1.7 eV, reviewed by Bourgoin and Lano, [4]. It is surprising that they are very close to classical results of Bemski, [5] for silicon monocrystals.

REFERENCES

1. J. Lomer, E. Mitchellin "Radiation Damage in Solids, IAEA Vienna, 1962.
2. V. van Lint et al., "Mechanism of Radiation effects in Electronic Materials, Vol. 1., Wiley, NY, 1980.
3. Z. Zimek et al., Nukleonika 17, 75, (1972).
4. J. Bourgoin, M. Lanoo, "Point defects in Semiconductors, II, Springe, Berlin, (1983).
5. G. Bemski, J. Appl. Phys., 30, 1195, (1959).

ACTIVATION OF SHALLOW As^+ IMPLANTS IN (100) SILICON BY INCOHERENT LIGHT ANNEALING

B. Canut , J. Pivot, [1]) D. Barbier , G. Chaussemy , J-F. Joly [2])

ABSTRACT

70 keV As^+ implants with various doses (2×10^{14} to 10^{16} cm^{-2}) in CZ boron-doped (100) silicon were rapidly annealed (< 12 s) in the 930-1175 °C temperature range, using an incoherent light annealing device, in order to form shallow N^+ layers. The samples were investigated by means of channeling techniques and electrical characterization. Outdiffusion or surface accumulation limits the As activation. The best N^+ layers were achieved from 4×10^{15} cm^{-2} implants annealed at 1100°C.

INTRODUCTION

Rapid Thermal Annealing (RTA) is now becoming the most attractive technique to remove implantation damage in silicon. The potential advantage of RTA over classical furnace annealing is to produce a near complete acitvation of implanted dopants without excessive spreading of the doping profile [1]. Besides, shallow and high dose As^+ implantation are required in the fabrication of the thin and highly doped emitter region of bipolar transistors. In this work, we realized N^+ layers in CZ Boron-doped silicon substrates by As^+ implantation followed by annealing with an incoherent light furnace. The As activation was controlled by Rutherford Backscattering Spectrometry (RBS) in the channeling geometry and by sheet resistance measurements. The aim of this paper was to correlate these measurements in order to test the N^+ layer characteristics as a function of the ion dose and the annealing temperature.

EXPERIMENTAL

The starting material was boron-doped CZ (100) silicon , 10 Ω.cm in resistivity. The wafers were implanted at room temperature with 70 keV As^+ ions with four doses (2×10^{14}, 10^{15}, 4×10^{15}, 10^{16} cm^{-2}). The RTA treatments were performed in a 30 kW halogen lamp furnace (ADDAX R1000 of AET, Meylan, France) under a continuous Ar gas flow. All the RTA experimental characteristics have been detailed in a previous work [2]. Five maximum plateau temperatures have been selected: 930, 970, 1025, 1100, 1175 °C. The annealing time within a temperature interval of less than 1% below these quoted maxima was 7 s ± 10%, except for 1175 °C (12 s). Fig.1 illustrates a typical RTA time-temperature cycle performed in the present work. The As activation has been studied by RBS experiments in the random and channeling geometries using 2 MeV $^4He^+$ ions and a detection angle of 150 °. Sheet resistance measurements were carried out by the four probe technique.

RESULTS

As an example, Fig. 2 shows the RBS spectra (randon and channeled) obtained on a Si sample implanted with 10^{16} As^+/cm^2 and annealed at 1100°C for 7s. The following parameters have been extracted from these spectra:

---

[1]) Laboratoire de physique de la matière, Institut National des Sciences Appliquées de Lyon, 20 Avenue Albert Einstein F69621 Villeurbanne cedex.

[2]) Département de physique des matériaux, Université Claude Bernard-Lyon I, 43 Bd du 11 Novembre 1918, F69621 Villeurbanne cedex.

i) the Si dechanneling yield (χ_{min}Si) measured behind the disordered region due to the implantation, ii) the As dechanneling yield (χ_{min}As) which is the ratio of the measured channeled and random As doses. The As substitutional fraction α was calculated with the usual formula [3]: $\alpha=(1-\chi_{min}As)/(1-\chi_{min}Si)$. The lowest As dose (2 x 10^{14} cm^{-2}) has not been investigatedby RBS experiments, and for the 10^{15} cm^{-2} implants only the χ_{min}Si values could be obtained due to the poor sensitivity of RBS in the As detection. The results are collected in table I in conjunction with the sheet resistance values. The decrease of χ_{min}Si when increasing the RTA temperature indicates a damage removal process in the implantation region. The best results were obtained for the 4 x 10^{15} cm^{-2} implants annealed at 1175°C. In these conditions χ_{min}Si reached the theoretical value of 3% corresponding to virgin single crystal Si.The χ_{min}Si values measured for the 10^{15} cm^{-2} implants were higher after annealing at any temperature (>8%). The As doses measured after annealing are in good agreement with the implantation values except for the highest temperature (1175 °C). At this temperature the ratios between the doses measured before and after annealing are 2.9 and 4.3 respectively for the 4 x 10^{15} and 10^{16} cm^{-2} implants. These results reveal that an As outdiffusion phenomenum significantly occured at temperatures higher than 1100°C. This outdiffusion phenomenum was also observed in Secondary Ion Mass Spectroscopy experiments (SIMS) carried out on samples implanted and annealed in the same conditions [4]. The As substitutional fraction α was rather low and significantly insensitive to the annealing temperature within the measurement accuracy. A slightly better substitutionality was however obtained for the 4 x 10^{15} cm^{-2} implants (mean value: 75%), whereas for the 10^{16} cm^{-2} implants the mean value of α was only 65%.

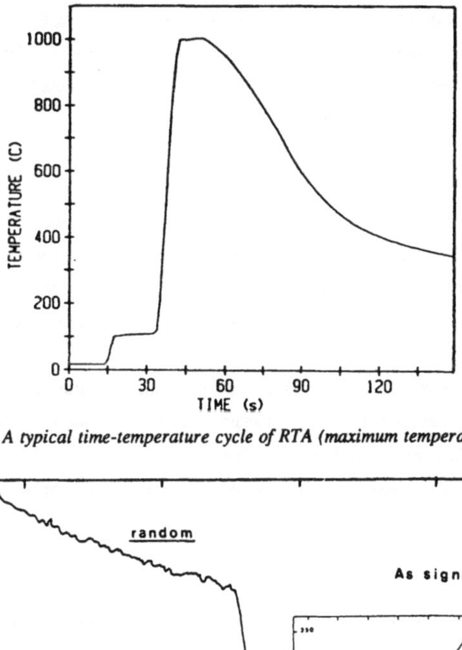

Fig.1: *A typical time-temperature cycle of RTA (maximum temperature 1025 °C)*

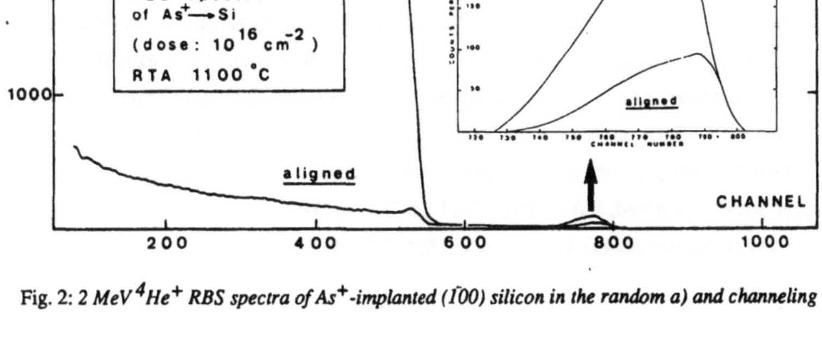

Fig. 2: *2 MeV $^4He^+$ RBS spectra of As^+-implanted (100) silicon in the random a) and channeling conditions b).*

RTA temperature:	as implanted	925	970	1025	1100	1175
DOSE :4,00E+15						
Si dechanneling yield:	17,30%	5,50%	5,10%	4,80%	4,40%	2,60%
As dechanneling yield:	100%	29%	29%	33%	32%	15%
As substitutionnal fraction:	0%	75%	75%	70%	71%	87%
sheet resistance	346	181	118	65	46	155
DOSE :1,00E+16						
Si dechanneling yield:	62,70%	8,60%	9%	6,60%	6%	6%
As dechanneling yield:	100%	44%	42%	33%	35%	42%
As substitutionnal fraction:	0%	61%	64%	72%	69%	62%
sheet resistance Ω		146	100	61	30	46

TABLE I: *RBS and sheet resistance measurement results obtained after annealing the 70 keV As^+ implants at various temperatures, as a function of the ion dose.*

Note that for such a dose implanted at 70 keV, the As solid solubility limit in Si is exceeded. In addition, an As pile-up phenomenum at the wafer surface was observed whenever no outdiffusion occured (annealing temperature below 1100°C), together with a poorer As substitutionality in the surface region (see Fig.2). According to Narayan, Holland and coworkers, the As nonsubstitutional fraction after RTA could be due to precipitation and/or clustering phenomena occuring during the damaged layer regrowth [5,6].
Furthermore, the sheet resistance of annealed samples dropped down when increasing the RTA temperature from 930 to 1100 °C. This feature must be related to a growing fraction of electrically active As after annealing at least up to 1100°C. For a given dose the sheet resistance value rose again when annealing the samples at 1175 °C.

CONCLUDING REMARKS

In this paper we have investigated the effects of both the annealing temperature and those of the implantation dose on the activation of shallow As^+ implants by incoherent light annealing for times lower than 12 s. The electrical activity measurements closely correlate the RBS observations, which suggests the following conclusions:

-Defect removal by RTA in 70 keV As^+ implants is better achieved for high dose implants, provided the solid solubility limit is not exceeded.

-A good compromise between a sufficient activation and a limited outdiffusion of the implanted As is obtained with annealing temperatures near 1100°C. The remaining fraction of unsubstitutional As seems to be mostly related to the accumulation of the dopants near the wafer surface.

- Within the experimental conditions tested in this work the 4 x 10^{15} cm^{-2} implant gave the best N^+ layers.

REFERENCES

[1] T.E. Seidel, C.S. Pai, D.J. Lischner, D.M. Maher, R.V. Knoell, J.S. Williams, B.R. Penumalli and D.C. Jacobson in *Energy Beam-Solid Interactions and Transient Thermal Processing* , ed. by D.K. Biegelsen, G.A. Rozgonyi, and C.V. Shank, Material Research Society Proceedings, vol. 35, p. 329 , MRS Pittsburgh (1985).

[2]J. Narayan in *Energy Beam-Solid Interactions and Transient,Thermal Processing*, ed. by J.C.C. Fan and N. M. Johnson, Material Research Society Proceedings, vol. 23, p. 335 , North Holland, New York (1984).

[3] W.K. Chu, J..W. Mayer and M. A. Nicolet in *Backscattering Spectrometry*, Academic Press, New York (1978).

[4] G. Chaussemy, B. Canut, D. Barbier, M. Remram, J-F. Joly, J. Pivot, and A. Laugier, European Material Research Society Proceedings, Journ. de Phys., to be published (1987).

[5] J. Narayan and O.W. Holland, J. Appl. Phys., 56, 10, p. 2913 (1984).

[6] O.W. Holland, J. Narayan, and D. Fathy, J. Appl. Phys., 59, 3, p. 905 (1986).-

ELECTRON BEAM HEAT SOURCE OF MODIFIED POWER DENSITY DISTRIBUTION

J.Felba*), K.Friedel*), K.Przybecki*)

An electron beam is the heat source of unusually high efficiency. It is therefore more and more widely used in materials processing where a phase transition takes place. In the thermal processes of surface mofification (e.g. annealing, hardening, plating), additional advantages of using the electron beam are: short interaction time, the possibility of depositing great amounts of heat into a selected area and the possibility of influencing the temperature distribution in the heated layer.

The deposition of energy into the heated surface can be realized by the irradiation of a material either with a scanned electron beam, or by means of a pulse electron beam covering simultaneously the whole selected area. In the latter case a glow discharge electron gun with cold cathode can be used. Such an electron gun makes use of the effect of glow discharge. At the moment when a voltage U of several or some scores kilovolts is applied between the anode (Fig.1) and cathode of the gun

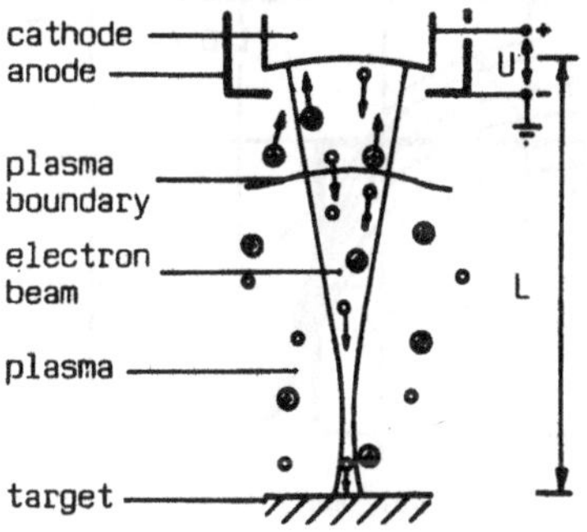

Fig.1 Main features of a simple glow discharge electron gun with cold cathode

(the gun being immersed in a gas atmosphere at a specific reduced pressure usually in the range 0.1 Pa to 10Pa), the ionization of the gas and then a discharge take place, which create plasma of high ion and electron concentration. Positive ions supplied by plasma are accelerated in a strong electric field and they bombard the cathode thus liberating electrons in a secondary emission process. The electron beam power is a function of the gas pressure, and of accelerating voltage and electrode geometry. A proper electrode structure and gun parameter selection can produce a variety of beam shapes.

For experimental purposes an electron gun with the electrode configuration having an axial symmetry was constructed, which generated

*) Institute of Electron Technology, Wrocław Technical University, ul. Grabiszyńska 97, 53-439 Wrocław, POLAND

a conical electron beam. With such beam geometry the change of working distance L results in the change of beam diameter on a target. With the change of the distance L, the beam power density distribution also changes (Fig.2).

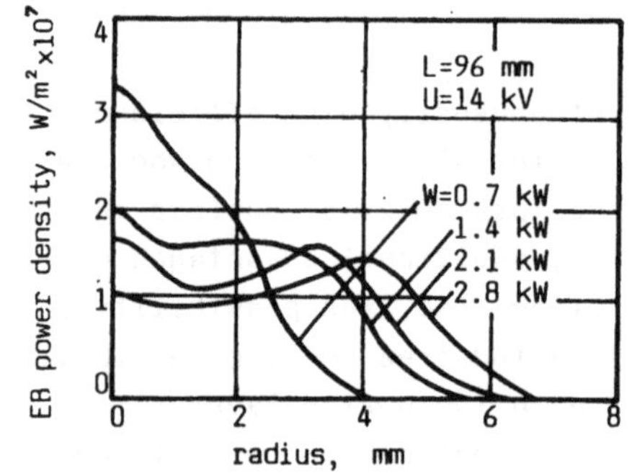

Fig.2 Experimental power density distribution of electron beam for different working distances L

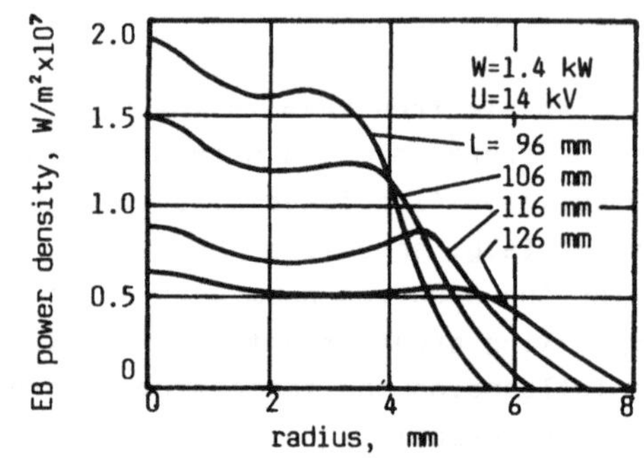

Fig.3 Experimental power density distribution of electron beam for different total beam power W

Another way of modifying the power density distribution is to change the beam power (Fig.3), leading to the change of diameter of the cathode

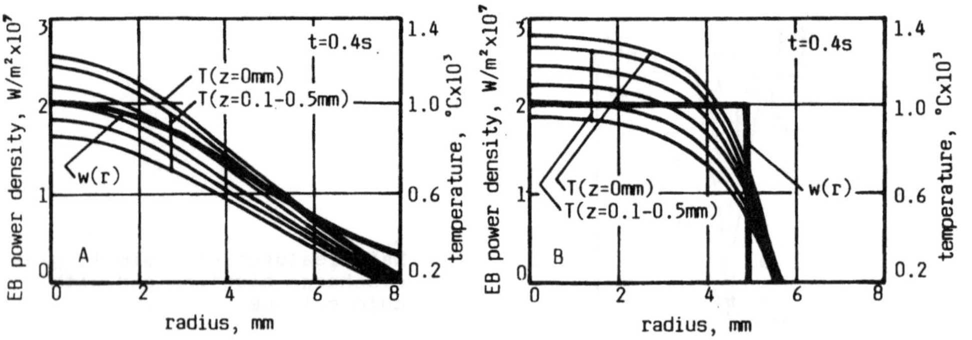

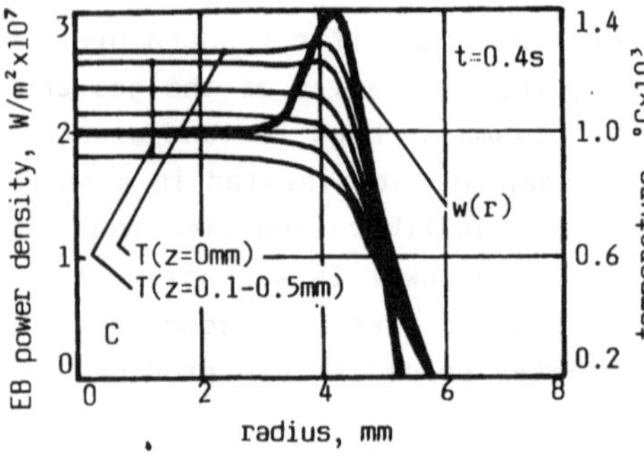

Fig.4 The temperature T distribution on the surface (z=0) and under the surface (z >0) for electron beams of different power density distributions w(r):
A - gaussian
B - rectangular
C - modified

emission area and beam convergence. Thus, practically, the selection of a proper cathode geometry, electron beam power and working distance L makes it possible to set the power density distribution of a heat source on target. Additionally, by the control of interaction time it is possible to obtain the required temperature distribution in the treated material.

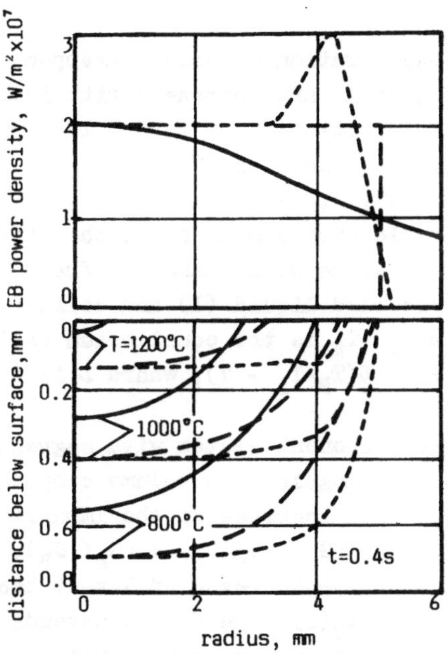

Fig.5 Selected power density distributions and corresponding isothermal lines

Surface modification of a material requires a determined temperature field in the treated zone resulting from the technology of specific materials processing and related with the power density distribution of a surface heat source and heat conduction. A numerical analysis of two-dimensional heat transfer for a flux of energy with optional distribution w(r) irradiating the material surface was carried out. Fig.4 presents exemplary calculated temperature distributions on the surface and in the undersurface layer of steel containing 0.55% C, by interaction time 0.4s of a surface heat source with different power density distribution. Selected isothermal lines for these distributions are shown in Fig.5. As it can be seen, the numerical analysis of the heat transfer makes it possible to modify the power density distribution of a surface heat source, which will be causing expected temperature distributions in the modified material, and which will enable the selection of the proper parameters and electrode configuration of the glow discharge electron gun with cold cathode.

NANOSECOND PULSE RECRYSTALLIZATION OF HIGHLY DOPED SILICON LAYERS

V.P.Popov, A.V.Dvurechenskii, Yu.A.Manjosov[1)]

Numerous experiments show that the concentration of solute trapped impurity atoms C_B and partition coefficient K' are increased with increase in interface velocity at rapid solidification /1, 4 to 6/. The crystal growth velocity V for a pure system is defined as /7/:

$$V = V_0 \exp(-H/kT_i)[1 - \exp(-\Delta G/kT_i)], \qquad (1)$$

where thermodynamic factor in the quadratic brackets accounts the difference in the forward and reverse jumps of atoms with change of free Gibbs energy $\Delta G = G_l - G_s$ between undercooled liquid (l) and solid (s) states at the interface temperature $T_i < T_M$ (T_M is the equilibrium melting temperature). Here ΔG is equal $\Delta G_v = L(T_M/T_i - 1)$, where L is the latent heat of crystallization.

In the case of solidification of a two-component system with segregation of solute atoms, it is possible to distinguish the volume component ΔG_v and the change of the surface energy with increase in the impurity concentration at the interface ΔC_B from ΔG: $\Delta G = \Delta G_v(C_B) + \Delta G_s(C_B)$. The dependence ΔG_v on C_B is determined from the decrease of latent heat $L(C_B) < L(0)$ and melting temperature $T_M(C_B) < T_M(0)$. The former depends on the mixing entropy, state density of electrons and phonons, lattice polarization near by ionized impurity atoms, the latter one may determine from the condition $\Delta G_v = 0$ at $V=0$ /5/.

The crystal growth velocity V at the steady-state conditions depends only on the temperature gradient in the solid phase ΔT_s and the latent heat L /4/:

$$V = K/\rho L \times \nabla T_s, \qquad (2)$$

where K is the thermoconductivity coefficient, ρ is the specific density. By reason $L(C_B) < L(0)$ with a glance to (2), $V(C_B)$ must be greater than V(0). An expression (1) gets simplified to $V = \text{const} \cdot L \Delta T$ at the small undercooling $\Delta T = T_M(C_B) - T_i \simeq 10$ K. Keeping in mind, that $V(C_B) > V(0)$, we get from (1), that $\Delta T(C_B) > \Delta T(0)$. In such a manner an undercooling of the melt is greater in the layer with greater impurity concentration and we may expect an increase in partition coefficient with increase of impurity content at identical impulse treatment.

If $K' \ll 1$, then a concentration of impurity is much less the thermodynamic limit of the solution C_B^c after passing the front of crystallization. The value of C_B^c is determined from the condition $L(C_B^c) = 0$ /3/. In the case of $C_B^s \ll C_B^c$ one may neglect a change in the latent heat of crystallization. The main effect of impurity concentration dependence must be connected with increase in the impurity content before the front of crystallization. An analysis of (1) shows that with about the same crystallization ve-

1) Institute of Semiconductor Physics, Academy of Sciences of the USSR, 630090 Novosibirsk, USSR

locity V, a low solubility impurity enrichment of the interface must increase ΔT, as the result of the decrease of surface energy between two phase. In this case we may see also in experiments an increase in partition coefficient K' with a decrease in interface temperature lowering the probability of a thermal emission of solute trapped atoms to the liquid phase /5/.

In fact, this behavior of K' we observed in (100) silicon wafers, implanted with different (one order of magnitude) doses of thallium atoms ($\varphi = 3 \times 10^{14}$ and 3×10^{15} cm^{-2}) and pulse annealed with ruby laser (τ_i = 20 ns, E_i = 1.5 J/cm^{-2}). The profiles of total and substitutional concentration of Tl atoms were gotten by RBS measurements with channeling and angular scanning near by $\{100\}$ and $\{110\}$ axis. The results are presented in Fig. 1. The maximum concentration of atoms in substitutional position C^s_{Tl} is observed at the depth ~120 nm and achieved $(3.5 \pm 0.5) \times 10^{18}$ or $(1.0 \pm 0.1) \times 10^{20}$ cm^{-3} respectively (the curves 1' and 2'), as the initial concentration (curve 3) differed in order of magnitude. As TEM measurements for the samples with greater dose showed, a cell formation and precipitation of impurity are observed at the depth lower 100 nm. The partition coefficient K' was not determined for these depths. At the greater depth, K' was calculated for the best fitting of the profiles of thallium atoms concentration in substitutional position (Fig. 1). The values of K' are equal to 0.1 and 0.4 for the lower and greater doses respectively.

We observed an increase in solute trapping of thallium atoms with introduction of other impurities to very high concentrations ($C^s_i \gtrsim 1$ at.%), such as B, P, As. This experiments are illustrated in Fig. 2. The laser treatment was the same as in Fig. 1, but some samples were doped before pulse recrystallization by 35 keV B^+ implantation. Boron concentration was ~4 at.% after laser annealing. Tl atoms concentration in lattice site (curve 2') was well over $(1.5-2.0) \times 10^{20}$ cm^{-3}, while in the layers with Tl atoms only (curve 1') - 1.0×10^{20} cm^{-3}. Solute trapping in this case is enhanced by increase of the growth velocity and the melt undercooling at double doping.

An undesirable effect at the increase in the second impurity concentration is the increase in the nonequilibrium trapping of the defects. High defect concentration f is measured by backscattering on displaced Si atoms and depends approximately linearly on the impurity content (0.4-4.0 at.%) for only As doping, not distorted the lattice. This effect is much more likely to be the same, as Tl enhanced trapping at double doping.

In order to compare these results with the prediction of Wood's theory /8/, we calculated $\Delta T(C^s_{As})$ and $V(C^s_{As})$ on the basis of the equations (1), (2) and theory of the regular solution for $L(C^s_{As})$ and $T_M(C^s_{As})$. V(0) was chosen 1.5 m/s by experimental conditions. Wood's dependence f(V) was corrected to $\Delta T(f)$ by (1). In Fig. 3 Wood's data are compared for (100) and (111) orientation (left and right dotted curves respectively) with the values $\Delta T(C^{S}_{As})$, converted to $\Delta T(f)$. A divergence is connected possibly with taken no to the account the lesser temperature in the model /8/.

Thus, we investigated the effects of the enhancement of the solute trapping of impurities and defects with the increase of the content of the same or another impurity at pulse liquid phase epitaxy. This behavior is opposite to that near by equilibrium condition /2/ and is accounted for by the increase of the melt undercooling and growth velocity in two-component system.

REFERENCES

(1) Aziz M.J., White C.W.: Phys.Rev.Lett. 57 (1986) 2675.
(2) Brice J. In: Atomic Diffusion in Semiconductors, Plenum Press, New York 1973.
(3) Cahn J.W., Corviel S.B., Boettinger W.J. In: Laser and Electron Beam Processing of Materials.-Acad.Press., New York 1980.
(4) Campisano S.U.: Appl.Phys.A 30 (1983) 195.
(5) Jackson K.A., Gilmer G.H., Leami H.J. In: ibid. /3/.
(6) Khaibullin I.B., Shtyrkov E.I., Zaripov E.M. et al.: Fiz.Tekhn.Poluprov. 11 (1977) 330-334.
(8) Wood R.F. In: Laser-Solid Interactions and Transient Thermal Processing of Materials. North-Holland, New York MRS 1983.
(7) Spaepen F., Turnbull D. In: Laser Annealing of Semiconductors, Acad. Press 1982.

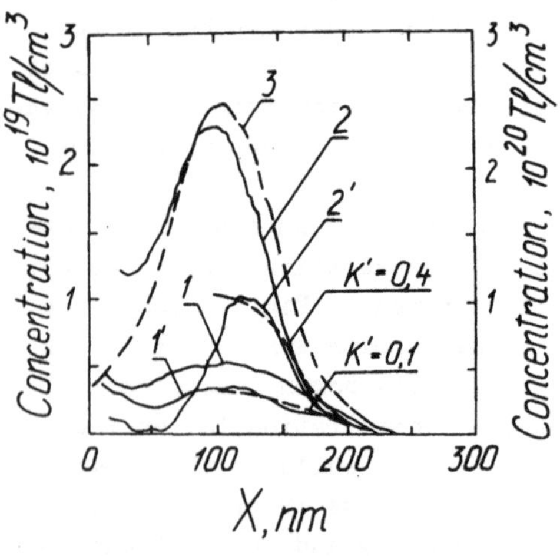

Fig. 1. Total (1,2), substitutional (1', 2') and initial (3) concentration profiles of Tl atoms after implantation (500 keV, $3x10^{14}$ (1), $3x10^{15}$ (2) cm^{-2}) and laser annealing.

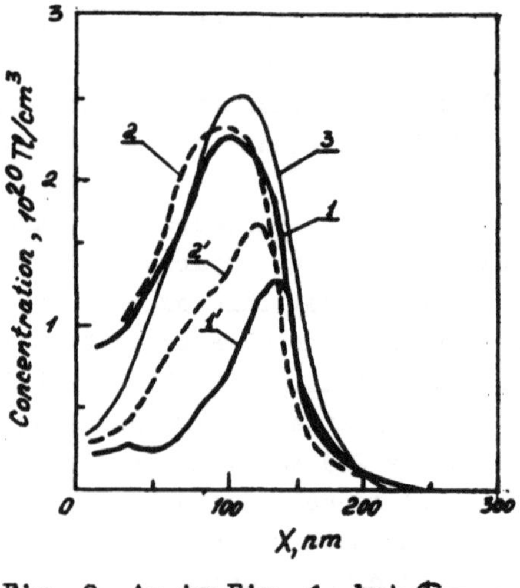

Fig. 2. As in Fig. 1, but Φ = $3x10^{15}$ (1); with B^+ implantation (35 keV, $2.5x10^{16}$ cm^{-2}).

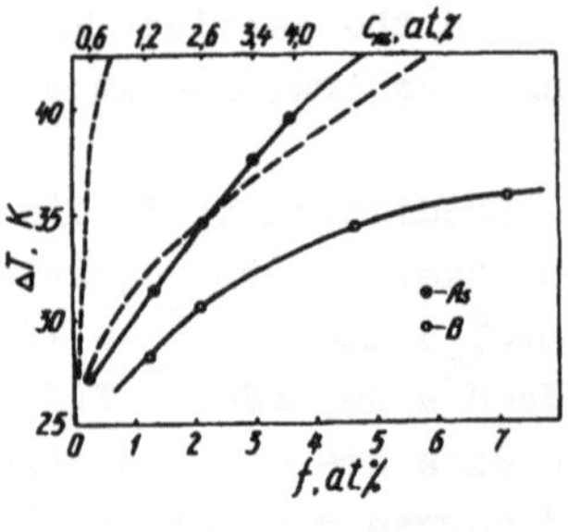

Fig. 3. The melt undercooling ΔT and part of displaced Si atoms f_s with increase in impurity content C_i^s.

THERMAL STRESSES IN SILICON WAFER DURING PULSED FLASH ANNEALING

V.A. Pilipovich, E.I. Gatskevich, V.L. Malevich[1)]

Pulsed annealing of ion-implanted semiconductor structures gives rise to a significant thermal stress which can reach critical values and lead to degradation of electrophysical properties of annealed structures. To determine optimal annealing conditions it is important to know distributions of temperatures and stresses depending on the thermal and energy characteristics of the anneal pulse.

This paper concerns the investigation of temperature and stress distributions in a silicon wafer subjected to flash annealing with a pulse of a xenon lamp with duration of 10^{-5} to 10^{-1} s.

The temperature field is calculated on the basis of a nonlinear heat flow equation using the finite-difference method. The temperature dependences of the specific heat and thermal conductivity of silicon are taken into account [1,3]. The wafer surface is supposed to be uniformly irradiated. For anneal times used, radiation and convection heat losses can be neglected. Hence, the problem can be treated as one-dimensional and temperatures can be considered as depending only on Z-coordinate (Z-axis normal to the wafer surface). The anneal pulse shape is taken as $I(t) = I_0[t/\tau \exp(1 - t/\tau)]^3$, where τ is the rise time to reach the peak power density I_0 [4]. The temperature distributions at the moment when the maximum temperature value on the irradiated surface is reached are shown in Fig. 1.

Thermal stresses are calculated in the plane-stress approximation. As follows from estimations [2], for the anneal pulse parameters specified the amount of wafer bending may be greater than its thickness h. This indicates that the small bending theory is not applicable to the case under study. Therefore, in contrast to ref. 1 we shall take into account stretching in the mid-plane of the wafer [6]. This problem is nonlinear and we shall use variational method to solve it.

The deformed state of the wafer is described by two functions, namely, $W(r)$ for bending and $u(r)$ for stretching in the mid-plane, where r is the radial coordinate. Then, the energy of the deformed wafer is described as

$$F = \frac{\pi E h^3}{12(1-\nu^2)} \int_0^R dr\, r \left[(\nabla^2 W)^2 - 2(1-\nu) \frac{1}{r} \frac{dW}{dr} \frac{d^2W}{dr^2} \right] +$$

$$+ \frac{\pi E h}{1-\nu} \int_0^R dr\, r \left[\left(\frac{du}{dr}\right)^2 + 2\nu \frac{u}{r} \frac{du}{dr} + \frac{u^2}{r^2} + \nu \frac{u}{r} \left(\frac{dW}{dr}\right)^2 + \right. \qquad (1)$$

$$\left. + \frac{du}{dr} \left(\frac{dW}{dr}\right)^2 + \frac{1}{4} \left(\frac{dW}{dr}\right)^4 \right] + \frac{2\pi M_T}{1-\nu} \int_0^R dr\, r\, \nabla^2 W -$$

1) Institute of Electronics, BSSR Academy of Sciences, Ya. Kolas St. 68/2, Minsk-90, 220841 (U.S.S.R.)

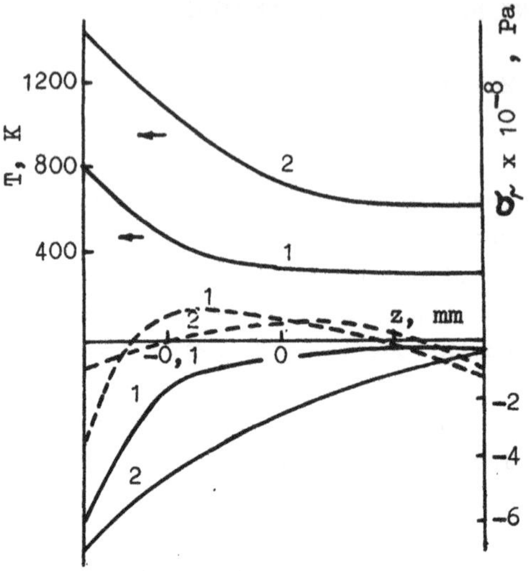

Fig. 1. Temperature and stress distributions for $r = 0$. Dashed curves correspond to the linear theory [2]. (1) $I_0 = 200$ kW/cm^2, $\tau = 50$ μs, (2) 50 kW/cm^2, 1000 μs.

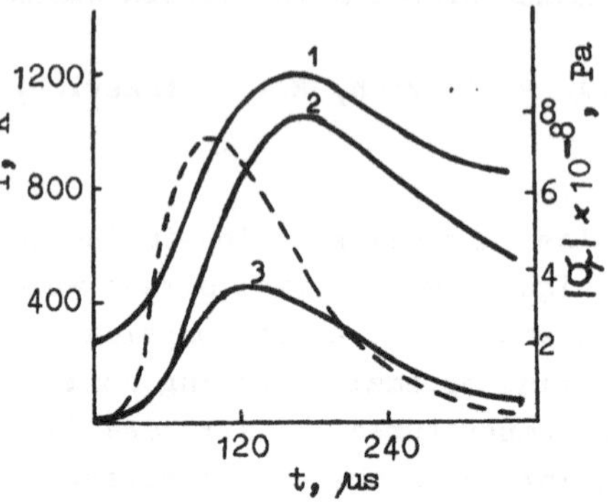

Fig. 2. Time dependences of temperature (1) and stress (2) with and (3) without stretching in the midplane taken into account. ($r = 0$, $z = -h/2$). ($I_0 = 150$ kW/cm^2, $\tau = 100$ μs). Dashed line corresponds to the anneal pulse shape.

$$- \frac{2\pi N_T}{1-\nu} \int_0^R dr\, r \left[\frac{du}{dr} + \frac{u}{r} + \frac{1}{2}\left(\frac{dW}{dr}\right)^2 \right],$$

where E is the Young's modulus, ν is the Poisson's ratio. The quantities M_T and N_T depend on the temperature distribution along z-coordinate and are expressed as

$$M_T = \alpha_T E \int_{-h/2}^{h/2} dz\, z [T(z) - T_0], \qquad N_T = \alpha_T E \int_{-h/2}^{h/2} dz [T(z) - T_0],$$

where $T_0 \simeq 300$ K is the initial temperature of the wafer corresponding to undeformed state, α_T is the linear thermal expansion coefficient. To find the extremum of the functional (1) we shall use the direct method. The boundary conditions for a freely-supported wafer are used [2]. The dependences of bending and stretching on the radial coordinate $W(r)$ and $u(r)$ taken as polynomials are

$$W(r) = ar^2 + br^3 + c, \qquad u(r) = \alpha r + \beta r^2 + \gamma r^3.$$

Substituting these functions into (1) and integrating over the wafer area, we derive the expression for the wafer energy as a function of the variables a, b, c, α, β, γ. Thus, the variational problem reduces to finding the conditional extremum of the function of many variables. By using method of Lagrange factors, we find the expression for α, β and γ as a function of a, b, and c. Substituting α, β and γ into the relationship for the energy (1) and, then, expressing b and c in terms of the variable a from the boundary conditions [2], we find the energy as a function of one variable a. This function has extremum for the value of a corresponding to the real root of the cubic equation

$$x^3 + C_1 p x^2 + (C_2 p^2 + C_3) x + C_4 p + C_5 p^3 = 0,$$
$$x = \frac{aR^2}{h}, \quad p = \frac{M_T R^2}{Eh^4}, \tag{2}$$

where the coefficients C_i depend on ν.

The tangential (σ_t) and radial (σ_r) components of stress are determined in the following way

$$\sigma_t = \frac{E}{1-\nu^2}\Big[\frac{u}{r} + \nu\frac{du}{dr} + \frac{\nu}{2}\Big(\frac{dW}{dr}\Big)^2 - \frac{z}{r}\frac{dW}{dr} - z\nu\frac{d^2W}{dr^2} -$$
$$- (1+\nu)\,\alpha_T\,[T(z) - T_0]\Big],$$

$$\sigma_r = \frac{E}{1-\nu^2}\Big[\frac{du}{dr} + \nu\frac{u}{r} + \frac{1}{2}\Big(\frac{dW}{dr}\Big)^2 - z\frac{d^2W}{dr^2} - \nu\frac{z}{r}\frac{dW}{dr} - \tag{3}$$
$$- (1+\nu)\,\alpha_T\,[T(z) - T_0]\Big].$$

Computations have been performed for [111]-oriented silicon wafers 350 µm thick and 6 cm in diameter. Other parameters were as follows: $\nu = 0.262$, $E = 1.29 \times 10^{11}$ Pa, $\alpha_T = 4.15 \times 10^{-6}$ K^{-1} [6].

Computation results (see Figs.1,2) indicate that stress is always maximum in the centre of the illuminated surface of the wafer ($r = 0$, $z = -h/2$). For anneal pulse energy density of about 100 J/cm^2 and duration of 10^{-4} to 10^{-3} seconds the computed stress is 2 to 5 times higher than the stress values found without stretching in the mid-plane taken into account. In contrast to the linear case where the tangential and radial stress components are equal and independent of r, these components take on different values (except for $r = 0$) and become dependent on the radial coordinate if stretching in the mid-plane is taken into account.

References

(1) Bentini, G.G., L. Correra: J. Appl. Phys. 54 (1983) 2057-2062
(2) Boley, B.A., J.H. Weiner: Theory of Thermal Stresses, Wiley, New York 1960
(3) Gurvich, L.V., I.V. Veits, V.A. Medvedev, et al.: Thermodynamic Properties of Individual Substances, Nauka, Moscow 1979 (in Russian)
(4) Impulse Light Sources, ed. by I.S. Marshak, Energija, Moscow 1978 (in Russian)
(5) Kontsevoj, Yu.A., Yu.M. Litvinov, E.A. Fattakhov: Plasticity and Strength of Semiconductor Materials and Structures, Radio i Svjaz, Moscow 1982 (in Russian)
(6) Landau, L.D., E.M. Livshitz: Theory of Elasticity, Nauka, Moscow 1965 (in Russian)

MULTIPLE PULSE IMPLANTATION DOPED LAYERS IN SILICON

Z.Werner, J.Piekoszewski, C.Pochrybniak, Z.Haratym, A.Sulik[1)]

A powerful, high-energy plasma pulse generated by plasma-injector type machines can reach an energy density sufficient to melt the surface layer of semiconductor in similar way as a laser or electron beam pulse. If such pulse contains dopant ions, it can dope the molten surface layer which next recrystallizes epitaxially. This approach to develop a single-step pulse implantation doping (PID) process has been extensively studied at our Institute over the last few years [2,3,5]. In the course of experiments it has become evident that there exist limitations of a single-pulse processing, consisting primarily in a limited concentration of dopant ions introduced by a single pulse (typically $10^{14}\div3*10^{14}cm^{-2}$). To explore a feasibility of introducing higher concentrations of ions by this technique, we investigated the properties of silicon layers doped by PID using multiple plasma pulses.

N-type silicon samples of (111) orientation and a resistivity of $4 \div 8$ Ωcm were subjected to 1 to 12 pulses of BF_3 plasma produced by JONOTRON'85 plasma pulse source [1]. The pulse energy was in the range $3.2 \div 5.7$ J/cm² and varied from pulse to pulse. The pulse duration is about 0.5 μs.

Carrier concentration and mobility were determined from Hall effect measurements on clover-shaped structures of Van der Pauw geometry. The concentration profiles were determined by sputtering the doped layer with 50 keV Ar^+ ion beam and measuring the resistance along a rectangular sample with coplanar electrodes attached to the surface. Spectral dependence of the photocurrent generated in the p-n photovoltaic junctions fabricated by multiple plasma pulse irradiation was measured on solar cell structures prepared on the samples in an identical way as those described in ref. [4].

Fig.1 presents the hole concentration and sheet resistance of silicon samples as functions of the number of BF_3 plasma pulses. It is seen that a sheet resistance below 10 Ω/sq. and a carrier concentration in excess of $7*10^{15}$ cm^{-2} can be obtained. The shaded region in the figure marked "linear" is a projection of the course of sheet resistance changes if each pulse produced equal increments of sheet conductance, with the spread of the values obtained after 1 pulse taken into account. The experimental data show that the effect of two pulses is far greater than the doubled effect of 1 pulse. This result is tentatively interpreted as follows. A single plasma pulse, apart from melting the silicon surface layer and introducing boron ions into the melt, deposits some boron

---

1) Institute for Nuclear Studies, 05-400 Swierk-Otwock, Poland.

atoms on the sample surface. These boron atoms arrive at the sample surface at a time when it is already solidified and they remain in electrically inactive form. The next pulse drives them into the molten layer and activates them electrically, thus enhancing the combined effect of 2 pulses.

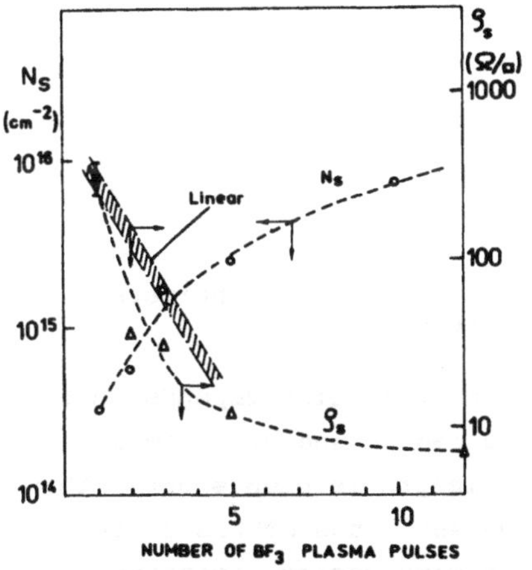

Fig.1. Surface concentration and sheet resistance of silicon samples as functions of the number of BF_3 plasma pulses.

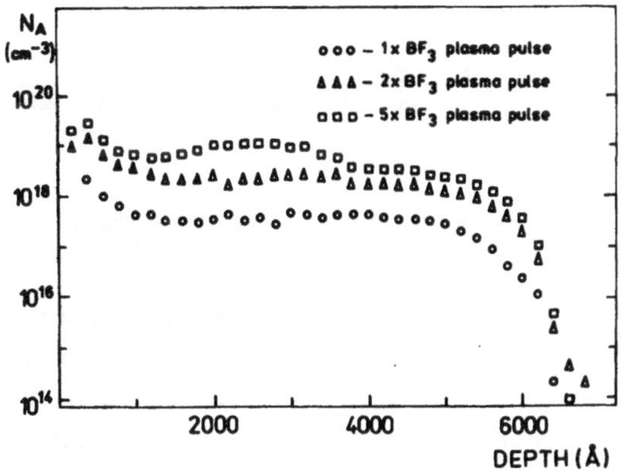

Fig.2. Carrier concentration profiles in silicon after 1, 2, and 5 pulses of BF_3 plasma.

The profiles of electrically active boron, presented in Fig.2 show the following features. Regardless the number of applied plasma pulses, all profiles extend to the same depth of about 6500 Å. The profiles remain essentially flat over most of the profile width. The effect of increasing the number of pulses is a gradual build-up of boron concentration. Again, the second pulse produces the greatest change of concetration. We interpret the obtained results as follows. The duration of the molten phase stage of the PID process is long enough for boron atoms to diffuse over the entire depth of the molten layer. Obviously, the profile terminates at a depth corresponding to the max. extension of the molten layer attainable at our experimental conditions, which turns out to be about 6500 Å. The successive pulses add more boron to this layer.

Fig.3. shows the external quantum efficiency of solar cells fabricated from the samples processed with varying number of plasma pulses. As the number of pulses grows, the short-wavelength part of the spectral characteristic deteriorates, while the long-wavelength part remains essentially unaltered. In the first approximation, the spectral characteristic of solar cell efficiency consists of two contributions; that of the front, heavily doped p-type layer, which dominates over the short-wavelength part, and that of lightly doped n-type substrate region, which dominates over the long-wavelength part of the characte-

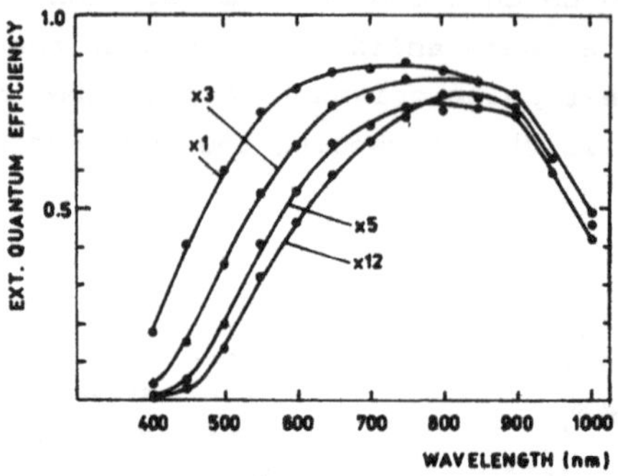

Fig.3. External quantum efficiency of silicon solar cells fabricated using varying number of BF_3 plasma pulses.

ristic. The magnitude of the front layer contribution depends on the diffusion length of minority carriers in this region and ultimately can drop practically to zero if the diffusion length is much shorter than the junction depth. With growing number of pulses, the diffusion length in the doped surface layer shortens due to growing doping concentration, contamination of the surface layer with heavy metal ions sputtered by plasma from the sample surrounding, and the build-up of the concentration of defects created by repeated fast recrystallization process.

Concluding, with increasing the number of BF_3 plasma pulses, the boron concentration in the doped surface layer of silicon grows monotonically. The boron profile remains essentially flat and extends to the same depth regardless the number of pulses. The minority carrier diffusion length in the doped layer deteriorates. Further work is in progress on better understanding of the processes involved in multiple plasma pulse doping technique.

REFERENCES

(1) Langner,J., et al.: this conference
(2) Piekoszewski,J., M.Gryzinski, J.Langner, Z.Werner: phys.stat.sol.(a) 67(1981)K163-K166
(3) Piekoszewski,J., Z.Werner, J.Langner, L.Jakubowski, C.Pochrybniak, A.Harasiewicz: Nucl.Instr.Meth. 209/210 (1983)477-482
(4) Piekoszewski,J., Z.Werner, C.Pochrybniak, J.Langner, A.Horodenski: this conference
(5) Werner,Z., T.Kozlowski, J.Piekoszewski, C.Pochrybniak, S.Gebalski, J.Langner. A.Sulik:Proc. EPM-84 Conf.(1984)147-151

PULSE IMPLANTATION DOPING - APPLICATION FOR PHOTOVOLTAIC JUNCTION FORMATION

J. Piekoszewski, Z. Werner, C. Pochrybniak, J. Langner, A. Horodeński+)

Pulse Implantation Doping (PID) technique makes use of the fact that the pulse of high-intensity plasma applied to the surface of the semiconductor can simultaneously melt and dope its near-surface layer if the kind, dose, energy of ions in the plasma and the pulse duration are appropriately chosen. PID is a genuine single-step doping process since in contrast to conventional implantation no further annealing of semiconductor is needed. In addition, there is an ease of scaling-up the dimensions of material processed in one step (over 100 cm^2) and of automation for high throughput ability. For these reasons, one of the areas where PID process can potentially be used is in the solar cell fabrication. In our previous papers by Piekoszewski et al. [2,3] using this technique with BF_3 plasma we demonstrated the feasibility of photovoltaic p+/n junction formation in n-type single-crystal silicon. The best conversion efficiencies of these junctions were reaching 6% on an 0.15 cm^2 area under AM1 illumination. In the present work we report new results of study in this field.

The samples were prepared of commercial n-type silicon of the resistivity of 4 to 8 Ωcm, (111) orientation, 3.6 cm^2 area and about 300 μm thick.

Solar cells were fabricated by the following sequence of principal operations:

1. formation of n+ layer on the back surface by evaporation of ≈200 A of Sb and alloying it into silicon by hydrogen plasma pulse,
2. formation of p+/n junction on the front surface by irradiation with BF_3 plasma pulses,
3. metallization of back and front contacts by evaporation of Ti-Pd-Ag layers and sintering at 500 °C for 15 min. The front-grid pattern consists of ten 1.3 mm spaced fingers and a contact bar,
4. removing the edge of the junction by masking and etching in CP4 for 12 sec.
5. AR coating by evaporization of 700 A SiO_x.

The BF_3 and hydrogen plasma pulses used to form p+/n/n+ structures were generated by a facility described by Langner at al. [1].

The doping conditions in this work were the following:

- energy densities E in the pulse: 3.5 to 6.4 J/cm^2, as measured by a set of calorimeters placed around the sample,
- ion source - sample distance: 25 cm giving the best homogeneity of

+)Institute for Nuclear Studies, PL-05-400 Swierk-Otwock, Poland.

sheet resistivity as compared to those obtained at 40 and 55 cm,

- number of pulses: single pulse of BF_3 to form p+/n and three pulses of hydrogen to form n+/n junctions.

Cells were tested by I-V measurements under an illumination of a tungsten halogen lamp adjusted to yield the specified AM1 short-circuit current in a calibrated reference cell. Table 1 gives the set of cell parameters grouped into two ranges of energy densities E at which p+/n junctions were formed. The range I corresponds to E from 3.5 to 4.6 J/cm² and the range II corresponds to E from 5.0 to 6.4 J/cm².

Table 1.

Sample No.	Energy density [J/cm²]	Isc [mA/cm²]	Voc [mV]	FF	η [%]
848		26.1	548	0.67	10.0
891		26.9	563	0.72	11.4
895	range I	27.5	553	0.70	11.1
898		28.7	543	0.66	10.7
900	3.5 - 4.6	29.3	555	0.67	11.4
901		28.7	544	0.66	10.8
941		25.9	562	0.74	11.2
842		25.1	560	0.64	9.4
844	range II	23.3	552	0.68	9.2
845	5.0 - 6.4	24.3	562	0.69	9.9
846		25.3	551	0.70	10.2
847		25.1	548	0.69	9.9

It is seen that in the range I the values of Isc are distinctly higher than in the range II; on average by 12%. This is consistent with the quantum efficiency (QE) versus wavelength data obtained for two cells representing both ranges of E as shown in Fig. 1.

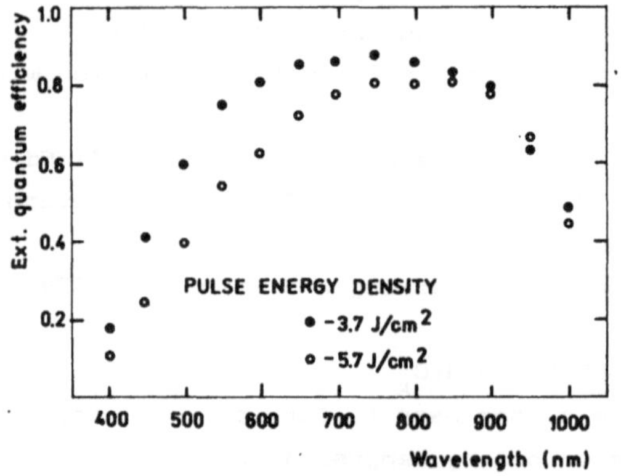

Fig. 1. External quantum efficiencies of two silicon cells processed by PID at different energy densities.

The junction depths (Xj) of these cells are 0.36 μm and 0.46 μm for E equal to 3.7 and 5.7 J/cm², respectively. This small difference in Xj cannot however account for the difference in QE over the whole wavelength range from 400 to 800 nm. The difference in Isc and QE observed between two ranges is presumably due to the difference in the minority carrier lifetimes τ in the emitter layers. Apparently, τ is smaller for the range II, i.e. for higher doping level and more violent thermal conditions of junction formation.

On the other hand, it is not clear at the moment why, on an average, FF and Voc in both energy ranges are the same whereas an average sheet resistance and thereby doping concentration of p+ layer differs in two ranges by a factor of 4.

The best cell obtained in the experiments with a single doping pulse has a conversion efficiency equal to 11.4 . Its I-V characteristic is shown in Fig. 2.

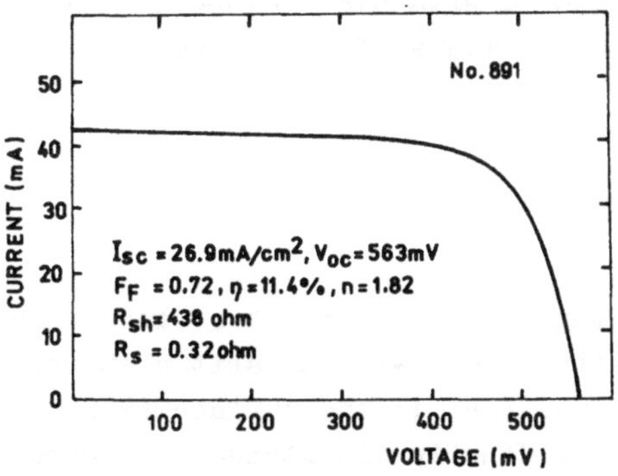

Fig. 2. AM1 illuminated I-V characteristic of silicon solar cell with p+/n junction formed by a single-pulse PID process.

It is obvious that resulting efficiencies are still too low to make PID process competitive with conventional technologies. However, in view of the facts that neither doping pulse parameters (ion energy distribution, intensity, time evolution) nor cell fabricating procedure are yet optimized. We feel confident that further improvement of cell parameters will be achieved in the near future. This expectation is additionally supported by efficiency of 13.3% observed in our laboratory when 3 doping pulses of BF_3 were used to form p+/n junction.

REFERENCES

1. Langner, J., et al., this conference.
2. Piekoszewski, J., Z. Werner, M. Gryzinski, J. Langner, J.Physique 43 (1982) 1353-1358.
3. Piekoszewski, J., Z.Werner, C. Pochrybniak, M. Gryzinski, J. Langner S. Gebalski, M. Klimkiewicz, A. Horodenski, Proc. Conf. Energy Pulse Modification of Semiconductors and Related Materials, vol. 1, pp. 142-146, Dresden, 1984.

EFFECT OF OXYGEN RECOILS ON THE DEFECT ANNEALING IN Si DURING RTA

D. Baither[1)], R. Koegler and E. Wieser[2)]

The structure of residual defects in silicon implanted and subsequently annealed with the aid of halogen lamps (RTA) was investigated by means of transmission electron microscopy. As implants through thin oxide layers are common to IC device processing, especially the influence of recoiled oxygen on the defect formation was considered.
At doses between 1×10^{15} and 2×10^{16} cm^{-2} arsenic ions were implanted into bare (100) silicon and into Si covered with an oxide layer.
The RTA treatment was carried out using a set of halogen lamps over the temperature range of 1050 °C to 1150 °C with heat pulses of 1 to 30 s. The TEM observations were performed by means of a high voltage electron microscope JEM 1000.

Implantation into bare silicon

At a dose of 1×10^{15} cm^{-2} annealing at 1000 °C for 1 s causes the formation of circular-shaped dislocation loops with a habit plan inclined and a Burgers vector (a/2 <110>) parallel to the surface (Fig. 1a). These loops slightly grow with their density simultaneously decreasing if the annealing time increases up to 3 s (Fig. 1b).

A higher dose of 5×10^{15} cm^{-2} leads to a considerable reduction in the defect density (Fig. 1c). The residual loops are elongated in the surface-parallel <110>-direction. Several loops have reached the surface and the remaining half loops are visible as straight lines. With increasing annealing time the small loops entirely disappear and the <110> orientation of the dislocation segments is improved yielding a regular network (Figs. 1d, e).

Further increasing the dose up to 10^{16} cm^{-2} implies no qualitative change in the defect structure (Figs. 1f, g).

Implantation through oxide layers

After implantation at a low dose of 2.5×10^{15} cm^{-2} circular loops described above become also visible, though their density is much lower

1) Institute for Solid State Physics and Electron Microscopy, Halle
2) Central Institute for Nuclear Research, Rossendorf

than after implantation at 1×10^{15} cm^{-2} into bare silicon (Fig. 2a).

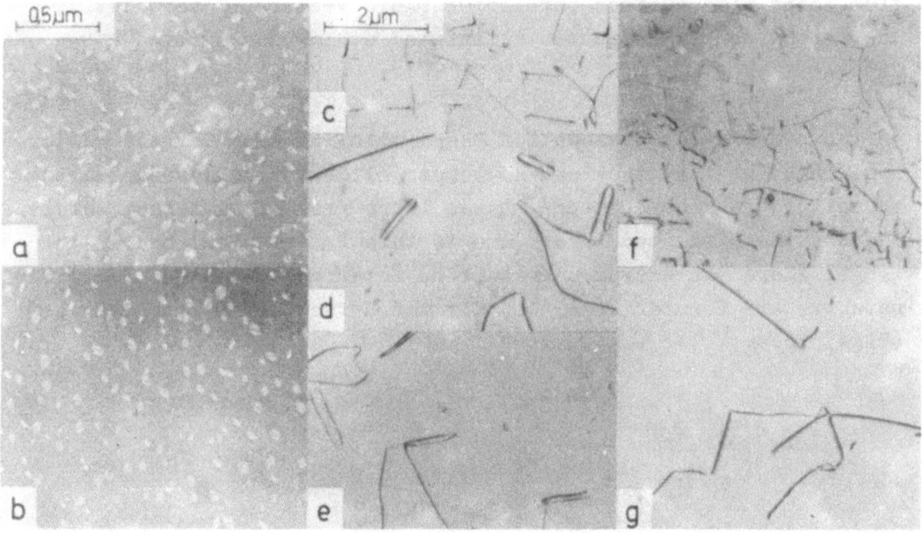

Fig. 1: As implantation into bare Si
a, b) $D = 10^{15}$ cm^{-2}; c - e) $D = 5.10^{15}$ cm^{-2};
f, g) $D = 10^{16}$ cm^{-2};
a, d, f) $T_a = 1100$ °C, $t_a = 1$ s
b, e, g) $T_a = 1100$ °C, $t_a = 3$ s
c) $T_a = 1055$ °C, $t_a = 30$ s

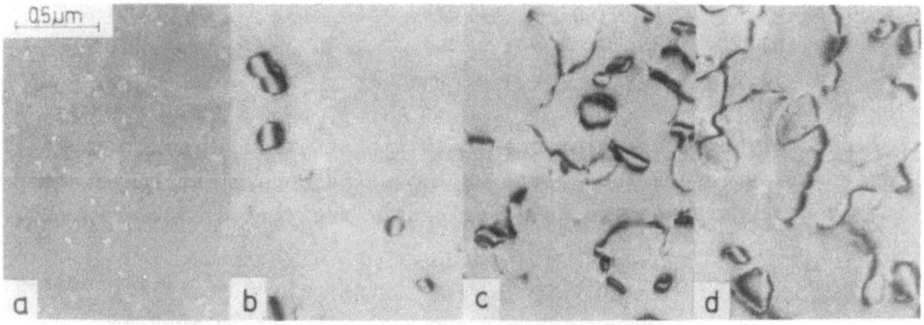

Fig. 2: As implantation into Si covered with SiO_2
$T_a = 1100$ °C, $t_a = 2$ s
a) $D = 2.5 \cdot 10^{15}$ cm^{-2}, b) $D = 5 \cdot 10^{15}$ cm^{-2},
c) $D = 1 \cdot 10^{16}$ cm^{-2}, d) $D = 2 \cdot 10^{16}$ cm^{-2}

At higher doses these loops entirely vanish and irregular-shaped loops with inclined Burgers-vectors arise. Their density grows with increasing dose (Figs. 2b - d).

The results confirm the assumption that the recoiled oxygen plays an important role in the defect formation /1/. Probably even the relatively low content after implantation through the native oxide layer /2/ influences the nucleation and growth of dislocation loops.

- Thus, the recoiled oxygen mainly restricts the growth of dislocation loops with surface-parallel Burgers vectors. Consequently, the defect concentration is diminished at low doses (Figs. 1 ↔ 2a).

- On the other hand, dislocation loops with inclined Burgers vectors are pinned so that they can no longer glide to the surface during annealing. As a consequence, these loops grow and interact before leaving the disturbed layer. Due to this interaction, beyond the range, where the amorph-/crystalline interface had been, irregular networks are formed (Figs. 4c, d); the defect density increases (Figs. 2c ↔ 1c - g).

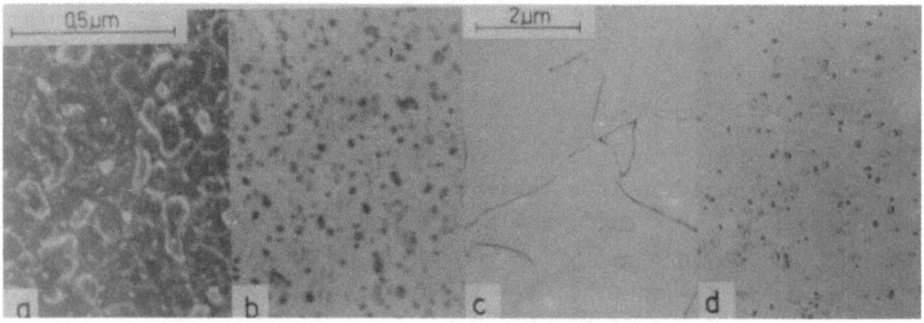

Fig. 3: As implantation into Si
$D = 5 \cdot 10^{15}$ cm^{-2}
a, c) bare Si; b, d) Si covered with SiO_2
a, b) T_a = 1125 °C, t_a = 1 s
c, d) T_a = 1225 °C, t_a = 1 s

- For high oxygen concentrations after implantation through thicker screen oxide layers the growth of the surface-parallel loops with inclined Burgers vectors is considerably restricted, too (Fig. 3).

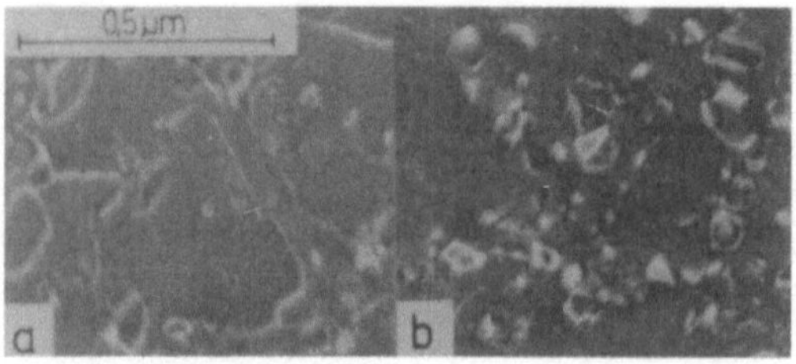

Fig. 4: As and O implantation into Si
a) 1.10^{16} As cm^{-2} + 7.10^{15} O cm^{-2}
b) 1.10^{15} As cm^{-2} + $1.5 \cdot 10^{16}$ O cm^{-2}

- The results mentioned above are supported by the investigation of oxygen-implanted specimens (Fig. 4).

/1/ D. K. SADANA, et al., Nucl. Instrum. Methods 209/210 (1983) 743 - 750

/2/ N. R. Wu, D. K. SADANA and J. WASHBURN, Appl. Phys. Lett. 44 (1984) 782 - 784

ANALYTICAL EVALUATION OF THE IMPURITY REDISTRIBUTION DURING POST-IMPLANTATION LASER ANNEALING

Florin Găiseanu[§], Marius Bâzu[§] and Lucian Gălăţeanu[§]

INTRODUCTION

A substantial redistribution of some impurities (such as B, P and As in Si) has been observed (5) after post-implantation annealing performed using a pulsed ruby laser. Althought the phenomena occuring during the laser annealing in the irradiated silicon are still a controversed topic, one generally admits (1,4,5) that this appreciable redistribution would be induced by the short-time pulsed annealing (10^{-7} - 10^{-8}s) into a silicon molten phase only. The purpose of this paper is to introduce an analytical model, discussed with respect to some available experimental results previously reported (5).

THEORETICAL MODEL

A diffusion coefficient D(x,t) depending on the spatio-temporal evolution of the laser-induced temperature T could be computed from a molten silicon model (1). However, such a model cannot provide the analytical required expressions. Defining an average diffusion coefficient D(t) with respect to the spatial coordinate, i.e. $D(t)=(1/x_j)\int_0^{x_j} D(x,t)dx$ (x_j-the junction depth) and furthermore by using the transformation $\theta=\int_0^t D(t)dt$, the simplest form of the diffusion equation $\partial C/\partial\theta = \partial^2 C/\partial x^2$ (were C is the impurity concentration) can be obtained. Moreover, taking into account that $\partial C/\partial x|_{x=0}=0$ would be a resonably fulfilled condition for the usual dopant impurities (6% agreement between the integral profiles before and after laser annealing for B and P and ±3% for As (4)), the following solution of the diffusion equation associated with the initial condition (Gaussian implanted profile) can be deduced:

$$C(x,\theta)=\frac{N}{2\sqrt{\pi}(2\Delta R_p^2+4\theta)}\left\{\exp\left[-\frac{(x-R_p)^2}{2\Delta R_p^2+4\theta}\right]\left[1+\mathrm{erf}\left(\frac{x/4\theta+R_p/2\Delta R_p^2}{1/4\theta+1/2\Delta R_p^2}\right)\right]\right. +$$

$$\left.+\exp\left[-\frac{(x+R_p)^2}{2\Delta R_p^2+4\theta}\right]\left[1-\mathrm{erf}\left(\frac{x/4\theta-R_p/2\Delta R_p^2}{1/4\theta+1/2\Delta R_p^2}\right)\right]\right\}, \quad (1)$$

were N is the implantation dose, R_p and ΔR_p the projected range and the standard deviation, respectively.

§ R&D Center for Electronic Components, str. Erou Iancu Nicolae 32B, Bucharest 72996, ROMANIA

PROFILE EVALUATION

As a validity check of relation (1), a comparison with some experimental data (5) obtained after the ^{11}B, ^{31}P and ^{75}As implantation in [001] Si wafers in a range of 35-100 keV and redistributed by pulsed laser annealing (λ=0.69 µm, the energy of incident beam E~1.5-1.7J/cm^2) were drawn (Fig.1). R_p and ΔR_p were calculated from an adequate polinomial approach (2).

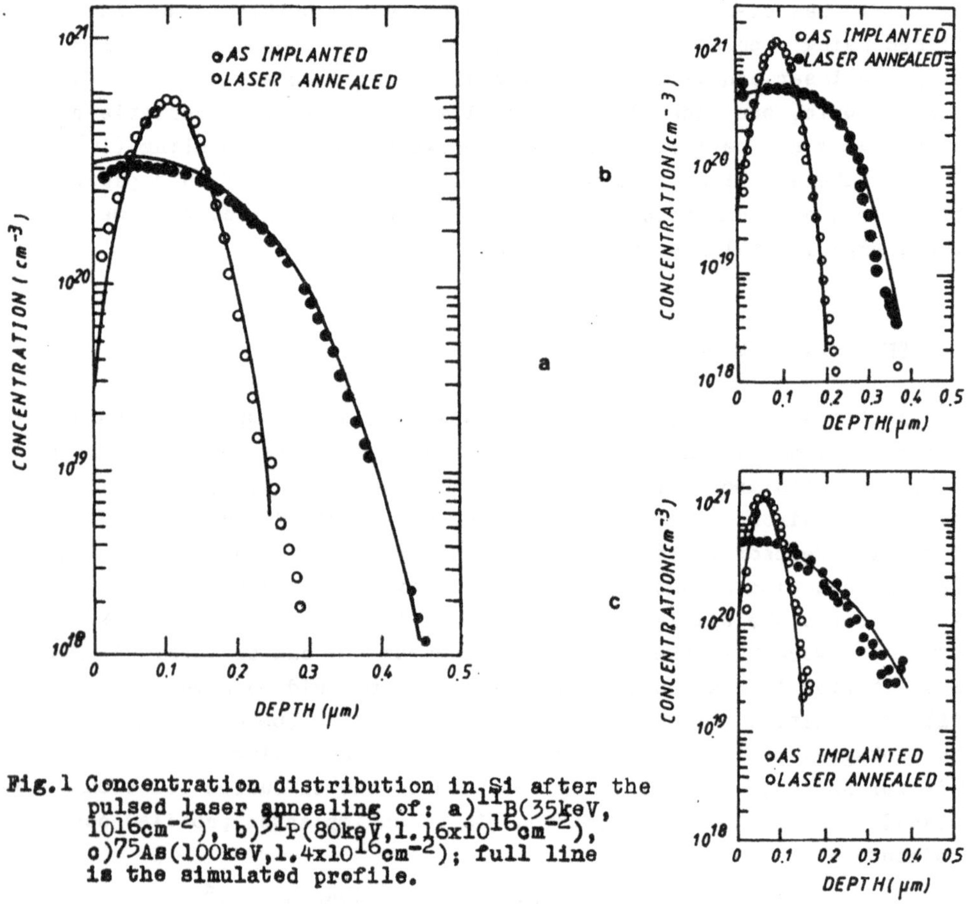

Fig.1 Concentration distribution in Si after the pulsed laser annealing of: a) ^{11}B(35keV, 10^{16}cm^{-2}), b) ^{31}P(80keV,1.16x10^{16}cm^{-2}), c) ^{75}As(100keV,1.4x10^{16}cm^{-2}); full line is the simulated profile.

As it can be seen from Fig.1, the agreement between the theoretical results (rel.(1)) and the experimental data is good if Θ=4.3x10^{-11} cm^2, 7x10^{-12}cm^2 and 10^{-10}cm^2 for B, P and As, respectively. From these data and taking into account the calculated time τ=230 ns during of which the Si implanted layer is in a liquid state (4), one may deduce the values of the effective diffusion coefficient D_{eff}=1.87x10^{-4}cm^2/s, 3x10^{-4}cm^2/s and 4.3x10^{-4}cm^2/s, respectively.

DISCUSSION

The typical deviations from the theoretically predicted profiles could come from: (i) the deviation from the initial Gaussian profile (a modified Pearson IV distribution would be adequate for B); (ii) segregations (lestes) phenomena; (iii) a dependence of D on C, intimately related to specific diffusion mechanisms at high concentrations (3); (iv) the motion of the molten front (4); (v) the specific temperature distribution induced by the laser incident beam. In connection with the items (v) (and (iv)), the region near the silicon surface would benefit of the highest temperature on the entire duration of the thermic excitation state of Si, so that an enhanced redistribution would be in this region stimulated. Because the low concentration range is to be exactly in an opposite situation, an impurity "accumulation" in the middle region of the profile (Fig.1b) would occur, depending on the particular values of D(T) for each impurity type. However, as it can be seen from Fig.1, the agreement between the theory (rel. (1)) and the experimental data can be appreciated as good, well justifying the approach proposed by rel. (1).

CONCLUSIONS

An analytical explicit expression relating the concentration redistribution during the laser annealing after the impurity implantation was derived from an adequate theoretical model. The comparison between the theoretical results and some experimental data shows a good agreement, allowing to appreciate that the possible deviations from the theoretical predictions could be considered as secondary effects.

REFERENCES

(1) Baeri, P., S.U.Campisano, G.Foti and E. Rimini: J.Appl.Phys.50(1979) 788-791

(2) Găiseanu, F. and M.Bădilă: Proc.of Microelectronic Conference, sept.4-6th (1984) 63-65

(3) Găiseanu, F.: J.Electrochem.Soc.132 (1986) 329C-331C

(4) Wang, J.C., R.F.Wood and P.P.Pronko: Appl.Phys.Lett.33 (1978)455-456

(5) White, C.W., W.H.Christie, B.R.Appleton, S.R.Wilson, P.P.Pronko and C.W.Maggie: Appl.Phys.Lett.33 (1978) 662-664

RAPID THERMAL ANNEALING : TEMPERATURE AND ACTIVATION UNIFORMITIES - COOLING SPEEDS AND MINIMUM DIFFUSION LENGTHS

J-M.DILHAC * and A.MARTINEZ *

ABSTRACT : Experimental evidence of the influence of thermal history and of gas pressure and flow, on temperature drop at periphery of a wafer, is given through thermal data and sheet resistance measurements. Cooling speeds are then presented and the related diffusion lengths are computed.

INTRODUCTION AND EXPERIMENTAL PROCEDURE

The temperature uniformity and the minimum diffusion lengths achievable through rapid thermal annealing, are essential in the area of semiconductor processing. We have measured or computed these features in the case of an annealer (ADDAX RV1000 3")(1) already presented elsewhere(2). 7-12 Ω.cm, P type silicon wafers are used in our experiments.

TEMPERATURE AND ACTIVATION UNIFORMITY

To measure the temperature drop, known to occur at wafer periphery(3), we have recorded the temperature reading of a thermocouple located 1 mm from the sample edge, and the output signal of a second thermocouple resting on the wafer center, whose temperature is kept constant. In the case of vacuum, this temperature is high enough to avoid measurements problems due to the wafer-thermocouple contact high thermal resistance(2). Fig.1 gives these temperatures versus time profiles in the case of a wafer heated in vacuum or at atmospheric pressure, the gaseous atmosphere being, in this case, first stagnant, and then N_2 being admitted at a flow rate of 5 l/min. The processing quartz chamber is either cold or hot (i.e., two cycles like the one recorded have been previously executed) at the beginning of the cycles. From the curves recorded at atmospheric pressure, it can be seen that the temperature drop i) is progressively reduced ii) is modified when N_2 is admitted iii) and that there is a memory effect related to the initial temperature of the processing chamber. The temperature drop is smaller under vacuum. On the other hand, boron implanted Silicon samples (2.10^{15}at.cm^{-2}, 35 kev) were annealed at atmospheric pressure and in vacuum, for different times and at various temperatures. The sheet resistances were then measured versus distance across the samples. A typical result is shown in Fig.2 : in the case of vacuum, the sheet resistance increase near the edges is smaller than in the case of atmospheric pressure processing. As already shown through thermocouple measurements, the temperature drop at the edge of a wafer is therefore smaller in vacuum, than at atmospheric pressure.

---

* Laboratoire d'Automatique et d'Analyse des Systèmes du CNRS, 7 avenue du colonel Roche, 31077 Toulouse CEDEX, FRANCE. (also INSA TOULOUSE).

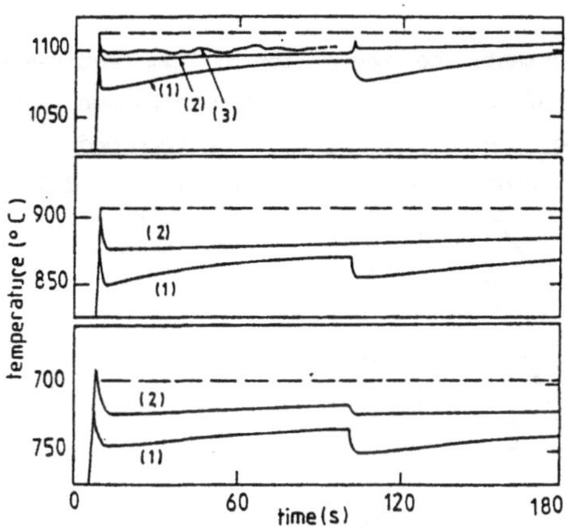

Fig.1 : Thermocouple temperature readings. Dashed curves : center temperature setpoint; solid curves : peripheric temperature, (1) atmospheric pressure, processing chamber initially cold, (2) atmospheric pressure, processing chamber initially hot, (3) vacuum, processing chamber initially cold. N_2 is started at t = 100 s.

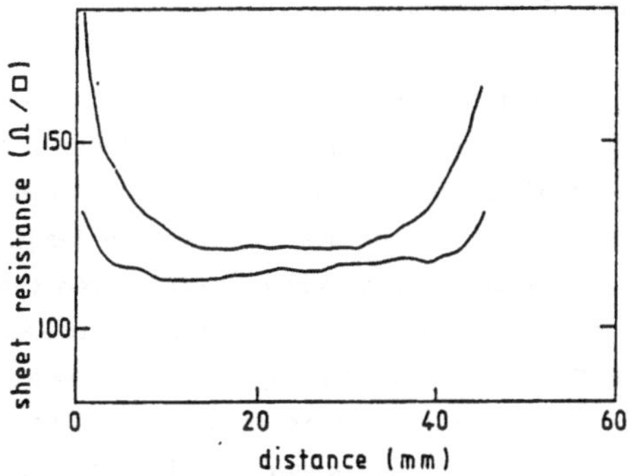

Fig.2 : sheet resistances as a function of distance across the silicon samples. Upper curve, atmospheric pressure ; lower curve, vacuum. Annealing : temperature 900°c, duration 2 min.

COOLING SPEEDS AND MINIMUM DIFFUSION LENGTHS

The ideal time-temperature profile generated in a rapid thermal annealer is rectangular. In practice, the ramp-up and ramp-down sections of the profile are not vertical, the plateau temperature may not be constant and the ramp-up may overshoot the temperature plateau(4). These deviations induce, by themselves, an additional impurity diffusion when a gradient is present. If an efficient closed loop control system is used, they may be avoided except for the ramp-down section (indeed, the cooling of a wafer through radiation and convection cannot be instantaneous) which therefore determines the smallest redistribution achievable through what has been called a "thermal spike"(5).
To estimate the diffusion length related to the ramp-down, consider the experimentally measurable parameter(6)

$$t_{eff} = \frac{1}{D(T_p)}\int_{T_a}^{T_p} D[T(t)]\,\frac{dt}{dT}\,dT$$

t_{eff} will be refered to as the effective diffusion time at temperature T_p (plateau temperature), T_a is the ambient temperature. The diffusivity D is assumed to be given by the Arrhénius expression $D(T) = D_{\circ}\exp(-E/kT)$ where E is the activation energy, k is the

Boltzmann constant and D_o is also a constant. The computation of t_{eff} requires the recording of the temperature-time cooling profile T(t), but since the ramp-down is almost linear in the early stages of the cooling, the only parameter to be recorded is the initial cooling speed. Fig.3 shows this speed versus plateau temperature, for 2 and 3 " wafers.

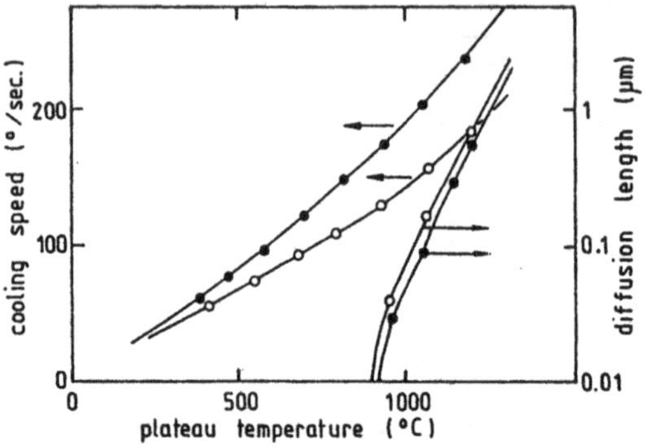

Fig.3 : experimental cooling speeds and calculated diffusion lengths (As in Si) as a function of plateau temperature. ● 2"wafer, ○ 3"wafer.

The different speeds are due to the different thicknesses of the wafers. The corresponding effective diffusion times are nearly constant vs T (0.18 and 0.23 $\pm$ 0.1 s respectively). They have been computed in the case of As in Silicon, since no enhancement due to RTA in the diffusion of As has been reported(7). The calculated diffusion lengths are also shown ($3\sqrt{D(T_p).t_{eff}}$), with $D(T_p)$ = 22.9 exp(-4.1 ev/kT) cm^2/s(8). From these data, it can be seen that when redistribution must be confined to 1 µm (typical production process), heat treatments up to about 1300°C are allowed. For typical development processes where redistribution must be confined to 0.1 µm, the limit is reduced to 1050°C, and for a redistribution of 0.01 µm the limit is 950°C. For such a small redistribution, the use of a thermal flux annealing technique may be necessary if high temperatures are needed(9).

CONCLUSION

In this paper, we have demonstrated the influence of two RTA equipment characteristics, the temperature uniformity and the cooling speeds on two properties of annealed wafers, i.e. the homogeneity of activation and redistribution lengths. Vacuum processing is recommended for high activation homogeneity, on the other hand, a means to compute the minimum redistribution achievable by a given RTA equipment has been given.
The authors would like to thank C.GANIBAL and F.MAROU for experimental assistance.

REFERENCES

(1) The ADDAX RV1000 3" model is manufactured by AET, ZIRST, chemin des malacher, 38240 Meylan, FRANCE
(2) J-M.DILHAC,C.GANIBAL, and A.MARTINEZ : Mat.Res.Soc.Proc.n°92,1987
(3) K.N.RITZ, M.DELFINO, C.B.COOPER III, and R.A.POWELL : J.Appl.Phys.,60(2),1986,p800
(4) J.D.FEHRIBACH,R.GHEZ, and G.S.OEHRLEIN : Appl.Phys.Lett.46(4),1985
(5) A.ARBEL, and M.NATAN : J.Appl.Phys.61(3),1987,p1209
(6) J.PHILIBERT : Diffusion et transport de matière dans les solides, Les Editions de Physique
(7) A.E.MICHEL : Mat.Res.Soc.Proc.n°52,1986,p3
(8) R.B.FAIR and J.C.C.TSAI : J.Electrochem.Soc.,122(12),1975,p1689.
(9) C.HILL : Mat.Res.Soc.Proc.n°13,1983,p381

THE INFLUENCE OF IMPLANTATION PARAMETERS AND ANNEALING CONDITIONS ON THE FORMATION AND PROPERTIES OF $MoSi_2$ LAYERS

Geza Valyi*, Christine Dehm*, Heiner Ryssel* and Wolfhard Möller**

ABSTRACT

$MoSi_2$ layers were fabricated by ion-beam mixing and subsequent annealing. The mixing was performed by arsenic ions, which is advantageous because the same process step can be used for silicidation and for doping of the underlying layer. The temperature during implantation was varied and various implantation energies (140-200 keV) were used in order to determine the effect of these parameters on the silicidation reaction. While the thickness of the silicide formed is proportional to dose and strongly dependent on implantation temperature, implantation energy is relatively unimportant in the energy range investigated. The results are in good agreement with dynamic Monte Carlo simulation of the mixing process.

1. INTRODUCTION

Metal silicides are of great importance in modern semiconductor devices. While the silicides of some noble or near-noble metals (Pt, Pd and Ni) can be used as ohmic contacts, refractory-metal silicides (i.e. the silicides of the metals belonging to groups IV, V, or VI of the periodic table), can be applied also as gates and interconnects, because of their low electrical resistivity and high temperature stability. The most important of the latter group are: $TiSi_2$, $TaSi_2$, $MoSi_2$, and WSi_2 [2].

Despite the immense work which has been put into the research of metal silicides and the great number of publications appeared in the last decade about this topic, there are still many problems which have not been resolved until now. One of these is the mechanism of silicide formation. This is not only a question of scientific interest, but it is also of practical importance, because the knowledge of the exact formation mechanism is inevitable for the optimum use of these materials.

In our work, we studied the formation of $MoSi_2$ layer by ion-beam mixing. Using this method, at first a layer of the respective metal is deposited on the surface of silicon (or polycrystalline silicon) and then ions are implanted through the metal layer into the silicon. The implantation of ions through the interface between metal and silicon

*Lehrstuhl für Elektronische Bauelemente, Universität Erlangen-Nürnberg
Artilleriestraße 12, D-8520 Erlangen, FRG

**Max-Planck-Institut für Plasmaphysik
Boltzmannstraße 2, D-8046 Garching/München, FRG

destroys the native oxide layer (which is always present at the surface of silicon) and the atoms of the metal and silicon are mixed [5-8]. Depending on the conditions, silicide is directly produced during implantation.

Ion-beam mixing is a very complex phenomenon. The following effects may contribute to the mixing process [7,8]:

The implanted ion loses its energy through collisions with the substrate atoms and comes to rest in less than about 10^{-13} seconds. Atomic displacements initiated by direct collisions are called "Ballistic Mixing". This prompt phase of ion mixing lasts less than 10^{-12} seconds.

The energy, which was deposited during the ballistic processes, is dissipated in the form of heat over the entire solid. If this thermalization process is faster than heat conduction in the material, regions with very high energy density will exist for a short time ($\leq 10^{-11}$ seconds). These regions are called "Thermal Spikes" [3]. Because the average energy of the atoms in the spike is often higher than the heat of melting of the substrate (approximately 1 eV/atom), the material of the spike is sometimes regarded to be in a liquid state. Since spikes exist only for very short periods, it is extremly difficult to study them. However, it is clear, that diffusion processes and chemical reactions will proceed much faster in the spikes.

After the energy is completely dissipated, there is a uniform temperature throughout the solid, but the system may remain in a non-equilibrium state with respect to variables other than temperature (e.g. defect concentrations). While the system returns to equilibrium, further atomic rearrangements may occur. These transport processes may last an hour, and because they are promoted by radiation-induced defects, they are called "Radiation Enhanced Diffusion".

Until now, we considered only physical effects. Since chemical reactions may also occur (and this is always true in the case of silicide formation), it is generally impossible, to describe ion-beam mixing in a simple manner. Which one of the above mentioned effects dominates may depend on the layers which are mixed, on the temperature during mixing, on ion dose, and even on the mass of the implanted ion.

In this paper, we present our results, compare them with those published by other authors and discuss the differences in respect to formation mechanism of the molybdenum silicide.

2. EXPERIMENTAL TECHNIQUES

For our experiments, <100> oriented, boron-doped silicon with a resistivity of 15 Ωcm was used. The wafers were etched prior to the molybdenum deposition in a 10% HF solution in order to minimize the thickness of the native oxide.

The ion-beam mixing was performed in a Varian DF 3000 implanter using arsenic ions. The dose was $5x10^{15}$ cm^{-2}. During the experiments with different implantation energies, the current was approximately 100 µA. In order to investigate the effect of implantation temperature, the current was varied between 50 and 1400 µA.

After ion-beam mixing, part of the wafers were annealed. Either a conventional annealing furnace, a rapid optical annealer Heatpulse 610, or a vacuum annealer Varian IA 200 was used. Annealing temperatures were varied between 450°C and 1100°C. The annealing time was between 5 minutes and 4 hours for thermal annealing, and between 10 seconds and 5 minutes for rapid thermal annealing.

The thickness of the silicide layers was determined using a Dektac IIA step height measuring apparatus after the unreacted molybdenum had been etched off by concentrated nitric acid. The sheet resistivity of the silicide layer was measured by a 4-point probe.

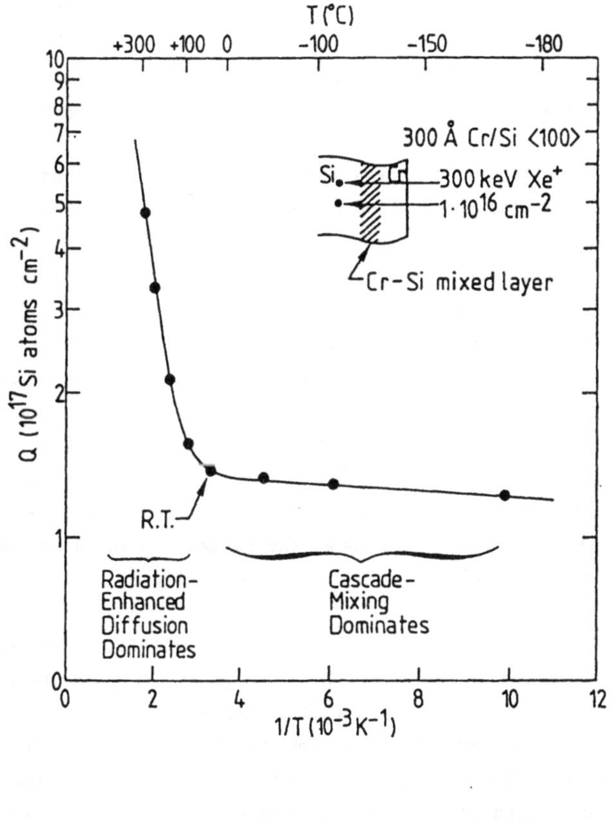

Fig.1. The amount of silicon atoms contained in the Cr-Si mixed layers, Q, vs reciprocal implantation temperature as given by Mayer et al. [8].

3. RESULTS AND DISCUSSION

The implantation of arsenic at different current densities resulted in different thicknesses of silicides formed through ion-beam mixing. This is caused by the current-density dependence of the wafer temperature.

The temperature dependence of the ion-beam mixing process can usually be described by a curve where the logarithm of the number of intermixed ions (lnQ) is plotted vs the reciprocal absolute temperature [7]. A typical example for this is shown in fig.1, for the case of mixing chromium with silicon [8]. The curve is characterized by two distinct temperature regimes. The linear dependence of ln(Q) on 1/T refers to an Arrhenius type process, and the activation energy can be calculated from the slope. Therefore, it is clear that one obtains two very different activation energies for the two regimes in fig.1, which refers to two different mixing processes. This phenomenon can be understood regarding the various processes contributing to ion-beam mixing, as described in the introducton part of this paper. At low temperatures the ballistic effects dominate, that's why the mixing process is relatively temperature insensitive, but at higher temperatures it becomes strongly temperature dependent because of enhanced diffusion or chemical reactions. The critical temperature, above which the mixing is temperature dependent, was usually found to be room temperature.

In order to obtain the lnQ vs 1/T curve for $MoSi_2$, the implantation current was varied between 50 and 1400 A. The temperature of the wafer was then calculated as a function of the heating due to the ion beam and the energy loss due to radiation [9]. The results are shown in fig.2. The thickness of the formed silicide layers shows the same temperature dependence as known from the literature for other silicides. However, the regime with the higher activation energy starts at about 400°C, not at room temperature as in the case of chromium silicide [8]. For this region, one can calculate an activation energy for the process of 0.32 eV. This value is much lower than those previously reported for the formation of $MoSi_2$ in thermal activated case. (E.g. Tu and Mayer [11] found it to be 3.2 eV, Guivarc'h et al. 2.4 eV [4], and Yanagisawa and Fukuyama 4.1 eV [13].) At 1000 μA, the complete molybdenum layer was converted into silicide. Therefore, there was no difference between the wafers implanted at 1000 and 1400 μA, and the result belonging to 1400 μA is not included in fig.2.

Shreter et al. investigated the formation of $NiSi_2$ and $CrSi_2$ and obtained similar results [10]. They found that the activation energy of $CrSi_2$ is lower in an ion-beam mixing process (0.2±0.1 eV) in comparison to a thermal annealing process (1.9-2.0 eV). They explained it by assuming the formation of thermal spikes. According to their model, one has to calculate with $T_0 + T$ instead of T in the Arrhenius equation, where T_0 is the temperature added to T by the spike energy. Thus, the thickness of the mixed layer can be given by the following equation:

$$d = c \cdot \Phi \cdot t_0 \cdot a_0 \cdot \mathrm{EXP}-(E_a/k(T_0+T)) \qquad (1)$$

where t_0 is the spike duration, Φ the ion fluence, a_0 is the effective area of the spike and E_a is the activation energy of the process. Calculating T_0, we applied the same method as Shreter et al. At first, we calculated the apparent activation energy, $E_{a,app}$, from the gradient of the curve in fig.2. It is 0.32 eV, as mentioned above. From eq.(1), the derivative of ln(d) vs 1/T is:

$$-E_a \cdot (k \cdot (1+T_0/T)^2)^{-1} \tag{2}$$

The measured gradient of the curve must be the same as the value correctly calculated from the activation energy:

$$-E_{a,app} \cdot (kT)^{-1} = -E_a \cdot (k \cdot (1+T_0/T)^2)^{-1} \tag{3}$$

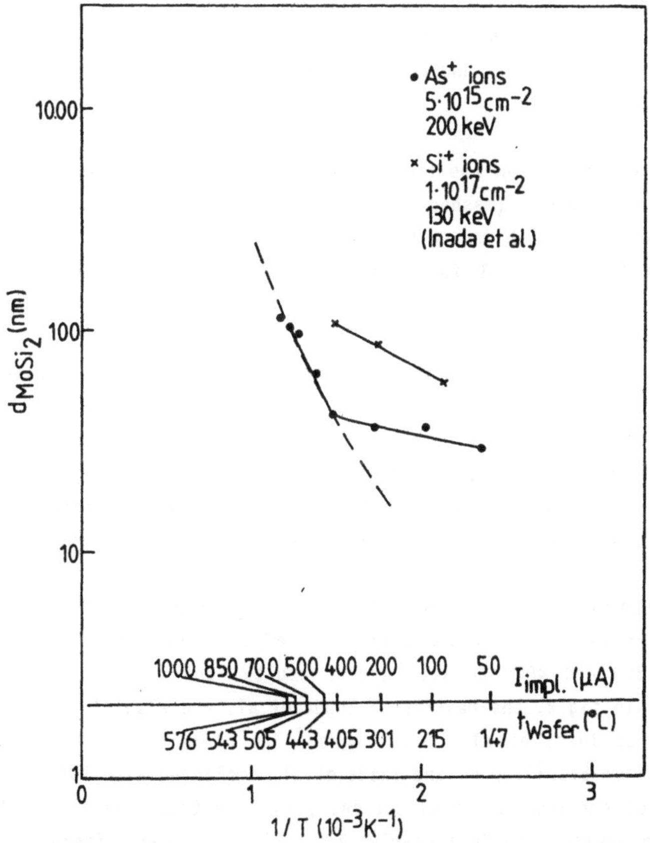

Fig.2. The thickness of $MoSi_2$ formed by ion-beam mixing with $5 \cdot 10^{15}$ cm^{-2} arsenic at 200 keV. The results of Inada et al. [5] are also shown. The dashed line is the curve we simulated using the spike model.

where the term on the left is the gradient of the strongly temperature dependent part of the curve in fig.2. We took for E_a the value given by Tu and Mayer, i.e. 3.2 eV, which is about the average of the values given in the literature. The ratio T_0/T is in our case 2.16, and T_0 is approximately 1670 K.

The spike area, a_0, can be calculated the following way [3]: If one assumes that the atoms within the spike must have at least an energy as high as the heat of melting of the substrate, their number is given by:

$$N = E \cdot \nu \cdot \delta H^{-1} \tag{4}$$

where E is the implantation energy ($2 \cdot 10^5$ eV), ν is the fraction of the energy lost in nuclear processes (0.8 for Si) and δH is the enthalpy of melting (0.7 eV/atom for Si). Thus, $N=2.3 \cdot 10^5$ atoms, and the spike area is approximately 320 nm^2.

The spike duration, t_0, is difficult to calculate exactly [3]. Therefore, we assumed for it a value of 10^{-11} sec. The factor c in eq.(1) can be calculated by substituting the measured value of d (43 nm) and the temperature which belongs to it (673 K) into eq.(1), and then solving for c.

Using these data, we could draw the theoretical lnQ vs 1/T curve for high temperatures as it can be seen in fig.2 (dashed line). This curve gives about a straight line between 830 and 650 K, which means that it is reasonable to use the slope of the measured curve in this interval for calculating T_0, and justifies the use of this model.

From these results, one can conclude that the model of Shreter et al. applies very well also to the ion-beam mixing of molybdenum silicide, which means that the silicidation reaction is promoted by thermal spike effects. As mentioned above, the silicidation reaction begins to be temperature dependent in most cases reported previously at room temperature, e.g in the case of $CrSi_2$, but we found that the critical temperature is about 400 °C. To find the cause of this difference needs further investigations.

The temperature dependence of the ion-beam mixing of molybdenum with silicon was also studied by Inada et al.[6]. They implanted 130 keV Si ions through 55 nm molybdenum into Si at 200, 300 and 400 °C. From their results, they calculated an activation energy of 0.09 eV. In fig.2, the results of Inada et al. are also given. Obviously, our results and those of Inada et al. do not agree. The cause of this difference may be that we used a dose of $5 \cdot 10^{15}$ cm^{-2}, and Inada et al. implanted with $10^{17} cm^{-2}$. As it was described in one of our previous publications [12], the thickness of the $MoSi_2$ layer formed by ion-beam mixing depends linearly on ion dose and the difference in dose may cause such a difference in layer thickness. Otherwise, the various ions which were used for mixing (As instead of Si) should have the the opposite effect because of the different masses. This is a good example for the complexity of the ion-beam mixing process.

Since the temperature dependence of the mixing can be well described by a thermal spike model, it is clear that the heating of the sample due to implantation results not in an annealing effect. To verify this, we performed also some post-implantation annealings in different rapid thermal annealers. We found that for annealing times of 10 seconds, only annealing at 1000 °C leads to further silicidation. On the other hand, temperatures between 400 and 600 °C during the implantation have a very strong effect on the mixing reaction, as we could show above. Experiments using various annealing conditions (rapid thermal annealing in vacuum or N_2 ambient and annealing in a conventional furnace in N_2 ambient) were also performed. We obtained the best results using rapid annealing in N_2, but vacuum annealing was almost as good. The lowest resistivity we obtained is 80 μΩcm, which compares well with published data of thermal silicide layers of 100 μΩcm [2]. Annealing in a conventional furnace resulted in layers with higher resistivity, because of oxidation effects.

In the results presented up to now, we dealt with the temperature dependence of the ion-beam mixing. We also performed some experiments to find its dependence on ion energy. We implanted arsenic with energies between 140 and 200 keV, but we did not find any strong dependence of the thickness of the mixed layer on the implantation energy. It is in good agreement with dynamic Monte Carlo simulation results of the mixing (fig.3), which does not show any change in the amount of the intermixed ions, either. Monte Carlo simulation can be used for simulating the ballistic mixing, but of course, it cannot take into account diffusion and chemical reactions in its standard form.

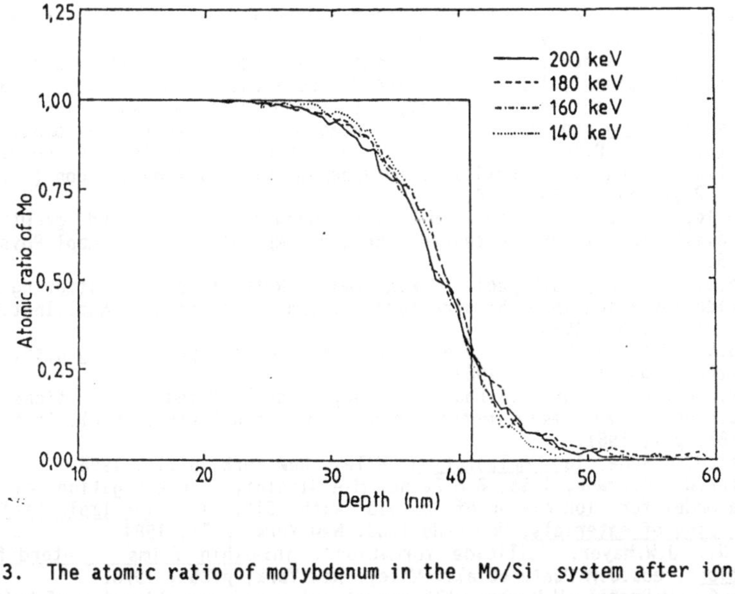

Fig.3. The atomic ratio of molybdenum in the Mo/Si system after ion-beam mixing using implantation energies of 140, 160, 180, and 200 keV as simulated by the TRIDYN computer code. Implantation dose: $5 \cdot 10^{15}$ cm^{-2}.

SUMMARY

We have studied the formation of $MoSi_2$ due to arsenic ion-beam mixing. In order to obtain information on the mechanism of the mixing, we examined the effect of implantation current (which determines the temperature during implantation) and implantation energy on the mixing process. Our investigations lead to the following conclusions:

- The temperature dependence of the $MoSi_2$ formation can be described by a curve of similar shape as that known for other silicides. However, the temperature dependent region begins at higher temperatures in the case of $MoSi_2$.
- The appearent activation energy of the process, calculated from the temperature dependent part of the ln(d) vs 1/T curve, is much lower than the activation energy which was determined for a pure thermal activation process (0.32eV vs 3.2eV). The difference can be explained using the spike model proposed by Shreter et al., which was originally applied for the Cr/Si and Ni/Si systems.
- Experiments using various implantation energies showed only slight variation of the thickness of $MoSi_2$. This is in good agreement with dynamic Monte Carlo simulation of the mixing process.

Acknowledgment

One of the authors, G.Valyi, had a fellowship from the Hanns-Seidel-Stiftung, Munich, and he is grateful for this support.

References

[1] Agamy,S.A., V.Q.Ho, H.M.Naguib, "As^+ implantation and transient annealing of $MoSi_2$ thin films", J.Vac.Sci.Technol. A3, p.718, 1985.

[2] Chow,T.P., A.J.Steckl, "Refractory metal silicides: thin film properties and processing technology", IEEE Trans. Electron Devices, ED-30, p.1480, 1983.

[3] Davies,J.A., "High energy density collision cascades and spike effects", in: Ion implantation and beam processing, Academic Press Australia, p.81, 1984.

[4] Guivarc'h, A., P.Auvray, L.Berthou, M.Le Cun, J.P.Boulet, P.Henoc, G.Pelous, A.Martinez: "Reaction kinetics of molybdenum thin films on silicon (111) surface", J.Appl.Phys. 49, p.233, 1978.

[5] d'Heurle,F.M., C.S.Petersson, M.Y.Tsai, "Growth kinetics and growth mechanisms for disilicide layers obtained through implantation", J.Appl.Phys., 53(12), p.8765, 1982.

[6] Inada,T., K.Kishi, S.Miyagi, H.Kakinuma, "Rutherford backscattering analysis of silicide formation in Mo-Si structures by ion implantation", Nucl.Inst.Meth. Phys. Res., 218, p.567, 1983.

[7] Lau,S.S., B.X.Liu, M.-A. Nicolet, "Ion mixing and phase diagrams", Nucl.Inst. and Meth., 209/210, p.97, 1983.

[8] Mayer,J.W., B.Y.Tsaur, S.S.Lau, L.S.Hung, "Ion-beam-induced reactions in metal-semiconductor and metal-metal thin film structures", Nucl. Inst. and Meth., 182/183, p.1, 1981.

[9] Ryssel,H. I.Ruge, Ion implantation, Wiley, New York, p.115, 1986.

[10] Shreter,W., C.Frank, T.So, B.M.Paine, M-A.Nicolet, "Investigation of a thermal spike model for ion mixing of metals with Si", in: Ion Implantation and beam processing of materials, North-Holland, New York, p.31, 1984.

[11] Tu,K.N., J.W.Mayer, "Silicide formation", in: Thin films - interdiffusion and reactions, eds.J.M.Poate et al., Wiley, New York, p.359, 1978.

[12] Valyi,G., H.Ryssel, W.Möller, "The fabrication and investigation of $MoSi_2$ layers", International symposium on trends and new applications in thin films, Strasbourg, 17-20 March 1987.

[13] Yanagisawa,S., T.Fukuyama: Reaction of Mo thin films on Si (100) surfaces, J.Elechtrochem.Soc. 127, p.1150, 1980.

LIQUID PHASE GROWTH OF $FeSi_2$

I.Kasko[1)], K.Gärtner and G.Götz[2)]

1. Introduction

Thermal annealing of iron films deposited on silicon (solid state reaction) results in the formation of Fe_3Si, FeSi and $FeSi_2$ ([1] and Refs. in there). At temperatures of 900–1100°C orthorhombic $FeSi_2$ as well as tetragonal $FeSi_2$ were found to grow epitaxially on <111> Si. In this paper the formation of $FeSi_2$ by irradiation with ns laser pulses (liquid phase reaction) is investigated.

2. Experimental and results

A Fe layer (40 nm) and an amorphous Si layer (20 nm) were e-beam-deposited onto a <100> Si substrate (a-Si/Fe/Si system). A second system was prepared (TH Karl-Marx-Stadt) by co-sputtering of Fe and Si onto a <111> Si substrate (a-$FeSi_2$/Si system). The resulting $FeSi_2$ layer of a thickness of 100 nm was shown to be amorphous. The two systems were irradiated with 40 ns pulses of a Q-switched Nd:glass laser.

Using 1.4 MeV He ions random and aligned backscattering spectra were measured before and after the laser irradiation with different laser energy densities. From these data the relative concentration of iron c_{Fe} and the minimum yield χ_{min} are determined as functions of the area density Nz. The results for the a-Si/Fe/Si system and the a-$FeSi_2$/Si system are depicted in Fig.1 and Fig.2, respectively.

As it can be seen by comparison of Fig.1a and Fig.2a the gradient of the Fe concentration is stronger in the case of the a-Si/Fe/Si system. The irradiation of the a-Si/Fe/Si system with laser energy densities of 0.85 and 1.43 Jcm^{-2} provides layers of mixed silicide phases with no preferential orientation ($\chi_{min} \approx 1$). After irradiation with 2.09 Jcm^{-2} the resulting layer is a bit more Fe rich than $FeSi_2$ with a remaining concentration gradient (Fig.1a). The corresponding minimum yield shown in Fig.1b indicates that the layer has grown epitaxially, however the crystalline quality is poor.

For the $FeSi_2$/Si system irradiated with a laser energy density of 0.99 Jcm^{-2} we find a buried epitaxial silicide layer (Fig.2b). The composition of this layer is near to the eutectic composition marked in

1) Lenin-University Minsk, Department of Physics, Lenin Prospekt 4, 220080 Minsk, USSR
2) Friedrich-Schiller-Universitat Jena, Sektion Physik, Max-Wien-Platz 1 Jena, DDR 6900

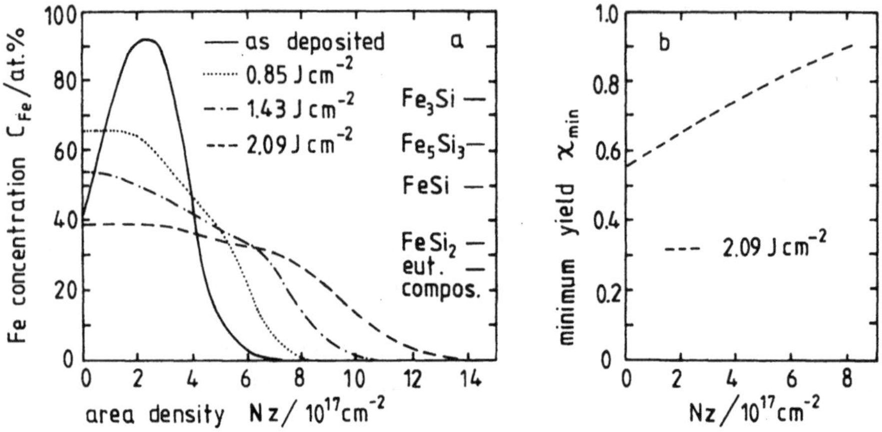

Fig.1. Depth dependence of the Fe concentration (a) and of the minimum yield (b) for the a-Si/Fe/Si system

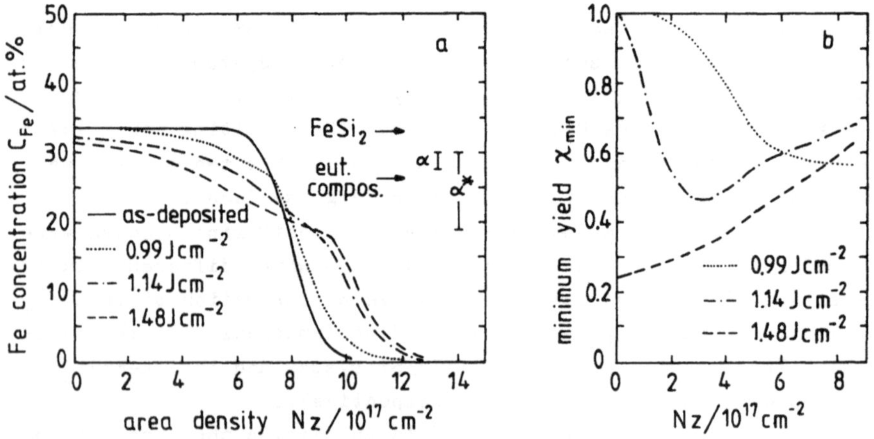

Fig.2. Depth dependence of the Fe concentration (a) and of the minimum yield (b) for the a-$FeSi_2$/Si system

Fig.2a (see also Fig.3). The remaining non-epitaxial surface layer has the composition of $FeSi_2$. With increasing laser energy density (1.14 Jcm^{-2}) the thickness of the buried epitaxial layer increases. The irradiation with a laser energy density of 1.48 Jcm^{-2} results in a completely epitaxial silicide layer (Fig.2b) with much better crystalline quality than that obtained from the a-Si/Fe/Si system. In spite of the relatively strong concentration gradient (Fig.2a) the silicide layer consists only of one phase. This was shown by TEM diffraction pattern which corresponds to that of the tetragonal $FeSi_2$.

3. Discussion

Investigations of the formation of $NiSi_2$ by irradiation with ns laser pulses [2] showed that the liquid phase formation of silicides can be understood qualitatively on the base of the equilibrium phase diagram. The phase diagram of the Fe-Si system is shown in Fig.3. Our interest is concentrated on the liquidus line L and the different $FeSi_2$ phases.

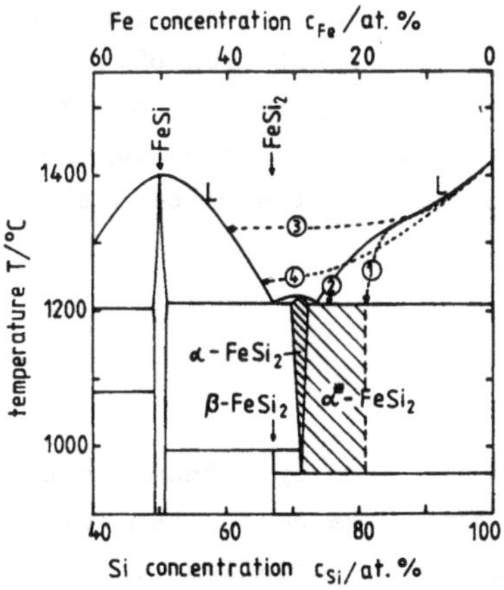

Fig.3. Phase diagram Fe-Si

Because of its low melting temperature the $FeSi_2$ layer of the $FeSi_2$/Si system melts continuously from the surface to the interface. For 1.48 Jcm^{-2} also a remarkable amount of Si is molten. The change of the Fe concentration in the melt at the interface (c_{Fe}^{I}) during the solidification is qualitatively given by the dotted line 1 in Fig.3.

First, Si recrystallizes epitaxially (proved by marker experiments) accompanied by an enhancement of c_{Fe}^{I}. After cooling below 1200°C a tetragonal non-equilibrium phase (α^*-FeSi) grows from the melt. The homogeneity region of this phase is much larger than that of the equilibrium α-$FeSi_2$ (see Figs.2a and 3). For lower laser energy densities only a small amount of Si is molten and the dotted line 2 in Fig.3 represents the solidification process. After epitaxial recrystallization of a small Si layer an epitaxial α^*-$FeSi_2$ layer grows until c_{Fe} reaches about 30 at.%. The solidification of the remaining melt provides a mixture of FeSi and $FeSi_2$ which does not grow epitaxially.

The situation is different for the a-Si/Fe/Si system where the melting starts from the a-Si/Fe interface and provides a strong gradient of c_{Fe}. Therefore the solidification starts with a rapid increase of c_{Fe}^{I} according to the dotted lines 3 or 4 in Fig.3. Then FeSi grows from a constitutional undercooled melt which can lead to interfacial instabilities and the formation of celluar structures. If the melt is undercooled in the whole region FeSi may partially grow till the surface and then $FeSi_2$ can grow epitaxially from the remaining more Si rich melt (2.09 Jcm^{-2}, dotted line 4).

[1] Cheng, H.C., T.R.Yew and L.J.Chen, J.Appl.Phys. 57 (1985) 5246-5250
[2] Kaschner, C., A.Witzmann, K.Gärtner and G.Götz, phys.stat.sol.(a) 94 (1986) 787-791

LIQUID PHASE GROWTH OF $NiSi_2$ AND $CoSi_2$

A.Witzmann, C.Kaschner, K.Gärtner, and G.Götz[1)]

1. Introduction

In 1982 Tung et al.[2] succeeded in the liquid phase formation of homogeneous, epitaxial $NiSi_2$ and $CoSi_2$ layers by ns-irradiation of mono- or disilicide layers on Si. In this paper the dependence of the resulting silicide layer on the laser energy density is reported for ns-irradiation of Ni and Co silicide layers and a model of the processes by means of the equilibrium phase diagram is proposed.

2. Experimental and results

80 nm CoSi, NiSi and $NiSi_2$ layers have been prepared by e-beam deposition of metal layers under a pressure below $1*10^{-6}$ Pa onto chemically cleaned <111>-Si and additional vacuum heating. A variation of the averaged silicide concentration between NiSi and $NiSi_2$ has been achieved by the implantation of Ni and Si ions in the NiSi and $NiSi_2$ layers, respectively. The irradiation has been performed with 40 ns pulses of a Nd:glass laser (λ = 1.06 μm) with energy densities between 0.5 and 3 Jcm^{-2}. The recrystallized films have been investigated by RBS and TEM. For the irradiation of CoSi layers a continuous increase of the silicide layer thickness with the laser energy density can be seen from the concentration profiles (Fig.1). At a laser energy density of 1.02 Jcm^{-2} the layer is completely molten, which is indicated by the crystallinity of the layer ($\chi_{min} \approx 70\%$). The deeper melting into the substrate for higher laser energy densities is accompanied by a broadening of the interface width. The minimum yield χ_{min} of the layers remains relatively constant (60...80%) for 1.33 Jcm^{-2} and above.

The situation is different for the irradiation of Ni silicide layers. In this case crystalline $NiSi_2$ layers are formed over a wide range of energy densities ($E_L \approx$ 1.4...2.3 Jcm^{-2}) as can be seen in Fig.2. For lower energy densities the melting depth is not sufficient to form a melt with $NiSi_2$ composition which results in inhomogeneous multiphase layers. At laser energy densities > 2.3 Jcm^{-2} the interface width and the fraction of Si in the crystallized layer increase, which is connected with a decrease of the crystalline quality.

---

1) Friedrich-Schiller-Universität Jena, Sektion Physik, Max-Wien-Platz 1 Jena, DDR 6900

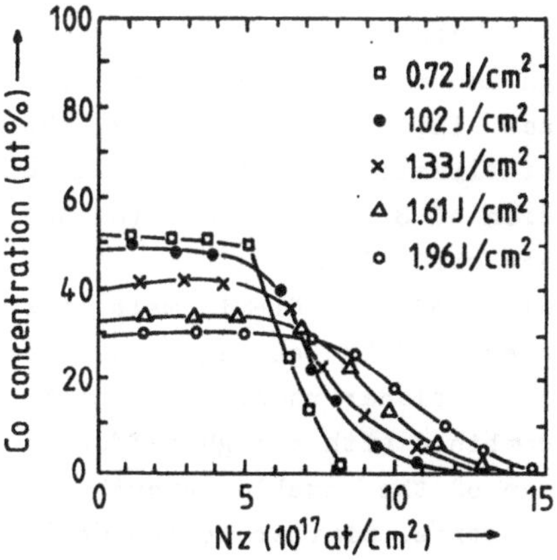

Fig.1. Depth profiles of the Co concentration of laser irradiated CoSi samples.

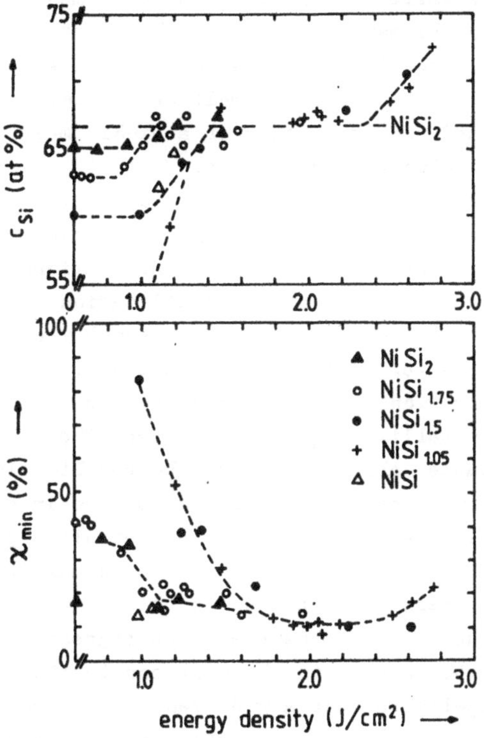

Fig.2. Averaged Si concentration and minimum yield χ_{min} of laser irradiated Ni silicide layers as a function of the laser energy density.

3. Discussion

The discussion of the melting and resolidification processes in the ns-regime is performed by means of the equilibrium phase diagram of the binary metal silicon system. The comparison of the systems Co-Si and Ni-Si reveals qualitative differences in their phase diagrams (Fig.3a and b) which causes a different behaviour of this systems after ns-irradiation.

The high melting temperature of CoSi in comparison to the considerably lower eutectic temperatures causes a melting of the layer starting from the interface and the formation of a melt with high concentration gradients (especially for the low energy densities). The irradiation of Ni silicides (NiSi...$NiSi_2$) causes a melting from the surface and the formation of a melt with low concentration gradients since the melting temperatures of the silicides and the eutectic temperatures are only slightly different.

According to the phase diagram the resolidification starts with the crystallization of Si from the melt. The existence of a layer of recrystallized Si has been confirmed by marker experiments in the case of the formation of $NiSi_2$. The recrystallization of Si continues until the eutectic concentration is reached at the interface liquid-solid in the case of a Co-Si melt or the peritectic concentration of $NiSi_2$ is reached in the case of a Ni-Si melt. Afterwards the corresponding disilicide recrystallizes.

In the Co-Si system at low energy densities the actual interface concentration is strongly determined by the high concentration gradient of the melt. Regions of constitutional supercooling exist causing the instability of the recrystallization front. Therefore the recrystallized silicide layer consists presumably of $CoSi_2$ columns surrounded by a mixture of silicides and silicon. At high laser energy densities the segregation of Co atoms at the interface liquid-solid during the recrystallization of Si causes the recrystallization front to become instable already during the resolidification of Si which leads to the observed decreased sharpness of the interface silicide-Si.

Because of the low concentration gradient of the Ni-Si melt the actual interface concentration is mainly determined by segregation. In the energy density region below $\approx 2.3\ Jcm^{-2}$ a planar recrystallization of the melt is possible. The segregation combined with a high diffusion of the atoms in the melt can be the reason of the stable growth of $NiSi_2$-layers over a wide range of energy densities. Similar to the Co-Si system a critical supercooling is reached for higher energy densities which causes the instability of the interface liquid-solid and the recrystallization of a cellular structure.

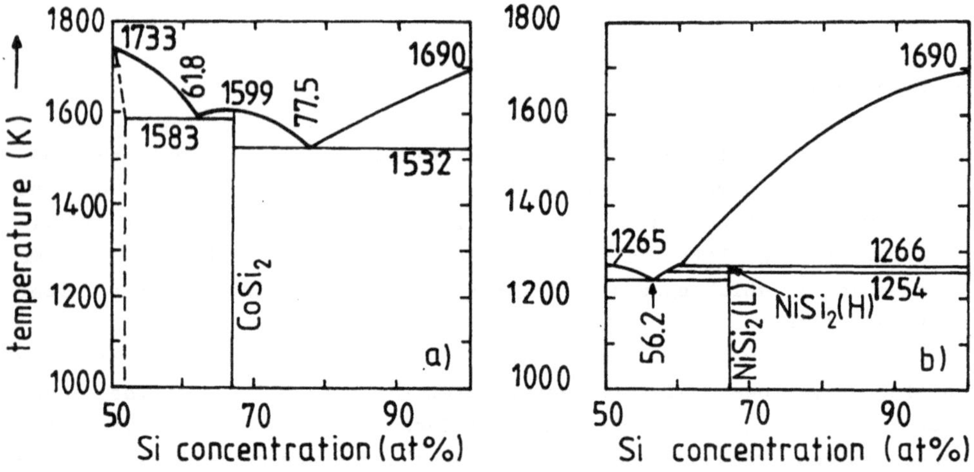

Fig.3. Phase diagrams of the systems Co-Si a) and Ni-Si b) (Ref.[1]).

[1] Hansen, M., K.Anderko: Constitution of Binary Alloys, Mc Graw Hill, New York, Toronto, London, 1958.

[2] Tung, R.T., J.M.Poate, J.C.Bean, J.M.Gibson, and D.C.Jacobson: Thin Solid Films 93 (1982) 77.

FORMATION OF EPITAXIAL $NiSi_2$ LAYER BY HIGH DOSE ION IMPLANTATION AND RAPID THERMAL ANNEALING

V.Yu.Petukhov[1)], I.B.Khaibullin[1)], M.M.Zaripov[1)], E.Wieser[2)], R.Groetzschel[2)]

The formation of the epitaxial Ni-disilicide layer on silicon by means of the reaction between the thin metallic nickel film deposited on silicon and substrate using the different treatments (furnace - (2,5), laser - (3,6), electron - (4) or ion-beam-annealings (1)) has been studied earlier. The application of incoherent light pulse annealing to produce nickel silicide has been studied by J.T.Lue (7). In this work, however, the surface morphology has been shown to be of a cellular structure: the cell walls are composed of a mixture of silicides; the cell material identified by TEM is polycrystalline silicon epitaxially regrown from the crystalline substrate, but epitaxial $NiSi_2$ has not appeared.

In previous work (8), we have produced $NiSi_2$ in the surface layer of silicon by the only high dose Ni^+-implantation, the disilicide being in the fine dispersed form.

In this paper we show the possibility of the epitaxial $NiSi_2$ synthesis by the high dose Ni-implantation and the following rapid thermal annealing (RTA).

The silicon single crystals of (111)-orientation were implanted with 40 keV and dose $2.5 \cdot 10^{17} Ni^+/cm^2$ at the room temperature. RTA was performed using the galogen-lamp technique for 20 s in the air atmosphere. The temperatures tested by the pyrometer were chosen at 600 °C, 750 °C or 900 °C. The effect of the high dose Ni^+-implantation and the following RTA on the depth distribution of implanted ions and the epitaxial regrowth was investigated by both random and channelling RBS experiments.

RBS spectra of 'as implanted' and 600 °C RTA samples practically coincided. Nickel and silicon concentrations and the thickness of the nickel-enriched layer (1) calculated from the RBS spectra are shown in the Table 1. For these two samples the Ni-enriched layer is placed immediately on the surface. Electron diffraction patterns of the surface layer of these samples revealed only the amorphous halo.

Unlike the first two, the spectra of the samples annealed at T_s = = 750 °C and 900 °C revealed the redistribution of Ni atoms into the depth

1) Kazan Phys.-Techn. Institute Academy of Sciences of the USSR 420029, Kazan, USSR.

2) Academie der Wissenscaften der DDR ZfK Rossendorf Postfach 19 Dresden DDR-8051

Table 1. Effect of RTA temperature of the sample, T_s on the thickness and composition of the Ni-enriched layer (I) and on the thickness of the amorphous layer (II)

RTA		As implan.	After RTA 600 °C	750 °C	900 °C
(II) Amorph. d_2	d_1, nm	35	38	56	55
(I) Ni_xSi_{1-x} d_1	d_2, nm	-	-	4	8
Si-substrate	Composition layer (I), x/(1-x)	$\frac{0.47}{0.53}$	$\frac{0.43}{0.57}$	$\frac{0.3}{0.7}$	$\frac{0.3}{0.7}$

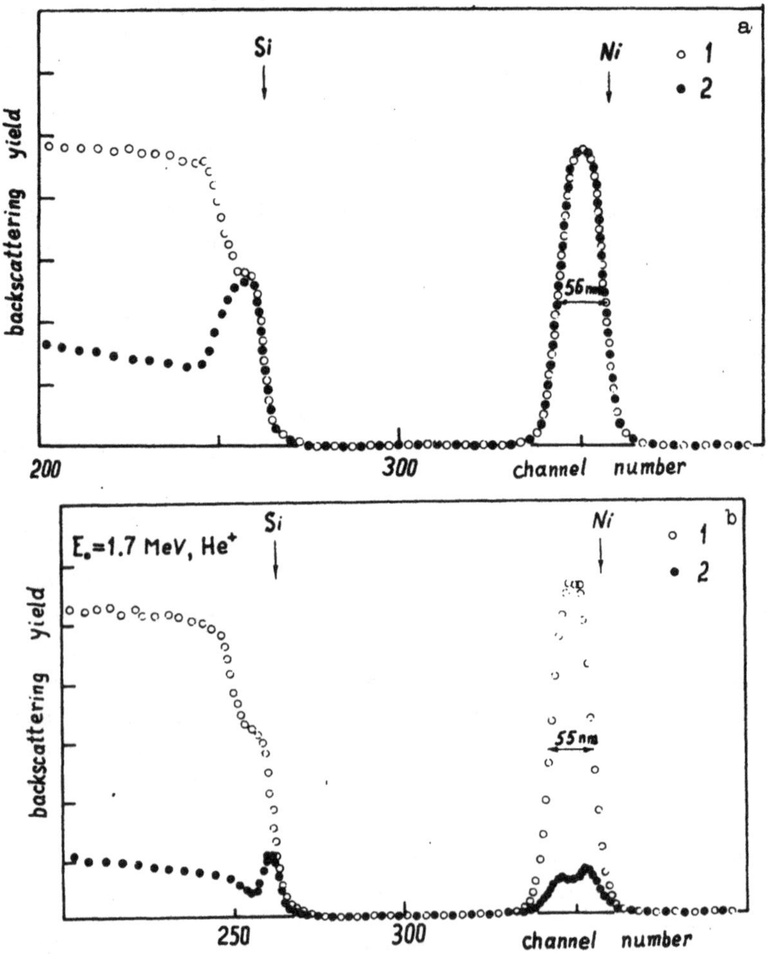

Fig.1. RBS spectra of Ni^+-implanted silicon after 20 s RTA at different temperatures of samples: a) T_s=750 °C; b) 900 °C (1 - random, 2 - channelled)

of substrate as it is seen in Fig.1. The thickness of the Ni-enriched layer increased up to 55-56 nm and the correlation of concentrations of Ni and Si atoms is equal to N_{Ni}/N_{Si} = 0.43 (Table 1). In both cases the Ni-enriched layers are not placed on the surface of samples: at the 750°C RTA the layer (I) is placed starting from the depth of 4 nm, at 900°C RTA - 8 nm. The channelling spectra (Fig.1a and 1b, curve 2) show that at the 750°RTA random and aligned curves practically coincide but at the high RTA temperature the yield of backscattered ions is essentially decreased. The minimum yield of the channelled He-ions backscattered from Ni-atoms in contrast to the random spectrum is equal χ_{min}= 11% testifying to the formation of the sufficiently good epitaxial layer. The composition of the formed silicide is near to the one of the stoihiometric compound $NiSi_2$ with a little abundance of Si. Electron diffraction patterns consist of the amorphous halo and weak point reflexes with Kikuchi lines. This reveals the existence of a thin amorphous layer (II) and epitaxial silicide layer (I) in the small depth of the sample. The amorphous layer (II) formed on the surface of the samples at 750°C, 900°C RTA (see Table 1) is the product of oxidation by atmospheric oxygen and it consists, apparently, of silicon oxide because the Si atoms, unlike the Ni atoms, are placed starting from the surface as it was mentioned above and, moreover, the spectrum of channelled He ions backscattered from surface Si atoms (Fig.1b, curve 2) reveals the resolved maximum while in the same spectrum the similar peak corresponding to Ni atoms is absent.

In summary, the epytaxial $NiSi_2$ layer on Si(111) implanted by high dose Ni^+ is formed during 20-second RTA at surface temperature 900°C although the disilicide composition is already achieved at 750°C RTA. The investigation of the epytaxial growth showed both the threshold behaviour and the absence of the influence of the diffusion process in our experiments.

References.

(1) Chen, L.J. et.al.: Appl. Phys. Lett. 40 (1982) 595-597.
(2) Chiu, K.C.R., et.al.: Appl. Phys. Lett. 36 (1980) 544-547.
(3) Grimaldi, M.G. et.al.: Appl. Phys. Lett. 43 (1983) 244-246.
(4) Harper, R.E. et.al.: Vacuum 34 (1984) 875-879.
(5) Ishiwara, H. et.al.: Surf. Sci. 86 (1979) 711-717.
(6) Kascher, C. et.al.: Phys. stat. sol. A 94 (1986) 787-791.
(7) Lue, J.T.: Solid State Electronics 26 (1983) 787-793.
(8) Petukhov, V.Yu. et.al.: Poverhnost. Fiz., Him., Meh. 2 (1985) 104-107.

REDISTRIBUTION OF IMPLANTED DOPANTS IN $MoSi_2$/POLY-Si STRUCTURES DURING SILICIDATION TEMPERING

E.Wieser[1], Ch. Weise[1], E.Vetter[2], and Th. Gessner[2]

Refractory silicides combined with a polysilicon layer to a polycide structure are of growing interest for MOS gate metallization schemes. Thereby, doping of the polysilicon sheet is mainly realized by implantation. The dopant redistribution during the high temperature processing, necessary in order to anneal the implanted layer and for silicide formation, is a very important process. It will influence sheet and contact resistance, the threshold voltage, and also the etching behaviour.

As dopants B, P and As are considered. The results of rapid thermal annealing (RTA) are compared with that one of furnace annealing.

The investigated sheet system consists of thermally oxidized (100) silicon (150 nm SiO_2) covered with 200 nm CVD polysilicon and 200 nm $MoSi_2$. The silicide was deposited by cosputtering of Mo and Si in the stoichiometric ratio. The As redistribution was also investigated on a structure of 100 nm $MoSi_2$ on 150 nm polysilicon.

Doping was carried out by the following implantations: B - 1×10^{16} cm^{-2}, 90 keV; P - 1×10^{16} cm^{-2}, 300 keV; As - 5×10^{15} and 2×10^{16} cm^{-2}, 150 keV, and 1×10^{16} cm^{-2}, 125 keV.

For annealing and silicidation furnace tempering at 900 and 1000 °C (30 min) in N_2 and rapid thermal annealing at maximum temperatures between 1000 and 1150 °C with a dwell time of 15s and a temperature rise time of about 10s in air were used.

The depth distribution of the dopants was determined by SIMS using O^- primary ions. The specific resistance of the $MoSi_2$ layer was between 90 $\mu\Omega$cm and 120 $\mu\Omega$cm for the investigated annealing treatments.

Fig. 1 shows As distributions characteristic for low average As concentration ($c_{av.}^{As} \leqq 10^{20}$ cm^{-3}). The dopant redistribution proceeds obviously in times as short as 15s. This statement is valid also for higher concentrations and the other investigated dopants. A nearly equal average As concentration in $MoSi_2$ as well as within the polysilicon is observed. A somewhat enhanced As concentration is found in the depth of the original As profile and at the interface $MoSi_2$/polysilicon. A tendency to surface segregation connected with an As loss can be stated.

Enhancement of the average As concentration results in an enhanced As content within the poly-Si layer, but not within the $MoSi_2$ as shown

---

1 - Central Institute of Nuclear Research Rossendorf, PSF 19, DDR - 8051 Dresden

2 - Technische Universität Karl-Marx-Stadt, PSF 964, DDR - 9010 Karl-Marx-Stadt

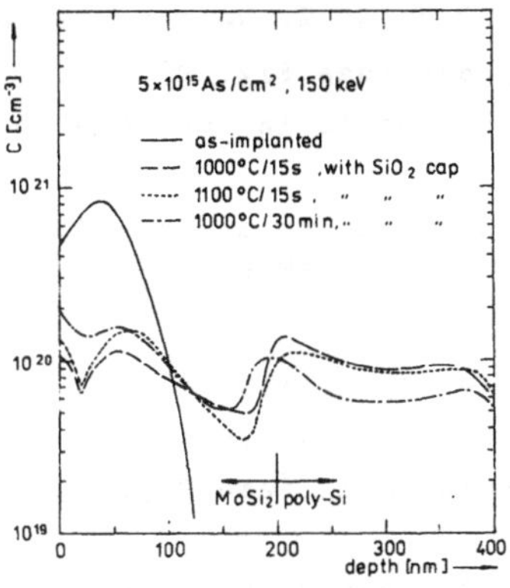

Fig.1: As distributions in 200 nm $MoSi_2$ on 200 nm poly-Si after different heat treatments

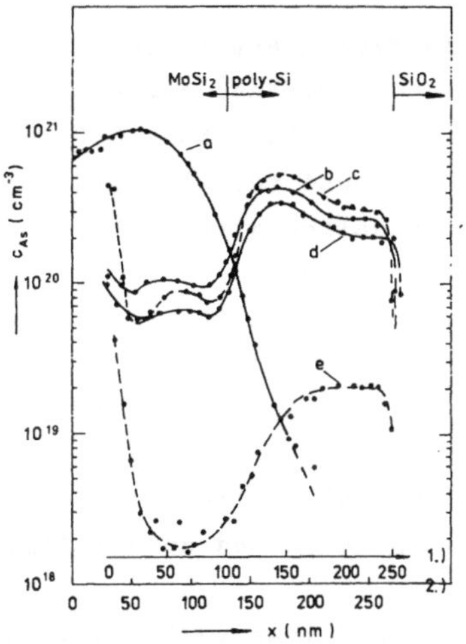

Fig.2: As distribution in 100 nm $MoSi_2$ on 150 nm poly-Si, annealing: a-as-implanted; b and c-1100 $^{\circ}C$/15s with and without SiO_2; d and e - 1000 $^{\circ}C$/30 min, with and without SiO_2

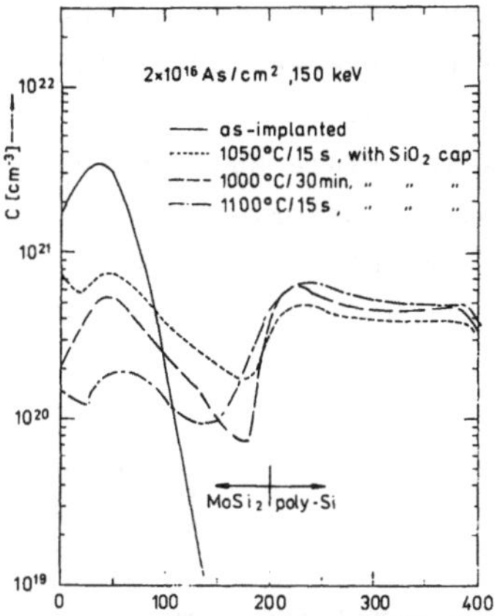

Fig.3: As distributions for high concentrations

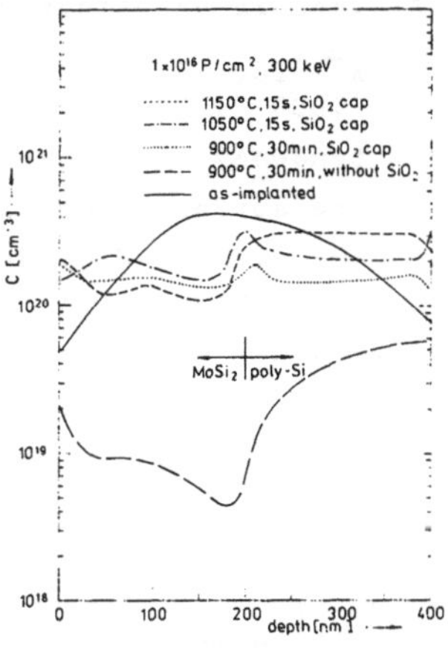

Fig.4: P distribution for medium concentrations

in Fig. 2. This behaviour is characteristic for medium average As concentrations (10^{20} cm^{-3} = $c_{av.}^{As}$ = 4 x 10^{20} cm^{-3}). It leads to the conclusion that a limited solubility of As in $MoSi_2$ of about 10^{20} cm^{-3} exists.

At $c_{av.}^{As}$ 4 x $10^{20} cm^{-3}$ (Fig. 2) the As concentration in the poly-Si layer corresponds to the equilibrium solubility at 1000 to 1100 $^{\circ}C$ ($c_{sol.}$ = 3-4 x $10^{20} cm^{-3}$). For high average As concentrations ($c_{av.}^{As}$=5x10^{20} cm^{-3}, Fig. 3) the As content exceeds the equilibrium solubility in $MoSi_2$ as well as in the poly-Si. The main part of the excess As is found in the depth of the original As profile. The amount of excess As decreases with increasing annealing temperature and/or time by As loss due to surface segregation.

The As loss by RTA with a SiO_2 cap amounts to 20 to 30 %. No remarkable difference is found for RTA with and without SiO_2 cap (see Fig. 2). By furnace annealing without cap layer the main part of As is lost (90 % according to Fig. 2e).

Fig. 4 shows P distributions after different annealing treatments for a medium average P concentration (1×10^{16} P/cm^2 = 2.5×10^{20} P/cm^3). The behaviour of P is very similar to that of As.

The solubility of P is higher in poly-Si than in $MoSi_2$. With rising annealing temperature the P concentration increase in the poly-Si, possibly by the enhanced solubility, and decreases in $MoSi_2$ (see Fig. 4a, 4b).

A remarkable P loss ($\sim$40 %) is already observed after furnace annealing with a SiO_2 cap at 900 °C/30 min. The P loss by furnace annealing without a cap layer is very high (> 90 %). RTA with a SiO_2 cap leads only to a small P outdiffusion ($\leq$ 15 %).

A comparison of RTA and furnace annealing of B implanted structures (Fig. 5 and 6) shows a higher B concentration within the poly-Si layer obtained by RTA. This is explained by the temperature dependence of the B solubility ($T_{RTA} > T_{furn.}$) as well as by the B loss observed after furnace annealing at 1000 °C/30 min with SiO_2 cap ($\sim$25 %).

A significant tendency to B segregation at the interface $MoSi_2$/poly-Si as well as at the surface (interface $MoSi_2/SiO_2$) is observed.

Furnace annealing without SiO_2 cap leads to a very strong B loss (> 90 % already at 900 °C/30 min). The B loss by RTA at 1000 to 1100 °C without cap layer is not serious ($\leq$ 20 %).

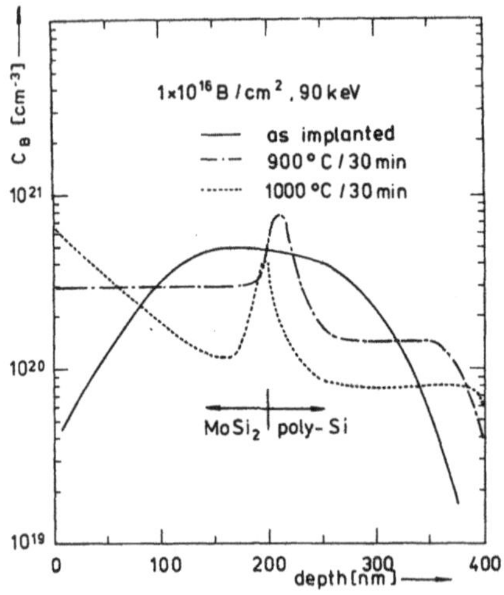

Fig.5: B distribution after furnace tempering

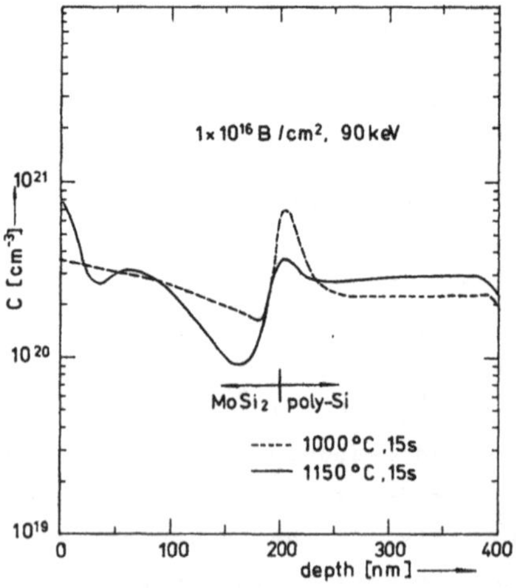

Fig. 6: B distribution after RTA

R.B.S. STUDIES OF WSi_2 FORMATION FROM Si/W/Si LAYERS BY W^+ ION MIXING FOLLOWED BY ANNEALING WHEN THE W FILMS CONTAIN OXYGEN

B.Canut*, M.Kadri+, J.Pivot*and A.Deneuville+

ABSTRACT

Formation of c-WSi_2 by ion mixing of Si/W/Si layers is shown to be very dependant on residual O in the W film. There is a mechanical mixing, but the associated diffusion is prevented or delayed. Formation of C-WSi_2 needs at 650°C additional Si diffusion. For higher annealing temperatures silicon and oxygen segregate on the top of the sandwich.

1. INTRODUCTION

Refractory metal silicides formed by metal-Si interdiffusion are interesting in V.L.S.I. Si microelectronic, to benefit of the self aligned technics but a low temperature of formation is wished to avoid doping atoms redistribution [1] . In this paper we study the formation of WSi_2 enhanced by ion mixing and the use of amorphous hydrogenated silicon in Si/W/Si sandwich. Usually, the W films contain a significant fraction of oxygen (here 6%). We used a Si encapsulating layer in order to avoid the W oxydation and to facilitate the WSi_2 formation which is prevented or delayed by the occurence of oxygen both at the Si/W interface or within the W film [2] .

2. PREPARATION AND EXPERIMENTAL TECHNIQUES

Thin films of a-Si:H (540 Å), W (110 Å), a-Si:H (150 Å) have been deposited succesively on c-Si by reactive cathodic sputtering of Si and W targets in a 20% H_2 / 80% Ar reactive mixture [3]. The sandwiches are ion mixed at room temperature by W^+ ions with 0,1 μA current up to 10^{16} ions/cm^2 fluence. Samples are then annealed in a classical Joule owen during one hour at 550, 650, 750, 850 and 1000°C .

R.B.S. is performed with 2 Mev $^4He^+$ ions and various incidence angles (φ = 0, 60 and 75°) , in order to increase the thickness resolution. The formation of crystalline WSi_2 is detected by X rays at glancing incidence ($\alpha < 0{,}2°$) .

3. RESULTS AND INTERPRETATION

The R.B.S. spectra of the various sandwiches : virgin (□) , just after implantation (0), with subsequent annealing at 850°C (Δ) are presented on Fig1.

* On the virgin sandwich (□) there is first a W peak nearly symetric and significantly below the channel corresponding to the surface atoms of W. This was expected from the preparation process of the sandwich. Assuming bulk densities, W and Si encapsulating thicknesses (respectively 100 Å and 150 Å) are deduced. The R.B.S. spectrum is then dominated by the Si signal with a hump followed by a step. The hump corresponds to the encapsulating Si layer and gives a top Si thickness of ≈150 Å in agreement with its nominal value. We clearly see on the Si step the oxygen bump. It is located in the W film below the surface. This result agrees with S.I.M.S. [3] measurements which give respectively [O] ≈ 6% atom and 0,08% atom in the W and Si films of such sandwiches.

* Just after W implantation (O) the R.B.S. spectrum is quite different. The W peak broadened with a larger tail toward

*Université Claude Bernard - LYON 1, Département de Physique des Matériaux (U.A. 172) ,43 Boulevard du 11 Novembre 1918 - 69622 VILLEURBANNE Cedex , France .

+C.N.R.S. Laboratoire d'Etudes des Propriétés Electroniques des Solides, Université Scientifique Technologique et Médicale de Grenoble , B.P. 166 , 38042 GRENOBLE , France .

the substrate. W atoms reach now the surface, and their concentration decreases slowly deep in the sandwich. Drastic changes have also occured on the Si signal. There is still Si atoms at the surface, but only a structure remains instead of the hump in a smooth Si variation where the bottom Si film interface has also disappeared. A new break appears on the Si signal approximatively at the edgė between the bottom Si film and the Si substrate. This confims a large intermixing of W and Si atoms in the sandwich and suggests it occurs from the surface to the c-Si substrate. We used a technic developped by one of us [4] to deduce the ratio α = [W] / [Si] from the Si spectrum when it is compared to what it would be for a pure Si sample. Results are presented on Fig.2 . Because the local density is usually unknown, the depth is scaled in $\mu g/cm^2$. The α values are very inaccurate near the surface and so the profiles are drawn only from a significant depth. We see on Fig.2 that just after implantation (*) there is no plateau in the profile; the [W]/[Si] ratio which is near 1 in the central part of the sandwich decreases slowly to zero at the c-Si edge. A final important point is the oxygen behaviour. We see on Fig.1 that oxygen has been spread in the whole sandwich. This confirms the previous S.I.M.S. [3] and infrared [5] measurements on these sandwiches.

* One hour annealing changes the W/Si profiles accross the sandwich and this is well shown on Fig.2 . Annealing at 550°C induced mainly a larger spread of W. There is still no plateau in the profile but α values are lower (≈ 0,33 in the central part). From X rays, this is the beginning of the crystalline tetragonal WSi_2 formation [5] . The W/Si ratio in the central part of the sandwich has decreased with increasing annealing temperature, which shows an increasing diffusion of Si in W assisted by the implantation defects.

* Annealing at 750°C broadens the plateau which is now around W/Si =0,4 . This is coherent with the increases of the c-WSi_2 signal (X rays) and of the Si-O signal (Infrared) . Some Si leave the central part of the sandwich to be oxidized closen to the surface.

* Annealing at 850°C result in the same behaviour with an increase near 0,5 of the W/Si ratio in the central part of the film.

* Annealing at higher temperature forms a SiO_2 layer on the sandwich surface which has been etched by HF, but SiO signal from the bulk sandwich remains. For instance Fig.2 shows the W/Si ratio after annealing at 950°C . The plateau remains and broadens but at a lower (≈ 0,35) W/Si ratio. This is interpreted as the appearance of amorphous SiO mixtures around the WSi_2 crystallites by Si diffusion from the bottom layer and partial oxydation within the sandwich. There is an oxygen motion toward the top of the sandwich. The oxygen accumulation in a superficial layer becomes specially obvious on the O and W signals at 850°C. This c-WSi_2 has to be contrasted with that occuring with the same W ion mixing with subsequent annealing of the same sandwich but with a negligible O content in W [6] . In this case amorphous stoechiometric WSi_2 is formed just after W implantation and the profile of the chemical species remains identical with subsequent annealing. C-WSi_2 is then formed by crystallisation at 350°C . It thus appears that ion mixing is a composite process. Besides the purely mechanical effects, there is a diffusion enhanced by the implantation defects which gives additional species redistribution. If the mechanical effect is always nearly identical, the induced diffusion may be, as here, significantly hindered by the contamination of the films.

4.CONCLUSION

The R.B.S. studies allow a description of the formation of crystalline WSi_2 by W ion mixing then annealing ofa a-Si:H / W / a-Si:H sandwich when the W film contains oxygen.

* The ion mixing is efficient and mechanically spreads O and W in the whole sandwich up to the surface, but oxygen prevents the associated diffusion.

* The formation of crystalline c-WSi_2 needs additional Si diffusion and occurs for a central composition W/Si ≈ 0,27 (instead of 0,5) around 650°C.

* With higher annealing temperatures (Ta > 850°C) both Si and 0 diffuse toward the surface to form a top SiO_2 layer.

Fig.1

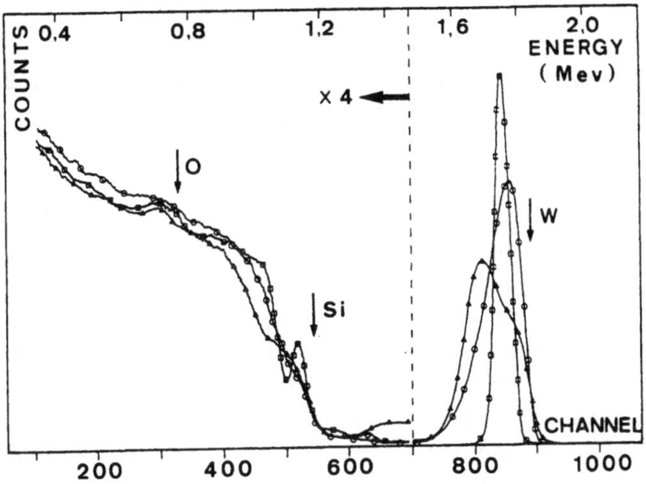

Fig.2

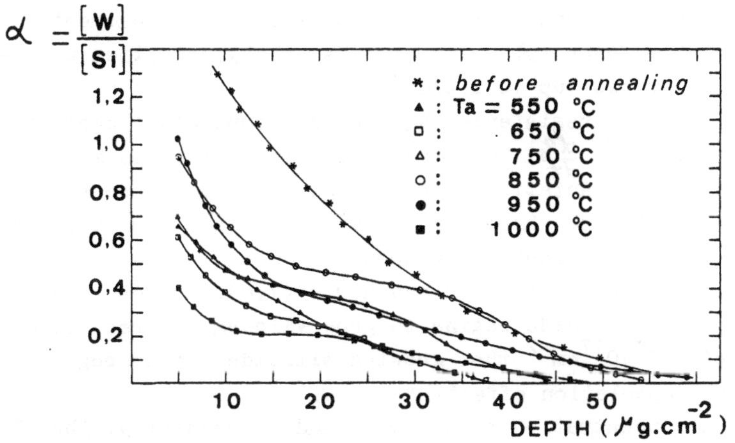

REFERENCES

[1] A. Deneuville : Ann. Chim.Paris, 11 (8) (1987) 603.

[2] D.J. Silversmith, D.D. Rothman and R.W. Mountain : Thin Solid Films, 93 (1982) 413 ; J. Lajizerowic Jr and G. Bomschil : Ibidem 124 (1985) 19.

[3] M. Kadri, A. Deneuville and M. Brunel : Proc. IPAT Brighton June (1987).

[4] D. Fleche, J.P. Gauthier, J. Pivot and J.A. Roger : J. Microsc. Spectrosc. Electron. 10 (1985) 219.

[5] M. Kadri, A. Deneuville and M. Brunel : Proc. E-MRS Strasbourg (1986).

[6] B. Canut, M. Kadri, J. Pivot and A. Deneuville : Proc. E-MRS Strasbourg (1987), to be published.

ELECTRICAL RESISTIVITY OF SILICON IMPLANTED BY HIGH DOSE Cr^+ IONS AFTER LASER ANNEALING

V.Yu.Petukhov[1], I.B.Khaibullin[1], N.V.Kurbatova[1], E.Wieser[2], R.Groetzschel[2]

Recently, it has been shown that the silicides of some transition metals can be formed immedeately during the high dose ion implantation (7,8). It is obvious that in this case the electrical properties of the surface layer are to be change not only because of the metal-silicon compound formation but also due to the influence of the irradiation on the conductivity. To remove the radiation damage pulse laser annealing is often successfully used (2).

In this communication the influence of the implantation dose as well as the energy of the following laser annealing on the electrical resistivity of the surface layer of Cr-implanted silicon are reported.

Cr^+-ions implantation was carried out in the 10 Ohm.cm p-silicon. at energies 40 or 60 keV and doses ranging from $0.5\cdot10^{17}$ to $4\cdot10^{17}$ ions/cm^2 at the room temperature. At the following treatments by 30-ns-pulses of the ruby laser (λ =0.69 μm) the energy densities on the samples were in the region up to 2 J/cm^2.

The sheet resistance, R_s and the temperature coefficient of the resistance, $\alpha = \frac{1}{R_s}\frac{dR_s}{dt}$, both measured using a 4-probe method, as a function of the implantation dose, Φ at different dose rate, j are shown in the Fig.1. The threshold behaviour of this observed dependencies can arise due to at least two reasons.

i) As it is usual for ion synthesis (7,8), at dose increasing, the microdispersed silicide begins to grow yilding the resistance decrease. At dose of $1.2\cdot10^{17}cm^{-2}$ the growing silicide grains begin to overlap that causing percolation effect.

ii) The concentration of Cr-ions and,consequently, the electrical properties of the surface layer can be saturated during the high dose implantation because of the sputtering effect.

To explain the obtained dependensies RBS and electron diffraction experiments were carried out. Backscattering spectra showed only small sputtering up to the dose of $2.5\cdot10^{17}$ cm^{-2}, the sputtered yield being 20% or less in respect to the implantation dose. At the latter dose value the correlation of Cr- and Si-concentrations was equal to N_{Cr}/N_{Si} =

[1] Kazan Phys.-Techn. Institute Academy of Sciences of the USSR 420029, Kazan, USSR.

[2] Academie der Wissenscaften der DDR ZfK Rossendorf Postfach 19 Dresden DDR-8051

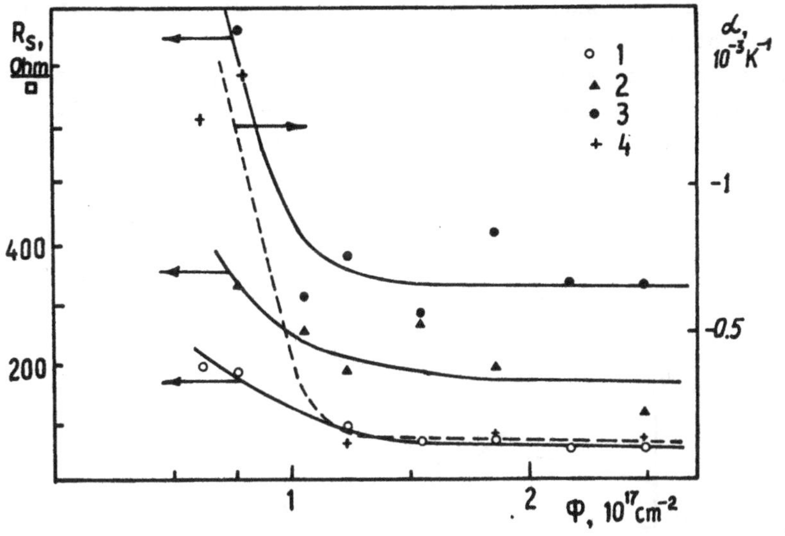

Fig.1. The sheet resistance, R_s and temperature coefficient of the resistance, α vs. dose implantation, φ for silicon implanted by Cr^+ (E=40keV) at the different dose rate, j: 1,4- $0.6\cdot10^{14}$, 2 - $1.5\cdot10^{14}$, 3 - $2.5\cdot10^{14}$ $Cr^+/cm^2 s$.

=0.46, the layer thickness being $\sim$ 50 nm. Additional electron diffraction investigations revealed the appearance of the polycrystalline chromium disilicide for the samples implanted with the dose $\sim 10^{17} cm^{-2}$ and higher and at $j > 1.5\cdot10^{14} cm^{-2}s^{-1}$.

It means that at low doses the process i) takes place. Then, only at the dose of $2.5\cdot10^{17}$ cm^{-2} the continuous silicide film is formed. The influence of the dose rate on the resistance consists in the formation of the micropore with the increase of j, resulting in the anomalously high swelling (7).

As one can see in Fig.1 the temperature coefficient, α is negative at all implantation doses investigated, while α for $CrSi_2$ (6) and 10 Ohm·cm - p-silicon crystal (1) at room temperature are to be positive. We beleive that such sign of the temperature coefficient can be caused by at least 3 reasons yilding negative α :

1) the near-surface silicon layer, disordered by implantation;

2) Cr-atoms act as donor impurity in Si (3), so they can partially compensate the shallow acceptor impurity but Cr-doped silicon also has the negative temperature coefficient of the resistance;

3) the boundaries of polycrystalline grains.

In this case, at the relatively low implantation dose before the percolation moment the contribution of the above mentioned 3 factors into the measured value of R_s is great but then at the dose $\varphi \gtrsim 1.2\cdot10^{17} cm^{-2}$ the influence of first two factors consists in the shunting effect. Of cause, the influence of grain boundaries is remained.

Thus, the measured resistance has to be determined by the synthesized grains of Cr-disilicide plus the amorphous Cr-doped silicon plus shunting of silicon substrate taking into account contact effects at interfaces of different phases.

To improve the electrical properties of samples the laser annealing was used. In Fig.2 the influence of laser energy density, E on the sheet resistance and the temperature coefficient are presented.

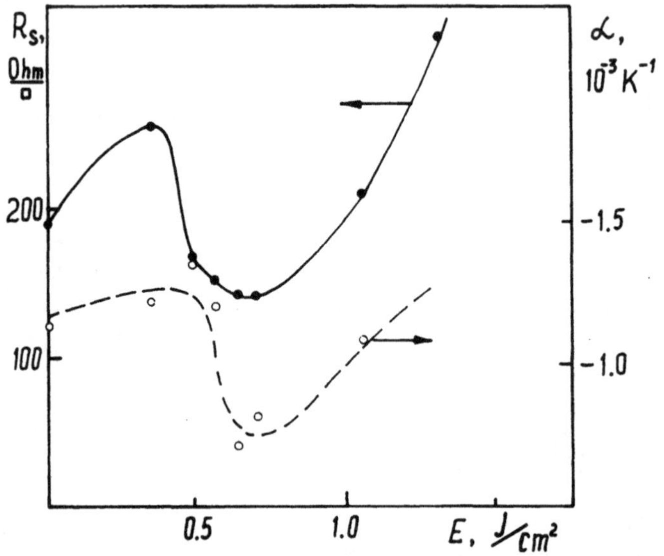

Fig.2. Sheet resistance, R_s and the temperature coefficient of the resistance, α vs. the density of the laser energy on the sample, E for silicon implanted by 40 keV Cr^+ with the dose of $2.5 \cdot 10^{17}$ cm^{-2} and $j=1.5 \cdot 10^{14} cm^{-2}s^{-1}$.

As RBSspectra and electron diffraction patterns demonstrated, the decrease of R_s and α up to the E ≅ 0.7 J/cm^2 was conditioned by the annealing of the radiation damage as well as by the partial crystallization of the amorphous silicon. The following increase of both curves at E > 0.7 J/cm^2 are due to the redistribution effect: almost continuous polycrystalline layer of $CrSi_2$, arising after $2.5 \cdot 10^{17} cm^{-2}$ implantation, decays, Cr-atoms diffuse into the depth of the silicon and after the finish of the laser pulse recrystallizes again. Moreover, the surface layer including the silicide was partially evaporated. Mean concentration of $CrSi_2$ on the surface of the sample falls down to ~5% and $CrSi_2$ spreads to the depth of several hundreds nm as it can be seen in Fig.3. At the same time the amorphous silicon layer homoepitaxially recrystallizes up to the surface.

Most surprising is the fact of very fast Cr-diffusion in silicon. The diffusion coefficient Cr in Si estimated from the RBS-spectrum was found to be about $10^{-2} cm^2/s$. Similar high value of the diffusion coefficient for zinc in silicon in the Zn-implanted Si after ns-laser annealing was found earlier (5). The clearing up of the diffusion mechanism requires additional investigations.

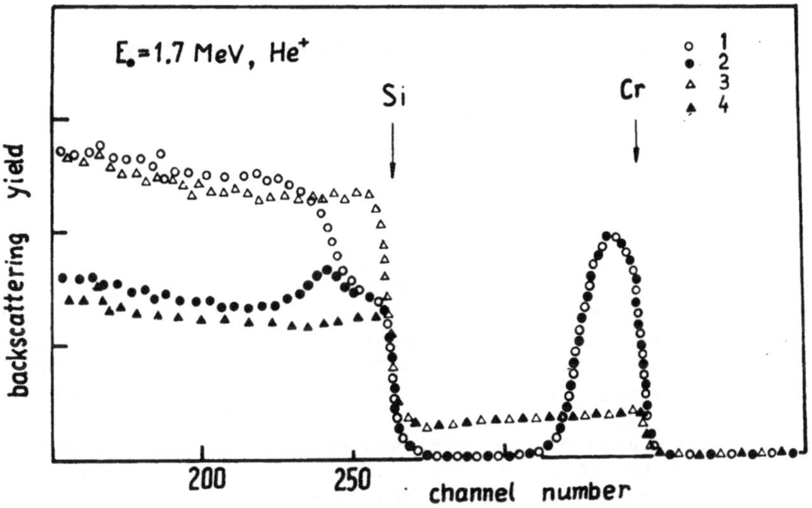

Fig.3. RBS spectra of silicon implanted by Cr^+ with $\Phi = 2.5\cdot10^{17} cm^{-2}$, $j=0.6\cdot10^{14} cm^{-2}s^{-1}$ curves: 1,2 - as implanted; 3,4 - after laser annealing with $E = 1.3\ J/cm^2$; 1,3 - random; 2,4 - aligned.

Conclusions.

1. At Cr^+-implantation in Si the chromium silicide can be formed, $CrSi_2$ being polycrystalline at $j \gtrsim 1.5\cdot10^{14} cm^{-2}s^{-1}$.
2. The following laser treatment at $E \cong 0.7\ J/cm^2$ leads to the improvement of crystalline structure of $CrSi_2$ and to the reduction both of the electrical sheet resistance and the temperature coefficient of resistance as well as the thermal stability of electrical properties due to the annealing of radiation damages.
3. Taking into account the determined thickness of the silicide layer the calculated resistivity of the conducting layer is equal $\rho = 7\cdot10^{-4}$ Ohm·cm corresponding to the reported data (4).

References.

(1) Bullis, W.M et al.: Solid-State Electronics 11 (1968) 639-646.

(2) Khaibullin, I.B. et al.: Rad.Eff. 36 (2976) 225-233.

(3) Lebedev, L.A., N.A.Sultanov: Fiz. Tekn. Poluprov. 4 (1970) 2208-2211.

(4) Murarka, S.P.: Silicides for VLSI applications, Academic Press 1983.

(5) Mustafin, T.N. et al.: Fiz. Tekn. Poluprov. 12 (1978) 1312-1317.

(6) Nishida, I., T.Sakata: J.Phys.Chem.Solids 39 (1978) 499-505.

(7) Petukhov, V.Yu., I.B.Khaibullin, M.M.Zaripov: Poverhnost. Fiz., Him., Meh. N2 (1985) 104-107.

(8) Petukhov, V.Yu. et. al.: Phys.stat.sol.(a) 96 (1986) 463-468.

CRYSTALLINE AND AMORPHOUS Cr-Si ALLOY FORMATION WITH ENERGETIC PULSES

E.D'Anna, G.Leggieri, A.Luches and M.Martino *

1. INTRODUCTION

The very fast heating and cooling rates reached in pulsed laser and electron beam irradiation of materials allow metastable phases to be obtained. Microcrystalline phases not obtainable by conventional furnace annealing were detected in metal/silicon systems /1/. Many glassy metal-silicon alloys have been obtained by nanosecond laser irradiation of various metal/silicon systems (Au/Si, Ag/Si, Pt/Si, Pd/Si, V/Si and Nb/Si) and many interesting features in the electronic properties of these alloys were found /2,3/.

We extensively studied the Cr/Si system by irradiating with laser and e-beam pulses both thin Cr films deposited on single crystal silicon (c-Si) and on evaporated silicon layers (a-Si).

2. EXPERIMENTAL APPARATUS

The first kind of samples consisted of c-Si wafers covered with Cr films 65 and 130 nm thick. The second consisted of c-Si wafers on which a 500 nm thick SiO_2 layer was formed . On this substrate we deposited first an a-Si layer and then a Cr film. Two different samples were prepared: Cr(60 nm)/a-Si(700 nm) and Cr(100 nm)/a-Si(200 nm), to have a Cr/Si atom ratio of 13/87 and 45/55, respectively, corresponding to the deepest eutectic compositions of the Cr-Si binary system. The samples were irradiated in vacuum with a 25 keV - 50 ns e-beam /4/ and in air with a Nd:glass laser (1.06 μm, 30 ns). Some samples were also irradiated with a KrCl excimer laser (222 nm, 15 ns).

Single pulse irradiation was always accomplished. The irradiated samples were analyzed by using Rutherford backscattering spectrometry (RBS) with 2 MeV He^+ particles. X-ray difractometry investigations were accomplished on some samples.

3. RESULTS AND DISCUSSION

The Cr/c-Si samples, submitted to e-beam irradiation /5/ , present evidence of reaction at current densities higher than about 1000 A/cm^2. In particular, the Cr(65 nm)/c-Si sample presents a $CrSi_2$ layer 70 nm thick at the Cr/Si interface after a 1000 A/cm^2 pulse. The layer thickness increases with beam fluence and the Cr film has completely reacted after a 1200 A/cm^2 pulse. RBS analysis shows formation of a $CrSi_2$ layer about 200 nm thick in this last sample. In the Cr(130 nm)/c-Si samples, a 145 nm thick $CrSi_2$ layer is formed after a 1000 A/cm^2 pulse. After a 1200 A/cm^2 pulse the silicide layer thickness is of about 260 nm and all the Cr film is consumed after a 1400 A/cm^2 pulse.

The same samples, submitted to KrCl laser irradiation /6/, react at quite low fluences, due to the strong coupling of ultraviolet radiation to the metal film. A 40 nm silicide layer is formed at the Cr-Si interface after irradiation of the Cr(65 nm)/c-Si sample at 0.35 J/cm^2. The reacted layer thickness is of about 90 nm in the sample irradiated at

---

* Universita' di Lecce, Dipartimento di Fisica, 73100 Lecce, Italy

0.60 J/cm^2, and the reaction is complete after a 0.9 J/cm^2 pulse. In the Cr(130 nm)/c-Si samples, a reacted layer about 80 nm thick is formed after a 0.45 J/cm^2 pulse, and a reacted layer about 160 nm thick after a 0.9 J/cm^2 pulse. From the RBS and x-ray diffraction spectra, a composition of the reacted layers close to the $CrSi_2$ silicide can be deduced.

Using the Nd:glass laser, in the Cr(65 nm)/c-Si samples reactions starts at a fluence of about 0.5 J/cm^2. Complete reaction is evident after a pulse of about 1.5 J/cm^2 (Fig. 1). At higher fluences a non uniform in depth composition is evident from RBS spectra.

More complex are the RBS spectra obtained after irradiation of the Cr/a-Si/SiO_2/c-Si samples. Submitted to e-beam pulses, the Cr(60 nm)/a-Si(700nm)/SiO_2(500 nm)/c-Si samples show Cr/Si intermixing after a 950 A/cm^2 pulse; the mixing is almost complete after a 1100 A/cm^2 pulse.

At higher fluences (1200 - 1400 A/cm^2) we observed an almost complete removal of the deposited materials. The SiO_2 substrate appeared as the sample surface. At an even higher fluence (1600 A/cm^2), intermixing of the deposited films was again evident. A 250 nm - thick layer with a composition close to the $Si_{67}Cr_{33}$ eutectic was observed (Fig. 2).

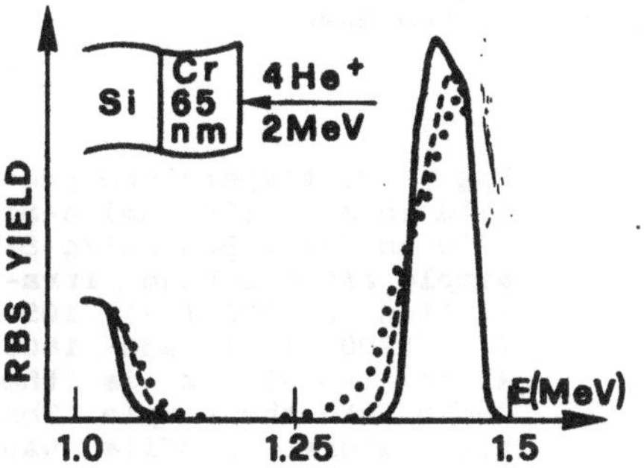

Fig. 1: RBS spectra of a Cr(65 nm)/c-Si sample; as deposited (-), after a 0.55 J/cm^2 (--) and a 1.0 J/cm^2 (..) pulse (Nd:glass laser).

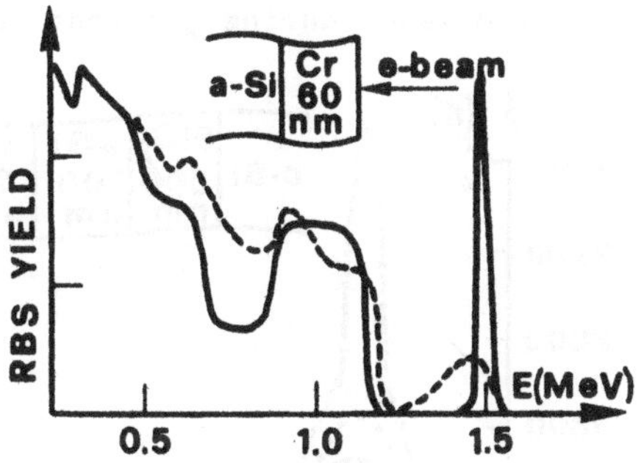

Fig. 2: RBS spectra of a Cr(60 nm)/a-Si(700 nm)/SiO_2(500 nm)/c-Si sample; as deposited (-) and after a 1600 A/cm^2 pulse (--).

The same trend was evident also in the Cr(100 nm)/a-Si(200 nm)/SiO_2(500 nm)/c-Si samples, at slightly different fluences in respect of the above samples, due to the different thickness of the deposited films. We noted that the sample submitted to a single 1500 A/cm^2 pulse presented an intermixed layer about 200 nm thick, with a composition near to that of the $Cr_{45}Si_{55}$ eutectic.

The same kinds of samples were also irradiated with the Nd:glass laser. At moderate fluences we noticed intermixing of the deposited materials. The composition of the reacted layers was close to the $CrSi_2$ phase. Only after a 1.1 - 1.2 J/cm^2 pulse, thin (150 - 180 nm) layers with an eutectic composition were obtained. It was not possible to obtain thicker layers by increasing the laser beam fluence, due to strong evaporation of the Cr film.

From x-ray diffraction analysis, it results that microcrystalline $CrSi_2$ silicide was always formed when a Cr film deposited on a c-Si wafer was irradiated. The same happens when only partial reaction is obtained in the Cr/a-Si/SiO_2/c-Si samples. When the Cr/Si reaction is almost complete, $CrSi_2$ nanocrystals are usually detected . They are embedded in an amorphous layer. Only when complete mixing occurs an amorphous layer is obtained. This means that it is quite impossible to obtain an amorphous Cr/Si alloy when the melt composition is not close to the eutectic composition. It must be also noted that when the energy fluence is sufficient to cause complete mixing of the Cr/Si deposited layers the resulting alloy does not adhere to the SiO_2 substrate and can be easily ripped off. But when the energy fluences are increased, the alloyed layer presents a good adhesion to the substrate.

The thermal effects produced into the samples by the e-beam and laser pulses were studied by numerically solving the heat diffusion equation /5,6/. As previously determined, intermixing starts when the Cr/Si interface reaches the lowest eutectic temperature of the binary Cr/Si system. Diffusion go on as long as the interface temperature is higher than the eutectic temperature. The growth rate of the reacted layer results to be about 1 m/s.

It must be noted that also the adhesion behaviour in the Cr/a-Si/SiO_2/c-Si samples can be understood when the temperature profile into the samples is considered (Fig.3). When the e-beam (laser) fluences are high enough to cause complete mixing of the Cr/a-Si films there is a temperature mismatch between the molten layer and the solid SiO_2 surface, which prevents adhesion. At higher fluences, when the temperature at the SiO_2 surface is close to the SiO_2 melting temperature (~ 2000 K), the molten layer wets the SiO_2 underlayer. This fact prevents ripping off of the alloyed layer during the fast cooling down of the sample.

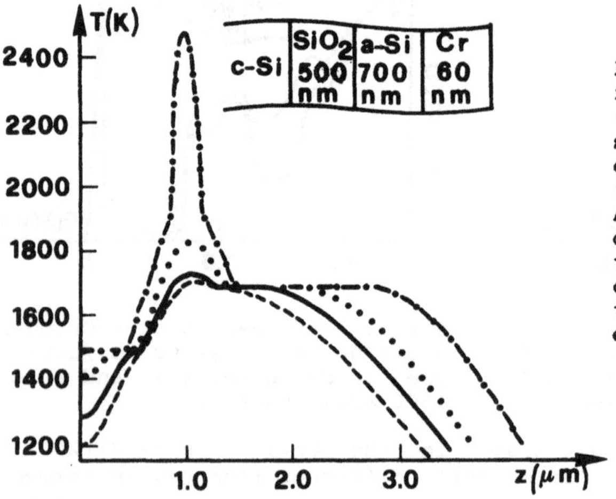

Fig. 3: temperature profile in a Cr(60 nm)/a-Si (700 nm)/SiO_2(500 nm)/c-Si sample after e-beam irradiation at 950 (--), 1050 (-), 1200 (..) and 1600 A/cm² (- • -). z is the depth into the sample. The temperature profile was calculated at the instant t=100 ns (end of the e-beam pulse).

The authors want to thank A.Drigo, G.Majni and S.Martelli for help in the experiments and discussion.
Work supported in part by Consiglio Nazionale delle Ricerche.

/1/ von Allmen,M., S.S.Lau, M.Maenpaa and B.Y.Tsaur: Appl. Phys. Lett. 37 (1980) 84
/2/ von Allmen,M., S.S.Lau, M.Maenpaa and B.Y.Tsaur: Appl. Phys. Lett. 36 (1980) 205
/3/ Huber,E. and M.von Allmen: Phys. Rev. B28 (1983) 2979
/4/ Luches,A., V.Nassisi, A.Perrone and M.R.Perrone: Physica 104C (1981) 228
/5/ D'Anna,E., G.Leggieri, A.Luches, G.Majni and G.Ottaviani: Thin Solid Films 136 (1986) 93
/6/ D'Anna,E., G.Leggieri, A.Luches, G.Majni, F.Nava and G.Ottaviani: Appl.Phys. A 40 (1986) 183

COMPUTER SIMULATION OF ION BEAM MIXING OF COBALT ON SILICON

I. R. Chakarov and D. S. Karpuzov[1)]

1. Introduction

Many theoretical works deal with recoil mixing of thin or thick layers on a substrate /2,3,4,10/. These studies were carried out within the assumption of isotropic angular distribution of moving low-energy cascade atoms. In the case the initial energy of primaries should be $E \gg E_o$, E_o being between 10 and 100 eV /11/. Any deviation from isotropic distribution of the cascade atoms affect the recoil mixing /5,6,9,12/ and the mixing efficiency will depend not only on the energy deposition (F_D) only but on the momentum deposition function (F_P) as well, especially near the surface. Such a situation is encountered for example in the dynamic recoil mixing (DRM), a process where a constant-thickness thin layer is bombarded by a energetic ion beam to obtain recoil mixing at the interface between the layer and the substrate /1/. Therefore the computer calculation methods are most suitable for analyzing DRM, especially if the full kinetics of the process is simulated.

2. Calculational method

The recoil mixing was simulated by using the binary-collision computer program MARLOWE /7,8/. The main assumptions include : a) taking into account the process kinematics only, i.e. without chemical driving forces, thermal spikes and radiation enhanced diffusion, b) the mixing is considered as a linear superposition of results originated by different primary ions and c) no composition or structural changes of both the film and the substrate are assumed during the bombardment.

The recoil mixing was simulated for Co-film on a Si-substrate. A typical run follows the trajectories of 200 to 600 incident Ar ions with energy of 10 keV and all originated cascade atoms in the target. The information concerning the transmission sputtered atoms, interstitials and vacancies created in each cascade is stored and later analyzed. Three basic structures were investigated. First, a thin cobalt film between 6 and 75 Å thick was bombarded to determine the transmission sputtering yield for these thicknessses. Second, a semiinfinitive Co-target was irradiated by 10 keV Ar^+ to determine the energy deposition or damage profile. Third, two layer structure was bombarded consisting of thin Co-film on Si-substrate with Co-layer thickness between 10 and 60 Å to examine the recoil mixing efficiency. In all cases Co-layer was amorphous and the Si-substrate was polycrystalline. The minimal energy above which the recoil target atom is incorporated in the cascade was 25 eV for Co and 15 eV for Si atoms. The same values were used as cut-off energies.

---

1) Institute of Electronics, Bulg. Acad. Sci., Boul. Lenin 72, Sofia 1784, Bulgaria

The latter determine the minimum energy of particles followed in the cascade. An atom or ion with lower energy is considered as stopped in a replacement or interstitial position. Surface binding energy was 4.4 eV.

3. Results

The transmission sputtering yield as a function of the Co-film

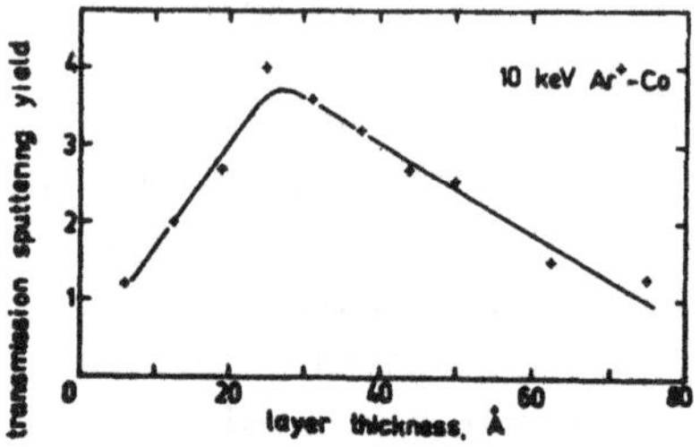

Fig.1. Transmission sputtering yield vs Co-film thickness

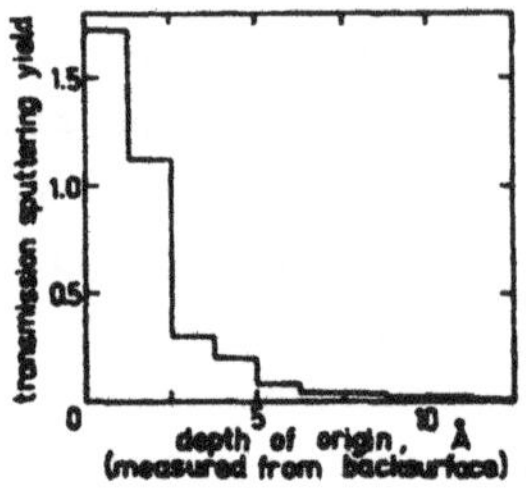

Fig.2. Depth of origin of transmission sputtered Co-atoms at layer thickness corresponding to maximum yield

thickness is represented in Fig.1. We have compared this function with the dependence of E_{xt}/E_t on the thickness. Here E_{xt} is the mean energy of transmitted ions related to their motion along the surface normal and E_t is their full energy. The comparison of the yield with E_{xt}/E_t shows that the maximum transmission sputtering yield corresponds to the maximum deposited momentum along that direction, P_x. Fig.2 illustrates the depth of origin of transmission sputtered Co-atoms, measured from the sputtered (back) surface.

In Fig.3 we summarize the results of computer simulations of recoil mixing of Co-layer with different thicknesses on silicon substrates. In addition, in Fig.4 the radiation damage profile in a thick cobalt target is plotted. The comparison of the results illustrated in Figs. 1, 3 and 4 indicates that the recoil mixing efficiency well correlates with the transmission sputtering yield and the deposited momentum, P_x, but not with the radiation damage profile or the deposited energy function, $F_D(x)$.

Concentration depth profiles of the recoil-transmitted Co-atoms through the interface into the silicon substrate and of Si atoms into the Co-film are plotted in Fig.5. The thickness of the cobalt layer chosen

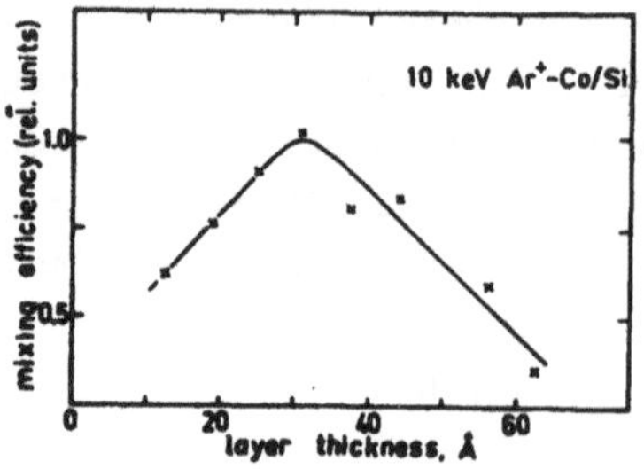

Fig.3. Mixing efficiency of Co/Si structure vs the thickness of the Co-layer

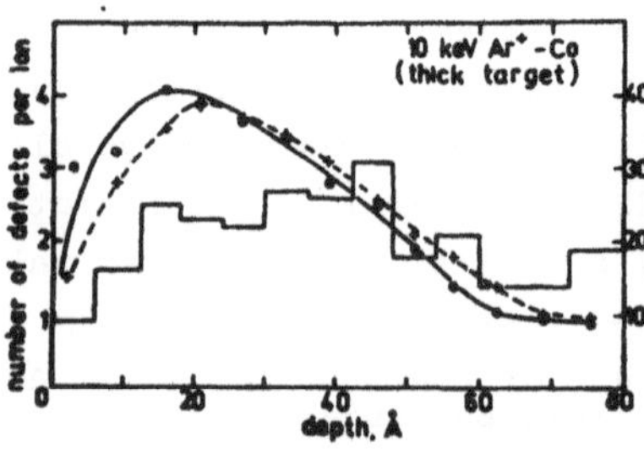

Fig.4. 10 keV Ar^+ bombardment of thick Co-target. Histogram- distribution of Ar^+, crosses- interstitials, points- vacancies

corresponds to the maximum mixing efficiency. Short- and long-range recoil mixing can be distinguished for Co-atoms in the Si-substrate. FWHM is only 5 Å. The cascade-mixed area of Si atoms in the cobalt is narrrower than the forward mixing. A comparison with the depth of origin of transmission sputtered Co-atoms for this case (Fig.2) demonstrates that only layers closest to the interface can contribute to the mixing. For instance, first four layers yield 95 % of the sputtered cobalt atoms, of which 77 % should be attributed to the first layer, 13% to the second and 3% to the third. Therefore a presence of a potential barrier at the interface different from that of binding energies of the host matrixes used in the present calculations would affect the concentration profiles of recoil-transmitted atoms /3,13/ since most atoms ejected from the first layer has low energies (their distribution was found to be close to E^{-2}). Generally the reduction of the barrier increases both the transmission sputtering yield and the mixing rate /3,13/. At high fluences one has to take into account also the modification of the composition near the interface.

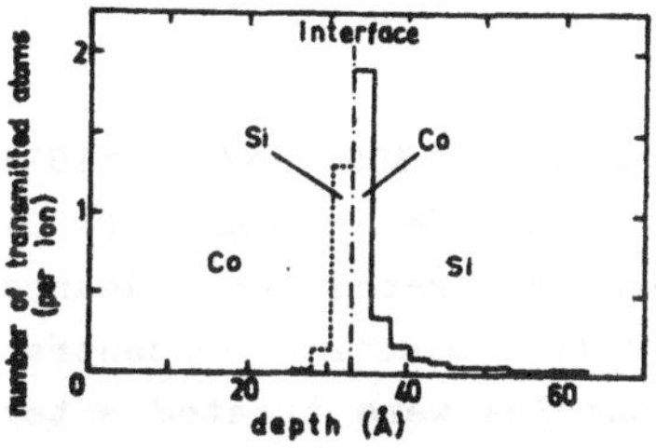

Fig.5. Depth profiles of recoil transmitted Co and Si

4. Conclusions

The present calculations illustrate the application of a BC computer model to atomic mixing at low energies and small layer thicknesses. In these circumstances the results can be hardly analyzed by using theoretical descriptions of F_D and F_P.

References

(1) Argyrokastritis,P., D.S.Karpuzov, J.S.Colligon. A.E.Hill and H.Kheyrandish : Phil.Mag. A49 (1984) 547
(2) Dzioba,S. and R.Kelly : J. Nucl.Mater. 76&77 (1978) 175
(3) Fisher,G., G.Carter and R.Webb : Radiat.Eff. 18 (1978) 41
(4) Littmark,U. : Nucl.Instrum.Meth. B7/8 (1985) 684
(5) Littmark,U. and W.O.Hofer ; Nucl.Instrum.Meth. 168 (1980) 329
(6) Littmark,U. and P.Sigmund : J.Phys.D - Appl.Phys. 8 (1975) 241
(7) Robinson,M.T. in : Sputtering by Particle Bombardment I, ed. by R.Behrish, Springer, N.Y.,1981
(8) Robinson,M.T. and I.M.Torrens : Phys.Rev. B9 (1974) 5008
(9) Roosendal,H.E., U.Littmark and J.B.Sanders : Phys.Rev.B26(1982) 5261
(10)Sigmund,P. : Phys.Rev. 184 (1969) 383
(11)Sigmund,P. : Rev.Roum.Phys. 17 (1972) 823, 969, 1079
(12)Sigmund,P. : See Ref.10
(13)Grötzschel,R., R.Klabes, U.Kreissig and A.Schmidt : Radiat.Eff. 36 (1978) 129

This work is sponsored by the Science Committee of PRB, Research contract No.107

I-V AND C-V STUDIES OF ION MIXED Au/GaAs (100) CONTACTS - EFFECT OF THE ANNEALING

Zs. J. Horváth[1], B. Pécz[1], E. Jároli[2]

Experimental

The I-V and C-V [6] characteristics of Ar^+ (400 keV, $1*10^{16}$ ion/cm^2) and Xe^{++} (700 keV, $1*10^{14}$ ion/cm^2) bombarded Au/GaAs (100) Schottky contacts have been studied. The Au was evaporated onto cleaned sulphur doped GaAs VPE layer with $1.4*10^{17}$ cm^{-3} free electron concentration at 150°C by resistance heating [7]. The samples were treated after the bombardment by consecutive heating for 10 min. at increasing temperatures in the range of 375-525°C in forming gas, while the reference sample without bombardment in temperature range of 250-425°C. 7-10 devices were studied in each sample. The deviation of the examined parameters was less then 10%.

I-V measurements

The reference sample showed normal Schottky I-V characteristics up to 350°C annealing. The barrier height decreased with increasing annealing temperature, as it is seen in Table 1 presenting the barrier height values obtained from the reverse I-V characteristics. After 350°C heat treatment the contacts have become near ohmic.

The forward I-V characteristics obtained for the bombarded contacts had continuous curvature (had no straight region) [7]. This indicates the presence of a nonlinear resistive layer in series connection with the Schottky contacts. Since the curvature of the I-V characteristics increased (the current decreased), therefore the resistance of the above layer also increased with increasing annealing temperature. The resistance of this layer seems to be higher for the devices bombarded by Xe^{++} ions, than for those bombarded by Ar^+ ions.

The barrier height values obtained from the reverse I-V characteristics of the bombarded devices (the forward ones had continuous curvature) are given in Table 2. The bombardment with Ar^+ ions decreased the barrier height much stronger than that with Xe^{++} ions. This is probably connected with the much higher dose of Ar^+ bombardment which might create much more defects, than the bombardment with Xe^{++} ions [7].

----

[1]Research Institute for Technical Physics of the Hungarian Academy of Sciences, Budapest, P.O.Box 76. H-1325, Hungary.
[2]Central Research Institute for Physics of the Hungarian Academy of Sciences, Budapest, P.O.Box 49. H-1525, Hungary.

Table 1: The barrier height obtained from reverse I-V characteristics and the capacitance at zero bias for reference devices without bombardment as a function of the annealing temperature

t, °C	ϕ_{bI}, eV	C_0, pF
-	0.72	34.2
250	0.64	-
275	0.59	35.5
300	0.55	36.0
325	0.54	37.0

Table 2: The barrier height obtained from reverse I-V characteristics and the capacitance at zero bias for Xe^{++} and Ar^{+} bombarded devices as a function of the annealing temperature

	Xe^{++} bombarded		Ar^{+} bombarded	
t, °C	ϕ_{bI}, eV	C_0, pF	ϕ_{bI}, eV	C_0, pF
-	0.67	4.12	0.47	19.25
375	0.69	4.49	-	-
400	0.70	4.53	0.51	14.55
425	0.69	4.28	0.53	-
450	0.68	4.10	0.52	12.90
475	0.68	3.82	0.53	11.45
500	0.67	3.80	0.54	7.25
525	0.67	3.80	0.56	6.10

The barrier height for Ar^{+} bombarded devices increased with increasing annealing temperature, while for Xe^{++} bombarded devices was nearly constant with a slight maximum at 400°C.

C-V measurements

Although the reference devices had good C-V characteristics up to 350°C heat treatment (see Table 1), those for the bombarded ones were poor, alike those in bad quality MIS structures. This indicates the presence of a rather thick interfacial layer with high interface state density.

It is very interesting that while the barrier height was much stronger decreased after Ar^{+} bombardment than after Xe^{++} bombardment, the capacitance at zero bias decreased after Xe^{++} bombardment much stronger, than after Ar^{+} one, as it is also indicated in Table 2. This is probably connected with the resistive layer mentioned above.

The annealing decreased monotonously the capacitance values at zero bias for Ar^{+} bombarded contacts, while these values for the Xe^{++}

bombarded devices had a slight maximum at 400°C, like the barrier height for these devices.

Conclusions

On the basis of the existing ideas [1-5,8-10] concerning the physical picture of Schottky contacts these experimental results may be explained qualitatively by the following model:

The ion bombardment creates an interfacial layer with high density of surface states at the Au-GaAs interface and a transition resistive layer located deeper in the GaAs. In this layer the free electron concentration is significantly reduced in comparison with the bulk. For Ar^{+} bombardment either the thickness of this layer or the concentration reduction in it is much less, than for Xe^{++} bombardment, but the interface state density seems to be higher. The heat treatment decreases the interface state density and increases either the thickness of the resistive layer or the concentration reduction in it.

Acknowledgements

The authors are indebted to J. Gyulai, I. Gyúró, I. Mojzes and M. Németh-Sallay for their support and the valuable assistance.

References:

(1) Card H.C., E.H.Rhoderick: J.Phys.D: Appl.Phys. 4 (1971) 1589-1601.
(2) Cowley A.M.: J.Appl.Phys. 37 (1966) 3024-3032.
(3) Hanselaer P.L., R.L.Van Meirhaeghe, W.H.Laflére, F.Cardon: Semicond. Sci. Technol. 2 (1987) 94-101.
(4) Horváth Zs.J.: Proc.Symp.Electron.Technol.'87, Sept.15-18, 1987, Budapest, Hungary, in press.
(5) Horváth Zs.J., I. Gyúró, M. Németh-Sallay, A. Nagy, B. Szentpáli: 15th School on the Physics of Semiconducting Compounds, April 21-25, 1986, Jaszowiec-Ustron, Poland.
(6) Horváth Zs.J., B.Pécz, P.Tüttő, E.Jároli, M.Németh-Sallay, I.Gyúró: Proc. 2nd Conf. on Physics and Technology of GaAs and other III-V Semiconductors, Sept.8-11, 1986, Budapest, Hungary, in press.
(7) Jároli E., B.Pécz, J.Gyulai, M.Fried, L.Petrás, E.Zsoldos, T.Lohner, I.Mojzes: Nucl.Instr.Meth.Phys.Res. B19/20 (1987) 767-772.
(8) McLean A.B., I.M.Dharmadasa, R.H.Williams: Semicond. Sci. Technol. 1 (1986) 137-142.
(9) Tseng H.-H., C.-Y.Wu: Solid-State Electron. 30 (1987) 383-390.
(10) Wu C.-Y.: J.Appl.Phys. 51 (1980) 3786-3789.

ION MIXING IN GLASSES

A.A.Deshkovskaya*, V.P.Komar**, I.V.Skornyakov**, G.P.Pimenova***, V.I.Ovchinnikov***

Four processes are used for glass modification with ion beams, namely 1)ion depositing, 2)ion etching, 3) ion implantation, 4) ion mixing.

Contrary to the electron and photon bombardment, in all these cases ion bombardment results in simultaneous change both in the composition and in the glass surface layer structure. These changes are most intensive in the case of ion mixing. In comparison with direct ion implantation, ion mixing is characterized by a number of peculiarities indicative of broad opportunities this method offers and its universality.

1) As an additive source there may be used various chemical compounds applied to the glass surface.

2) When using one and the same type of bombardment particles a good deal of different type additives may be introduced into the glass surface layer simultaneously.

3) A high concentration of applied additives in the glass surface layer may be achieved while primary ion dose is comparatively low.

When using ion mixing method in glass production the above-mentioned factors become particularly important because contrary to the semiconductors, glass is liable to be modified substantially only if a considerable concentration of additives (tens of atomic percents) is introduced.

In the paper discussed is ion mixing used to change optical and mechanical properties of silicate glass of various compositions(quartz, pyrex, alkalialumosilicate).Various coatings,namely films (Al,Cu,Fe,C, Ti,Jn,Ho,Y,Yt) produced by means of magnetron or thermal spraying as well as salt solutions of the same metals were applied to plane-parallel glass samples. To obtain maximum modification effect the metal film thickness was chosen so that maximum of bumping collision liberated energy for the given metal fell on metal-glass interface, i.e. d≈Rpd or Rpd<d<Rp, where d- a film thickness, Rpd- a mean projective range of bumping collision liberated energy, and Rp- a mean projective range of an additive in the given film. The samples were subjected to argon and antimony ion bombardment through the applied film layers at room temperature. Ion energy- 100-300keV, dose- 10^{15}-10^{16}ion/cm^2, current density during bombardment- 1-5mcA/cm^2.

*Minsk radioengineering institute, P.Brovka,6,220069,Minsk, USSR
**Institute of physics,Academy of sciences of BSSR,Lenin pr.,70,220002, Minsk,USSR
***Byelorussian republican powder metallurgy scientific & production corporation,Platonov,21,Minsk,USSR

The research of optical properties of samples modified by ion mixing was carried out on the basis of reflectance IR spectra obtained on Perkin-Elmer-180 spectrophotometer. Glass mechanical properties (Vickers microhardness, crack resistance) were studied with the help of a method similar to the one described in (1). In the reflectance IR spectra of the glass modified by ion mixing there was observed a shift in Si-O-Si valence oscillation bands to lower frequencies. In the case of mixing the shift was bigger than in the case of direct ion implantation(Fig.1). The largest shift under equal experimental conditions is observed during ion bombardment of two-layer coating (Ho-Y). The condition of the applied film (a metal or its compound) affects the frequency shift value $\Delta\nu$ too. For the metal salt films $\Delta\nu$ is higher than for the pure metal films which probably is the result of larger structural disturbance caused by simultaneous implantation of atoms of different sort (e.g.iron and chlorine of $FeCl_3$salt) into the glass. Preliminary diffusion facilitates coating atom implantation into the glass. It was noted that ion mixing shows differently in the glasses of different composition.

Such a band shift is usually observed when modifier atoms are introduced into the glass during its synthesis and indicates an increase in the degree of silicon-oxygen network depolymerization. Thus, it may be assumed that in the given case the implanted additives play the role of modifiers - they build into the glass network increasing the number of non-bridging oxygens. This probably leads to rearrangement of interaction between short-range order areas and changes interaction of tetrahedron structural element normal oscillations in the glass surface layer.

Mechanical properties of the glass modified by ion mixing were studied by means of crack resistance tests with varying loads on the indenter. A number of cracks in the doped samples was found less than in those not doped. Fig.2 shows the effect of glass hardening as a result of carbon ion mixing by argon.Not doped samples were found to have bigger spread in microhardness data. Such sample hardness variation results in lower mechanical resistance of the glass and increases possibility of crack formation. Ion mixing flattens microhardness variations creating a uniform transition layer thus improving crack resistance. Better hardening is to be accounted for the formation of a considerably changed glass layer where glass structural network depolymerization under the impact of ion bombardment is accompanied by simultaneous complex and radical formation participating in which are implanted additives the interaction of which during the process of atomic rearrangement may result in higher silicon-oxygen framework binding degree.

Formation of a disturbed layer of considerable thickness is well proved by electron-microscope pictures of sample cross spalling surface after C+ and Al+ implantation by mixing (Fig.3). The different spalling pattern in the doped and not doped areas is obvious: in the not doped area it is like that of plastic materials whereas in the doped area it has a form of steps or long narrow slots along the ion beam path.

Thus, it was determined that as a result of implantation into the glass of recoil atoms (bombarded layer atoms), argon and antimony primary ions, chain mixing and radiation stimulated diffusion in the glass surface layer there is formed a layer different in its composition, structure and properties from the bulk of the glass. This is proved by the IR reflectance spectra, surface mechanical properties data and results of electron-microscope research.

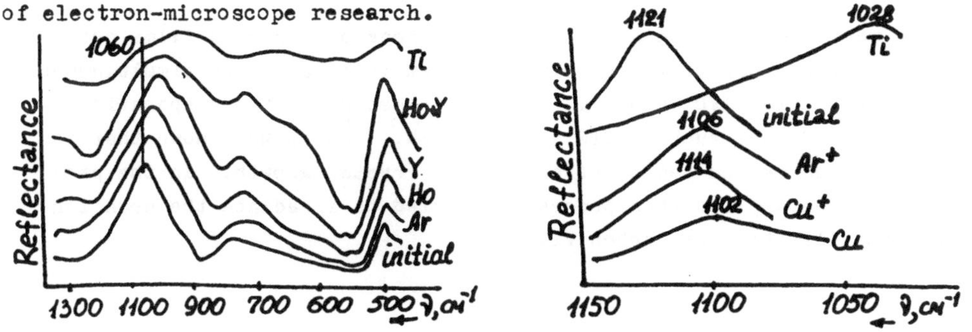

Fig.1. IR reflectance spectra of alkalialumosilicate (a) and quartz (b) glasses with different coatings prior and after argon ion bombardment.

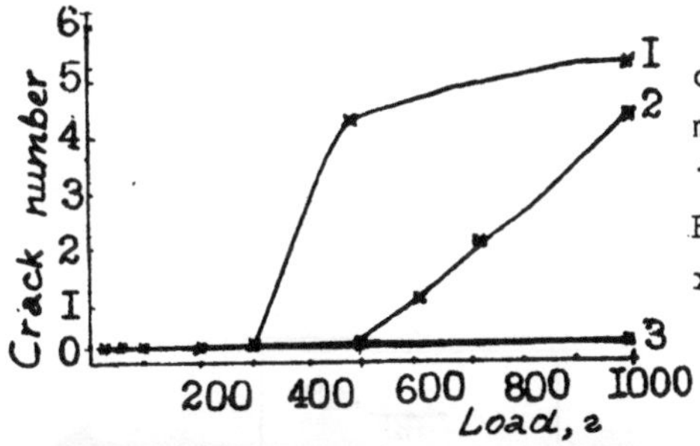

Fig.2. Crack number as a function of load on the indenter for alkalialumosilicate glass
1- initial; 2- carbon mixing by argon, E=200keV, $D=6.2\cdot10^{15}cm^{-1}$; 3- carbon mixing by argon, E=300keV, $D=6.2\cdot10^{15}cm^{-1}$

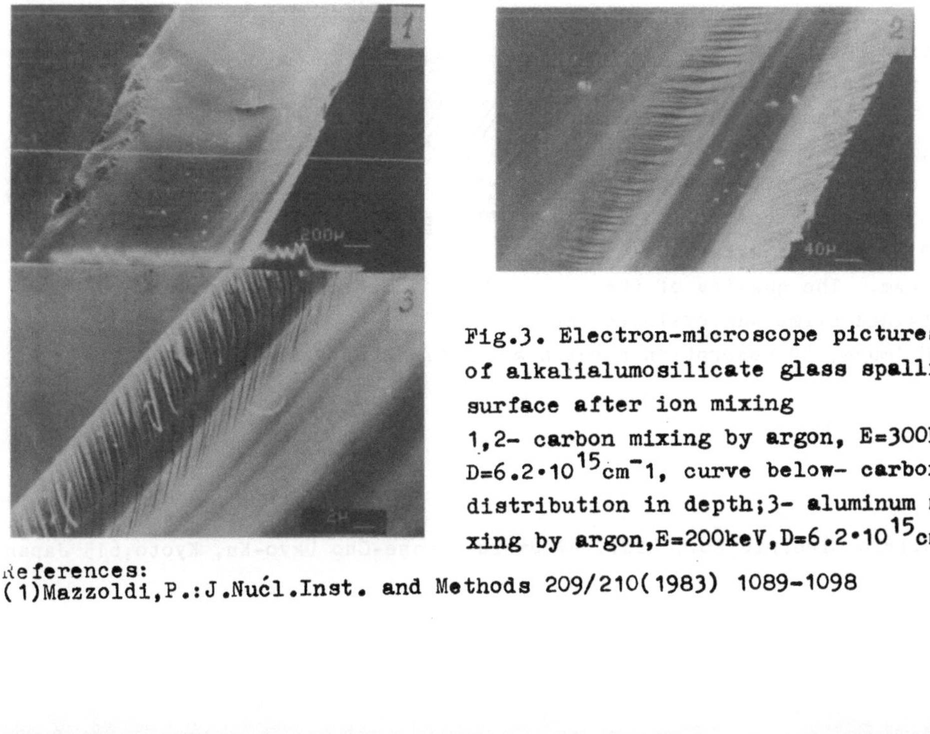

Fig.3. Electron-microscope pictures of alkalialumosilicate glass spalling surface after ion mixing
1,2- carbon mixing by argon, E=300keV $D=6.2\cdot10^{15}cm^{-1}$, curve below- carbon distribution in depth; 3- aluminum mixing by argon, E=200keV, $D=6.2\cdot10^{15}cm^{-1}$

References:
(1) Mazzoldi, P.: J. Nucl. Inst. and Methods 209/210 (1983) 1089-1098

ON THE FORMATION OF CARBON FILMS BY ION BEAM ASSISTED METHOD

Kiyoshi Ogata, Yasunori Andoh and Eiji Kamijo[1)]

1. INTRODUCTION

We report on an original preparation process of carbon films onto low temperature substrate, namely, ion beam irradiated and simultaneous vapour deposition(IVD method). We have investigated the preparation for ceramic films, such as TiN and cubic BN,[(1),(2)] and it was shown that ion beam assisted method had an effect on the crystal growth. In the present work, we applied this process and tried to irradiate the neon ion for crystallization of carbon films.

2.EXPERIMENTAL PROCEDURE

A schematic diagram of IVD machine is shown Fig. 1. The details of this machine have been described elsewhere(3).

Carbon was evaporated by electron beam bonbardment on Si substrate and neon ion was irradiated simultaneously. The acceleration voltage of neon ion was changed in the range of 200-1000v. The deposition rate of carbon was kept at the rate 1.3A/sec constantly, and beam current was changed. And the ratio of Ne/C was in the range of 2.5-50%. The substrate was controlled within 100 degrees by water cooling system. The quality of the obtained films was evaluated by SEM image, IR-absorption spectra and RAMAN-scattering pattern.

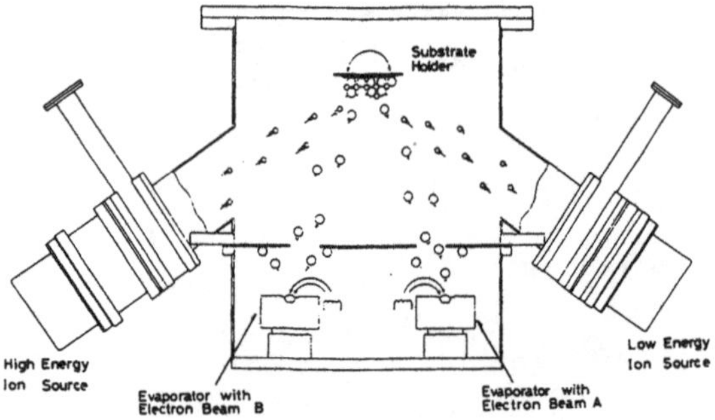

Fig. 1. Schematic diagram of IVD machine

1)Nissin Electric Co., Ltd., 47 Umezu-Takase-Cho Ukyo-Ku, Kyoto,615 Japan

3.EXPERIMENTAL RESULTS

Figure 2 shows the relation between the fraction of crystallization and ratio Ne/C, that was obtained by IR-absorption measurement. The value of vertical axis uses the ratio of peak intensity for spectra indicated in the vicinity of $2200cm^{-1}$, that agreed with TYPE-II diamond structure, and $1000cm^{-1}$ of TYPE-I.[4] And so the increase of this value means to grow well for crystalline diamond.

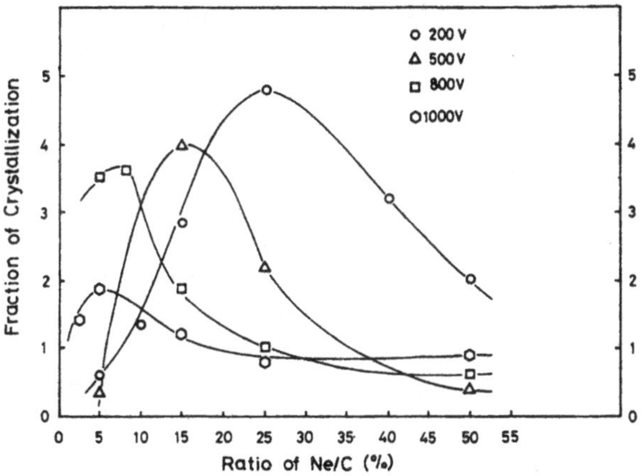

Fig. 2. Variations between the fraction of crystallization and ratio Ne/C

The figure indicated tendency to increase and decrease in proportion to the change of Ne/C, independent of ion beam energy. Especially, the films prepared with accelerated ions at 200v and the ratio Ne/C of 25% indicated the optimum condition for crystallization. Therefore, we evaluated in detail this film by using of SEM image and RAMAN-scattering patterns.

Figure 3 shows a SEM image of this sample and they appear nealy spherical and rectangular particles. In addition, the result of RAMAN-scattering for evaluating a crystal structure is shown Fig. 4. This spectrum indicates a clear peak at about $1333cm^{-1}$ and broad peak to higher region. The former agrees with cubic diamond of sp^3 structure and another due to a graphite or carbon,[5] and these facts show that this film is diamond-like carbon.

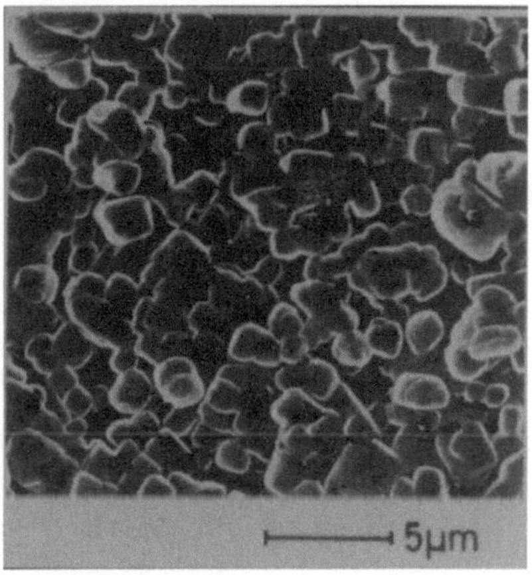

Fig. 3. SEM image of carbon film (accele. voltage: 200v, Ne/C: 25%)

In the other conditions of Ne/C ratio, the status of crystallization varied even though the energy of ion beam was fixed. And then the large part of films were seemed to compose of amorphous carbon and/or graphite phase from the external view of SEM image. The results of RAMAN-scattering and TED patern were content with the characteristic of Fig. 2. Thus the crystallization of carbon is able to be controlled by experimental terms.

4.CONCLUSIONS

We have investigated diamond films prepared by neon ions at low acceleration voltage 200-1000 v and the simultaneous carbon-vapour deposition. The results show that the status of crystalline growth was followed by the ion beam energy and ratio Ne/C. And we could be obtained the diamond crystallization under the low temperature by ion beam assisted method, making use of the optimum conditions.

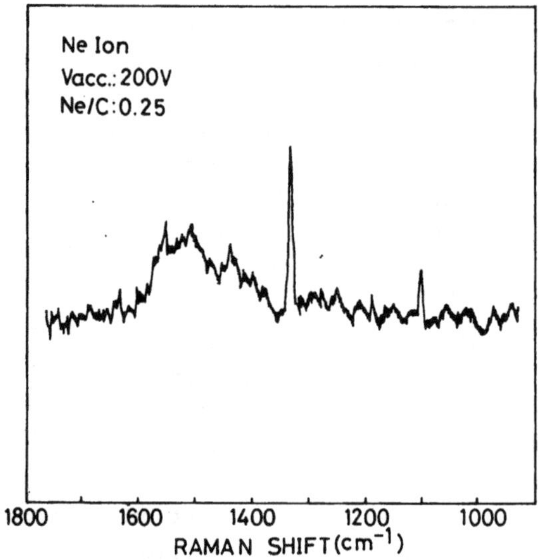

Fig. 4. RAMAN-scattering pattern (accele. voltage: 200v, Ne/C: 25%)

5.REFERENCE

(1)M. Satou, Y. Andoh, K. Ogata, Y. Suzuki, K. Matsuda, and F. Fujimoto: Jpn. J. Appl. Phys. 24 (6) (1985) 656-660

(2)Y. Andoh, K. Ogata, Y. Suzuki, E. Kamijo, M. Satou and F. Fujimoto: Nucl. Instr. and Meth. B19/20 (1987) 787-790

(3)Y. Andoh, Y. Suzuki, K. Matsuda, M. Satou and F. Fujimoto: Nucl. Instr. and Meth. B6 (1985) 111-115

(4)J. R. Hardy, S. D. Smith: Phil. May., 6,1163 (1961)

(5)S. A. Solin, A. K. Ramdas: Phys. Rev. , B1 (1970) 1687

ION MIXING IN Cu/Au MULTILAYERED THIN FILMS

K.Hohmuth, R.Grötzschel and B.Rauschenbach[1)]

The ion beam mixing technique is used to produce stable and metastable crystalline or amorphous phases. In the field of this technique, many experimental and theoretical investigations have been carried out to determine the mixing process as a function of initial composition, temperature, and irradiation conditions /1/.
In following, results are reported here on investigations of the mixed Cu/Au multilayers by transmission electron microscopy, Auger electron spectroscopy, and Rutherford backscattering.
Copper and gold layers of thickness ≥ 7 nm were deposited on substrates at room temperature in the high vacuum. The evaporation process of the multilayers took placed without interruption of vacuum. Between 4 and 10 different layers were deposited, resulting in a total thickness of order of 2 x R_p, where R_p is the mean projected range of the noble gas ions. The thickness of the copper or gold layers were adjusted such that the average film composition is Cu_xAu_{100-x} with x = 90, 85, 75, 65, 50, and 35 for Xe^+-ions irradiation and with x = 70, 50, 40, and 30 for Ar^+-ions irradiation. The specimens were bombarded at room temperature with Ar^+- ions or Xe^+- ions at energies from 100 up to 300 keV and to fluences of 5×10^{15}, 1×10^{16} and 2.5×10^{16} ions/cm^2, respectively. The redistribution of the thin layers in the depth by ion beam mixing have been measured by Rutherford backscattering technique (RBS) and Auger electron spectroscopy (AES). The crystallographic structure and morphology of the Cu/Au multilayers before and after mixing were investigated by high-voltage electron microscopy (HVEM) and transmission high-energy electron diffraction (THEED); for experimental details see Refs. /2//3/.
Some important results are summarized as follows:

(1) A pronounced increase of the grain size could be measured as a result of ion beam mixing /3/.

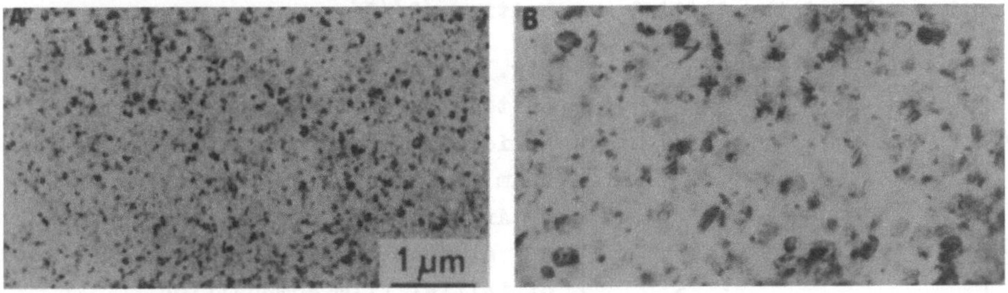

---

1) Central Institute for Nuclear Research Rossendorf
GDR-8051 Dresden, P.O. 19

Fig.1: HVEM-micrographs of multilayered Cu/Au thin films before (A) and after Ar^+-ion beam mixing with 1 x 10^{16} ions/cm^2 (B) and 2.5 x 10^{16} ions/cm^2 at room temperature (energy: 200 keV)

(2) Uniformly mixed Cu/Au layers were obtained by implantation with Xe^+-ions to fluence of 5 x 10^{15} ions/cm^2 and Ar^+-ions to fluence of 1 x 10^{16} ions/cm^2 (Fig. 2) /2/.

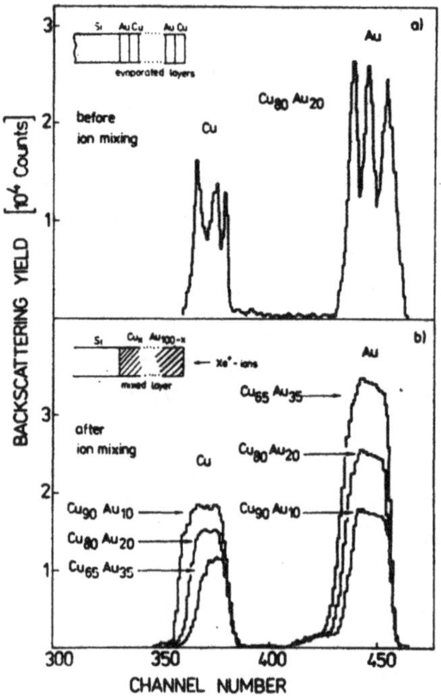

Fig.2:
He^+-Rutherford backscattering spectra of multilayered Cu/Au thin films taken before and after 220 keV Xe^+-ions bombardment to a fluence of 1 x 10^{16} ions/cm^2 at room temperature Spectra showing the discrete layered structure before irradiation ($Cu_{80}Au_{20}$) and the formation of uniformly mixed Cu/Au layers by ion beam mixing ($Cu_{65}Au_{35}$, $Cu_{80}Au_{20}$; $Cu_{90}Au_{10}$).

(3) Copper-rich and gold-rich solid solutions are formed by ion beam mixing at room temperature for a fraction of copper x ≥ 70, where the Cu-rich solid solution is the dominant phase /2//3/.

(4) Solid solutions with a middle composition could be found for x ≤ 65 and x ≥ 30 of the initial composition /2//3/.

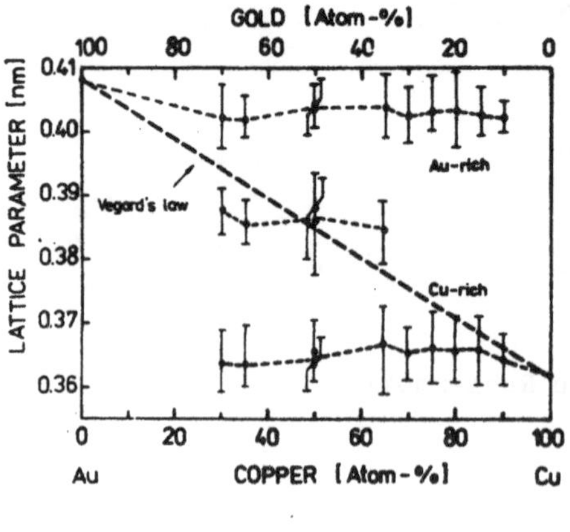

Fig.3:
Lattice parameters of metastable solid solutions formed by Xe^+-ion beam mixing (solid circles, •) and by Ar^+-ion beam mixing (open circles, o) at room temperature as a function of initial film composition.

(5) The copper-rich solid solution is performed at low implantation doses and also the gold-rich solid solution is produced with increased dose for the initial composition $x > 70$ /3/.

(6) At initial composition $x < 70$, a strong tendency could be determined to form solid solutions with a middle composition in addition to the copper-rich and gold-rich solid solutions. This tendency is increased with increasing of implantation fluence /3/.

(7) The formation of solid solutions can be predicted to use the free energy diagram as function of concentration and ion fluence. The formation of a solid solution with a middle composition is thermodynamically favoured over copper-rich or gold-rich compositions (characterized by a minimum of the free energy) /4/.

(8) Formation of solid solutions induced by ion beam mixing can be interpreted in terms of collisional mixing and chemical parameters /3//4/.

(9) Post-annealing (temperature < 380 °C) of thin $Cu_{50}Au_{50}$ layers mixed with xenon ions leads to an order-disorder transformation. It is formed a superlattice structure CuAu I of the $L1_0$-type /3/.

(10) The ordering of atoms on the lattice sites of the disordered parent is accompanied by a distortion of the lattice of sites itself. The distortions produce dislocation loops, which are characterized by a special arrangement /3/.

References

/1/ Proceedings on conference Ion Beam Modification of Materials, Catania 1986, published in Nucl. Instr. Meth. B 19/20 (1987)

/2/ B.Rauschenbach, K.Hohmuth, E.Richter, and R.Grötschel, Nucl. Instr. Meth. B 15 (1986)692

/3/ B.Rauschenbach, Appl. Phys. A 40 (1986) 47

/4/ B.Rauschenbach, phys. stat. sol. (a) 94 (1986) 409

A MODEL OF SIMULTANEOUS OR ALTERNATING ION IMPLANTATION IN FILMS AND COATINGS BEING DEPOSITED

Iskanderova, Z.A., Radjabov, T.D., Leiderman, R.Yu., and Tukfatullin, F.K.+)

Many interesting investigations and applications of ion beam assisted deposition such as the effect on the film growth and structure, increase in the film sorptive capacity, heterostructure formation, synthesis of compounds and improved surface coatings, immobilization of radioactive gases, etc /1-3,5-10/ have been known. The distribution and retained dose of the implanted impurity are the most important for the final result in many cases. This process is mainly realized in two ways: 1 - simultaneous and 2 - alternating deposition and ion bombardment using both the ion sources and discharge. At 1 the presence of thermo- and radiation-enhanced processes is very likely because of continuous irradiation of the layers being condensed. That is why our model 1 takes the diffusion into account.

1. Equations and boundary conditions in a coordinate system with the origin at the film-substrate interface are of the form

$$\left(\frac{\partial}{\partial t} - D_1\frac{\partial^2}{\partial x^2}\right)C(x,t) = qF(x+Vt;Vt), \quad -Vt < x < 0, \tag{1}$$

$$\left(\frac{\partial}{\partial t} - D_2\frac{\partial^2}{\partial x^2}\right)C(x,t) = qF(x+Vt;Vt), \quad 0 < x < \infty, \tag{2}$$

$$D_1\frac{\partial}{\partial x}C(-Vt,t) = uC(-Vt,t) + \sum_s K_s\left[C(-Vt,t)\right]^s, \tag{3}$$

$$D_1\frac{\partial}{\partial x}C(0-0,t) = D_2\frac{\partial}{\partial x}C(0+0,t) = H\left[\kappa C(0+0,t) - C(0-0,t)\right], \tag{4}$$

$$C(\infty,t) = 0, \quad C(x,0) = 0. \tag{5}$$

Here $C(x,t)$ is the impurity concentration, atom·cm^{-3}; D_1 and D_2 are the diffusion coefficients in the film and substrate, respectively; q is the ion current density; $V=u-\upsilon$, u is the deposition and υ is the sputtering rate; $F(x+Vt;Vt)$ is the depth distribution function of the implanted ions, which is parametrically dependent on the film thickness Vt. (In the case of polykinetic implantation there is need to carry on an averaging according to the energetic and angular distribution of ions).

In equation (3) K_s are the constants of surface reaction rates with the kinetics of arbitrary orders. Equation (4) is the most general (linear) condition at the interface. If the current densities are small enough and/or if the maximal implantation depth is lesser than Vt, then

+) Central Designed Technological Office for Research Device Engineering, Institute of Electronics, Tashkent, USSR

the parameters κ and H^{-1} are the same as the ratio of equilibrium concentrations (κ) and (at $\kappa=1$) of diffusive resistance (H^{-1}). The above mentioned assumptions are also necessary for the film-substrate interface to consider to be sharp, neglecting ion mixing.

Appel transformation /4/

$$\xi = \frac{x}{Vt} + 1, \ \tau = -\frac{D_1}{V^2 t}, \ \psi(\xi,\tau) = \frac{V}{q}\left(\frac{V^2 t}{D_1}\right)^{1/2} exp\left(\frac{x^2}{4Dt}\right) C(x,t), \tag{6}$$

where $D=D_1$ at $-Vt<x<0$, $D=D_2$ at $0<x<\infty$ transforms (3) and (4) into conditions on "surfaces" $\xi=0$, $\xi=1$. The transformed equations (1) and (2) in this case keep the form of Fickian equations. An exact analytical expression for Green function may be obtained in the case of kinetics of the first order ($\kappa_1 \geqslant 0$, $\kappa_s = 0$ at $s \neq 1$) and $H \to \infty$ or $C(-Vt,t)=0$ and at arbitrary H. In a general casethe problem may be reduced to a non-linear integral equation in $\psi(0,\tau)$. We do not stop here at these calculations. Some of the particular cases have been considered earlier in /6,7/.

2. The process of alternating condensation and ion implantation (neglecting diffusion) is described in a moving coordinate system by

$$\left(\frac{\partial}{\partial t} + V(t)\frac{\partial}{\partial x}\right) C(x,t) = q(t)\, F(x), \tag{7}$$

where

$$V(t) = -v, \ q(t)=q \ \text{at} \ m(T_1+T_2)<t<m(T_1+T_2)+T_1 \ (m=0,1,\dots,n),$$
$$V(t) = u, \ q(t)=0 \ \text{at} \ m(T_1+T_2)+T_1<t<(m+1)(T_1+T_2) \ (m=0,1,\dots,n-1) \tag{8}$$

T_1 and T_2 are the durations of bombardment and deposition, respectively. (The differences in a stopping power and coefficicnts of the film and substrate sputtering are neglected; the ion bombardment is assumed to be the initial and final stage of the process).

Equation (7) must be solved under conditions $C(\infty,t)=0$, $C(x,0)=0$. To the final moment of the process

$$C(x, t=nT_1+nT_2+T_1) = \frac{q}{v}\sum_{m=0}^{n} \theta(x+mvT_1-muT_2) \int_{x+mvT_1-muT_2}^{x+(m+1)vT_1-muT_2} dy\, F(y), \tag{9}$$

where $\theta(x)=1$, if $x \geqslant 0$, $\theta(x)=0$, if $x<0$; the film thickness is $d=n(uT_2-vT_1)$. For an exponential range distribution $F(x)=(1/r)\,exp(-x/r)$, where r is the average implantation depth (small energies or irradiation in a discharge),

$$C(x,nT_1+nT_2+T_1) = \frac{q}{v} e^{-X} \cdot \frac{1-e^{-b}}{1-e^{-a}} \cdot \begin{cases} exp(a[X/a]) - e^{-a}, & n > X/a > 0, \\ e^{na} - e^{-a}, & X/a > n, \end{cases} \tag{10}$$

where $X=x/r$, $a=(uT_2-vT_1)/r$, $b=vT_1/r$, $[z]$ is the integral part of a number z. At Gaussian $F(x)$ (ion beam irradiation, average energies)

$$C(x,nT_1+nT_2+T_1) = \frac{q}{2v}\sum_{m=0}^{n} \theta(X-ma)\left[erf\, \rho(X-1+b-ma) - erf\, \rho(X-1-ma)\right], \tag{11}$$

where $\rho = r/(\sqrt{2}\,\Delta r)$, Δr is the straggling.

Analysis of (9) and results of calculations (fig. 1) show that at b =const the concentration grows with a decrese in a and at

$an > x/r \gtrsim 2 \div 3$ oscillates with a period a. If $a \ll 1$, then

$$C(x, nT_1 + nT_2 + T_1) = C_0 + \frac{1}{2} C(0, T_1)\, \eta(x/ra) + O(a), \qquad (12)$$

where $C_0 = \int_0^{\infty} dx\, C(x, T_1)/(ra)$, $\eta(z) = 1 + 2[z] - 2z$.

In this case the concentration fluctuates with an amplitude $\frac{1}{2} C(0, T_1)$ near the mean value C_0. "Sawtooth" oscillations in reality will be smoothing both by internal processes in the sample (diffusion, ion mixing) and by an inertia in external actions (electrical, mechanical, etc), i.e. discontinuity of ion bombardment, movement of the sample.

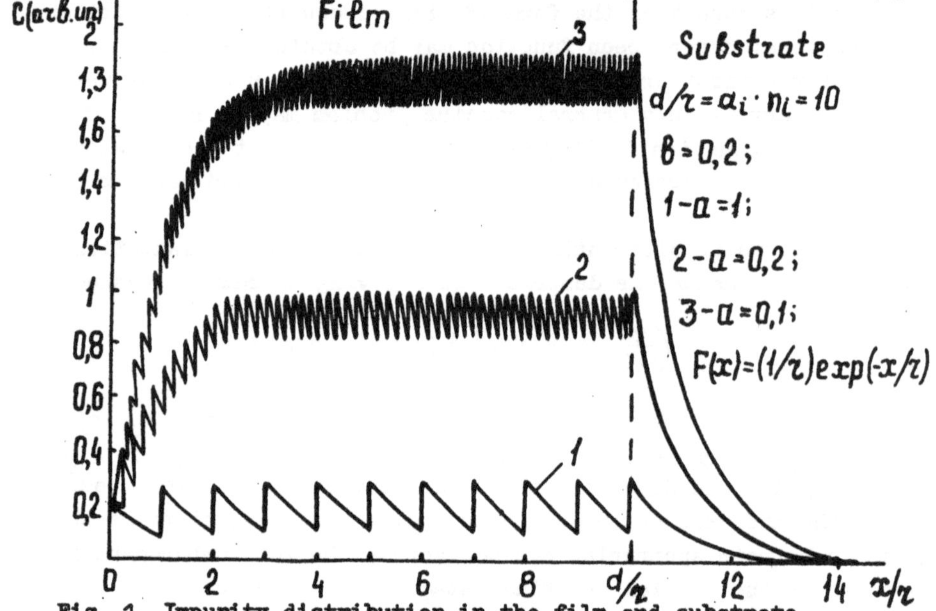

Fig. 1. Impurity distribution in the film and substrate at alternating film deposition and ion bombardment.

(1) Iskanderova, Z.A., L.F.Lifanova and T.D.Radjabov: Vacuum 32 (1982) 269-275

(2) Harper, J.M.E., J.J.Cuomo, R.J.Gambino and H.R.Kaufman: Nucl.Instr. Meth. B7/8 (1985) 886-892

(3) Kant, R.A., B.D.Sartwell, I.L.Singer and R.G.Vardiman: Nucl.Instr. Meth. B7/8 (1985) 915-919

(4) Miller, W., Jr.: Symmetry and Separation of Variables, Addison-Wesley, Massachusetts 1977

(5) Pranevicius, L.I. and Yu.I.Dudonis: Modification of Solid Properties by Ion Beams. Vilnius: Mokslas 1980 (in Russian)

(6) Radjabov, T.D. and Z.A.Iskanderova: Journal of Technical Physics 48 (1978) 1957-1965 (in Russian)

(7) Radjabov, T.D., Z.A.Iskanderova, V.G.Dyskin and E.O.Arutjunova: Journal of Technical Physics 51 (1981) 1219-1226 (in Russian)

(8) Schiller, S., U.Heisig and K.Goedicke: J.Vac.Sci.Technol. 12 (1975) 858-864

(9) Weissmantel, C.: Thin Solid Films 58 (1979) 101

(10) Whitmell, D.S.: Rad.Eff. 53 (1980) 209-218

INFLUENCE OF HIGH ENERGY PARTICLES ON THIN FILM FORMATION BY LPVD

W. Pompe, M. Bobeth[1] and Yu.V. Trushin[2]

The production of thin films by means of laser pulse vapour deposition (LPVD) is characterized by some peculiarities compared to other deposition methods. For an incident laser power density of 10^8 W/cm^2 and a pulse duration 10^{-8} s the following deposition parameters can be expected: a very short deposition time per laser pulse $t = 10^{-6} - 10^{-4}$ s and a high deposition rate $R \lesssim 10^{26}$ m^{-2}s^{-1}. The time between pulses is usually considerably larger than the deposition time (typical pulse frequencies 1 - 10^3 Hz). The particles impinging on the substrate have energies in the region 1 - 10^3 eV. An amount of about $\lesssim 10$ % of the particles may possess energies $\gtrsim 100$ eV.

In the present work the influence of the special deposition procedure of LPVD on the initial kinetics of film formation is investigated. At first kinetic effects are considered on the basis of the following rate equations

$$\dot{n}_1 = R - n_1/\tau - 2\left[s_{11}Dn_1 + s_{R1}R\right]n_1 - \left[s_{1z}Dn_1 + s_{Rz}R\right]n_z$$
$$\dot{n}_z = \left[s_{11}Dn_1 + s_{R1}R\right]n_1 \qquad (1)$$

describing the time dependence of the mean surface density of single atoms (adatoms) $n_1(t)$ and the cluster density $n_z(t)$ including all clusters with the number of atoms $z \geq 2$. According to the high deposition rate, dissociation and migration of clusters are neglected and their coalescence can be omitted in the initial stage of deposition. Different from usual models we have considered the direct impingement $2s_{11}R(t)n_1$ in the nucleation rate, which becomes important at low temperatures and high deposition rates. The deposition rate is modelled by the step function $R(t) = R_o$ for $0 < t < t_p$ and zero otherwise. The adatom surface diffusion constant D and the desorption time τ are assumed to be given by $D = D_o \exp(-E_{sd}/kT)$ and $\tau = \nu_o^{-1} \exp(E_{des}/kT)$ with the activation energies E_{sd} and E_{des} ($D_o = 10^{-7}$ m^2/s, $\nu_o = 10^{12}$Hz).

---

1) Zentralinstitut für Festkörperphysik und Werkstofforschung der AdW der DDR, Dresden, 8027, GDR

2) Physical Technical Institute of the Academy of Sciences of the USSR, Leningrad, 194021, USSR

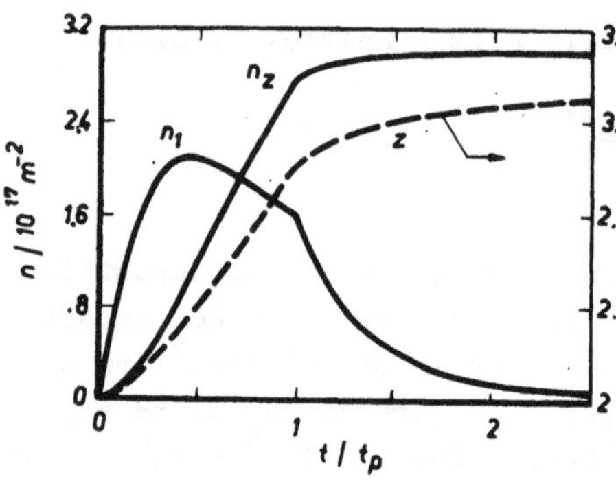

Fig. 1

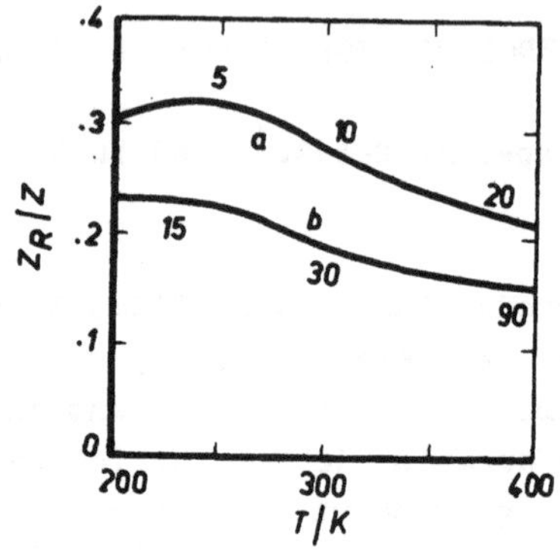

Fig. 2

Since we mainly deal with a highly non-stationary process the depletion of adatoms in the vicinity of clusters is neglected and the capture numbers are estimated by geometrical considerations: $s_{Rz}=a_o^2\pi(1+\sqrt{z/\pi})$, $s_{1z} = 2\pi(1+\sqrt{z/\pi})$ and s_{R1}, s_{11} analogously with z = 1. z is the mean number of atoms in a cluster and a_o means the characteristic lattice constant of the adatoms.

The solution of (1) is plotted in Fig. 1 (E_{des} = .6 eV, E_{sd} = .3 eV, T = 240 K, $R_o = 10^{23}$ $m^{-2}s^{-1}$, $t_p = 10^{-5}$ s, $a_o = 3 \cdot 10^{-10}$ m). At low temperatures a great number of adatoms still exists after the deposition pulse, and therefore considerable nucleation and growth takes place also between single pulses, whereas at higher temperatures the adatom density becomes zero immediately. The mean number of atoms per cluster z increases roughly linear with time. The ratio z_R/z of the mean number of particles z_R joined to a cluster by direct impingement and the total number z is plotted in Fig. 2 at a coverage $\pi a_o^2 z n_z$ = .2 (parameters as in Fig. 1 except E_{des} = .9 eV, $R_o = 3 \cdot 10^{21}$ (b); 10^{23} $m^{-2}s^{-1}$ (a)). The numbers at the curves denote the cluster size z. The adatom density between pulses goes to zero. The ratio z_R/z describes, at least for z > 10, in a rough manner the kinetically controlled cluster shape as a function of the deposition conditions. At higher temperatures z_R/z decreases with temperature. Due to a faster diffusion the mean cluster density decreases connected with a relative diminishing of direct impingement. Surprisingly a maximum of z_R/z is found for higher deposition rates, which may be caused by a dominating cluster growth between deposition pulses at low temperatures. Estimating the height of the clusters by $h/a_o \approx 1/(1 - z_R/z) \approx 1.5$ we find layer growth to occur. Depending on the mean cluster size this corresponds to atomic rough layers for small z (high R, low T) and to a roughness on the scale of the cluster sizes for z > 10 (small R, high T), respectively.

To our knowledge it is not clear until now to what extent the presence of high energy particles in the laser plasma influences the formation of epitaxial layers as a post-nucleation phenomenon. These particles may affect film formation in several respects: sputtering, defect production, local activation of desorption and migration. Because of the great number of particles with energies of 20 - 200 eV in the LPVD plasma we deal here with the influence of highly mobile interstitials which migrate to the surface leading to an additional mass flow from the bulk $R_b(t)$ modelled as $R(t)$ by a step function with the duration $t_i = d^2/D_i$ (d ... implantation depth of high energy particles, D_i ... diffusion constant of interstitials) supposing the deposition time of high energy particles to be smaller than t_i. Recombination of interstitials and vacancies can be neglected for not too high defect concentration N_d, if $4\pi a_o d^2 N_d < 1$. Let us suppose the formation of clusters of epitaxial as well as of random structure with certain probabilities. They are described by the densities $n_{ze}(t)$ and $n_{zr}(t)$, respectively. To rearrange random clusters into epitaxial ones an amount of energy is necessary which, besides thermal activation, may be provided by directly impinging particles or by interstitials diffusing from the bulk to the surface. These processes lead to the following equations

$$\begin{aligned}
\dot{n}_1 &= R + R_b - n_1/\tau - 2[s_{11}Dn_1 + s_{R1}(R + R_b)]n_1 \\
&\quad -[s_{1ze}\,Dn_1 + s_{Rze}(R + R_b)]n_{ze} - [s_{1zr}\,Dn_1 + s_{Rzr}(R + R_b)]n_{zr} \\
\dot{n}_{ze} &= [\alpha s_{11}Dn_1 + s_{R1}(\beta R + \gamma R_b)]n_1 + s_{Rzr}(\mu R + \mu_b R_b)n_{zr} \qquad (2) \\
\dot{n}_{zr} &= [(1-\alpha)s_{11}Dn_1 + s_{R1}((1-\beta)R + (1-\gamma)R_b)]n_1 - s_{Rzr}(\mu R + \mu_b R_b)n_{zr}
\end{aligned}$$

The relative coverage of epitaxial clusters v_e after one deposition pulse is shown in Fig. 3 as a function of the interstitial flow (parameters as in Fig. 1 and $\alpha = \beta = \gamma = \mu = .1$, $\mu_b = 1, .5, .2$ from top to bottom, $t_i = 10^{-5}$ s). It can be seen that defect production can improve epitaxial growth if the supposed transformation of random into epitaxial clusters by interstitials is sufficiently effective ($\mu_b \gtrsim .1$). A substantiation of model (2) requires of course further detailed microscopic investigations as an estimation of the transition probabilities $\alpha, \beta \ldots \mu_b$.

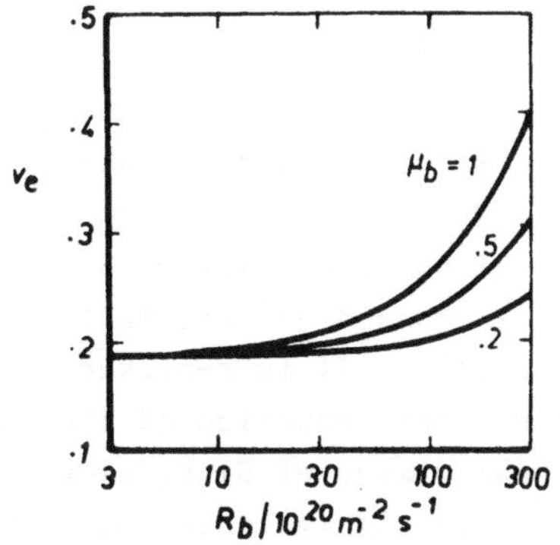

Fig. 3

MODIFICATION OF SURFACE PROPERTIES OF TOOL STEEL BY ION BEAM MIXING

E. Hensel, R. Grötzschel, U. Scholz and E. Richter[1)]

Introduction

During the last few years the application of ion beam techniques for the improvement of surface properties of metals such as hardness, fatigue, wear and corrosion has become more and more important. Besides direct implantation, ion beam mixing (ion-irradiation-induced material transport in solids) of surface layers on metals is a promising method. A good description of the mechanisms for ion beam mixing has been given by Wang /2/.

It is the purpose of our investigations to produce Si_3N_4 hard coatings fast bonded with tool steel substrates by ion beam mixing. From this combination we expect an improvement of hardness, wear and corrosion resistance.

Experimental

- Tool steel: 90MnV8 (0.85 % C, 0.28 % Si, 2.0 % Mn, 0.017 % P, 0.011 % S, 0.22 % Cr, 0.08 % Ni) and 210Cr46 (2.102 % C, 0.32 % Si, 0.30 % Mn, 0.022 % P, 0.018 % S, 11.31 Cr, 0.16 % Ni), mechanically polished
- Deposition of 46, 51 and 115 nm Si_3N_4 by reactive sputtering onto steel
- Ion implantation: 10^{15}, 10^{16} and $5 \cdot 10^{16}$ Kr^+ cm^{-2}, 130 keV and 10^{16}, $5 \cdot 10^{16}$ and 10^{17} Kr^+ cm^{-2}, 330 keV. The implantation energy was chosen in such a way to locate most of the implanted ions in the Si_3N_4/steel interface region
- Chemical etching in 20 % HCl, 50 °C, 15' and 60'
- Measurements: Wear (pin on disk without lubrication), Tally step, Rutherford Backscattering Spectroscopy (RBS) 1.2 MeV $^4He^+$

Results and Discussion

By Kr^+ implantation into Si_3N_4/steel a good adhesion of the surface coating on steel could be obtained. In the unimplanted case exfoliation of the Si_3N_4 layer around Vickers diamond identations took place. But after irradiation with krypton no exfoliation was found. Wear measurements (fig. 1) showed an important decrease of the wear volume (factor 5) after implantation. It is remarkable, that no significant difference of wear exists after reduction of the coating layer thickness from 115 to 46 nm. Ion beam mixing of Si_3N_4/steel also leads to an increase of corrosion resistance. At implanted Si_3N_4/90MnV8 the corrosion rate decreases by

---

[1)] Zentralinstitut für Kernforschung, Rossendorf, Bereich 2, PF 19, Dresden, 8051, DDR

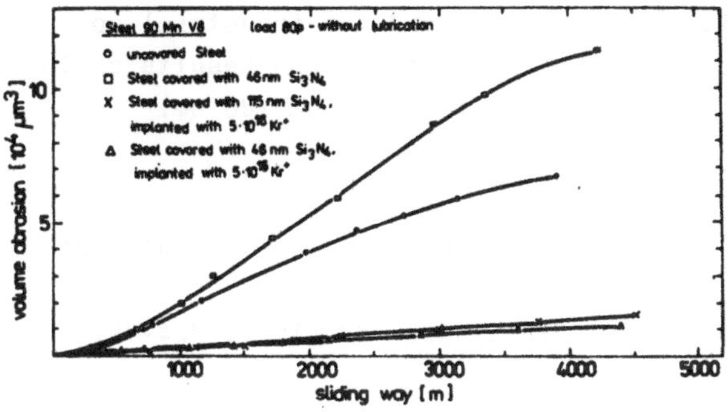

Fig. 1
Wear diagram

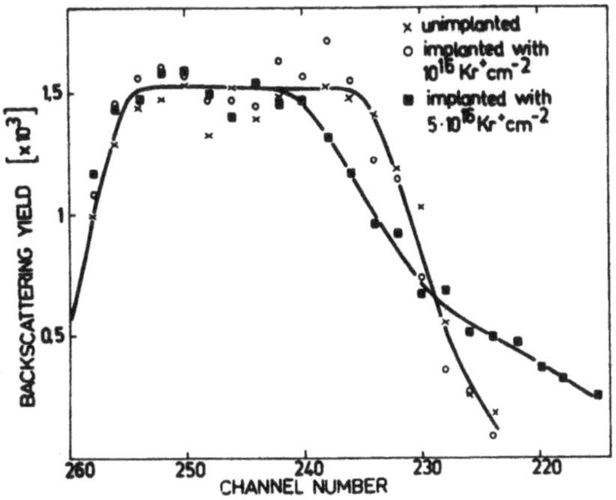

Fig. 2
RBS spectra of the Si profiles in 115 nm Si_3N_4/90MnV8, unimplanted and implanted with different Kr^+ doses, 330 keV

50 % after an etch time of 15'. After 60' etching no difference between uncovered steel and implanted Si_3N_4/steel was found. At irradiated Si_3N_4/210Cr46 the corrosion rate decreases by ≈60 % after 15' etch time, and still rests of the original surface exist. Also after 60' a decreased corrosion rate of ≈ 50 % was established. Clearly, there is an additional influence of the different steel alloy content on the etch behaviour of the investigated samples.

The effects after ion implantation mentioned above are caused by atomic mixing at the coating layer/substrate interface. The migration of Fe into Si_3N_4 of different thickness was shown by RBS measurements. Fig. 2 demonstrates a broadening of the Si distribution in the transition region to the steel substrate after ion irradiation with different doses. The improvement of surface properties by atomic mixing occurs only above a definite ion dose ($> 10^{16}$ cm^{-2}) and is limited by blistering of the implanted krypton (10^{17} cm^{-2}) in the Si_3N_4 layer /1/. The wear and corrosion measurements showed an increased resistance still in a depth of several microns, contrary to the atomic mixing being only in a range of few tens nanometers. An explanation could be a migration of nitrogen and silicon with the wear and etch front. This problem will be investigated by ESCA measurements in the next time.

Conclusions

Ion beam mixing of Si_3N_4/steel is a promising method for industrial applications needing both wear and corrosion resistance. Good results could be obtained already with a Si_3N_4 coating layer thickness < 100 nm.

References

/1/ Hensel, E., R. Grötzschel, E. Richter: 4. Meeting on Nucl. Anal. Meth., Dresden 1987, Proceedings Vol. 1, S. 92-99

/2/ Wang, Z.L.: Nucl. Instr. and Meth. in Phys. Res. B2 (1984) 784-791

COMPUTER SIMULATION AND EXPERIMENTAL MEASUREMENTS OF RECOIL IMPLANTATION OF GOLD INTO SILICON

K. Paprocki, I. Brylowska, W. Szyszko, D. Jaworska and M. Subotowicz*)

Introduction

Let us evaporate a thin film of a particular material on the surface of the substrate and let us bombard this structure by the ion beam of the e.g. noble gas ions. In the interaction of the ion beam with atoms of thin film these atoms can acquire an energy sufficient to pass the film-substrate interface. This process will be called recoil implantation. The following characteristics of the recoil implantation: the recoil yield and the depth distribution of recoiled atoms were intensivelly investigated theoretically [1-3] and experimentally [1,4-6]. The recoil yield was assumed as the ratio of introduced atoms number from a thin film to number of implanted ions.

The aim of this paper was to determine the recoil yield and depth distributions of Au atoms concentration in the case of the Au/Si system implanted by Ar ions. The experimental concentration profiles were obtained by using remaing radioactivity measurements (RM) and from analysis of the computer simulated Rutherford back-scattering spectra (RBS). The experimental recoil yield were calculated as the integral of concentration profile divided by dose of Ar ions. These experimental data were compared with the theoretical ones calculated by computer simulation of the recoil process in Au/Si system applying Monte-Carlo method.

Experimental

Thin 25-29 nm Au films evaporated in vacuum on the monocrystalline silicon were implanted with the 120 keV Ar ions doses of 10^{13}-10^{15} ions/cm^2. The energy of Ar ions was chosen in such a way that the thickness of the Au film was equal to the projected range R_p. This condition gives the maximum efficiency of recoil process. After recoil implantation the residual Au films were etched chemically. A part of the samples was activated in the neutron flux $7x10^{13}$cm^{-2} for 10 h. Depth distributions of Au atoms were measured by use of the RBS (JINR in Dubna, USRR) and RM techniques.

---

*) Institute of Physics, Maria Curie-Sklodowska University, Pl. M. Curie -Sklodowskiej 1, 20-031 Lublin, Poland

Results

In Fig. 1 are presented experimental and theoretical depth distributions of Au atoms introduced in Si by the recoil technique using of Ar ions. Computer simulated (CS) depth distributions lie below experimental profiles. CS profiles were calculated by the Monte-Carlo method for 50000 Ar ions.

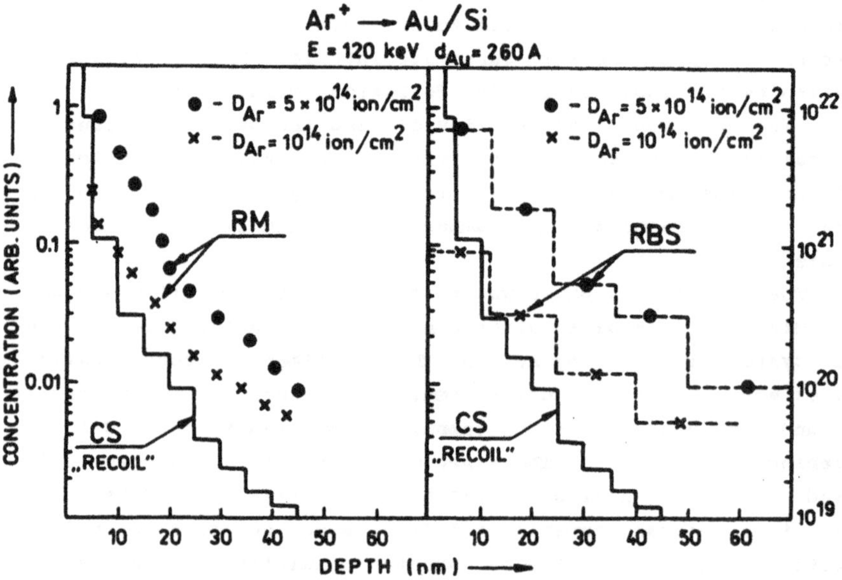

Fig. 1. Depth distributions of Au recoil atoms in Si introduced by Ar ions (●, × RM, -●- -×- RBS). Histograms (CS) represent the computer simulated distributions for dose 5×10^{14} ions/cm^2.

It was accepted two-body collision model. The motion of the Ar atoms was simulated in the Au-film only. In the calculations the displacement energy E_d and the certain threshold energy E_c were selected as $E_d = E_c = 20$ eV. Au atoms, which acqire an energy larger than E_d, have been included into the collision cascade and move until their energy decrease to E_c. It follows from the Fig. 1 that the accepted model of the binary collisions does not describe satisfactorily the experimental data. The influence of the annealing at 600° C on the Au recoil profile is shown in Fig. 2.

We did not notice the change of the shape of Au profiles after the heat treatment.

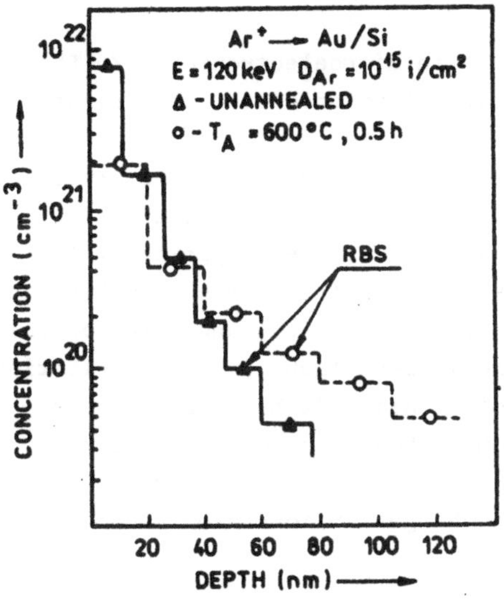

Fig. 2. Depth distribution of the Au recoil atoms in Si implanted by Ar ions.

Conclusions.

1. The experimentally (RBS) determined recoil yield was 4.2 and theoretically (CS), was 2.8. Both values are greater than unity. It means that recoil implantation is more efficient process than direct implantation.

2. The model of the simple binary collisions is not sufficient. There should be taken into the account the diffusion and implantation dynamics processes.

3. The anneling at 600° C does not influence on the shape of the depth distributions.

The paper was supported by the CPBP-01.08.C.1.3 and partialy by CPBP-01.06.1-16.01.

References

1. R. S. Nelson: Rad. Eff. 2, 47, 1969
2. A. Gras-Marti: Phys. Stat. Sol. (a) 76, 621, 1983
3. I. Abril, R. Garcia-Molina, A. Gras-Marti: Phys. Stat. Sol. (a) 95, 766, 1986
4. K. Paprocki, I. Brylowska: Rad. Eff. Let. 87, 63, 1985
5. B. Maillot, M. Bruel: Nucl. Instr. Meth 209/210, 707, 1983
6. M. Hautala: Nucl. Instr. Meth. B13, 357, 1986

ION BEAM MODIFICATION OF THIN POLYIMIDE FILMS

E.Venkova[3)], K.Kostov[2)], S.Georgieva[1)], I.Katardjiev[1)], D.Karpuzov[1)],
Ts.Marinova[2)], P.Kirova[4)]

1. Introduction

Polyimides are finding increased use in microelectronics /2/. This includes applications as fabrication aids (photoresists, planarization layers in multilevel photoresist schemes and ion implantation masks) as well as passivants/interlevel insulators among the others. Of crytical importance for the high-resolution ion-implant mask, especially in reactive-ion-etch mode is the stopping power of the polyimide at the desired ion energy, the dimensional stability of the etch structure and the degradation of polymer under high implant doses along with possible effects of temperature rise during the implantation. On all fabrication-aid applications the polyimide is removed after it has performed its task, the removal being normally accomplished in an oxygen plasma.

We examine here how the ion-oriented applications and properties interact, by studying the composition changes in ion-beam irradiated polyimide films.

2. Experimental

The polyamide was synthesized in the Central laboratory for polymers and the Institute of microelectronics. Two thin layers were spinned on silicon or metal-covered cital substrates and cured to obtain good-quality polyimide films (in terms of electrical and adhesion properties). The final film thickness was about 2 μm.

The samples were irradiated by Ar^+ (10,40 and 80 keV), Kr^+ (40 and 80 keV), Bi^+ or Cr^+ (40 keV) to fluences ranging from $2x10^{10}$ to $2x10^{16}$ ions/cm^2 at normal incidence. The implant was usually rastered over an area of a few sq.cm. and uniformity is estimated to be within 5 %. Most of the low-energy Ar^+-implantations were carrried out in the UHV chamber of the electron spectrometer ESCALAB Mk II (VG Scientific) with a residual gas pressure of $1x10^{-8}$ Pa. For the XPS measurements a MgK$_\alpha$ excitation source ($h\nu$=1253.6 eV) was used, the total instrumental resolution being ~1 eV as measured for the FWHM of the Ag $3d_{5/2}$ photoelectron peak. RBS analysis of the samples was carried out using a well-collimated beam of 2 MeV He^+, typical beam fluence was between 5 and 20 μC. Particles scattered through an angle of 168° were detected at an analyzer resolution of 5 keV/channel.

3. Results and discussion

Figures 1 and 2 illustrates the RBS spectra for Ar^+ and Bi^+ implanted polyimide films respectively. Fig.1 demonstrates that the Ar^+ bombardment

1) Institute of Electronics, Bulg.Acad.Sci.,Sofia 1784, Bulgaria
2) Institute of General and Inorganic Chemistry
3) Institute of Microelectronics, Sofia 1784
4) Central Laboratory for Polymers,Bulg.Acad.Sci., Sofia 1040

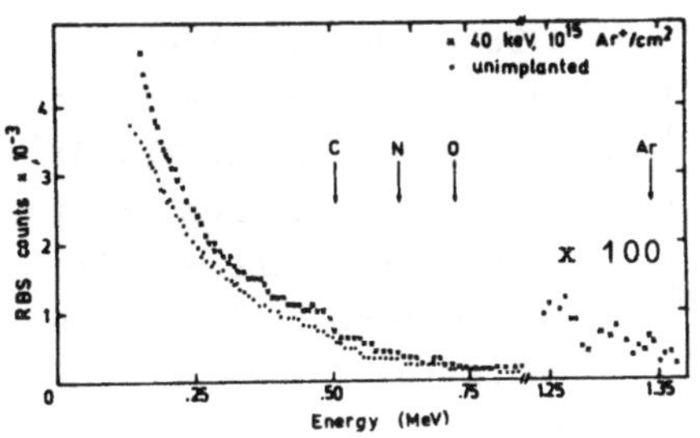

Fig.1. RBS spectra of virgin and Ar^+ implanted polyimide/cital

does not introduce additional peaks in the spectra. The Ar-peak itself is not clearly pronounced even at fluences of 10^{16}. Its behaviour with fluences shows that not all Ar is accumulated in the target, so it probably diffuses out partly. The same picture but with much better resolved implant peak is observed for higher energy Ar^+ and especially for Kr^+. Similar RBS spectra were reported for 120 keV implantation of both Ar^+ and Cu^+ in polyimide /4/, where migration and aggregation of Cu-atoms to form tiny islands in the surface region was also pointed out. In Fig.2 the spectra for Bi^+ implantation are shown which are quite close to that measured for Cr^+ in shape. The difference with Ar and Kr is clear: (a) much better resolved implant peak and (b) well defined peak in the region of O and N. To our knowledge the second peak is not reported in the literature although it was found for other polymers (see /3/ for pristine). Since N and O bonds are easily scissed by ion irradiation this peak appearence is probably associated with oxygen trapping after the implantation. We want to mention here that the RBS spectra were measured normally a few days after the implantation and a few samples analyzed about 9 months later did not show a difference. The estimated projected ion ranges for KAPTON polyimide (density 1.5 g/cm^3) are: 60 nm and 100 nm for 40 and 80 keV Ar^+, respectively, 41 and 64 nm for 40 and 80 keV Kr^+ and 30 nm for 40 keV Bi^+ /1/.

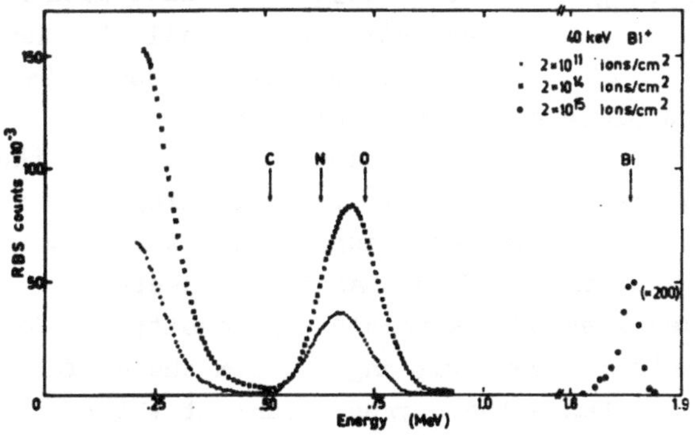

Fig.2. RBS spectra of Bi^+ implanted PI/cital

The C(1s) photoelectron peak of untreated polyimide layer deposited onto metal-covered cital substrate shows a complex structure with one strong peak at 285 eV (Fig.3a). This is a contribution from the carbon in benzene- and possibly CH_2-groups. The peak at 289.5 eV is assigned to carbon in C=O groups. The atomic ratio of the elements is [C]:[O]:[N]=11.7:2:1. After a bombardment with 40 keV Ar^+ at $2 \times 10^{15} cm^{-2}$ the peak related to C=O groups disappears (Fig.3b). The intensities of N(1s) and O(1s) peaks decrease with 60 and 30 % respectively. The higher value for oxygen is due to the exposure of the bombarded surface to air prior to inserting it in the UHV chamber of electron analyser.

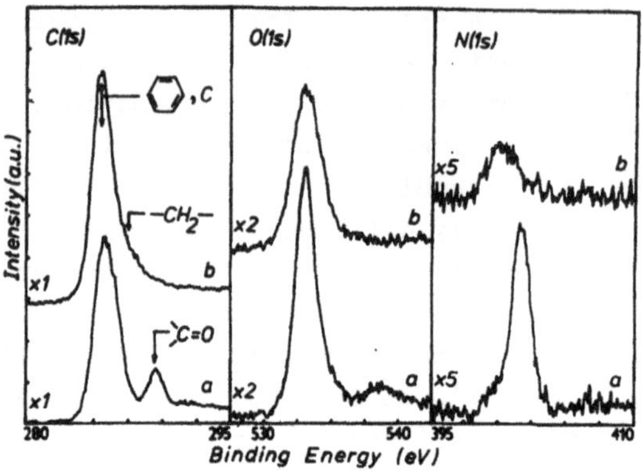

Fig.3. C(1s)-, O(1s)- and N(1s)-photoelectron spectra of polyimide on cital: a) untreated, b) after a bombardment with 40 keV argon ions at dose $2x10^{15}$ cm^{-2}

To prove this the polyimide layer on silicon was bombarded by 10 keV Ar^+ to dose of 10^{13} ions/cm^2 in the UHV chamber of the spectro spectrometer. Prior to the irradiation the ratio [C]:[O]:[N] ratio was found to be 11.8 : 1.9 : 1.which is close to the ration given above for polyimide on cital. After the bombardment in the chamber the XPS spectra show a ratio of [C]:[O]:[N]=11.8 : 0.2 : 0.3. These data with no exposure to air indicate that as a result of the bombardment C=O and C-N bond breaking takes place and N and O are desorbed. The results described above suggest that after the bombardment with Ar^+ the surface is relatively enriched with carbon due to bond breaking.

The XPS experiments with different take-off angles show that C=O groups have lower concentration at the untreated surface than in the deeper layers. This was observed as decrease of the relative intensity of 289.5 eV peak with decreasing the photoelectron exit angle (from the surface).

The Bi^+ implantation did not change the oxygen amount at the surface (~15 Å): it is desorbed but bound partly in a BiO_x state.The XPS measurements also indicate that after the bombardment with Bi^+ the concentration of N decreases similar to the irradiation by argon beam.

4. Conclusions

(a) The untreated surface is depleted of C=O groups within 5-10 Å. Deeper layers are characterized by stoichiometric ratio of [C],[O]and[N].

(b) The ion bombardment causes breaking of C=O and C-N bonds accompanied by desorption of O and N in the case of Ar^+ and N mainly in the case of Bi^+ (and probably Cr^+) irradiation.

(c) Exposure to air result in trapping of oxygen in the ion-irradiated polyimide.

References

(1) Biersack,J.P.: in Ion Modification of Insulators, ed. by P.Mazzoldi and G.Arnold, Elsevier, N.Y. 1987, p.1
(2) Senturia, S.D.: in Proceedings of ACS, Anaheim, Sept. 1986.
(3) Venkatesan,T., T.Wolf, D.Allara, B.J.Wilkens, G.N.Taylor: J.Appl. Phys. 56 (1984)
(4) Youshida,K. and M.Iwaki; Nucl.Instr.Meth. B19 (1987) 878-881

This work is sponsored by the Science Committee of PRB, Res.Co Research contract No.107.

STUDY OF ELECTRICAL AND OPTICAL PROPERTIES OF FUSED QUARTZ IMPLANTED BY HIGH DOSE Cr^{+}- AND Fe^{+}-IONS

V.Yu.Petukhov[1], I.Z.Ilaldinov[1], I.B.Khaibullin[1], K.Wollschläger[2]

It is well known that at the high dose implantation the formation of fine dispersed phase in the surface layer of solids is possible. If dielectric is chosen as a matrix and implantation is carried out by metal ions then under the definite implantation conditions the formed metallic inclusions start to coalesce which must essentially influence the electrical and optical properties of the surface layer. Such structures are interesting as a model system in order to check and to study some peculiarities of the percolation effect.

In this report the preliminary results of electrical and optical investigations of fused quartz implanted by Fe^{+}- or Cr^{+}-ions are presented.

The slices of fused quartz were implanted at room temperature with 40 keV ions and doses varied up to $4\cdot10^{17}cm^{-2}$.

Fig.1 shows the electrical resistance of a silicon oxide film vs. the dose of implanted Fe-ions. Similar curve was also obtained for Cr-implanted samples. For both systems investigated it was found that at

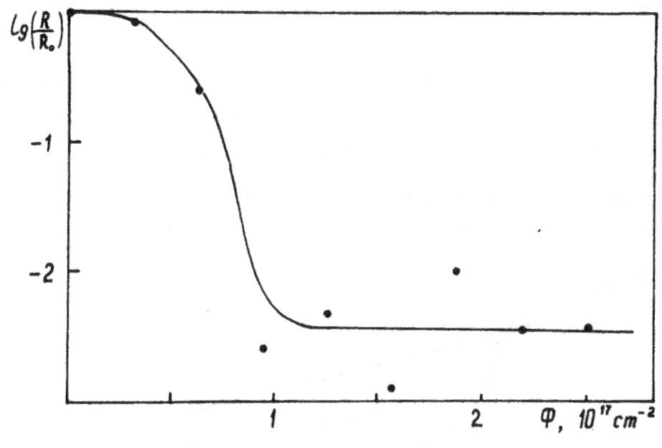

Fig.1. Electrical resistance of implanted fused quartz vs. implantation dose of Fe^{+} ions.

[1] Kazan Phys.-Techn. Institute Academy of Sciences of the USSR 420029, Kazan. USSR.

[2] Akademie der Wissenschaften der DDR ZfK Rossendorf Postfach 19 Dresden DDR-8051

some implantation doses, dependent on the ion flux density, the sharp decrease in the resistance took place. Such threshold behaviour could arise when percolation moment sets in. However, the sign of the measured temperature coefficient of resistance was negative, not showing the metallic conductivity, as it was usual at the formation of the dispersed metallic film in SiO_2 (1). Moreover, a more considerable decrease in resistivity, over more than 6-8 orders of magnitude (1), was to be expected in this case.

To reveal the reasons of the observed behaviour the RBS, RHEED, X-ray and SEM experiments were performed. RBS-spectra showed the maximum concentration of metal atoms to be $2\cdot10^{22}cm^{-3}$ at implantation dose $2.5\cdot10^{17}cm^{-2}$, the thickness of layer being about 40 nm. Although, as one can see, the sputtering is significant and concentration of metal atoms on the surface is large enough ($1.8\cdot10^{22}cm^{-3}$), nonmetallic conductivity predominates. SEM micrographs did not show any metal clusters on the sample surface and electron diffraction patterns consisted only in amorphous halo.

Thus, we can suggest that we are not able to see the percolation effect at the chosen implantation conditions because, due to sputtering metal concentration is less than 30 vol% while in (1) a very rapid change in resistance has been observed when metal concentration was about 35-40 vol % . Therefore, the measured resistance is determined by that of dielectric SiO_2 interlayer between small metal clusters. This corresponds to the conclusion of Neugebauer and Rairden (2), which states that the resistivity of cermet films is determined principally by extent of volume dilution of metal by SiO_2 but not by the nature of the metal.

Optical measurements were performed in the visible and near IR regions. The transmission spectra of Fe-implanted sample, shown in Fig.2 (curve 1,2), revealed the decrease of transmission, which usually takes place after the metal implantation into silica glasses.

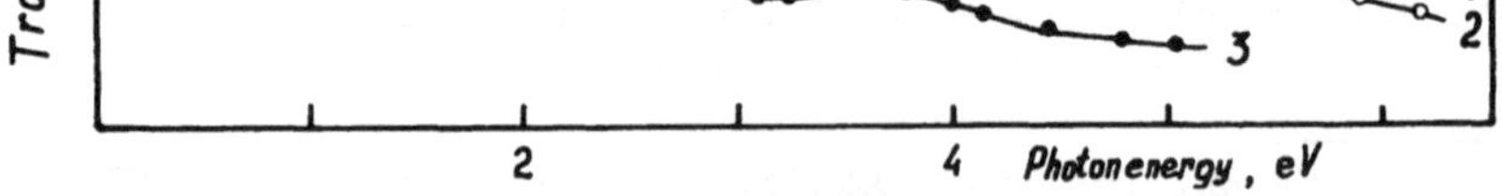

Fig.2. Transmission spectra of fused quartz implanted by Fe^+-ions with doses: 1 - 10^{17}, as implanted; 2 - $2.5\cdot10^{17}$, as implanted; 3 - $2.5\cdot10^{17}cm^{-2}$, annealed at $900^\circ C$ x 10 minutes.

Similar spectrum was observed for fused quartz, implanted by Cr-ions. The decrease of transmission for 'as implanted' samples is caused by conduction electron absorption in small metallic clusters, as well as by the generation and accumulation of radiation defects.

The following thermal annealing showed the essential changes in optical transmission spectra, the changes being dependent on the implanted metal. For both systems investigated ($SiO_2 \leftarrow Fe^+$ and $SiO_2 \leftarrow Cr^+$) the subsequent heat treatment led to the partial annealing of the defects mentioned above. The spectrum for Fe-implanted sample revealed the appearance of at least two new absorption lines ($\lambda \approx 280$ and 420 nm) (Fig.2, curve 3), resulting in the orange colour of sample, while for Cr-implanted - only one ($\lambda \approx 300$ nm) peak was observed. The appearance of these absorption lines took place at annealing temperatures higher than 600 °C for both cases and could be connected with the chemical reaction between the metal and substrate material. This suggestion was confirmed by X-ray measurements which showed the formation of metal oxides, for example α-Fe_2O_3 in $SiO_2 \leftarrow Fe^+$ system. It must be mentioned that oxide were formed in the near-surface layer and metal diffusion was not observed, as RBS spectra showed.

In summary, it should be pointed out that to study the percolation effect by high dose implantation other implantation conditions must be chosen providing the maximum metal concentration in the near-surface layer at least 40 vol % .

References

(1) Perkins, J.G.: J. Non-Cryst. Sol. 3 (1972) 349-364.

(2) Neugebauer, C.A., J.R.Rairden: Compt. Rend. Conf. Intern. Sur des Couches Minces, Cannes, 1970 (SFITV, Paris) p. 331.

ION BOMBARDMENT INDUCED MODIFICATION OF LaB_6

G. Oertel and W. Henrion [1)]

LaB_6 single crystals are used as cathodes with high brightness and long life. The origin of this investigation was the need to optimize the performance parameters of these cathodes. We found that the emission characteristics of these cathodes are influenced by contaminations. Auger electron spectroscopy (AES) results, received with an equipment PHI model 590 in the derivative spectra mode, showed that mainly oxygen and carbon on the emitting 100-surfaces of the crystals reduced the emission current density and stability of the cathodes. For surface cleaning we bombarded the crystals with Ar^+-ions of energies between 1 and 5 keV. After ion bombardment up to steady state conditions the colour of the crystals had changed from red to grey-blue. Figure 1 shows the colour appearance, that is the spectral distribution of the product of the relative spectral sensitivity of human eye and the re-

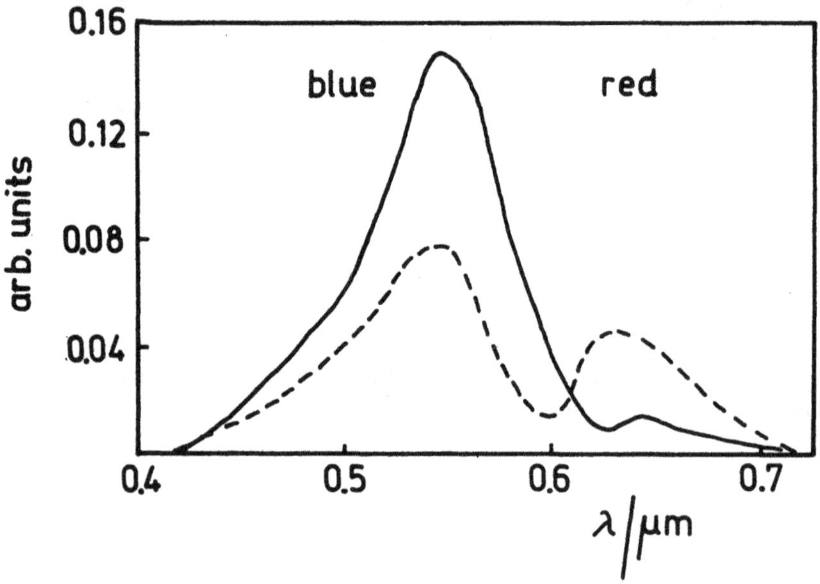

Fig. 1: Colour appearance of LaB_6 before (---) and after (——) ion bombardment

1) Central Institute of Electron Physics of the Academy of Sciences of the GDR, Rudower Chaussee 5, 1199 Berlin, GDR

flectivity of the sample before and after ion bombardment.
Looking for an explanation of this colour change we made a quantitative analysis by AES. The molar fractional contents C_i are given by /1/

$$C_i = \frac{I_i S_i^{-1}}{\sum_i I_i S_i^{-1}} \qquad (I)$$

I_i are the Auger intensities of La or B in LaB_6. For calibration the sensitivity factors S_i were established from an in situ fracture surface referencing to I_{Ag} (351) for 3 keV primary electrons. The bulk stoichiometry of the crystals was ascertained by chemical analysis to be B/La = 5.9 (5.8 for polycrystalline material). The measured intensities of La 78, La 625 and B 179 are shown in Table 1. The values for in situ fractured crystals are extrapolations to zero coverage of the adsorption curves from the residual gas atmosphere. The total pressure was about 10^{-7} Pa. The intensities of the Ar^+-ion bombarded surfaces were measured after a dynamical equilibrum of the sputter etching and modification processes was reached.

Table 1: Auger intensities

	fractured crystal	Ar^+-bombarded surfaces polycryst.	100	110	111
La 78	46.3	69	71	71	71
La 625	18	24	26	24	22
B 179	72.5	72	71	71	71

Insertion of these intensities and the resulting sensitivity factors $S_{La\ 78} = 0.60$, $S_{La\ 625} = 0.23$ and $S_{B\ 179} = 0.16$ in equation (I) gives for the low energy La peak a composition of about LaB_4. Goebel et al. reported /2/ that for 100 eV proton bombardment also the modification to LaB_4 occurred. For the high energy La Auger electrons we get larger B/La ratios (Table 2).

Table 2: B/La ratios for Ar^+-bombarded LaB_6 surfaces

	polycryst.	100	110	111
La 78, B 179	3.9	3.8	3.8	3.8
La 625, B 179	4.3	3.9	4.3	4.6

In a simple model, we can assume that the bombarded LaB_6 crystals are covered by a thin LaB_4 layer of thickness d. The low energy Auger electrons inform mainly about the composition of this thin layer, whereas the high energy Auger electrons have a larger information depth. The contributions of the LaB_4 layer (I^I_{625}) and the underlying LaB_6 crystal

(I^{II}_{625}) to the signal are /1/

$$I_{La\ 625} = I^{I}_{La\ 625} + I^{II}_{La\ 625} \qquad (II)$$

$$I^{I}_{La\ 625} = I^{LaB_4}_{\infty, La\ 625} \left[1 - \exp\left(-d/\lambda \cdot \overline{\cos\varphi}\right)\right] \qquad (III)$$

$$I^{II}_{La\ 625} = I^{LaB_6}_{0, La\ 625} \exp\left(-d/\lambda \cdot \overline{\cos\varphi}\right) \qquad (IV)$$

with $\overline{\cos\varphi} = 0.74$ /3/. The attenuation length λ for the La 625 Auger electrons is 2.1 nm, calculated by an empirically deduced equation for inorganic compounds /4/. Insertion in (II) gives for d the values in Table 3.

Table 3: Thicknesses d of the LaB_4 layer

surface	polycryst.	100	110	111
d(nm)	1.9	3.3	1.9	1.0

References

(1) Klaua, M. und G. Oertel, in: Festkörperanalyse mit Elektronen, Ionen und Röntgenstrahlen, VEB Deutscher Verlag der Wissenschaften, Berlin 1980, 295

(2) Goebel, D. M., Y. Hirooka and T. A. Sketchley, Rev. Sci. Instrum. 56 (1985) 1717

(3) Seah, M. P., Surf. Sci. 32 (1972) 703

(4) Seah, M. P., and W. A. Dench, Surf. Interface Anal. 2 (1980) 53

PYROLYTIC LASER DEPOSITION OF TUNGSTEN

T. Szörényi[1], K. Piglmayer and D. Bäuerle[2]

1. INTRODUCTION

Laser-induced chemical vapor deposition (LCVD) allows maskless single-step patterning of materials. The technique is applicable both to planar substrates, and three--dimensional fabrication. Although extensive work has already been devoted to LCVD demonstrating that practically all metals, semiconductors and insulators of interest for semiconductor device fabrication can be deposited by lasers [1], few systematic works have been published on the effect of deposition conditions on deposited film quality [2]. The present report summarises our first systematic study on pyrolytic laser deposition of W by hydrogen reduction of WF_6.

2. DEPOSITION OF W

2.1. Optimization of gas partial pressures

In laser-induced pyrolytic deposition of W by hydrogen reduction of WF_6 as described by the reaction:

$$WF_6 + 3\,H_2 \rightarrow W\downarrow + 6\,HF\uparrow\,, \qquad (1)$$

the proper choice of the partial pressures of H_2 and WF_6 has been found to be of fundamental importance. With increasing laser power P the amount of excess H_2 should increase: for P well below 100 mW and low partial pressures of WF_6 $p(H_2)/p(WF_6)>5$ is sufficient, whereas ratios of $p(H_2)/p(WF_6)\geq 50$ are necessary to ensure stable and reproducible deposition for laser powers in excess of $\approx$100 mW and higher partial pressures. Consequently, the pressure domains appropriate for practical applications are limited to approx. 1-20 mbar WF_6 and 50-1000 mbar H_2; minute WF_6 partial pressures result in very weak deposition, and for $p(WF_6)>20$ mbar hydrogen pressures above 1 bar would be necessary [3].

2.2. Deposition of spots

Scanning electron micrographs of a series of spots deposited from an admixture of 5 mbar WF_6 and 500 mbar H_2 on fused silica covered with a 700 Å thick layer of sputtered W are shown in Fig.1. The 647.1 nm TEM_{00} beam of a Kr^+ laser was focused onto the substrate surface (perpendicular illumination) by a combination of a simple lens and a mic-

1) Research Group on Laserphysics, Hungarian Academy of Sciences, H-6720 Szeged, Dóm tér 9., Hungary
2) Angewandte Physik, Johannes-Kepler-Universität Linz, A-4040 Linz, Austria

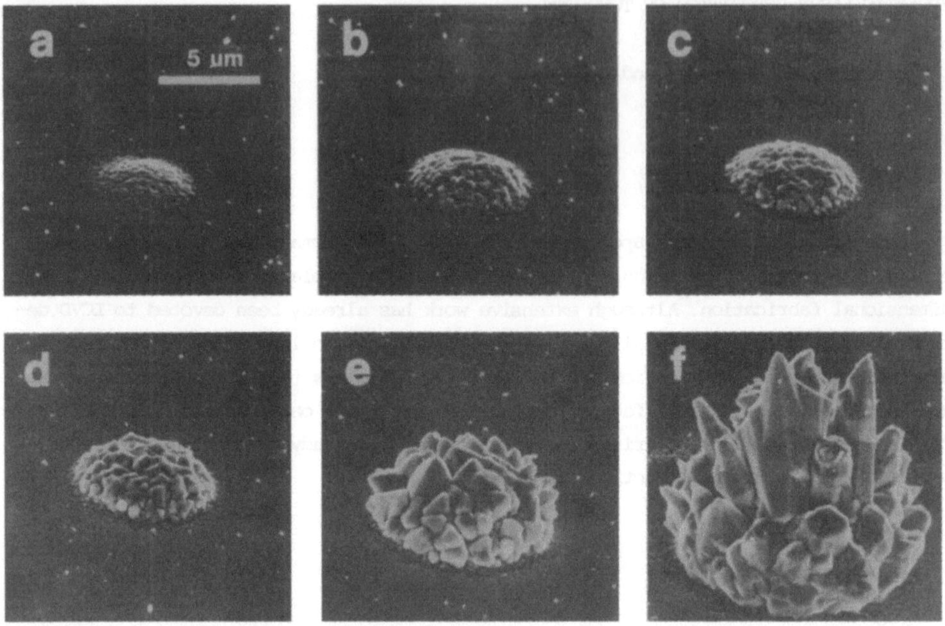

Fig.1. Time evolution of W spots. The laser illumination time increases from a) to f) with: a) 0.02 s, b) 0.05 s, c) 0.07 s, d) 0.1 s, e) 0.2 s, f) 0.5 s. P=130 mW.

roscope objective (f=17 mm). With increasing laser illumination time both the diameter of the spots and their height/diameter ratio increases. The microstructure is polycrystalline, with fine grains as deposition commences (Fig. 1a-c), then becomes more and more coarse textured (Fig. 1d,e). For longer illumination times the lateral growth practically stops and axial growth prevails (Fig. 1f). For laser powers from 30 to 200 mW the spot diameter first increases exponentially with time and then saturates at a given value which depends on the laser power. The exponential dependence suggests a thermally activated behaviour. For fused silica substrates, covered with a-Si, the time evolution of the process shows the same character with somewhat shorter saturation times at somewhat larger diameters.

2.3. Direct writing

Scanning electron micrographs of typical W stripes deposited from 5 mbar WF_6 diluted in 400 mbar H_2 on fused silica covered with 1200 Å a-Si are shown in Fig. 2.

Fig. 2. Scanning electron micrographs of tungsten stripes deposited with 647.1 nm Kr^+ laser light of P = 146, 132, 120, 107 and 97 mW (left to right). Scanning speed: 100 µm/s.

The characteristic cross-section of the stripes is a well defined mesa type with a width to height ratio of ≈10. With increasing WF_6 partial pressure the cross-section changes to semicircular with a ratio of ≈5 for $p(WF_6)$=10 mbar. Both the width and the thickness increase approximately linearly with increasing laser power [3]. The width is, within the accuracy of the measurements, independent of both the WF_6 and the H_2 partial pressures. The thickness increases with increasing WF_6 pressure but is independent of the amount of H_2 (provided that appropriate $p(H_2)/p(WF_6)$ ratio is used: cf. chap. 2.1). For the two types of substrates the results were qualitatively identical.

Electrical conductivity data measured on W stripes deposited from 5 mbar WF_6 diluted in H_2 at pressures ranging from 100 to 800 mbar are compiled in Fig. 3. The strongly decreasing resistance with increasing power is due to the increasing cross-section of the stripes. The slightly increasing resistivity may be accounted for by stripe texture which becomes more pronounced with increasing laser power. We note that these resistivity values are the best reported for LCVD of W from WF_6+H_2 and they are only a factor of 1.3 to 2.3 larger than those of the bulk material.

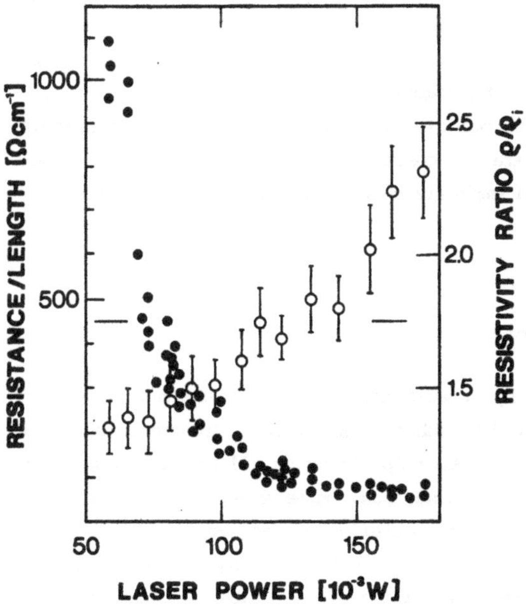

Fig.3. Electrical resistance per unit lenght (left scale) and resistivity ratio (right scale) of W stripes as a function of processing laser power. A value of ρ_i=5.33 μΩcm has been used for bulk W.

3. CONCLUSIONS

Excellent stability and reproducibility in pyrolytic laser deposition of W is achieved by optimum control of the process paramters. Illumination times between ≈0.01 s and 0.1 s result in patterns of excellent morphology using $p(WF_6)$=5 mbar and $p(H_2)\geq200$ mbar. Stripes of 1.5 to 15 μm widths and 0.1 to 3 μm thicknesses can be deposited applying laser powers between 30 and 200 mW, the resistivities of which are very close to those of the bulk material.

REFERENCES

[1] For most recent reviews see: a) Bäuerle, D.: Chemical Processing with Lasers, Springer Series in Materials Science 1, Springer, Berlin, Heidelberg 1986
b) Bäuerle, D., K.L. Kompa and L.D. Laude (eds.): Laser Processing and Diagnostics II, Les Editions de Physique, Les Ulis 1986

[2] Houle, F.A.: Appl. Phys. A41 (1986) 315-330

[3] Zhang, G.Q., T. Szörényi and D. Bäuerle: J. Appl. Phys. 62 (1987) 673-675

PULSED AND cw LASER SYNTHESIS OF III-VI AND IV-VI COMPOUND FILMS

K. Bali, T. Szörényi and I. Hevesi[1)]

1. INTRODUCTION

Laser chemical processing (LCP) is a challenging alternative for producing thin films both locally with lateral dimensions down to the submicrometer range and in large area form [1]. Laser-induced synthesis - i.e.: mixing of sandwich films consisting of alternating layers of the constituents by laser irradiation - is a unique technique to prepare compound semiconducting films [2]. The reactions involved, and, consequently, the properties of the synthesized films are principally determined by the laser irradiation conditions. In the following we compare the characteristics of cw and pulsed (submicrosecond) laser processing with special emphasis on the complexity of the process under high power density irradiation.

2. SAMPLE PREPARATION AND PROCESSING

Compound films have been synthesized by irradiating In/Se and Ge/Se multilayer structures vacuum evaporated onto glass or NaCl substrates (overall thickness 100-150 nm) with the scanned focused TEM_{00} beam of an Ar^+ laser operating on all green lines, or one single pulse from a Q-switched ruby laser: λ=694 nm, max. pulse energy 2 J, pulse length 40 ns at FWHM. The beam of this latter laser was spatially homogenized [3]. The fluence Φ on the sample surface was changed with calibrated filters and using homogenizing rods of different diameters. Due to errors in the calibration of the pulse energies and inaccuracies in the determination of the spot dimensions the resulting uncertainty in quoted fluence values is maximum 15 %.

3. cw LASER SYNTHESIS

By focusing high power ion lasers (I up to 10^7 W/cm^2) illumination times as short as nanoseconds should be sufficient to ensure energy densities above threshold (10^{-2}-1 J/cm^2; constant for such short pulses [2]), nevertheless in practice only irradiation (dwell) times longer than $2\cdot10^{-5}$s [4] have been realized by chopping or scanning the laser beam [2]. This low power processing favours, in general, crystallization: in fact, a great variety of polycrystalline films of surprisingly well defined stoichiometry have been prepared using cw lasers [2]. Amorphous film formation is also possible when the illumination time is sufficiently short as observed in the Ge-Se system for

1) Research Group on Laserphysics, Hungarian Academy of Sciences, H-6720 Szeged, Dóm tér 9., Hungary

extremely high scanning speeds [4]. The energy threshold for transformation is well defined; in some cases, e.g. in the Ge-Se system, several thresholds were reported resulting in different phases [4]. As followed by optical spectra of InSe and In_2Se_3 thin films the properties of the synthesized compounds very sensitively depend on process parameters just above the threshold [5]. In each case cw laser synthesis is a pure and relatively simple process during which the whole material completely transforms (practically without material loss) preserving the initial composition.

4. PULSED LASER SYNTHESIS

In order to assess the more complex processes involved in high power pulsed laser processing of such structures let us consider the case of Ge-Se system. Very recent results of Canivez et al. on excimer laser processing of Ge/Se multilayer films indicated selective etching of the Ge layer (and no mixing) or compound formation on the surface depending on the sequence of the constituents [6]. In order to decide whether thermal-or photo--effects are dominant in this strange behaviour we processed Ge-Se sandwiches of nominal composition of $GeSe_2$ with a ruby laser.

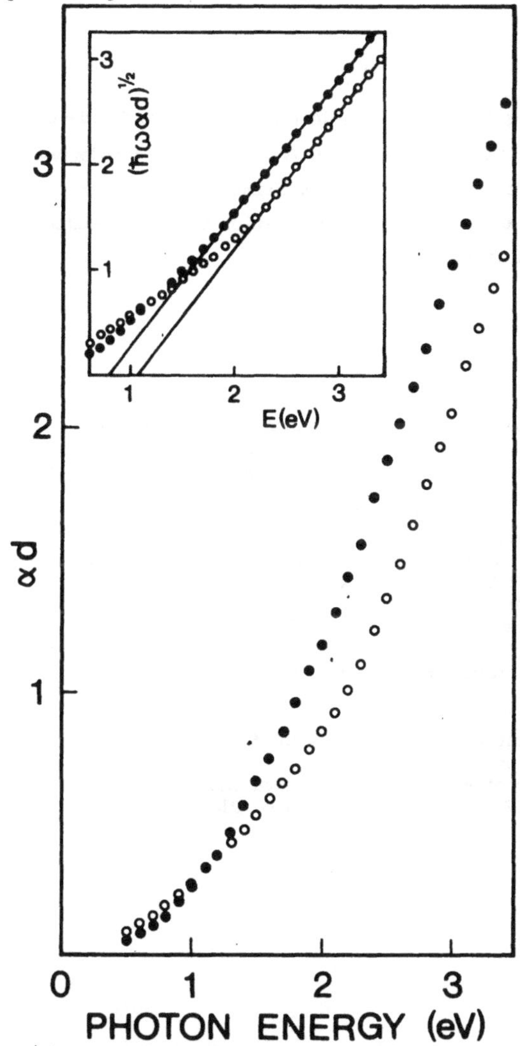

Fig.1. Optical absorption spectra of ruby laser processed Ge-Se films. Φ=300 mJ/cm^2:O and 610 mJ/cm^2: •

For fluences up to approx. 60 mJ/cm^2 no visible changes were observed on Ge/Se/glass substrate structures (thicknesses 300 Å and 740 Å, respectively). At around 80 mJ/cm^2 a complete ablation of the Ge layer took place, while the underlying Se layer remained practically undamaged. Further increase in the fluence resulted in gradual thinning of the selenium film. The fact that ruby laser pulses of energy density very near to that of excimer processing [6] but in terms of photon energy well below the photoelectric threshold of both constituents can also produce ablation suggests that the electronic model put forward in ref. [6] is not applicable. Our finding that pulses of fluences of the same order of magnitude produce ablation of thin single layers of Ge, Ti, Cr, V and Al from glass substrates clearly indicates that the role of underlying Se layer is not decisive.

For samples of the opposite sequence of the constituents the threshold lies somewhat below 300 mJ/cm^2. Above this energy density two simultaneous processes compete depending on the laser fluence: (i) compound formation and (ii) partial removal of selenium or that of the formed compound. This latter phenomenon can be visualized experimentally by another adjacent blank glass plate in front of the multilayer structure (typically 0.1-0.3 mm). This is a realization of pulsed laser deposition [7,8 and references therein]. The compound films formed are amorphous as determined by electron microscopy. The analysis of αd data calculated from measured transmittance and reflectance (Fig.1) in terms of indirect transitions (inset in Fig.1) gives energy gap values close to 1 eV indicating that the composition of the synthesized material is near to GeSe [9], in agreement with the Se-loss described above.

5. CONCLUSIONS

Synthesis of compound semiconducting films is possible both with cw and pulsed lasers. While cw processing preserves the initial composition and so offers a relatively simple and pure method for the production of excellent quality films, pulsed laser processing is much more complex because the partial or complete ablation of the constituents or that of the compund, competes with synthesis.

REFERENCES

[1] For most recent reviews see: a) Bäuerle, D.: Chemical Processing with Lasers, Springer Series in Materials Science 1, Springer, Berlin, Heidelberg 1986
b) Bäuerle, D., K.L. Kompa and L.D. Laude (eds.): Laser Processing and Diagnostics II, Les Editions de Physique, Les Ulis 1986

[2] Laude, L.D., M. Wautelet and R. Andrew: Appl. Phys. A40 (1986) 133-143

[3] Cullis, A.G., M.C. Webber and P. Bailey: J. Phys. E 12 (1979) 688-690

[4] Antoniadis, C., L.D. Laude and P. Pierrard: In Energy Beam-Solid Interactions and Transient Thermal Processing, eds.: V.T. Nguyen, A.G. Cullis Les Editions de Physique, Les Ulis 1985 pp. 387-390;
Laude, L.D.: Progr. Cryst. Growth Charact. 10 (1985) 141-158

[5] Szörényi, T., K. Bali, G. Radnóczi and L.D. Laude: In [1]b pp. 157-161

[6] Canivez, Y., A. Jadin, R. Andrew, L.D. Laude and M. Wautelet: In [1]b pp. 153-155

[7] Metev, S.: In [1]b pp. 143-152

[8] Gaponov, S.: In Laser-Assisted Modification and Synthesis of Materials, ed.: S. Metev Sofia University, Sofia 1985 pp. 216-229

[9] Abrikosov, N.K., V.F. Bankina, L.V. Poretskaya, L.E. Shelimova and E.V. Skudnova: Semiconducting II-VI, IV-VI and V-VI Compounds, Plenum, New York 1969

MODIFICATION OF ELECTROCHROMIC WO_3 FILMS BY NANOSECOND LASER PULSES

B.A. Budkevich, I.A. Ges, S.P. Zhvavyj, G.D. Ivlev, V.A. Pilipovich and I.M. Romanov[1)]

The reversible electrochromic effect in thin WO_3 films manifests itself in a change of absorption of the visible light resulting from application of external electric field. This effect offers ample opportunities for using this material in light modulators and information displays. However, the use of WO_3 films in optoelectronics is hindered by two basic limitations, i.e. low colouration/bleaching speed and short life-time of such films.

This paper reports the results of investigation of electrochromic properties of WO_3 films irradiated with laser pulses of nanosecond duration. Amorphous films 0.5 to 1 μm thick were prepared by vacuum evaporation onto quartz substrates. Three laser irradiation modes were chosen: A (λ = 694 nm, τ = 80 ns), B (347 nm, 70 ns) and C (337 nm, 10 ns).

To study the electrochromic properties the films were placed into a standard electrochromic cell filled with an acid electrolyte.

With cathodic polarization of the WO_3 films placed into the cell electrochromic effect arises [1] which consists in the occurrence of a broad absorption band in the visible and near-infrared regions. Laser treatment results in a 3 to 6-fold increase in the colouration/bleaching speed (V) (see Fig. 1) and a 1.2 to 1.5-fold increase in the electrochromic efficiency.

As seen from Fig. 2, nonirradiated films obtained under the conditions mentioned above loose their electrochromic properties after about a month of storage in air, whereas colouration/bleaching speeds of laser-treated films remain practically unchanged.

Investigation of optical spectra of WO_3 films has shown that laser treatment results in a substantial scattering of light in the samples. SEM analysis of the film surface and measurements of the reflectivity dynamics during laser pulse action indicate that diffuse scattering of light is caused by partial evaporation of the film and the emergence of a developed microrelief on the film surface that remains after evaporation.

The development of the microrelief leads to a change in the conditions of charge flow across the WO_3-electrolyte interface and to increased current density and, hence, colouration/bleaching speeds in the films.

Structural analysis of the WO_3 films by TEM, electron diffraction (in transmission and reflection modes) as well as data of time-resolved reflectivity measurements have revealed that under laser pulse irradiation thin (about 0.1 μm) near-surface layer of the film that remains on the

1) Institute of Electronics, BSSR Academy of Sciences, Ya. Kolas St. 68/2, Minsk-90, 220841, U.S.S.R.

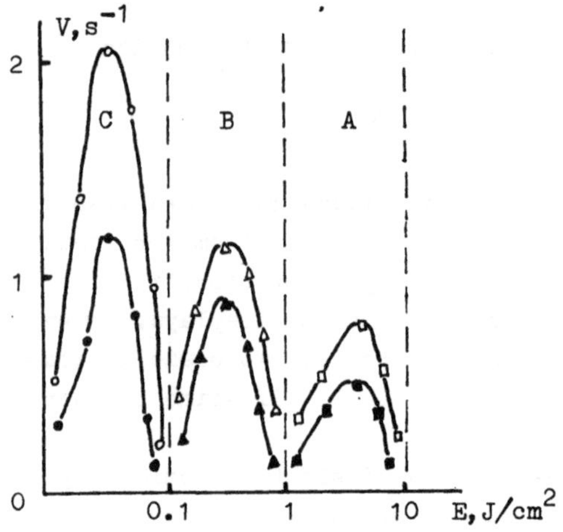

Fig. 1. Colouration (●, ▲, ■) and bleaching (○, △, □) speeds for different irradiation modes: A (■, □), B (▲, △), C (●, ○).

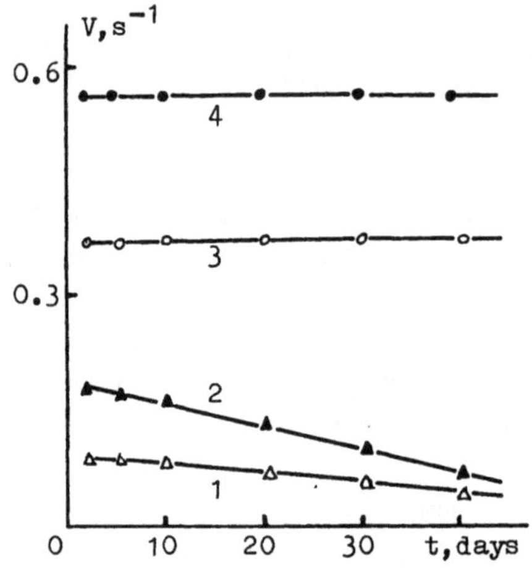

Fig. 2. The effect of laser irradiation on the speed of response of WO_3 films 1 μm thick under storage in air. (1,3) colouration speed, (2,4) bleaching speed of (1,2) nonirradiated and (3,4) irradiated films in the A mode for = 0.2 J/cm².

substrate after evaporation crystallizes into a nonstoichiometric phase $WO_{2.9}$ with a tetragonal lattice. No structural changes have been detected at the film depth. This experimental evidence agrees well with the results of the computational modelling of the kinetics of heating of WO_3 film with nanosecond laser pulses. It follows from computations that the process of the film heating and cooling is characterized by a high temperature gradient, and phase changes "solid-liquid" and "liquid-vapour" can proceed simultaneously owing to closeness of WO_3 melting and evaporation temperatures. The possibility of WO_3 crystallization under such conditions is confirmed by experiments. The average grain size of the polycrystal thus formed is about 0.1 μm which is anomalously large for solid-phase crystallization.

Theoretical and experimental data on the effect of laser radiation on WO_3 films allowed us to qualitatively analyze the causes of changes in their electrochromic efficiency.

The polycrystalline $WO_{2.9}$ phase occupies the lesser part (about 10-20%) of the irradiated film volume. This can explain the fact that in absorption spectra only a slight shift of the absorption band maximum is observed in the coloured state and there is no shift of the fundamental absorption edge typical of crystallized WO_3 films. $WO_{2.9}$ crystals like all nonstoichiometric tungsten oxides are of blue colour. However, in the absorption spectra of irradiated uncoloured WO_3 films no absorption edge is observed before the fundamental absorption edge which also supports the validity of our deduction about the crystallization of the surface layer rather than the entire film.

It is known [2] that in stoichiometric WO_3 all WO_3-octahedra are joined to each other by their vertices, and in WO_{3-x} ordered shear structures are formed with certain crystallographic orientations. It has been found that in the regions of crystallographic shear WO_6-octahedra are joined by their edges [3] . In the tetragonal lattice of nonstoichiometric tungsten oxides, strong distortions of WO_6-octahedra are observed [4] voids and mesa-pores of pentagonal cross-section are formed which have large effective surface and may promote higher electrochromic efficiency as compared with other oxide phases with a more ordered structure. The mesa-pores act as penetration sites for protons and increase transport of H^+-ions through the crystal which leads to increased colouration speed of the WO_3 films.

In conclusion, under nanosecond laser treatment of WO_3 films, phase changes due to thermal effect of absorbed radiation are observed. This results in the formation of a developed microrelief on the film surface. At the same time, initial quasi-amorphous phase of WO_3 changes to a polycrystalline $WO_{2.9}$ phase that is thermodynamically more stable. The film ageing rate decreases and the conditions for charge transfer through WO_3 bulk and across the WO_3-electrolyte interface are improved. As a result, colouration/bleaching speeds and electrochromic efficiencies are increased.

References

(1) Deb, S.K.: Phil. Mag. 27, N4 (1973) 801

(2) Gado, P., A. Modueli: Acta Chem. Scand. 19 (1965) 1514

(3) Ljima, S.: J. Sol. State Chem. 14 (1975) 52

(4) Kolong, R.: Nonstoichiometry, Mir Publishers, Moscow 1974 (in Russian)

HARDENING OF CEMENTED CARBIDES BY LASER PULSE IRRADIATION

B. Schultrich, H.-J. Scheibe, G. Keßler, M. Ermrich, H. Müller,[1]
W. Hauffe[2]

Laser induced surface modification of metals is well-known practice now. For more brittle ceramic-like materials possible favourable effects are in general overshadowed by intense thermal cracking and other damages in the surface region /1/. But a deeper understanding of these processes may throw another light on the high-temperature interaction of the components of the material and reveal possibilities for avoiding damage.

In this respect composite materials of the hardmetal type are especially promising, where a hard and brittle refractory phase (e.g. WC) is combined with a ductile metallic binder (e.g. Co alloyed with W and C). Apart from their practical importance as wear resistant material, WC-Co cemented carbides represent an outstanding model system. Firstly, by changing the cobalt content the brittleness and the creep behaviour may be varied within a wide range between ceramic-like and metallic-like mechanical behaviour. Secondly, the combination of a high and a lower melting component (WC about 2700 °C, Co binder about 1350 °C) may cause special interaction effects related to liquid phase sintering . Thirdly, its hard phase undergoes a high temperature phase transition from hexagonal to cubic structure at about 2500 °C /5/, setting unambigous temperature marks.

The investigated WC-Co hardmetals were produced along the usual powder metallurgical lines with cobalt volume fractions from 10 to 30% and averaged WC grain sizes from 1 to 9 μm. The materials were applied with different surface states: as received from sintering, ground or polished. They were irradiated by a neodymium-glass laser (wave length 1.05 μm) in free generation (energy about 5 J, impulse length about 1 ms) and with Q-switch (energy about 0.5 J, impulse length about 10 ns), as well. Mean power densities, averaged over the laser spot were up to 10^5 W/cm^2 and 10^{10} W/cm^2, respectively. For structural investigations the original inhmogeneous energy distribution over the laser spot was preferred, which allowed a direct comparison of different energy densities. For hardness test samples an optical homogenizer was used. The absorbitivity depends on composition and surface state. Therefore a calorimetric method for direct measurement of the real energy input was installed. Furthermore the number of repeated laser pulses is of marked influence. They may retard or promote further structural changes depending on material state

1) Zentralinstitut für Festkörperphysik und Werkstofforschung, Helmholtzstr. 20, Dresden, 8027, GDR

2) Technische Universität Dresden, Mommsenstr. 27, Dresden, 8027, GDR

and irradiation conditions. For these investigations we succeeded in direct in-situ observation of surface changes in the LASEM-equipment, where the laser irradiation is carried out within a scanning electron microscope /8/.

At lower power densities the surface temperature does not exceed the transformation point or melting point of the tungsten carbides (about 2500 and 2700 °C, respectively) and processes are concentrated in the lower melting binder phase. Rising the energy input, at first selective removal of binder material occurs, caused by local melting and evaporation. Similar to the liquid phase sintering, the molten binder will solve some carbide. Under special conditions, dendritic solidification structures are developed due to the high cooling rate. (A similar picture was observed after electron irradiation /4/.) Apart from the indirect influence via the binder , the WC grains are not altered. At some higher power densities an increasing fraction of the carbide grains is dissolved and is partially reprecipitated during cooling as fine secondary carbides within the binder. The joints between the carbide crystals, originally forming a connected skeleton, are broken up and the remaining grains, now isolated, can be easily displaced in their molten surrounding.

Finally, the whole surface layer is temporarily molten and has completely transformed. Instead of the initial structure with polygonal WC crystals (of about 1 µm in this case), a wavy surface with irregularly bounded regions of about 10 µm size appears. The nature of this layer is best revealed by investigation of a section. Due to preparational problems at such near-surface regions, a special ion-cutting technique was applied instead of usual mechanical sectioning /3/. The following conclusions can be drawn for Q-switched irradiation: 1) The transformed layer is restricted to a depth of 1 to 2 µm. 2) The boundaries, seen on the top surface, are not traces of surface cracks but represent small binder films separating larger tungsten carbide aggregates. This suggests that during cooling the solidification front of the carbide push away the melt (enriching with binder) until meeting the neighbouring front. 3) The interface beneath the transformed structure is marked by pores caused by extracted binder. (Similar effects are known from heat treatment of TiC coated cemented carbides.) Beyond the pore borderline an additional zone appears. Here the grains have retained their original form, but larger amounts of WC have dissolved and are integrated in the binder now.

The pecularities of structural changes by high power irradiation can be clearly demonstrated at inhomogeneous laser spots. A dark appearing area marks the high-temperature region where intense melting has occured. Localized simultaneous X-ray analysis with a position sensitive detector proves that in this region under unswitched irradiation the originally hexagonal α-WC has completely been replaced by W_2C and the high-temperature cubic ß-WC_z (z = 0.65 according to the measured lattice constant) in the case of unswitched irradiation. This shows that 1) the high temperature state is quenched in (no decay of ß-WC_z into α-WC and W_2C has been found) and 2) some decarborization must have occured by precipitation of free carbon.

The kind of the carbide transformation depends on the original carbon distribution and the diffusional time for its separation during the molten state. Thus for Q-switched irradiation (with much smaller pulse time) there is less carbon diffusion and ß-WC_z and α-WC is produced during the hot stage instead of ß-WC_z and W_2C. Furthermore, starting from a fused carbide αWC - W_2C, the cubic ß-WC_z overwhelmingly arise from W_2C where less carbon diffusion is necessary /2/.

Hardness represents a characteristic property for wear resistant and tool materials. It can markedly be rised by low power and high power irradiation, as well. In the first case increase of hardness is caused by solution and dispersion hardening of the binder phase, oversaturated with W and C during the high-temperature stage. Hardness increments of 5 GPa and more were achieved (e.g. rise from 18 to 24 GPa for a WC-6%Co hardmetal). In this low power irradiation regime, damages can be avoided by not to high energy densities, which are yet sufficient for the solution-precipitation effects /6/. In the high power regime we succeeded in suppression of thermal cracking by reducing the pulse time /7/. Then, similar increase of hardness by 5 GPa were proved. In the deeper layers the same solution-precipitation mechanism should work, whereas in the transformed surface region grain refinement and reduction of binder content is assumed to dominate.

(1) Darwish,A.H. and S.P.Andersen: Quatrieme Symposium Europeen de Metallurgie des Poudres, Grenoble, 1975, p. 5.

(2) Ermrich,M.: Thesis, Akademie der Wissenschaften der DDR, Berlin, 1987.

(3) Hauffe,W.: Thesis B, Technische Universität Dresden, 1978.
Hauffe,W.: Elektronika (1985) Nr. 3/4 1.

(4) Roebuck,B. and E.G.Bennett: 11th Plansee Seminar, Reutte, May 1985.

(5) Rudy,E. and J.R.Hoffmann: Planseeberichte Pulvermetallurgie 15 (1967) 174.

(6) Ryžov,E.V., V.I.Averčenkov, V.V.Naduvaev, O.A.Gorlenko and E.N.Frolov: Sverchtverye materialy (1985) /Nr.4/ 9.

(7) Scheibe,H.-J, B.Schultrich, W.Pompe and M.Ermrich: Proceedings 5th Internat. Conf. on Lasers and their applications, Oct. 1985, Dresden, GDR. p. 192.

(8) Schultrich,B. and K.Wetzig: J. Mater. Sci. 19 (1987).

TRANSIENT NUCLEATION IN LASER PLASMA DEPOSITION OF THIN FILMS

A. RICHTER [+)]

The pulsed laser plasma method has been applied for deposition of thin films and turned out to be a powerful technique for synthesis of polycomponent materials and structures of unique properties many of which cannot be obtained in another way (1,2). Peculiarities of the deposition method with relevance to the film structure are high deposition rates $R > 10^{15}$ $cm^{-2}s^{-1}$, high heating and cooling rates up to 10^{10} Ks^{-1}, definite material supply by pulse like material removal, detailed time stages and the aggregation of clusters with different size, frequency, electrical charge and energy $E_{cl} = 10 \ldots 200$ eV.

Thin film formation is a complex process of nucleation, growth and coalescence of islands to reach continuity. Growth mode, that is 3-dimensional island or 2-dimensional layer growth, and structure of the deposit mainly depend on the mobility of particles on the substrate, the cluster distribution in the laser plasma, the microstructure of substrate and interface as well as the supersaturation determined by the material transport of the laser plasma (3,4). An analysis of the deposition process with respect to the mentioned peculiarities necessarily requires the investigation of transient nucleation, kinetics of layer growth and a bond energy description of the nucleus.

Kinetics of aggregation of clusters in the plasma and of nucleation and growth on the substrate is described by a set of coupled differential equations with different boundary conditions and a specification of D^+ and D^- related to the process considered:

$$\frac{dZ(i,t)}{dt} = D^+_{i-1}(t)Z(i-1,t)-D^-_i(t)Z(i,t)-D^+_i(t)Z(i,t)+D^-_{i+1}(t)Z(i+1,t)$$
$$= I_{i-1}(t) - I_i(t).$$

The cluster densities $Z(i,t)$ depend on time and cluster size i, D^+_i and D^-_i are the frequencies of a single atom attachment to and detachment from a cluster, respectively. Taking into account specific cluster bond energies known from band structure LCAO-x_α-calculations (5), size dependent sticking coefficients as well as time dependent collision rates, a numerical simulation of kinetical formation of clusters is possible:

$$\Delta Z(i,t) = (I_{i-1} - I_i)\,\Delta t.$$

+) Zentralinstitut für Festkörperphysik und Werkstofforschung der AdW der DDR, Helmholtzstr. 20, Dresden 8027, GDR

This expression does not include the alteration in cluster population during the sufficiently small time intervall Δt, and is true only in the limit $\Delta t \rightarrow 0$.

Initial temperature and density of the laser plasma determine the frozen-in cluster distribution after the adiabatic expansion of the plasma. As can be seen in fig.1, a characteristic relative cluster stability due to bond energies and thus a preformation of a phase in the plasma is obtained for a special set of parameters mainly determined by the laser power density and the length of the laser pulse. Ion mass spectroscopy yields an experimental evidence for such a cluster distribution (6).

Similar considerations for nucleation and growth on the substrate taking into account an initial cluster distribution result in a growth diagram distinguishing between 3-dimensional island and 2-dimensional layer growth as well as amorphous films controlled by the deposition rate and mobility of particles on the substrate. High values of supersaturation permit to prepare crystalline films of a few monolayers with a low surface roughness. However, very large deposition rates result in an extremly quick filling of the layer without any possibility of the atoms to arrange in crystalline positions such that amorphous films are obtained.

In conclusion, the plasma composition characterizes supersaturation, preformation of nuclei and cluster energy which are essential parameters in the kinetics of crystalline layer growth. Cluster energy and large deposition rates cause high lateral mobility and fluidity of the material after impact upon the surface. Therefore, the peculiarities of the LPVD controlled by the power density and the interaction time of the laser beam with the solid can be used to prepare films with a definite microstructure.

REFERENCES

(1) W. Pompe et al. Wiss. Ber. ZFW Nr. 26 (1984) 87,
W. Pompe et al. Thin Solid Films 144 (1986) 77.

(2) S.V. Gaponov, Proc. Int. School Laser-Assisted Modifications and Synthesis of Materials, ed. S. Metev (Sofia University Press, 1985) p.216.

(3) D. Kashchiev, J. Cryst. Growth 40 (1977) 29.

(4) A. Richter and W. Pompe, Proc. ICALS, Prague 1987, to be published.

(5) G. Seifert, ZfI Mitt. (Leipzig) to be published (1987).

(6) H.-J. Dietze and S. Becker et al., Isotopenpraxis 22 (1986) 6.

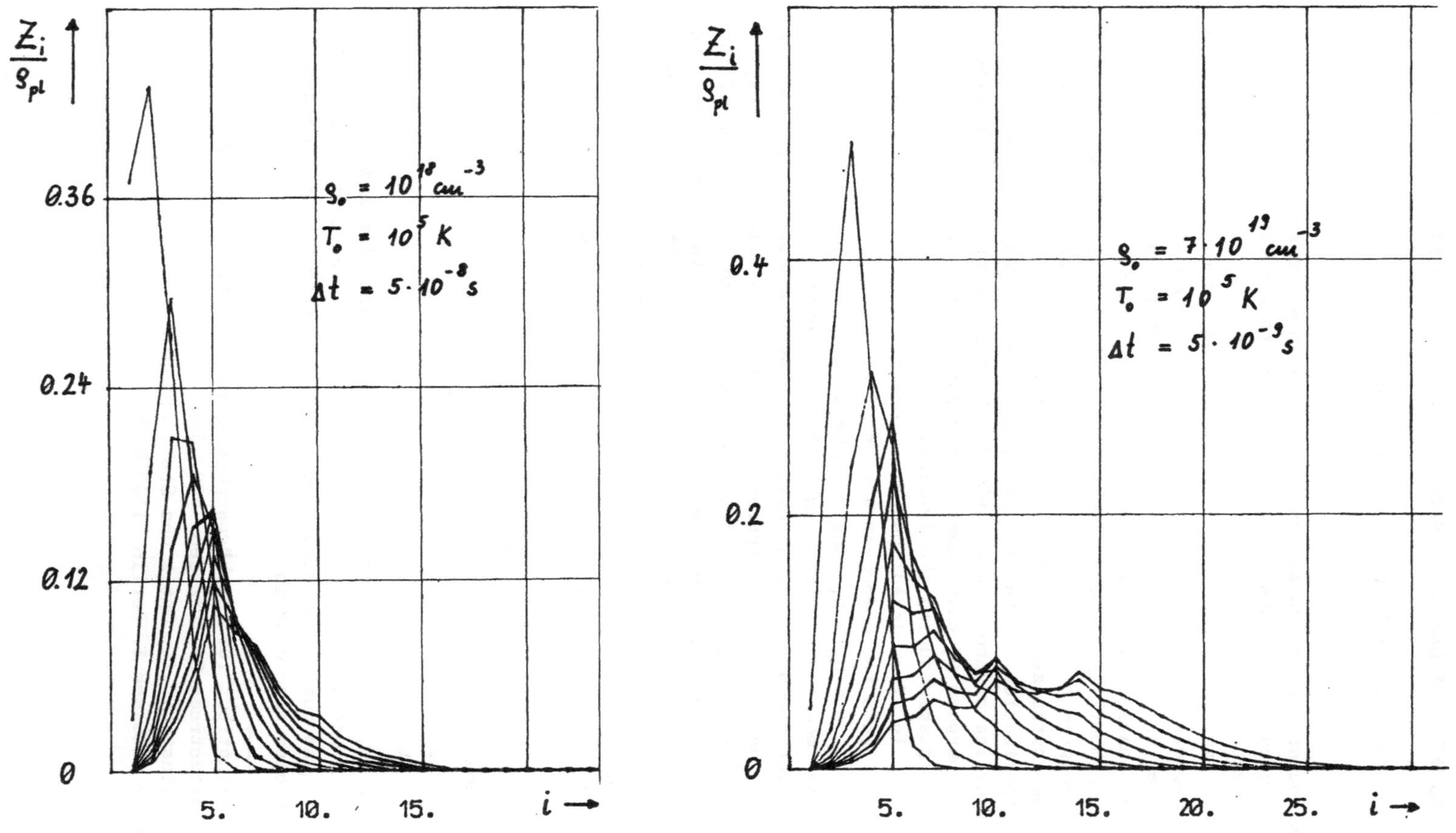

Fig.1: Time dependent cluster size distribution in the laser plasma with different initial plasma parameters

THEORETICAL MODEL OF LASER GENERATED PLASMA FOR THIN FILM DEPOSITION

S.Völlmar[1], Th. Schlegel[2], W.Pompe[1], W.Brunner[2]

In laser pulse vapour deposition the intensive evaporation of a solid target is used as a plasma source. In order to deposit thin films of definite crystalline structure and chemical composition on a substrate, the energy spectrum of the plasma should be formed in a definite manner. Especially, particle energies in the range from 10 to 100 eV seem to be appropiate for the formation of high-quality thin film structures characterized by even thickness, good adhesion to the substrate, high mobility of deposited particles during film formation. For instance, compare the paper by A. Richter (2). Therefore the irradiation has to be optimized with respect to creation of a suitable plasma energy distribution. Intense target evaporation can be expected for energy densities of about 10 to 1000 J/cm^2 from a Q-switched Nd-glass-laser. The purpose of this paper is the description of the development of the laser plasma during the interaction with the high energy pulse.
The problem is formulated using a one-dimensional hydrodynamical description. In a two-component system of electrons and ions (m_e, m_i - their masses, and n_e, n_i -their numbers per volume) the mass conservation, the momentum balance as well as the energy balance must be fulfilled. The computer code describing the hydrodynamical behaviour is based on a "one fluid/ two temperatures" model with velocities $v=v_e=v_i$ and temperatures $T_e \neq T_i$ (1). In order to describe the interaction of the laser pulse with the generated plasma the inverse bremsstrahlung mechanism is discussed as the dominating process. Additional ionization or recombination processes are neglected, therefore the ionisation state q is held constant.
The equations are: continuity of electrons and of ions

$$\frac{\partial n_e}{\partial t} + \frac{\partial}{\partial x} n_e * v = 0 \qquad \frac{\partial n_i}{\partial t} + \frac{\partial}{\partial x} n_i * v = 0 \tag{1}$$

momentum conservation:

$$m_i n_i \left(\frac{\partial v}{\partial t} + v * \frac{\partial v}{\partial x} \right) = - \frac{\partial}{\partial x} \left(p_i + p_e + \tau_i + \tau_a \right) . \tag{2}$$

The symbol τ_i denotes the viscosity of heavy particles and τ_a is an artificial viscosity for avoiding instabilities in computations. The pressure is the sum of the pressures of the electron gas and the gas of heavy particles according to the ideal gas law.

[1] Zentralinstitut für Festkörperphysik und Werkstofforschung
Helmholtzstraße 20, Dresden, 8027, GDR

[2] Zentralinstitut für Optik und Spektroskopie
Rudower Chaussee 5, Berlin, 1199, GDR

Energy conservation of ions:

$$\frac{3}{2} n_i \left(\frac{\partial T_i}{\partial t} + v* \frac{\partial T_i}{\partial x} \right) + (p_i+\tau_i) \frac{\partial v}{\partial x} = \frac{3}{2} n_e*(T_e-T_i)/\tau_{ei}, \qquad (3)$$

where τ_{ei} is the energy relaxation time between electrons and ions due to the coulomb collisions. Energy conservation of electrons:

$$\frac{3}{2} n_e \left(\frac{\partial T_e}{\partial t} + v* \frac{\partial T_e}{\partial x} \right) + (p_e+\tau_a) \frac{\partial v}{\partial x} = - \frac{3}{2} n_e*(T_e-T_i)/\tau_{ei} + \frac{\partial}{\partial x} K_e \frac{\partial T_e}{\partial x} + W. \qquad (4)$$

K_e is the electron thermal conductivity. W is the power absorbed by unit volume of plasma from the irradiated laser beam through the inverse bremsstrahlung process. The system of equations is solved in the Lagrangian formalism with 100 spatial steps and time steps in the order of 1-10 ps. As results, the distributions of the temperature, the density, and the velocity near the target surface are obtained.

Fig. 1 Temperature versus time for

Carbon Molybdenum

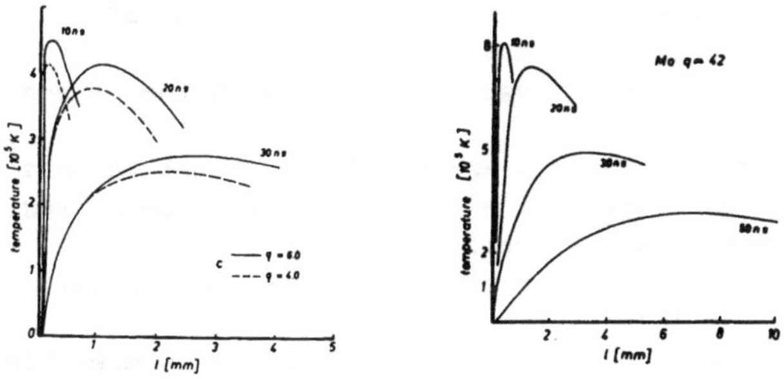

Fig. 2 Velocity versus time for

Carbon Molybdenum

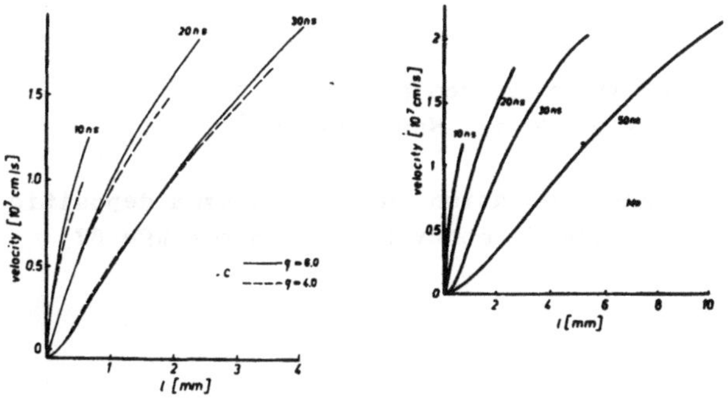

Fig.3 Density versus time for

Carbon Molybdenum

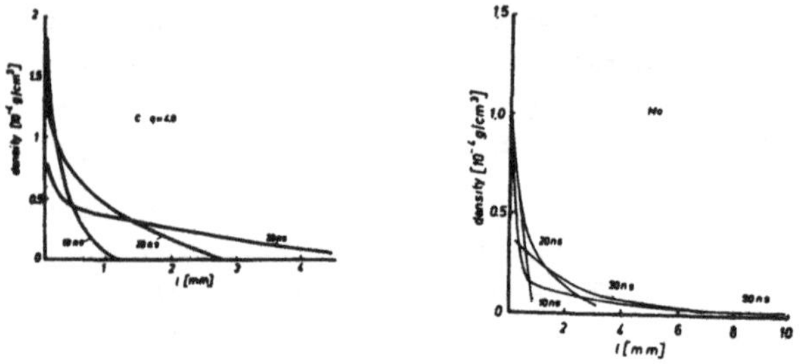

In the following, the effect of such a laser pulse on carbon plasmas with averaged ionizations q=4 and 6, and molybdenum plasmas with q=42 is considered. As parameters of the irradiation we have the wavelength 1.06 µm, the pulse energy 100 J/cm^2 , a triangular pulse shape with duration 20 ns with the maximal power density of 10^{10} W/cm^2.

We can draw the following conclusions:

- The plasma energy is in the range of 50 eV, with the considered parameters of irradiation.
- Electrons and ions temperatures are nearly equal. At the end of the laser pulse the temperature of electrons in the corona region is about 3% higher then that of the ions.
- The energy distribution of the plasma from carbon target varies only weakly with the degree of ionization.
- Carbon: The assumed degree of ionization is compatible with the calculated temperature, e.g. the temperature comes up to 10% of the energy of ionization.
- Molybdenum: In this analysis the degree of ionization exceeds the calculated temperature.

literature

(1) Hora, H.: "physics of laser driven plasmas", John Wiley & Sons, New York 1981

(2) Richter, A.: "Transient nucleation in laser plasma deposition of thin films", contribution to the EPM'87

EXPERIMENTAL SET-UP FOR LASER PULSE VAPOUR DEPOSITION IN UHV

R. Wünsch, H.-J. Scheibe, H.J. Weiß, W. Resch [1)]

Multiple thin-film layers of two or more different materials have a very wide range of applicability and provide. Laser induced physikal vapour deposition (LPVD) under ultra high vacuum innovative concepts of fabricated multilayer structures. Our set-up consists of an UHV stainless steel chamber and a Q-switched Nd-glass-laser or CO_2-TEA-laser was used to deposit the layers. A block diagramm shown in Fig. 1. The schematic diagramm of the vacuum system is given in Fig. 2. The sputer ion pump and titanium getter pump produced base pressure of 10^{-7} Pa range. The substrate for deposition as well as the targets consisting of the substances to be deposited are plased on a substrate-target holder.

Whenn focussing pulses of a Nd-glass or CO_2-laser on the target, with power densities $\gtrsim 10^6$ Wcm^{-2}, plasma flares emanate from the target. Their size and shape are such that the deposit in apparently homogeneous over a large substrate area, provided that the mutual position of target and substrate are appropriately chosen.

Technical Data

- vacuum system

eff. pumping speed	for chemicaly active gases	1000 ls^{-1}
	for inert gases	80 ls^{-1}
bakeable		250° C/150° C
basic pressure		10^{-7} Pa

[1)] AdW der DDR,
Zentralinstitut für Festkörperphysik und Werkstofforschung,
Helmholtzstraße 20, Dresden, 8027
DDR

- substrate-target holder

target holder	2 targets of various materials
substrate holder	Substrate 3" diameter, thickness 4 mm max.
the target to substrate distanze	10 mm ... 240 mm
- incident density up to	10^9 Wcm^{-2}

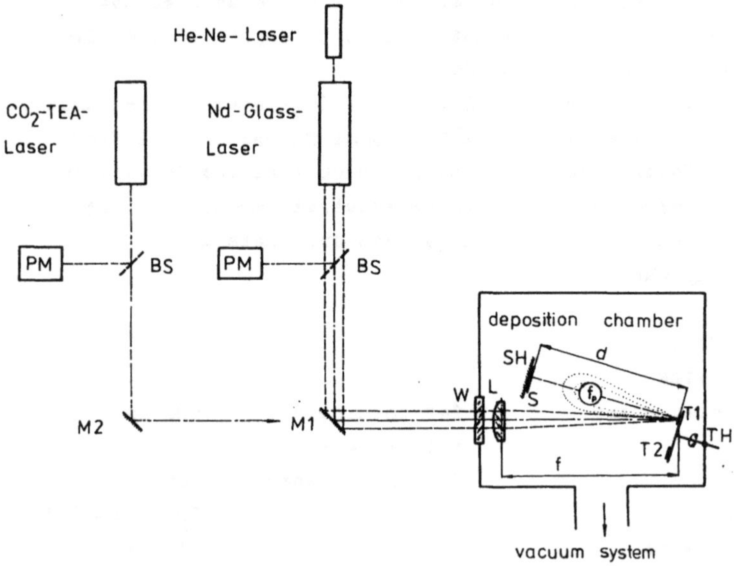

f_P Plasma flares
SH Substrate holder
S Substrate material
L Lens
W Window
M Mirror
d Distanze between target and sub.
TH Target holder
T1 Target material 1
T2 Target material 2
PM Power meter
BS Beamsplitter

Fig. 1. Block diagramm of the set-up

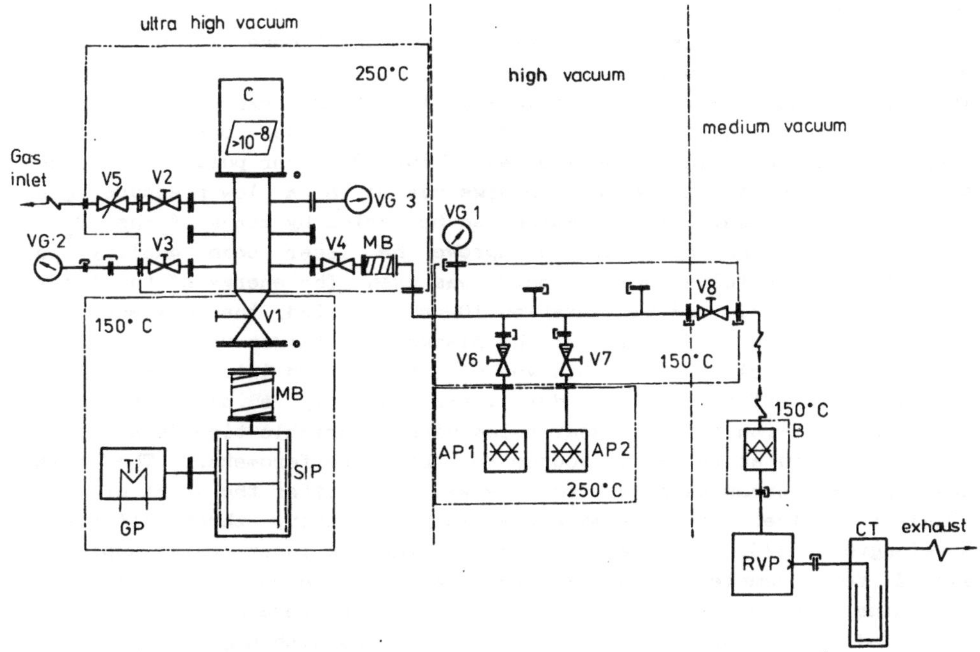

C	UHV Chamber	V1	Plate valve
/ /	Total pressure /Pa/	V2...V4	Metal angle valves
SIP	Sputter ion pump	V5	Gas dosing valve
GP	Titanium getter pump	V6...V8	Angle valve
AP	Adsorption pump	VG1	Vacuumgauge of the Pirani prinziple
RVP	Rotary vane pump	VG2	Cold-cathode ionization gauge
CT	Condensate trap	VG3	Hot-cathode ionization gauge
B	Baffle		
MB	Metall bellows		

Fig.2. Schematic diagramm of the vacuum system

CHARACTERIZATION OF THE LPVD-PROCESS BY MICHELSON INTERFEROMETRY

W.Ross[1], H.Beyer[1], S.Völlmar[2]

In laser pulse vapour deposition the intense evaporation of a solid target is used as a source of particles deposited later on a substrate. In order to get a deeper insight into the strong non-linear interaction process of the high laser power with a solid target as well as the generated plasma, a direct experimental observation of the propagating plasma cloud is necessary.

The experimental setup is shown in Fig. 1. In these experiments LPVD-plasmas are produced by focussing CO_2-laser pulses onto a carbon target under an incident angel of 45°. The CO_2-laser pulses have been generated with a transverse discharge plasma in a low-pressure gas mixture. The laser pulse duration can be varied by means of the CO_2-content in the gas mixture. To improve the laser beam quality, a resonator with a folded-beam geometry was used. The energy of the laser pulses was 1 Joule and the duration 10 µs. To study the influence of the pressure on the formation of the plasma, the target was placed into a stainless steel vessel. This vessel was inserted into one arm of a Michelson interferometer. A dye laser pumped by a nitrogen laser synchronised with the CO_2-laser pulse with a variable time delay up to 1 ms was used as the light source for the interferometer. The time delay for all examples presented here was 10 µs after the start of the CO_2-laser pulse. To determine the plasma density, three different wavelengths (420, 570, and 650 nm) of the dye laser were applied. Fig. 2 for example shows the interference patterns, obtained with Rhodamine (570 nm). In all cases with pressures above 50 Torr a laser supported detonation wave (LSD) exist,e.g. zones with higher density in the front and a rarefaction array behind the front (2,3).

The deviations from the rectilinear patterns contains information concerning the index of refraction n of the plasma (7). In a cylindrical symmetry the integral equation for calculating from each line is

$$z(x)=2\int_x^R [n(r)-1]*r*dr/\sqrt{r^2-x^2}\ , \qquad (1)$$

where n(r)-1 is the refractivity difference, r the radial cylindrical coordinate, x a direction perpendicular to the symmetry axis, z(x) is the deviation from the straight line in units of the wavelength λ/2, and R the outer radius of the plasma.

[1] Zentralinstitut für Elektronenphysik, Hausvogteiplatz 5/7, Berlin, 1086,GDR

[2] Zentralinstitut für Festkörperphysik und Werkstofforschung Helmholtzstraße 20, Dresden, 8027,GDR

Fig.1 Experimental setup

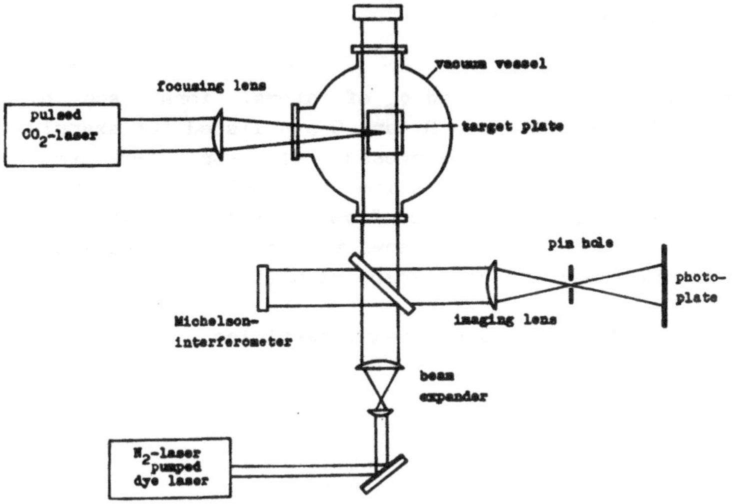

Fig. 2 Interference patterns obtained with Rhodamine.

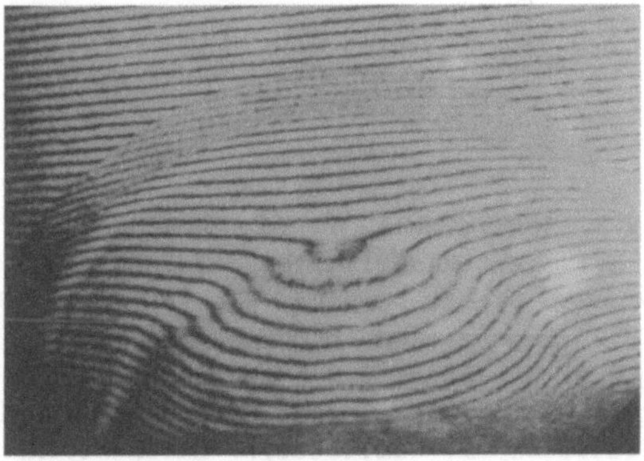

The fast registration of the values z(x) is carried out with the help of the image processing system robotron A6471. The solution of the equation (1) by the Abel inversion (4,6)

$$n(r)-1= -\frac{1}{\pi}\int_r^R dx \frac{\partial z(x)}{\partial x} \frac{1}{\sqrt{x^2-r^2}} \qquad (2)$$

is analysed with a numerical procedure on a computer SM4. The density of heavy particles (atoms and ions) and electrons can be derived from interference patterns obtained with different wavelengths.

For the ions and atoms the equation

$$n-1=(A+B/\lambda^2)*(\varrho_0+\varrho_i) \quad (3)$$

holds, and for the electrons

$$n-1=-C\varrho_e*\lambda^2 \quad (4)$$

with the densities ϱ_0, ϱ_i, and ϱ_e of atoms, ions, and electrons, respectively. The parameters A,B, and C are listed for example in (6). The interference patterns at a pressure of 1 atmosphere are analysed mathematically. As an illustration the calculated densities are represented in Fig.3 for a distance from the target surface of about 4 mm. The low density of electrons shows that the degree of ionisation is very small in this case (to be compare with (5)).

Fig. 3 Densities of electrons and heavy particles

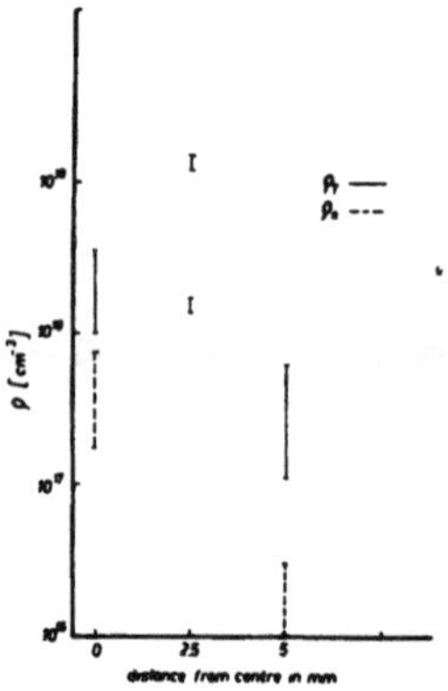

<u>literature</u>

(1) Barkhudarov,E.M.,V.R. Berezovskii, T. Ya. Chelidze, G.V.Gelashvili, W.O.Mdivnishvili, M.I. Taktakishvili, N.L. Tsintsadze, Internat. Conference on plasma physics, Nagoya ,April 1980

(2) Бархударов,Э.М., В.Р. Березовский, М.О.Мдивнишвили, М.И.Тактакишвили, Н.Л. Цинцадзе, Т.Я. Челидзе: Письма b журнал технической физики том 10 (1984), 1178-1181

(3) Cornolti F., A. Giulietti, D. Giulietti,M. Lucchesi, M. Vaselli: Optics Communications 57(1986), 249-253

(4) Kümmel, F.: Tabellen zur Abelschen Integralgleichung , Veröffentl.des Phys.-Techn. Instituts der DAW Nr. 1 (1960)

(5) Lachambre J.L., C.R. Neufeld, Phys Fluids 24 (1981) 2336-2344

(6) Neumann,W.: "Spectroscopic Methodes of Plasma Diagnostic" in Progress in plasma and Gas Electronics Vol1, ed. R. Rompe,M.Steenbeck, Akademie Verlag Berlin 1975

(7) Schmitz,G., Z. für Physik 126 (1949),1-12

ON DEFECTS AND STRESSES IN THIN FILMS

A. Richter[1)], A.E. Romanov[2)], W. Pompe[1)] and V.I. Vladimirov[2)]

During deposition of thin films an interface between substrate and deposit is generated, the structure of which is very important for growth, adhesion and thickness of the film (1,2,3). Thermal defects as surface vacancies and adatoms result in a roughness of the interface and a decrease of the anisotropy of the specific surface energy. Dislocations are non-thermal defects correlated with the incoherent interface. However, especially screw dislocations are also centres of nucleation on a surface and promote 3-dimensional island growth. Deposition techniques as e.g. the pulsed laser plasma deposition method (LPVD) (4) additionally generate defects, mainly vacancies and interstitials, due to high deposition rates and a significant number of particles of an energy $E > 20$ eV. The aim of this paper is to investigate the stress field and the elastic energy of defects and interfaces in order to correlate defect formation with structural properties of thin films.

Different mechanisms cause internal stress within the film:

(a) Crystalline films with a misfit $\varepsilon_i = (a_s - a_f)/a_f$ between the atomic lattice constants of the substrate a_s and the deposite a_f possess an incoherent interface. The elastic stress field depends on the thickness of the film h, the density of dislocations and deformation ε.

(b) Temperature changes during the deposition create a thermal misfit with a deformation

$$\varepsilon_{th} = \int_{T_0}^{T_1} (\alpha_s - \alpha_f)\, dT$$

where $\alpha_{s/f}$ are the thermal expansion coefficients of the substrate and the film. This mechanism is important for deposits with high melting temperature and deposition at high substrate temperature T_s.

(c) Grain growth in thick films is described with a deformation $\varepsilon_g = (\Delta\text{-a})(D_0{}^{-1}\text{-}D^{-1})$. D_0 and D are the average distances between grain boundaries before and after grain growth, respectively, and $(\Delta\text{-a})$ is the relaxation length of the interface. This description can be applied to the corresponding case of amorphous-crystalline transformations within the concept of medium range order, too.

(d) Growth mode and layer coverage (5) determine compressive as well as tensile stress in thin films. Cap-shaped nuclei possess compressive stresses $\sigma \approx \gamma/r$, increasing with the surface energy γ and decreasing with the droplet radius r. Small nuclei with a large curvature contain stresses of about 10^8 Pa. However, if

[1)] Zentralinstitut für Festkörperphysik und Werkstofforschung der AdW der DDR, Helmholtzstr. 20, Dresden 8027, GDR

[2)] A.F. Ioffe Physico-Technical Institute, Academy of Sciences of the USSR, Leningrad 194021, USSR

layer coverage is performed, compressive stress is reduced, because islands glue together. This process is connected with significant changes of tensile stress at points of coalescence. Two-dimensional layer growth mode is characterized by the material parameter $\Delta\gamma = \gamma+\gamma_g-\gamma_s$, where γ, γ_g and γ_s are the surface energies of the deposit, interface and substrate, respectively. Stress within the film is important for layers of a few atoms, since $\sigma \approx \Delta\gamma/h$. Due to the sign of $\Delta\gamma$, a good adhesion of the film is observed for $\Delta\gamma < 0$.

Defect formation or rearrangement of defects takes always place if the system can relax into an energetically favourable state. Particles with a sufficiently high energy can induce transformations of the growing defect structure during the deposition process and thus stimulate the relaxation of the defect network. Since the elastic strain fields are of long range order, the interaction energy between defects and the surface is of special importance to estimate the critical thickness of the deposited layer connected with the formation of a dislocation network. The elastic energy density of such a system contains contributions of the homogeneous elastic strain parallel to the film surface, a thickness independent term and the interaction term of dislocations with the free surface:

$$E = 2\frac{1+\nu}{1-\nu}\mu h\varepsilon_i^2 + \frac{\mu b}{4\pi(1-\nu)}(\varepsilon_i-\varepsilon) - 2\mu\frac{1-\nu}{1-2\nu}(h-b)(\varepsilon_i-\varepsilon)\varepsilon_i - \frac{\mu}{2(1-\nu)}(\varepsilon_i-\varepsilon)^2\Phi(\varepsilon,h).$$

μ is the shear modulus, ν the Poisson ratio, b the value of the Burgers vector, $\Phi(\varepsilon,h)$ a function of strain and film thickness (6). Fig. 1 shows the change of the energy density with increasing film thickness. The formation of dislocations is energetically favourable for $h \approx 300b$ with $\varepsilon_i = 10^{-3}$. The critical film thickness and the remaining strain ε is approximately given by

$$h^* = b/(8\pi\varepsilon_i)*\ln \varepsilon_i^{-1} \qquad \text{and} \qquad \varepsilon = b/(16\pi h)*\ln \varepsilon_i^{-1}.$$

The internal stress in the film is a quantity which gives information on the defect density. Here we want to distinguish between a row of dislocations of length 2L with 2L >> h and 2L = h. The component σ_{yy} of the stress tensor of a row of dislocations is given by

$$\sigma_{yy} = \frac{2\mu\varepsilon_i}{\pi(1+\nu)}\left(\arctan\frac{y-L}{h} - \arctan\frac{y+L}{h} + \frac{h(y+L)}{h^2+(y+L)^2} - \frac{h(y-L)}{h^2+(y-L)^2}\right) .$$

A comparison between the stress field of the defect layer $\sigma_{yy}+\sigma_{\varepsilon_i}$ with $\sigma_{\varepsilon_i} = 2\mu(1-\nu)\varepsilon_i/(1-2\nu)$ and the characteristic stress τ to generate dislocations gives the following results:

- $-\tau \gg \frac{1}{2}\sigma_{\varepsilon_i}$ no formation of defects,
- $-\tau \ll \frac{1}{2}\sigma_{\varepsilon_i}$ formation of a row of dislocations with 2L >> h,
- $-\tau \lessapprox \frac{1}{2}\sigma_{\varepsilon_i}$ periodical arrangement of dislocations with 2L = h and large distances between dislocation crowds.

This means, if the internal stress is very large, the system will relax in the formation of a dislocation row, whereas only crowds of dislocations are generated at medium values of internal stress.

In conclusion, we can state that a definite defect formation reduces the remaining internal stress within the thin film and favours a regular growth of the interface. Therefore, the generation of defects in a region close to the surface during deposition can influence the thin film structure in a positive manner.

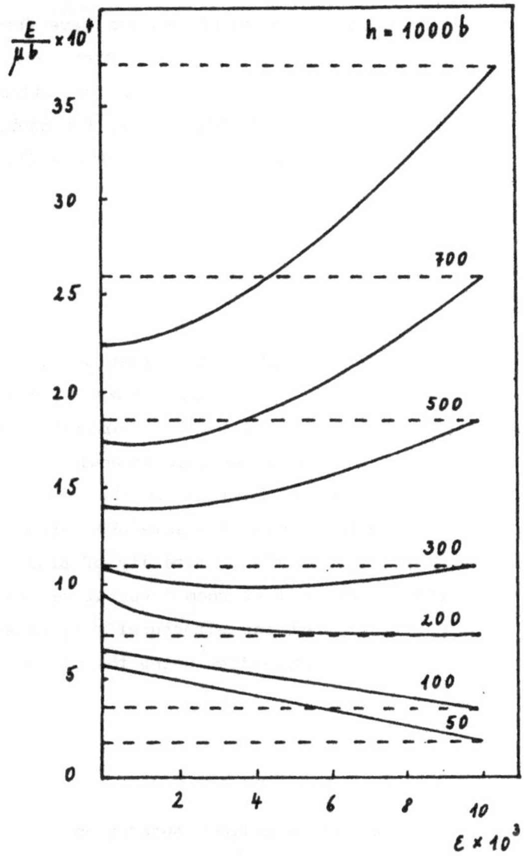

Fig. 1: Energy density via elastic strain for a film with initial misfit $\varepsilon_i = 10^{-3}$ at different film thickness h. —— system with dislocations, - - - without dislocations.

References

(1) J.M. Blakely, Surface Physics of Materials, Academic Press, New York, 1975.

(2) H. Trinkaus, Vorlesungsmanuskripte 17. Ferienkurs 1986, Kernforschungsanlage Jülich, p. 377.

(3) H. Ullmaier, Vorlesungsmanuskripte 17. Ferienkurs 1986, Kernforschungsanlage Jülich, p. 401.

(4) W. Pompe et al., Proc. 2nd Conf. Trends in Quantum Electronics, ed. by A.M. Prokhorov and I. Ursu, Springer Verlag, Berlin 1986, p. 331.

(5) A. Richter, ZfI-Mitt. (Leipzig) in print 1987

(6) V.I. Vladimirov, M.J. Gutkin and A.E. Romanov, Preprint "Relaxation of internal stresses in heterogeneous systems", to be published in Solid State Physics 1987, in Russian.

ESR STUDY OF SiN FILMS PREPARED BY ECR PLASMA CVD METHOD

T. IZUMI[+], M. SHIBUYA[+], T. HIRAO*, T. KAMADA* and T. MATSUMORI[+]

Two kinds of ESR (electron spin resonance) centers have been observed in silicon nitride (SiN) films prepared by the ECR (electron cyclotron resonance) CVD method. One had a g-value of 2.0055 which is identical to that of silcon dangling bonds in amorphous Si. The other had a g-value of 2.0042, due to the combination of Si dangling bonds with nitrogen neighbors. The properties of these ESR centers, which are sensitive to SiN preparation condition, are discussed.

1. Introduction

SiN films prepared by plasma CVD method have been widely used in Si integrated circuits as a good passivation material. This method, however, has several disadvantages, since it repuires the specimens to be heated up to approximately 300°C, and produces an uneven deposition reaction on the specimens so that molecules, atoms and radicals which do not sufficiently decompose tend to be contained in the deposited films.[1)] As a new alternative to this deposition method, the ECR plasma CVD method has become the object of researchers' attention, since it has the possibility of producing high quality thin films at low gas pressure (10^{-4} Torr) and at room temperature; and does not require any thermal reaction. We have investigated the structural properties of SiN films which were prepared by several types of film deposition using the ESR method.

2. Experimental

SiN films of 1 μm were deposited on silicon substrates by using an ECR plasma CVD apparatus (Microwave frequency=2.45 GHz, magnetic flux density=875 G). Deposition gases were introduced through two respective inlet systems into the plasma chamber and into the specimen chamber. The sample preparation conditions are summarized in Table 1. ESR measurements were performed using an X-band spectrometer at room temperature. The g-value, the line width and the signal intensity were determined by using signals of Mn^{++} or DPPH as the calibration reference. The IR spectra were measured with a Hitachi 260-10 spectrometer. Isochronal annealing was carried out in flowing dry argon in the temperature range of room temperature to 900°C.

3. Results and Discussion

Figs. 1a and b show typical examples of the ESR spectra in SiN films which were

+Dep. of Electronics, Fac. of Engineering, Tokai University, Hiratsuka, Kanagawa, Japan
*Central Resarch Laboratories. Matsushita Electric Industrial Co.,Ltd., Moriguchi, Osaka, Japan.

prepared by the ECR plasma CVD method. Two kinds of paramagnetic damage centers were observed in this figure. One had a g-value of 2.0055, which is identical to that of the silicon dangling bond in amorphous Si. The other had a g-value of 2.0042 which is an unknown damage center. The total density of Si dangling bonds was $\sim 10^{18}/cm^3$. Morimoto et al observed an ESR signal with g=2.0030 and ΔH_{pp}=13 G for amorphous $Si_{1-x}N_x$:H films which were prepared by r.f. glow discharge decomposition.[2] It was found that the g-value increases from 2.0029 to 2.0050 as x decreases from about 0.6 to zero. Hobbs et al have also observed an ESR signal with g=2.0039 which is due to Si dangling bonds with nitrogen neighbours in the high-dose N implanted silicon.[3] Hence, a variety of the g-value between 2.0050 and 2.0030 depends on the changes in the amount of nitrogen in the vicinity of Si dangling bonds. Therefore, it seems that the ESR signal with g=2.0042 is due to Si dangling bonds with nearest neighbor N atoms.

Gas flow dependence on the ESR center;

Fig. 2 shows the variation of the ESR signal intensity and the g-value as a function of the gas flow-ratio $r=SiH_4/N_2$. The spin density takes a minimum value of r=0.4 and increases below and above this value. The g-value shifts from 2.0042 to 2.0050 as the introducing gas flow ratio increases from 0.4 to 0.6. The observed infrared absorption spectra as a function of introducing gas flow ratio is shown in Fig. 3, where three distinctive absorption peaks are seen, corresponding to Si-N bond (and 835 cm^{-1}), Si-H bond (at 2175 cm^{-1}) and N-H bond (at 3350 cm^{-1}), respectively. Especially, the Si-H bond spectrum at 2175 cm^{-1} is observed only in the case of the gas flow ratio in the range of 0.6 to 0.8. This means that the defect structure due to the Si dangling bond observed at larger r is the same as that of a-Si:H. At the introducing gas flow ratio of 0.2, the Si-H bond spectrum at 2175 cm^{-1} was not observed, and the N-H bond was observed. The N-H bond peak increased with decreasing r. This means that hydrogen atoms link together nitrogen dangling bonds dominantly, since the binding energy of an N-H bond (93.4 kcal) is larger than that of an Si-H bond (70.4 kcal). At or near the stoichiometric composition (r=0.4), the density of Si dangling bonds was minimal.

From the above results, a Si-dangling bond with g=2.0055 is a dominant defect in SiN films for r>0.4. On the contrary, a Si-dangling bond with nearest neighbor N atom (giving g=2.0042) is a dominant defect in SiN films for r<0.4.

Annealing properties;

Fig. 4 shows the isochronal annealing curves of the specimens under the deposition condition described in Table 1. The isochronal annealing curve line of SiN film prepared by the conventional plasma CVD method is also shown with a solid line. Although the spin density in No.13 is much larger than the other, it decreased abruptly with an increased annealing temperature and reached about $7X10^{17}/cm^3$ at about 400^oC. ON the other hand, no change was observed in the spin density of No.29, even with increased annealing temperature of room temperature to 400^oC. The spin density of the other samples was on the order of $8X10^{17}/cm^3$ after anneal at 400^oC. The g-value shifted to 2.0055 which is identical to the g-value of a-Si films.

When the SiN films anneal at 400°C, H-atoms and N-atoms might migrate in SiN networks to reduce Si-dangling bonds. The annealing in the temperature range above 400°C to 900°C did not change the spin density of dangling bonds except for No.13. The spin density of No.13 which was prepared by a low microwave power of 50 W increased abruptly and reached to $4X10^{18}/cm^3$ at the 900°C annealing. Clearly, more work is needed in order to clarify the annealing mechanism of SiN films by the ECR method.

4. Conclusion.

The observed ESR signal in SiN films prepared by the ECR plasma CVD method was interpreted in terms of Si dangling bonds in SiN. One had a g-value of 2.0055 which is identical to that of Si dangling bonds in amorphous Si. The other had a g-value of 2.0042 which is due to Si dangling bonds with nitrogen neighbours. Intensities of these ESR centers depend on the preparation condition of SiN films.

References

1) S. Matsuo and K. Kikuchi; Jpn. J. Appl. Phys. 22 (1986) L210.
2) A.Morimoto, Y.Tsujimura, M.Kumeda and T.Shimizu; Jpn.J. Appl. Phys. 24 (1985) 1394.
3) A.Hobbs, R.C.Barklie, PLF Hemment and K.Reeson; J.Phys. C: Solid State Phys. 19 (1986) 6433.

Preparation Condition for Si-N Films Deposition

(1) ECR Plasma CVD

SAMPLE No.	GAS FLOW (SCCM) SiH_4	N_2	GAS PRESSURE (Torr)	MICROWAVE POWER (W)
5	6	15	8.3×10^{-4}	100
6	6	15	8.3×10^{-4}	150
8	3	15	8.3×10^{-4}	150
9	9	15	8.3×10^{-4}	150
10	12	15	8.3×10^{-4}	150
13	6	15	8.3×10^{-4}	50
21	8	12	8.3×10^{-4}	150
29	6	15	8.3×10^{-4}	500

(2) Plasma CVD

SAMPLE No.	GAS FLOW RATIO (SiH_4/N_2)	GAS PRESSURE (Torr)
3	3.0	0.1~5

Table 1

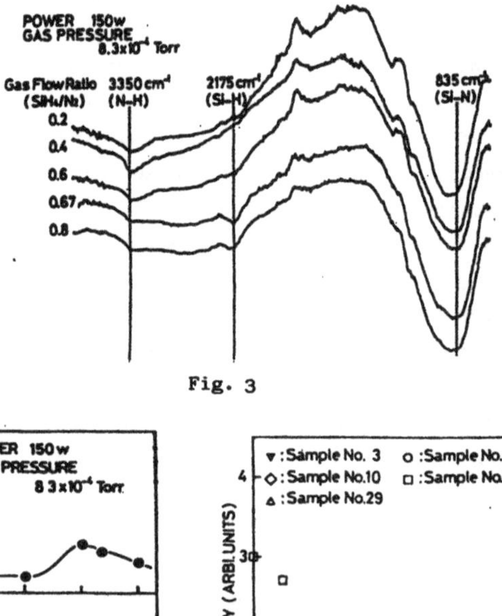

Fig. 3

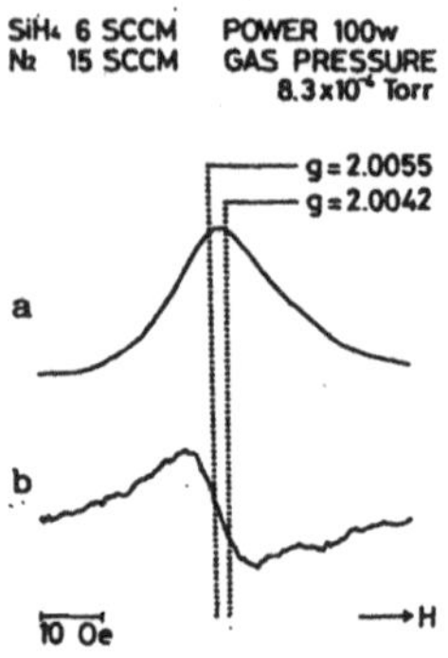

Fig. 1

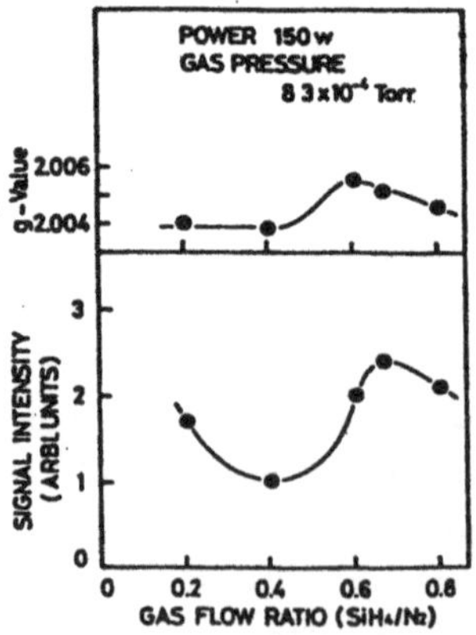

Fig. 2

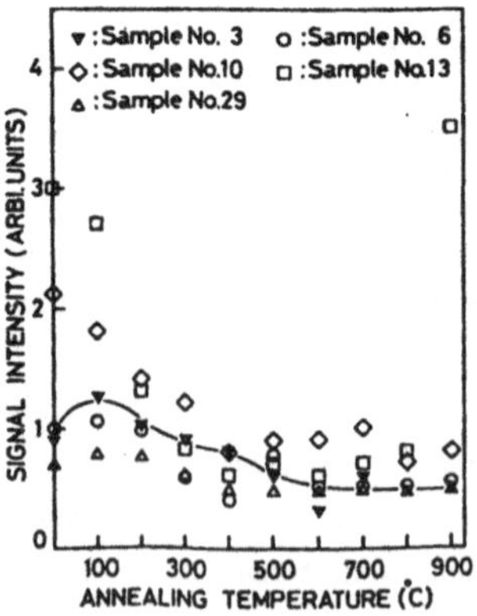

Fig. 4

ELECTRICAL PROPERTIES OF SILICON FILMS ON SiO_2 FORMED BY EXPLOSIVE CRYSTALLIZATION

N.Winkler, W.Andrä, E.Glaser, G.Andrä[1)]

1.Introduction

In the last years explosive liquid phase epitaxy (ELPE) has been investigated in silicon [1,2]. ELPE can be used to form a laminated crystalline layer on SiO_2. To characterize these SOI-layers Hall-effect and resistivity measurements were carried out and the Hall-mobilities and carrier concentrations were determined.

2.Experimental

The ELPE-process was realized in 400 nm lp-CVD-Si films on 350 nm SiO_2 layer on bulk silicon or in lp-CVD-Si amorphized by silicon implantation ($1*10^{16}$ cm^{-2}, 300 keV). Before or after the crystallization phosphorus ($N_D = 1*10^{13}$ cm^{-2}...$1*10^{15}$ cm^{-2}) and in one case arsenic ions ($1*10^{15}$ cm^{-2}) were implanted into the films at 100 keV. The ELPE was realized on the whole wafer by flash lamp pulses (pulse length τ = 1.5 ms). The layers implanted after crystallization were annealed at a temperature of 920 K for 30 min in N_2-atmosphere. These heat treatment don't cause an additional bulk nucleation [3]. For the Hall-effect and resistivity measurements test structures (Fig.1) were prepared by means of photolithography. To measure electrical parameters as a function of the distance from the ignition point the process was ignited by a pulse of a Q-switched Nd:glass laser with a pulse length τ_i = 50 ns [4] and a lateral region of a length $l \approx 700\,\mu m$ and a width $d = 20\,\mu m$ was crystallized.

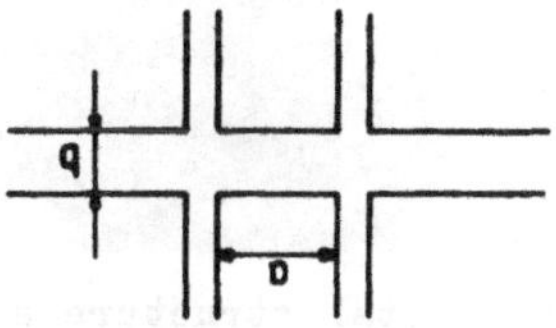

Fig.1. Test structure used for the Hall-effect and resistivity measurements ($a=50\,\mu m$, $b=30\,\mu m$).

3.Results and Discussion

Fig.2 shows the measured Hall-mobilities μ_s as a function of the

1) Friedrich-Schiller-Universitat Jena, Sektion Physik, Max-Wien-Platz 1 Jena, DDR 6900

carrier concentration N_s. Silicon implanted ELPE layers show larger crystal lamellae than not Si^+ implanted layers (Fig.3) and higher values of μ_s and lower values of the sheet resistivity ρ_s are obtained.

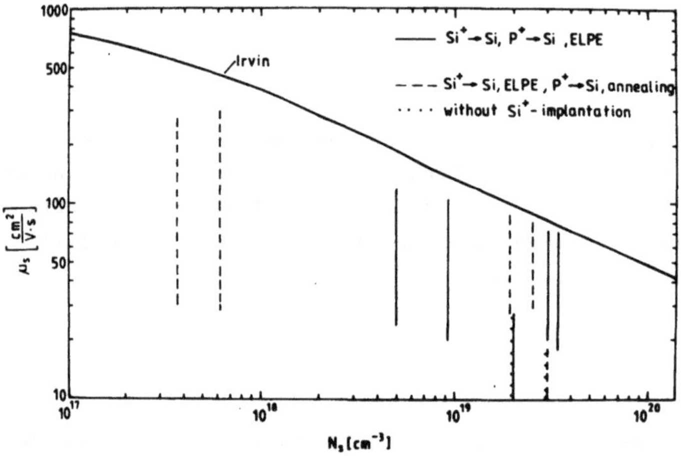

Fig.2. Hall-mobility μ_s as a function of the carrier concentration N_s compared with the mobility for single crystal silicon by Irvin [5]

a) b)

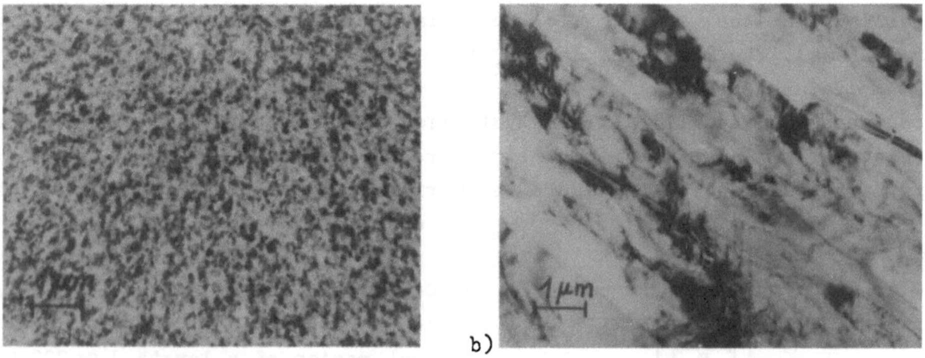

Fig.3. Dark-field TEM micrographs of typical crystal structure after ELPE of a lp-CVD-Si-layer (a) and a Si^+ implanted lp-CVD-Si-layer (b)

Fig.2 shows a large spread of the μ_s-values measured for a fixed carrier concentration. This can be explained by the fact that ρ_s and μ_s are functions of the angle α between the direction of the crystal lamellae and the current direction in the test structure (Fig.4). If α increases the number of grain boundaries intersecting a given current direction becomes larger and higher values of ρ_s are obtained. For a fixed angle α the spread of ρ_s and μ_s additionally is a function of the size of the crystal lamellae. A comparison with the data for polycrystalline silicon with the same sheet parameters shows that electrical properties comparable to that of polycrystalline silicon are obtained if the direction of the crystal lamellae is perpendicular to the current direction. If the lamellae are parallel to the current direction the electrical properties of the layer are not significant influenced by the grain boundaries [6]. For decreasing α the increase of μ_s is caused by the elongated size of the crystal lamellae formed by ELPE. In the case of the best conditions ($\alpha \approx 0$) for $N_s \geq 2.5*10^{19}$ cm^{-3} the measured mobility amounts about 90% of that of monocrystalline silicon [5]. At these high N_s nearly identical μ_s for samples implanted befor or after the ELPE process are measured.

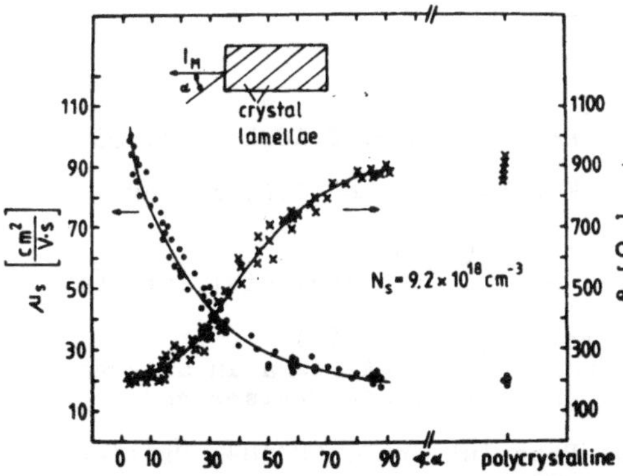

Fig.4. μ_s and ρ_s as a function of the angle α between the direction of crystal lamellae and the current direction in the test structure
• Hall mobility μ_s x sheet resistivity ρ_s

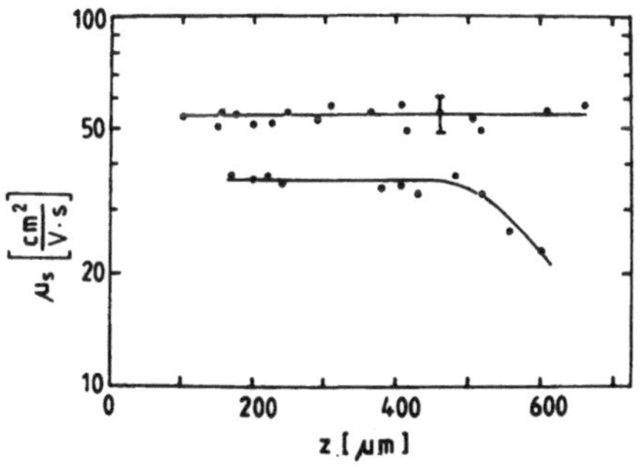

Fig.5. Hall-mobility μ_s as a function of the distance z from the ignition point
• P^+- implanted
o As^+- implanted

Fig.6. Bright-field TEM micrograph of a region after ELPE showing a polycrystalline region at the edge

Fig.5 shows μ_s for P^+ or As^+ implanted layers with $N_s = 2.5*10^{19}$ cm^{-3} as a function of the distance z from the ignition point (samples implanted before ELPE).The direction of the crystal lamellae is equal to the current direction (α = 0) in the test structure. For the P^+ implanted samples μ_s proves to be nearly independent of the distance z within a large distance range ($\approx$ 650 μm). But in the As^+ implanted structures μ_s decreases and ϱ_s increases correlated with a decrease of the crystallite size [7]. For distances z > 450 μm the decrease of the crystallite size and of μ_s becomes especially significant.

All the test structures to measure the electrical properties as a function of the distance z show polycrystalline regions ($d_{poly} \approx 2\,\mu$m) at the edges of the structures (Fig.6). For the P^+-implanted samples this polycrystalline region is the reason of the low μ_s compared to the best results of μ_s ($\alpha \neq 0$) in Fig.2.

[1] Geiler, H.-D., E.Glaser, G.Götz, and M.Wagner: J.Appl.Phys. 59, 3091 (1986)
[2] Auvert, G., D.Bensahel, A.Perio, V.T.Nguyen, and G.A.Rozgonyi: Appl.Phys.Lett. 39, 724 (1981)
[3] Köster, U., phys.stat.sol. (a) 48, 313 (1978)
[4] Wagner, M., H.-D.Geiler, G.Andrä, and G.Götz: phys.stat.sol. (a) 83, K1 (1984)
[5] Irvin, J.C., Bell.Syst.Techn.Journ. 41, 387 (1962)
[6] Nishimura, T., K.Sugahara, S.Kusunoki, and Y.Ahasaka in Extended Abstracts of the 17th Conference on Solid State Devises and Materials, Tokyo, 1985, p.147
[7] Andrä, G., E.Glaser, and G.Götz: Proc.Int.Conf. EPM'84, Dresden, ed. by K.Hennig (Dresden 1985) p.492

FORMATION OF BURIED SILICON NITRIDE AND OXYNITRIDE LAYERS IN SILICON BY ION BEAM SYNTHESIS

W. Skorupa, K. Wollschläger, M. Voelskow (1); H. Bartsch (2); J. Albrecht, J. Götz (3)

The formation of buried compound layers by ion beam synthesis to produce SOI (silicon on insulator)-structures for modern VLSI-systems has become one of the most promising methods. Recently, we concentrated our research work on the formation of buried silicon nitride and oxynitride layers /1-3/. Some new activities in this field will be shortly reported in this paper.

Most investigations were done at silicon substrates (n, ⟨100⟩, 3-5Ωcm) which were implanted at an energy of 330 keV and a target temperature of 500 °C with nitrogen doses in the range of $0.9\text{-}1.5 \cdot 10^{18}$ cm^{-2} and then annealed at 1200 °C, 5 h, N_2 /1/. By implanting a dose of $1.2 \cdot 10^{18}$ cm^{-2} in this way a stoichiometric silicon nitride layer with a silicon top layer nearly free of defects will be produced. Previously, the nitrogen profile was deduced by RBS. To reach profiles with higher depth resolution a scanning AES probe (beam diameter ≤ 0.5 μm) was used to evaluate the profile at a bevel (≤1°). The advantage of this method is to avoid destruction of the chemical compound by sputtering as in conventional AES profiling. In Fig. 1 profiles measured at a nearly stoichiometric layer before and after annealing are shown. Both, after implantation and annealing a layered structure of the silicon nitride compound (0.4 - 0.7 μm) is visible which is more distinct after the anneal step. An overstoichiometric peak as found by RBS in the center of such layers after annealing /1/ was not seen. This is due to outgassing of nitrogen at the bevel surface. Furthermore, after annealing the steepness of the profile tails between silicon nitride and silicon is twice as high as before and comparable to that one between a silicon nitride surface layer deposited by rf-sputtering and

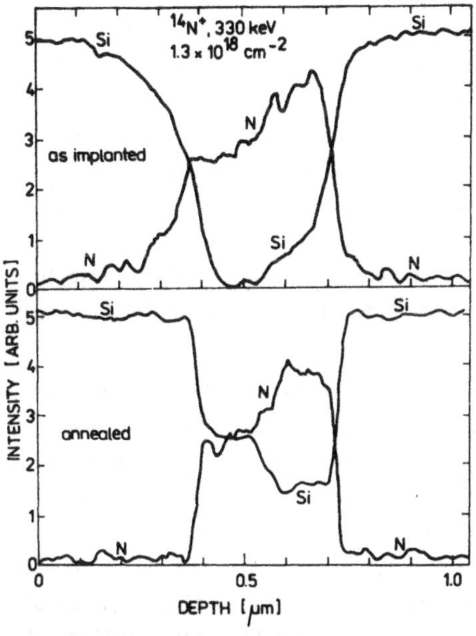

Fig. 1
AES depth profiles of silicon after high dose nitrogen implantation and annealing at 1200 °C, 5 h, N_2

(1) Central Institute for Nuclear Research Rossendorf, Academy of Sciences of the GDR, PF 19, DDR-8051 Dresden

(2) Institute for Solid State Physics and Electron Microscopy, Academy of Sciences of the GDR, Halle

(3) ZMD Dresden, GDR

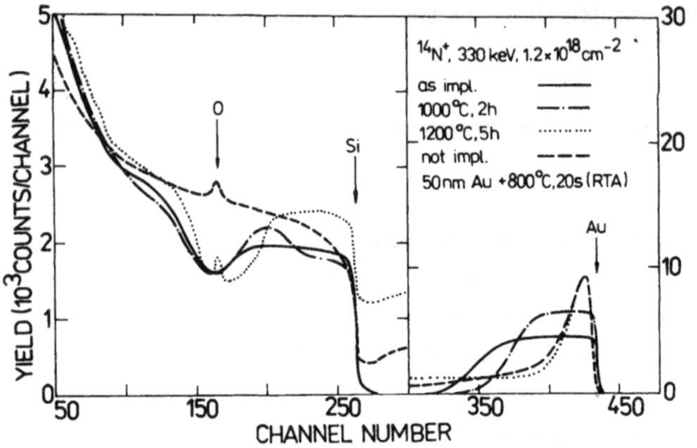

Fig. 2

RBS spectra after high dose nitrogen implantation, annealing at different temperatures, gold evaporation and gold diffusion by rapid thermal annealing

the silicon top layer. This surface layer was used as a surface marker.

Recently it was shown that in epitaxially thickened silicon layers above buried silicon nitride relatively high minority carrier lifetimes were found /4/. From further lifetime measurements over a wide nitrogen dose range it was suggested that heavy metal gettering in the buried layer system may be the cause for this effect /5/. For this end and also to prove the diffusion inhibiting properties of silicon nitride we investigated the diffusion of gold in such systems. Firstly, gold was evaporated and diffused into an as-implanted (a), a 1000 °C/2 h annealed (B) and a 1200 °C/5 h annealed (C) SOI-structure with a stoichiometric layer. From the RBS spectra in Fig. 2 it is visible that gold was redistributed only within the silicon top layer for the cases A and B. In the not implanted case and case C gold diffuses through the silicon nitride layer far into the semiconductor bulk. This is due to the amorphous silicon nitride structure in the cases A and B and the polycrystalline structure in the case C. The presence of grain boundaries in the latter case leads to diffusion along them and so no diffusion inhibition takes place. In a second experiment a lower gold amount of 10^{16} cm^{-2} was implanted and diffused. By RBS gold trapping was found immediately above the amorphous compound layer of case B where a precipitation rich dislocation network is present as deduced from XTEM investigations. For case C no trapping took place. This means that heavy metals are first of all gettered at silicon defects near to the compound layer. Annealing at 1200 °C leads to the elimination of such defects /1/ and so no or negligible gettering is possible.

In the literature it is the prevailing opinion that for the formation of SOI-structures with buried layers implantation energies $\geq$ 150 keV should be necessary. We could show that buried compound layers with monocrystalline silicon top layers may be formed already at an energy of 60 keV /2/. This should be of importance for stacking of buried layers. In connection with this work silicon oxynitride layers were also produced. They were synthesized by implanting half of the total dose of $6.2 \cdot 10^{17}$ cm^{-2}, as necessary for nearly stoichiometric compounds, with each of nitrogen and oxygen. The implantation sequence of these two elements was also varied. The resulting layer types will be further called SiNO and SiON. It is important for the understanding of the microstructure of such a layer system to know the nitrogen and oxygen profiles. RBS, being the most usual method in buried layer research, is unable to separate these profiles due to the low mass difference of nitrogen and oxygen. We have used Elastic Recoil Detection with Bragg peak-spectroscopy to evaluate the profiles /3/. In Fig. 3 this is shown for the as-implanted case

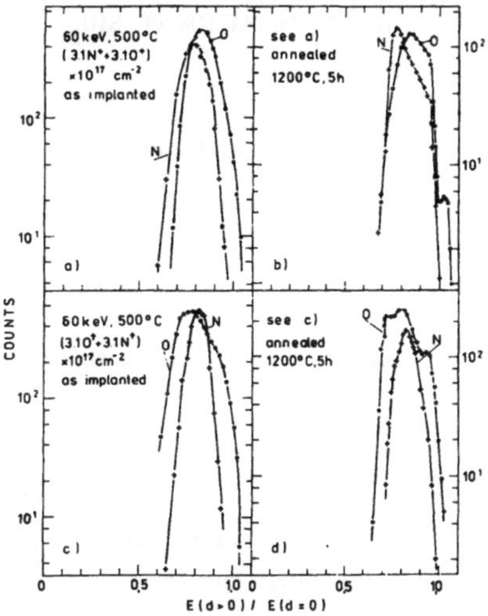

Fig. 3

Oxygen and nitrogen profiles of silicon after implantation at 60 keV and annealing. The profiles were measured by Elastic Recoil Detection (33.6 MeV $^{35}Cl^{+}$-ions) with Bragg peak-spectroscopy. The surface is located at E(d≫0)/E(d=0) = 1.0, the depth increases going left from this value (E - energy of the particles recoiled from a depth d)

and after annealing at 1200 °C/5 h. After implantation for SiNO (Fig. 3a) the nitrogen profile lies somewhat deeper which is caused by the higher stopping power of nitrogen rich silicon as compared with pure silicon. Annealing leads to a piling up of nitrogen at the back interface (Fig. 3b). For SiON the nitrogen profile is rather centered within the oxygen profile (Fig. 3c). At the surface side of the latter profile a shoulder is visible which is more distinct after annealing (Fig. 3d). Generally, after annealing the profile tails are steeper pointing to narrower interfaces. This was revealed by XTEM. These and other facts are discussed in more detail elsewhere /2,3/.

REFERENCES

/1/ Skorupa, W., K. Wollschläger, U. Kreissig, R. Grötzschel and H. Bartsch: Nucl. Instr. Meth. B19/20 (1987) 285

/2/ Skorupa, W., K. Wollschläger, M. Voelskow, K.-H. Heinig and H. Bartsch: Proc. Europ. Mat. Res. Soc. Mtg., Straßbourg, June 1987

/3/ Skorupa, W., K. Wollschläger, R. Grötzschel, J. Schöneich, E. Henschel, R. Kotte, F. Stary, H. Bartsch and G. Götz: Proc. Int. Conf. Rad. Eff. in Insulators (REI 4), Lyon, July 1987

/4/ Skorupa, W., U. Kreissig, E. Hensel and H. Bartsch: Electron. Lett. 20 (1984) 426

/5/ Skorupa, W., H. Oertel and H. Bartsch: Electron. Lett. 22 (1906) 1062

MODELING OF ^{18}O TRACER STUDIES OF THE OXYGEN REDISTRIBUTION DURING FORMATION OF SiO_2 LAYERS BY HIGH DOSE IMPLANTATION

H. U. Jäger [1], J. A. Kilner [2], R. J. Chater [2], P. L. F. Hemment [3], R. F. Peart [3] and K. J. Reeson [3]

The formation of buried SiO_2 layers by high dose oxygen ion implantation is increasingly being studied because of its expected importance in silicon-on-insulator (SOI) technology (1).

The evolution of the oxygen depth distributions during implantation from a Gaussian-like to a flat topped distribution has been modeled by several workers (2, 4). In calculating the distributions these authors take into account both the internal oxidation by diffusion of excess oxygen and those other effects such as target swelling, sputter erosion of the surface and changes in the stopping power. All of these factors become important during high dose implantation.

The tracer investigations of Kilner et al. (3) provided more detailed information on the mass transport processes occuring in silicon during the formation and growth of the buried oxide. In this experiments a high dose of ^{16}O and a lower dose of ^{18}O ions were implanted. After both implantations the depth profiles of the two oxygen isotopes were measured using secondary ion mass spectrometry (SIMS).

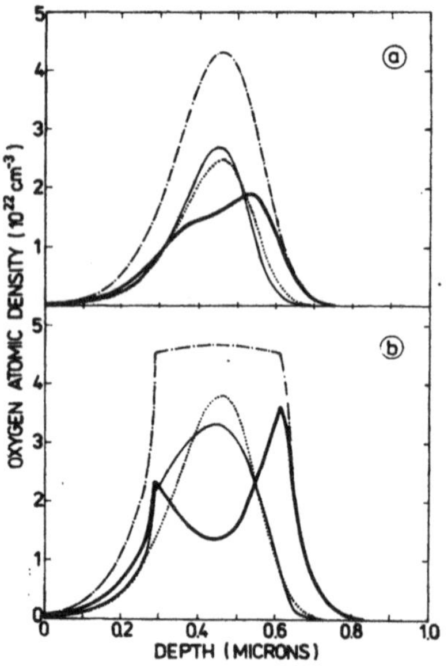

Fig. 1. Modeling of the redistribution of oxygen in silicon during two successive high dose implantations of 200keV O^+ ions. The two doses implanted are assumed here to be equal and amount to $D=0.6\times10^{18}$ cm^{-2} (a) or 1.0×10^{18} cm^{-2} (b). The oxygen depth profiles shown refer to:
(i) single implantation to dose D (--------),
(ii) final distribution of the initially implanted material (━━━━),
(iii) distribution of the secondly implanted material (———), and
(iv) total oxygen distribution after the two implantations (—·—·—).

[1]) Zentralinstitut für Kernforschung, Postfach 19, DDR-8051 Dresden, GDR
[2]) Department of Metallurgy and Materials Science, Imperial College, London, SW7 2BP. U.K.
[3]) Department of Electronic and Electrical Engineering, University of Surrey, Guildford, GU2 5XH. U.K.

It is the aim of the present contribution to examine the phenomena associated with layer formation in the light of the data of Kilner et al. For this purpose we generalized the model explained in ref. (2) by taking into account the implantation of a second oxygen isotope. In calculating the ^{16}O and ^{18}O profiles expected after the two successive implantations, one of the main conclusions made by Kilner et al. (3) is adopted: viz during implantation the incremental dose of oxygen migrates freely through the preformed SiO_2 layer and in doing so it freely exchanges with the matrix oxygen within the layer.

The first model predictions shown (fig. 1) are for the somewhat fictitious situation that the two doses implanted are equal to each other. For this case the consequences of our approach become most apparent. The results for the implantation of an understoichiometric dose are shown in fig. 1(a). Here the second implantation causes a redistribution of the initially implanted oxygen for which the concentration profile asymmetrically broadens and peaks at increasing depths. This effect is due to target swelling. The decrease of mean ion range and straggling with increasing amounts of implanted material favours the modification of this profile to an asymmetric one. When the total implanted oxygen dose is high, so that a broad buried SiO_2 layer is formed (fig. 1(b)), the computed final distributions of the two isotopes differ even more markedly. The assumption of a perfect exchange between migrating and bonded oxygen leads to the result that during the second implantation the initially implanted isotope diffuses preferentially out of

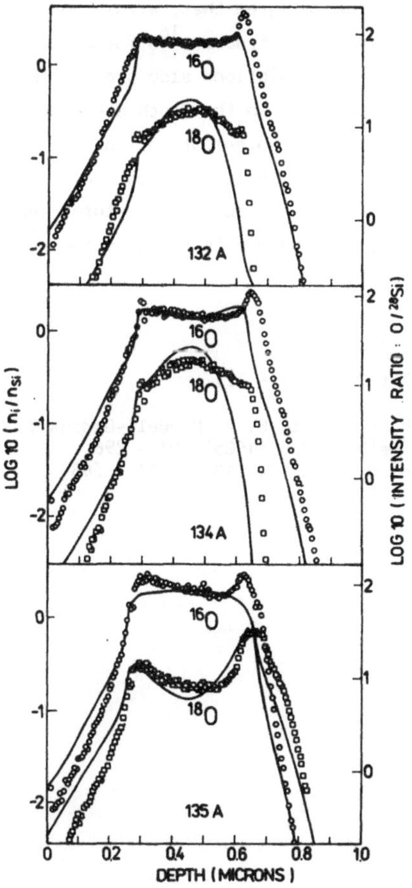

Fig. 2. ^{16}O and ^{18}O depth profiles in as-implanted samples. The SIMS profiles (right-hand scale) measured in samples 132A, 134A and 135A of Kilner et al. (3) are compared with corresponding model predictions (left-hand scale); for details of the implantations, see table 1.

Table 1. Implantation history of the reference samples (3)

Sample	First implantation			Second implantation		
	Species	Dose (cm^{-2})	Energy (keV)	Species	Dose (cm^{-2})	Energy (keV)
132A	$^{16}O^+$	1.8×10^{18}	200	$^{18}O^+$	2×10^{17}	210
134A	$^{16}O^+$	1.65×10^{18}	200	$^{18}O^+$	3.5×10^{17}	210
135A	$^{18}O^+$	4×10^{17}	210	$^{16}O^+$	1.8×10^{18}	200

the oxide layer and is accumulated at the interfaces. Only the distribution of the finally implanted isotope is peaked inside the oxide layer. Because of target swelling and changes in the stopping power the centres of the two oxygen distributions do not coincide. The ions of the secondly implanted isotope come essentially to rest in the upper part of the oxide layer, where they diffuse towards the front interface. Therefore their concentration decreases strongly near the back interface.

In fig. 2 we compare the SIMS results of Kilner et al. with our computed ratios of oxygen to silicon concentration. Examination of fig. 2 shows that a striking similarity is achieved between the calculated and measured depth profiles. The only essential exception is the peak in the SIMS data at the lower SiO_2 interfacial zone which was already stated (3) to reflect most probably the microstructure at the interface rather than chemical composition. In all cases experiment and calculation agree that, within the buried layer, the depth distribution of the initially implanted tracer (^{16}O for samples 132A and 134A, ^{18}O for 135A) shows concave curvature. The calculations also predict that the second isotope to be implanted should show convex curvature in the depth distribution in this same region. The experimental evidence is in good agreement for the ^{18}O distribution in samples 132A and 134A. For sample 135A ^{18}O was implanted first followed by ^{16}O. The ^{16}O distribution for this sample shows, if anything, concave curvature. The explanation of that is not yet clear but could possibly also be due to the changing nature of the SIMS signals in the interface region.

References

(1) Hemment, P. L. F.: Mat. Res. Soc. Symp. Proc. 53 (1986) 207 - 221
(2) Jäger, H. U.: contrib. 8.6 to this conference
(3) Kilner, J. A., R. J. Chater, P. L. F. Hemment, R. F. Peart, E. A. Maydell-Ondrusz, M. R. Taylor and R. P. Arrowsmith: Nucl. Instr. Meth. B 7/8 (1985) 293 - 298
(4) Maydell-Ondrusz, E. A. and I. H. Wilson: Thin Solid Films 114 (1984) 357 - 366

IMPROVED MODELING OF OXYGEN DEPTH PROFILES IN HIGH DOSE OXYGEN-IMPLANTED SILICON

H. U. Jäger [1])

The evolution of the oxygen depth profiles during high dose oxygen implantation into silicon has been modeled by Maydell-Ondrusz et al. (10) and Jäger (6). The basic assumptions of these two approaches are similar. Both models take into account the change of ion range profile with changing target composition, the (vertical) widening of the target by the collected oxygen and the surface sputtering. If the implanted oxygen exceeds the SiO_2 stoichiometry, initially at the profile maximum, the excess oxygen is assumed to diffuse readily to substoichiometric regions, i.e. towards surface and bulk, thereby oxidizing silicon in the interface region to SiO_2. In this way the shape of the oxygen depth profile changes with increasing dose from Gaussian-like to flat topped.

In order to describe these mass transport processes in the target, we solved one-dimensional mass continuity equations, where a source term $\gamma(z,t)$ depending on depth z and time t describes the oxygen ions that come to rest in a volume element per unit time (6). The resultant oxygen concentration profiles $n_O(z,t)$ allow to predict the critical ion dose for the formation of a buried SiO_2 layer, the thickness and depth position of this layer, and the amount of oxygen impurities in the silicon surface layer of as-implanted samples. If the model is extended by postulating an extensive exchange between diffusing and bonded oxygen in the buried SiO_2 layer, it is also possible to analyse recent ^{18}O tracer experiments (9, 4). In a foregoing investigation (7) we have done this using the experimental data reported by Kilner et al. (9) and have found that a good agreement could be generally attained, but our theoretical profiles had obviously an incorrect slope in the lower SiO_2/Si interface region, i.e. at the bulk side of the buried SiO_2 layer. The same problem arises if such model predictions are compared with the new data of Chater et al. (4) (fig. 1, dashed lines).

One crucial parameter of our phenomenological model which can affect this slope is the function $\gamma(z,t)$. This function is the ion deposition rate per volume element. It depends on time only due to the changing target composition and can be built up using ion range data from literature. Up to now it has been assumed to be a simple Gaussian for the pure silicon target (i.e. at low ion doses). When a distinct buried oxide layer has been already formed it has been assumed to consist of parts of three different Gaussians which have been put together by the method of dose matching (1). The parameters needed in doing so are the effective ranges and stragglings of oxygen ions in silicon and SiO_2. They have been taken from ref. (2).

It is well known that the tails of theoretical ion range distributions are essentially changed if higher moments of these distributions are taken into account. For example, the third moment (skewness) of a 200keV O^+ distribution in Si or SiO_2 is predicted (2) to be approximately -1.0. For an implantation energy of 1MeV which has been also used in studying buried oxide formation (11) the skewness would be even about -2.15. Such large neg-

[1]) Zentralinstitut für Kernforschung, Postfach 19, DDR-8051 Dresden, GDR

Fig. 1. Oxygen depth profiles obtained after a 200keV $^{16}O^+$ implantation to a dose of 1.8×10^{18} cm^{-2} and after a subsequent 250keV $^{18}O^+$ implantation to a dose of 3×10^{17} cm^{-2}. The SIMS profiles given refer to the right-hand scale and have been reported by Chater et al. (4) (sample 139A). They are compared with theoretical results obtained in the framework of our former (- - -; refs. (6, 7)) and present (——) approach.

ative skewness values will have the effect of increasing the function $\tau(z,t)$ in the surface region ($z \rightarrow 0$) and of decreasing it on the bulk side ($z \rightarrow \infty$).

It is the aim of the present contribution to show how a more suitable ion range distribution affects the computed oxygen depth profiles. For this purpose we repeat essentially our foregoing investigations (6, 7), but use now Pearson type IV distributions instead of Gaussians for constructing the function $\tau(z,t)$. The effective range, straggling and skewness values are taken from the new tables of Burenkov et al. (3). These data differ from those given in ref. (2) since Burenkov et al. introduced an improved, empirically determined electronic stopping cross section for light ions such as oxygen. All other parameters of our calculations are determined as indicated previously (6).

Fig. 1 shows the theoretical oxygen depth profiles computed in the framework of our foregoing and present approaches, respectively, and compares them with experimental results. Fig. 1 (and also fig. 2 of ref. (8)) demonstrates that the introduction of Pearson type IV ion range distributions leads to a substantially better agreement of theoretical predictions with experimental data. The slope of the oxygen concentration in the lower SiO_2/Si interface region is increased in such a way that it coincides practically with experiment. The simultaneous decrease of the slope near the upper Si/SiO_2 interface, i.e. in the surface region, is altogether still in line with the experimental data.

Thus we can conclude that our present phenomenological model reproduces the oxygen depth profiles of as-implanted samples fairly well. The discrepancies, which appeared originally (7) in the wings of the distributions and which have been mentioned also in a recent review of Hemment (5), can be removed if more realistic ion range distributions (3) are used.

References

(1) Amaratunga, G. A. J., K. Sabine and A. G. R. Evans: IEEE Trans. Electron. Devices ED-32 (1985) 1889 - 1890

(2) Burenkov, A. F., F. F. Komarov, M. A. Kumakhov and M. M. Temkin: Tablitsy Parametrov

Prostransvennogo Raspredeleniya Ionno-implantirovannykh Primesei, Izdatel'stvo BGU im. V. I. Lenina, Minsk 1980
(3) Burenkov, A. F., F. F. Komarov, M. A. Kumakhov and M. M. Temkin: Prostranstvennye Raspredeleniya Energii, Vydelennoi v Kaskade Atomnykh Stolknovenii v Tverdykh Telakh, Energoatomizdat, Moskva 1985
(4) Chater, R. J., J. A. Kilner, P. L. F. Hemment, K. J. Reeson and J. R. Davis: Nucl. Instr. Meth. B 19/20 (1987) 290 - 293
(5) Hemment, P. L. F.: Mat. Res. Soc. Symp. Proc. 53 (1986) 207 - 221
(6) Jäger, H. U.: Nucl. Instr. Meth. B 15 (1986) 748 - 751
(7) Jäger, H. U., J. A. Kilner, R. J. Chater, P. L. F. Hemment, R. F. Peart and K. J. Reeson: to be published in Thin Solid Films
(8) Jäger, H. U., J. A. Kilner, R. J. Chater, P. L. F. Hemment, R. F. Peart and K. J. Reeson: contrib. 8.5 to this conference
(9) Kilner, J. A., R. J. Chater, P. L. F. Hemment, R. F. Peart, E. A. Maydell-Ondrusz, M. R. Taylor and R. P. Arrowsmith: Nucl. Instr. Meth. B 7/8 (1985) 293 - 298
(10) Maydell-Ondrusz, E. A. and I. H. Wilson: Thin Solid Films 113 (1984) 357 - 366
(11) White, A. E., K. T. Short, J. L. Batstone, D. C. Jacobson, J. M. Poate and K. W. West: Appl. Phys. Lett. 50 (1987) 19 - 21

THEORETICAL INVESTIGATION OF THERMALLY ACTIVATED AND ION BEAM INDUCED LATERAL SOLID-PHASE EPITAXY

H.-J. Müller and K.-H. Heinig[1)]

I - INTRODUCTION

Because of the low transformation temperature as low as about 600°C, the solid-state transformation is of increasing interest for 3-d integrated ULSI circuit production. A promising technique for the fabrication of monocrystalline Si layers in SOI-technology is the so-called lateral solid-phase epitaxy (LSPE) of thin amorphous Si-layers, starting from a seed window within the underlying SiO_2 layer (formed e.g. by LOCOS) /1/.

The LSPE is limited by nucleation and growth of randomly oriented crystallites (known as primary crystallization PC) ahead of the LSPE frontier. When the epitaxial frontier does impinge with such a growing crystallite, a local stopping of the LSPE is induced. If the density of crystallites becomes large enough, the local stopping leads to a complete blocking of the growing LSPE frontier. The fundamental characteristic of the LSPE is the free epitaxial growth length from the seed to the region of blocking (LSPE length).

Normally, the solid phase transformation is induced by furnace annealing for some hours at low temperatures (about 600°C). Both the LSPE and the PC are thermally activated. The growth rate of the LSPE is comparable to the radial growth velocity of the poly-Si grains /2/. Therefore, the mean LSPE length is of the order of the mean grain radius. An enhancement of the ratio (LSPE length)/(mean grain radius) can be achieved by a strong increase of the LSPE growth rate with respect to the poly-Si growth velocity. Such an increase can be realized, if the LSPE is induced by a laterally scanned focused ion beam (LSFIB) combined with a thermal activation at 500°C /3/. The microscopic growing conditions within such a LSFIB region are not yet well understood, but an estimation of the upper limits of the macroscopic LSPE length caused by the competitive process of PC can be given.

The aim of this paper is to present results of a theoretical investigation of the LSPE, carried out by the help of computer experiments.

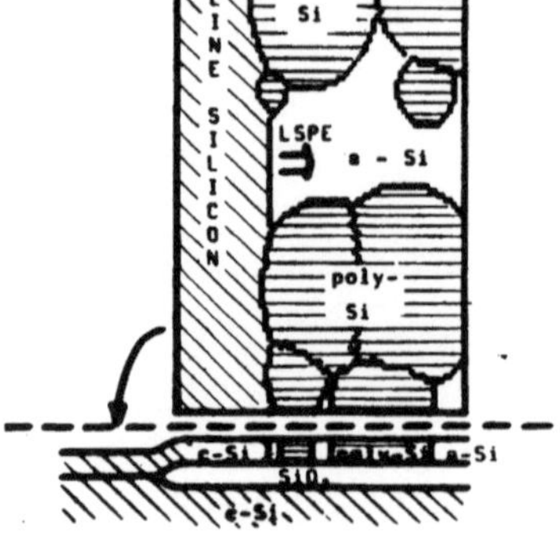

Fig. 1: computer plot during the transformation process together with a scheme of the layer system and the seed window; the underlying substrate acts as a seed; the competition between LSPE and PC can be seen.

---

1) Zentralinstitut für Kernforschung, Rossendorf, DDR-8051 Dresden, PF 19, GDR

II - MODEL

A semiempirical computer simulation (CS) has been proven to be a useful way for a comprehensive study of static and kinetic features of the polycrystalline formation /4/. The principles of that CS have been extended to describe the complex competition between PC and LSPE in thin films /5/. The LSPE has been assumed to start from a linearily shaped seed. In Fig. 1 a snap shot picture of the transformation is given.

The present CS is gouverned by three fundamental parameters, which are the rates for

(i) the nucleation of randomly oriented crystallites J_n

(ii) the radial growth velocity of the nucleated grains V_p and

(iii) the epitaxial growth rate of a monocrystalline Si layer V_e.

The values of these macroscopic parameters are determined by the external conditions during the transformation. Thus it becomes possible for the first time, to simulate a wide range of different external conditions (temperature, doping, layer quality or beam treatment) within a unified formalism only by varying these three parameters. It will be demonstrated at two examples in the following section.

III - RESULTS

The computer experiments have been carried out in two steps. At first, to select basic features of the LSPE, calculations were made at different ratios between nucleation and growth of the PC. These ratios are controlled by the transformation temperature (see /4/). To realize sufficient statistics, at every choosen temperature 10 transformations have been simulated.

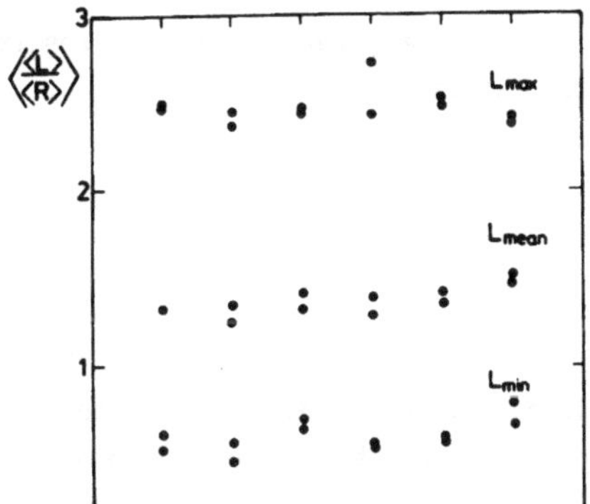

Fig. 2: calculated minimal, mean, and maximal LSPE length vs. temperature

undoped thermally activated LSPE

Thermal activation influences the LSPE as well as the grain growth. Their rates can be assumed to be equivalent (V_e=V_p). Otherwise, the temperature influences the ratio of nucleation and growth. Its dependence on the transformation temperature results in a variation of the absolute value of the epitaxial length and the mean grain radius R_p of the formed polycrystalline structure.

The main result of the investigation of thermally activated LSPE consists in the statement that the ratio between the epitaxial length (the averaged minimal L_{min}, maximal L_{max} as well as the mean L_{mean} length) and the corresponding mean radius of poly-Si is <u>independent of the transformation temperature</u> (Fig. 2). That means that the ratio is not influenced by the values of the nucleation and the growth rate. These ratios have been estimated to be

$$L_{min}/R_p = 0.5 \quad ; \quad L_{max}/R_p = 2.5 \quad ; \quad L_{mean}/R_p = 1.3 \; . \qquad (1)$$

The sligth increase of the dots in Fig. 3 at higher temperatures is due to the bad geometrical resolution of the CS in this temperature range.

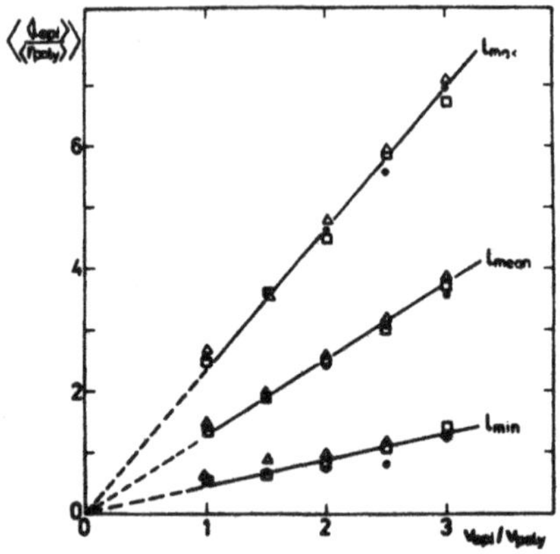

With respect to the found scaling behaviour it is obvious that the characterization of the LSPE length should be based on the mean grain radius R_p. It has been established to be the natural length unit of the LSPE.

<u>ion beam induced LSPE</u>

It can be assumed that the LSPE growth rate V_e is equal to the velocity of the LSFIB,

Fig. 3: influence of an enhanced LSPE growth rate on the normalized LSPE length

which is swept perpendicular to the LSPE frontier /3/. The width of the LSFIB has been neglegted. For this reason, no influence of the ion beam on the nucleated grains has been included into our model.

To simulate different sweep velocities, the ratio of V_e/V_p has been varied. The results are given in Fig. 3. Three features are worth to be discussed:

(i) the relation between V_e/V_p and L/R_p is independent of the temperature,

(ii) the LSPE length changes linearly with the LSPE growth rate, and

(iii) the maximal, the mean, and the minimal LSPE length are influenced differently.

The slopes of the curves in Fig. 3 are given by

$$m_{min} = 0.43 \quad ; \quad m_{mean} = 1.25 \quad ; \quad m_{max} = 2.32 \qquad (2)$$

A quantitative comparison with experimental data is not possible, because in experiment the LSPE is at present limited not only by PC, but also by strong crystal damage induced by the LSFIB /3/.

The theoretical results give a qualitative insight in the competition of LSFIB induced LSPE and PC. The basic role of the ratio between the LSFIB seep velocity and the radial grain growth velocity for the enhancement of the LSPE length has been demonstrated.

<u>References</u>

/1/ H.Ishiwara, A. Tamba and S. Furukawa: Appl. Phys. Lett. 48 (1986) 773-775.
/2/ H. Yamamoto, H. Ishiwara and S. Furukawa: Jap. J. of Appl. Phys. 25 (1986) 667-672.
/3/ T. Kanayama, H. Tanoue and M. Komuro: Jap. J. of Appl. Phys. 26 (1987) L84-L86.
/4/ H.-J. Müller and K.-H. Heinig: Nucl. Instr. & Meth. 22 (1987) 524-527.
/5/ H.-J. Müller and K.-H. Heinig: Proc. E-MRS-Conf. 1987, Strasbourg, to be published.

MATHEMATICAL MODELLING OF THE CAPACITANCE CHARACTERISTICS AND THE THRESHOLD VOLTAGE OF A MOS STRUCTURE ON A THIN SILICON SUBSTRATE

N. Sieber[1], G. Otto[1]

A sucsessfull application of SOI films in the microelectronic requires a knowledge of the electrical properties of the recrystallized silicon (r-Si) as well as of the interface r-Si/SiO_2. MOS structures fabricated in thin silicon layers on SiO_2 have electrical properties which are typically influenced by the charge coupling between front and back silicon interfaces /1,2 /. Thus, a precise determination of electrical properties of these structures by CV measurements such as free carrier concentration, generation lifetime, threshold voltage etc. can be affected by the silicon thickness and the charge at the interface recristallized silicon/underlying insulator.

It is the purpose of this paper to give a general solution for the dependence of the space charge capacitance characteristic and the threshold voltage of a thin - based MOS structure on various properties of the back interface.

The semiconductor capacitance can be calculated using the approximate charge analysis /3/. For the case of a p-type semi-infinite semiconductor with a fully ionized homogeneous impurity concentration N_A under equilibrium condition the normalized electric field is given by

$$F(u,u_F) = \left[e^{u_F}(e^{-u}+u-1)+e^{-u_F}(e^{u}-u-1)\right]^{1/2} \qquad (1)$$

where u and u_F are the normalized electrostatic and bulk Fermi potentials. In the case of thin Si films where the thickness is comparable with the Debye length L_D of the semiconductor the normalized electric field is strongly dependent on the potential of the back surface u_{S2}. For the three potential distributions shown in Fig. 1 the normalized electric field can be writen in the form

$$F_A(u,u_o,u_F)=\left[e^{u_F} (e^{-u}-e^{-u_o} +u-u_o)+e^{-u_F} (e^{u}-e^{u_o} -u+u_o)\right]^{1/2} \qquad (2)$$

$$F_B(u,u_{S2},F_{S2},u_F)= \left[e^{u_F} (e^{-u}-e^{-u_{S2}} +u-u_{S2})+e^{-u_F} (e^{u}-e^{u_{S2}} -u+u_{S2}) + F_{S2}^2\right]^{1/2} \qquad (3)$$

$$F_C(u,F_o,u_F) = \left[e^{u_F} (e^{-u}+u-1)+e^{-u_F} (e^{u}-u-1)+ F_o^2 \right]^{1/2} \qquad (4)$$

Here u_o is the zero field normalized potential, F_{S2} is the normalized electric field of the back surface and F_o is the normalized electric field for u=0 (x=d_1, see Fig. 1).

For a given semiconductor thickness d_{Si} the values u_o, F_{S2}, F_o can be determined by fitting the integrals

---

1 - Central Institute for Nuclear Research Rossendorf, Academy of Sciences of GDR, PSF 19, DDR-8051 Dresden

$$\left| \int_{u_o}^{u_{S1}} du/F_A(u,u_o,u_F) + \int_{u_o}^{u_{S2}} du/F_A(u,u_o,u_F) \right| = d_{Si}/L_{Di} \tag{5}$$

$$\left| \int_{u_{S2}}^{u_{S1}} du/F_B(u,u_{S2},F_{S2},u_F) \right| = d_{Si}/L_{Di} \tag{6}$$

and

$$\left| \int_{u_{S2}}^{u_{S1}} du/F_C(u,F_o,u_F) \right| = d_{Si}/L_{Di} \tag{7}$$

where L_{Di} is the Debye length of an intrinsic semiconductor.

The calculation of the low and high frequency capacitance was performed for the case with accumulated back surface ($u_{S2} \geqslant 0$). With the electric field F determined using (1) - (4) for each surface potential u_{S1} the capacitance characteristics can be calculated in analogy to the bulk semiconductor

$$C_{sc} = C_{sco} \frac{dF}{du_{S1}} , \tag{8}$$

where C_{sco} is the Debye capacitance.

The relationship between the voltage applied to the gate V_G and the surface potential u_{S1} is given by

$$V_G = \frac{kT}{q} \left[u_{S1} + \mathrm{sign}(u_{S1}) \frac{\varepsilon_{Si} \cdot d_{ox}}{\varepsilon_{ox} \cdot L_{Di}} \cdot F \right] \tag{9}$$

Here ε_{Si}, ε_{ox} are the permittivities of the semiconductor and of the oxide, d_{ox} is the oxide thickness. In (9) we do not take into account the existence of the work function difference, oxidecharge and surface states at the front and back interface.

The expression for the high frequency capacitance of the thin semiconductor can be approximated in the three ranges of u_{S1} as given below

a) Accumulation range ($u_{S1} \leqslant 0$)

$$C_{sc} = C_{sco} \left[\sinh(u_{S1} - u_F) + \sinh u_F \Big/ F(u_{S1}, u_F) \right] , \quad L_D \leqslant 2d_{Si} \tag{10}$$

$$C_{sc} = C_{sco} \left[dF_A(u_{S1}, u_o, u_F) / du_{S1} \right] , \quad L_D > 2d_{Si} \tag{11}$$

where L_D is the Debye length of the accumulation layer

b) Depletion range ($0 \leqslant u_{S1} \leqslant u_F + 6$)

$$C_{sc} = C_{sco} \left[\left\{ F_o \, dF / du_{S1} + 1/2 \, e^{u_F} (1 - e^{-u_{S1}}) \right\} / F_{C1}(u_{S1}, F_o, u_F) \right] , \tag{12}$$

where

$$F_{C1} = \left[e^{u_F} (e^{-u_{S1}} + u_{S1} - 1) + F_o^2 \right]^{1/2} \tag{13}$$

c) Inversion range ($u_{S1} \geqslant u_F + 6$)

$$C_{sc} = C_{sco} / \left[(C_{sco}/C_{sc})_{u_{S1} = u_F + 6} + \int_{u_1}^{u_{S1}} du/F_C(u, F_o, u_F) \right] \tag{14}$$

where u_1 is determined so that the continuous transition is achieved going from range b to range c.

The expression for the threshold voltage V_{TH} under all possible steady-state charge conditions at the back surface can be derived using equations (2), (3), (4) and (9)

$$V_{TH} = \frac{kT}{q}\left[(u_{FB} + 2\ln N_A/n_i + \mathrm{sign}(u_{S1})\ \frac{\varepsilon_{Si}\cdot d_{ox}}{\varepsilon_{ox}\cdot L_{Di}}\cdot F\right], \qquad (15)$$

where u_{FB} is the normalized front-gate flatband voltage.

The calculated low frequency and high frequency capacitance characteristics of a n-type substrate ($N_D = 10^{15}$ and $10^{16}/cm^3$ and with $u_{S2}=0$ are shown in Fig. 2, 3 and 4. The decreasing base thickness results in an increasing minimum normalized capacitance C_{min} for both the low frequency and the high frequency characteristic. For the same thickness the increase of C_{min} is inversely proportional to the substrate doping. This is due to the fact that the reduced back side field is inversely proportional to the substrate thickness and the doping concentration. The back side field affects also the width of the low frequency capacitance characteristic at the voltage corresponding to the depletion and inversion range. Consequently, for the surface potential $u_{S1}=2u_F$ the threshold voltage V_{TH} can significantly differ from the conventional ones. Indeed, the dependence of V_{TH} on the substrate thickness (Fig.5) illustrates the threshold voltage shift for different conditions at the back surface.

These results show that the determination of V_{TH} requires precise knowledge of the back side potential and of the thickness of the thin substrate. Furthermore, the determination of the doping concentration from the capacitance-voltage characteristic has to be performed in the accumulation region because C_{min} is considerably influenced by the thickness and the potential at at the back side surface of thin silicon films.

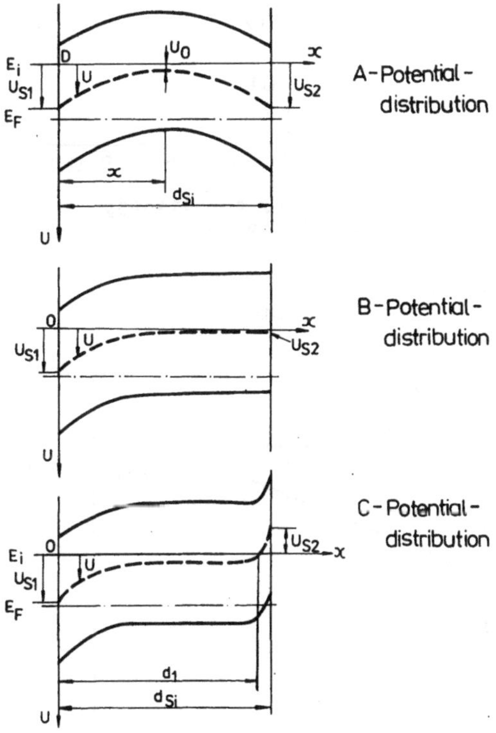

Fig.1: Three typical kinds of potential distribution in a thin semiconductor

References

/1/ Jerhot,J., Phys. stat.sol(a), 96 (1986) K 95.

/2/ Lim H.-K. and Fossum J.G., IEEE Trans.Electron.Dev. ED-30,10(1983)1244

/3/ Nicollian E.H., and Goetzberger, A., Appl. Phys. Lett. 7 (1965) 216

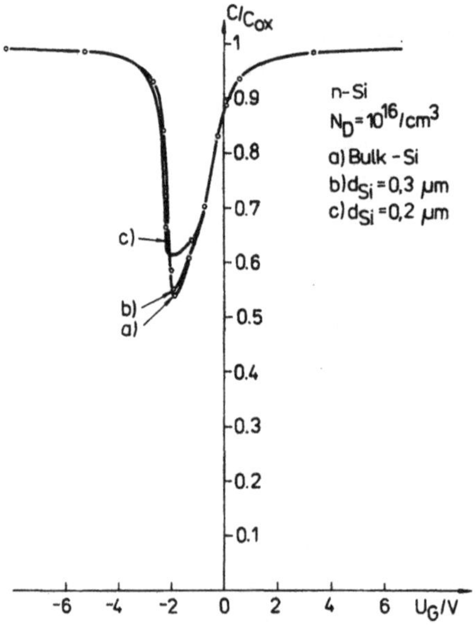

Fig. 2: Low frequency capacitance characteristic ($N_D = 10^{16}\ cm^{-3}$)

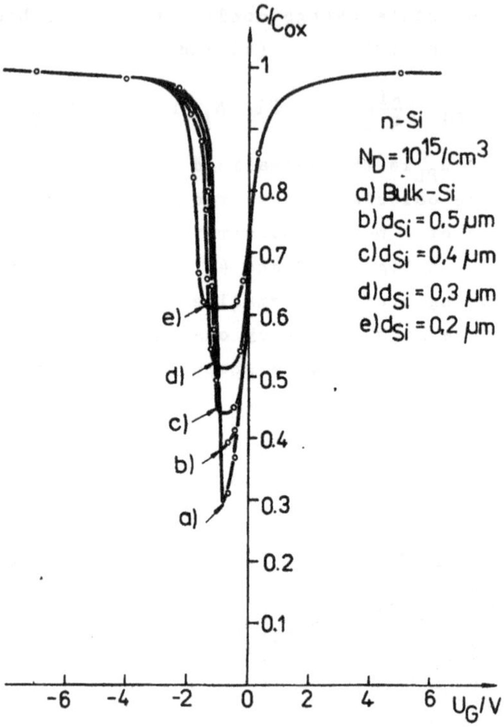

Fig. 3: Low frequency capacitance characteristic ($N_D = 10^{15}\ cm^{-3}$)

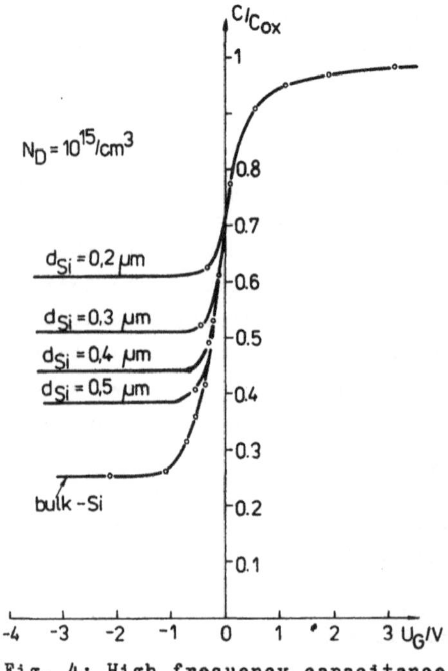

Fig. 4: High frequency capacitance characteristic ($N_D = 10^{15}\ cm^{-3}$)

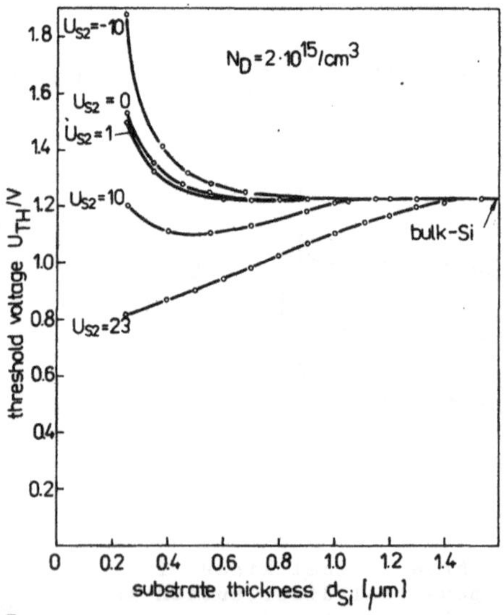

Fig. 5: Dependence of the threshold voltage on the substrate thickness for different back-side potential

INFLUENCE OF THERMAL GRADIENT ON THE RECRYSTALLIZATION OF THICK POLYCRYSTALLINE SILICON ON SiO_2

H.H.Richter*; B.Tillack*; R.Banisch*; H.Andrae**; W.Weinelt**

1. Introduction

In the last few years the creation of large area monocrystalline Si films on insulating layer (SOI) has been the subject of numerous investigations with the aim of producing device- worthy material for electronic applications /1/. Particularly thick monocrystalline Si films are of interest because it is possible to form dielectrically isolated areas, e.g. for high voltage devices.

The aim of this paper is to explain some experimental results /2/ which were received by zone melting recrystallization (ZMR) carried out using irradiation of a scanned halogen lamp or a cw CO_2-laser. The application of the two irradiation sources allows a large variation of the melting zone geometry which has a great influence on the thermal gradient at the liquid/ solid interface. As consequence, this quantity seems to be significant for the origin of defects in the recristallized film like experimental results demonstrate. This arguments are supported by simple mathematical model calculations.

Tab.1 compares experimental parameters which influence the thermal gradient at the solid/ liquid interface during the ZMR.

Tab.1 Experimental parameters

	laser	halogen lamp
minimum width of the melting zone	80 µm	1 mm
preheating temperature	~ 1300 K	~ 1550 K
scan velocity	0,2...1,0 mms^{-1}	0,5...4 mms^{-1}

Fig.1 shows the sample structure applied for the experiments and for the mathematical calculations.

2. Remarks to the mathematical model

The numerical results are based on FDM approximations of the following version of a mathematical model describing phase transition processes:

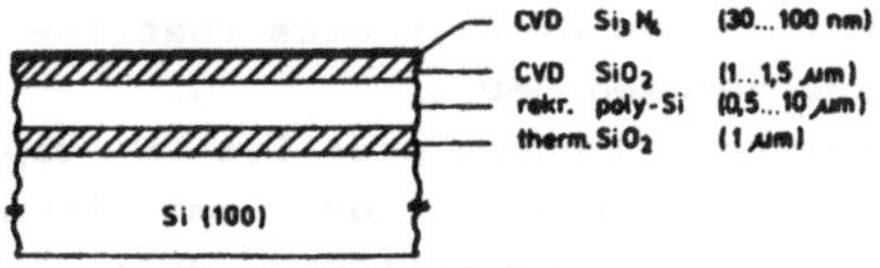

Fig.1 Sample structure

* Institut fuer Halbleiterphysik der Akademie der Wissenschaften der DDR, Frankfurt (Oder) 1200, DDR, PF 409.

** Technische Universitaet Karl-Marx-Stadt, Karl-Marx-Stadt 9010, DDR, PF 964.

- The temperature and (liquid) phase fraction fields are determined by a system of two partial differential equations (heat conduction and phase transition equations).
- The instationary temperature field is one- dimensional (in the space variable corresponding to the scan velocity direction), i.e. this temperature is to be understood as the mean value with respect to the thickness of the Si film.
- The enthalpy- type formulation of the heat balance contains irreversible phase transitions due to undercooling.
- The second differential equation determines (locally) the rate of the variation of the phase fraction, where values $v \in (0,1)$ of this quantity correspond to mushy (2-phase) regions.
- Convection due to hydrodynamics in the melting zone and the occurence of strains in the solid phase due to thermal elast(oplast)icity are not considered.

For the numerical analysis and related results see for instance /3/.

3. Results and discussion

ZMR of thick (up to 10 µm) polycrystalline Si on SiO_2 without crystallization seed gives Si films with very large monocrystalline areas separated by grain boundaries (Fig.2).

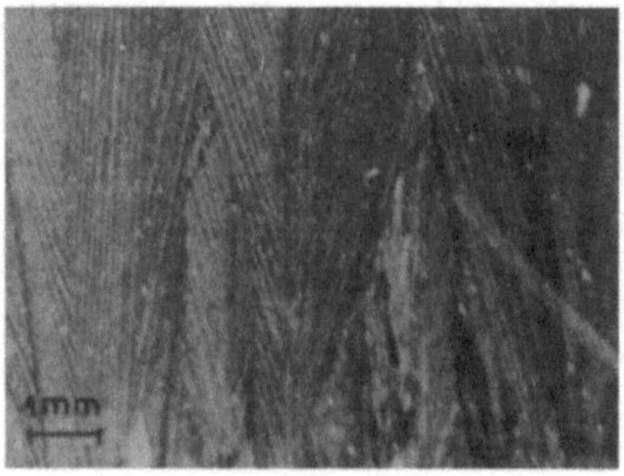

Fig.2 Optical micrograph of a recrystallized Si film, 10 µm thick (Schimmel etched)

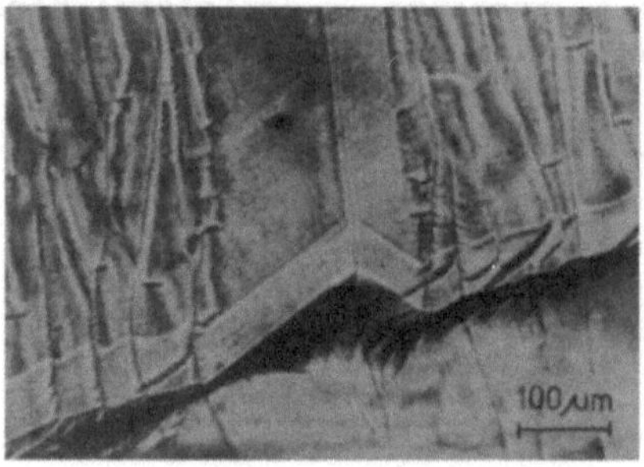

Fig.3 Surface of a recrystallized area showing the facetted interface, film thickness 10 µm

The main defects are subgrain boundaries (SGB's) within the grains. The detailed mechanism of the formation of SGB's is not yet fully understood. Fig.3 shows the origin of SGB's at the interior corners of the facetted solid/ liquid interface and demonstrates that the size of the facets determines the spacing between SGB's.

Consider the calculated temperature and phase profiles in dependence on the film thickness for laser as well as halogen lamp irradiation (Fig.3). The results can be discussed in the following way:

During the ZMR with halogen lamp there is a wide range where the temperature is approximately equal to the melting temperature, i.e. in this area both phases exist simultaneously. By increasing the Si film thickness the maximum temperature decreases and thus the temperature gradient at the melting front, too. During the laser ZMR the higher thermal gradient at the interface is reflected by a very sharp temperature profile, i.e. the 2- phase range is very narrow. Numerical and experimental results show a good correspondence to each other.

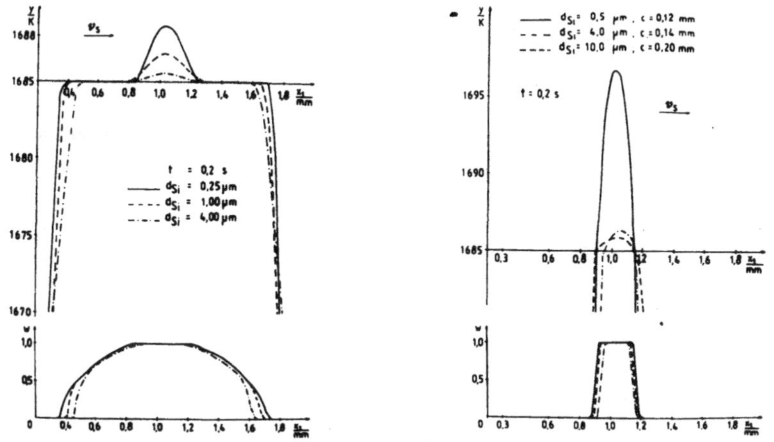

a) b)

Fig.4 Calculated temperature and phase profiles for different film thicknesses, w= fraction of the liquid phase, w= 0 solid, w= 1 liquid; a) Halogen lamp b) Laser

Based on the concept of the facetted interface /4/ one can conclude that these facets must be orginated in the 2- phase region. Therefore an extended 2- phase range must cause great spacings between two adjoining SGB's. That also means: Great spacings can only arise if a sufficient large 2-phase region exists. Fig.5 shows the experimental confirmation of this conclusions. For samples irradiated by halogen lamps the spacing between two SGB's is 2 to 3 times larger as for samples of comparable thickness recrystallized with laser.

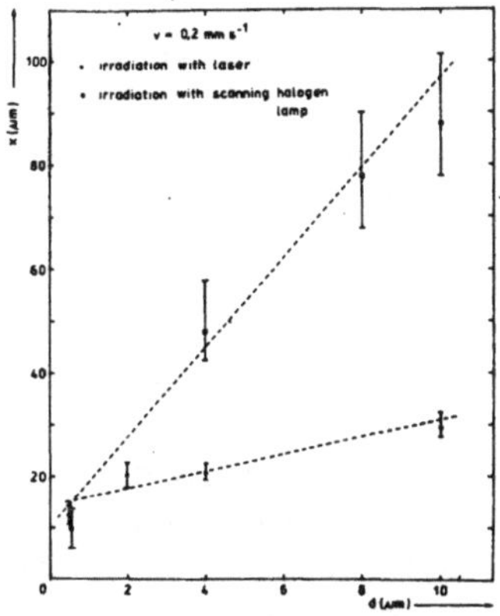

Fig.5 Spacing (x) between SGB's as function of film thickness (d)

In summary, the simple mathematical model can explain some experimental results qualitatively. Other calculated temperature profiles in dependence on various quantities will be published elsewhere /5/.

References

/1/ Jastrzebski, L.: RCA Rev. 44 (1983) 250.

/2/ Tillack, B.; H.H. Richter; R. Banisch : this conference.

/3/ Andrae, H.; U. Streit; W. Weinelt: Wiss. Inf. THK 60a, Karl-Marx-Stadt.

/4/ Geis, M.W. et. al.: J. Electrochem. Soc. 129 (1982) 2812.

/5/ Richter, H.H.; H. Andrae; B. Tillack; W. Weinelt: in preparation.

ELECTRICAL CHARACTERIZATION OF THICK SOI-FILMS

R. Banisch*, B. Tillack*, B. Hunger*, P. Barna**,
J. Gyulai**, M. Adam* und V. Schiller*

1. Introduction

Silicon-on-Insulator (SOI) is a promising material for device applications /1/. Most of the work done to date concentrated on preparation and characterization of thin (few hundred nm) SOI-structures. SOI-films in the thickness range of few µm or more are also under investigation /2, 3/, but there is only poor information on properties of this material.

In the present paper characteristics of electrical test structures prepared in thick SOI-films will be discussed.

2. Experimental

About 4 µm thick SOI-films were obtained by a ZMR (zone melting recrystallization)-process using a halogen lamp strip heater. The recrystallization was carried out without an external seed. Experimental conditions are presented elsewhere /3/. The doping level of the SOI-layer is 1×10^{15} P cm^{-3}.

In this material p- and n-channel MOS transistors were prepared by a standard p-well Al-gate CMOS-technology and also diodes by a bipolar process.

After electrical measurements, Al and SiO_2 layers were removed and the surface of the SOI-wafers was etched to delineate (sub-) grain boundaries. For comparison, test structures were prepared on a single-crystalline Si wafers.

3. Results and Discussion

Material characteristics. The singlecrystalline SOI-films showed (100) orientation /5/. Dominating defects were subgrain boundaries (SB) but, in contrast to thin SOI-films, the distance between SB-s was about 80 µm and their branching feature was lost. A detailed discussion of crystallographic characterization of this material is presented elsewhere /3, 4/.

CMOS-Transistors. Typical output characteristics of n- and p-channel tansistors with channel length of 5 µm and 4 µm, respectively, prepared in SOI-films are shown in Fig. 1. A summary of transistor properties is given in Tab. 1. The inverse subthreshold slope is about 10-15% higher than for the transistors prepared in monocrystalline Si-wafers. The leakage current is in the 10^{-12} A level for all transistors. Channel mobilities, µ, for the SOI-films are obtained from the drain current measurement /10/.

* Institut fuer Halbleiterphysik der Akademie der Wissenschaften der DDR, 1200 Frankfurt (Oder), PF 409

** Central Research Inst. Physics, H-1525 Budapest, PO Box 49, Hungary
Work partially supported by Hung. Acad. Sci. OTKA grant No. 1794

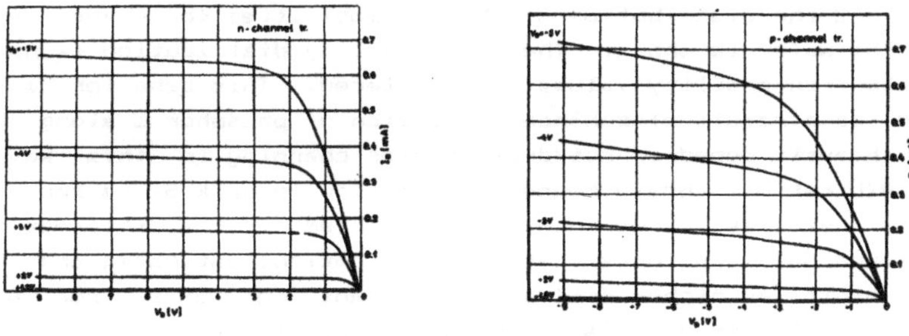

Fig.1:Output characteristics of n-and p-channel transistors

Tab. 1: Characteristics of the CMOS transistors

	SOI		"bulk"	
	p-ch.	n-ch.	p-ch.	n-ch.
inv. subthreshold slope (mV/dec)	90	165	80	145
gate breakdown voltage (V)	23,0		25,8	
threshold voltage	- 1,25	1,05	-1,00	1,25
channel mobility (cm^2/Vs)	250	550	350	850

In the SOI-material the channel mobilities reach values of 65-70% of the reference wafer.

Also in the case of thin SOI-films, carrier mobilities were reduced as a result of interaction with SB-s. In the thick SOI-films, the density of these defects is low /3/ and there were many transistors with SB-free channel region. The lowering of the mobility in this case is believed to be due to the relatively high surface roughness

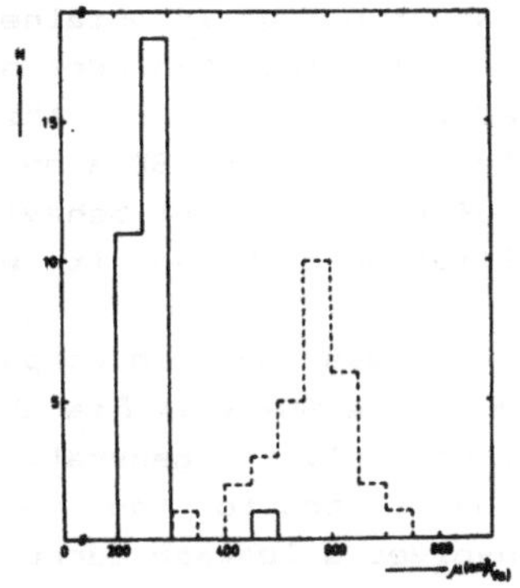

Fig.2:Channel mobility distribution; full line: p-channel transistor, dashed line: n-channel transistor

(up to 100 nm) resulting from the deposition of the poly-Si.

In Fig. 2 the distribution of channel mobilities for scans over 3 lines on the wafer is shown. The hole mobility distribution is narrow, but the electron mobility values are scattered. This behavior may be understood as a result of enhanced diffusion of phosphorus along SB-s from the heavily doped source/drain region changing in this way the effective W/L ratio. Grain boundary diffusion in bulk Si is known from the early sixties, but now this problem is of interest for recrystallized SOI-structures. Later investigations estimated that P and As /6/ exhibit an higher diffusion constant for SB's. For B, no significant influence of SB's on the diffusion constant was found /7/. REM-investigations on Sirtl-etched pn-junctions prepared in this material also show diffusion but only for P /8/.

pn-diodes measurements. Emitter-basis (EB)- and basis-collector (BC)-diodes of an npn-transistor were used. Breakdown behavior and reverse leakage current of the diodes was investigated and also the correlation of SB-s with these properties. In Fig.3 we see the influence of the SB-s on the breakdown voltage (U_B). If a SB crosses the EB-diode, the breakdown voltage is reduced and leakage current is generated well below U_B. On the other hand ,in case of BC-diodes,no influence of SB-s can be found.

The value of the U_B for the SB-free SOI area in comparison with the reference wafer is the same for EB-diodes. For BC-diodes, the U_B,is about 60% higher in the SOI-film. This finding,observed also by Tsaur is related to the buried SiO_2 layer.

The leakage current of the diodes was measured at 2 V reverse voltage. In the approximation for one-sided abrupt juction the minority carrier lifetime is calculated /11/. The results are summarized in Tab. 2.

Tab. 2: Minority carrier lifetime τ (s)

	SOI without SB	SOI with SB	reference preparation
EB-diode	3×10^{-8}–3×10^{-7}	2×10^{-11}–$1{,}5\times10^{-7}$	6×10^{-8}–4×10^{-7}
BC-diode	3–4×10^{-6}	3×10^{-7}–3×10^{-6}	3×10^{-6}–2×10^{-5}

For τ in the EB-diode a value of about 0,1 µs is obtained but a SB causes a striking decrease. In contrast for the BC-diodes no significant reduction of τ due to SB was observed. In agreement with this result CV-measurements /9/ indicate very little influence of SB-s on the minority carrier lifetime. The explanation of the different behavior of the two diodes is also the SB-enhanced diffusion of P as it was discussed earlier.

While in the EB-diode no difference in τ was found in comparison with the reference material for BC-diodes in the SOI-film τ is about 5 times lower. This finding indicates that not only the generation of minority carriers in the depletion region contributes to the reverse leakage current. A possible explanation involves a leakage current path due to the second Si-SiO_2-interface.

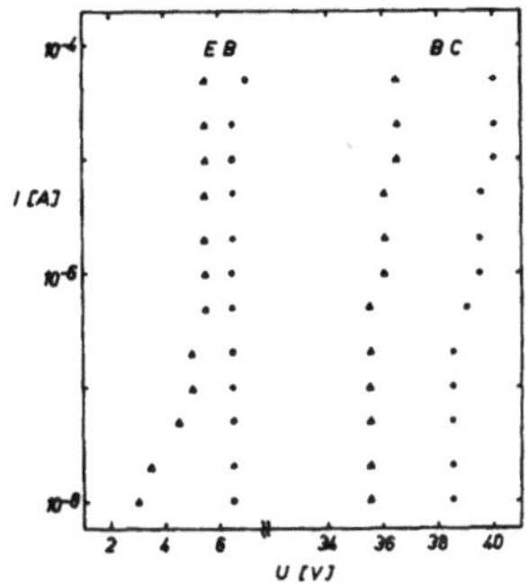

Fig.3: Breakdown behavior of the EB-and BC-diode; • without SB; ▲ with SB crossing the junction

4. Conclusions

Electrical properties of thick SOI-films prepared by ZMR were investigated. From test structures it was established that carrier mobilities, as well as minority carrier lifetime reach values comparable to these of the reference singlecrystalline Si. The subgrain boundaries themselfes do not influence these parameters very much only in combination with other effects - i. e. outdiffusion of P.

The results show clearly that this SOI-material is usable for unipolar as well as for bipolar devices. Comparision of thick films with thin SOI-structures shows, that the density of SB-s is much lower. Therefore in applications an increase of layer thickness can be recommended together with elimination of SB-s.

5. References

/1/ Jastrzebski L., RCA Review 44 (1983) 250

/2/ Pfeiffer L., G. K. Celler, T. Kovacs, Mc. D. Robinson, APL 43 (1983) 1048

/3/ Tillack B., P. Moeck, R. Banisch, R. Reinboth, H. H. Richter, M. Voelskow, J. Matthaei, F. Wieser, phys. stat. sol.(a) 94 (1986) 871

/4/ Tillack B., R. Banisch, H. H. Richter, this conference

/5/ Atwater H.A., H. I. Smith, C. V. Thompson, M. W. Geis, Mater. Lett. 2 (1984) 269

/6/ Baumgart H., H. J. Leamy, L. E. Trimble, C. I. Doherty, G. K. Celler in "Grain Bondaries in Semiconductors", Eds. G. E. Pike, C. H. Seager, H. J. Leamy, North Holland, New York 1982, p. 311

/7/ de Graff H.C. in "Polycrystalline Semiconductors" Ed. G. Harbeke, Springer Series in Solid State Science, 1985, p. 170

/8/ Borchardt S., R. Banisch, unpublished results

/9/ Tsaur B.-Y., J. C. C. Fan, M. W. Geis, APL 41 (1982) 83

/10/ Sze S.B. "Physics of Semiconductor Devices" Wiley, New York 1969, chap. 10

/11/ ibid, chap. 3

A MODEL FOR NITROGEN REDISTRIBUTION PROCESSES IN HIGH DOSE NITROGEN-IMPLANTED SILICON

E. Sobeslavsky, H. U. Jäger, W. Skorupa, K. Wollschläger[1)]

1. Introduction

SOI structures formed by high dose implantation of reactive ions, such as O^+ or N^+, into silicon have been shown to be an attractive material for microelectronic devices /2,3/. In this study we present a model of the evolution of nitrogen depth profiles during high dose N^+-implantation into silicon and compare it with RBS results from samples implanted with 20 keV N^+-ions, see Fig. 1. Nitrogen implantation differs from the better known high dose oxygen implantation. Unlike the as-implanted oxygen concentration the nitrogen profiles do not saturate if more than a critical dose N_D^{crit} has been implanted, at which the nitrogen to silicon ratio begins to exceed that of the compound.

The model used in the present investigation is a modification of a model originally developed to describe high dose O^+-implantation into silicon /1/. It takes into account volume swelling and the change in ion range profiles due to the formation of Si_3N_4. In target depths where the nitrogen to silicon ratio exceeds that of Si_3N_4, we assume that the excess nitrogen causes an additional volume swelling which broadens the profile. In this region the nitrogen is also allowed to diffuse both to the surface and to the bulk in order to form silicon nitride. We also take into account the sputtering process on the target surface.

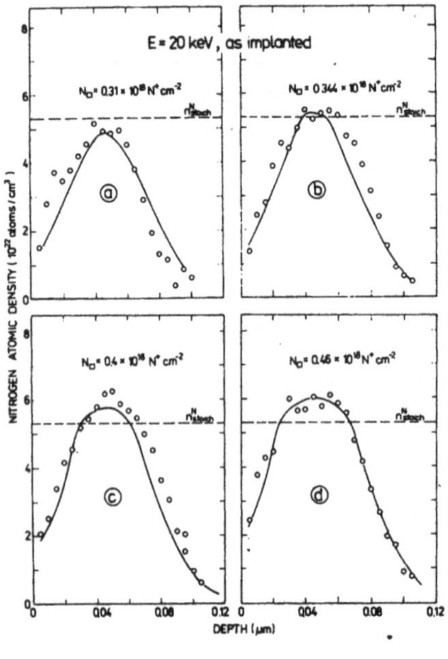

Fig. 1

Implantation of 20 keV N^+-ions into silicon. Comparison of the experimental RBS curves of the nitrogen concentration (•••) with theoretical profiles (—). The broken line (---) indicates the stoichiometric nitrogen concentration for Si_3N_4 with a density of ρ = 3.18 g cm^{-3}.

2. Physical and mathematical details

We consider the evolution of the nitrogen concentration profiles in terms of the atomic number density of implanted nitrogen $n^N(z,t)$, t is the implantation time, z means the depth in the target. The basic equations are the continuity relations both for the implanted nitrogen species and the silicon atoms of the target

$$\frac{\partial n^N(z,t)}{\partial t} + \frac{\partial j^N(z,t)}{\partial z} = \nu(z,t) \qquad (1)$$

$$\frac{\partial n^{Si}(z,t)}{\partial t} \quad \frac{\partial j^{Si}(z,t)}{\partial z} = 0 \, . \qquad (2)$$

The terms $j^N(z,t)$, $j^{Si}(z,t)$ stand for the currents in the target modifying the ion range pro-

1) Zentralinstitut für Kernforschung, DDR - 8051 Dresden, PF 19

files. The source term $\nu(z,t)$ gives the number of nitrogen atoms implanted per unit volume and time. Equ. (1), (2) together with an appropriate volume condition and the requirement that the synthesized Si_3N_4 must not segregate, allows us to express $j^N(z,t)$ in terms of $n^N(z,t)$ and to compute $n^N(z,t)$ by a step by step solution of equ. (1), cf. /1/. Here we have to distinguish two cases. The nitrogen concentration firstly reaches the stoichiometric level of the nitride, n^N_{stoich}, at a critical dose $N^{crit} \approx (4\cdot\sqrt{2\pi}\cdot A/M(Si_3N_4))\cdot \Delta R_p(N\rightarrow Si)\cdot \rho(Si_3N_4)$, where $\Delta R_p(N\rightarrow Si)$ is the straggling, $\rho(Si_3N_4)$ and $M(Si_3N_4)$ denote the density and the molecular weight of silicon nitride, respectively. $A=6.023\cdot10^{23}(Mol)^{-1}$ means the Avogadro number. In target depths where the nitrogen concentration is below n^N_{stoich} the flow density $j^N(z,t)$ can be shown to be

$$j^N(z,t) = n^N(z,t)\cdot\left[C(t) + \Delta\Omega\cdot\int_0^z dz'\,\nu(z',t)\right] \qquad (3)$$

Here we assume that every bonded atom requires a volume

$$\Delta\Omega = (1/4)\cdot\left[\Omega(Si_3N_4) - 3\cdot\Omega(Si)\right], \qquad (4)$$

where $\Omega(Si)$ and $\Omega(Si_3N_4)$ are the volumes of a Si-atom and a nitride molecule, respectively. The term $C(t)$ is determined by the boundary conditions.

In target depths where the nitrogen concentration exceeds the stoichiometry level n^N_{stoich} the flow density reads analogously

$$j^N(z,t) = n^N(z,t)\left[j^N(z_{stoich},t)/n^N(z_{stoich},t) + \Omega^N_{exc}\cdot\int_{z_{stoich}}^{z} dz' \quad (z',t)\right]. \qquad (5)$$

At $z = z_{stoich}$ the profile intersects the threshold value $n^N(z,t) = n^N_{stoich}$. The term Ω^N_{exc} means the volume an excess nitrogen atom. The amount of Ω^N_{exc} follows by a fit procedure of the evaluated nitrogen profiles to the experimental data. So, putting $j^N(z,t)$ into equ. (1) we are able to compute $n^N(z,t)$.

3. Results and discussion

The model involves a lot of parameters, they are listed in the following table.

Parameter	Silicon		Si_3N_4
ρ (g cm^{-3})	2.33		3.18
R_p (Å)	506		354
ΔR_p (Å)	247		152
Y		0.2	
D_{eff} (cm s)	-		10
Ω^N_{exc}, cm^3	-		$1.7\cdot10^{-23}$

The values of the range and straggling parameters and also the silicon and nitride densities are taken from literature. The sputtering parameter Y is fitted to the experimental profiles, it is somewhat lower than the value known from literature. A comparison between the calculated and measured profiles shows the diffusion coefficient to be negligible.

The most remarkable parameter is the volume of an unbonded nitrogen atom Ω^N_{exc}. The evaluated profiles fit the experimental nitrogen concentration curves if we take a volume of $\Omega^N_{exc} = 1.7\cdot10^{-23}$ cm^3. This corresponds to an excess nitrogen density of

$\rho(N^N_{exc}) = 1.36$ g cm^{-3}, which is higher than that for solid nitrogen ($\rho(N^N_{sol}) = 1.03$ g cm^{-3}).

With this input data our model fairly well reproduces the experimental nitrogen concentration profiles in the considered case of nitrogen ions implanted into silicon with an energy of E = 20 keV, see Fig. 1. Our studies suggest that both the synthesized silicon nitride and the unbonded excess nitrogen within the nitride layer are in a high density, high pressure phase.

References

/1/ H.-U. Jäger et al.: Thin Solid Films 123 (1985) 159-169

/2/ U. Kreißig et al.: Thin Solid Films 100 (1983) L25-L28

/3/ K.J. Reeson et al.: Radiat. Effects 99 (1986) 71-81

PHYSICAL MODEL, PROPERTIES AND OPTOELECTRONICAL APPLICATION OF HYDROGEN IMPLANTED SOI MOSFET'S

S. Wagner, F. Neubert, Th. Otto[1]) U. Kreißig, J. Schöneich[2])

The investigation of active electron devices, realized in thin polycristalline silicon layers on insulating substrates (pc-SOI-MOSFET's) has importance for the developement of foundations both of threedimensional integration and optoelectronical macro-chip displays /1,2/. Implantation of atomic hydrogen plays an important role on the improvement of the electron device behaviour. It is the aim of this letter to study the influence of hydrogen implantation to the properties of the material and polycristalline-silicon MOSFET's.

N- and p-channel enhancement and deep depletion mode MOSFET's, as well as resistors, gate capacitors, inverters and ring oscillators have been investigated, which were fabricated in polycristalline silicon layers, produced by LPCVD on an insulating layer, grown by thermal oxidation of silicon wafers, and 350 nm and 1 µm thick, respectively. TEM-investigations /Fig.1/ showed the dimensions of grain size in polycristalline silicon to be 200 to 350 nm in the average, doping concentration of n- and p-silicon are $7*10^{15}$ and 10^{18} cm^{-3}, respectively. The photomicrograph /Fig.2/ of a pc-SOI-MOSFET cross section reveals a large inhomogenity of the silicon thickness and an undulating structure of the 100 nm SiO_2 gate insulator. The implantation of atomic hydrogen was carried out with complete prepared wafers.Implantation dose was 10^{16} cm^{-2} and energy 235 keV. It was followed by a 400 °C annealing process in hydrogen athmosphere for 30 minutes. Threshold voltages and channel mobility were found out by a linear regression method from the transfer current-voltage dependences. In Fig.3 the logarithmic transfer current-voltage characteristics of a p-inversion channel transistor before and after hydrogen implantation are represented.

The threshold voltages of all transistors (p-channel: -7V ± 1V, n-channel: +17V ± 1V with channel length 10 µm and Drain-Source voltage 5V) after hydrogen implantation moves to more negative amounts, for p-channel transistors by 20 to 70 %. The subthreshold slope considerably enlarges. The channel mobility of unpassivated p-channel transistors with channel length between 10 and 50 µm is constant and amounts to 5 cm^2/Vs (with V_{DS} = 5V). If channel length is less than 10 µm the mobility grows and reaches 18 cm^2/Vs in the maximum. After implantation this amount grows by 30...90 % and in short-channel transistors (L = 1 µm) reaches

[1]) Technische Universität Karl-Marx-Stadt, Sektion Physik/Elektronische Bauelemente, 9010 Karl-Marx-Stadt, PF 964, GDR

[2]) Zentralinstitut für Kernforschung der AdW der DDR, Rossendorf, 8051 Dresden, PF 19, GDR

about 60 cm^2/Vs. Leakage current, measured at $V_{GS} = 0$, $V_{DS} = 5$ Volts, L = 10 µm, amounts to 100 pA/µm and after H^+-implantation diminishes to 0,2...2 pA/µm. As a consequence the current relation I_{ON}/I_{OFF} at $V_{DS} = 5$ Volts before and after hydrogen implantation grows from $10^4...10^5$ to $10^7...10^8$. The properties of unpassivated n-channel transistor correspond to those of p-channel types (disregarding the high threshold voltage, caused by the high doping concentration),but after the H^+-implantation we have to point out a deterioration of parameters. The leakage currents increase by $10^4...10^5$ because at 5 Volt drain-source voltage breakthrough begins. All unpassivated deep depletion transistors (DDT) are self-closed. The transfer characteristic of n-channel DDT corresponds to those of the p-channel inversion type transistors, disregarding sign and amount of the threshold voltage. After H^+-implantation the leakage current of these transistors increases in a drastic manner. The properties of the corresponding p-channel transistors do not remarkable change, with the threshold voltage as an exception, if H^+-implantation is carried out. Ring oscillators with 29 inverter stages are well working and allowed the destination of the gate-delay time. At 30 Volts these times with unpassivated and H^+-implanted devices amounted to 10 ns and 7,5 ns, respectively. The power-delay product runs to 5 pJ with and without implantation.

An interpretation of these results is based upon the following physical model of the polycristalline silicon:

1. All grains of the polycristalline silicon are of uniform dimension, which is much larger than the thicknes of the disordered zone at the grain boundaries;
2. The doping atoms are homogeneously distributed, without segregation at the grain boundaries;
3. There exist surface states at the grain boundaries both of acceptor- and donor types with a uniform energetic distribution;
4. The current flux over the potential barriers takes place by thermionic emission.

Starting from this base we calculated the Fermi-levels and the barrier potentials at the grain boundaries which are located perpendicular to the silicon-insulator interface and mainly control the current flux. With regarding the geometrical form of the devices and the influence of the electrical contacts we can describe the working principles and properties of pc-SOI-MOSFET's as follows:

In the **"off-region"** the leakage current can be very small because of the self depleted state of the polycristalline layers caused by the grain boundary charges, but is effectively influenced by the possibility of the existence of channels at the backside. High densities of traps and a corresponding large field-enhanced generation current in the high-field zone at the drain /3/ lead to the so called "bath-curve" of the logarithmic transfer characteristic. The passivation of boundary states by hydrogen therefore must effectively diminish the leakage current.

In the **subthreshold region** any variation of gate voltages requires recharging of surface states which leads to a low slope of the current-voltage characteristic. Hydrogen passivation of these states counteracts this disadvantage.

The **threshold voltages** are higher than the corresponding single crystal amounts because of the high surface state density. The potential barriers at the grain boundaries, which originally repel the majority carriers, with growing gate voltages are diminished, but then renew remarkable amounts with oposite sign just before reaching the threshold voltage and therefore the effective mobility of minority carriers, which build up the inversion channel, is not much different from the corresponding majority carrier amount. At the same reason the inversion condition at the grain boundaries requires larger gate voltages than the interior grain region, thus producing a threshold voltage-region instead of a single amount. The hydrogen passivation of boundary states weakens this effect.

In the **active region** of the characteristic the effective mobility strongly depends on the gate voltage, which continuously deforms the potential barrier smaller because of the strongly growing inversion-charge density in the vicinity, so the barrier high diminishing, although the grain-boundary charge continues moderately to grow. The mobility also depends on the drain voltage, which deforms the barriers, and on the channel length, which determines the number of grain boundaries.

Based on these conceptions of the model and using the experimental results of transistors, resistors and gate capacitors, the influence of the hydrogen implantation upon different parameters of the polycristalline silicon was determined and collected in the following table. It is evident, that the experimental results confirm the effects mentioned above but moreover lead to the conclusion that the H^+-implantation, because penetrating the gate oxide, causes a considerable increase of the fixed positive effective surface charge at the SiO_2-Si boundary.

It has been stated out /4/, that hydrogen implanted pc-SOI-MOSFET's are suitable to allow the realisation of a liquid-crystal matrix display including the integration of the whole peripherical electronic circuitry up to the volume of more than 10.000 pixels with the condition of using the conventional TV picture-changing frequency. If the frequency is lower, as for instance at displays of computers, displays with considerable more pixels and larger areas can be realized.

The authors thank Mrs.Quäck and Mr.Skalaud from ZMD for the preparation of the electron micrograph in Fig.2.

References

1. Malhi S.D.S. et al. IEEE Trans.El.Dev., vol. ED-32, No.2(1985) 258-281
2. Oana Y. SID Intern. Symp., Dig. of Tech.Pap., San Franz.(1984) 312-315
3. Rodder M.S., Antoniadis D.A. IEEE Trans.El.Dev., vol.ED-34, No.5 (1987) 1079-1083
4. Otto Th. Thesis A, TU Karl-Marx-Stadt, 1987

Table

Parameters in pc-Si $N_D = 7 * 10^{15}$ cm^{-3}	unpassivated	after implantation of H^+ $D = 10^{16}$ cm^{-2}
resistivity, ohm*cm	$5,2 * 10^5$	6,3
averidge carrier concentration, cm^{-3}	$2,2 * 10^{12}$	$5 * 10^{15}$
majority carrier mobility, cm^2/Vs	5,5	210
potential barrier, V	0,16	0,07
density of recharging acceptor states at the grain boundaries, cm^{-2}	$8,4 * 10^{11}$	$1,6 * 10^{11}$
density of recharging states at the gateoxide-silicon boundary, cm^{-2}	$2,2 * 10^{12}$	$4,4 * 10^{11}$
density of fixed positive effective charge at the SiO_2-Si boundary, cm^{-2}	10^{11}	$1,5 * 10^{12}$

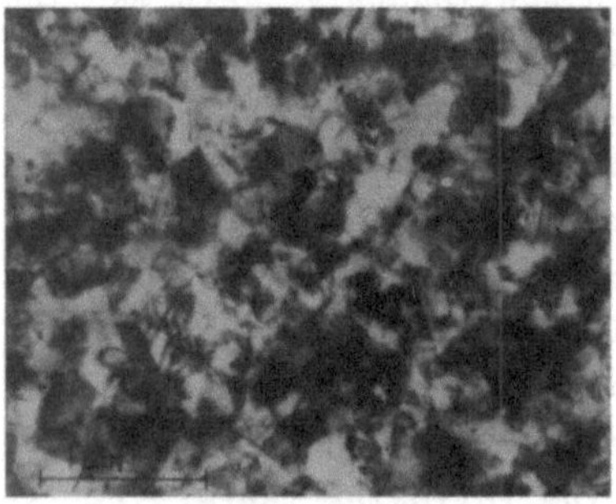

Fig.1 TEM-photomicrograph of a polycrystalline silicon layer

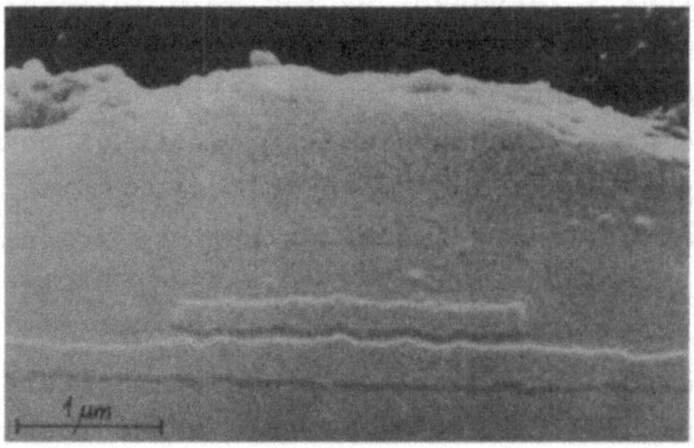

Fig.2 Cross-section of a pc-SOI-MOSFET

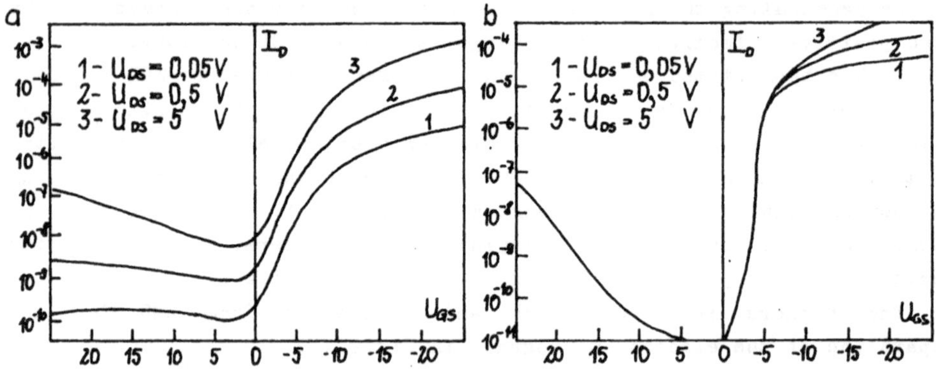

Fig.3 Logarithmic transfer characteristic of a p-channel transistor before (a) and after (b) hydrogen implantation

PULSE LASER INDUCED IGNITION OF EXPLOSIVE LIQUID-PHASE CRYSTALLIZATION OF AMORPHOUS SILICON

M.Wagner, G.Andrä, E.Glaser, G.Götz[1] and H.Bartsch[2]

1. Introduction

Liquid-phase explosive crystallization of an amorphous silicon layer results in the formation of a laminated crystalline layer consisting of longish crystallites (~100 μm x 1 μm) whose long axis are parallel to the spreading direction of the explosion front [1]. The electrical parameters of the crystallized layers are dependent on the orientation of the lamellas with respect to the direction of the carrier movement [2]. Thus, for an application of explosively crystallized silicon layers in microelectronic technologies the crystallization direction has to be aligned according to the layout of the chip. In addition, good electrical parameters of the layers can only be achieved, if the explosion starts at a well-defined temperature, that means after a certain time-delay after the beginning of the heating pulse providing the substrate temperature necessary for the explosion process. Both claims can be well satisfacted by using a second pulse namely a ns-laser pulse for the ignition of the crystallization. Such a double pulse technique provides a suitable method for controlling the moment of the beginning of the crystallization process as well as the direction of the phase front movement [3].

2. Experimental

The amorphous silicon layers under investigation were produced by UHV-evaporation of 350 nm Si on oxidized (200 nm SiO_2) silicon substrates. For reaching the spreading temperature of a liquid-phase explosion front (~1200K) either the pulse of a free-running Nd-glass laser or that of a set of Xe-flash lamps was applied. When providing the spreading temperature by the pulse of the free-running laser the back-sides of the samples were irradiated in order to prevent spontaneous ignitions of explosions due to the speckles in the laser beam [4]. With a certain time delay the heated amorphous layer is irradiated with the focused beam of a Q-switched Nd-glass laser causing the explosion to start.

1) Friedrich-Schiller-Universität Jena, Sektion Physik, Max-Wien-Platz 1 Jena, DDR 6900

2) Institut für Festkörperphysik und Elektronenmikroskopie der AdW der DDR, Weinberg 2, PF 250, Halle, DDR 4010

3. Results

Fig.1 shows the outer region of the spot where the ignition pulse hitted the sample. As can be seen in the optical micrograph (a) explosion process started within a well defined narrow zone of the ignition spot. The symbols in the figure mark characteristic regions of the ignition spot. The typical crystal structures of these regions are demonstrated by the TEM-micrographs (b)...(d). Region A is characterized by entire melting of the amorphous material. After solidification crystallites of 0.2...0.7 μm dimension can be found. Assuming a solidification velocity of 10...15 ms^{-1} in the strongly undercooled melt the existence time of the liquid can be estimated to be in the order of 100 ns.

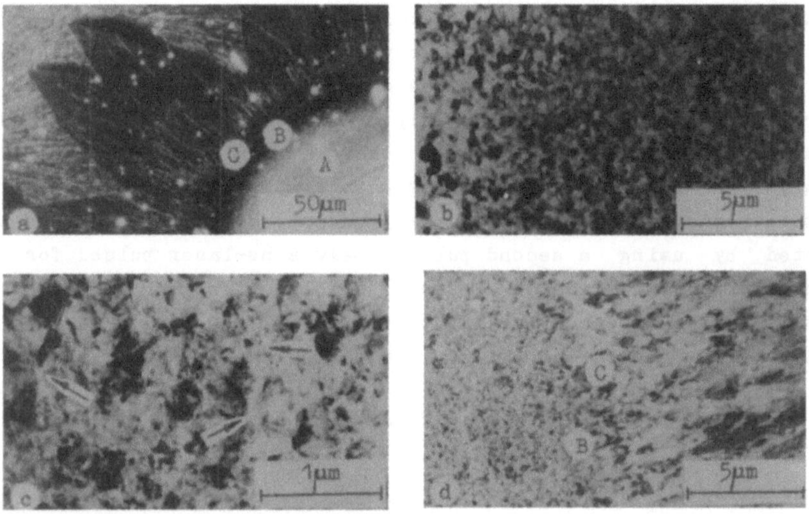

Fig.1. Optical (a) and TEM micrographs of the outer region of an ignition spot. The parts (b), (c), (d) of the figure show TEM micrographs of the regions A, B, C, respectively, marked in part (a)

In region B there were slightly lower temperatures compared with region A. Therefore, in this region there were produced isolated melt seeds which turn out to be the starting points of a multitude of exlosion processes (see the arrows in Fig.1c). Due to the high density of the melt seeds generated the explosion fronts cannot spread over long distances but they die out due to collisions with other melt seeds or explosion fronts in the neighbourhood. In the farther distance of the centre of the ignition spot the TEM-analysis reveals a decreasing density of melt seeds (region C). This region turns out to be the actual starting zone of the large-area explosive crystallization leaving long crystalline lamellas behind. From an analysis of the dimension of the transition region between complete melting of the amorphous material and the actual starting zone of the explosion process and an estimation of the temperature profile caused by the ignition pulse it can be concluded that this transition region corresponds to a temperature interval in the order of 10K.

A reduction of the intensity of the ignition pulse causes the region A and farther also region B to disappeare. Finally, the explosive crystallization starts at single isolated points. The TEM analysis of such an ignition point clearly reveals that the explosion starts at longish nuclei of about 50x250 nm^2 dimension (see the arrow in Fig.2).

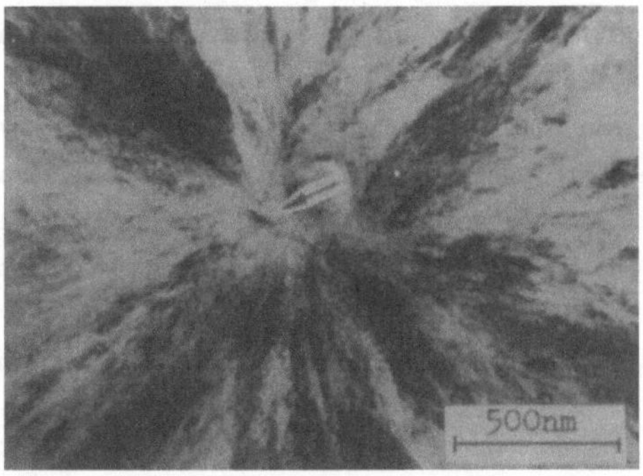

Fig.2. TEM-micrograph of a point-like ignition centre

Within the time interval available (much less than 1 ms at sufficient high temperatures during irradiation) crystallites of such dimensions cannot result from solid phase crystallization. They must be a result of local melting in connection with a momentary resolidification of the amorphous material. The results gained from our experiments lead to the following model of the ignition of an explosion process:

Due to the additional fast heating caused by the irradiation with a further energy pulse in the amorphous material there arises an isolated tiny molten region. This molten region is highly undercooled with respect to the crystalline state of silicon. Therefore, inside of such a "melt-droplet" a crystalline nucleous is immediately formed which can be still detected by TEM after the irradiation of the sample. The heat locally liberated due to the formation of this nucleous causes a further melting of the surrounding amorphous material. A quantitative estimation shows that the amount of liberated heat would be enough to produce an intermediate liquid layer with a thickness of about 40 nm between the crystallite and the surrounding amorphous silicon. In this way the starting configuration for a liquid-phase explosion process [1] is provided.

[1] Geiler, H.D., E.Glaser, G.Götz, and M.Wagner: J.Appl.Phys. 59 (1986) 3091-3099

[2] Winkler, N., W.Andrä, E.Glaser, G.Andrä: these proceedings

[3] Wagner, M., E.Glaser, G.Andrä, and G.Götz: Proc. of the EPM, Dresden 1984, 485-491

[4] Wagner, M., H.D.Geiler, G.Andrä, and G.Götz: phys.stat. sol (a) 83 (1984) K1

COMPUTER SIMULATION OF GRAIN BOUNDARY CONFINEMENT AT ZONE MELTING RECRYSTALLIZATION

K.-H. Heinig and H.-J. Mueller [1)]

1. Introduction

It is generally recognized that the silicon-on-insulator (SOI) technology is attractive for high-density CMOS-integrated circuits of the next generation. In order to take full advantage of this technology, device-quality single-crystalline silicon layers on insulating substrates must be available.

One approach to forming SOI layers is zone melting recrystallization (ZMR) by lasers /1/, halogen lamps /2,3,4/ or graphit strip heaters /5/. In ZMR a hot zone (usually a narrow line) of molten Si is scanned at a velocity v over a thin poly-Si film on an oxidized wafer. At the rear edge of the molten zone the film recrystallizes monocrystalline.

Besides of topographic imperfections /6/, the principal microscopic defects in zone melting recrystallized SOI films are low-angle grain boundaries, briefly subboundaries, which form a network rougly aligned to the scan direction. The subboundaries can provide a path for enhanced dopant diffusion, which can cause shorting of source and drain in MOS transistors /7/, or they can locally enhance the etching rate.

For this reason, in recent years several techniques for controlled melting and recrystallization have been proposed to eliminate or localize subboundaries. This has been achieved by an undulation of thickness /8/, reflectivity /9/ or thermal conductivity /4,9/ of the film structure. The method has been named subboundary confinement or also subboundary entrainment.

The major problem in implementing subboundary confinement is the lack of a clear picture of the mechanism of the formation of subboundaries. Geiss and co-workers /5/ developed a qualitative model for subboundary formation. A computer simulation which is based on this model has been carried out by Pfeiffer et al. /10/. Their computer graphics reproduces the main features of subboundary pattern in zone melting recrystallized silicon layers. Encouraged by their success, that model has been generalized in order to describe subboundary confinement.

2. The model

Up to now enough experimental results have been obtained in order to provide assumptions for the formulation of a model of subboundary pattern formation /10/ and, as it will be shown, for subboundary confinement: After ZMR the film tends to have (i) (100) texture and (ii) in-plane <100> direction within +/- 25° of the zone-motion direction /8,11/. (iii) The melt-solid growth interface consists of alternating (111)

---

1) ZfK Rossendorf, DDR-8051 Dresden, PF 19, GDR

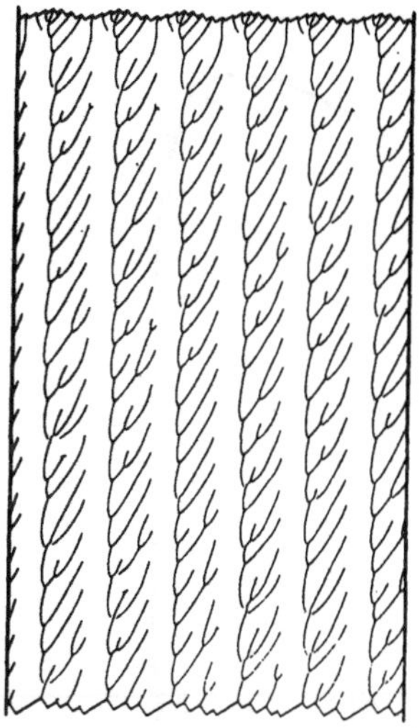

Fig. 1. Computer-generated plot of a subboundary pattern for the case of subboundary confinement due to a temperature undulation transversal to the scan direction (from top to bottom). The tilt between scan direction and in-plane <100> direction is 20°, the scan velocity v=0.035 (a.u.).

and (11$\bar{1}$) facets. (iv) Subboundaries originate from the interior corners of the faceted growth front. (v) The growth rate of each facet is purely controlled by the nucleation of new monolayers on the crystal plane. (vi) The nucleation rate is zero at the melting isotherm and increases linearly with supercooling. (vii) If a tip of fast growing facets overtakes the melting isotherm, then the tip melts and two additional facets form. (viii) A transversal undulation of the temperature field causes a subboundary confinement.

Based on these assumptions and a sinoidal temperature isotherm perpendicular to the scan direction, we derived a system of coupled, nonlinear differential equations for the length of the facets /12/ which describes the evolution of the faceted solidification front and, especially, of the interior corners of that front, i.e., of the subboundaries. The parameters of the model are the scan velocity, the angle between the scan direction and the <100> direction of the recrystallized film, the temperature gradient at the liquid-solid interface and kinetic factors /12/. Furthermore, the amplitude, the wavelength, and the phase of the sinoidal temperature isotherm can be varied.

Systems of typically 20...2000 coupled differential equations have been solved numerically using a Runge-Kutta procedure. To avoid an influence of artificial boundary conditions on the evolution of the faceted growth front, periodical boundary conditions have been implemented. The initial structure of the faceted front has no influence on the results and has been choosen arbitraryly.

3. Results and discussion

Without an undulation of the temperature isotherm , the system of differential equations becomes identical to that of Pfeiffer et al. /10/, i.e., it describes the evolution of a faceted growth front gouverned by a straight melting isotherm. The

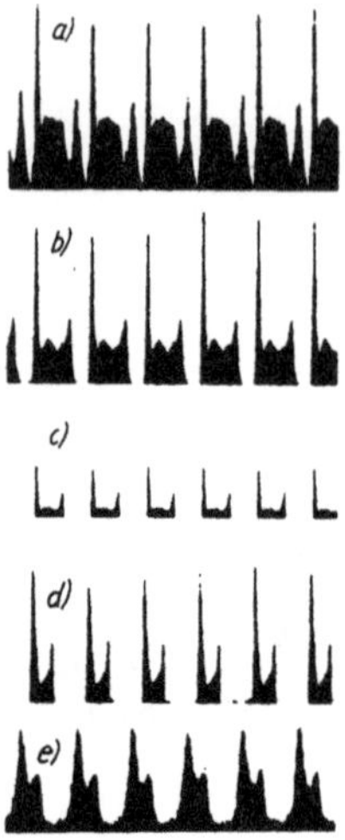

Fig. 2. Density of subboundaries perpendicular to the scan direction for different scan velocities v (a.u.) and a tilt between scan and <100> direction of 20°. v=
(a) 0.0125; (b) 0.025;
(c) 0.06; (d) 0.15; (e) 0.155

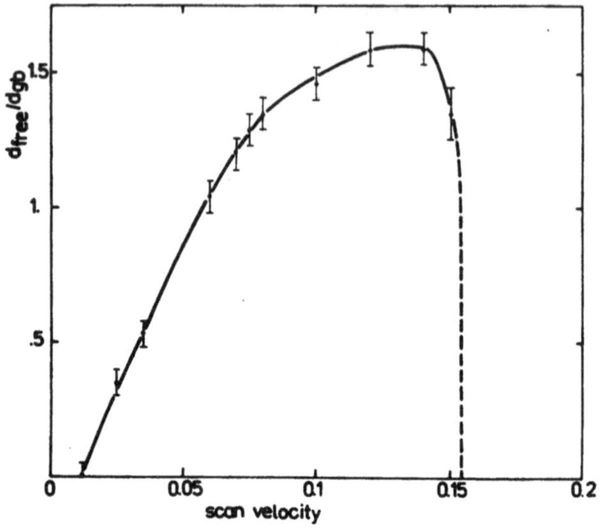

Fig. 3. Ratios between the widths of stripes with and without subboundaries versus the scan velocity (a.u.).

numerical results have been proven to be equivalent. No subboundary confinement is observed, the subboundaries distribute randomly like a network over the wafer.

According to Geiss and co-workers /8/ and others /4,9/, an undulation of the melting isotherm perpendicular to the scan direction leads to subboundary confinement. At first we investigate the case where the in-plane <100> direction is parallel to the scan direction /12/. For this case it was found that subboundary confinement occurs only in a narrow range of scan velocities. In that range the wavelength of the undulation of the isotherm is approximately equal to the mean distance between subboundaries. The permitted deviation of +/-50% from the optimum velocity, which has been found recently /8/, is in good agreement with this computer simulation.

Here we consider the case where the in-plane <100> direction is not parallel to the scan direction. The temperature isotherms are undulated perpendicular to the scan direction. In this case, subboundary confinement can be achieved even if the natural distance between subboundaries is much smaller than the wavelength of the temperature undulation (Fig. 1). Between the regions with subboundaries remain stripes free of grain boundaries.

Fig. 2 shows computer generated plots of the density of subboundaries perpendicular to the scan direction calculated at different scan velocities. For a wide range of scan velocities we found, in-between defective stripes of width d^{gb}, regions of width d^{free} which are free of subboundaries.

In Fig. 3, the ratios d^{free}/d^{gb} have been plotted versus the scan velocity. It can be seen that there is an optimum scan velocity where the defect-free areas are much wider then the defective one. For the considered tilt between the in-plane <100> direction and the scan direction as well as the amplitude of the temperature undulation it is demonstrated that the fraction of the subboundary-free area amounts to about 60 % (Note that this is no optimum case.).

In conclusion, our computer simulation predicts that a tilt between the in-plane <100> direction of the crystalline layer and the scan direction improves the grain boundary confinement during zone melting recrystallization.

REFERENCES

/1/ G.K. Celler and L.E. Trimble, J. Phys. 43 (Colloque C1) (1982) C1-353

/2/ T.J. Stultz and J.F. Gibbons, Appl. Phys. Lett. 41 (1982) 827

/3/ M. Voelskow, J. Matthaei, K.-H. Stegemann, D. Gerisch and S. Mutze, Proc. Int. Conf. Energy Pulse Modification of Semiconductors, Ed. K. Hennig, ZfK-555 (1985) 402

/4/ M. Hoand, D. Dutartre and D. Bensahel, Proc. Mat. Res. Soc. Symp., Boston, Dec. 1985 (preprint)

/5/ M.W. Geiss, H.I. Smith, B.-Y. Tsaur, J.C.C. Fan, D.J. Silversmith and R.W. Mountain, J. Electrochem. Soc. 129 (1982) 2813

/6/ C.K. Chen. M.W. Geis, B.-Y. Tsaur, R.L. Chapman and J.C.C. Fan, J. Electrochem. Soc. 131 (1984) 1707

/7/ B.-Y. Tsaur, Proc. Mat. Res. Soc. Symp., Boston, Dec. 1985, (preprint)

/8/ M.W. Geis, H.I. Smith and C.K. Chen, to appear in J. Appl. Phys. (preprint)

/9/ I.N. Miaoulis and B.B. Mikic, J. Appl. Phys. 59 (1986) 1658

/10/ L. Pfeiffer, s: paine, G.H. Gillmer, Wim van Saarloos and K.W. West, Phys. Rev. Lett. 54 (1985) 1944

/11/ M.W. Geis, H.I. Smith, B.-Y. Tsaur, J.c.c. Fan, E.w. Maby and D.A. Antoniadis, Appl. Phys. Lett. 40 (1982) 158

/12/ K.-H. Heinig ,H.-J. Mueller, G. Otto, and K. Hennig, J. de Physique EMRS-XII (1987) 185

STUDY OF STRUCTURE FORMATION BY THE MELT INSTABILITY OF SILICON AT LIGHT HEATING

K.-H. Heinig and H.-J. Mueller[1)]

1. Introduction

In the pioneering work of Bosch and Lemons [1] it was found that silicon-on-insulator films, when heated by a radiant flux, melts inhomogeneously. Later on, it has been shown [2,3] that, during optical heating of bulk silicon, this break up of the surface into small regions of solid and liquid occurs too. Thus, this process is apparently quite general. The photograph of Fig. 1 has been taken after irradiation of a (100)-oriented wafer of bulk Si with a Nd-YAG laser pulse of about 1 ms duration [5]. The transiently molten regions are visible due to the surface roughening at resolidification. Other examples of inhomogeneous melting can be found in refs. [4,6].

The inhomogeneous melting arises naturally when a jump in reflectivity at the semiconductor-metal melting transition implies a negative feedback mechanism [6]. For silicon, the reflectivity of the melt is 0.7 compared to 0.3 of the solid. As long as the temperature of silicon is below melting temperature T_m, the temperature increases proportional to the incident light flux. Raising the light flux slightly from a steady state just below the melting point will create a small amount of liquid. Since the liquid reflects more light, less power will be absorbed. A further increase of the light flux increases the portion of the molten surface, but not the amount of absorbed power. Thus, for a wide range of light intensity, the surface will remain partly solid and partly liquid. Only if the absorbed power is sufficient to melt the surface completely, the temperature will continue to rise with increasing light flux.

For optical heating of silicon, the formation of several distinct and reproducible microscopic patterns of solid and molten regions with various degrees of long-range order has been observed. These patterns show "slush" zones, meanders (Fig. 1) and lamellae [4]. Existing models explain the structure formation in terms of the interference of incident and surface scattered beams of coherent light (see e.g. [4] and references therein). However, this theory cannot explain (i) the occurence of lamellar, inhomogeneous melt regions at heating with incoherent light and (ii) that lamellae often align with crystallographic axes and not with polarization (Fig. 1, [9,14]).

In this paper we propose a model which requires no external field like coherent light for pattern formation and symmetry breaking. It will be shown that the unique melting kinetics leads to spontaneous pattern formation which can , of course, align with any external field.

2. The model

In our model, the pattern formation reflects the competition of two driving forces [9]. On the one hand, the unusual thermal situation [7] due to the jump in reflectivity causes nucleation and/or melting in overheated silicon and nucleation and/or recrystallization in undercooled liquid, respectively, which tends to produce fine-grained "slush" of solid and liquid. On the other hand, the interfacial energy between liquid and solid areas favours the growth of large solid and liquid regions.

---

1) Zentralinstitut fuer Kernforschung Rossendorf, DDR-8051 Dresden, PSF 19, DDR

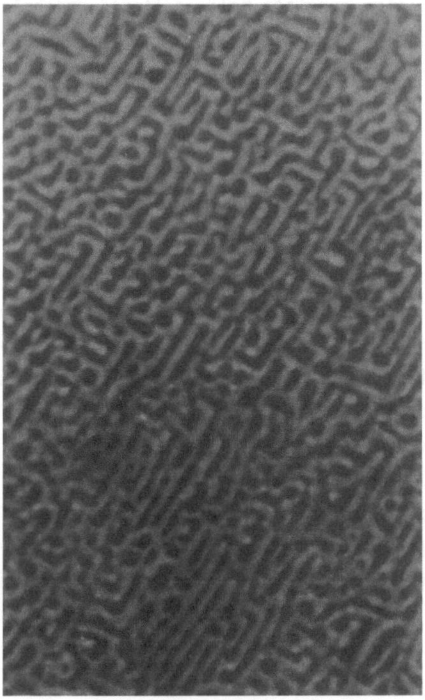

Fig. 1. Photograph of (100)-oriented bulk silicon taken after irradiation by a pulsed Nd-YA laser (1 ms)[5]. The light areas were molten.

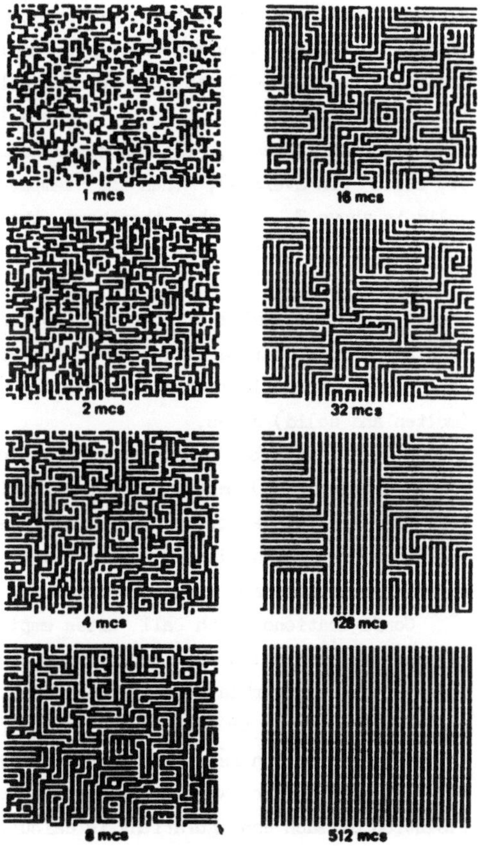

Fig. 2. Computer simulation of the evolution of the melt morphology if nucleation events are permitted. mcs - Monte Carlo step

Starting from phenomenological optical and thermodynamical laws, this competition can be modeled by a short-range interaction between solid and liquid. The evolution of pattern by this short range interaction has been simulated by a two-dimensional cellular automaton (CA) [15]. The CA consists of a square lattice of 4096 sites having periodical boundary conditions, with each site carrying one of two possible values. In analogy to crystal growth simulations [16,17] it is assumed that empty and occupied sites of the lattice of the CA represent blocks of atoms of solid and liquid, respectively. To simulate Boltzman statistics, values of sites are updated in a random sequence. 4096 alternations of sites are called one Monte-Carlo step (mcs). A site with its eight nearest neighbours represents a configuration. The evolution code of the CA determines for each of the 512 configurations which value the central site attains after an evolution step. For this class of CA 2^{512} different evolution codes exist.

Now we determine the evolution code of our CA. According to the classical transition state theory [19], the driving force for melting and solidification is the difference between the Gibbs free energies of the liquid and solid phases,

$$G_{sl} = L|T - T_c|/T_c \tag{1}$$

with L the latent heat of fusion and $T_c = T_m + a\ E_i$ is the actual melting temperature. It differs from the equilibrium melting temperature T_m by that amount of interfacial energy E_i which a curved or rugged interface has in excess of a plane interface. For a curved interface, E_i is given by the well-known Gibbs-Thomson relationship [20]. Thus, we obtain

$$G_{sl} = a_1 |T - T_m| + a_2 E_i. \quad (2)$$

If, for instance, the inhomogeneous melt region consists half of liquid and half of solid (in average), then the "temperature of configurations" with more than 50% solid is above T_m. On the other hand, the temperature of configurations with less than 50% solid is below T_m. Thus, it can be assumed that within a configuration of the CA

$$T - T_m = const*(D_m - D), \quad (3)$$

where D is the number of occupied (molten) sites in the configuration and D_m is the mean number of occupied sites averaged over the whole lattice. The interfacial energy of a configuration is proportional to the number of edges S between occupied and empty (molten and solid) sites,

$$E_i = const*S. \quad (4)$$

Now, it will be predicted that a block of atoms (i.e., the central lattice site of a configuration) melts, if the Gibbs free energy of the resulting configuration is less than the initial one. Otherwise, that central lattice site remains unchanged. An analogous argument holds for solidification. Choosing a fixed ratio for the constants a_1 and a_2, we have an unique procedure for the evaluation of the evolution code.

Configurations with all sites empty or all sites occupied have been treated separately. The solid and molten regions corresponding to these configurations are, of course, highly overheated and undercooled, respectively. However, for these configurations the solid or melt cannot grow, phase transformation can occur by nucleation only. As nucleation requires an especially high activation energy, two cases must be distinguished. Here we investigate the case that nucleation is allowed. Then, the evolution of such configurations is calculated as desribed above. In the second case, no nucleation is permitted. That situation, where melting must start from a precursor seed, has been studied elsewhere [9].

3. Results and discussion

Fig. 2 shows a computer plot of the evolution of the melt morphology, which has been simulated by the cellular automaton described above. Nucleation in completely molten or solid configurations is allowed. It can be seen that during the first Monte-Carlo step melting starts randomly at many sites (nucleation stage). After a short time (2 mcs to 4 mcs) an ordering process in the melt morphology can be observed. The patterns become very similar to that of Fig. 1. In this experiment, due to the short laser pulse, we have in the initial stage a strong superheating of the solid. Thus, melting can start at many nucleation sites. During the laser pulse little time is available for ordering processes. Therefore, a melt morphology like that obtained after two or four Monte-Carlo steps is quenched in.

If nucleation is inhibited and "free" boundary conditions are implemented, similar patterns grow from a liquid seed [9]. It can be thought that at small overheating melting starts in the middle of the laser spot which should be the hottest point. Then, for cw laser irradiation, there is enough time to compensate for overheating by growth of the molten area.

The alignment of the lamellae occurs at random in any (of the possible) directions. As in the Heisenberg model, where an isotropic short-range interaction produces antiferromagnetism with long range order, we have in our model a spontaneous symmetry breaking and formation of alternating solid and liquid lamellae. In contrast to existing theories [4], our approach requires no external source for symmetry breaking. If there is any external anisotropy present (e.g., polarization of coherent light, crystal orientation etc.), it acts like an external field within the Heisenberg model, i.e. the spontaneously formed internal field will align with the external one.

Biegelsen has shown [8] that the presence of such a mobile, inhomogeneous melt region is one prerequisite for crystal orientation control in SOI-layer production. If there are (100) precursor crystallites in the SOI-layer, the formation of liquid and solid lamellae leads to (100) texture [10]. It should be noted that non-radiant heating (e.g., electron beam) does not produce inhomogeneous melting and does not result in (100) texture [8]. Egami et al. [11] have observed that the liquid lamellae align with Si pattern edges of a graphoepitaxy structure. In zone melting recrystallization, this melting behaviour has also a drastic effect on the quality of the recrystallized films [12]. And finally, very recently, Im et al. [13] have observed inhomogeneous melting at the rear edge of the molten zone in zone melting recrystallization experiments. This melt instability has a strong influence on the formation of low-angle grain boundaries and protrusions. These examples show that a deeper understanding of the melt instability is important for a further improvement of SOI-layer production.

In conclusion, we suggest that in silicon heated by a radiant flux (i) the formation of lamellae is quite general, and (ii) the patterns form spontaneously. External anisotropies (e.g. interferences) are not the driving force for pattern formation, although the alignment with any existing anisotropy occurs easely.

REFERENCES

[1] M.A. Bosch and R.A. Lemons, Phys. Rev. Lett. 47 (1981) 1151
[2] G.K. Celler, McD. Robinson, L.E. Trimble, and D.J. Lischner, Appl. Phys. Lett. 43 (1983) 868
[3] K.-H. Heinig, in: Proc. Int. Conf. EPM 84, Dresden,Ed. K. Hennig, ZfK-555, Dresden 1985 (p.265)
K.-H. Heinig, M. Voelskow, J. Matthai, A. Zetzsche, and C. Treutler,ibid.(p.280)
[4] J.S. Preston, H.M. van Driel, and J.E. Sipe, Phys. Rev. Lett. 58 (1987) 69
[5] H.-D. Geiler and K.-H. Heinig, to be published
[6] W.G. Hawkins and D.K. Biegelsen, Appl. Phys. Lett. 42 (1983) 358
[7] K.A. Jackson and D.A. Kurtze, J. Cryst. Growth 71 (1985) 385
[8] D.K. Biegelsen, 2nd Int. Workshop on Future Electron Devices, SOI Technology and 3D Integration [FED SOI/3D Workshop], March 19-21, 1985, Shuzenji (p.31)
[9] K.-H. Heinig and H.-J. Mueller,Proc.EMRS-Conf.1987 Meeting,Strasbourg,June 1987
[10] D.K. Biegelsen, L.F. Fennell, and J.C. Zesch, Appl. Phys. Lett. 45 (1984) 546
[11] K. Egami, M. Kimura, T. Hamaguchi, and N. Endo, MRS Symp.Proc.Vol.35 (1985) 669
[12] D. Dutartre, Appl. Phys. Lett. 48 (1986) 350
[13] J.S. Im, C.V. Thompson, and H. Tomita, preprint, Massachus. Inst.of Techn.(1987)
[14] M. van Allmen, W. Luthy, and K. Affolter, Appl. Phys. Lett. 33 (1978) 824
[15] for a review see: S. Wolfram, Rev. Mod. Physics 55 (1983) 601
[16] S. Wolfram, preprint "Cellular Automata and Condensed Matter Physics", The Institute of Advanced Study, Princeton (1985)
[17] N. Packard and S. Wolfram, J. Stat. Phys. 38 (1985) 901
[18] N. Packard, preprint "Cellular Automaton Models for dendritic crystal growth",The Institute of Adv. Study, Princeton (1985)
[19] J.Y. Tsao, P.S. Peercy, M.O. Thompson, and M.J. Aziz, Phys. Rev. B (1987)
[20] W. Kurz and D.J. Fisher, Fundam. of Solidification(Trans.Tech.Publ., 1984) p.13

High Throughput CO_2 Laser Recrystallization for 3D integrated Devices

K. Haberger, P. Panish, R. Buchner, H. Steinberger[+] and R. Bower*

The crystallization of a layer of silicon deposited on an insulator has become the focus of intense research, holding the hope of producing good quality large-crystal silicon suitable for vertically integrated circuits. The possible advantages of such 3 dimensionally integrated circuits would be the increase of packing density without further reduction of the lateral device dimensions as well as the combination of different technologies and semiconductor materials.

Most of the recrystallization work for 3 dimensional circuit integration has been done utilizing argon laser raster scanning systems [1]. Due to the long processing time required, more than 30 minutes for a 4"wafer, the system throughput is limited as is the maximum allowable substrate temperature due to electrical dopant deactivation and diffusion of the devices already processed in the bulk silicon. Moreover, the energy conversion efficiency of visible lasers is very restricted. Zone melt techniques using high throughput graphite strip heaters, on the other hand, have been used to produce high crystalline quality silicon layers for SOI applications [2]. This method is not ideally suited to the 3 D field due to the extreme heating of the substrate involved.

A good compromise between these two methods could be the use of a high power CO_2 laser which allows full wafer processing using a sweeped-line zone melt system. This system offers the high power density required to maintain a temperature difference between the molten poly-Si layer and the substrate, as well as imposing the high throughput rate of a zone processing system due to the fact that the wafers may be processed in a normal atmosphere. The danger of contaminating the recrystallized polysilicon is avorded through the used protective absorptive capping layers similar to those used in traditional zone melt system [3].

[+] Fraunhofer Institut für Festkörpertechnologie
D - 8000 München 60, West Germany
*Department of Electrical Engineering
University of California Davis
Davis, California 95616

The configuration of the CO_2 laser zone melting system we have constructed for thin layer poly-SI recrystallization is shown in Fig. 1. The beam of a 1 kW CO_2 laser, which has a diameter of 23 mm, is reflected onto the wafer surface with a rotating polygonal mirror. In addition to the scanning provided by the rotation of 36 facet on the polygon wheel the beam is focused to a line by cylindrically machining the facet surface to a focus length of 250 mm. The polygon wheel is made of precision machined and polished aluminium. No optical corrections such as f-Θ lenses have been utilized. Due to the rotation speed of the polygon an effective line focus is formed with a length of 15 cm and a width of roughly 200 µm yielding a power densitiy on the order of 10^4 W/cm^2. The wafer is preheated to 1100°C and moved perpendicular to the focused scan line at speeds between 0.5 and 10 mm/sec. The melting can be monitored by a He-Ne laser reflectivity measurement system. Utilizing this equipment the surface of a full 3" wafer can be recrystallized within a few minutes.

The absorption of the 10.6 µm radiation of the CO_2 laser in polysilicon is, even at elevated temperatures, rather poor. Therefore, a capping layer of SiO_2 deposited over the polysilicon has to be used. In this manner, the heat applied to the capping layer flows from the top surface down. Heat transport is retarded due to the heat capacity of the upper layers. To ensure the efficient energy coupling to the polysilicon the following layer structure compatible to conventional silicon technology is used: A 0.5 µm low temperature LPCVD oxide and a planarization layer are deposited over the previously processed n-channel bulk MOS devices. Next an 0.5 µm poly-Si layer is deposited at a temperature of 615°C followed by a 1.3 µm LPCVD oxide capping layer. This thick capping layer serves to strongly absorb the 10.6 µm laser line. In addition, for some wafers the SiO_2 capping layer has been implanted with nitrogen at doses up to 5 x 10^{16} cm^2 to increase the absorption at the CO_2 wavelength.

Besides basic investigations on the absorption and melting as a funtion of the preheating temperature, rotationg speed of the polygon wheel as well as the mechanical scan velocity, we have fabricated p-channel MOS transistors in the recrystallized layer [4]. A cross section view is shown in fig. 2. The effect of the fabrication process on devices in the silicon substrate as well as in the recrystallized polysilicon layer has been characterized through electrical measurements. These measurements give a channel mobility of up to 60 cm^2/Vsec for the boron doped MOSFETs in this layer under optimum conditions. A characteristic curve is shown in fig.2. The main goal of this early investigation was to test the feasibility of a fast 3 D laser recrystallization method. With this rather crude equipment, the incident laser power was poorly regulated resulting in localized overheating of the wafers. We have observed void formation due to the contraction of silicon upon melting. We hope to overcome this well-known agglomeration problem through the use of proper surface melting techniques.

Another problem common to all zone melt systems is wafers warpage and distortion which in this case is on the order of 1 µm/cm dependent on the substrate orientation.

The authors would to thank for the technical support of Martin Bleier, Anke Heidenreich and Bernhard Rose. This project was supported with funds from the West German ministry of Research and Development.

[1] Y. Akasaka, Proceedings of the IEEE, 74 (1986) pp 1703
[2] M.W. Geis et.al., Electron Devices Letters - 7 (1986) pp 41
[3] B.Y. Tsaur et.al., IEDM Proceedings, (1981)
[4] R. Buchner, K. Haberger, P. Seegebrecht and P. Panish, Proceedings ESSDERC, Bologna (1987)

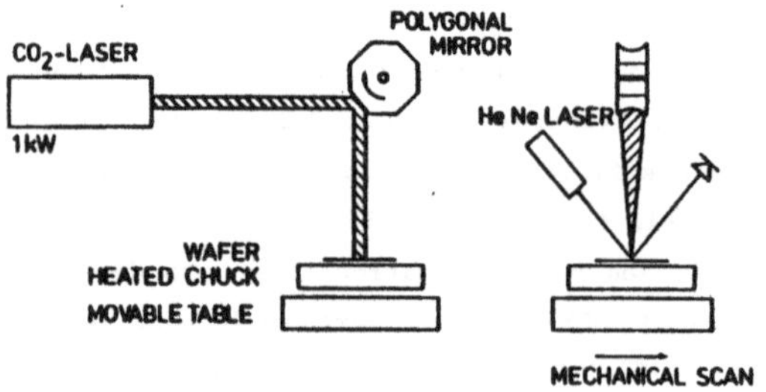

Figure 1: Schematic representation of the CO_2 laser recrystallization system utilized in the experiments showing components and direction of wafer motion

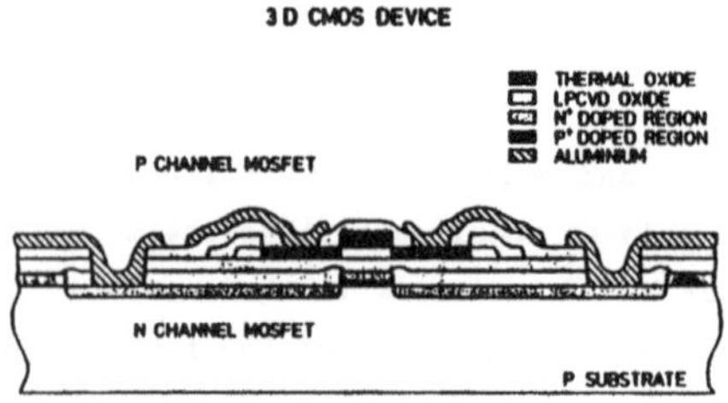

Figure 2: Schematic cross section of a 3 D MOS device

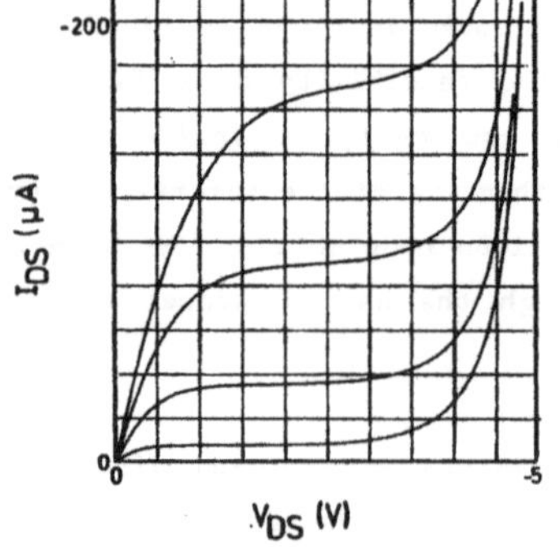

Figure 3: Characteristic curves of a p - channel MOS transistor processed in the recrystallized silicon layer

FORMATION OF BURIED NITRIDE LAYERS BELOW $NiSi_2$ BY ION IMPLANTATION

J.Danilowitsch, K.Gärtner, G.Götz and B.Weber[1)]

1. Introduction

Recently, it has been shown that the synthesis of silicon-on-insulator (SOI) structures formed by high dose ion implantation of nitrogen or oxygen followed by a high temperature annealing process has attracted increasing attention (see for example refs.[1-3]). Silicide-on-insulator systems will also be of interest for the fabrication of isolated conducting structures in three-dimensional integrated circuits. Up to now no investigations are reported in literature concerning this topic.

This paper presents some preliminary results of the formation of a Si_3N_4 layer between a monocrystalline $NiSi_2$ layer and single crystalline silicon substrate by high dose implantation of N_2^+ into a $NiSi_2$/Si-substrate system.

2. Experimental

A 120 nm thick $NiSi_2$ layer was produced by vacuum annealing (400°C, 15 min and 800°C, 15 min at 10^{-4} Pa) of a layer system consisting of 30 nm amorphous Si and 35 nm nickel e -beam deposited onto <111> Si-substrate. N_2^+ ions were implanted with an energy of 300 keV and a current density of < 5 μA cm^{-2} to a dose of $8*10^{17}$ N cm^{-2}. The samples were heated during implantation to 300, 450, and 500°C, respectively, by mounting on a heated copper plate. The composition and crystalline quality of the layer were investigated by Rutherford backscattering spectrometry (RBS) combined with channeling techniques using 1.4 MeV He ions and a backscattering angle of 170°. The resistivity of the silicide layer was measured by means of the van der Pauw method.

3. Results and discussion

Fig.1 shows the backscattering spectra aligned in <111> direction for the different implantation temperatures T_i, a random spectrum (the random spectra for all three values T_i are equal), and a simulated random spectrum for comparison assuming a 200 nm thick homogeneous Si_3N_4 layer between the $NiSi_2$ layer and the silicon substrate.

---

1) Friedrich-Schiller-Universität Jena, Sektion Physik, Max-Wien-Platz 1
Jena, DDR 6900

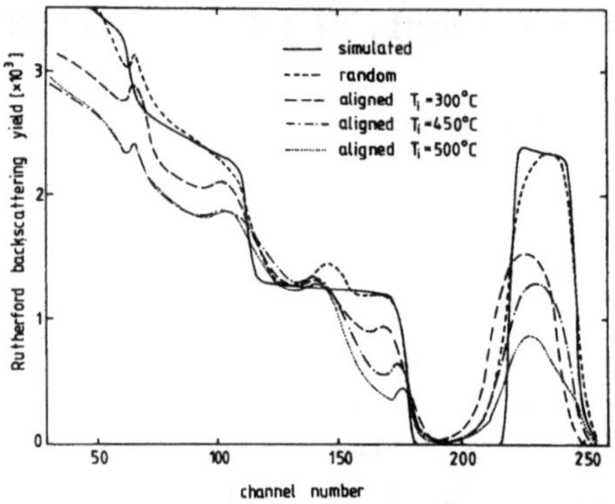

Fig.1. Backscattering spectra (1.4 MeV He^+) in random and aligned direction of 120 nm epitaxial $NiSi_2$ on <111> silicon substrate after implantation with 300 keV N^+ ions at 300, 450, and 500°C. The full line represents a simulated random spectrum of a 120nm $NiSi_2$/ 200nm Si_3N_4/Si-substrate system.

The minimum yields and the resistivities of the $NiSi_2$ layer before and after implantation are listed in Table 1. The crystalline quality of the $NiSi_2$ layer before implantation ($\chi_{min} \approx 10\%$) is poorer than the best values known from literature ($\chi_{min} \approx 3\%$) but it is sufficient for these preliminary experiments. The comparison of the measured and the simulated random spectrum shows that the composition of the silicide layer corresponds exactly to that of $NiSi_2$. The implanted layer is not composited homogeneously. The distribution of the nitrogen atoms corresponds approximately to the calculated (from ref.[5]) implantation profile which is shown in Fig.2. The maximum nitrogen concentration reaches the value of Si_3N_4. In Fig.1 the height of the silicon part of the implanted layer is equal in minimum to the height of the simulated Si_3N_4 part. This is in accordance with the results of the literature (see for example refs.[3,5]) for high dose N^+ implantation in silicon, where the stochiometric composition is reached in the maximum of the nitrogen distribution at a dose of about $0.8 \cdot 10^{18}$ N cm^{-2}. X-ray diffraction measurements indicate that some α-Si_3N_4 has already formed.

Table 1

Mean Rutherford backscattering minimum yield χ_{min} and resistivity ρ of the silicide layer before and after nitrogen implantation at different implantation temperatures.

quantity	before implantation	after implantation at RT	300°C	450°C	500°C
χ_{min} (%)	10	100	80	55	30
ρ (/uΩcm)	31		110	54	40

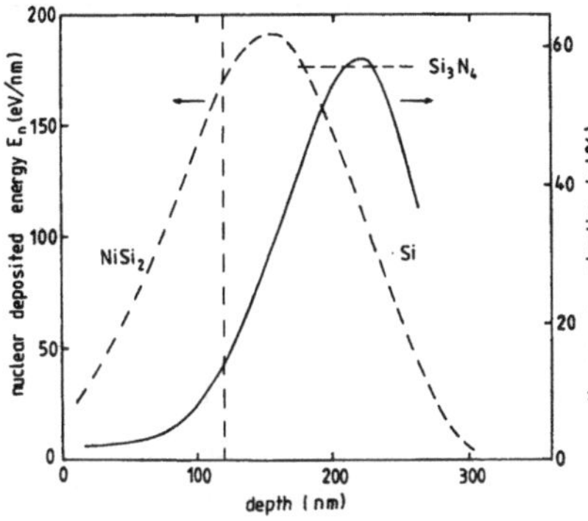

Fig.2. Nuclear deposited energy and nitrogen concentration calculated according to ref.[5] for implantation of 150 keV N^+ into $NiSi_2$ as function of the depth

A considerable amount of nuclear deposited energy (Fig.2) is located in the $NiSi_2$ layer which is responsible for strong radiation damage in this layer. Information about the damage for the different T_i is given by the χ_{min} values in Table 1. For implantation at room temperature the silicide layer is completely amorphized (from a channeling point of view).

Ishiwara et al. [6] found that the amorphization dose for $NiSi_2$ is about three times that for silicon under argon irradiation. Therefore, in the case of nitrogen implantation the amorphization dose is expected to be $< 2*10^{16}$ cm^{-2} which is significantly below the dose used. With increasing T_i the mean minimum yield and the resistivity of the silicide layer decrease steadily (Fig.1, Table 1). For $T_i = 500^oC$ the resistivity is not far from the initial value but the minimum yield is still 30%. In the case of oxygen implantation in silicon Golanski et al. [7] found an optimum value of $\chi_{min} \approx 15\%$ at $T_i \approx 550^oC$. If there exists also an optimum value of χ_{min} for N^+ implantation in a $NiSi_2$ system it has to be at $T_i \gtrsim 500^oC$.

The experiments must be continued for higher implantation temperatures and higher nitrogen doses using high quality silicide layers. The formation of the nitride layer by high temperature annealing has to be investigated.

[1] Pinizzotto, R.F., Mater. Res. Soc. Symp. Proc. 27(1984)265.
[2] Hemment, P.L.F., Mater. Res. Soc. Symp. Proc. 53(1986)207.
[3] Reeson, K.J., Nucl. Instr. & Meth. B 19/20(1987)269.
[4] Hemment,P.L.F., R.F.Peart, M.F.yao, K.G.Stephens, R.J.Carter, J.A.Kilner, D.Meekison, G.R.Booker and R.P.Arrowsmith, Appl. Phys. Lett. 46(1985)952
[5] Burenkov, A.F., F.F.Komarov, M.A.Kumakhov and M.M.Temkin, Spatial distribution of energy deposited in atomic collision cascades in solids, Energoatomoisdat, Moscow 1985.
[6] Ishiwara,H., K.Hikosaha and S.Furukawa, Appl.Phys.Lett.32(1977)23.
[7] Golanski,A., A.Perio, J.J. Grob, R.Stuck, S.Maillet and E.Clavelier, Appl.Phys.Lett. 49(1986)1423.

EMISSION CHARACTERISTICS OF EUTECTIC ALLOY LIQUID METAL ION SOURCES FOR FOCUSED ION BEAM SYSTEMS

F.Machalett, R.Mühle, I.Stiebritz [1)]

1. Introduction

In recent years there has been a growing interest in liquid metal ion sources (LMIS) to obtain fine focused ion beams of high intensity with potential applications such as ion beam lithography, maskless ion implantation and focused ion beam sputtering [1,10,11]. For these applications various kinds of ion species are required which are partly delivered by liquid metal ion sources [2].

But unfortunately this type of ion sources is applicable only to elements having a relatively low melting point and a low vapour pressure at this point [3]. For this reason silicon and germanium ions, which are often used as dopants for GaAs, cannot be generated by the ionization of the pure elements in LMIS. Therefore an eutectic alloy with gold is generally emloyed as source feed material because of its good wetting properties and the relatively low eutectic temperature of 362°C and 356°C [7,8] for silicon and germanium, respectively.

In the present paper investigations on mass spectra of Au-Si and Au-Ge alloy LMIS are presented. The influence of the alloy composition (only Au-Si) and of the total source current on the beam composition was investigated. The eutectic point of the Au-Si alloy was redetermined on account of the published contradictory values [6,7].

2. Experiment

The ion source and its operation parameters are described in [9]. Mass distributions of the emitted ions were investigated by means of a magnetic mass spectrometer, also described in [9]. Three different compositions of Au-Si alloys ($Au_{94}Si_6$, $Au_{82}Si_{18}$, $Au_{69}Si_{31}$) and the eutectic composition of the Au-Ge alloy ($Au_{73}Ge_{27}$) has been used as source material.

3. Results and discussion

To find out the lowest temperature of complete melting of the Au-Si alloy we carried out a thermoanalysis and determined the eutectic

1) Friedrich-Schiller-Universität Jena. Sektion Physik, Max-Wien-Platz 1 Jena, DDR 6900

composition to be (18.0 ± 0.5) atomic percent of silicon and the temperature of the solidus line to be T_E = (362 ± 4)°C. These values are in good correspondence with those given by Gerlach and Goel [6], but differ from those given by Hansen [7] (31 atomic percent of silicon).

The observed mass spectrum of the Au-Si alloy and the table of the corresponding ion flux fractions are shown in [9]. The influence of the source feed composition on the relative intensity of the main ion species is represented in Fig.1. The increase of the silicon fraction in the source material causes an increase of the proportion of silicon ions and of the Au_nSi compounds too. But the silicon fraction x cannot be increased excessivly because this is connected with an increase of the liquidus line in the T-x phase diagram.

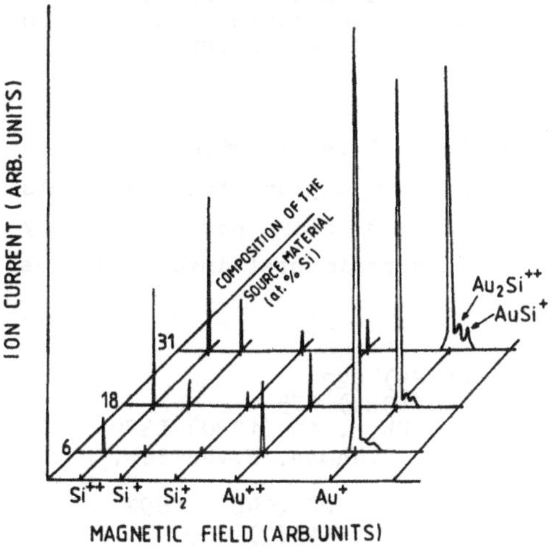

Fig.1. Main ion species of the Au-Si mass spectrum in dependence on the material composition

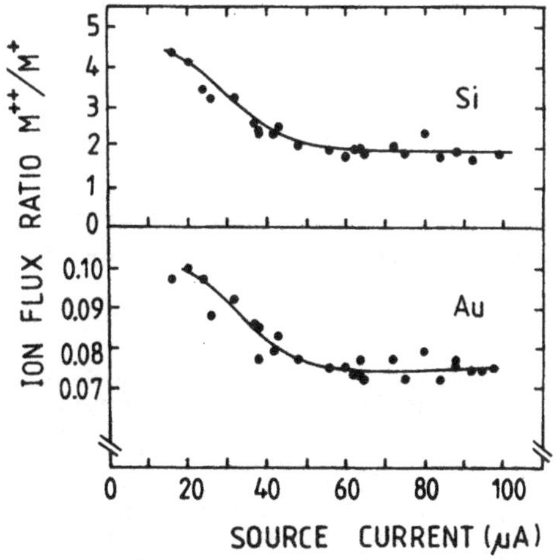

Fig.2. Ratios of double to single charged ion fluxes as a function of the source current ($Au_{82}Si_{18}$ - LMIS)

It is shown in [9] that for increasing source current the Si^+ and $AuSi^+$ fractions increase while the Au^+ fraction decreases and those of Si^{++}, Si_2^+, Au_2Si^{++}, Au^{++} are nearly constant. From these measurements the ratios of double and single charged ions, shown as functions of the source current in Fig.2, are determined. As can be seen, the two ratios decrease with increasing current. These results are in good correspondence with those obtained by Gamo et al.[5].

The observed mass spectrum of the $Au_{73}Ge_{27}$ alloy is represented in Fig.3. It is similar to the Au-Si mass spectrum in [9], but differs from the Au-Ge mass spectrum obtained by Ishitani et al.[8]. They had not found the peaks of the $AuGe^{++}$, Au_2Ge^{++}, Au_2^+, Au_2Ge^+ and Au_3^+ cluster ions. Instead of the Au_2Ge^{++} peak they found Ge_3^+. The influence of the source current on the ratio of Ge^{++}/Ge^{+} is shown in Fig.4.

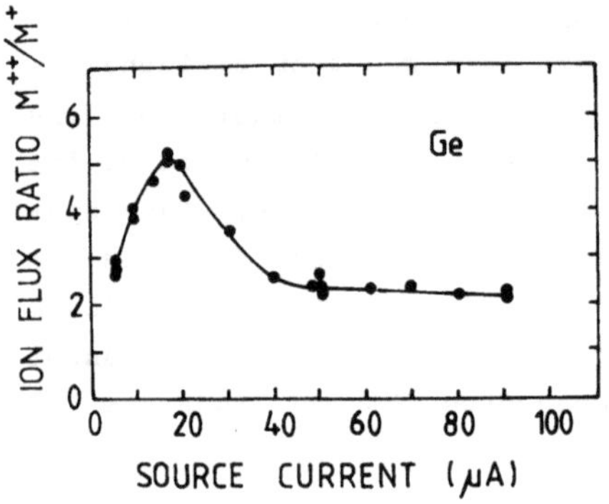

Fig.3. Mass spectrum of the $Au_{73}Ge_{27}$ ion source at a source current of 50 μA

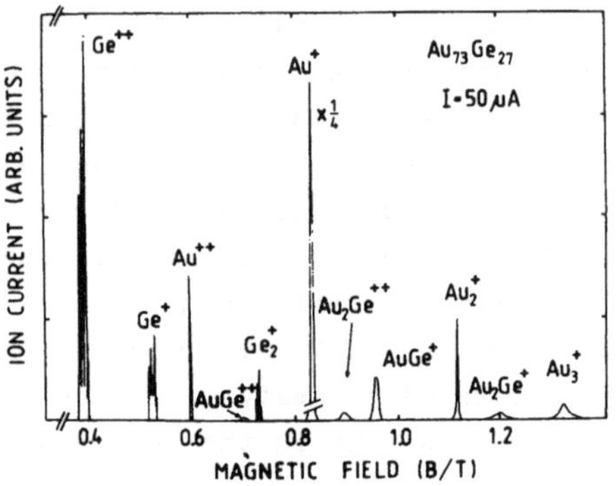

Fig.4. Ratio of double to single charged ion fluxes as a function of the source current

Between 15 and 20 μA source current a maximum of the function could be found. At present we can't give a satisfactory explanation for this maximum, because the emission process is generally not completely understood. That's why further theoretic and experimental investigations are necessary.

[1] Adesida, I., Nucl.Instr.Methods B7/8 (1985) 923

[2] Broughton, D. and R.Clampitt, Vacuum 34 (1984) 275

[3] Gamo, K., T.Ukegawa, S.Namba, Jpn.J.Appl.Phys. 19 (1980) L379

[4] Gamo, K., T.Ukegawa, Y.Inomoto, Y.Ochiai, S.Namba, J.Vac.Sci. Technol. 19 (1981) 1182

[5] Gamo, K., T.Matsui, S.Namba, Jpn.J.Appl.Phys. 22 (1983) L692

[6] Gerlach, W., B.Goel, Solid-State Electron. 10 (1967) 589

[7] Hansen, M., Constitution of binary alloys, Mc Graw-Hill Book Company, Inc., New York, Toronto, London, 1958

[8] Ishitani, T., K.Umemura, H.Tamura, Nucl.Instr.Methods B218 (1983) 363

[9] Machalett, F., R.Mühle, I.Stiebritz, to be published in J. of Phys.D

[10] Wagner, A., Nucl.Instr.Methods 218 (1985) 355

[11] Wang, V., J.W.Ward, R.L.Seliger, J.Vac.Sci.Technol.19 (1981) 1158

CHARACTERISTICS OF AN ExB MASS SEPARATOR FOR FOCUSED ION BEAM SYSTEMS

R.Mühle, F.Machalett, I.Stiebritz [1)]

1. Introduction

In focused ion beam systems ExB type mass filters are often employed for the separation of the different ion species, delivered by the liquid metal ion sources [1,5]. While its mass resolution is considerably lower than that of a sector magnet, the use of this velocity filter leads to a simple optical configuration because of the rectilinear path of the separated ions. The ion optical properties of the ExB type mass separator are described in detail by Seliger [4].

In this paper, the characteristics of such a filter type are reported, which is constructed for ion energies up to 200 keV [3].

2. Experiment

A schematic diagram of the ion optical column is shown in Fig.1. It consists of the following elements: ion source, condenser lens, ExB mass separator and Faraday cage. The emitted ions are accelerated by the extractor voltage to energies between 5 and 25 keV and are focused by the condenser lens on the separation aperture. The length of the field region and of the drift space are 50 and 70 mm, respectively. Separation apertures with diameters of 1, 0.5 and 0.1 mm were used. The magnetic field of the strength of 0.1 T was realized by a Maniperm-ceramic magnet. The ions were collected by a Faraday cage. The separation properties of the filter were tested with AuSi and AuBe alloys. The obtained mass spectra are shown in Fig.2 and 3. The atomic composition of the alloys and the values of source current I_s and extractor voltage U_{ex} are indicated. Using a 0.5 mm diameter separationn aperture, a mass

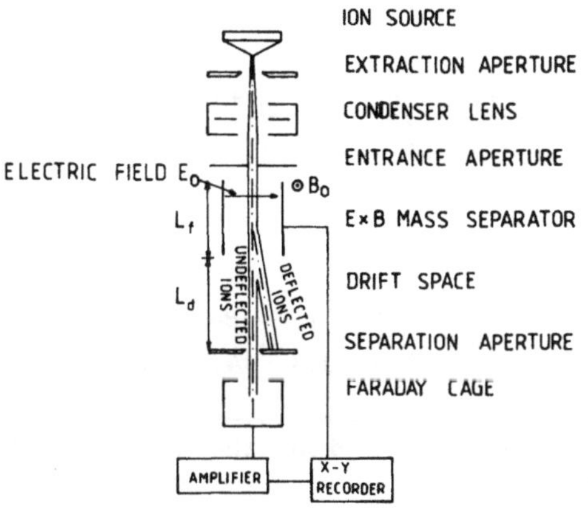

Fig.1. Ion optical column

1) Friedrich-Schiller-Universität Jena, Sektion Physik, Max-Wien-Platz 1 Jena, DDR 6900

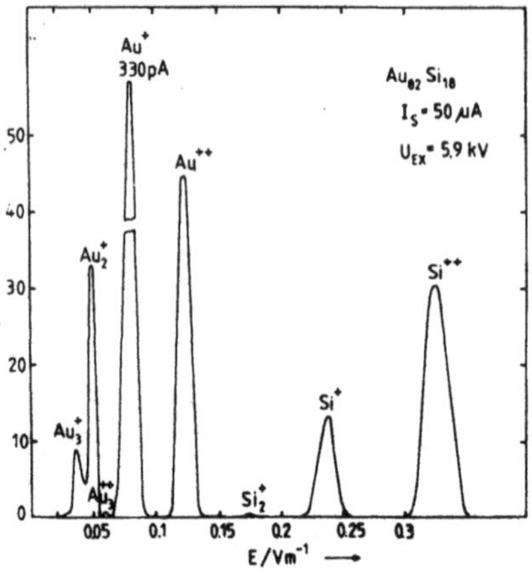

Fig.2. Mass spectrum of the $Au_{82}Si_{18}$ alloy

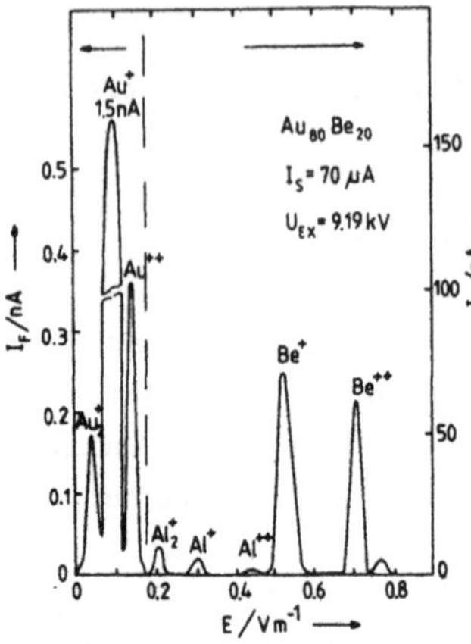

Fig.3. Mass spectrum of the $Au_{80}Be_{20}$ alloy

resolution of about 30 was obtained. The silicon ions could be separated. The observed Al ions in the AuBe mass spectrum arise from the corundum ampoule where the alloy was molten. The separated ion species and the related ion flux fractions are the following:

- $Au_{82}Si_{18}$

ion species	Au_3^+	Au_2^+	Au^+	Au^{++}	Si_2^+	Si^+	Si^{++}
ion flux fraction (%)	2.7	10.0	62	6.7	0.12	7.3	11.2

- $Au_{80}Be_{20}$

ion species	Au_2^+	Au^+	Au^{++}	Al_2^+	Al^+	Al^{++}	Be^+	Be^{++}
ion flux fractions (%)	7.3	65.2	6.8	1.2	0.8	0.4	14.4	3.9

The obtained values for the AuSi alloy correspond with those measured with a magnetic mass spectrometer [2].

[1] Ishitani, T., K.Umemura and H.Tamura, Nucl.Instr.Methods 218 (1983) 363-367
[2] Machalett, F., R.Mühle and I.Stiebritz, to be publ. in J.Phys.D
[3] Matthies, A., Diplomarbeit, FSU Jena, 1987
[4] Seliger, R.L., J.Appl.Phys. 43 (1972) 2352-2357
[5] Shiokawa, T., P.H.Kim, K.Toyoda, S.Namba, T.Matsui and K.Gamo, J.Vac.Sci.Technol. B1 (1983) 1117-1120

CHANNELING EFFECTS AT LOW-ENERGY ION IMPLANTATION

J. Bollmann, A. Mertens, H. Klose [1)]

Introduction

In many cases low-energy ions are used to modify the semiconductor surface (e.g. high-dose ion implantation for preparing of Schottky- or ohmic contacts, of silicide layers, sputter cleaning and etching, or ion implantation for very shallow active layers). With decreasing ion energy the critical angle for scattering of ions into an axial/planar channel increases. The influence of channeling on the depth distribution of implanted ions can not be neglected, especially in the concentration range below about $1x10^{15}cm^{-3}$. At present SIMS is the commonly used technique for detecting depth distributions of foreign atoms in solids. Depending on the element under study only in special cases the detection limit is below $10^{15}cm^{-3}$. For electrically active shallow dopants depth distributions can be detected by differential CV-method. In this case the depth resolution is some Debye-lengths (proportional to the square root of the inverse actual detected concentration), i.e. measurements below $10^{14}cm^{-3}$ will become senseless. Especially during dechanneling, the energy loss of channeled ions causes the creation of vacancy- as well as dopant-coupled point-like defects, often resulting in deep levels within the forbidden gap. The detection by defect-spectroscopic methods combined with depth profiling (DDLTS) /1/ opens the possibility to measure very low concentrations with high accuracy. An upper limit is given by the donor density N_d of the semiconductor, while the lower one is about $10^{-4}N_d$. Combining SIMS and DDLTS depth distributions over a very wide range of concentration can be measured. Additionally, it is possible to study the connection between dechanneled atoms

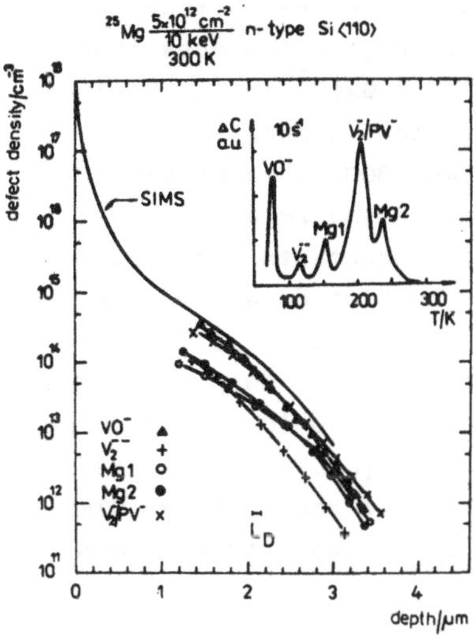

Fig. 1: DLT-spectrum and depth distributions of defects due to Mg-ion implantation in Si

1) Humboldt University , Department of Physics
Invalidenstraße 110, Berlin, 1040, GDR

and the damage produced by them, especially in the channel tail, shown in Fig. 1.

Experiments

Magnesium ions were implanted into 2-5 Ohmcm n-type silicon of different crystal orientation. Some samples could be aligned within 0.1° to the ion beam direction or misaligned into a random direction by RBS-analysis (implanted in a 350-kV-implanter). The most samples were implanted in a 25-kV-implanter with not so good possibility of sample alignment, but the differences of profile tail depths are within of the measuring accuracy (≈0.1 μm). The ion energy was varied in the range from 3 to 25 keV and the ion dose from $5x10^{12}$ to $8x10^{16}cm^{-2}$. The implantation temperature was 77 K and 300 K, respectively. After the implantation Schottky-contacts were prepared by high dose Ag- or Mg -ion implantation /2/ or by gold evaporation.

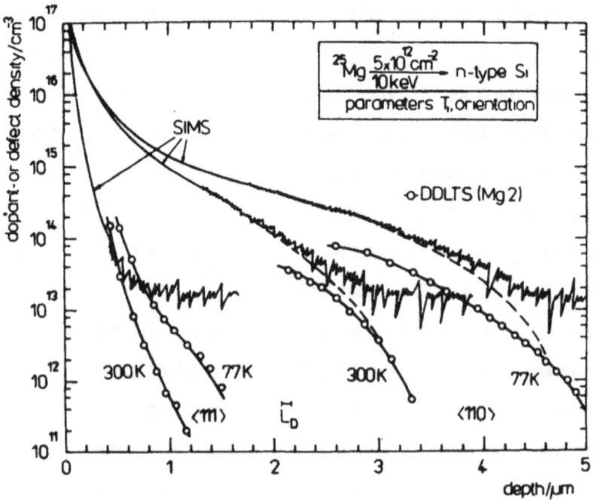

Fig. 2: Mg-depth distributions in Si, measured by SIMS and DDLTS

Measurements of deep level depth distribution were performed with a high sensitive look-in type DLT-spectrometer using a double correlation technique (DDLTS). An emission rate range from 10^5 to 0.2 s^{-1} could be chosen. For some samples SIMS-measurements of the dopant distributions were performed with a CAMECA IMS 3 f. [1)]

Results

A typical DLT-spectrum of deep levels due to mg-ion implantation and the depth distributions of Mg-correlated deep levels as well as vacancy -coupled defects detected by means of DDLTS are shown in Fig. 1. The depth resolution is about one Debye-length (L_D), determined by the donor density N_d. In Figs. 1 and 2 a comparison between the density of deep levels and the density of implanted Mg-ions (SIMS) is also repre-

1) We wish to thank Mr. Gericke for performing the SIMS-measurements.

sented. Obviously, the Mg-correlated defect density is roughly identical to the dechanneled one, and the produced vacancy concentration is about of the same order of magnitude. As Fig. 2 shows, the depth distributions are strongly affected by the crystal orientation and the implantation temperature. The detected distributions follow the expected trend of decreasing penetration depth with increasing temperature due to thermal vibrations of the "channel walls".
With increasing ion dose the amount of ions comming to rest in the tail increases up to a limit of $\gtrsim 10^{14} cm^{-2}$ (Fig. 3). Above this value amorphisation is achieved, and the destroyed surface layer prevents further channeling into the profile tail.

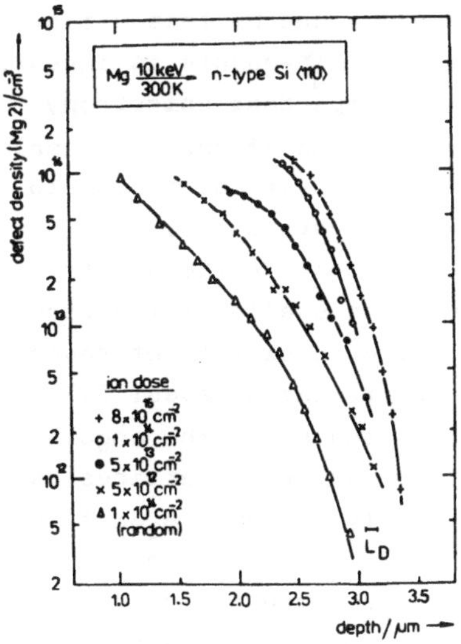

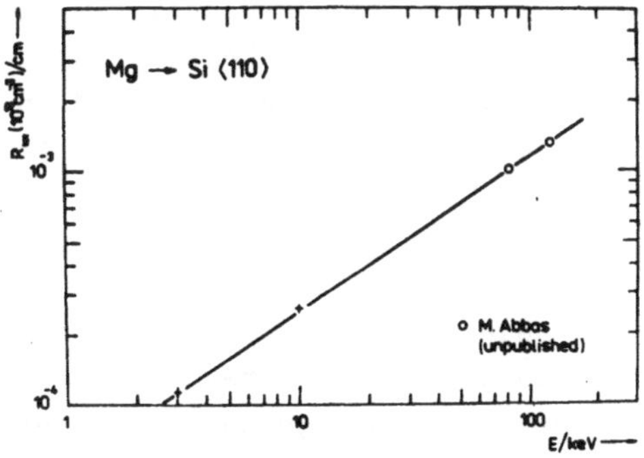

Fig. 4: "Maximum" range R_{tot} of Mg in Si versus ion energy

Fig. 3: Defect (Mg 2) depth distributions in Si dependent on ion dose

Fig. 4 shows a plot of the "maximum" range R_{tot} defined by taking the depth in the tail region at an arbitrary density ($1x10^{13} cm^{-3}$) versus the ion energy.[1] Using the semiempirical fit (electronic stopping) $R_{tot} = A\, E_o^{\,s}$ one will get values of $s = 2/3$ and of the extrapolated electronic stopping power S_e ($1.5x10^8$ cm/s) = $4x10^{-13}$ eVcm^2/atom.

References

/1/ Lefevre, H., M. Schulz: Appl. Phys. 12 (1977) 45.

/2/ Bollmann, J., H. Klose and A. Mertens: phys. stat. sol. (a) 97 (1986) K95.

1) We are most grateful to our colleagues Dr. Kerkow for carry out special aligned implantations and Mr. Abbas for two unpublished data.

POSSIBILITIES OF THE TDPAC METHOD FOR THE DIAGNOSIS OF RADIATION DAMAGE AND ANNEALING PROCESS IN IMPLANTED SEMICONDUCTORS

S. Unterricker[1)]

Nuclear methods are of growing importance in the diagnosis of implanted semiconductors. In particular nuclear spectroscopic methods like Mössbauer effect, perturbed angular correlations (TDPAC) and muon spin rotation (μSR) can yield valuable informations concerning the structure of the microscopic environments of implanted atoms /1, 2/. Thereby cases are of special interest where nuclei of implanted atoms act as probes. By TDPAC the hyperfine interaction of certain radioactive probe nuclei which causes a precession of nuclear spins and by this a fast rotation of the emission characteristics of nuclear γ-quanta is investigated. The precession is reflected in the time dependent anisotropy $A_2G_2(t)$. From the time spectra $A_2G_2(t)$ distributions of the quadrupole frequencies, statements to the symmetry of probe environments and to the fractions of probes at different sites can be evaluated.

The quadrupole frequencies which are proportional to the electric field gradients = efg's only respond to the first shells of the probe environments therefore they can yield a microscopic picture of these environments. We distinguish the following characteristic cases: (i) for purely cubic surroundings $A_2G_2(t)$ is time independent or the angular correlation is unperturbed, (ii) tetragonal or hexagonal lattices as well as neighboured single defects cause strictly periodic $A_2G_2(t)$ pattern, (iii) frequency distributions with a most probable frequency $\hat{\omega} = 0$ (random defects, radiation damage) produce a monotonic decay of $A_2G_2(t)$ up to the hard core (h.c.) value, and (iv) for frequency distributions with a maximum at $\hat{\omega} > 0$ we get a more or less damped version of the second case. $A_2G_2(t)$ decays in an oscillating manner to the h.c. value.

An essential problem is the implantation of radioactive probes into the solids. Preferably we used the recoil implantation after nuclear reactions. The main advantage of this method is the simultaneous production and implantation of radioactive probes. Naturally, the number of systems which can be prepared by this method is limited.

We have investigated Si, the $A^{III}B^{V}$ compounds InP, InAs and InSb, the $A^{II}B^{VI}$ compounds CdS, CdSe, CdTe and ZnSe and ternary semiconductors like $CdSiP_2$, $CdGeP_2$, $CdSnP_2$, ...

Some results

^{111}In (daughter ^{111}Cd) implanted n-type Si shows wide frequency distri-

1)Bergakademie Freiberg, Sektion Physik, B.-v.-Cotta-Str. 4, Freiberg 9200, DDR

butions $P(\omega)$ with a maximum at $\hat{\omega} = 0$ (Fig. 1). By a thermal treatment the decay of $A_2G_2(t)$ will be flattened and $P(\omega)$ becomes smaller. As verified by Fig. 1, a nearly complete annealing can be obtained. For p-

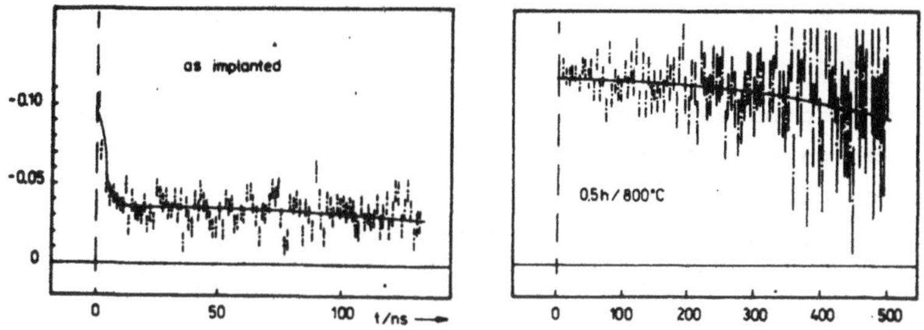

Fig. 1: $^{111}In(^{111}Cd)$ implanted n-type Si

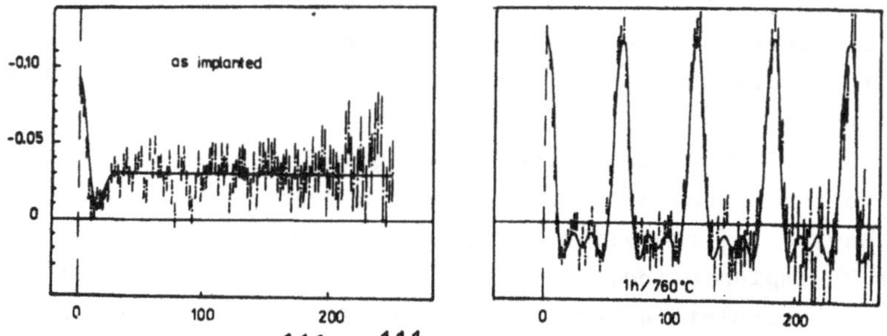

Fig. 2: $^{111}In(^{111}Cd)$implanted $CdSiP_2$

type and intrinsic Si such a complete annealing of probe environments is impossible /3/. This behaviour can be explained, after /3/, by the concentration of free electrons in the conduction band. These electrons can prevent $^{111}In \rightarrow {}^{111}Cd$ decay after-effects.

A complete annealing is also possible in the cases of $^{111}In(^{111}Cd)$ as well as $^{118}Sb(^{118}Sn)$ implanted InP. But the two probes behave different because In is a host atom and Sb an impurity in InP. In the first case the annealing starts with a homogenization and after this a cancellation of lattice defects whereas in the latter the decrease of quadrupole frequencies and the increase of annealed probe environments take place simultaneously.

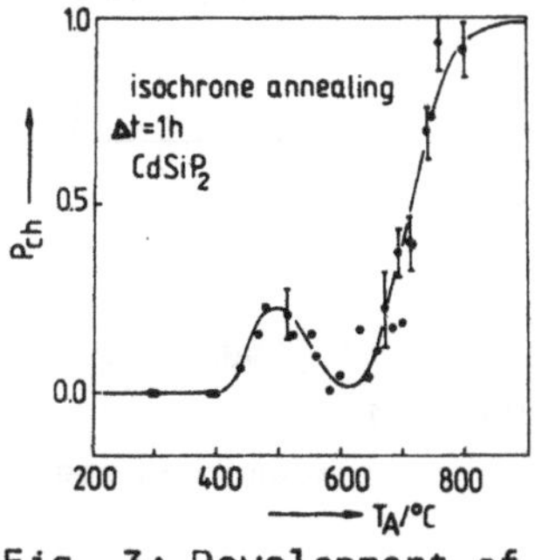

Fig. 3: Development of chalcopyrite environments in $CdSiP_2$

Fig. 2 shows $A_2G_2(t)$ spectra for as implanted and annealed $CdSiP_2$ which has a noncubic lattice structure. Again the implantation produces highly disordered regions. But now $P(\omega)$ has its maximum at $\hat{\omega} > 0$. The entirely periodic time spectrum of the annealed sample demonstrates that the efg is axially symmetric and with the quadrupole coupling constant ν_Q the A-site position of the In-probe in the chalcopyrite structure ($\bar{4}2m$) could be decided. Drastic changes of the time spectra occur in a sequence of isochrone annealing steps. The part p_{ch} of probes in chalcopyrite environments versus the annealing temperature T_A is represented in Fig. 3. In the heating progress we see an inverse annealing (T_A = 500...600 °C) and after this a steep second annealing stage.

Discussion

In the as irradiated samples a considerable part of probes is positioned in highly disordered (amorphized) regions. The maximum of the corresponding frequency distributions $P(\omega)$ is at $\hat{\omega} = 0$ for semiconductors with cubic structures, for such with noncubic structures $\hat{\omega} > 0$ it is valid. By the bond bending model which is based on the conception of a distorted tetrahedral coordination around the atoms (probes) we can understand this experimental fact. The bond directions of the four next neighbours (n.n.) are bent in a random manner thereby the n.n. arrangement of the ordered structure is most probably. According to this model $P(\omega)$ has its maximum at $\hat{\omega} = 0$ for cubic structures and $\hat{\omega} \sim 2\pi \nu_Q$ for noncubic structures.

The annealing behaviour of the investigated semiconductors is closely connected to their bonding properties. As a measure for extent and annealing behaviour of radiation damage we use the part p_d of probes which are, after RT implantation, localized in defect clusters and the temperature T_A^c for complete annealing (see Table). The different bond properties of the substances are reflected by the spectroscopic ionicity f_i of Phillips /4/. If f_i increases p_d and T_A^c decrease like other material properties which characterize the bond strength. The ternary $A^{II}B^{IV}C_2^V$ compounds behave similar to the $A^{III}B^V$ substances.

	p_d in %	T_A^c in °C	f_i /4/
A^{IV} Si	100	650	0
$A^{III}B^V$ InP	90, In-111 70, Sb-118	650	0.42
$A^{II}B^{VI}$ CdS	33	300	0.69
$A^{I}B^{VII}$ CuBr	0 (EPR) /5/	RT	0.74
$A^{II}B^{IV}C_2^V$ $CdSiP_2$	100	720	0.38

References

/1/ Weyer, G. et al.: Phys. Rev. B 11 (1980) 4939

/2/ Deicher, M. et al.: Nucl. Instr. Meth. Phys. Res. B 13 (1986) 499

/3/ Wichert, Th. et al.: Mat. Sci. Forum 10-12 (1986) 551

/4/ Phillips, J. C.: Bonds and Bands in Semiconductors, Academic Press, New York, London 1973

/5/ Goltzene, A. et al.: Nucl. Instr. Meth. Phys. Res. B 1 (1984) 427

QUADRUPOLE-SIMS-MEASUREMENTS OF IMPURITIES IN SiO_2/Si-STRUCTURES

H. U. Marquardt, U. Gunst, and B. Müller [1)]

1. Introduction

The redistribution of impurities in Si or poly-Si and especially on the interface SiO_2/Si in the process of oxidation is a very important problem in the microelectronic technology (1, 5). It appears that quadrupole based equipments have a advantage for the investigation of SiO_2/Si-structures, if there is no possibility for the compensation of the by ion bombardment produced charge on the insulator surface. We have developed a small quadrupole based SIMS equipment for depth profiling measurements with Ar^+-bombardment, electronic gating, and O_2-jet (3, 4). The redistribution of phosphorus on the SiO_2/Si interface due to the oxidation (wet) of phosphor-doped poly-Si-layers under variation of process conditions, and first chlorine profiles were studied in our system.

2. Methodical and experimental aspects

Electrical nonconducting samples requier a special sample preparation and/or measuring method to compensate the charge on the sample or to eleminate the influence of the charge. There are mainly two effects influencing the results of the measurements:

1) The charge built up on insulators during the analysis causes primary a strong intensity-loss of the measured secondary ions. The reason is the shifting of the energy distributions about the amount of charge of the sample surface.
2) A further effect consits in the diffusion of mobile ions in the insulator, which results from the electrical field in it.

We were able to compensate during the measurements the shifting of the energy distributions through continuous variation of the sample-potential, as is also shown in Fig. 2. We changed this potential in such a kind, that the measured $^{28}Si^+$- respectively $^{16}O^+$-Intensities were constant. With the help of a

1) PSF 327, DDR - 6300 Ilmenau, GDR.

simple model we have evaluated the field intensity in the SiO_2 from the potential shifting (3).

The enhancement of mobile ions on the interface SiO_2/Si is not avoided through the variation of the sample-potential. As our measurements have shown the maximum of intensity of this ion kind lies on the interface. We used this fact in the cases, in which the measurements were done with oxygen-inlet in order to identify the interface. For the correction of the sputtering rates we used a very simple model assuming a linear change of the sputtering rates with the sputtertime. The measurements were taken up at a primary ion energy of 4 keV respectively 3 keV and primary ion fluxes between 3 and 40 µA/cm² (scanning ion beam). In the case of the phosphorus measurements we worked with oxygen inlet to increase the secondery ion yields. For the transmission of secondary ions into the quadrupol-system an electronical window was used. The estimation of the sputtering rates tooke place by means of known layer thickness, crater profile and depth measurements with the devices "Taly Step" and "OPM", and the sputtertime.

3. Results

3.1. Measurements of chlorine profiles

Chlorine profiles in thermal silicon oxides prepared in O_2/HCl ambient (3% HCl) at 1373 K have been measured. Typical chlorine profiles for pos. and neg. secondary ions are shown in Fig. 1 and Fig. 2.

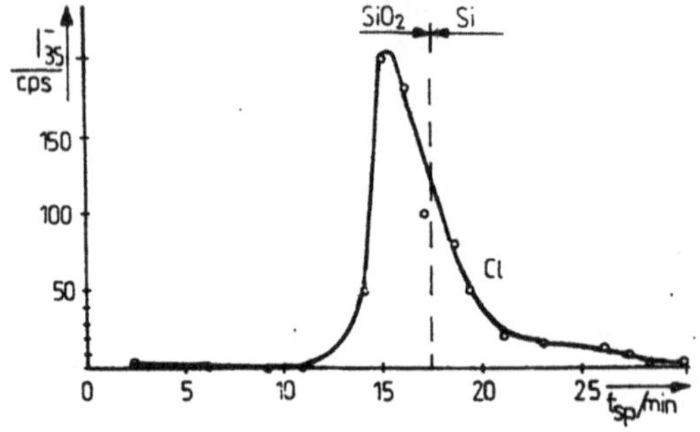

Fig.1 Cl-depth profile (neg. secondary ions) of SiO_2 and Si (70.1 nm gate oxide, sample potential constant V_t = 0 V, Ar^+ bombardement only).

The oxides represent standard processing conditions. As in other works carried out in this direction (2, 6) Cl piled up in the SiO_2 in a region less than 20 nm from the SiO_2/Si-interface.

The value for the Cl-peak halfwidth (100 nm for pos. and 140 nm for neg. Cl-ions) shows a good agreement with (2).

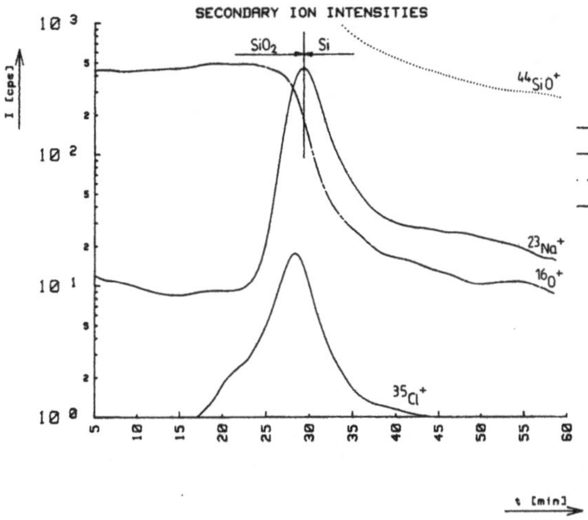

Fig.2 SIMS depth profile of SiO_2 and Si (70.1 nm gate oxide, measured without charge compensation, sample potential variable V_t = (0...30) V, Ar^+-bombardment only).

3.2. Measurements of phosporus profiles

Besides gate oxides also thermal oxides from poly-silicon have many uses as mask oxides for poly-Si pattering, or as an intermediate dielectric. The control of the phosphorus content in the oxide is important in all cases because the poly-Si structure, kind and concentration of impurities, and oxidation parameters influence the redistribution of the impurities on the SiO_2/Si-interface and in the SiO_2 bulk. Therefore we have oxidized (wet,under different parameters) and ivestigated high-doped LPCVD poly-Si (Tab. 1). Fig. 3 shows a typical sample geometry (the poly-oxide thickness was 40 nm up to 110 nm) and measured Na- and P-distribution for sample 2.

sample no.	oxide-thickness nm	oxidation-time min		oxidation-temp. K	q_{SIMS}	q_{AES}
1	101	10	2)	1223	-	1.3
2	76	20	2)	1123	1.8	1.5
3	45	60	1)	1023	1.5	1.8
4	89	30	2)	1123	2.7	2.2
5	92	420	2)	1123	1.2	1.2
6	59	60	2)	1023	-	2.1

Tab. 1: List of samples with parameters of wet oxidation and pile up effect measured with SIMS and AES (3), temp. of bubbler 1) 297 K and 2) 363 K, q is ratio of the dopand concentration in the maximum to the bulk($q = c_{Pmax}/c_{Pbulk}$).

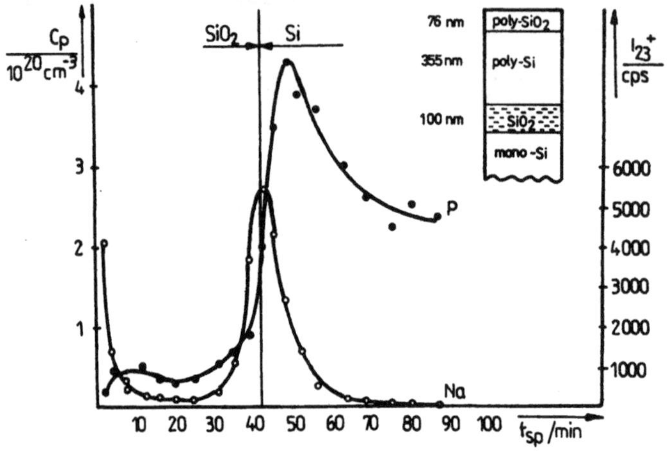

Fig. 3: P and Na depth profile of poly-SiO_2 and poly-Si, without charge compensation, sample potential variable, Ar^+-bombardment and O_2-backfilling.

In every cases of 1223 K we could not find phosphorus in the oxide ($10^{19}/cm^3$).But for 1023 K and short time oxidation (wetness high) we measured posphorus concentrations up to $10^{20}/cm^3$. For 1023 K we could detect phosphorus in the oxide in all cases. The pile up effect measured with SIMS and AES is shown in Tab. 1 (3). There are a qualitative agreement between SIMS and AES results, beyond it we have also in the tendency a good agreement with the literature (1, 5). The measurements will be in progress with better instrumental parameters (new ion gun).

References

(1) Chang,C.C., T.T.Sheng, T.A.Shankoff: J.Electrochem.Soc. 5 (1983) 1168-1171

(2) Deal,B.E., A.Hurrle, M.J.Schulz: J.Electrochem.Soc. 12 (1978) 2024-2027

(3) Marquardt,H.U., Th.Stauden: Dissertation, Ilmenau Institute of Technology 1986

(4) Müller,B., H.U.Marquardt, Th.Stauden: Proceedings of the 7th Czchoslovak Conference on Electronics and Vacuum Physics, Slovak Academy of Science (1985) 377-386 and 405-413

(5) Schwarz,S.A., R.W.Barton, C.P.Ho and C.R.Helms: J.Electrocem.Soc. 5 (1981) 1101-1106

(6) Tsai,H.L., S.R.Butler, D.B.Williams, H.W.Kraner and K.W. Jones: J.Electrochem.Soc. 2 (1984) 411-418

HIGH TEMPERATURE ION SOURCE FOR IMPLANTATION PURPOSES

K. Kiszczak, A. Latuszyński, D. Mączka,[+] A. Podgórski*

INTRODUCTION

The modification of surface properties of solids by means of ion beam method has met with growing interest both in research laboratories and in industry. This situation produces some specific requirements which newly constructed ion implanters, and mainly their sub-assemblies - ion sources, must fulfill.

The production of ion beams of a number of elements, sometimes with very different physical-chemical properties, makes it necessary to equip implanteres with several various types of ion sources, or compels one to look for some universal solutions. Fundamentally, it is possible only in the case of the generation of ions of stable elements, when a smaller ionization efficiency is permissible and the essential thing is to obtain properly strong and stable ion beams.

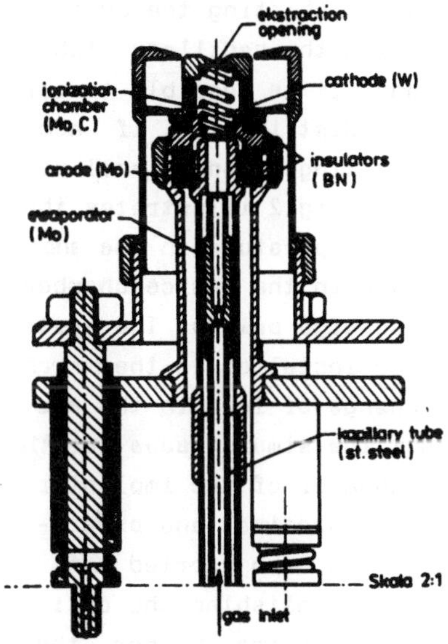

Fig.1 Discharge chamber and mechanical construction of the vacuum part of the source.

This paper presents the structure and performance characteristics of a plasma ion source with the so-called hollow cathode discharge. Its construction has been the development of the sources worked out in the Institute of Physics, MCSU, and in the Joint Institute for Nuclear Research in Dubna for ISOL systems[(1-4)]. The changes introduced have aimed at an increase of ion currents, especially during the ionization of hard-volatile elements. For example, with Al, Ti, Mn, we always get currents of the order of 200 μA with current densities ~ 0.5 mA/mm^2.

The model presented in this paper has been, for many years, used in the ion implanter 300 keV in the Institute of Physics, MCSU, Lublin, and also with the implanter MPB 202RP or BALZERS in IEMT,

[+]Institute of Physics, M. Curie-Skłodowska University, MCSU, Lublin, Poland
*Institute of Electronic Materials Technology, IEMT, Warsaw, Poland

Warsaw, in which, as its parameters were more favourable, it replaced a factory-made source of DANFYSIK[5].

CONSTRUCTION AND SOURCE CHARACTERISTICS

Fig.1 shows the mechanical construction of the vacuum part of the source. The electrodes of a discharge chamber are made of molybdenum and the spiral cathode of wolfram wire; d = 0.5 - 0.8 mm.

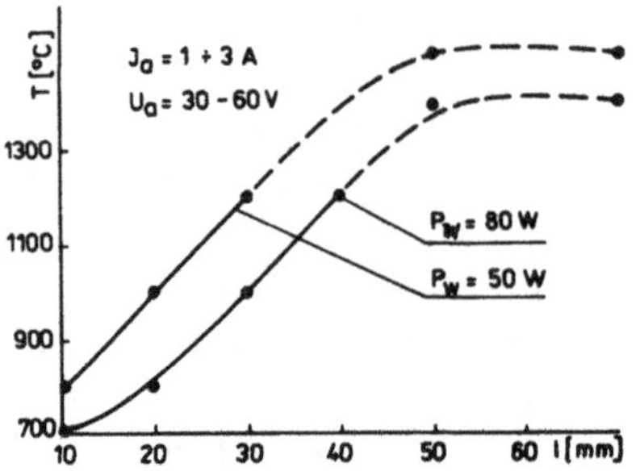

Fig.2 Temperature distribution along l the anode cylinder and the source discharge chamber for different powers emitted within the region of the discharge.

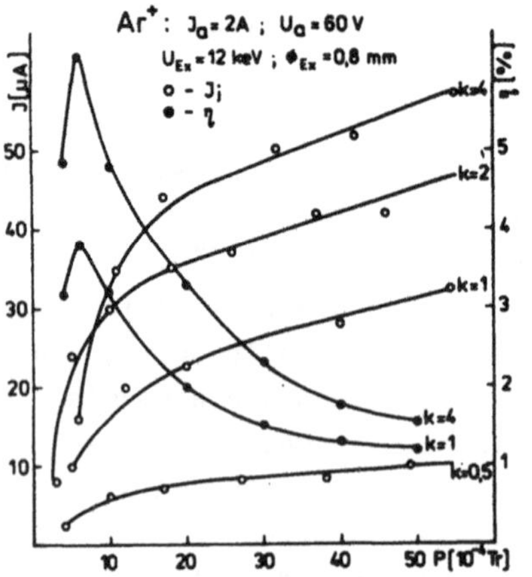

Fig.3 The dependence of ion currents J emitted from the source and of ionization efficiency η on pressure **p** in the discharge chamber for different values of the parameter characterizing the emission channel, **k**.

A system of compression springs, applied here, eliminates the negative effects of thermal expansion of particular elements of the source and ensures satisfactory air-tightness of the discharge chamber. Ionized substance is put into the evaporator in the cylindrical anode and is evaporated with heat supplied from the discharge chamber. A suitable temperature of the evaporator is chosen by means of shifting the evaporator, together with the capillary tube within the anode. It is possible to correct temperature distribution, if necessary, by installing an additional resistance furnace. Fig.2 illustrates the distribution of temperature in the anode hollow region and in the source chamber, for various discharge powers. The system of vacuum valves installed in the source body enables change of load in the evaporator without the simultaneous aeration of the source chamber of the implanter.

Extensive theoretical and experimental research has been carried out with the aim of establishing the optimum geometry of the discharge chamber, the ion extraction unit, and also the operating conditions of the source in the process of ionization of various substances. Fig.3 shows some of the results of this research, i.e. the dependence of ion currents emitted from the source J, as well as in source efficiency η , on pressure in the discharge chamber, with regard to the geometric factor of the emission

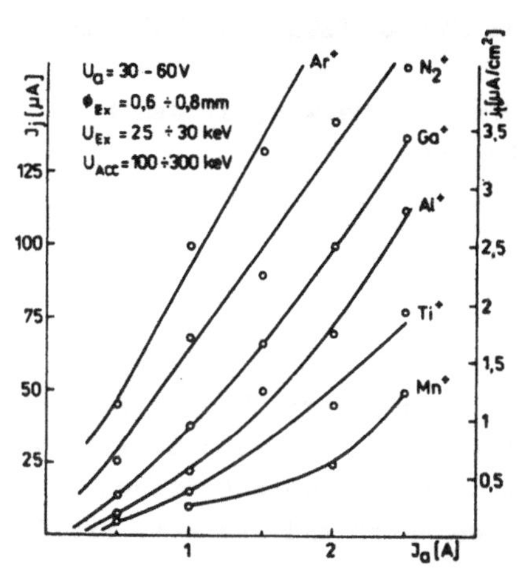

Fig.4 Ion currents J_j and ion current densities j_+ on target surface of chosen elements as a function of source discharge current, J_a.

channel. The parameter describing the emission channel has been established as the ratio of its radius **R** to its length L; ($k = \frac{R}{L}$). The dependencies presented in Fig.3 can be described by approximate formula:

$$J_+ = Cen_+ \sqrt{\frac{kT}{2\pi M}} S_e\left(\frac{R}{L}\right)$$

where: n_+ is ion density in the plasma, T - temperature, **k** - the Boltzmann constance, **M** - the mass of the gas atom, $S_e(\frac{R}{L})$ - area of the plasma meniscus. **C** = const.

In the process of generating ions of solids, pure elements and their compounds were used as output material, chlorination with CCL_4 carrier the source volume was also applied. Fig.4 shows obtained ion currents of various elements as a function of source discharge current. The results of research permits the determination of the following technological--operating paraments of the source model presented in this paper:

TABLE 1. Ion source parameters

1. discharge current; J_a=0-3A	2. discharge voltage; U=30-60V
3. cathode current; J_k=26-40A	4. magnetic field; B=20mT
5. evaporator temp.; t=500-2000^{0}C	6. ionization efficiency; $\eta \approx 6$ %
7. gas consumption: Q=10-30 cm^3Tr/min	8. density of the ion beam ~ j = 0.5mA/mm^2
9. cathode life-time: 0.6mm, t ≳ 25 h	10. emission channel geometry; Ø=0.8mm, R/L ≳ 2
11. minimal working pressure in the discharge chamber; p≥5·10^{-4}Tr	

This work was partly supported by the CPBP 01.06 programme of the University of Warsaw.

REFERENCES

(1) Sidenius, G: Proc. 7th Int. EMIS Conf. Marburg, 1970
(2) Latuszyński, A., D. Mączka. Yu. Yushkievich, K. Kiszczak: Vacuum 36 (1986) 263
(3) Mączka, D., A. Latuszyński, R. Kuduk, J. Partyka: Nucl. Instr. and Meth. B21 (1987) 521
(4) Latuszyński, A., D. Mączka and Y. Yushkievich, Proc 4th Int. Conf. on Ion Impl. Berchtesgaden, FRG (1981) 106
(5) DANFYSIK, Catalogue, 1980

A NEW CONSTRUCTION OF ION IMPLANTER AT THE SLOVAK TECHNICAL UNIVERSITY

Š. Bederka, I. Bello, V. Dúbravcová, J. Grňo, R. Harman, J. Janík,
J. Kral, V. Kremničan, J. Mráz, A Ritomský 1/
J. Ďurčanský, I. Grellneth, V. Illich 2/

Introduction

An ion implanter was developed and constructed at the Department of Microelectronics, Slovak Technical University. This machine will be used for research and technology development in microelectronics. In this paper, the construction of this ion implanter and its some preliminary characteristics are described.

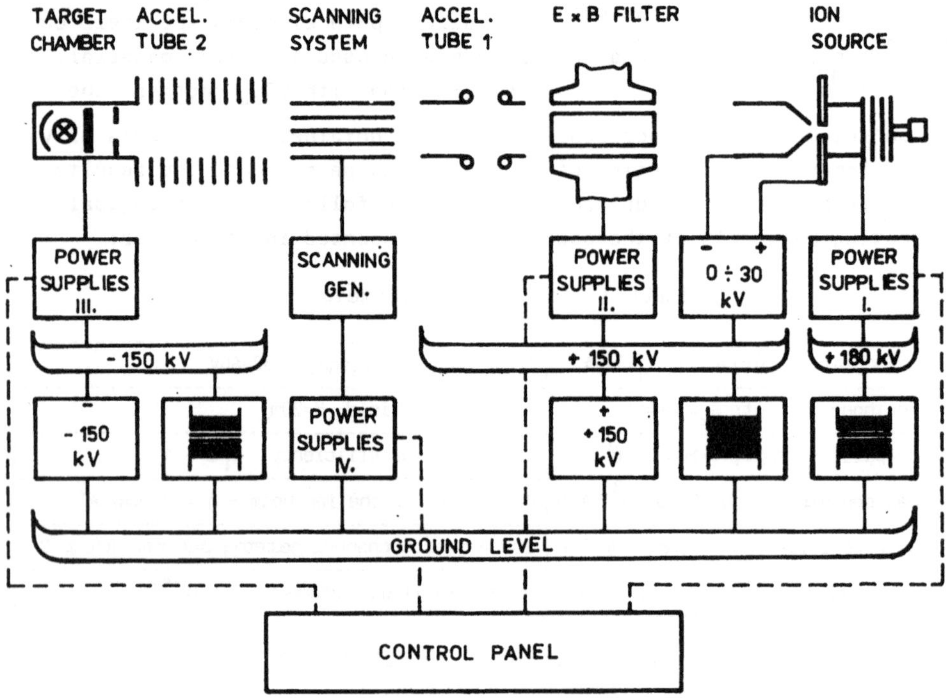

Fig. 1. Block diagram of the ion implanter

1/ Microelectronics Department Slovak Technical University, Mlynská dolina 812 19 Bratislava.

2/ Development Laboratories and Workshops, Slovak Technical University, Vazovova 5, 880 43 Bratislava.

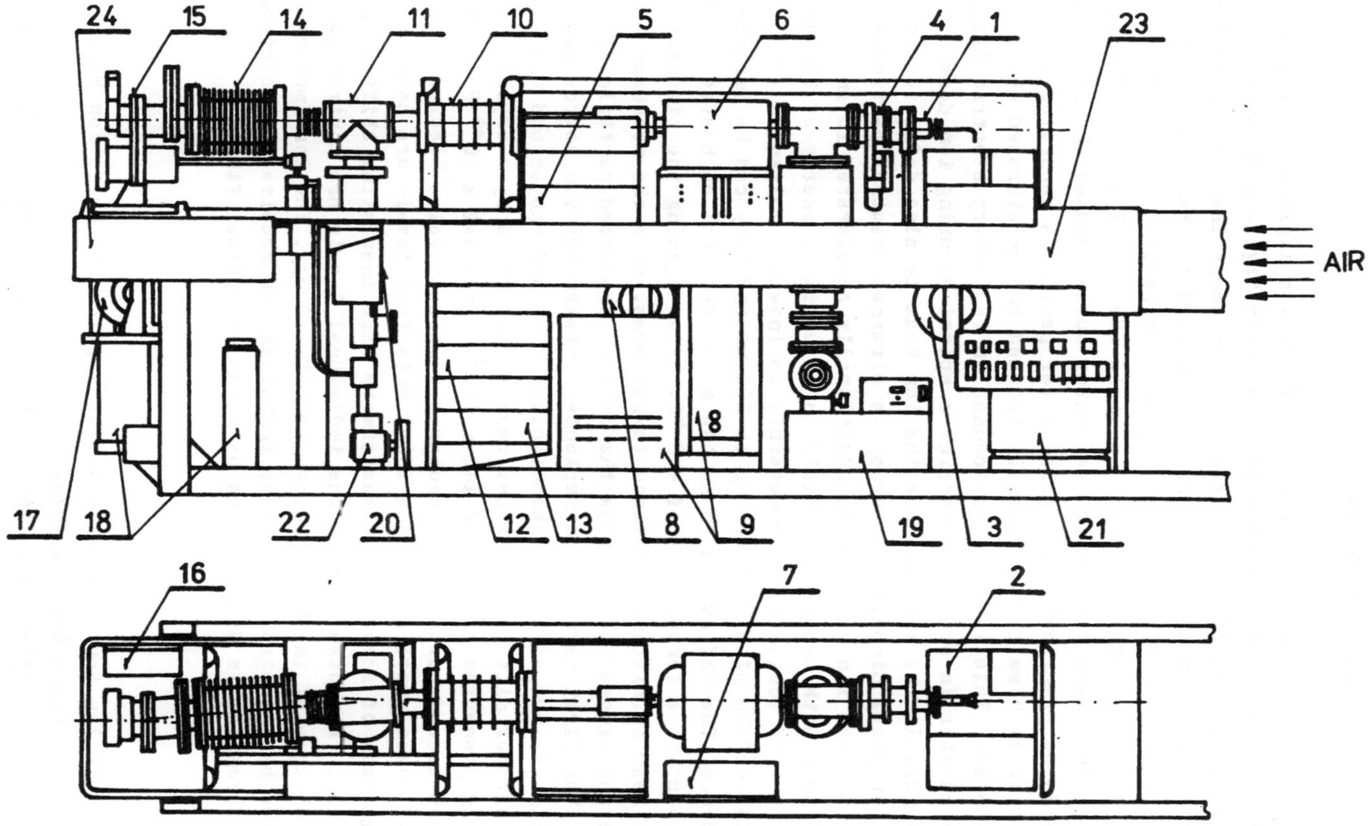

Fig. 2. Overal view of the ion implanter design

1,2 - ion source and its power supplies, 3,8,17 - insulating transformers, 4 - extraction chamber, 5,6,7 - mass separator and its power supplies, 9,18 - HV power supplies, 10,14 - accelerating tubes, 11,12 - scanning chamber and generator, 13 - power supplies, 15,16 - target chamber, 19,20,21,22 - vacuum pumps, 23,24 - insulating tables.

Description of Ion Implanter System

A construction concept and function of our ion implanter are evident from Fig.1 and the arrangement is clear from Fig.2. Ion species are created in a hollow cathode ion source /1/ from where they are extracted by electric field. Then they are separated in an E x B mass separator with energy from 25 to 30 keV. In the separation region, the maximal magnetic induction and maximal electric field are 0.72 T and $2x10^5$ V/m, respectively. The plate lens in the separator outlet causes the ion beam crossection to be practically a circle. After passing a separator slit with a diameter of 3 mm the ion beam is focused and accelerated up to an energy of 180 keV. This ion beam is scanned in the rotary electric field of the scan chamber in order to describe a spiral securing ion dose homogeneity /2/. In the scan chamber the ion beam is also deflected by 5° from the original direction in order to eliminate neutral particles. Already the scanned ion beam can be additionally accelerated in an accelerated tube with homogeneous field to an ultimate energy of 330 keV for single charged ions. There is possible to insert simultaneously 25 semiconductor wafers of a diameter up to 100 mm and they can be implanted in sequence. The wafers placed on the metal pads can be heated during implantation up to a temperature of 400 ^oC. After passing the circular orifice ion beam is hitting a defined target area. From the measured beam charge and exposed target area the dose of implanted particles is determined. The first part of ion implanter between the ion source and the separator slit is pumped by a turbomolecular pump and the second one from the separator slit to the target chamber is pumped by a cryogenics pump. Working pressures are around $1 \div 2 \times 10^{-3}$ Pa and $1 \div 2 \times 10^{-4}$ Pa in the first and the second parts of the equipment, respectively. Electronic circuits and power supplies on individual potential levels are connected to mains by means of insulating transformers with insulating ability up to 200 kV /3/. The potentials of the individual levels are controlled by output voltages of the HV supplies /150 kV - 2 mA/. The ion implanter is controlled by a control unit, that transmits and receives the control and back information in digital form by means of optical fibers.

References

/1/ The Danfysik Ion Source Model 911, Description and Operation Manual, Danfysik Jillinge - Denmark 1975.

/2/ Bederka, Š., Kremničan, V., Višniar, M.: Proc. of 7-th CS Conf. on Electronics and Vacuum Phys., Bratislava 1985, s. 990

/3/ Mráz, J., Klimeš, M., Plávka, J.: Elektronik, Vol. 40, No 5 /1982/ 131 /In slovak/

THERMAL PROCESSES UNDER INTERACTION OF FOCUSED ION BEAMS WITH SEMICONDUCTOR CRYSTALS

I.A.Glushtsov, V.I.Korzjuk, Y.A.Podlipko, A.I.Urbanovitch[1)]

Usage of focused ion beams (FIB) in microelectronics is one of the most promising trends allowing to simplify VLSI production technology and increase their integration level. The use of FIB is connected with application of liquid metal ion sources in ion-focusing systems. The advantages of such sources make them applicable in such developing at present fields using FIB as ion beam submicron lithography [1], maskless implantation [2,3], analytical scanning ion-mycroscopy [4], maskless etching [5]. However, up to the present many questions concerning the usage of such beams have not been solved yet, in particular, the problem of thermal heating of materials exposed to radiation. The calculation of temperature fields in Si wafers and in a series of materials exposed to irradiation by ion beams with submicron diameter has been carried out in this work. The model used represents the case when ion beam diameter is much less than the wafer thickness. Calculating the temperature fields the temperature changes have been taken into account not only in radial direction over the crystal surface, but also into the depth of the crystal. Let ion be normally incident at the boundary of semi-infinite medium z=0. Strictly speaking, it is necessary to allow for a finite thickness of the target d. In real experimental conditions, however, $d \sim 10^{-4}$ m and ion-beam diameter was $2a \sim 10^{-6}$ m and less. Because of this the medium could be considered as semi-infinite. As a result of interaction of the beam with the target the main part of the energy that is released by ions retarding in the crystal is turned into the target heating. Let the function describing this heat released at the boundary z=0 be given as Gaussian that characterizes the density distribution of incident ions in the beam:

$$A(x,y,t) = \mu \cdot I_0(t) \cdot e^{-\frac{x^2+y^2}{a^2}} \tag{1}$$

where μ $(0<\mu<1)$ is the factor allowing for the energy fraction of ion beam transforming into heat; a and I_0 are a halfwidth of the beam and an intensity in the center of the beam respectively.

Part of the energy which is released as heat would be radiated at the boundary z=0 from the target surface according to the Stefan-Boltzmann law. The energy losses due to convection and dissipation have been ignored as a negligible quantity. The linear approximation for heat irradiation depending on temperature has been used. Based on this linear dependence the temperature distribution T(x,y,z,t) in the target would

1) Belorussian State University Minsk, Prospect Lenin 4, Minsk 220080, USSR

obey the equation:

$$\rho C \frac{\partial T}{\partial t} = \chi_0 \cdot \nabla^2 T \tag{2}$$

with initial conditions:

$$T\big|_{t=0} = T_0 = const \tag{3}$$

and boundary conditions:

$$\chi_0 \frac{\partial T}{\partial z} - 4\sigma T_0^3 (T - T_0)\big|_{z=0} = A(x, y, t) \tag{4}$$

where σ is Stefan-Boltzmann's constant; ρ, C, χ_0 are density, specific heat capacity and thermal conductivity of the target respectively; T_0 is ambient temperature. The solution of the task (2-4) could be written inthe form:

$$T(r, z, t) = T_0 + \frac{\mu a^2}{\sqrt{\rho C \pi \chi_0}} \cdot \int_0^t \frac{I_0(\varphi) \cdot H(z, 0, t-\varphi) \cdot e^{-\frac{r^2}{a^2 + \frac{4\chi_0}{\rho C}(t-\varphi)}}}{(t-\varphi)^{1/2} \cdot \left[a^2 + \frac{4\chi_0}{\rho C}(t-\varphi)\right]} d\varphi \tag{5}$$

where:

$$r^2 = x^2 + y^2 \tag{6}$$

$$H(z, 0, t-\varphi) = e^{-\frac{\rho C z^2}{4\chi_0 (t-\varphi)}} - \frac{4\sigma T_0^3}{\chi_0} \cdot \int_0^\infty e^{-\frac{4\sigma T_0^3}{\chi_0}\omega - \frac{(z+\omega)^2 \rho C}{4\chi_0 (t-\varphi)}} d\omega \tag{7}$$

So as to analyse the solution (5) it is convinient to go to dimensionless variables:

$$\xi = \frac{r}{a}; \quad \tau = \frac{t}{t_0}; \quad \ell = \frac{z}{a}; \quad \eta = \frac{\varphi}{t_0} \tag{8}$$

and designations:

$$t_0 = \frac{\rho C a^2}{4\chi_0}; \quad \Theta = \frac{T - T_0}{T_0}; \quad \gamma = \frac{4\sigma T_0^3 a}{\chi_0}; \quad F(\tau) = \frac{\mu \cdot a \cdot I_0(\tau)}{2\sqrt{\pi}\, \chi_0 T_0} \tag{9}$$

$$H(\ell, 0, \tau-\eta) = e^{-\frac{\ell^2}{\tau-\eta}} - \gamma \int_0^\infty e^{-\gamma y - \frac{(\ell+y)^2}{\tau-\eta}} dy \tag{10}$$

Based on (8-10) and (5) we receive:

$$\Theta(\xi, \ell, \tau) = \int_0^\tau \frac{F(\eta) \cdot H(\ell, 0, \tau-\eta) \cdot e^{-\frac{\xi^2}{1+\tau-\eta}}}{(\tau-\eta)^{1/2} \cdot (1+\tau-\eta)} d\eta \tag{11}$$

For simplicity we assume that ion beam intensity is constant with time, i.e. $F(\eta)$=const. For this case a set of universal curves $\Theta(\xi)$ (fig.1) and $\Theta(\ell)$(fig.2) has been received that defined temperature changes in the crystal in radial and axial directions to the incident ion beam for different times τ. Quantitative estimates for temperature fields in the crystals of silicon, SiO_2 and organic materials used in the processes of ion-beam lithography are given in the table. The calculations were carried out for FIB that are employed in practice in the processes of maskless implantation and submicron ion lithography.

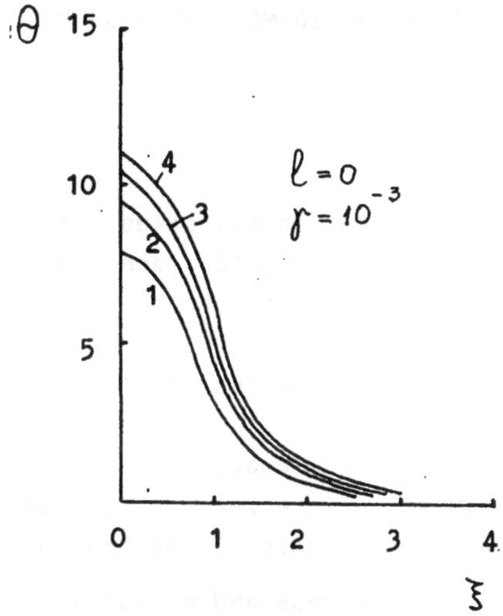

Fig.1. Temperature profile over the crystal surface when $F(\eta)$=const=5. Radiation time: 1-τ=1, 2-τ=2, 3-τ=3, 4-τ=4.

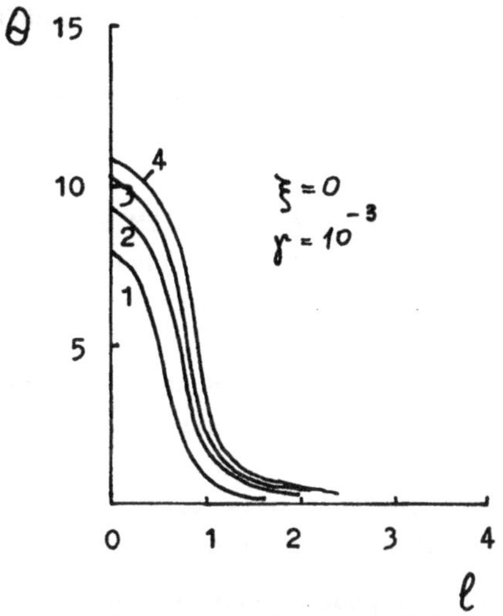

Fig.2. Temperature profile in depth of the crystal when $F(\xi)$=const=5. Radiation time: 1-τ=1, 2-τ=2, 3-τ=3, 4-τ=4.

Table. Heating temperature of materials in the center of incident ion beam during irradiation by ion beam with energy 100 keV.

material	current in the beam, A	beam diameter, μm	temperature, °C
Si	10^{-7}	0,1	825
		0,5	33
SiO_2	10^{-8}	0,1	195
		0,5	8
organic coating	10^{-9}	0,1	295
		0,5	12

References:

(1) Stangl B., Mitterauer J., Ruedenauer F.G., Marowsky G.: J.Vac.Sci.Technol.B3(1985)477-480
(2) Tamura M., Shukuri S., Moniwa M., Default M.: Appl.Phys.A39(1986) 183-190
(3) Hart R.R., Anderson C.L., Dunlap H.L., Seliger R.L., Wang V.: Appl. Phys.Lett.35(1979)865-867
(4) Levi-Setti R., La Marche P.H., Lam K., Shields T.H., Wang Y.-L.: Nucl.Instr.Meth.Phys.Res.218(1983)368-374
(5) Berry I.L., Caviglia A.L.: J.Vac.Sci.Technol.B1(1983)1059-1061

MASS AND ENERGY DISTRIBUTION MEASUREMENTS OF AuSi-LIQUID-METAL-ION SOURCE

U. Kreißig, A. Kahn[1)], F.G. Rüdenauer, W. Steiger[2)]

Liquid-metal-ion sources (LMIS) are now being actively investigated due to their enormous potential advantages for focused-ion-beam (FIB) technology such as maskless ion implantation, microfabrication and microscopies (SIM, SIMS).

The versatility of LMIS has been greatly increased by the use of alloy rather than pure elements.

The source emission characteristics related to mass and energy analyses are essential not only to estimate the particular ion-species FIB-size due to chromatic aberration, but also to consider the ion emission mechanism.

The present paper describes an experimental study of mass and energy analyses of ions emitted from AuSi-alloy LMIS at total emission currents below 10 µA.

A double-focusing four-sector mass spectrometer (1) was employed to investigate the ions emitted from the LMIS. The aparatus was adjusted to have a mass resolving power of $m/\Delta m \simeq 500$ and an energy resolution of $\Delta E = 4$ eV at $E = 2$ keV. Mass spectra were obtained by keeping the emitter potential at a constant value and sweeping the magnetic field. The energy analyses was carried out by scanning the emitter potential while keeping both the electrostatic and magnetic fields constant. As the emitter potential is swept to higher values, ions with increasingly lower initial energies, i.e. with a higher energy deficit. are able to pass the spectrometer.

A conventional needle-type LMIS (2) was employed throughout these experiments. The LMIS consisted of a tungsten emitter tip spot-welded to a tungsten hairpin filament and the extractor electrode. The emitter tip was electrochemically etched to an apex radius of about 4 µm. After the ion emitter had first been flushed of impurities by heating it to 1500 °C in a vacuum, it was wetted by the $Au_{82}Si_{18}$ (at%) alloy. A small amount of the alloy was put at the end of the hairpin filament and the emitter and resistively heated up to the melting temperature of the AuSi-alloy (≃360 °C).

The wetted emitter tip was set at a gap of 0.2 mm from the extractor electrode with 1.5 mm-diameter aperture.

Heating the emitter and applying a high voltage of about 6 to 8 kV between the extractor and the emitter the ions start to emit from the Taylor cone formed by stretching out the molten alloy with an intense electric field. The emitter temperature could be adjusted with the filament current. The

---

1) Central Institute for Nuclear Research, Rossendorf, GDR-8051 Dresden, POB 19

2) Austrian Res. Ctr. Seibersdorf, A-2444 Seibersdorf, Austria

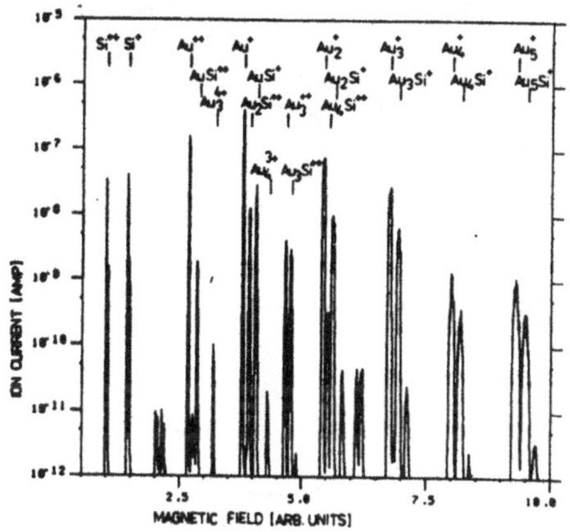

Fig. 1

Mass spectrum of $Au_{82}Si_{18}$ alloy LMIS at 4.8 μA total emission current

emitted ions were decelerated to 2 keV and focused at the entrance slit of the spectrometer.

Mass analysis measurements showed that the AuSi-LMIS emitted singly and doubly charged ions of the alloy components and also multiatomic, multi-charged and molecular ions at about 5 μA total emission current (Fig. 1). The Au^+-peak is in reality a mixture of Au^+ and Au_2^{++} since Au_2Si^{++}-ions were observed.

Field evaporation is the main ion generation process and field evaporation and postionization produce the multiply charged ions (3).

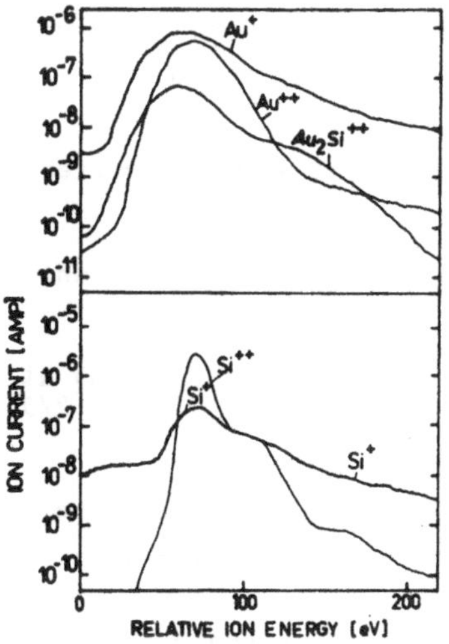

Fig. 2

Energy spectra of most prominent species at 4.8 μA total emission current

Some singly charged atoms and molecules are also generated by vapourization and subsequent ionization of neutrals or by charge exchange processes between ions and neutrals. Energy distributions of the mainly emitted ions are shown in Fig. 2. It's obvious that atomic ions and even-charged atomic ions have a smaller energy width than molecular and odd-charged species respectively. Furthermore the energy width of Si-ions are smaller than that of Au-ions. We conclude: the lower the mass-to-charge ratio of the ion the lower its energy width. In some energy spectra (e.g. Si^{++}, Au_2Si^{++}) a low energy shoulder, i.e. a somewhat higher energy deficit, was observed. This may be a contribution of postionization process several tens of Å away from the emitter surface.

Fig. 3 and Fig. 4 show the full width of half maximum (FWHM) and the peak positions of the energy distributions depending on total emission current. The increase of the FWHMs with increasing current (Fig. 3a, 4a) follows a 2/3 power law. This dependence has been predicted by Knauer (4) assuming that field emission is nearly collision free. An alternative broadening process due to steady coulomb repulsion among the emitted particles is proposed by the same author.

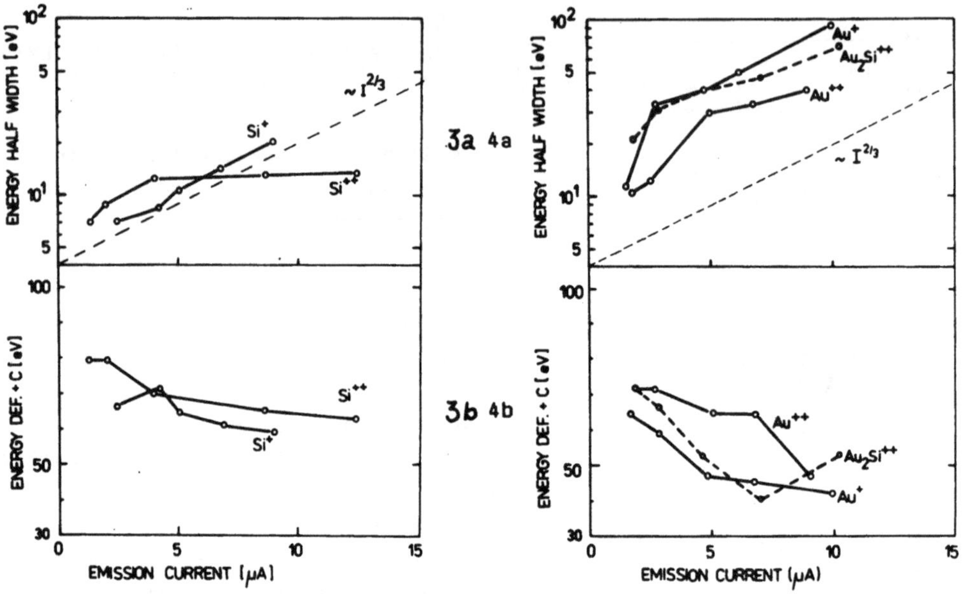

Fig. 3a, 4a
FWHM of the energy distributions in dependence of total emission current

Fig. 3b, 4b
Relative energy deficit of ions in dependence of total emission current

The energy width of Si-ions (Fig. 3a) are generally lower, about 7 eV at 3 μA emission current. The current dependence is corresponding with $i^{2/3}$ except for Si^{++}-ions above 5 μA.

Generally the current dependence of energy broadening isn't fully understood, e.g. also fluctuations of the liquid metal surface in the apex region contribute to it.

The energy deficit could be determined only in a relative scale (Fig. 3b, 4b). According to Swanson and Kingham (3) the difference between the energy deficits of singly and doubly charged ions have to be of the order of $I_2-\Phi$, where I_2 is the second ionization potential and Φ the surface work function. This corresponds to $\simeq$16 eV for Au- and $\simeq$12 eV for Si-ions. The energy deficits of the atomic ions decrease slightly with increasing emission current. This points to an increasing and field-dependent activation energy as the field evaporation mechanism changes to more thermal field evaporation at higher emission currents.

References

(1) F.G. Rüdenauer and W. Steiger: Jpn. J. Appl. Phys., Suppl. II, Pt. 1 (1974) 383

(2) F.K. Naehring and A. Kahn: Proc. Int. Conf. Electron Beam Technologies, Bulgaria, Varna 26 May - 2 June 1985, p. 523

(3) L.W. Swanson and D.R. Kingham: Appl. Phys. A41 (1986) 223

(4) W. Knauer: Optik 59 (1981) 335

CLUSTER IONS IN LASER MASS SPECTRA OF WO_3/C TARGETS

H.-J. Dietze[1], S. Becker[1], W. Pompe[2]

For the diagnostics of laser plasmas, a knowledge of the fraction, and of the abundance distribution of positively and negatively charged ions contained therein is also of importance.
This holds particularly true for the task of producing thin films by laser pulse vapour deposition (LPVD) from the laser plasma.
The analyses of ion distribution in laser plasma was performed with a double focussing mass spectrograph with Mattauch-Herzog geometry (type MX 3301, SKB Leningrad). Parameters of the laser ion source, designed and built at the ZfI, are described in /1/.
The target material chosen was a mixture consisting of 50 weight-% each of WO_3 and high purity graphite, as well as tungsten carbide produced by a melting process.
At a power density of roughly 5×10^8 W . cm^{-2} a multiplicity of ion clusters could be observed in the laser plasma of WO_3/C targets.
When the cluster ion formation of the WO_3/graphite system is compared to the of the WC target, the low intensity of the detected WC_n^+ and WO_n^+-ions for the compact tungsten carbide is obvious (see Fig. 1).
The observed abundance distribution of negative C_n^- cluster ions corresponds to the distribution measured and discussed by FÜRSTENAU et al. /2/.
In the laser plasma, negatively charged C_n^- ions are formed at higher intensities than positively charged spezies ($C_2^+/C_2^- = 1,2 \cdot 10^{-2}$)
WO_3^- ions are formed with the high intensities in the laser plasma by resonance capture

$$WO_3 + e^- \longrightarrow WO_3^-$$

Where electron energie are between 0 and 1 eV with regard to the formation of WO^- and WO_2^- ions, a dissociative resonance capture process can be assumed, e.g.

$$WO_3 + e^- \longrightarrow WO_2^- + O$$

Electron energies in this process are between 2 - 15 eV. New types of clusters $WC_nO_m^-$, also occur in the negativ ion mass spectra, were detected.

---

[1] Central Institute of Isotope and Radiation Research, Academy of Science of the G.D.R., Leipzig

[2] Central Institute of Solid State Physics and Material Science, Academy of Science of the G.D.R., Dresden

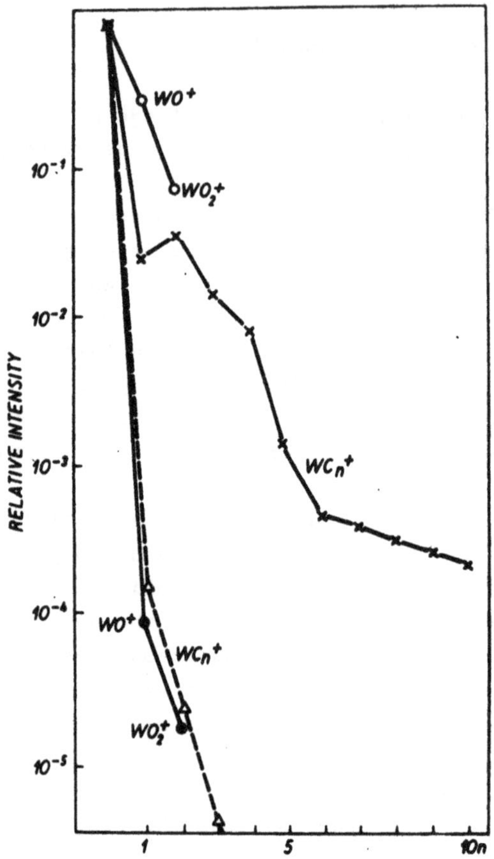

Fig. 1: Comparison of cluster abundances in the laser plasma of a WO_3/C and a compact tungsten carbide target.

Literature:

/1/ H.-J. Dietze, S. Becker, Fresen. Z. Anal. Chem. 321 (1985) 490

/2/ N. Fürstenau, F. Hillenkamp, Intern. J. Mass Spectrom. Ion Phys. 37 (1981) 135

ANALYSIS OF Ar ION BEAM ETCHING (IBE) INDUCED SURFACE DAMAGE OF GaAs BY MEANS OF ^{64}CU ADSORPTION / AUTORADIOGRAPHY AND RBS / CHANNELING

A. Schindler[+], R.Fechner[+], R. Flagmeyer[++], D. Panknin[+++], F.Bigl[+], G. Otto[++]

With the advance of GaAs electronic and optoelectronic devices, GaAs dry etching techniques have become increasingly important in realizing fine pattern delineation. However, any ion bombardment induces radiation damage on the surface of the materials processed.

In this paper studies of Ar+ ion bombardment damage of GaAs (100) surfaces by means of ^{64}Cu adsorption and autoradiographic detection and RBS/channeling are presented.

Ar ion beam etching was performed at R.T. using a Kaufman-typ broad beam ion source (Ar^+ ions, working pressure $2*10^{-2}$ Pa, ion energy 0.5 to 2.5 keV, ion dose $1*10^{13}$ to $5*10^{16}$ cm^{-2}, ion incident angle with respect to the surface normal 0°.

The ^{64}Cu adsorption was carried out at R.T. in an aqueous solution (1mg Cu^{II}-ions, 10 ml 1n HCl and 15 Mbq ^{64}Cu in 100 ml H_2O) for 5 minutes. The autoradiographs were taken with the samples in contact with the film for 1 to 10 hours. The blacking curves were measured by a photometer. The quantitative Cu concentration measurements were made by autoradiographic imaging of calibration standards on the same film.

RBS measurements were carried out with 1.2 MeV He^+ ions using a backscattering angle of 95° (5° grazing exit geometry). With the given detector resolution power of 15 keV the surface depth resolution was better than 4 nm.

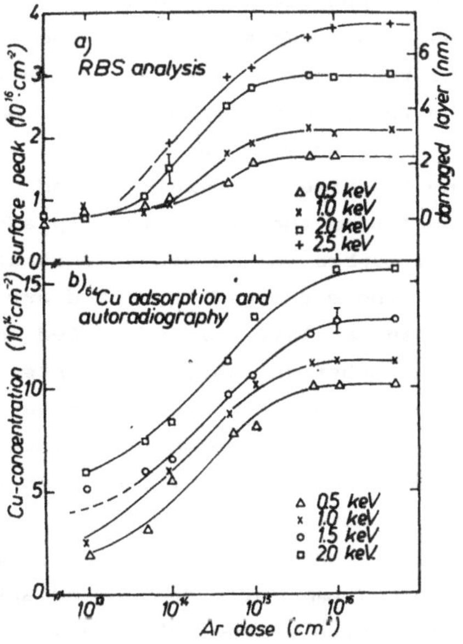

Fig. 1
Total amount of crystal damage measured by (a) surface peak intensities of <100> aligned RBS spectra and (b) adsorbed Cu concentration on Ar+ ion beam etched GaAs surfaces as a function of ion dose and energy. (The curves are only guides for the eye.)

In Fig.1a the total amount of damage as a function of the Ar^+ ion dose is shown for four ion energies. The RBS-surface peak intensity increases

---

+ Zentralinstitut für Isotopen- und Strahlenforschung Leipzig der AdW der DDR, Permoserstr. 15, Leipzig, 7050

++ Karl-Marx-Universität Leipzig Sektion Physik, Linnestr.5, Leipzig,7010

+++ Zentralinstitut für Kernforschung Rossendorf der AdW der DDR,PF 19, Dresden, 8051

with the ion dose and begins to be saturated at a dose of 1..3*10^{15} cm^{-2}. This is in good agreement with the measurements by Williams (11).
The behaviour of the amount of adsorbed Cu as a function of the ion dose and the energy , shown in Fig.1b, is in qualitative agreement with the RBS results.
The absolute Cu concentrations adsorbed on the etched surfaces range between 2*10^{14} and 1.6*10^{15} atoms cm^{-2}. On the non-etched areas of the samples the concentration is less than 6*10^{13} atoms cm^{-2}. Due to this low value which corresponds to only about 10% of the topmost monolayer atom density of GaAs (6.3*10^{14} cm^{-2}) the ^{64}Cu adsorption technique is very sensitive to damage at low ion doses, in contrast to RBS/channeling.
The copper concentrations at saturation damage for all etch energies are larger than the monolayer equivalent , the maximum being 2.7 times larger at 2.5keV ion energy.
The comparison of the behaviour of the adsorbed copper with the damage thickness behaviour as a function of the ion energy for saturation ion doses in Fig.2 shows good agreement.
The experimental results lead to the conclusion that the primarily adsorbed/ chemisorbed (6) Cu atoms penetrate into the whole damaged layer and are gettered there. This is supported by the high substitutional and interstitial diffusion even at low temperatures (4) and by the high solubility of copper in defect and/or vacancy rich GaAs (3).
The autoradiographically measured copper concentrations are integrated ones from all depths because the attenuation of the ^{64}Cu radiation can be neglected in our case. We measured an attenuation coefficient of 0.16μm^{-1} in GaAs.

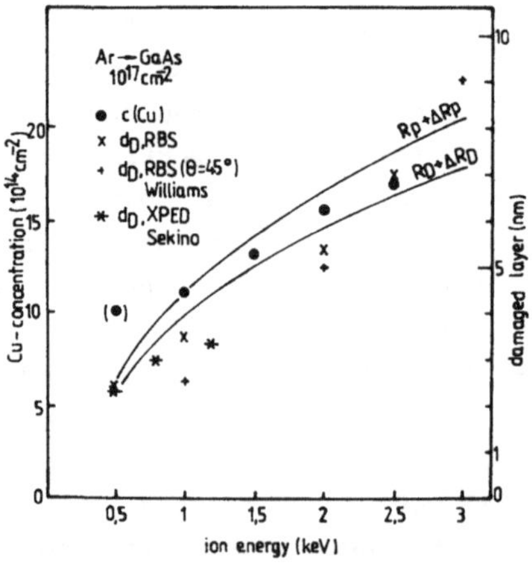

Fig. 2
Damaged layer thickness d_D (right hand scale) and adsorbed Cu concentration c (left hand scale)at the saturation Ar ion dose (>10^{16} cm^{-2}) in dependence of the ion energy. Also shown are the theoretical curves of the maximum penetration depth of the ions ($R_p+\Delta R_p$) (2) and of the damage ($R_D+\Delta R_D$) (10), the projected ranges plus the standard deviations of the ranges, and d_D measurements by RBS and XPED from Williams(11) and Sekino et al.(9).

Depth profile measurements were carried out in order to check the assumption of the distribution of the copper atoms in the damaged layer. The sample preparation was done by very low energy ion beam etching (0.2 keV Ar^+) on the damaged and Cu adsorbed GaAs surface which gave bevelled sections with an inclination angle of about 2*10^{-4} degree (Fig. 3 inset) (1). Autoradiographic registration of the ^{64}Cu distribution at the bevelled sections and differentiation of the blacking curves in connection with the mechanical profile measured by the Talystep method give the copper concentration depth profile with a resolution better than 2 nm. Fig. 3 shows the depth distributions for four different ion energies and for the undamaged sample part. The measurements clearly show that the copper is distributed throughout the damaged layer.

There are very deep tails in the profiles, far deeper than the predicted and observed thicknesses of the heavily distorted layers. This suggests that enhanced diffusion of IBE-induced defects takes place in GaAs.

Similar results on defect migration have also been obtained for Ar etched Si (7) by SIMS, for Ar etched GaAs an InP by PL (12) and for Cs bombarded GaAs by AES (5).

There are obvious problems with the used profile measurement technique because damage caused by ion beam etching has to be measured by ion beam etching depth profiling. Absolute concentration profiles cannot be reliably measured in this case but contrasts between different conditions of damage production are easily registered.

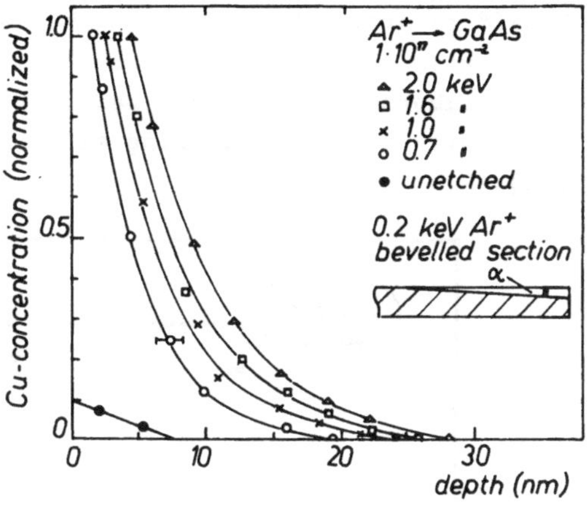

Fig. 3
Depth distribution of adsorbed ^{64}Cu on Ar^+ ion beam etched GaAs (100). The profiles are normalized to unity at the maximum. (The curves are only guides for the eye.)

A first attempt was made to recover the disordered surface layer by means of rapid thermal annealing (8).

The best results up to now have been achieved with 0,4 sec halogen lamp annealing at temperatures around 400 °C. As measured by RBS the damage recovered completely only at ion doses below the monolayer equivalent dose were damage saturation is not be reached. At the damage saturation level about 60 % the disorder still remains.

R e f e r e n c e s

(1) Bigl,F., A. Schindler, J. Flachowsky, K. Freyer and H.-Chr. Treutler: Int. Conf. Ion Beam Mod. of Mat., Budapest 1978, p.1743

(2) Burenkov,A.F., F.F. Komarov, M.A. Khumachov and M.M. Temkin: Tabellen-Parameter der räumlichen Verteilung implantierter Ionen, Minsk 1980 (in russian)

(3) Furukawa,Y. and C.D. Thurmond: J.Phys. Chem. Solids 26, 1535 (1965)

(4) Hall,R.N. and J.H. Racette: J. Appl. Phys. 35, 379 (1964)

(5) Homma,Y., H. Okamoto and Y. Ishii: Jap.J.Appl.Phys. 24, 934 (1984)

(6) Larrebee,G.B.: J. Electrochem. Soc. 108, 1130 (1961)

(7) Mizutani,T., C.J. Dale, W.K. Chu and T.M. Mayer: Nucl. Instr. Meth. Phys. Res. B7/8, 825 (1985)

(8) Schindler,A., R. Fechner, D. Panknin, R. Flagmeyer, F. Bigl, G. Otto : to be published

(9) Sekino,Y., N. Owari, M. Kudo and Y. Nihei: Jap.J.Appl.Phys. 25, 38 (1986)

(10) Sigmund,P. and J.B. Sanders: Int. Conf. Appl. Ion Beams Semicond. Technol., Grenoble 1967 p. 215

(11) Williams,R.S.: Sol. State Comm. 41, 153 (1982)

(12) Yuba,Y., K. Gamo, H. Toba, Xi Guan He and S. Namba: Jap.J.Appl. Phys. 22, 1206 (1983)

A MeV ION PROBE AT TU PRAGUE

M.Setvak, Z.Hulek, J.Voltr, Z.Haba, R.Salomonovic, A.Sibl, J.Kral [1]

An analytical equipment using ion beam from a 2.5 MeV Van de Graaff generator has been finished and put in operation at Technical University of Prague. The whole equipment is placed in a two floor heavy laboratory originaly built for accelerator and nuclear physics with thick walls and ceiling made of reinforced monolithic concrete (Fig. 1).

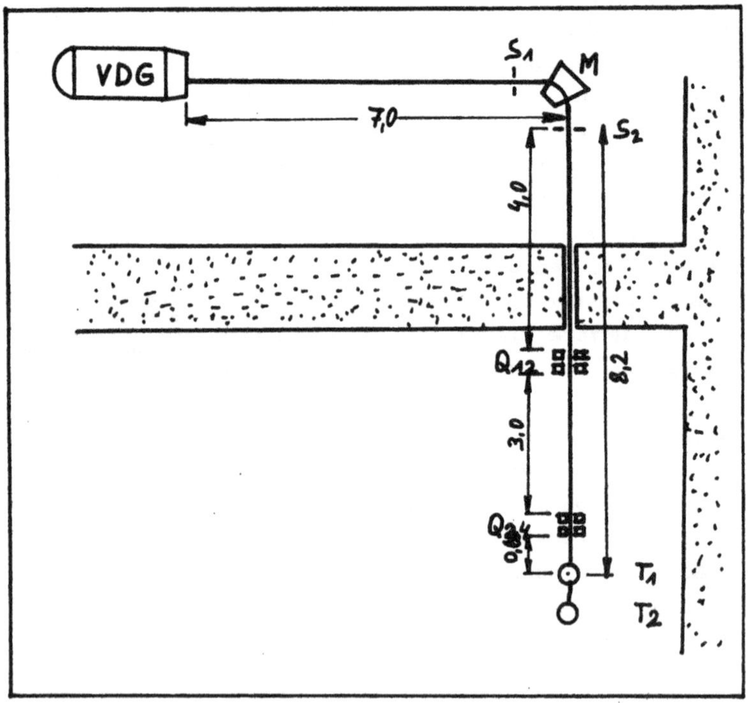

Fig. 1. General view of the system, the length are in meters. VDG - High Voltage Eng. Model A-2500 Van de Graaff generator, $Q_{1,2,3,4}$ - Bruker B-QD 50/100-5 quadrupole doubletts.

A vertical analytical channel has been built using two quadrupole doubletts for compression of the ion beam to small diameter (1). To prevent mechanical vibrations from outside when using a microfocused beam the whole system is connected to the thick ceiling which divides the two laboratories and is a part of heavy concrete monobloc. A computing program TRANSPORT (2) has been used for computing the optics of the system. An orthomorfic system of focussing consisting of two quadrupole doubletts has been designed to focus the beam to a small dia-

[1] Technical University of Prague, Faculty of Nuclear Science and Physical Engineering, Brehova 7, 115 19 Prague, Czechoslovakia.

meter (Fig. 2). With this system a demagnification of the beam in the slit S_2 behind the bending magnet M on the target is about M = 0.1. The bending magnet is a double focussing Rogowski shaped dipole with mass-energy product of 10 amu.MeV. Two target chambers were built, one for scattering experiments, the second - T_2 is a carousel chamber for PIXE analysis.

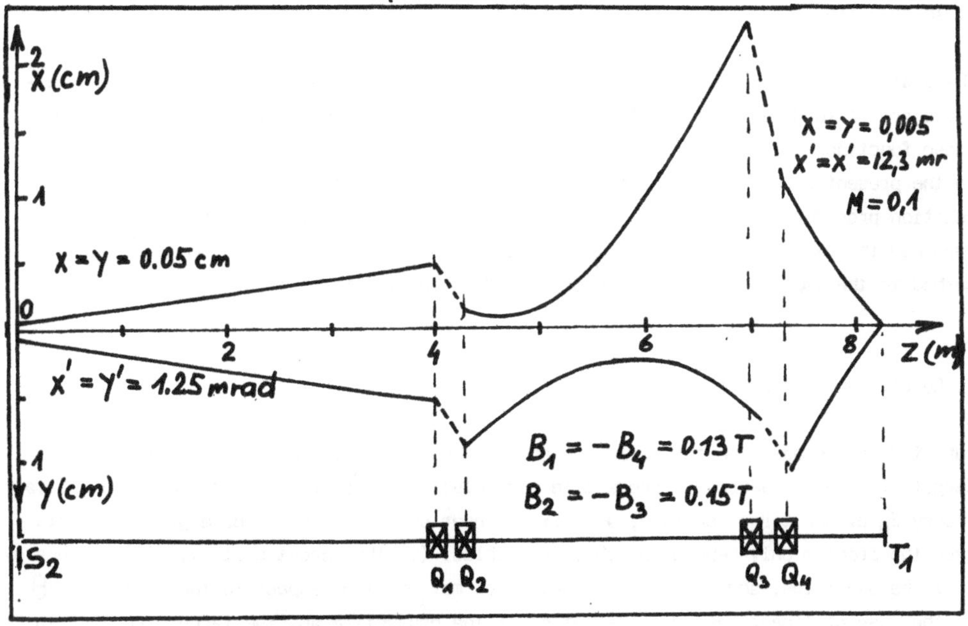

Fig. 2. Ion beam envelopes between the slit S_2 and the target T_1 computed by TRANSPORT.

The vacuum system of the ion probe has been designed and constructed as a clean (oil free) system with the limit pressure in the 10^{-5} Pa range.[3] The beam line from the Van de Graaff switching magnet to the experimental chamber has a length of about 12 m. It consists of the elements fabricated mostly from a stainless steel tube of the diameter 70 mm. The beam line is closed on the both sides with gate valves and pumped with three sputter ion pumps having 125 l/s pumping speed each. The flanges (according to ISO standards) are sealed with viton gaskets. During the running in periode when the beam line is sometimes open to the atmosphere a turbomolecular pump is used instead of the sputter ion pump. The experimental chamber is pumped by a turbomolecular pump , (Balzers TPU 330), equipped with a vibration dumper. A computer control is being developped for the whole vacuum system.

References:

(1) M.Setvak: Dissertation Thesis, TU Prague, 1980 (in Czech)
(2) K.L.Brown et al.: TRANSPORT - a computer program for designing particle beam transport system, Stanford, 1977
(3) Z.Hulek et al.:The vacuum system of the analysing channel of a Van de Graaff accel., Proc. 7th EVF Conf., Bratislava 1985

COMPUTER STUDIES OF THE STATISTICAL DISTRIBUTION FUNCTIONS OF PARTICLES MOVING IN COLLISION CASCADES

M. Posselt, G. Otto [1)]

1. Introduction

The statistical distribution functions of the particle species moving in collision cascades are of fundamental theoretical interest. There are two alternatives to determine these functions: (i) the solution of Boltzmann equations, and (ii) computer simulations. In the present work the TRIM Monte Carlo simulation program (1)(5)(3) is used. Our calculation procedure is similar to the method of Hautala and Whitlow (2)(4) who have determined the directional distributions of the recoiling target atoms. We apply our method to the implantation of phosphorous and antimony into silicon.

2. Results and discussion

Our studies are based on the following assumptions: (i) stationary irradiation of the target, (ii) plane source of projectile particles with normal incidence and with initial energy E_0 at the target surface, and (iii) linear cascades. Then, for a given particle type the problem only depends on three variables: (i) the target depth x, (ii) the energy E of the particles, and (iii) the direction of motion with respect to the x-axis, $\cos\theta$. In the case of linear cascades the motion of the projectile and the target recoils can be simulated by so-called static computer codes. For our purposes we employ the TRIM program. To study the statistical distribution functions we have added the following procedure to the TRIM code. In the target certain marker planes are defined. For a given species the number of particles dJ penetrating the marker plane at x during dt with (E,dE) and ($\cos\theta$, $d\cos\theta$) is logged

$$dJ = j(x,E,\cos\theta)\, dE\, d\cos\theta\, dt \qquad (1)$$

$j(x,E,\cos\theta)$ is the differential flux at x. The number of particles moving at time t in (x,dx) with (E,dE) and ($\cos\theta$, $d\cos\theta$) is given by

$$dF = f(x,E,\cos\theta)\, dE\, d\cos\theta\, dx = (j(x,E,\cos\theta)/v\cdot|\cos\theta|)\, dE\, d\cos\theta\, dx \qquad (2)$$

$f(x,E,\cos\theta)$ is the momentum distribution at x. The functions j and f are very similar. For our purposes we call j and f the statistical distribution functions. In the following we want to limit our considerations to the function $j(x,E,\cos\theta)$. Instead of particle numbers we can define probabilities by normalization. For example,

[1)] Academy of Sciences of the GDR, Central Institute for Nuclear Research, Rossendorf, Postfach 19, DDR-8051 Dresden, GDR

$$dJ_N = j(x,E,\cos\theta)dE\, d\cos\theta / \left(\iint j(x,E,\cos\theta)dE d\cos\theta\right) = j_N(x,E,\cos\theta)dE d\cos\theta \quad (3)$$

is the probability that a particle penetrates the marker plane at x with (E,dE) and ($\cos\theta$,$d\cos\theta$). The three-dimensional distribution $j_N(x,E,\cos\theta)$ can be reduced to the two-dimensional distributions

$$(\partial \tilde{J}_N(x,E)/\partial E) = \int j_N(x,E,\cos\theta)d\cos\theta, \quad (\partial \tilde{J}_N(x,\cos\theta)/\partial \cos\theta) = \int j_N(x,E,\cos\theta)dE \quad (4)$$

Figs. 1-4 show the energy and directional distributions of the projectiles for 50 keV P^+ and Sb^+ implantation into silicon. The distributions have been calculated for characteristic depths. It is evident that the peak of the energy distribution must shift from E_0 at x=0 to E_f at sufficiently high values of x. E_f is the final energy in the simulation of the particle motion. We have chosen E_f=15 eV. The energy distributions of antimony and phosphorous are different. In the first case the slowing down of the projectile is always dominated by the nuclear stopping which is the result of many individual elastic scattering events between the projectile and the target atoms. This leads to important energy fluctuations. Therefore, the system loses the memory of the monoenergetic initial condition at relatively small depths. In the second case the electronic stopping is not negligible. It can be assumed to act as a continuous braking force. Therefore, the slowing down of the phosphorous projectiles is more deterministic than in the case of antimony. The directional distributions for phosphorous and antimony show a strong anisotropy. For our examples it can be shown that the fluctuations in the direction of motion are in the same order of magnitude. Therefore, in contrast to the energy distributions the directional distributions are very similar. We have also studied the energy and directional distributions of the target recoils. Two cases have been considered: (i) The histories of the particles have been followed until the displacement energy of the target (E_f=15 eV). (ii) Only the recoils with energies $E \geq E_f$=5 keV have been simulated. In both cases there are very marked peaks in the energy distributions at E_f, i.e. the great majority of the particles are always the recoils with the lowest energies. The thin high energy tails of the energy distributions are the remaining memory of the energy distributions of the projectiles. Figs. 5 and 6 show the directional distributions of recoils for the Sb^+ implantation. The distributions of the particles with energies $E \geq E_f$ = 5 keV show a stronger anisotropy than the distributions of the recoils with $E \geq E_f$ =15 eV. This is clear since the distribution of the high energy recoils is more influenced by the directional distribution of projectiles. The influence decreases with decreasing energy and with decreasing generation of the cascade recoils. The isotropic limit is given by the dashed lines. With increasing marker depth the deviation of the distributions from the isotropic limit increases slightly. For the phosphorous case the directional distribution of the recoils is very similar to that of the antimony case. This is due to the fact that the directional distributions of the projectiles are nearly identical.

3. Conclusions

A method to calculate the statistical distribution functions of all particles moving in ion-beam-induced collision cascades has been developed. For the two examples the energy and directional distributions of the projectiles and the recoils have been discussed. Further applications of our procedure and a more detailed discussion will be published elsewhere.

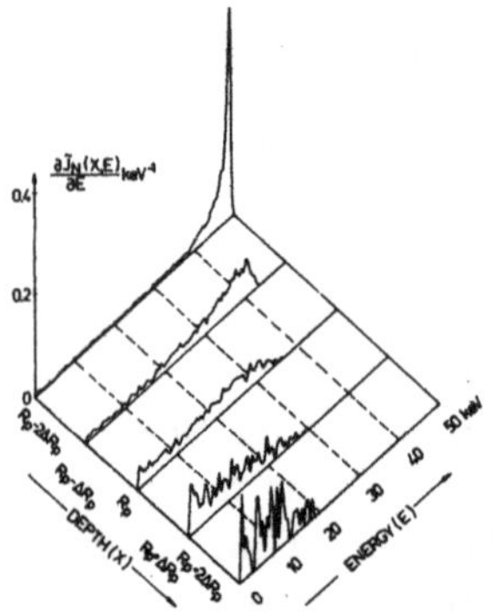

Fig. 1 : 50 keV P^+ into Si
(projectiles)

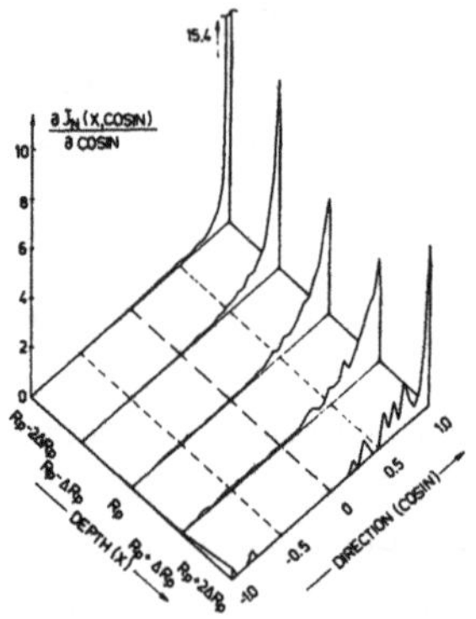

Fig. 2 : 50 keV P^+ into Si
(projectiles)

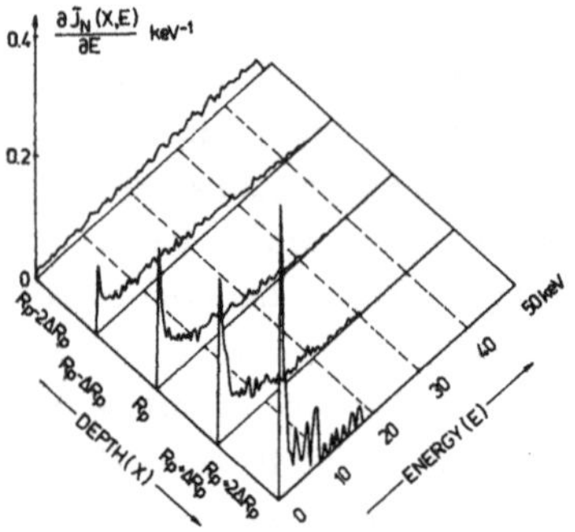

Fig. 3 : 50 keV Sb^+ into Si
(projectiles)

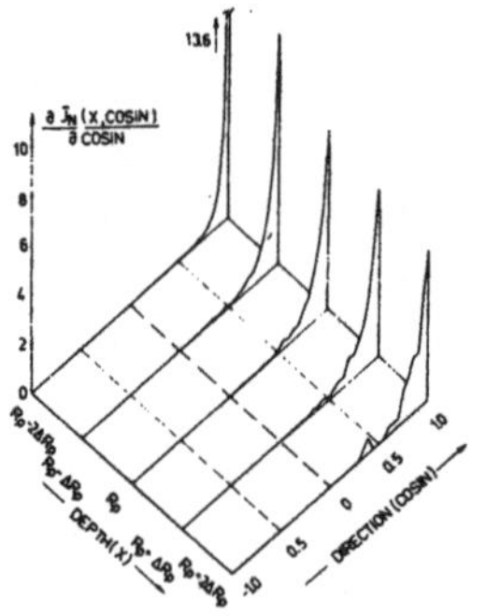

Fig. 4 : 50 keV Sb^+ into Si
(projectiles)

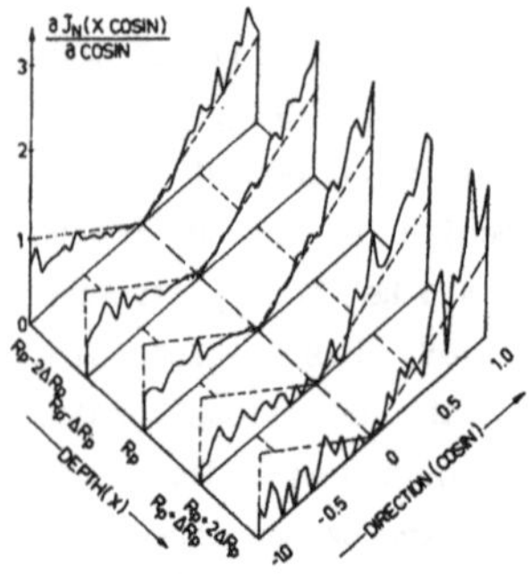

Fig. 5 : 50 keV Sb^+ into Si
(recoils)

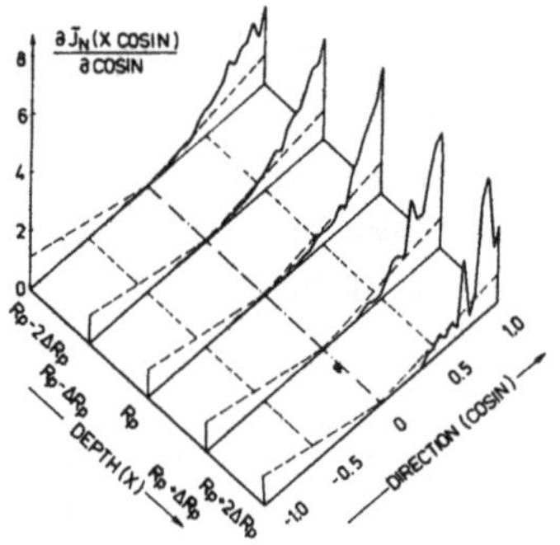

Fig. 6 : 50 keV Sb^+ into Si
(recoils)

References

(1) Biersack, J. P. and L. G. Haggmark: Nucl. Instr. and Meth. 174(1980) 257
(2) Hautala, M. and H. J. Whitlow: Nucl. Instr. and Meth. B6(1985) 460
(3) Posselt, M. and J. P. Biersack: Nucl. Instr. and Meth. B15(1986) 20
(4) Whitlow, H. J. and M. Hautala: Nucl. Instr. and Meth. B18(1987) 370
(5) Ziegler, J. F., J. P. Biersack and U. Littmark: The Stopping and Range of Ions in Solids, Pergamon Press, New York 1985

MONTE CARLO CALCULATIONS OF HIGH DOSE ION BOMBARDMENT ACCOUNTING FOR DYNAMIC TARGET MODIFICATION

A. Schönborn, N. Hecking and E. te Kaat+)

A dynamic Monte Carlo code simulating high dose ion implantation of materials, accounting for sputtering, stopping power changes due to deposited ions and knock-on processes has been developed. Results for silicon bombardment with 150 keV nitrogen ions in the dose range from 7×10^{16} cm^{-2} to 1×10^{18} cm^{-2} are discussed.

I. INTRODUCTION

Computer simulations are widely used in studies of ion bombardment of materials to calculate the distributions of implanted ions, radiation damage and other quantities. In most Monte Carlo codes a native target is assumed for all simulated ion tracks, which is a good approximation in case of low ion fluence. In some recent applications of ion implantation, for example formation of buried insulating layers for SOI semiconductor devices, or improvement of corrosion resistance and wear behaviour of metals, a high concentration of impurity atoms is needed. If high doses are implanted, however, stopping power and volume changes result from deposited ions. The target surface is eroded by sputtering, and implanted atoms may be further displaced in collision cascades. To account for these high dose effects a dynamic Monte Carlo code has been developed. It is based on a static simulation program described in /1/.

II. METHOD

In the program "HIDOS" ion trajectories are computed in binary collision approximation. The differential cross-section, as suggested by Winterbon et al. /2/ or Kang et al. /3/, is used to calculate the free pathlength between successive collisions with target atoms, the energy transfer and scattering angles. The electronic energy loss is calculated according to LSS theory, including experimentally determined correction factors. The target is divided into slabs whose individual compositions are altered by deposited ions. In slabs consisting of more than one element, Bragg's rule is applied to estimate the electronic stopping. The type of the nuclear collision partner is determined

+) Institute of Physics, University of Dortmund, POB 500500, D-4600 Dortmund 50, FRG statistically from cross-sections and concentrations.

To account for recoil implantation and sputtering, the complete collision cascade is simulated. All primary and secondary recoil atoms are followed according to /4/ until their energy falls below a cut-off energy. Recoil atoms reaching the surface with an energy exceeding the planar surface potential /5/ are considered as sputtered particles. Since the local density is allowed to relax, deposition, sputtering and recoil motion result in thickness changes of the slabs.

III. RESULTS AND DISCUSSION

As an example, a simulation of 150 keV nitrogen ions implanted into silicon with doses up to 1×10^{18} cm^{-2} is presented. Similar energies and doses are used in SOI-technology to form a buried silicon nitride layer /6/. In the simulation nuclear collisions were calculated using the differential cross-section of Kang et al. /3/. The correction factors of the electronic stopping power were: 1.43 for N in Si /7/, 1.09 for Si recoils in Si (derived from MeV Si - Si implantations), 1.0 for Si in N and N in N (correction factors unknown). The silicon surface binding energy used was 7.8 eV /8/. Several thousands of cascades were simulated, but wiggles may still occur in the profiles due to statistical inaccuracy.

The results of the simulation are shown in fig. 1 and 2. The projected ion range decreases almost linearly with increasing ion dose. At the highest dose which leads to a relative fraction of 59 % nitrogen in the profile maximum this shift amounts to 6 % (22 nm). Detailed studies show that it is mainly caused by sputtering resulting in a surface erosion of 24 nm. An additional decrease is caused by

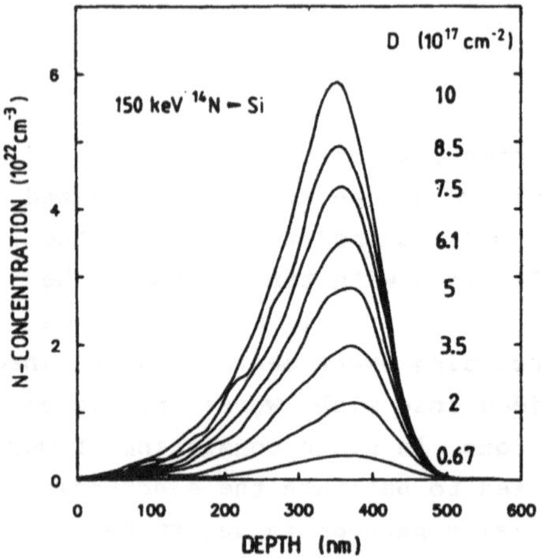

Fig. 1: Calculated depth profiles of N ions in Si for different ion doses D.

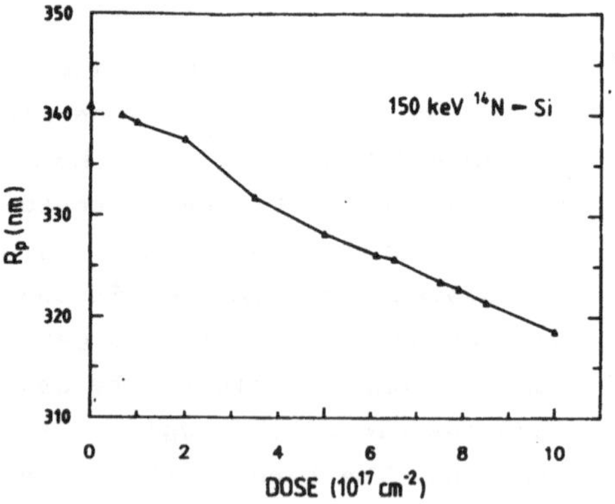

Fig. 2: Dose dependence of the projected ion range R_p of N ions in Si.

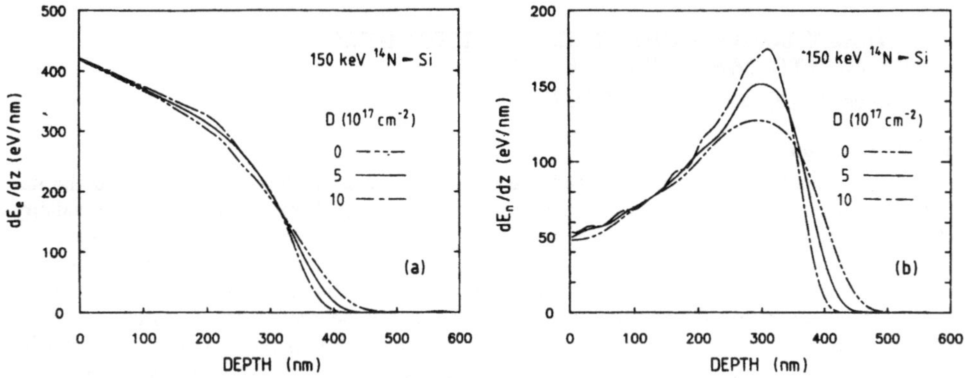

Fig. 3: Calculated stopping power profiles for N ions after implantation of 0, 5 and 10 x 10^{17} cm^{-2} N into Si.
a) Electronic stopping dE_e/dz, b) nuclear stopping dE_n/dz.

increasing stopping powers (fig. 3) due to deposited nitrogen and cascade mixing. Within the cascades silicon recoils are transported farther into the target than the lighter nitrogen recoils, whose motion is nearly isotropic.

The shape of depth profiles is found to be nearly independent of ion dose because broadening by cascade mixing and sputtering on the one hand and contraction by increased stopping powers on the other hand compensate each other in the above example.

IV. SUMMARY

The influence of sputtering, range shortening due to composition changes and motion of recoil atoms on the depth distribution of 150 keV nitrogen ions in silicon has been studied by dynamic Monte Carlo simulation. The projected ion range is shortened, whereas the shape of profiles remains unchanged. In general, direction and relative amount of individual high dose effects depend on the ion-target combination and on energy. Quite pronounced profile distortions are to be expected in specific cases.

REFERENCES

/1/ H.F. Kappert, K.F. Heidemann, D. Eichholz, E. te Kaat and W. Rothemund: Appl. Phys. 21 (1980) 151
/2/ K.B. Winterbon, P. Sigmund and J.B. Sanders: Mat. Fys. Medd. Dan. Vid. Selsk 37, No. 14 (1970)
/3/ H.J. Kang, E. Kawatoh and R. Shimizu: Jap. J. Appl. Phys. 23 (1984) L262
/4/ W. Moeller and W. Eckstein: Nucl. Instr. Meth. B2 (1984) 814
/5/ M.T. Robinson in Sputtering by Particle Bombardment I, Ed. R. Behrisch, Springer (Berlin 1981)
/6/ K.J. Reeson: Nucl. Instr. Meth. B19/20 (1987) 269
/7/ U. Bußmann: Dipl. Thesis, University of Dortmund (1984)
/8/ H.H. Andersen: Appl. Phys. 18 (1979) 131

ENERGY LOSS OF CHANNELED HELIUM IONS BY MONTE CARLO SIMULATION

Andrzej DYGO and Andrzej TUROS[1]

1. Introduction

A great number of experimental and theoretical investigations has been devoted to the important problem of energy loss of channeled particles. The usual way of estimating the validity of a particular model of channeled particle stopping power was to compare the energy loss of the best channeled particles calculated for different crystal orientations with appropriate experimental data. However, the measured energy distributions of particles emerging from crystals are strongly influenced by experimental resolution factors and the precise determination of energy loss of the best channeled particles is quite difficult. Moreover, the energy loss of the best channeled particles alone cannot yield full information on the mechanism of the channeled particles energy loss.

These problems can be solved by calculation of energy distributions for different orientations of the same crystal axis with respect to the beam direction. The computer simulation by means of the Monte Carlo method has been used for the purpose. This report presents the results of the Monte Carlo simulation for 2 MeV He-ions traversing a $\langle 100 \rangle$ Si single crystal. The calculated distributions are compared with the experimental data reported recently by Jin and Gibson [5].

2. Three component model of energy loss

It is generally accepted to express the random stopping cross section as a sum of the stopping due to core and valence electrons: $\epsilon_r = \epsilon_c + \epsilon_v$. Basing on the random phase approximation of Bohm and Pines ϵ_v component can be divided into two contributions [1]: $\epsilon_v = \epsilon_{v1} + \epsilon_{v2}$, where

$$\epsilon_{v1} = \frac{4\pi Z_1^2 e^4}{m_e v^2} Z_v \ln \frac{2m_e v v_F}{\hbar\omega_p} \tag{1}$$

is the stopping due to single-particle excitations of the valence electrons and

$$\epsilon_{v2} = \frac{4\pi Z_1^2 e^4}{m_e v^2} Z_v \ln \frac{v}{v_F} \tag{2}$$

is the stopping due to plasmon excitations. In the above formulae Z_1 is the atomic number of the incident ion, v its velocity, m_e is the electron mass, Z_v is the number of valence electrons per crystal atom, v_F is the Fermi velocity, $\hbar\omega_p$ is the plasmon energy.

In order to calculate local i.e. position-dependent stopping cross section $\epsilon(\vec{r})$ of an ion moving through the crystal the ϵ_c component is assumed to be proportional to the local core electrons density $n_c(\vec{r})$, and accordingly ϵ_{v1} and ϵ_{v2} are proportional to the local valence electrons density $n_v(\vec{r})$ and to the average valence electrons density n_v, respectively. Thus one obtains

$$\epsilon(\vec{r}) = \frac{n_c(\vec{r})}{n_c}\epsilon_c + \frac{n_v(\vec{r})}{n_v}\epsilon_{v1} + \epsilon_{v2}, \tag{3}$$

where n_c is the average core electrons density, ϵ_{v1} and ϵ_{v2} are given by formulae (1) and (2), respectively, and $\epsilon_c = \epsilon_r - (\epsilon_{v1} + \epsilon_{v2})$. The above formulation is consistent with the random stopping ϵ_r. The ion energy dependence enters through factors $\epsilon_c(\epsilon_r)$, ϵ_{v1} and ϵ_{v2}.

In the calculations presented below the Roothan-Hartree-Fock atomic wave functions [2] were used in order to obtain local core and valence electrons densities $n_c(\vec{r})$ and $n_v(\vec{r})$. Actually, to calculate energy lost in a single ion-atom collision, electronic densities integrated along the ion trajectory are needed. These can be found assuming that the ion trajectory is a straight line.

Also the Lindhard two component model of channeled ion stopping [6] was used for a comparison. It can be written in the following way:

$$\epsilon(\vec{r}) = \frac{n_e(\vec{r})}{n_e}\alpha\epsilon_r + (1-\alpha)\epsilon_r, \tag{4}$$

where $n_e(\vec{r})$ is the local electronic density (core and valence), n_e is the average electronic density and the constant $\alpha \geq 1/2$. The atomic Independent Particle Model (IPM) potential [4] was used to obtain electronic density $n_e(\vec{r})$. The value of the constant α was assumed to be equal 1/2.

[1] Institute for Nuclear Studies, Hoża 69, 00-681 Warsaw, Poland

3. Monte Carlo simulation results

Modified binary collision model was used in the Monte Carlo simulation program. The influence of the crystal atoms on the ion trajectory was taken into account up to the maximum impact parameter of 2 Å. The deflection function was determined by the classical scattering integral using IPM ion-atom potential [3]. The root-mean-square amplitude of the thermal vibrations of the crystal atoms was taken according to the Debye theory ($\theta_D = 543\,°K$ for Si). The electronic multiple scattering according to Lindhard [6] was included as an option. In the calculations of the energy loss the three component model (eq. (3)) as well as the Lindhard model (eq. (4)) were used.

The calculations by means of the Monte Carlo program were performed using the input parameters corresponding to the experimental conditions of Jin and Gibson [5]. In this experiment 2 MeV He-ion beam was collimated to 0.01° divergence and traversed ⟨100⟩ Si single crystal of 8800 Å thickness (the thickness uniformity was ±5%). The fraction of the beam scattered from 150 Å gold film was registered by the silicon detector. By changing the tilt angle θ of the ⟨100⟩ axis with respect to the beam direction, the angular dependence of the particle energy loss in the crystal was determined. The results were presented as a ratio $R(\theta)$ of the most probable energy loss for a given angle θ and the random energy loss and are reproduced in fig. 1.

The calculated He-ion energy distributions for different tilt angles θ according to the three component model of energy loss are shown in fig. 2. The calculation was made for a uniform 8800 Å thick ⟨100⟩ Si perfect crystal and perfectly collimated monoenergetic 2 MeV He-ion beam. The electronic multiple scattering was included although the influence of this process was found to be small. The scanning through different θ angles has been performed in the plane rotated 13° off {100} plane.

The distributions in fig. 2 do not only show changes of the maximum position, but significant changes of the distribution shape itself are observed. To make a comparison with the experimental $R(\theta)$ values, the influence on the distributions of the crystal thickness nonuniformity as well as scattering on the gold film and detector resolution were also taken into account. $R(\theta)$ values obtained from the simulated distributions are plotted in fig. 1 along with the experimentally determined ones. Also shown in this figure are the $R(\theta)$ values calculated for the Lindhard model of energy loss (electronic multiple scattering was not included in this case). As seen in fig. 1 the calculated results are quite similar for the two models of energy loss. Both sets of points show distinct discrepancy with the experimental results for larger tilt angles. Since the experimental points clearly tend to a level below unity for $|\theta| > 1.5°$, it seems that the normalisation of the experimental data was incorrect. The agreement between the experimental results and calculated ones for $|\theta| < 0.5°$, although not worse than about 10%, is rather qualitative than quantitative. However, it should be noted that $R(\theta)$, being defined by the most probable energy loss, is not the best parameter to represent the energy loss data. The simulated distributions for θ near ±0.4° indicate that there exist two components of a similar height in these distributions. Therefore it would be more meaningful to compare directly calculated and measured distributions.

4. Conclusions

It has been demonstrated that it is possible to reproduce by Monte Carlo simulation technique experimental data for channeling energy loss—measured for different incident directions of 2 MeV helium beam with respect to ⟨100⟩ Si axis—with an accuracy better than about 10%. The $R(\theta)$ values calculated for the three component model and the Lindhard model of energy loss are quite similar. Larger differences are found to occur in the calculated energy distributions.

It can also be concluded that the experimental data require verification because of apparently wrong normalization. The more insight can be gained by measuring the energy distributions along different crystallographic directions and by comparing the experimental results with those of the Monte Carlo simulations.

References

[1] Appleton, B. R., C. Erginsoy and W. M. Gibson: Phys. Rev. **161** (1967) 330

[2] Clementi E. and C. Roetti: At. Data Nucl. Data Tables **14** (1974) 177

[3] Dygo, A. and A. Turos: Nucl. Instr. Meth. **B18** (1987) 115

[4] Green, A. E. S., D. L. Sellin and A. S. Zachor: Phys. Rev. **184** (1969) 1

[5] Jin, H. S. and W. M. Gibson: Nucl. Instr. Meth. **B13** (1986) 76

[6] Lindhard, J.: Kgl. Danske Videnskab. Selskab., Mat. Fys. Medd. **34** (1965) No. 14

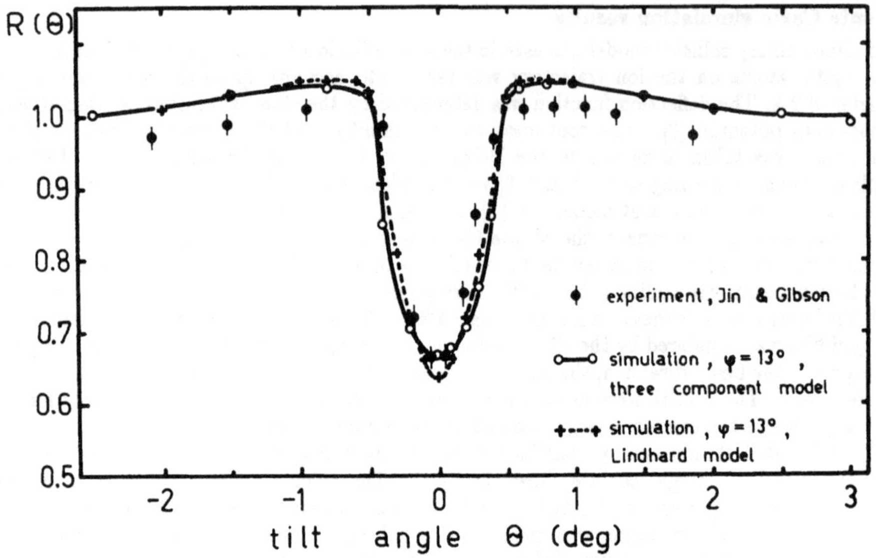

Fig. 1. Ratio $R(\theta)$ of the most probable energy loss and the random energy loss as a function of ⟨100⟩ tilt angle θ.

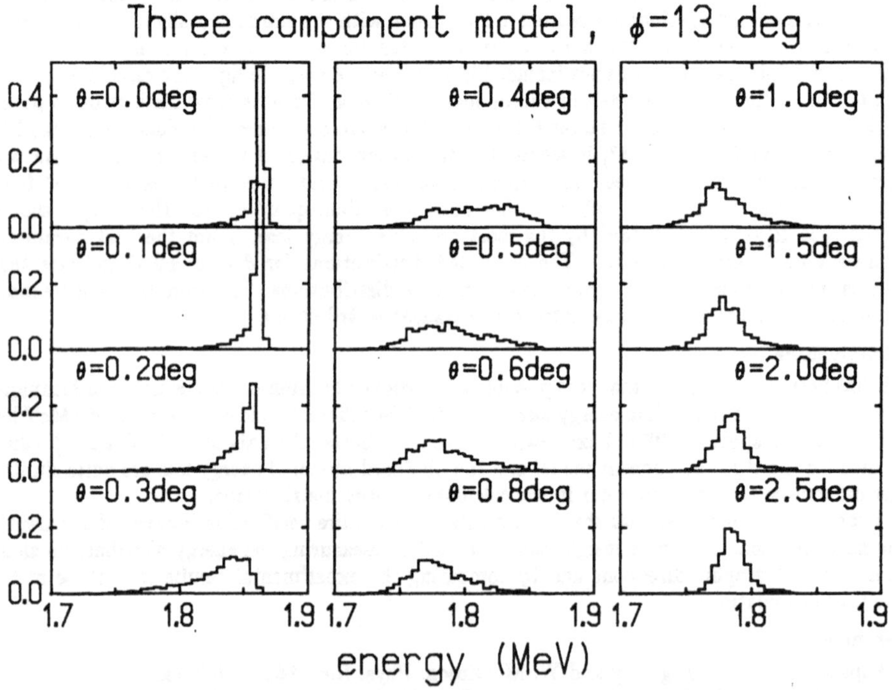

Fig. 2. Monte Carlo simulated energy distributions of 2 MeV He-ions after traversing 8800 Å thick ⟨100⟩ Si single crystal for different tilt angles.

TRANSMISSION SPUTTERING OF THIN GOLD FILMS: A COMPUTER SIMULATION STUDY

W.Szyszko,K.Paprocki,J.Sielanko*

1. Introduction.

Over the last decade considerable interest has been shown in the modification of material properties by ion implanted multicomponent targets, for example metal-semiconductor structures. In this method, atoms from the top layers are sputtered onto a semiconductor substrate and a shallow implanted layers are produced at the interface. The transmission sputtering yield can be calculated either from Sigmund formulae [1] or by a computer simulation technique [2-4].

Several measurements were performed on transmission sputtering yields of thin gold layers bombarded with noble gas ions [4-7]. The aim of our calculations was to study the transmission sputtering yields, energy and angular distributions and possible differences for the thin Au films bombarded by Ar,Kr and Xe ions. The calculated forward sputtering yields were compared with the measured total number of gold atoms implanted into Si by low dose ($5*10^{14}$ ions/cm^2) noble gas ion irradiation of Au/Si structures.

2. Computer model.

The computer simulations were performed within the binary-collision approximation. For the calculations the Monte-Carlo procedure (described in details in [8]), was modified so that now the whole collision cascade could be followed. In order to reduce the computer time (CPU) the trajectory of recoiled Au atom was followed if its maximum range, approximated by $R_{max}=3.6*\sqrt{E}$, was larger than its distance from the Au surface. A similar cut-off value of R_{max} was also used by Biersack and Eckstein [9]. In the simulations the Au film was assumed as amorphous and the maximum impact parameter was chosen 0.162 nm [2,10].

The calculations were performed in two ways:

(i)- no surface binding energy (U_o=0.0eV) at the Au back side was used, cut-off energy (E_c) and displacement energy (E_d) of gold atoms were selected as $E_c = E_d$ = 20 eV,

(ii)-a planar surface binding energy model was used,i.e. U_o=3.8 eV [11], E_c=3.8 eV and E_d=7.6 eV.

In both cases about 5000 trajectories of incoming ions were simulated.

3. Experimental technique.

Thin gold foils were evaporated in vacuum on silicon crystals. The thickness of Au films (Δx) varied from 25 to 30 nm. Au/Si samples were

* Institute of Physics,M.Curie-Skłodowska University,20-031 Lublin,Poland

implanted to the dose $5*10^{14}$ ions/cm^2 of 120 keV Ar^+, 200 keV Kr^+ and 300 keV Xe^+. After the recoil implantation, gold films were chemically removed and the number of Au atoms was studied by RBS technique.

4. Results and discusion.

The calculated transmission-sputtering yields Y_o and Y_{sb} are shown in table 1 together with the experimentally determined recoil yields (R). Y_o and Y_{sb} data were calculated using the models (i) and (ii) respectively.

TABLE 1. Transmission sputtering yield of gold (Y_o, Y_{sb} and R in [atoms/ion])

Ion	Energy [keV]	Δx [nm]	Y_o	Y_{sb}	R
Ar^+	120	26	3.3	9.1	4.2
Kr^+	200	29	7.6	19.2	7.2
Xe^+	300	26	14.6	25.9	15.0

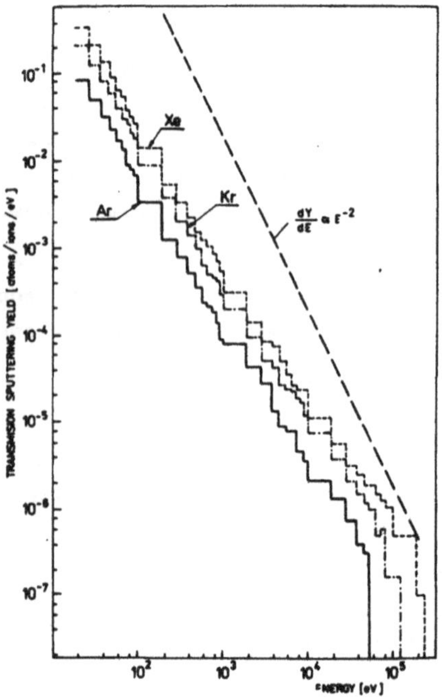

Fig. 1. Energy distribution of Au atoms sputtered with Ar, Kr and Xe ions.

The calculations were performed for the different values of E_d and E_c and the best fit to the measured recoil yields (R) was obtained for the values $E_d=E_c=20$eV with surface binding energy at the interface $U_o=0$ (model (i)). Also Karpuzov et al.[4] observed a good correlation of experimentally determined yields of gold atoms into silicon substrate with calculated yields if Au atoms crossed the interface withought energy loss. In the paper [4] both values were selected as $E_d=E_c=35$ eV for 10-20 keV Ar bombardment. In reality the U_o value can varied with the prolonged irradiation of the sample. In the transport theory [1] the sputtering yield is inversly proportional to the surface binding energy U_o. This assumption corresponds to our (ii) model (Y_{sb} calculations). The values of Y_{sb} are larger of a factor of 2-3 than Y_o yields. In our opinion model (ii) may be useful for the transmission sputtering study of Au atoms from thin gold layers into the vacuum. Therefore the values of Y_{sb} could be compared with the experimentally determined transmission sputtering yields. Ecker and Merkle [12] determined the transmission- sputtering yield Y=5 for 200 keV Ar bombardment of 25 nm Au film. The Y value calculated by Sigmund for 100-200 keV Ar^+bombardement of 25 nm Au films was equal $Y \simeq 15$ [1]. The transmission sputtering yield obtained in this work for 120 keV Ar irradiation was equal $Y_{sb} = 9.1$.

The energy spectra of transmission-sputtered Au atoms with energies larger than 20 eV are shown in Fig.1. The dashed line indicates a E^{-2} distribution.

The depth from origin and the angle of ejection distributions of forward sputtered Au atoms are compared in Fig.2. As it can be expected $\simeq$90% of sputtered atoms originate from within the last few atomic layers (Fig.2a). The calculated angle of ejection distributions are close to the cosine distribution with a small increase in intensity for the emittance angle $\varphi \geq 50^0$. Almost the same angular and depth distributions from origin for Ar and Xe ions let us conclude that cascade recoils dominates in forward sputtering of Au foils studied in present paper and primary recoils of gold atoms could be neglected for low dose noble gas ion bombardement.

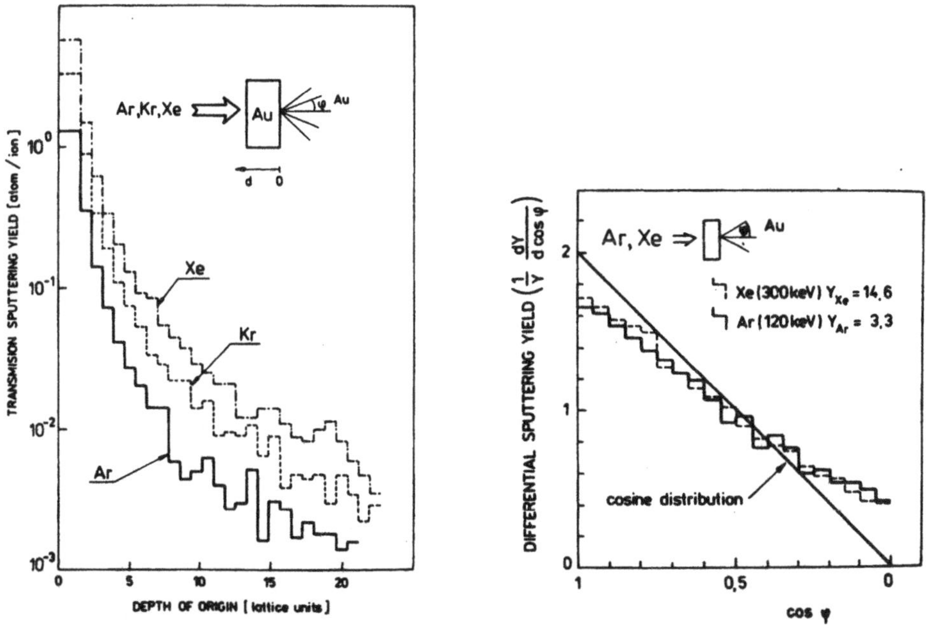

Fig.2. Depth from origin (a) and angle of ejection (b) distributions of transmission sputtered Au atoms.

This work was supported by CPBP 01.06. and CPBP 01.08.C1.3 research programes.

References.

1.P.Sigmund Phys.Rev. 184 (1969) 383
2.M.Hautala Nucl.Instr.and Meth. B13 (1986) 357
3.P.Argyrokastritis,D.S.Karpuzov,J.S.Colligon,A.E.Hill,H.Kheyrandish Phil.Mag. A49 (1984) 547
4.D.S.Karpuzov,J.S.Colligon,H,Kheyrandish,A.E.Hill Nucl.Instr. and Meth B6 (1985) 474
5.H.L.Bay,H.H.Andersen,W.O.Hofer,O.Nielsen Appl.Phys. 11 (1976) 289
6.H.L.Bay,H.H.Andersen,W.O.Hofer Radiat.Effects 28 (1976) 87
7.K.H.Ecker Radiat.Effects 23 (1974) 171
8.J.Sielanko,W.Szyszko Surf.Sci. 161 (1985) 101
9.J.P.Biersack,W.Eckstein Appl.Phys. A34 (1984) 73
10.M.Bister,M.Hautala,M.Jantti Radiat.Effects 42 (1979) 201
11.Ch.Steinbruchel Appl.Phys. A36 (1985) 37
12 K.H.Ecker,K.L.Merkle Phys.Rev. B8 (1978) 1020

MODELLING OF NONEQUILIBRIUM PHASE TRANSITIONS INITIATED BY SHORT LASER PULSES IN SILICON

D.Stock and H.-D.Geiler[1)]

INTRODUCTION

Irradiation of semiconductors with short laser pulses can result in ultrafast melting and solidification of thin surface layers. The strong nonequilibrium character of the occuring phase transitions requires the introduction of interfacial kinetics during melt-in and regrowth into modeling [1,2]. Assuming that silicon transforms continuously and heterogeneously during melting and freezing the kinetics of the motion of the liquid-solid interface can be described completely by an interface response function v(T) which is a central point of interest in modeling of highly nonequilibrium melting and solidification. In this paper we will investigate theoretically different estimates of the physical parameters of the v(T) relationships for crystalline Si (c-Si) and amorphous Si (a-Si) according to currently known experimental constraints. Therefore, heat-flow calculations including the interfacial kinetics were carried out and a comparison with recent experimental results concerning pulsed-laser melting of a-Si is made [3].

FORMULATION OF THE MODEL

The model is based on the solution of the heat-flow equation which is coupled with the interface response function v(T) at the liquid-solid interface. The surface region of the sample is modeled as a slab of finite thickness d extending in positive x-direction. Assuming a 1D-moving-boundary problem our approach can be summarized as follows:

$$\rho \partial_t h = \partial_x [K \partial_x T] + \nu(x,t,T) \qquad (1)$$

$$v(T_i) = v_o e^{-Q/kT_i} \left[e^{-\Delta S_{sl}/k} - e^{-\Delta H_{sl}/kT_i} \right] \qquad (2)$$

$$\partial_x T = 0, \quad T(d,t) = T_o, \quad T(x,0) = T_o \qquad (3)$$

Eq.(1) is the heat-conduction equation in terms of the local balance of enthalpy h(T) which involves the latent heat of the phase transition.

---

1) Friedrich-Schiller-Universität Jena, Sektion Physik, Max-Wien-Platz 1 Jena, DDR-6900, GDR.

Moreover the enthalpy formalism allows easily to treat the interfacial undercooling or overheating. The enthalpy may change with time due to heat flow with the thermal conductivity K and due to energy deposition ν from the laser pulse [4]. The interface response function $v(T_i)$ given by Eq.(2) provides the interface velocity v in response to the local temperature T_i at the solid-liquid interface according to the rates of thermally activated forward and backward jumps across the interface. The kinetic prefactor is determined by an attempt velocity v_o and the activation enthalpy Q. The thermodynamic quantities ΔS_{sl} and ΔH_{sl} are the entropy of fusion and the enthalpy of fusion, respectively. The boundary conditions and the initial temperature distribution are given by Eq.(3) where T_0 is the substrate temperature. The Eqs.(1-3) were solved numerically using a finite-difference method.

In the case of silicon each of the transitions, c<-->l and a<-->l, can be described by a relation according to Eq.(2). The curves should have a crossing point at a velocity of about 15 ms^{-1} which is known as the amorphization threshold for <100> Si [5]. Fig.1 shows v(T) for crystalline and amorphous Si in the case of two different sets of parameter values.

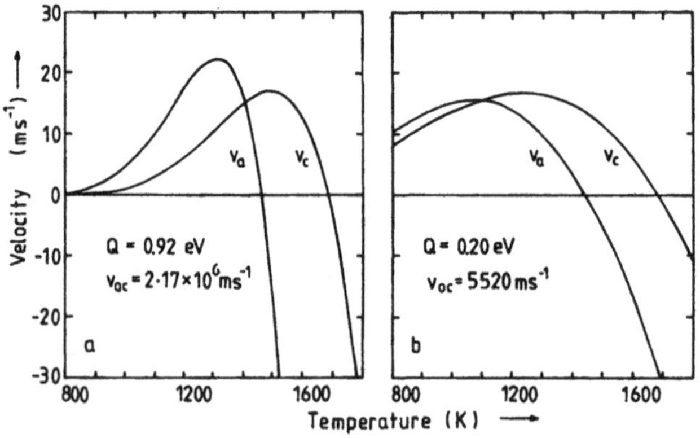

Fig.1. Interface response functions v(T) according to Eq.(2). The parameter values correspond to (a) $v_{oc} = 2.17\times10^6$ m/s and Q = 0.92 eV; (b) v_{oc} = 5520 m/s and Q = 0.20 eV.

The parameter values used in Fig.1a correspond to experimental constraints resulting from explosive crystallization of a-Si [6], i.e. (i) $\max\{v_c\} \simeq 17$ m/s and (ii) at this temperature a-Si should melt with nearly the same velocity. The parameter values of Fig.1b were derived by Tsao et al. [7] using transient-conductance measurements which provided $v_a(T_{mc}) = -29$ m/s. As can be seen from Fig.1 this constraint is not consistent with the above from explosive crystallization. The following thermodynamic data were used: ΔH_{cl} = 0.525 eV, ΔH_{al} = 0.385 eV, T_{mc} = 1685 K, T_{ma} = 1460 K (Fig.1a) rsp. 1442 K (Fig.1b).

RESULTS AND DISCUSSION

The results in Fig.2 are for a semiinfinite a-Si layer irradiated with 7.5 ns pulses from a frequency-doubled Nd:YAG laser (λ = 532 nm) corresponding to the experimental conditions [3].The curves of maximum melt depth versus laser energy density show different slopes in dependence of the used set of parameter values of v(T).

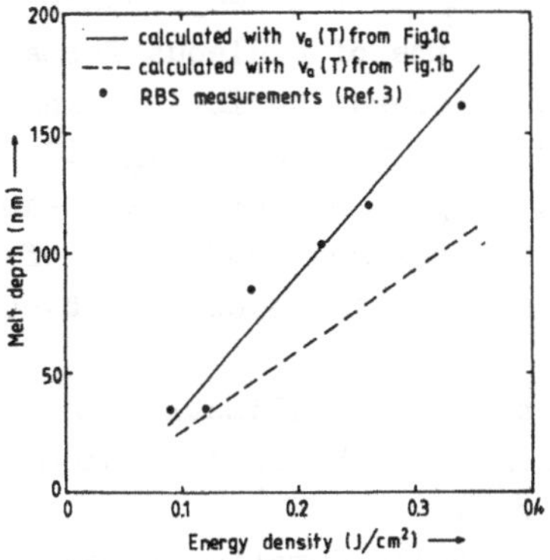

Fig.2. Maximum melt depth versus laser energy density;results from model calculations using different parameter values for v(T) in comparison with RBS measurements (solid circles) of Bruines et al.[3].

As can be seen the model calculations using the parameter values of v(T) consistent with the explosive crystallization of a-Si (see Fig.1a) provide a quite good correspondence with the experimentally determined maximum melt depth from RBS. The lower slope of the dashed curve which we obtained with the parameter values estimated by Tsao et al. is apparently caused by the lower slope dv_a/dT for melting (see Fig.1). With increasing laser energy density the difference in the maximum melt depth increases due to the increase of initial overheating. We conclude that further theoretical and experimental investigations are still necessary in order to clarify the $v_a(T)$ and $v_c(T)$ relationships for silicon.

[1] Wood, R.F. and G.A.Geist, Phys. Rev. B34 (1986) 2606
[2] Geiler, H.-D., D.Stock, G.Götz and K.Hennig, Proc. E-MRS Strasbourg 1986 153
[3] Bruines, J.J.P., R.M.P.van Hal, H.M.J.Boots, W.Sinke and F.W.Saris, Appl. Phys. Lett. 48 (1986) 1252
[4] Geiler, H.-D., K.Hehl and D.Stock, phys.stat.sol.(a) 78 (1983) 193
[5] Cullis, A.G., N.G.Chew, H.C.Webber and D.J.Smith, J. Cryst. Growth 68 (1984) 624
[6] Andrä, G., H.-D.Geiler, E.Glaser, G.Götz, M.Wagner and K.H.Heinig, Nucl.Instr.Meth. B19/20 (1987) 571
[7] Tsao, J.Y., M.J.Aziz, M.O.Thompson and P.S.Peercy, Phys. Rev. Lett. 56 (1986) 2712

INTERMEDIATE CRYSTALLIZATION OF AMORPHOUS SILICON LAYERS AT NANOSECOND PULSED LASER ANNEALING

V.Yu.Balandin, A.V.Dvurechenskii, L.N.Aleksandrov[1)]

The melting point of the amorphous material T_A is lower than that of the crystalline material T_M. The liquid silicon is undercooled immediately after the a-Si melting. In the event of the heating rate at that time $\dot{T} < (T_N-T_A)/\tau(T_N)$, provided $T_N > T_A$, where T_N is the temperature of maximum nucleation rate, $\tau(T)$ is the waiting time of nucleous formation, the intermediate crystallization (IC) can occure. During heating crystallites are formed and then are melted as T rises above T_M. IC realizes in the maximum heat generation in a sample. So for laser beam IC takes place at the surface of the sample. IC changes conditions of light reflection and absorption and hence strongly influences on subsequent crystallization dynamics.

In this paper we discuss the conditions of IC realization and it's role in a crystalline structure formation. The physical problem and numerical model were described elsewhere /1/.

As our calculations show IC realizes if pulse duration Γ exceeds some value (~170 ns). The dynamics of phase transformations in the a-Si layer at different energy densities E are shown in Fig. 1. For given experimental conditions the a-Si layer melting occures as $E > 0.25$ J/cm^2. Figure 1a shows that some time after a-Si melting crystallites are formed near the surface. The latent heat release at nucleation provides the further melt front penetration and, as the temperature rises, causes the nucleous crystal growth. After liquid front reaches c-Si substrate, there are two crystal growth fronts moving toward each other. Defects in amorphous-crystalline near-boundary region should remain (dotted line).
The slush-zone realizes as energy density is increased (Fig. 1, b). In this annealing regime the auto-oscillations of the melt and crystallites fraction is possible because of reflectivity changing at phase transition. When $E > 1$ J/cm^2, the slush-zone complitely melts and IC realizes. It is interesting to note, as IC occures, the monocrystalline structure is restored (Fig. 1, c). However, when E reaches ~ 2.3 J/cm^2 the polycrystalline layer is formed. We can explain this feature by a changing of the heat generation profile. The higher the energy density, the less is the existing time of the near-surface polycrystalline layer (Fig. 1, c-f). The heat production in the sample by the laser beam falls due to the life time increasing of the high-reflectivity liquid phase. It leads to the nucleation in the depth of the molten layer. The following energy density increasing once more produces the monocrystalline layer, as it is shown on Figure 1, e-f.

Since we made calculations not only for the fixed pulse duration, but

1) Institute of Semiconductor Physics, Academy of Sciences of the USSR, 630090, Novosibirsk, USSR

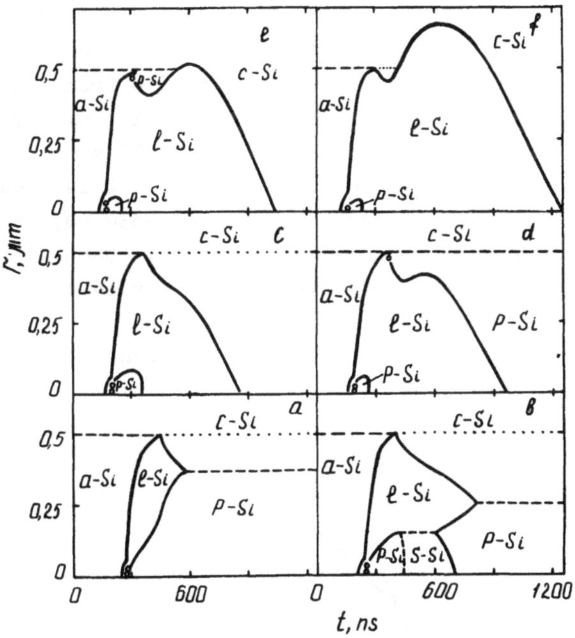

Fig. 1. Melt, crystal growth, and nucleation fronts position as a function of time pulse ruby laser annealing of a-Si, pulse duration 300 ns for E, J/cm^2: 0.8 (a); 1.2 (b); 1.8 (c); 2.4 (d); 2.8 (e); 3.4 (f). Solid curves -melting and crystal growth; circles-homogeneous nucleation; dotted lines-boundaries between region with different structures; points-remaining defects of a-Si-c-Si boundary.

also for different Γ, to characterize the annealing processes, we introduce a diagram. The diagram of the typical structural transformation processes realization versus the pulse duration and energy density is shown on Figure 2. Points with coordinates E and Γ lying in regions surrounded by curves correspond to processes having some particular features. So, a-Si does not undergo the transformation in region 1. Region 2 corresponds to the a-Si melting and melt reamorphization. Regions 3, 6, 7 are distinguished by a direction of crystallization front movement: in 3 - from the depth toward the surface; in 6 - from the surface toward the depth; in 7 - toward each other. Region 8 is remarkable for the slush-zone formation. IC realizes as E-Γ point lies in region 5 or 9. Monocrystal forms in 5, polycrystal forms in 9. It is clear to see that there are two E intervals provided Γ = 220-330 ns, in which polycrystalline (or mono-) layer realizes. In general, the polycrystalline structure is formed as E-Γ point lies between solid curves 'A' and 'B'.

Experimental results /2/ on reflectivity data apparently confirms our investigation on IC.

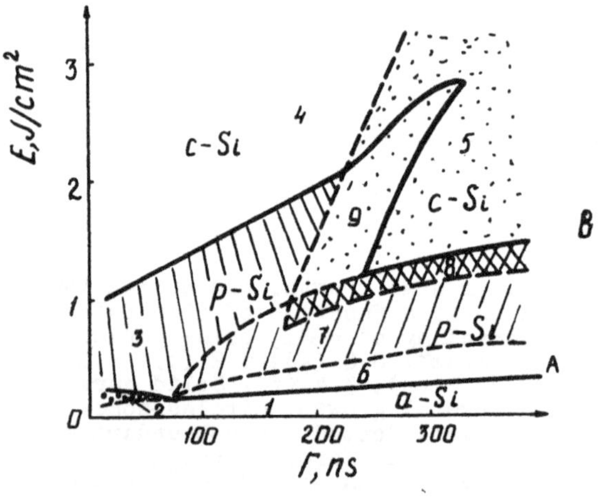

Fig. 2. Diagram of the typical structural process realization at the ruby laser pulse annealing of a-Si (0.5 /um) - c-Si structure. T_A=1200 K.

REFERENCES

(1) Balandin V.Yu., Dvurechenskii A.V., Aleksandrov L.N. Poverkhnost´ (1986) N1, 53-60.

(2) Ivlev G.D. Pis´ma v Zh.TF 8(1982), 468-472.

INVESTIGATIONS ON THE AMORPHOUS/CRYSTALLINE PHASE TRANSITION IN ION-IMPLANTED SILICON BY MEANS OF CROSS-SECTIONAL TEM IMAGING

H. Bartsch, D. Hoehl, P. Werner[1)]

The formation and recrystallization processes of amorphous(a) silicon layers by means of ion implantation are of fundamental interest, they have also consequences on the application in VLSI technology /1-9/. TEM cross-sectional(cs) imaging yields information on the geometry, the structure, and the interface kinetics of ion-implanted layers. Cs-TEM imaging first has to characterize the as-implanted state, as it is shown in Fig.1 (left). It becomes clearly visible that after ion implantation a heavily disturbed transition layer is formed between the implanted layer and the substrate. Buried layers are bound by two of

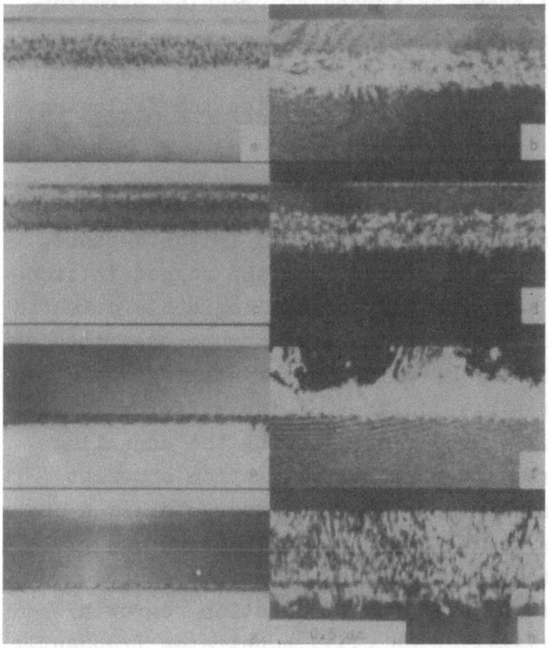

Fig.1: Si^{++}-implanted silicon ($E=140keV$, $T_i=RT$, <111>) as implanted(left) and annealed ($T_a=1070K$, $t_a=30min$)
(a,b): $D=5x10^{14}cm^{-2}$
(c,d): $D=10^{15}cm^{-2}$
(e,f): $D=5x10^{15}cm^{-2}$
(g,h): $D=10^{16}cm^{-2}$.

these regions the thickness of which depends on the implantation parameters and influences the regrowth kinetics. Experimental results show a decrease of the thickness of the transition layer with increasing ion dose, increasing ion mass, and decreasing target temperature. The thinnest transition layers (15...20 nm) were observed after implantation at 77 K. The diffraction contrast images (as in Fig. 1), however, do not show the crystallographic nature of the disturbed transition layer because of the limited resolution. For this reason high-resolution TEM (HRTEM) studies were carried out. Fig. 2 shows the result taken from a buried amorphous layer created by As^+-implantation at room temper-

1) Institute of Solid State Physics and Electron Microscopy, Academy of Sciences of the GDR, Weinberg 2, DDR-Halle/S., 4050

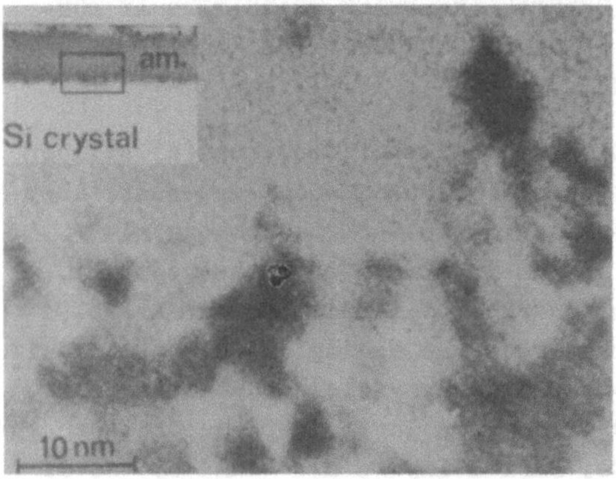

Fig. 2: Cs-HRTEM image of the transition region of a buried a-layer after As^{+}-implantation into <100>-Si ($E=130keV$, $T_i=RT$, $D=10^{14}cm^{-2}$).
Inset: whole layer system at low magnification.

ature (RT). The amorphous phase shows an irregular granular structure (bright regions), whereas the structural image of the crystalline(c) parts (darker regions) show regular bright spots caused by the atomic strings of the lattice. The transition region includes the following structural details:

- isolated a-clusters in the c-substrate,
- alternating a/c portions in the rugged interface region,
- isolated c-clusters (of substrate orientation) in the a-phase.

For this reason the epitaxial regrowth starts from the rugged intermediate layer, not from a smooth homogeneously crystallographic oriented interface. Considering the HRTEM image, this layer may be described as an arrangement of polyeders built up of different crystallographic planes. Due to the different growth velocities of the different crystallographic planes /2, 6/ defects are created during the annealing process of this layer. Therefore, after recrystallization a defect ribbon is very often arranged separately below the recrystallized layer (e.g. Fig. 1h, Fig. 3). In this region the defect formation can probably be minimized and simultaneously the rugged interface can be smoothed by preheating at lower temperatures (incubation regrowth) /2, 5/. The epitaxial recrystallization on <111> substrates frequently forms growth-induced microtwins, the density of which decreases with

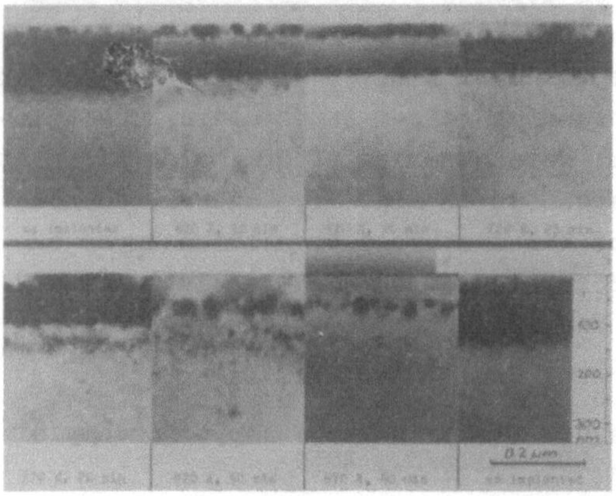

Fig. 3: Annealing sequence of an in situ heating experiment carried out in an HVTEM using the same material as for Fig. 2.

the distance from the interface. The epitaxial regrowth on <100> substrates, however, mostly forms interstitial dislocation loops in the implanted layer during annealing.

Annealing buried amorphous layers causes the regrowth to start from both interfaces /3/. In order to study the interface kinetics during recrystallization TEM in situ heating experiments are very instructive /1, 7/. Fig. 3 represents in situ annealing of a cs-sample, carried out in a heating cartridge of a high-voltage electron microscope (HVTEM). The thickness of the a-layer shrinks with increasing temperature. After annealing above 770 K the layer is epitaxially recrystallized. A double layer of dislocation loops is formed. The lower defect ribbon is caused after annealing the lower transition region, the other one is formed approximately in the depth of the maximum radiation damage (about 50 nm) owing to the excess interstitials. The surface layer, however, becomes free of extended defects due to the short distance of the moving interface from the surface (15...40 nm) and thus to the enhanced possibility of point defect interaction during regrowth.

A special situation turns out if the covering surface layer of a buried amorphized layer remains unclosed, particularly when the remaining crystalline parts become very small. In this case these crystallites can act as nuclei to crystallization in concurrence with epitaxial regrowth. Fig. 1f shows the case where very small nuclei of $3x10^8 cm^{-2}$ in density remain near the surface and act as nuclei to dendritic growth. The dendritic-grown crystallites stop the moving interface of the epitaxially recrystallizing layer in the half-depth of the layer. The density and the dimension of the nuclei, however, are too small to enable their detection in the as-implanted state of the layer (Fig.1e). We can therefore conclude that amorphization by implantation must be carried out by choosing parameters which guarantee a continuously amorphous layer up to the surface, or a buried amorphous layer covered with a closed monocrystalline surface layer, respectively.

__References:__

/1/ Bostanjoglo, O., F.E. Endruschat and W. Tornow: Materials Research Symp. Proc. Vol. 71 (1986) 345-350

/2/Csepregi, L., J.W. Mayer and T.W. Sigmon: Appl.Phys.Lett. 29 (1976) 92

/3/ Jones, K.S. and S. Prussin: MRS Symp.Proc. Vol. 71 (1986) 173-178

/4/ Ling,P.,N.R.Wu and J.Washburn: MRS Symp.Proc. Vol.71 (1986) 179-182

/5/ Maher, D.M. et al.: Semiconductor Silicon 1986, Pennington N.J.: The Electrochem. Soc. 1986

/6/ Narayan, J.: J. Appl. Phys. 53 (1982) 8607

/7/ Parker, M.A., T.W. Sigmon and R. Sinclair: MRS Symp. Proc. Vol. 62 (1986) 311-322

/8/ Turnbull, D.: MRS Symp. Proc. Vol. 51 (1986) 71-82

/9/ Williams, J.S.: MRS Symp. Proc. Vol. 51 (1986) 83-98

NUMERICAL MODELLING OF SILICON SAMPLE MELTING BY HIGH-INTENSITY ION-BEAM PULSE

M. Rabiński *

In the Pulse Implantation Doping process [4] the high-intensity ion beam transports the ion dose necessary to dope a semiconductor, and simultaneously it supplies energy sufficient to melt the surface layer of the sample. The thermal analysis of this phenomenon bases on the solution of so called Stefan problem of phase change. The appropriate mathematical model is given by two heat conduction equations for liquid and solid respectively, and by the heat-flow balance on the free boundary between the phases, with additional condition fixing the interface temperature.

For the numerical solution of Stefan problem the enthalpy and heat capacity methods are commonly used. This techniques reformulate the model system into a single equation by introducing the enthalpy as dependent variable or modifying the specific heat. The main disadvantage of this approach is that a highly nonlinear function is introduced, hence the explicit schemes usually have to be employed with all the consequences of their conditional stability.

To overcome the computational disadvantages of such methods, the moving heat-source formulation [2,3] has been used in a one-dimensional numerical program for the investigation of melting, surface evaporation and resolidification processes in a silicon sample exposed to the high-intensity ion-beam pulse. The mathematical model is given by the heat-conduction equation with the temperature dependent conductivity λ. The latent heat L absorbed (or liberated) during the change of state is treated as a moving heat sink (or source) located at the liquid/solid interface. The introduced dimensionless temperatures θ for the solid and liquid phases are defined as

$$\theta_s(x,t) = (T_s(x,t) - T_m)/\ \Delta T \tag{1}$$

$$\theta_l(x,t) = (T_l(x,t) - T_m)/\ \Delta T \tag{2}$$

where subscripts s and l denote solid and liquid phase respectively, T_m is the melting point temperature and ΔT is an arbitrary reference temperature difference. Thus the Stefan problem can be transformed into the following heat conduction equation

$$c\rho\ \partial\theta/\partial t = \frac{\partial}{\partial x}(\lambda\ \partial\theta/\partial t) + q + \gamma \tag{3}$$

where

$$\gamma(x,t) = \rho L\ v(t)\ \delta(x-x_o)\ /\ \Delta T \tag{4}$$

* *Institute for Nuclear Studies, 05-400 Swierk-Otwock, Poland.*

is the moving heat source at the phase change front, c is the specific heat, ρ denotes density, q is the yield of the inner source (e.g. ion energy deposit), v is an interface velocity in the direction of a unit vector pointing into the liquid phase, δ denotes the Dirac delta function, x_o is the location of the interface.

A finite-difference implicit method of the Crank-Nicolson type has been used for numerical solution of the conduction equation. This scheme is unconditionally stable and assures the heat flux conservation during the calculations [1] (thus the energy balance is also conserved). As consequence relatively long time steps can be used. For the determination of the moving source position and intensity a special computational procedure has been developed. The selfconsistency of the model should be achieved by some iterations. A special treatment has been proposed for the acceleration of convergence: because of the lack of the latent heat effects in the first guess the interface velocity (thus the moving source intensity) is initially overestimated, therefore the heat source is introduced gradually to smooth oscillations. Although the computational algorithm of the described program is more sophisticated than that of the enthalpy methods, the computer time for the problem solution is significantly shorter because of relatively longer time steps which can be used.

The phase change during the evaporation is treated analogous to melting but after each time step the evaporated material is released, thus the sample boundary changes. The ion-beam parameters determine the time-dependent heat flux at the sample boundary, or during the more detailed analysis the yield of the inner source results from the ion-energy deposition in the material. Also the radiation of sample surface according to the Stefan-Boltzman law can be included, but this effect was found negligible.

To illustrate the application of the above described program it was applied to the analysis of the thermal phenomena in a silicon sample exposed to 3.5 J/cm^2 250 ns pulse (see Fig.1). The results are presented in figures allowing correlate the temperature distributions, liquid/solid front position and evolution of sample boundary due to the evaporation. The influence of several beam parameters (such as the pulse shape, ion energy, etc.) on the sample melting depth is investigated in order to achieve the highest PID process technological efficiency.

References:

[1] Kalitkin,N.N.:Computational Methods (in Russian), Nauka, Moscow 1978
[2] Lightfoot,N.M.H.: Proc. London Math. Soc. **31** (1929) 97-119
[3] Ozisik,M.N.: J. Heat Transfer **100** (1978) 370-371
[4] Piekoszewski,J., M.Gryziński, J.Langner and Z.Werner: J. Physique **43** (1982) 1353-1358

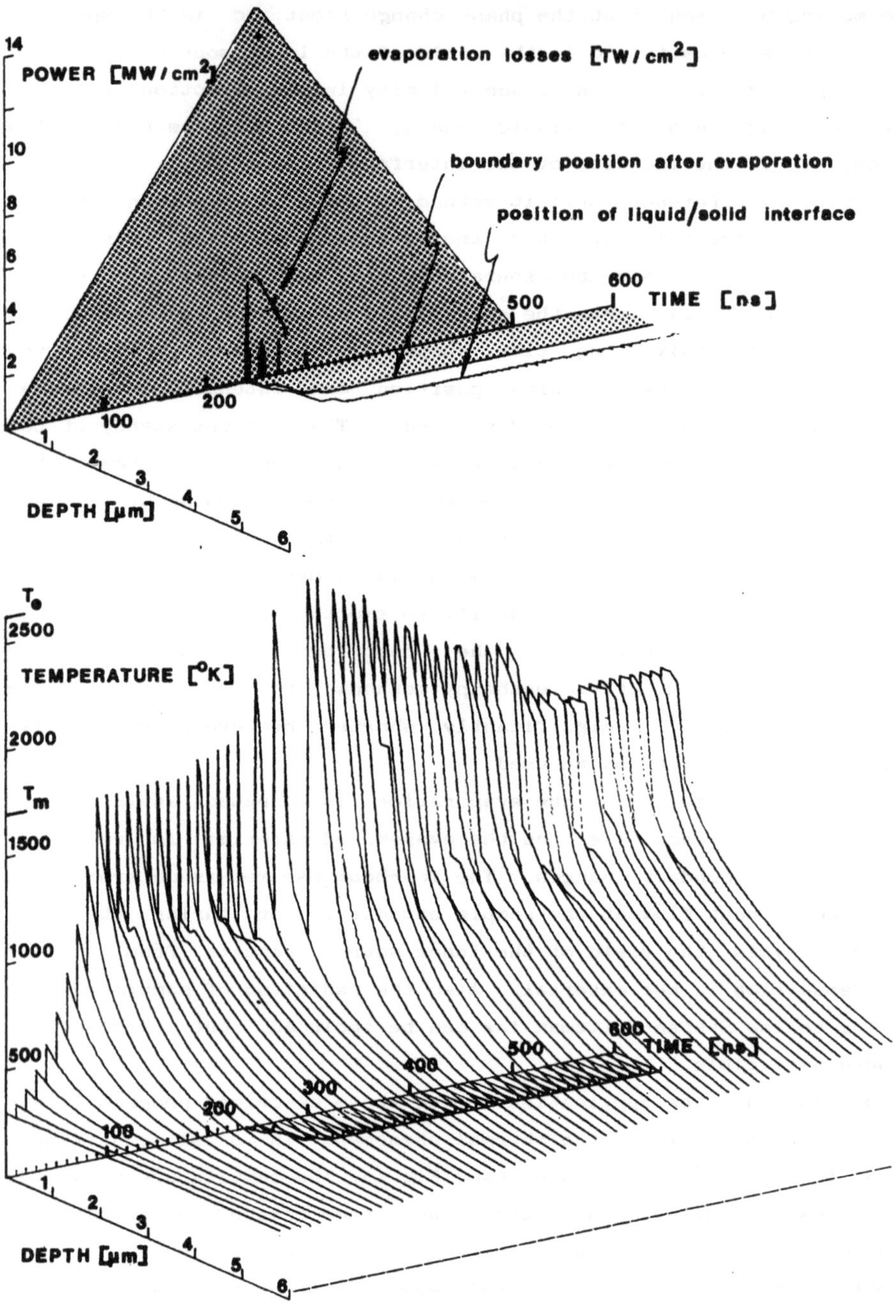

Fig.1. Temperature distributions in the surface layer of a silicon sample exposed to $3.5 J/cm^2$ 250ns pulse. Silicon parameters: λ_s=88/T cal/cm*s*K; λ_l=0.14 cal/cm*s; T_m= 1690 K; L_m= 430 cal/g; T_e = 2628 K; L_e = 3648 cal/g; c= 0.2 cal/g*K; ρ_s = 2.2 g/cm^3; ρ_l= 2.46 g/cm^3. In the case presented the significant part of the incident energy is removed with silicon vapor.

SIMULATION OF NANOSECOND LASER ANNEALING OF SILICON ALLOWING FOR CRYSTALLIZATION OF SUPERCOOLED MELT

S.P. Zhvavyi and O.L. Sadovskaja[1)]

This paper presents the results of computational simulation of the dynamics of nanosecond laser annealing of amorphous silicon layer (a-Si) of thickness $d = 0.1\,\mu m$ on a single-crystal (c-Si) substrate. It is assumed that the phase transitions a-Si ⟶ highly supercooled melt (sl-Si) and c-Si ⟶ melt (l-Si) occur at temperatures T_a = 1483 K and T_m = 1683 K, respectively, and in the region of the sl-Si phase the formation of fine-grained polycrystal (p-Si) proceeds on already available nucleation centres. The problem of heating and melting of silicon by laser radiation ($\lambda = 0.69\,\mu m$, τ_0 = 70 ns) with the crystallization of sl-Si taken into account was formulated as follows:

$$\rho\cdot[c(T)+\mathcal{L}(T)]\frac{\partial T}{\partial t} = \frac{\partial}{\partial x}\left[k(x,T)\frac{\partial T}{\partial x}\right] +$$

$$+[1-R(T)]\cdot\alpha(x,T)\cdot\frac{E(t)}{\tau_0}\exp\left[-\int_0^x \alpha(x',T)dx'\right]+\rho L_c\frac{\partial\Psi(x,t)}{\partial t}, \qquad (1)$$

$$T(x,t=0)=T_0\,, \quad \left.\frac{\partial T}{\partial x}\right|_{x=0}=0, \quad T(x\to\infty,t)=T_0\,,$$

where

$$\mathcal{L}(T)=\begin{cases}\theta(d-x)\cdot\{L_a\delta(T-T_a)-[1-\Psi(x,t)]\cdot L_c\delta(T-T_m)\}+L_c\delta(T-T_m), & \text{for } t<t^* \\ [1-\theta(d-x)\,\Psi(x,t)]\cdot L_c\delta(T-T_m), & \text{for } t\geq t^*\end{cases}$$

$$\theta(x)=\begin{cases}0, & \text{for } x\leq 0, \\ 1, & \text{for } x>0,\end{cases}$$

ρ, $c(T)$ and $k(x,T)$ are the density, specific heat and heat conductivity factor, respectively, $R(T)$ and $\alpha(x,T)$ are the reflectivity and absorptivity, respectively, $E(t)$ and τ_0 are the energy density and duration of laser pulse, T_0 is the initial temperature, L is the specific heat of phase transition, $\Psi(x,t)$ is the volume fraction that crystallizes by a time t after crystallization onset, t^* is the time for the melt to reach the maximum depth.

The last term in the right side of eqn. (1) stands for the power of the heat sources during crystallization of sl-Si. For heterogeneous nucleation

1) Institute of Electronics, BSSR Academy of Sciences, Ya.Kolas St. 68/2, Minsk-90, 220841 (U.S.S.R.)

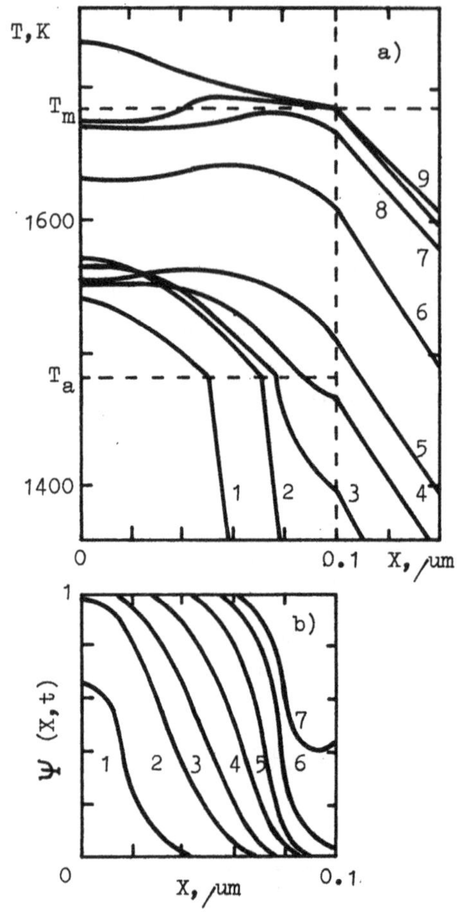

Fig. 1. (a) Computed temperature profiles and (b) crystallized fraction of sl-Si for E = 1.4 J/cm^2 at different time instants (ns): (1) 48.7, (2) 50.9, (3) 52.0, (4) 53.2, (5) 54.4, (6) 55.6, (7) 60.4, (8) 61.6, (9) 62.8.

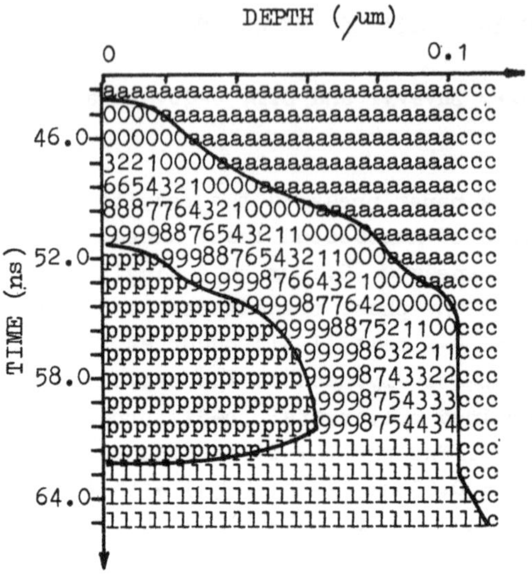

Fig. 2. Changes in aggregate states of silicon in the initial stage of nanosecond laser annealing.

mechanism the relationship for $\Psi(x,t)$ takes the form [5]:

$$\psi(x,t) = 1 - \exp\left\{-\frac{4\pi}{3} N_0 \left[\int_{\tau}^{t} V(t')\,dt'\right]^3\right\}, \tag{2}$$

where $V(t) = A \cdot \exp\left(-\frac{E_a}{kT}\right) \cdot \left[1 - \exp\left(-\frac{L_c}{kT_m} \cdot \frac{T_m - T}{T}\right)\right]$

E_a = 1.22 eV [1], N_0 is the density of crystallization centres. In computations N_0 was taken to be 10^{18} cm^{-3} which corresponds to the fine-grained polycrystalline phase with typical grain size of ~ 100 Å [3]. The value of the coefficient A = $5 \cdot 10^6$ cm/s was obtained by comparing the calculated values of maximum temperature and time of existence of the melt with experimental data [4]. In a two-phase region consisting of sl-Si and p-Si, all parameters of the problem were determined as follows:

$$a(x,t) = [1-\psi(x,t)]\, a_e(x,t) + \psi(x,t)\cdot a_c(x,t) \qquad (3)$$

It was also assumed that $R(T)$ is determined by the aggregate state of the surface layer.

Eqn. (1) together with eqns. (2) and (3) were solved numerically. Fig. 1 shows temperature profiles and crystallized fractions of the sl-phase at different instants of time and Fig. 2 depicts changes in the aggregate states of silicon over the depth of amorphized layer in the initial stage of annealing. Each symbol in the figure denotes state of a layer of thickness Δd = 40 Å: "a" corresponds to (a-Si), "p" – fine-grained polycrystal (p-Si), "c" – (c-Si), 1 – (1-Si) for $T \geqslant T_m$; N corresponds to the first significant digit (after decimal point) of the crystallized fraction $\psi(x,t)$. As seen in the figures, after formation of sl-Si the melt front movement into the sample is concurrent with formation of a fine-grained p-Si (curves 1,2). Evolution of the latent heat of phase transition (L_c) results in increased temperature in the near-surface region and an increase in the rate of melt front movement from ~6 m/s to ~ 10.5 m/s. However, as the melt front moves and, hence, the a-Si layer becomes thinner, the thermal gradient in this layer rises and its thermal resistance decreases (k_a= 0.02 W/cm·K $\ll k_c$[2]). At the same time, formation of the p-Si layer on the surface leads to a significant decrease in the absorptivity in this region ($\alpha_c = 3\cdot 10^3 cm^{-1} \ll \alpha_e = 10^6 cm^{-1}$). Therefore, the basic heat source, due to laser light absorption in the liquid phase and crystallization of sl-Si, moves from the surface to the sample depth. All these simultaneous processes slow down the melt front rate to about 4 m/s. Starting from formation of the p-Si layer on the surface the reflectivity takes on the value $R_c < R_e$, the absorptivity of silicon substantially increases, which leads to a sharp rise in temperature. Heating is accompanied with further penetration of heat source deep into the sample (curves 5-7), the temperature profile maximum being now not on the surface but at the depth x_1, where $\alpha(x_1) \approx \alpha_e$. Heating of the sl-Si to the temperature T_m results in p-Si melting which starts from the depth of ~0.06 µm rather than from sample surface. The melt front propagates at a rate up to 40 m/s to meet laser beam, i.e. from the sample depth to the surface. Only after the melt reaches the surface its front will again travel into the sample.

References

(1) Andra, G., H.-D. Geiler, G. Gotz, K.H. Heinig, H. Woittennek: Phys. Stat. Sol. 74(a), N2 (1982) 511-515

(2) Lowndes, D.H., R.F. Wood and J. Narayan: Phys. Rev. Lett. 52, N7 (1984) 561-564

(3) Narayan, J., C.W. White: Appl. Phys. Lett. 44, N1 (1984) 35-37

(4) Pilipovich, V.A., G.D. Ivlev, V.V. Zhidkov, V.L. Malevich: Pisma v Zhurn. Tekhn. Fiz. 9, N10 (1983) 594-598

(5) Skripov, V.P., V.P. Koverda: Spontaneous Crystallization of Supercooled Liquids, Nauka, Moscow 1977 (in Russian)

PHASE TRANSITIONS INDUCED BY NANOSECOND LASER HEATING OF AMORPHIZED SILICON

G.D. Ivlev and V.V. Zhidkov[1)]

The peculiarities of phase transitions induced by nanosecond laser irradiation of semiconductors have been the subject of many investigations. However, only few experimental studies [5,6] have been performed using the pyrometric method for measuring temperature in the irradiated zone. This can be partly explained by the fact that optical pyrometry is sometimes viewed [1] as being insufficiently reliable as applied to conditions of heating with short laser pulses. However, the surface temperature T of semiconductor can be found in such conditions by its thermal radiation (TR) which has been shown in [5,6]. By using optical pyrometry we have studied thermal processes taking place in monocrystal (c-Si) and amorphized (a-Si) silicon under the action of individual nanosecond (about 10^{-7}s) pulses of a ruby laser.

To prepare a-Si, wafers of c-Si(111) were implanted with P^+ions (70 keV, $2\cdot10^{15}cm^{-2}$). The a-Si layer thus formed was about 0.1 μm thick. Temperature of the silicon surface was measured by (1) the intensity of TR in a narrow spectral range and (2) the ratio of intensities of TR in two spectral regions. Calibrated photometric lamp was employed. In the first case, the spectral transmittance of optics of a photomultiplier-oscilloscope measuring system had a maximum for λ_o= 510 nm and its FWHM was $\Delta\lambda$= 60nm. In the second case, two photomultipliers were used to monitor quasimonochromatic fluxes of TR ($\Delta\lambda$= 10 nm, λ_o= 535 and 430 nm). Also, probe radiation (λ= 1.06 μm) reflected from silicon was monitored [4]. The probe beam was focused to the central zone ($\sim$1 mm in size) of the irradiated region 2.5 mm in diameter. The latter was also sighted by a pyrometric sensor. Nonuniformity of the laser pulse energy over the heated region was ±5-10%.

The obtained results indicate that for the laser energy density $W \geqslant 1.5$ J/cm^2 (the annealing threshold of the implanted layer) the maximum temperature of the silicon surface T_{max} exceeds the c-Si melting point T_{mc} = 1683 K and the melt exists during the time $\tau > 100$ ns. The TR of the melt having enhanced reflectivity reaches maximum intensity (Fig.1) with a delay relative to the laser pulse peak. It should be noted that by limiting high initial value of the pyrometric signal (using fast diode) fed to the oscilloscope and by increasing the sensitivity of the measuring system, we could observe a plateau which pointed to the constancy of the melt temperature during epitaxial crystallization. After crystallization is completed the TR intensity decreases abruptly and the reflectivity of the semico-

1) Institute of Electronics, BSSR Academy of Sciences, Ya.Kolas St. 68/2, Minsk-90, 220841, (U.S.S.R.)

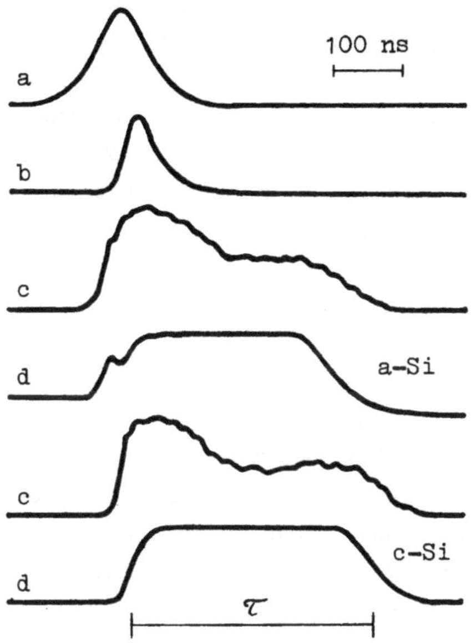

Fig. 1. Oscillograms of (a) laser pulse, (b,c) TR of silicon, and (d) reflected probe radiation. Oscillogram "c" was obtained using a fast diode. $W \simeq 2.6$ J/cm^2.

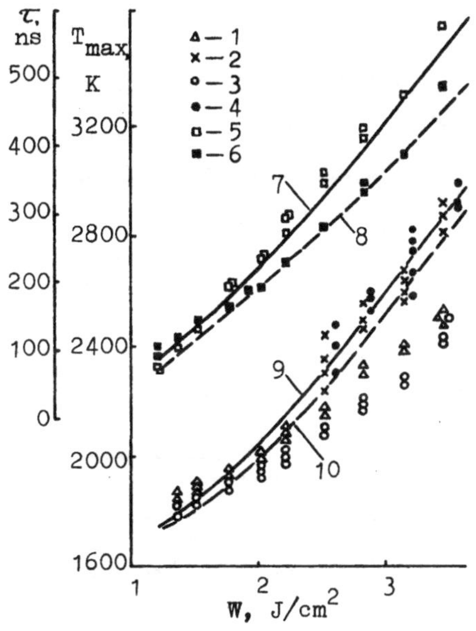

Fig. 2. Comparison [7] of experimental (1-6) and calculated (7-10) dependences of τ and T_{max} on W for c-Si (1,2,5,7,9) and a-Si (3,4,6,8, 10). T_{max} was found by the intensity of narrow-band TR (1,3) and by the ratio of TR intensities (2,4).

nductor falls off to the level characteristic of c-Si. The pyrometric measurements show that when the melt crystallizes $T \simeq T_{mc}$. It is worth noting that the established experimental dependences of T_{max} and τ on W agree with (Fig.2) the respective calculated dependences [7]. T_{max} was also measured with the value of W corresponding to the melting threshold of c-Si (1 J/cm^2). It was found that its value as well as the melt surface temperature during crystallization are, in fact, not different from T_{mc}. At the same time, the result was obtained that indicates a reduced melting temperature of a-Si, T_{ma}. The magnitude of T_{ma} was measured by the intensity of TR in the initial stage of a-Si melting for $W = 0.5 \pm 0.1$ J/cm^2. Starting from a-Si melting the TR intensity reached 20% of that found during crystallization of the melt with the temperature T_{mc} (plateau on oscillograms in Fig.1). Such a difference in the TR intensities corresponds to $T_{mc} - T_{ma} \simeq 200$ K if a certain difference in the spectral emissivities of silicon in the two cases is taken into account. Experimental value of T_{ma} was found to be 1484 ± 28 K provided that the plateau level corresponds to T_{mc}=1683 K. Because of a-Si melting at a reduced temperature, a metastable liquid phase is formed that may undergo intermediate crystallization in the course of laser annealing, which was observed in our experiments. The intermediate crystallization and subsequent melting of a fine-grained polycrystal give rise to a nonmonotonic increase in the reflectivity and a specific temporal structure of scattering of the probe radiation [3].

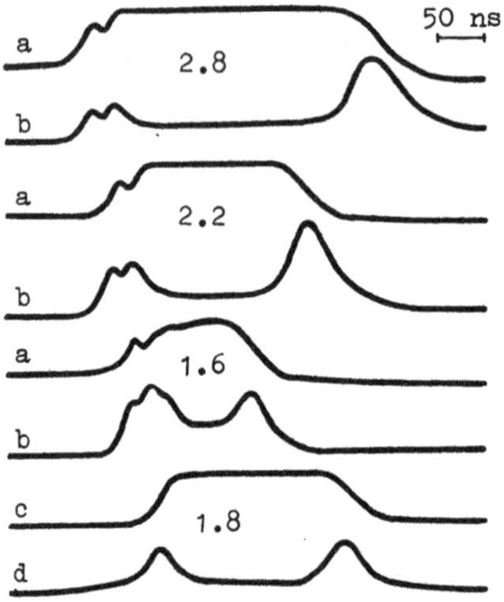

Fig. 3. Oscillograms of (a,c) reflected and (b,d) scattered probe radiation from silicon (see [3]). W is given in units of J/cm^2.

Such behaviour of the reflectivity and scattering in the process of laser annealing was not observed in the case of c-Si (Fig. 3), i.e. when the above-mentioned phase transitions did not take place. It is appropriate to note here that the possibility of intermediate liquid-phase crystallization during nanosecond heating of a-Si is also confirmed by the results of computational simulation [2]. A relatively weak scattering of the probe beam observed when the reflectivity increases to a quasi-stationary maximum and, then, decreases is explained by optical nonuniformity of silicon surface which consists of solid and liquid phases differing substantially in their physical properties. A continuous melted layer scatters insignificantly which is seen from the oscillograms (Fig. 3). The results obtained indicate a nonuniform behaviour of the phase transitions which, in particular, is due to microscale nonuniformity of the spatial and temporal structure of the laser radiation acting on silicon.

In conclusion, our investigation has shown that the process of nanosecond laser annealing of implanted silicon layers is not a simple sequence of the two phase transitions, namely, melting and liquid-phase epitaxy. Amorphization of semiconductor by ion implantation leads, in the course of, further annealing, to the occurrence of a supercooled melt in which intermediate crystallization may develop.

References

(1) Arutjunov, E.N., S.Yu. Karpov, Yu.V. Kovalchuk, et al.: Pisma v Zhurn. Tekhn. Fiz. 11 (1985) 921-924

(2) Balandin, V.Yu., A.V. Dvurechenskii, L.N. Aleksandrov: Phys. Stat.Sol. (a). 93 (1986) K105-K109

(3) Ivlev, G.D.: Fiz. Tekhn. Poluprov. 19 (1985) 2188-2190

(4) Ivlev, G.D.: Pisma v Zhurn. Tekhn. Fiz. 8 (1982) 468-472

(5) Kemmler, M., G. Wartmann, D. von der Linde: Appl. Phys. Lett. 45 (1984) 159-161

(6) Pilipovich, V.A., G.D. Ivlev, V.V. Zhidkov, V.L. Malevich: Pisma v Zhurn. Tekhn. Fiz. 9 (1983) 594-598

(7) Pilipovich, V.A., V.L. Malevich, G.D. Ivlev, V.V. Zhidkov: Inzhenernofiz. Zhurn. 48 (1985) 306-312

THEORETICAL TREATMENT OF THE IGNITION OF EXPLOSIVE LIQUID PHASE CRYSTALLIZATION PROCESSES

D.Wolff and H.D.Geiler[1)]

1. Introduction

The lateral explosive liquid phase epitaxy (ELPE) process in amorphous silicon (a-Si) layers on insulator is one of the well established autocatalytic crystallization phenomena [1,2]. In the following the ignition of the ELPE is considered. Therefore, the fully time-dependent heat flow control (HFC) problem coupled with the kinetic interface control (KIC) equations is investigated.
As reported from Wagner et al. [3] the ELPE starts from distinguished points and then radially spreads from its ignition centre (Fig.1). That ignition centre is a crystallite which is nucleated within the melt nucleus in a superheated a-Si layer. After the ignition process the ELPE becomes stable and moves with velocities of about 16 m/s over long distances [4]. Contrary to this experimental fact of a fixed velocity for all spreading conditions, i.e. for all substrate temperatures T_s, the steady state solutions of the ELPE problem yield a relatively wide spectrum of possible phase front velocities (8...16 m/s [4]) depending on T_s.

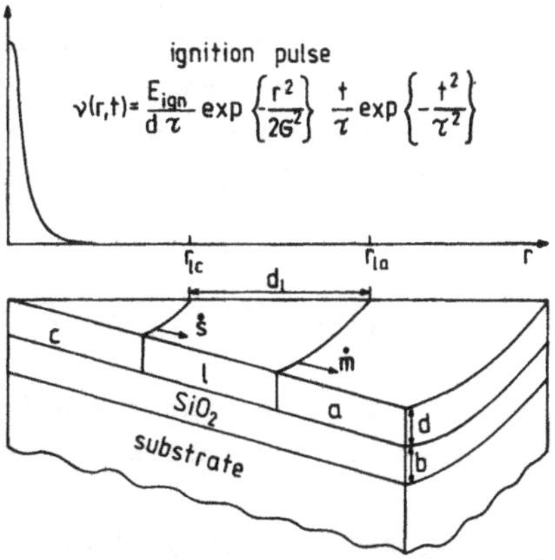

Fig.1. Schematic presentation of lateral ELPE.

2. The model

In cylindrical coordinates the enthalpy balance (HFC) in the Si layer is given by:

$$\partial_t h = \lambda_j \partial_r^2 T + \lambda_j \partial_r T / r + c\alpha(T_s - T) + \nu(r,t) \qquad (1)$$

(For a symbol list see Tab.1.) On the right hand side the first term describes the heat flow in the same way as in the one-dimensional (1D) model. The second one is its correction with respect to the cylindrical

1) Friedrich-Schiller-Universitat Jena, Sektion Physik, Max-Wien-Pl. 1, Jena, DDR 6900

Table 1.

symbol	meaning	value
λ_j	heat flow coefficient in phase j (W/m/K) j=c,l,a	35 (c-Si) 66 (l-Si) 3.61 (a-Si)
λ_{ins}	heat flow coefficient in the insulator (W/m/K)	0.704
d	thickness of the a-Si layer (nm)	360
b	thickness of the insulator (SiO_2) layer (nm)	400
E_{act}	activation energy (eV)	1.22
c	heat capacity of Si ($J/K/m^3$)	$2.2x10^6$
T_{ma}	melting temperature of a-Si (K)	1350
T_{mc}	melting temperature of c-Si (K)	1685
L_c	latent heat c<-->l (J/cm)	3856
L_a	latent heat a<-->l (J/cm)	2933
v_o	(m/s)	608000

symmetry. The third term refers to the heat losses into the substrate and includes the heat loss parameter α ($\alpha=\lambda_{ins}/b/d/c$). $\vartheta(r,t)$ expresses the enthalpy introduced into the system by a spike in the beam of a free running Nd-glass laser which heates up the Si layer to T_s. The enthalpy may change by changing the temperature as well as by phase transitions within the layer at the interfaces being at r_{1a} and r_{1c}, respectively:

$$\partial_t h=c\partial_t T+\dot{m}L_a\delta(r-r_{1a})-\dot{s}L_c\delta(r-r_{1c}) \quad (2)$$

Equ. (2) includes the phase front velocities at the melting front $\dot{m}$ as well as at the crystallization front $\dot{s}$ (KIC [1]):

$$\dot{m}=v_o\exp\{-E_{act}/(kT)\}[1-\exp\{L_a/(NkT_{ma})(1-T_{ma}/T)\}] \quad (3)$$
$$\dot{s}=v_o\exp\{-E_{act}/(kT)\}[1-\exp\{L_c/(NkT_{mc})(1-T_{mc}/T)\}] \quad (4)$$

The melting front consumes for further melting of a-Si most of the energy which is liberated at the crystallization front . Only a small part (about 30%) is available to compensate the heat fluxes into the crystalline silicon (c-Si), the a-Si and into the substrate, as well as to enlarge the the melt volume, i.e. the phase front distance d_1 after the ignition.
At t=0 the energy of a typical Gaussian laser spike (E_{ign}=0.08...0.3 J/cm^2, τ =0.1 ns, σ=0.25 um) at $T=T_s$ is incorporated into the a-Si layer and causes the melting of a-Si. When the melt temperature decreases below the c-Si melting temperature T_{mc}, a nucleation event appears in the centre of the molten phase and the ELPE begins.

3. Results and discussion

The temporal evolution of the interface distance d_1 depending on T_s is shown in Fig.2. After 80 ns and 400 ns in the case of T_s=1130 K and T_s=1200 K, respectively, the steady state d_1 values d_1^{SS} of the 1D model [1] are reached. At T_s=1320 K even after 500 ns, i.e. about 10 μm away

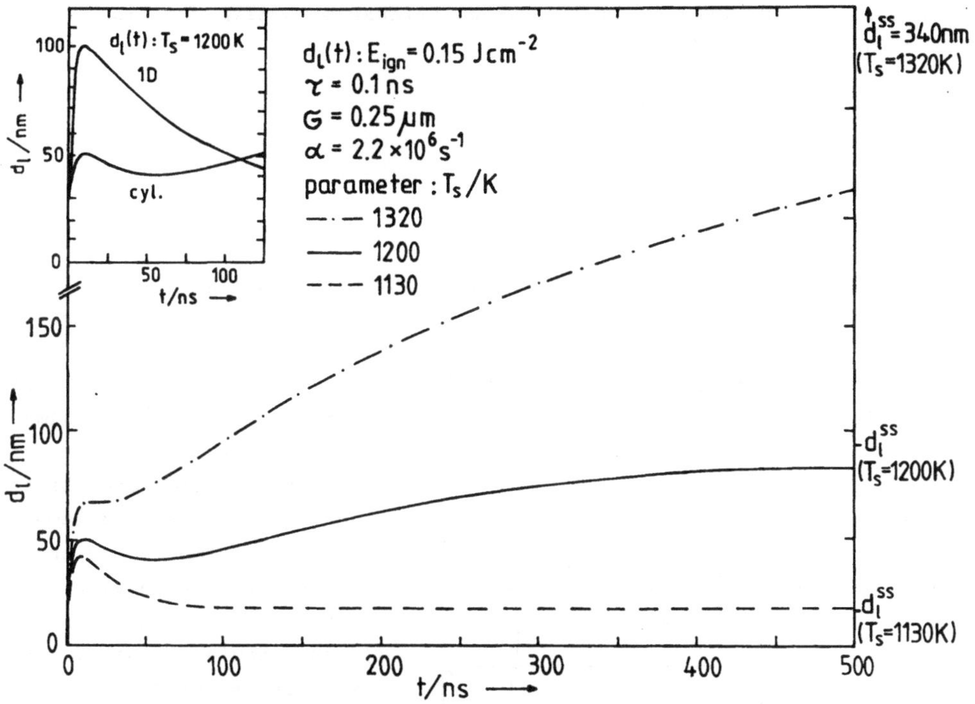

Fig.2. Temporal evolution of the phase front distance

from the ignition centre (r=0) d_l^{SS}=340 nm is not reached yet. d_l grows with a speed of about 0.15 m/s.
Generally, larger values of d_l^{SS} require longer times to reach the steady state. During these long evolution times the fronts are running with d_l essentially smaller then d_l^{SS}, but with velocities well above the correspondent v^{SS}. Hence, the realization of v^{SS} below 10 m/s requires evolution times longer than 1 µs. So, one may measure only the larger non-steady state values before competitive processes will stop ELPE [3].
To compare the long time solutions of (1-4) with solutions of the 1D model, it is necessary to know the influence of the 1/r-term in (1). The inserted picture shows the difference between the two models. Estimating the distance from the centre at which the influence of the 1/r-term is comparable to or less than the losses into the substrate:

$$r > \max_{j=c,l,a} \{db(\lambda_j/\lambda_{ins})(\partial_r T/(T-T_s))\}, \qquad (5)$$

one may find 50 µm, which correlates to ELPE run times of about 2.5 µs.

[1] H.D.Geiler,K.H.Heinig: Energy Pulse Modification of Semiconductors and related Materials,ed.by K.Hennig,ZfK Rossendorf,1985,p.466
[2] H.D.Geiler,E.Glaser,G.Götz,M.Wagner: J.Appl.Phys. 59(1986)3091
[3] M.Wagner,G.Andrä,E.Glaser,G.Götz,H.Bartsch: these proceedings
[4] G.Andrä,H.D.Geiler,E.Glaser,G.Götz,M.Wagner,K.H.Heinig: Nucl.Instr. Meth. B19/20(1987)571

MODELLING OF THE FACETED GROWTH DURING LATERAL EXPLOSIVE CRYSTALLIZATION OF AMORPHOUS SILICON

H.-D.Geiler and G.Götz[1)]

1. Introduction

Explosively crystallized thin films of amorphous silicon on SiO_2 or crystalline Si-substrate represent a typical crystallographic structure characterized by long single-crystalline lamellae separated by low angle grain boundaries [1,2,3]. Essentially all of the recent work in explosive crystallization (EC) of silicon on insulators (SOI) is aimed at minimizing the number of these subboundaries and broadening the crystalline lamellae. In the same manner as in the case of zone melting recrystallization (ZMR) of SOI [4] the knowledge about the lateral interface evolution is the key to solve the problem. Two questions must be answered before:

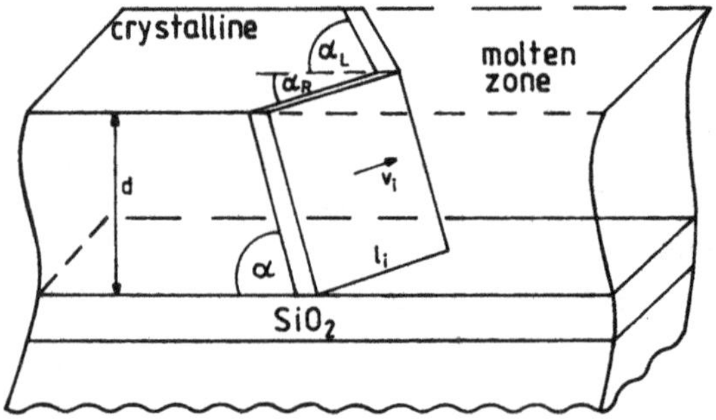

Fig.1. Scheme of lateral growing sharp interfaces in SOI-systems

(i) Which type of the melt/-crystalline (m/c) interface does exist under the condition of strong undercooling (temperature at the m/c-interface $T_{ci} = 1550$ K [7]) and following which interface kinetics does dominate?

(ii) In which manner the high temperature gradient laterally and vertically (gradT > 10^6 Kcm^{-1}) does influence the interface evolution?

2. Temkin model of the (m/c)-interface

The interface type (sharp or rough) depends on the change of the Gibbs free energy ΔG_n for building a new crystalline layer labeled n on the (n-1) layers crystallized before. Following the idea of Temkin [6] that the interface with the minimum ΔG will be established, the atomic fraction of occupied sites in the n-th layer c_n can be calculated by a variation of ΔG versus c_n. Fig.1 illustrates the geometrical situation

1) Friedrich-Schiller-Universität Jena, Sektion Physik, Max-Wien-Platz 1 Jena, DDR 6900

in the case of lateral growth. The vertical position of the new layer needs a separate thermodynamical consideration of the boundary regions adjecent to the substrate and the superstrate [8]. Excluding enhanced nucleation in these boundary regions their energy is scaled by the simple geometrical relation using the lattice spacing λ in the layer (see Fig.1): $r = \lambda \sin\alpha / d$. For large relative undercooling $\Delta T/T_{mc} > 10^{-1}$ and $r < 10^{-3}$, which is representative for EC in Si, the change of the specific surface energy and the suface excess bonding energies can be neglected. Hence, the classical Temkin formula is gained by summing up over all solid and liquid layers formed during lateral EC:

$$\Delta G/(NkT) = L/(kT)\,\Delta T/T_{mc}\sum_{n=-\infty}^{\infty} c_n + \varepsilon/kT\sum_{n=-\infty}^{\infty} c_n(1-c_n) - \sum_{n=-\infty}^{\infty}(c_n - c_{n+1})\ln(c_n - c_{n+1}) \quad (1)$$

with $\varepsilon = \xi \cdot L$, $\xi = 0.5$ for <100>,<110> and $\xi = 0.75$ for <111>
(latent heat per atom L= 0.525 eV, melting temperature of crystalline Si T_{mc}= 1685 K, N - total number of atoms per layer).
Then the minimum of $\Delta G/(NkT)$ is governed by the equation:

$$(\xi - \Delta T/T_{mc})L/(kT) - 2\xi L/(kT)c_n + \ln((c_n - c_{n+1})/(c_{n-1} - c_n)) = 0 \quad (2)$$

with $c_{-\infty} = 1$ (full crystallized) and $c_{\infty} = 0$ (full molten).
The numerical solution of (2) supplies the following results to answer the question (i): The <111> interface remains sharp up to the roughening transition of nonactivated growth at a temperature T_r= 1535K. This transition occurs in the case of the <100> interface at 1679 K.

3. Evolution of a faceted interface

The results presented in Chp.2 imply the existence of a <111> faceted interface for EC similar to the predictions given in [4,5] for ZMR and small undercooling. Yet, the interface is controlled not only by the kinetics but also by the heat flow. Because the lateral temperature gradient into the melt at the interface dominates ($\nabla T > 10^{-8}$) over the gradients in the other directions, the normal velocity of growth of a facet v_i is governed by $\varkappa_m \vec{\nabla} T = \vec{v}_i(T_{ci}) \cdot L$ with the heat conductivity in the melt $\varkappa_m$. This parallelism of v_i with the lateral gradient in the melt determines the geometrical position of the <111>-facet in the layer with $\alpha = 90^o$ (see Fig.1). From the crystallographic constrains follow the situation drawn in Fig.2. The plug build by the <111>-facets is <110> oriented that means no <100> fit to the substrate or superstrate. The consequence is a large strain in the layer, which relaxes after the crystallisation [9]. Assuming nucleation controlled growth one gets by use of the steady state interface induced nucleation rate I_o:
$v_i = \lambda^2 \cdot l_i \cdot d \cdot I_o(T_{ic})$.

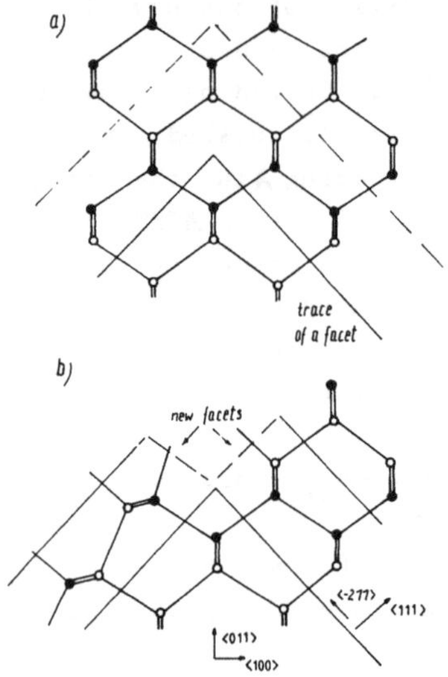

Fig.2. View on the (110)-plane with interface facets perpendicular on it forming the plug. a) case of ordinary growth, b) case of twin formation on the left facet causing the birth of a new pair of facets

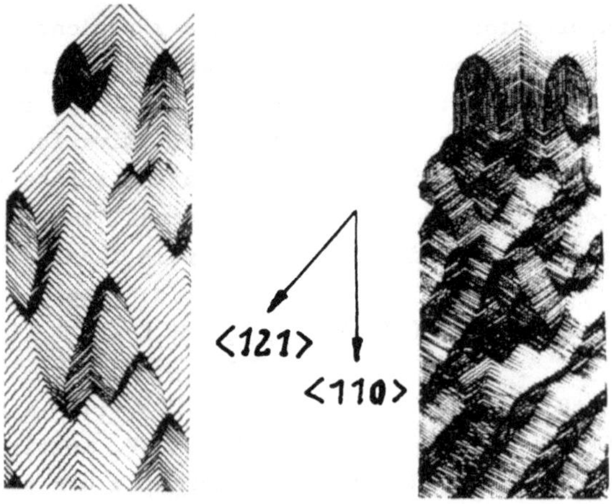

Fig.3. Computer plots ot the evolution pattern of faceted interfaces with mainly <110>-orientation (left part) and <112>-orientation (right part) of the lamellae

In the case of EC the interface temperature is constant along the faceted interface and represents an isothermal line. The competition between the facets is caused by the dependence of v_i on the length l_i. New facets were born by twin generation if v_i exceeds a critical v_c (see Fig.2b). A facet dies if their both neighbours collide. The evolution of the faceted interface is computed by the transformation of the coordinates of each facet step by step with the transformation formula given in [4]. Two typical results are shown in Fig.3a,b for lamellae oriented in <110> and <112> directions, respectively. Subboundaries should be formed at the ends of each plug of facets caused by twins and relaxed stress as explained in [10].

[1] Geiler, H.D., E.Glaser, M.Wagner, G.Götz, J.Appl.Phys.59(1986)3091
[2] Andrä, G., E.Glaser, this conferece
[3] Andrä, G., H.D.Geiler, E.Glaser, G.Götz, M.Wagner, K.H.Heinig, Nucl. Instr. and Meth. B 19/20 (1987) 571
[4] Pfeiffer, L., S.Paine, G.H.Gilmer, W. v. Saarlos, K.W.West, Phys. Rev.Lett. 54 (1985) 1944
[5] Pfeiffer, L., A.E.Gelman, K.A.Jackson, K.W.West, Proc. MRS-Meeting 1986, paper A6.1
[6] see for instance: P.Bennema, G.H.Gilmer in Crystal Growth: an introduction, Ed. by P.Hartman, North Holland Publ. Co. 1973
[7] Wolff, D., H.-D.Geiler, this conference
[8] see for instance: R.B.Griffith in Phase Transitions in Surface Films, Ed. by J.G.Dash, J.Ruvalds, Plenum Press, 1980
[9] Bostanjoglo, O., R.P.Tornow, W.Tornow, Ultramicroscopy (1987) in press.
[10] Baumgart, H., F.Phillipp, Proc. MRS-Symp. Vol. 35 (1085) 593

LASER PULSE-INDUCED PHASE TRANSFORMATIONS VISUALIZED BY NANOSECOND-EXPOSURE ELECTRON MICROSCOPY

O. Bostanjoglo*), R. P. Tornow*), W. Tornow*)

Introduction

Extremely large power densities are easily generated by focused laser pulses which are therefore very suitable for material processing, microstructuring and annealing. In order to fully exploit the properties of the laser a thorough knowledge is mandatory of the processes triggered within the material on the micrometer/nanosecond scale. This paper presents a time-resolved technique, providing for the first time electron micrographs with exposure times down to 15 ns taken within a Transmission Electron Microscope. The short exposure TEM was applied to visualize intermediate stages of laserpulse-induced phase transitions in thin films.

Experimental procedure

Fig. 1 shows a commercial TEM extended for short exposure investigations by installing an electron beam deflector, a stage for laser pulsing the specimen and adding a pulsed image intensifier below the final image plane. A pulse from a Q-switched frequency-doubled Nd:YAG laser (25 ns FWHM, TEM_{00} mode) is used to trigger fast phase transitions in the thin film specimen. The probing electron beam is switched on the specimen for a total observation time of several 100 ns only, in order to avoid radiation damage. A bright-field image or diffraction diagram is conventionally generated during this electron illumination period. The detector is a proximity focussing diode type image converter, consisting of a voltage pulsed microchannel plate and an Al-coated YAG:Ce single crystal scintillator at a constant accelerating voltage. The image appearing on the scintillator is picked up by a television set.

Electron beam and pulse laser are aligned with the help of a green He-Ne laser, observing the specimen lightoptically with a telescope.

The microchannel plate is pulsed for two reasons. Firstly, very short exposure times, down to 10 ns, are easily realized by exploiting the exponential gain/voltage characteristic. Secondly, the channel plates were found to survive pulsed voltages exceeding the safe DC values by a factor of two /1/. Thus, gain and maximum input current could be increased by two and four orders of magnitude, respectively, above the values realized in DC operation. The exposure voltage pulse is generated by a Marx cascade generator, based on avalanche transistors, with a switch-off unit. The generator is triggered during or, with a preselected delay, after the laser pulse. Since the avalanching transistors are heavily stressed by the MCP capacitance they were cooled with liquid nitrogen to increase their life time /2/.

Replacing the image intensifier by a photomultiplier tube the course of a laser-induced phase transition can be traced for microseconds with a time resolution of 3 ns in areas down to 0.3 µm in diameter /3/.

---

*) Optisches Institut der Technischen Universität Berlin
D-1000 Berlin 12, Straße des 17. Juni 135

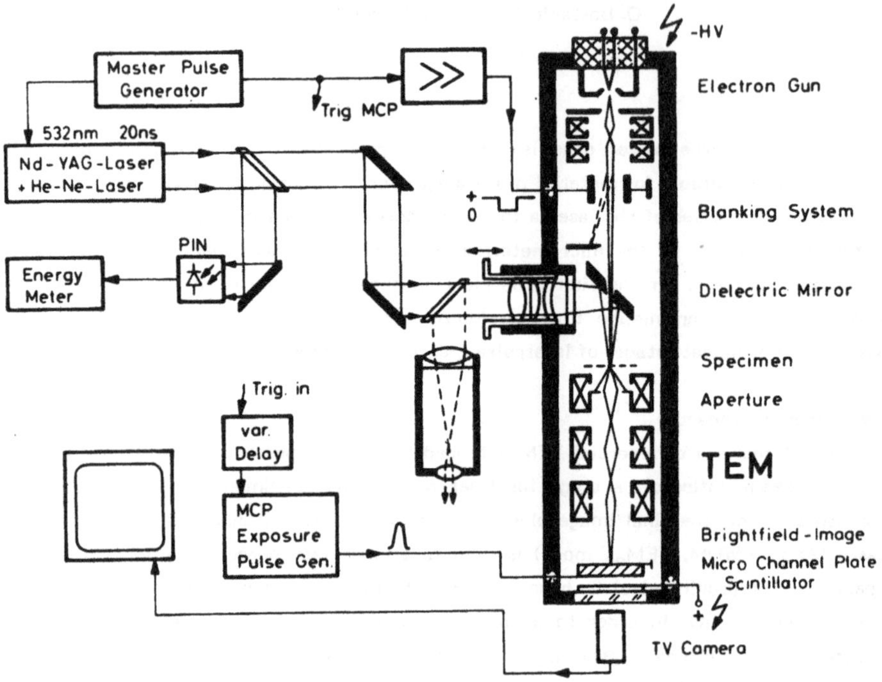

Fig.1 Nanosecond-exposure Transmission Electron Microscope for time-resolved investigations of laserpulse-induced phase transitions

Applications

The nanosecond-exposure TEM was applied to laser-induced explosive crystallization in amorphous Germanium films. Fig. 2 shows a series of typical intermediate stages during the phase transition, for microseconds after the initiating laser pulse of 25 ns FWHM. The pictures were taken at neighbouring areas of the same specimen under identical laser pulse conditions. The center of the laser focus was located below the lower left corners of the pictures. The upper micrographs a-d were taken with an exposure time of 15 ns at different points of time after the laser pulse, as indicated in the upper right corners. Below, pictures a'-d' present the final texture of the corresponding laserpulsed areas, photographed with a conventional exposure time of 0.5 s. About 100 ns after the laser pulse a ring-shaped crystalline phase nucleated outside the laser focus. Then, crystallization proceeded radially for about 300 ns with a velocity of $\approx$ 10 m/s (2a-2c). This value agrees with values directly measured with the 'integral' time-resolving TEM technique, using a photomultiplier as detector /3/. Radial crystal growth comes to an end $\approx$ 400 ns after the laser pulse, and it is only after $\approx$ 1 µs that crystallization continues, but now in a spiral growth mode (2d), lasting for $\approx$ 5 µs. The appreciable delay between end of radial and beginning of spiral growth coincides with the time it takes the temperature to rise due to heat diffusing from the laser focus accross the 'radial' crystals.

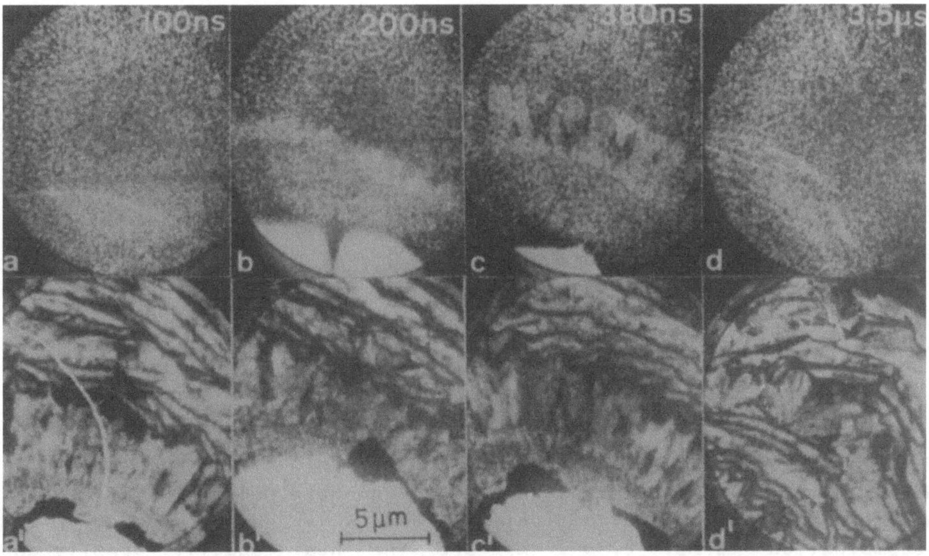

Fig.2 Electron micrographs of transient stages during laserpulse-induced explosive crystallization of an amorphous Ge-film. Exposure time 15 ns in a-d. Laser pulse energies are equal for a - d ($\approx$0.1 µJ).

The method presently is limited by shot noise in the electron image, which should be appreciably reduced by installing a laserpulse-driven electron gun /4/.

Acknowledgement

The work was partly supported by the Deutsche Forschungsgemeinschaft.

References

/1/ O. Bostanjoglo, R. P. Tornow, W. Tornow: Scanning Microscopy Suppl. 1 (1987) 197-203

/2/ O. Bostanjoglo, W. Tornow: Semiconductor Science and Technology 2 (1987) 175-176

/3/ O. Bostanjoglo, E. Endruschat, W. Tornow: Mat. Res. Soc. Symp. Proc. Vol. 71 (1986) 345-350

/4/ O. Bostanjoglo, F. Heinricht: to appear in J. Phys. E.: Sci. Instr. 1987

THERMAL PULSED ANNEALING WITH ACCOUNT OF PLASMA EFFECT

V.Yu.Balandin, A.V.Dvurechenskii, L.N.Aleksandrov

There are two alternative models of the a-Si pulsed annealing. The thermal annealing model is based on the supposition that the energy of light beams immediately transforms to lattice heat. The plasma model is described on the hypothesis that the structural trasformations proceed due to the bond weakening in a presence of the dense electron-hole plasma. The temperature change is neglected.

In this paper we account the electron-hole plasma influence on the phase trasformation kinetics through the change free energy δG of the system. It is supposed that $\delta G \sim \varepsilon_p$, where ε_p is the plasma energy per volume unit.

The mathematical description of the model is based on following equations /1,7/:

$$\partial N/\partial t = \nabla (D_p \nabla N) + B(r,t) - R(r,t)$$

$$c\rho\, \partial T/\partial t = \nabla (\kappa \nabla T) + \Delta \varepsilon_p R(r,t) + Q(S)$$

$$S = S(T, N)$$

Here N(r,t) is the concentration of electron-hole pairs; D_p is the ambipolar diffusion coefficient; B and R are the generation and recombination terms; c, ρ, k, are the heat capacity, density and thermal conductivity; $\Delta \varepsilon_p$ is the average energy transformed to the lattice at a recombination act; S is the logical phase transition function /1/. As aresult of the dense electron-hole plasma formation, the temperature and enthalpy of a-Si fusion are decreased:

$$T_A^*(r,t) = T_A - \varepsilon_p(r,t)\, T_A / H_A$$
$$H_A^*(r,t) = H_A - \varepsilon_p(r,t)$$

The melting of a-Si in some local volume goes on as $T(r,t) > T_A^*(r,t)$.

We have estimated the average life time of an electron-hole pair at Auger recombination with the formula:

$$\tau_e = N/R = N(1 + Na)^2/CN^3 \simeq a^2/C = 10^{-7}\ \mathrm{s}$$

where $a = e^2 h^2/(2\, \varepsilon\, \varepsilon_o kT_e\, m_e^*\, E_g) \sim$ 10-20 cm^3 at the average temperature of plasma $kT_e \simeq 0.5$ eV /3,7/. We have taken $E_g \simeq 0.3$ eV due to decreasing of band gap at the high density of the plasma /4/. $C \simeq 10^{-31}$ cm^6/s /2/. Consequently,

$$\varepsilon_p(r,t) = h\nu \int_{\tau_e} B(r,t')\, dt'$$

i.e. the heat generation term in the heat conductivity equation becomes τ_e late comparing to the function of the light absorption.

Institute of Semiconductor Physics, Academy of Sciences of the USSR, 630090 Novosibirsk, USSR

The experimental situation, described in /5/ is modelled. We have taken T_A =1500 K, H_A =1600 J/g. Plasma diffusion is neglected. The dynamics of the phase transformation during the action for different energy densities E is shown on Fig. 1.

For E low enough one can see the self- sustaining crystallization. (Fig.1,a). Crystallization at the surface of the sample occurs some earlier than in the depth and the buried liquid layer is formed. This phenomenon seems to be was found experimentally /6/. As E is increased

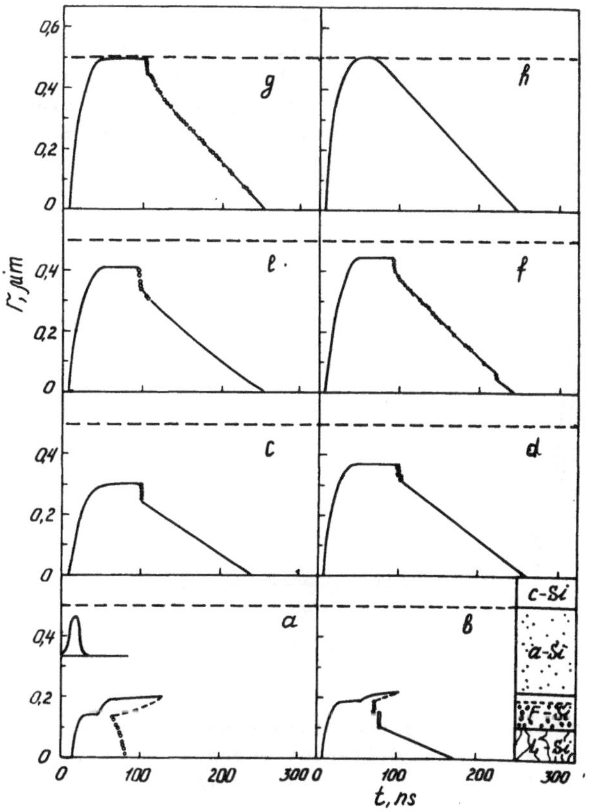

Fig.1. Melt-nucleation (.) and growth fronts positions as a function of time for E, J/cm²: 0.2 (a), 0.3 (b), 0.5 (c), 0.6 (d), 0.9 (e), 1.5 (f), 1.6 (g). Insert: ruby laser pulse shape (12 ns) and post annealing struture.

the fine- and large-grained polycrystalline layers (F-Si and L-Si) is realized (Fig.1, b-d). F-Si is formed by a homogeneous nucleation process, L-Si is due to crystal growth. Only the a-Si melting occurs during the laser beam action. Fig. 2 shows the decreasing of the melting temperature due to plasma effect at the surface. It is well seen that the plasma effects essentially at the pulse duration less than 50 ns.

As E is further rised (0.9-1.6 J/cm²) a new effect in F-Si - L-Si correlation occur: the poly-Si becomes advantageously F-Si (Fig.1,e,f). We called this phenomenon the effect of the a-Si thermal resistance. It is due to the lower a-Si thermal conductivity in comparison with crystalline one. The remaining a-Si layer restrains the heat flow to the c-Si substrate. At low E when the a-Si layer is thick enough the temperature rise due to latent heat of crytallization is higher then in the case of higher E when the a-Si laer is much thinner. The heat flow is more intensive then the heat generation and the crytallization of the melt proceeds at the strong undercooling where nucleation plays the main role. The effect of the a-Si thermal resistance do not depend on plasma because the crystallization takes place after plasma relaxation. It is worths to note that at E= 0.2-0.8 J/cm² there are good agreement with experimental data /5/.

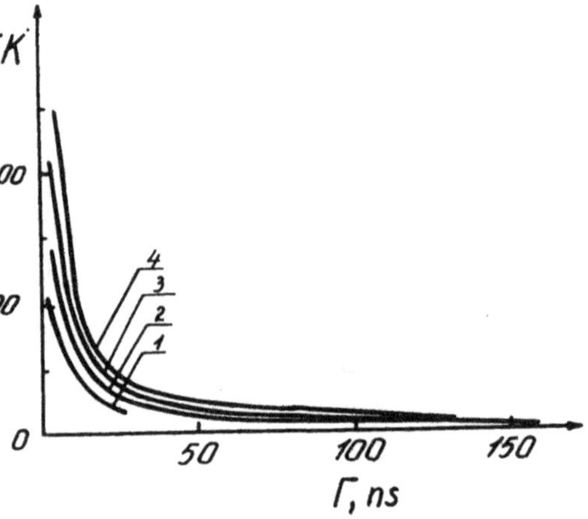

Fig.2. Surface melting temperature decreasing vs pulse duration for different energy densities, J/cm²: 0.2 (1), 0.3 (2), 0.8 (3), 2 (4).

REFERENCES

(1) Balandin V.Yu., Dvurechenskii A.V., Aleksandrov L.N. Poverkhnost', 1 (1986) 53-60.
(2) Dziewior J., Schmid W. Appl. Phys. Lett., 31 (1977) 346-348.
(3) Hang A. Solid Sate Electron., 21 (1978) 1281-1284.
(4) Kapaev V.V., Kopaev Yu.V. Mikroelectronika, 12 (1983) 499-511.
(5) Narayan J., White C.W. Appl. Phys. Lett., 44 (1984) 35-37.
(6) Thompson M.O. et al. Phys. Rev. Lett., 52 (1984) 2360-2363.
(7) Rimini E., Baeri P., Russo G. Proc. Intern. Conf. on Energy Pulse Modification of Semicond. Dresden,1984,ed by K.Hennig,p.3-13.
(8) Yoffa E.J. Phys. Rev.B, 21 (1980) 2415-2425.

AMORPHOUS SILICON MELTING TEMPERATURE VERSUS SELF-SUSTAINING CRYSTALLIZATION KINETICS

O.A.Kuljasova, V.Yu.Balandin, A.V.Dvurechenskii, L.N.Aleksandrov +)

It is well known that the melting of amorphous semiconductors, in particulary a-Si, is the first order phase transition, which occurs at a definite temperature T_A depending on structure. As it was shown in some publications, T_A has value from 1100 K to 1550 K. The T_A uncertainty one can understand by the uncertainty in structure which depends on the way of preparation, the storage conditions and other factors. The lateral propagation of the self-sustaining crystallization (SSC) through the a-Si film depends on the temperature and duration of the preliminary furnace annealing before the ignition of SSC /3/. An analysis of SSC propagation versus the a-Si structure dependent temperature T_A was made in /4/ where a full list of references is given.

The aim of this paper is to establish the relation between the distance of the SSC propagation and the melting temperature of a-Si that allows to define T_A experimentally from the measurement of the crystallized region.

Usually about 0.5 μm amorphous layers were used at the pulsed modification of semiconductors. That thickness is comparable with absorption lengthes of laser and electron beams. But in this case the SSC propagation into the sample is restricted by the interface. For the lateral SSC propagation there are many factors which hardly can be considered in caculation procedures. We have taken for investigation thick amorphous layers (the thickness of the a-Si layer is greater than the absorption length) laying on the monocrystalline silicon substrate.The calculation were made for 2 and 5 μm a-Si layers. The modeled situation correspondes to a 5 ns ruby laser annealing. In this case we deal with an onedimentional problem, described by the heat diffusion equation and the logical phase transitions equation /1,2/. The problem was solved by the finite-difference method.

The crystal growth rate versus the melt temperature T_M at the crystal liquid interface is expressed by a formula /2/:

$$V(T) = C \exp(-\Delta U/kT)\ (1 - \exp(\Delta H(T_M - T)/kT_M T))$$

where ΔU is the activation energy per atom and ΔH is the enthalphy of liquid-crystal transition.

We have obtained the following results. The melt-crystallization front movement is shown in Fig. 1.It is well seen that the laser energy density of 0.5 J/cm^2 leads to melting of about 0.4 μm of the a-Si layer. Then the homogeneous nucleation initiates the crystal growth toward the surface and SSC into the a-Si layer. The SSC was

---

+) Institute of Semiconductor Physics, Academy of Sciences of the USSR, 630090, Novosibirsk, USSR.

found to depend on T_A. The generated heat flows to the substrate only. The SSC develops as the crystallization energy release (Q_C) exeeds the latent heat of a-Si fusion (Q_A) and heat dissipation (D) so $Q_C - Q_A > D$. The heat conductivity of a-Si is less than that of c-Si. The remaining a-Si layer provids a heat resistance. As the SSC is penetrating to the a-Si layer the heat flow is increasing. As soon as $Q_C - Q_A < D$ the process decayed and then the SSC stopped, not reaching the c-Si substrate (Fig.1,a).

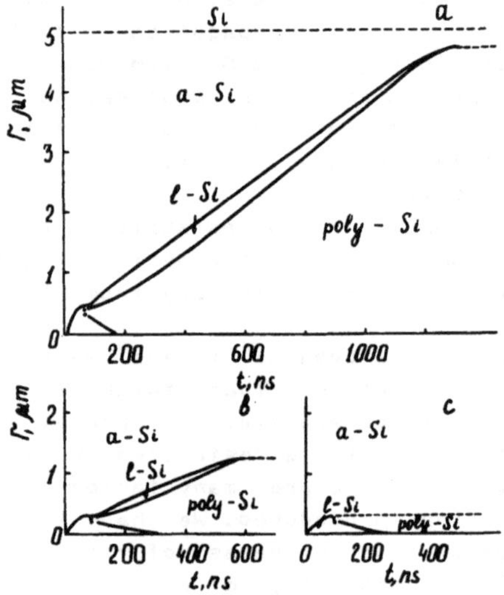

Fig.1. Melt-crystallization front and nucleation position as a function of time for 5 μm a-Si layer at T_A =1200 (a), 1400 (b), and 1600 K. Energy density is 0.5 J/cm².

To increase T_A one can make an oven annealing. The value of T_A depends on the stationary annealing temperature and its duration. At that time Q_A increases proportional to T_A. Consequently the heat flow from the crystallization to the melting front decreases as the temperature gradient in the molten zone falls down. As a result the SSC proceeds with the velocities less than in previous case and decayes faster (Fig.1,b).

As T_A exeeds 1600 K the SSC doesn't take place (Fig.1,c) because the condition $Q_C - Q_A < D$ is not fulfilled. The dependence of the thickness of the crystallized area on the a-Si melting temperature is stronger for the thick original amorphous layer (compare a and b in Fig. 2).

The SSC phenomenon in thick amorphous layers can be used to develop an experimental techniques to study a-Si properties, particularly the a-Si melting temperature and the structural transformation kinetics in a solid state.

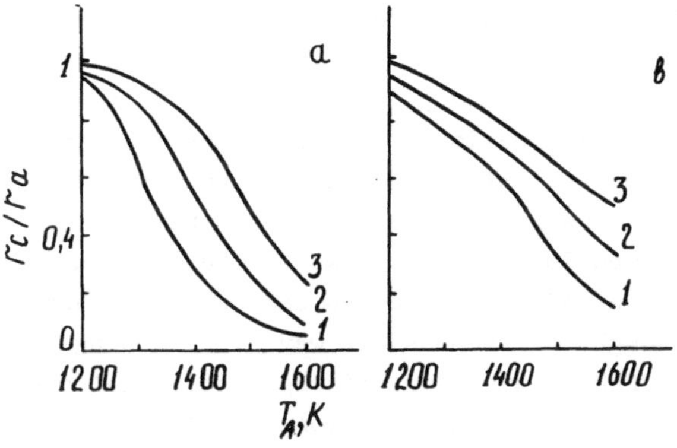

Fig.2. The ratio of crystallized area thickness r_c to the original thickness of the amorphous layer r_a as a function of a-Si melting temperature for 5 (a) and 2 μm (b) of the a-Si thickness. Pulsed energy densities are 0.3 (1), 0.5 (2), 1 J/cm² (3).

REFERENCES

(1) Balandin V.Yu., Dvurechenskii A.V., Aleksandrov L.N., Zn. Tekhn. Fiz. 56 (1986) 807-810.

(2) Balandin V.Yu., Dvurechenskii A.V., Aleksandrov L.N. Proc. Intern. Conf. on Energy Pulse Modification of Semicond.,Dresden, 1984, ed. by K. Hennig, p. 473-479.

(3) Bostanjoglo O. Phys. Stat. Sol.(a), 70 (1982) 473-481.

(4) Götz G. Proc. Intern. Conf. on Energy Pulse Modification of Semicond., Dresden, 1984, ed. by K. Hennig, p.423-439.

LONGITUDINAL-SPREAD CHARACTERISTICS OF RADIATION DEFECTS IN TELLURIUM-IMPLANTED GERMANIUM

M.Kalitzova[1)], N.Pashov[1)], D.Karpuzov[2)], P.Werner[3)], H.Bartsch[3)] and G.Vitali[4)]

1. Introduction

Ion bombardment induced disorder in Ge has been extensively studied by transmission electron microscopy (TEM). In addition high-resolution transmission electron microscopy (HRTEM) was used to study the atomic structure of the damage regions arising in the isolated atomic cascades. In these investigations information about the lateral-spread /1,3/ characteristics of the radiation defects is obtained by using plane-view specimens. Even more important is however the longitudinal spread distribution of the radiation induced damage in implanted materials. The most direct although not the easiest way of obtaining the depth dependent data is by using TEM observations of cross-sectional specimens. A particular valuable information at atomic scale can be obtained when HRTEM analysis of cross-sectional specimens (XHRTEM) is carried out.

In the present work the depth distribution of the damage created in Te^+-implanted germanium as well as the atomic structure of the defect clusters arising in isolated cascades were determined by using XHRTEM. The results on depth distributions are correlated with the simulated data obtained by MARLOWE computer code.

2. Experimental and computer techniques

The cross-sectional specimens were prepared for XHRTEM analysis by cutting (110) slices from (111) Ge (see Fig.1a) implanted at room temperature with 1.3×10^{12} Te^+/cm^2, 46 keV. The procedure was similar to that described in /6/. The final thinning employed two Ar^+ beams bombarding simultaneously both sides of a sandwich specimen. The samples were examined in a JEM-100C electron microscope equiped with a pole-piece for ultrahigh resolution, operated at 100 keV. The specimen position in the EM colomn is shown in Fig.1b. The crystal lattice imaging was carried out by using tilted three-beam dark-field mode /7/.

The computer code MARLOWE /4/ is a program simulating the development of atomic collision cascades by ion impact. It uses a binary-collision approximation to construct the trajectories of all energetic particles moving in a crystalline target. This code was used to calculate the depth distribution of the disorder created by 40 keV Te^+ in ⟨111⟩ Ge.

---

1) Institute of Solid State Physics and
2) Institute of Electronics,
Bulg. Academy of Sciences, Boul.Lenin 72, Sofia 1784, Bulgaria
3) Institute of Solid State Physics and Electron Microscopy,
Academy of Sciences of the GDR, Weinberg 4020 Halle(S), GDR
4) University of Rome "La Sapienza", Department of Energetics and GHSM of CHR, Via A.Scarpa 14, 00161 Rome, Italy

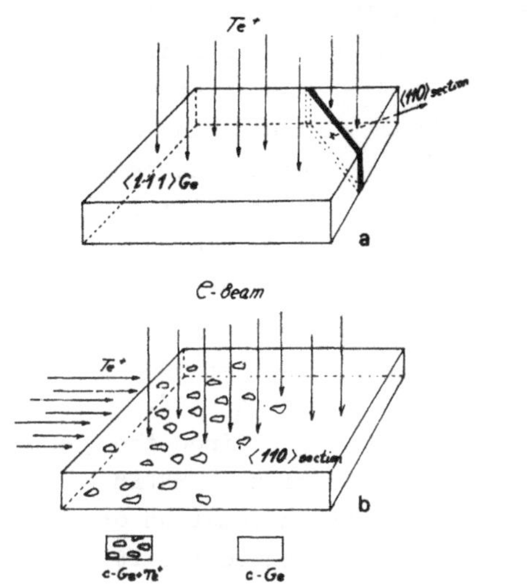

Fig.1. Schematics of : a)cutting of (110) slices from (111) Ge implanted with Te^+ ; b)positioning in the EM column

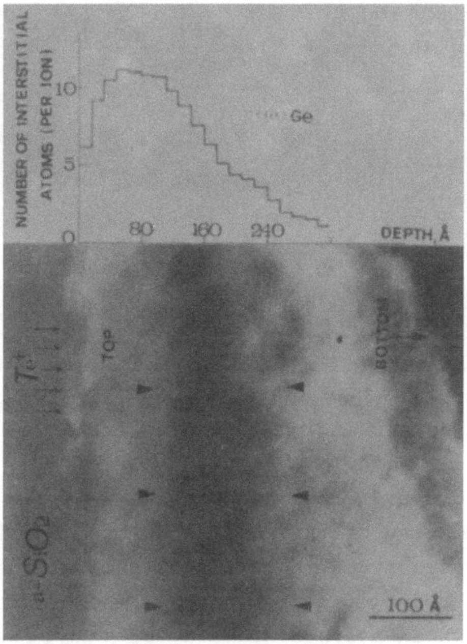

Fig.2. a) XHRTEM along ⟨110⟩; b)computer-simulated damage profile

3. Results and discussion

The micrograph in Fig.2 presents three-beam dark-field lattice image along ⟨110⟩ of (111) Ge bombarded with Te ions. This XHRTEM image shows the depth distribution of the radiation induced damage. A band with dark contrast (indicated by triangle marks) at a distance of 120-130 Å from the implanted surface having width of 160 Å is clearly demonstrated. In the dark-field mode used such darker contrast corresponds to accumulation of damage clusters, see /1/. The narrow well-expressed dark band along the crystalline edge is connected with local deformations of the thinned specimen and is not related to the effects of Te^+ implantation. In Fig.2 the experimental results are compared with a simulated depth distribution of the disorder induced in ion-bombarded (111) Ge with the same depth scale as in the micrograph. The calculated profile appears to be narrower and closer to the surface than the experimental one. The calculated distribution can also be affected by defect diffusion, not taken into account in the simulations. Better agreement should be expected if a shorter screening length in the interatomic potential is used (about 20 % reduction was found to be reasonable, cf. /2/).

Fig.3 illustrates a XHRTEM image of isolated damage clusters situated deeper than the wide maximum of the defect distribution, see for instance A-, B- and C-marked clusters in Fig.3. They appear as a dark central spot surrounded by a diffuse "grey zone" /5/. An analogy can be found between this appearence and the observed contrast of the damage clusters in the plane-view HRTEM /1/. In the analysis of HRTEM pictures such a contrast was explained as due to a spatial separation of vacancies from

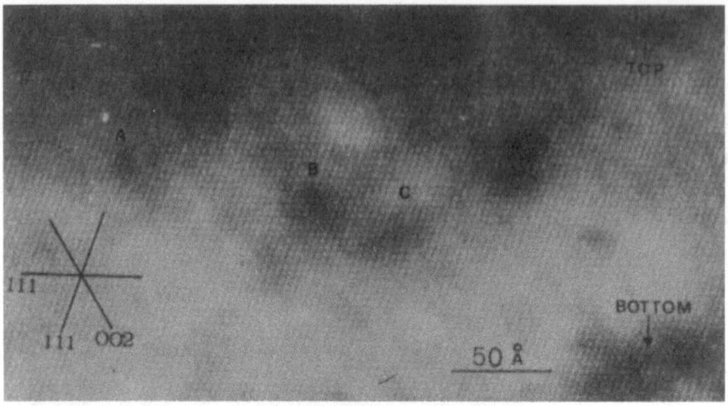

Fig.3. XHRTEM along ⟨110⟩ of isolated damage clusters in ion implanted (111) Ge

interstitials as a result of the collisional cascades. The existence of such separation of vacancies from interstitials around a Te^{+}-produced cluster was also shown by MARLOWE simulations of the damage /1/.

Since the quality of the XHRTEM pictures is not good enough we tried to obtain more information on the geometry of the grey zones by comparing the data estimated for 25 clusters. On the basis of such statistics the following conclusions can be drawn : (a) The vacancy reach region is approximately oval in shape with an average diameter of 4 nm ; The grey zones are asymmetric with tendency for elongation in ⟨001⟩ direction. The total mean dimension of the clusters in this direction is 11 nm while for all other directions it smaller than 9 nm. The asymmetry observed could be explained by a possible influence of the crystal lattice on the distribution of point defects,in this case interstitials.

The formation of visible grey zones still needs a better understanding and combine HRTEM and XHRTEM investigations might be helpful in resolving this problem.

References

(1) Kalitzova,M., D.S.Karpuzov and N.K.Pashov : Phil.Mag. 51 (1985) 373
(2) Kostic,S., J.J.Jimenez-Rodriguez, D.S.Karpuzov, D.G.Armour and G.Carter : Nucl.Instrum.Meth. B2 (1984) 182
(3) Parsons,J.R. and C.W.Hoelke : Phil.Mag. 21 (1970) 1105
(4) Robinson,M.T. and I.M.Torrens : Phys.Rev. B9 (1974) 5008
(5) Ruault,M.O., J.Chaumont and H.Bernas : Nucl.Instr.Meth. 209 (1983) 351
(6) Sheng,T.T. and R.B.Markus : J.Electrochem.Soc. 127 (1980) 737
(7) Zakharov,H.D.,M.Pasemann and V.N.Rozhanskij : Phys.Stat.Sol. (a) 71 (1982) 275

ANODIC OXIDATION OF Si AS A LOW TEMPERATURE OXIDATION METHOD FOR PASSIVATION OF SEMICONDUCTOR DEVICES

G. Mende[1)], M. Deutscher[1)], H.-J. Döring[2)], H. Flietner[3)] and B. Schmidt[1)]

1. INTRODUCTION

Thermal oxidation of Si is the most commonly used technique for the passivation of semiconductor devices. Anodic oxidation of Si /8,3/ on the other hand, has been much less explored from this point of view. As a low temperature method it should be applied especially advantageously in such cases where diffusion of impurities into the silicon and or a change of high-resistivity Si starting material must be avoided, e.g. during passivation of nuclear radiation detectors /4,9/ or electron detectors /5/. The smoothing action of anodic oxidation is speculated to be the reason for the high breakdown field strength of anodic oxide films on polysilicon /6/. Besides it can be attractive for producing of ultra thin oxide layers /1,7/.

In this paper we describe conditions for producing of MOS structures using anodic silicon oxide and investigate the influence of conventional and short light pulse annealing technique on these structures by high-frequency capacitance-voltage measurement. After irradiation by a 20-keV electron beam the samples were measured again.

2. EXPERIMENTAL

Polished n-type⟨100⟩oriented silicon substrates (5 Ohm cm) were cleaned and anodized in a special galvanic cell /2/. A 0.04 N solution of potassium nitrate in ethylene glycol containing about 0.35 vol % (per cent by volume) served as the electrolyte. The current density was 7 mA/cm^2. As the film grew, so the potential difference across it increased. When this reached a predetermined value - depending on the desired oxide film thickness - the voltage was maintained at this value for 30 min. "As-formed" anodic oxide films contain strained Si-O bonds, SiOH and SiH groups, absorbed water, glycol oxidation products, and ionic impurities (e.g. Na^+ ions). Therefore anodically grown silicon oxide must be cleaned and annealed.

MOS structures were produced by Al evaporation onto the oxide followed by dot photolithography. Afterwards the samples were annealed in a N_2 stream at 450 °C for 30 min.

Thermal SiO_2 is grown in a dry O_2 stream at 1000 °C followed by in situ tempering in N_2 for 20 min.

The electron irradiation dose was $3.6 \cdot 10^{-5}$ As cm^{-2}. No voltage was supplied to the sample during irradiation.

3. RESULTS AND DISCUSSION

Flatband voltage U_{Fb}, fixed oxide charge density $N_{eff,Fb}$ surface state density in midgap $N_{st,Mg}$ and mobile ionic charge density N_m of thermally and anodically grown MOS struc-

1) ZfK Rossendorf, Postfach 19, DDR-8051 Dresden
2) VEB Carl Zeiss Jena
3) ZIE Berlin

tures are practically identical, provided that anodic oxide films were annealed by furnace in a N_2 stream at 400 °C for 20 min (Table 1).

The minority carrier lifetime of detectors with high-resistivity starting material ≥10 kΩcm) does not change after anodic passivation under such conditions. Therefore the dark current of such detectors is very low compared with thermally passivated detectors /5/. However, in some type of electron detectors the anodic oxide film is irradiated by electrons, which generates an enhancement of the dark current caused by a change of electrophysical properties of the Si/SiO_2 interface.

Table 1

Electrophysical properties of $Si/SiO_2/Al$ structures

oxide type	U_{Fb} /V/	$N_{eff,Fb}$ /10^{10} cm^{-2}/	$N_{st,Mg}$ /10^{10} eV^{-1} cm^{-2}/	N_m /10^{10} cm^{-2}/
anodic	-0.18 ± 0.04	-3.0 ± 0.26	4.2 ± 0.8	1.8 ± 0.1
thermal	-0.27 ± 0.08	-0.9 ± 0.7	1.0 ± 0.2	2.6 ± 0.6

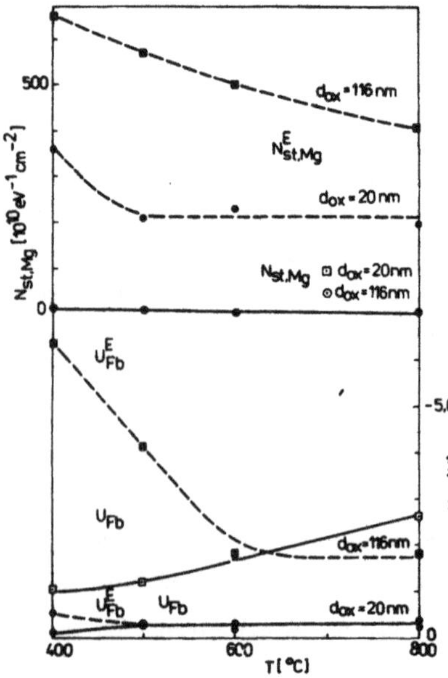

Fig. 1

The surface state density in midgap and flatband voltage of MOS structures using anodic silicon oxide films (cleaned in deionised water) before ($N_{st,Mg}$ respectively U_{Fb}) and after irradiation ($N_{st,Mg}$ respectively U^E_{Fb}) as a function of annealing temperature with oxide film thickness as parameter

The surface state density in midgap and flatband voltage of MOS structures before and after irradiation as a function of annealing temperature with oxide film thickness as parameter are shown in Fig. 1.

Best radiation resistance of relatively thick oxide films (d_{ox} = 116 nm) can be obtained after annealing at 800 °C, but the best radiation resistance of thin oxide films (d_{ox} = 20 nm) is already produced after annealing at 500 °C. After post-anodization cleaning with $HCl-H_2O_2-H_2O$ and annealing the radiation resistance of anodically and thermally grown MOS structures is practically identical /7/.

As the resistivity and the minority current carrier lifetime of high-resistivity starting material (≥ 10 k cm) changes after annealing treatment by means of a furnace above 500 °C, annealing treatment by short light pulses is investigated.

In Fig. 2 is represented the influence of halogen lamp annealing (t = 16 s) in air after and without furnace annealing at 400 °C in N_2 for 20 min on the electrophysical properties of the Si/SiO_2 interface before and after ir-

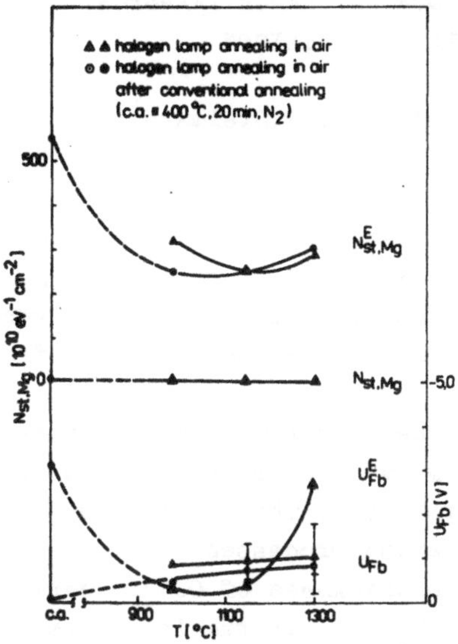

Fig. 2

The surface state density in midgap and flatband voltage of MOS structures using anodic silicon oxide films (cleaned in $HCl-H_2O_2-H_2O$) before ($N_{st,Mg}$ respectively U_{Fb}) and after irradiation ($N^E_{st,Mg}$ respectively U^E_{Fb}) as a function of halogen lamp annealing in air (16 s) after and without furnace annealing in N_2 at 400 °C for 20 min

radiation. Short light pulse annealed unirradiated oxide films show the same tendency with raising of temperature like furnace annealed oxide films. The smallest surface state density at relatively low flatband voltage can be reached after halogen lamp annealing at 1150 °C. Best radiation resistance can also be obtained after halogen lamp annealing at 1150 °C for 16 s. There are practical no differences between such annealed anodic oxide films and thermally grown SiO_2 films /7/.

4. CONCLUSIONS

1. At 400 °C annealed anodically grown SiO_2 films are comparable in quality to thermally grown SiO_2.
2. In radiation hardening technique thermal SiO_2 can be substituted by anodic SiO_2, which was annealed either conventionally at 500 - 800 °C for 20 min or with the help of a halogen lamp at 1150 °C for 16 s.

REFERENCES

/1/ Gaspard, F., A. Halimaoui et al.: Rev. Phys. Appl. 22 (1987) 65
/2/ Mende, G. and G. Küster: Thin Solid Films 35 (1976) 215
/3/ Mende, G.: J. Electrochem. Soc. 127 (1980) 2085
/4/ Mende, G., K.-D. Butter and B. Schmidt: Thin Solid Films 102 (1983) 65
/5/ Mende, G. et al.: Physik der Halbleiteroberfläche 14 (1983) 211
/6/ Mende, G. and J. Wende: Thin Solid Films 142 (1986) 21
/7/ Mende, G. et al.: Proceedings of the International Conference E-MRS 1987 (Strasbourg, France)
/8/ Schmidt, P.F. and W. Michel: J. Electrochem. Soc. 104 (1957) 230
/9/ von Borany, J., G. Mende and B. Schmidt: Nucl. Intrum. Meth. 212 (1983) 489

SELF-ORGANIZATION PHENOMENA DURING CHARGE TRANSPORT IN BORON DOPED POLYCRYSTALLINE SILICON

V.A.Dorosinetz, N.A.Poklonski, V.A.Samujlov, V.F.Stelmakh[1)]

The properties of the open nonlinear dynamical systems have been exhibited in a wide range of semiconductor materials, such as Ge /1/, GaAs /2/, InSb /3/, CdS /4/. Conductivity instabilities leading to generation of periodical and chaotic oscillations with a low number of freedom degree have been investigated /5/.

Infra-low-frequency current oscillations accompanied with space-time structures in the form of charge density waves were observed in thin polycrystalline silicon layers for the first time. These phenomena possess the main features of the self-organization processes.

The polycrystalline silicon films with thickness of 0.4 μm were deposited by low pressure decomposition of SiH_4 onto thermally oxidized monocrystalline silicon. The films were doped by ion implantation of B^+ with energy 30 keV. The irradiation doses were $1.3\cdot10^{12}$–$1.9\cdot10^{15}$ cm^{-2}. The mean grain size was ~110 nm. The planar resistance structures were investigated. These structures had the aluminium contacts marched onto the p^+-regions.

The undamped, infra-low frequency oscillations of electric current with frequency from $5\cdot10^{-3}$ to $2\cdot10^{-1}$ Hz were observed in the films with impurity concentration less than $2\cdot10^{17}$ cm^{-3} in VC-characteristics in quasi-static regime in high electric fields under $E \geqslant 10^4$ V/cm and T=300 K. The oscillation amplitude compiled some decades of percentage from the stationary value of current. A 1024-point fast Fourier transformation was applied to determine power spectra P(f) of time-current signals I(t) (Fig.1-3). From these spectra one can determine the degree of signal unharmonity. For periodical oscillations (Fig.1) the power spectrum consist of the main frequency and some harmonics multiple of it.

For the elucidation of oscillation nature oscillation phase pattern between two potential contacts was registered. The phase pattern form was close to ellipse. This fact is the evidence of space charge discontinuity /6/, i.e. charge density waves. These waves move slowly along the sample under the influence of electric field. The trap states in the polycrystalline silicon are in the regions of high disordering, i.e. on the grain boundaries. As a result we have a qualitatively new kind of wave charge transport, exactly the barrier recharging waves.

The experimental estimation of the wave density rate compiles $2\cdot10^{-3}$ cm/s. We can estimate the effective time of one barrier recharging $\tau\approx5\cdot10^{-3}$ s, with consideration of mean grain size in polycrystalli-

1) Physics Departament of the Byelorussian State University
Lenin Avenue 4, Minsk, USSR

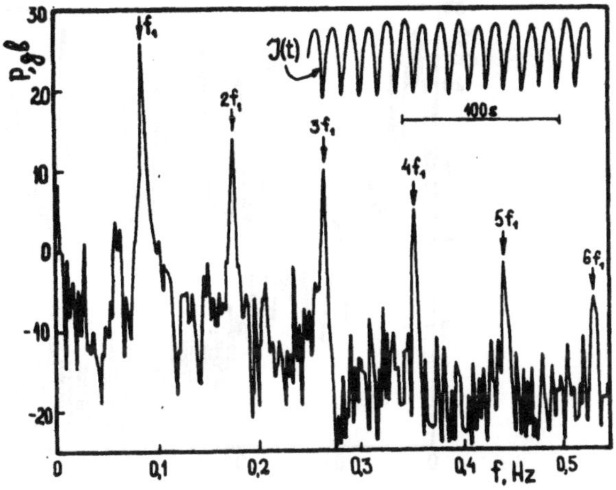

Fig.1. Power spectrum P(f) and time-current signal I(t) of periodical oscillations. E ≃1.05 V/cm.

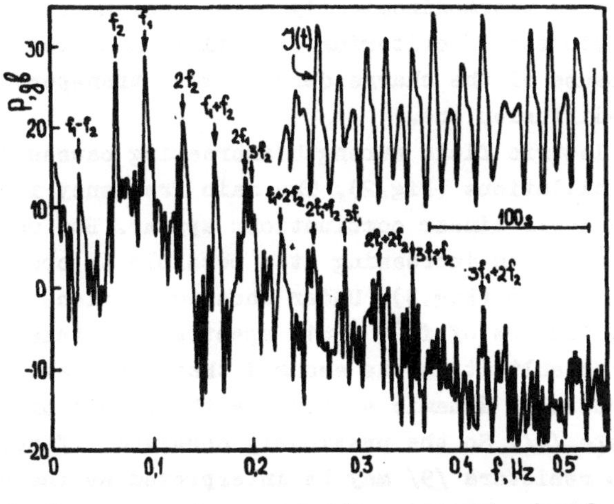

Fig.2. Power spectrum and time-current signal of quasi chaotic oscillations. E ≃1.15 V/cm.

ne silicon. If the probable value of the hole trapping coefficient onto deep levels in the band gap is $\gamma \simeq 10^{-8}$ cm^{3}s^{-1} and the mean hole concentration is $p \simeq 10^{11}$ cm^{-3}, τ has a good correlation with the value of relaxation time of the trap filling degree $\tau = (\gamma \cdot p)^{-1}$.

The estimation of an average one-barrier bias voltage magnitude, corresponding the current oscillation existence critical bias, was shown to be more than effective barrier height - electron charge ratio. From the previously presented materials here it may be concluded that the charge density waves generation takes place under the mobile charges in the grains and the localized states on the grain boundaries in polycrystalline silicon interaction. And the whole system is self-orga-

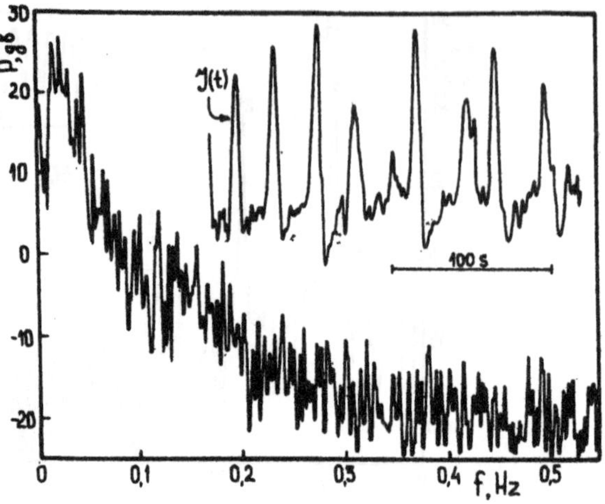

Fig.3. Power spectrum and time-current signal of chaotic oscillations. E $\simeq$ 1.25 V/cm.

nizing because of it satisfies the main criteria of self-organization /7/. System is thermodinamically open, nonlinear, over-critical and the Prigogine-Glansdorff criterium of a minimum entropy generation is completed because of the charge density wave transport is more ordered than the drift-diffusion one.

The electric field strength increasing causes the quasi chaotic current oscillations (Fig.2). The main frequency increases. Another frequency and their linear combinations appear. But under the great electric field strength increasing it's possible to create the chaotic current oscillations (Fig.3). Under these conditions the power spectrum takes on the feature of $f^{-\alpha}$ ($\alpha \simeq 1$) spectrum. To take into account this experimental result it may be assumed that the creation of the 1/f-noise in open nonlinear dinamic systems is the result of the order-chaos transition there /8/. So the previously observed 1/f-noise in polycrystalline silicon resistors /9/ may be interpreted as the high frequency tail of the chaotic oscillations.

/1/ Bumeliene S.B., Pozhela Yu.K., Pyragas K.A., Tamasevicius A.V.: Physica 134B (1985) 293-298
/2/ Marakas G.N., Porod W., Johnson D.A., Ferry D.K.: Physica 134B (1985) 276-280
/3/ Seiler D.G., Littler C.L., Justice R.J.: Phys. Lett. 108A (1985) 462-464
/4/ Gulyaev Yu.V., Rzhanov Yu.A., Balkarei Yu.I., Golik L.L., Elinson M.I., Pakseev V.E.: Proc. Int. Symp., Pushchino 1983, 64-67
/5/ Pyragos K.A., Pozhela Yu.K., Tamasevicius A.V., Ulbikas Yu.: Fiz. Tekhn. Poluprov. 21 (1987) 545-548
/6/ Zhdanova N.G., Kagan M.S.: Fiz. Tekhn. Poluprov. 15 (1981) 168-170
/7/ Haken H.: Advanced synergetics, Springer-Verlag, Berlin 1983
/8/ Bonch-Bruevich V.L.: Dokl. AN SSSR 278 (1984) 335-337
/9/ De Graaf H.C., Huybers M.T.M.: J. Appl. Phys. 54 (1983) 2504-2507

LUMINESCENCE OF CUBIC BORON NITRIDE
IMPLANTED WITH HIGH ENERGY IONS

A.M.Zaitsev, V.S.Varichenko, A.A.Melnikov, V.F.Stelmach[1)]

Recently the radiation optical centers (RC-centers) giving rise Mössbauer-type spectra have been discovered in cubic boron nitride (cBN) /1/. This fact stimulates the application of optical techniques for the studies of the irradiated material. The investigation of ion implanted cubic boron nitride with cathodoluminescence (CL) seems to be the most preferable. In the report we represent some results concerning CL properties of cBN implanted with B, Ne, Co and Xe ions with energies within the range 22÷130 MeV and doses up to $2x10^{17}$ cm^{-2}. The depth distribution of the luminescent centers was obtained by the ion sputtering technique.

The dominating feature of CL spectra of the high energy implanted samples are RC-centers which reveal stronger electron-phonon coupling as compared with these centers observed in electron irradiated or implanted with conventional (tens and hundreds of keV) energy ions samples. The irradiation with hevier ions causes the increase of the electron-phonon coupling and as a result the broadening of the zero-phonon lines and their phonon replicas. This behaviour of RC-centers may be explained by more effective coupling of the corresponding electron transitions with long-wave phonon modes what is probably due to more desorderly crystal structure.

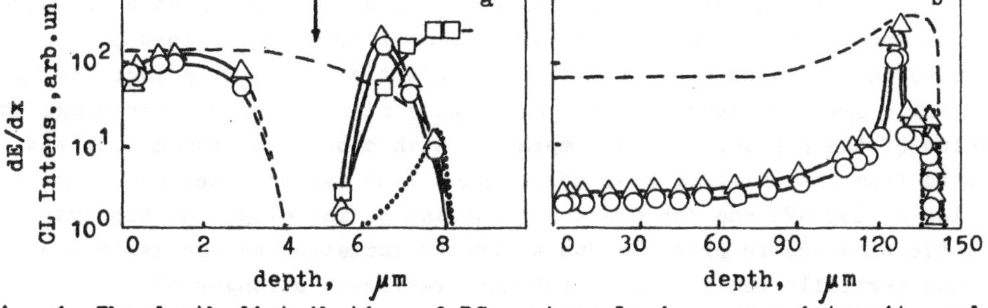

Fig. 1. The depth distribution of RC-centers luminescence intensity and the stopping power in cBN implanted with a) Co-ions, 64 MeV, $6x10^{14}$ cm^{-2}, b) B-ions, 92 MeV, $3x10^{13}$ cm^{-2}; O RC-1, △ RC-2, ‒‒ $(dE/dx)_e$, ··· $(dE/dx)_n$.

The depth of RC-centers penetration coincides well with the theoretically predicted strugglings of the ions /2/ so the existing theory accounts the action of the electron stopping mechanisms rather well. However the distribution profile of RC-centers intensity through the

1) Physics Department of the Byelorussian State University, Leninskii prospekt 4, 220080 Minsk, USSR

sample depth does not follow the electron stopping power magnitude (see Fig. 1). One can conclude that the defect production rate is not formed by the electron stopping completely. The most evident distinction can be observed in the layer which is situated just before the implanted one. Indeed this layer is characterized by sharp abruptness of luminescence intensity (marked with arrow, fig. 1), what results from extremely high defect concentration and therefore from low probability of radiative recombination of the luminescence centers. The weakening of the radiation centers luminescence in the mentioned layer strongly depends on the sort of ions and regime of the implantation. The sharp bondaries of the layer point to the dominance of some resonance-type defect production mechanisms. The most probable explanation of the fact seems to be the ion-explosion mechanism /3,4/ resulting in strong ionization of the lattice due to Auger-process. The condition of Auger ionization proceeding is coincidence of moving ion velocity with that of deep shell electrons of matrix atoms. The calculation of the mean value of these velocities in the layer under consideration reveals good aqreement. In the surface layers of the samples where the ions velocity was higher the thermal defect production mechanism /5/ apparently takes place. Since the energy passing to the lattice is proportional to the electron stopping power, so the defects distribution profile is qualitative coinciding with the stopping power curve. Such accordance was observed in cBN implanted with 90 MeV boron ions.

One of the most interesting features of the luminescence of cBN implanted with high energy ions is considerable shift of RC-centers zero-phonon lines to the long wave length spectral region. One can explain the fact as the straight conformation of strong hydrostatic pressure appearance in the irradiated layer. The most prominent pressure is observed in the layer subjected to the highest electron stopping. The hydrostatic pressure magnitude becomes higher with ion mass increasing. The value of the pressure in cBN implanted with cobalt and xenon ions estimated from the shift of RC-2 zero-phonon line (about 6 meV and 10 meV) makes up 2÷3 GPa and 4÷5 GPa correspondingly. The origin of the hydrostatic pressure is probably due to tracks formation by energetic ions. Being partially ionic crystal cBN must represent stronger efficiency in the tracks production /6/ as compared to pure covalent matrix.

Coming from the existing idea of the track features one can propose that the layer of cBN subjected to the strong electron stopping of ions possesses very heterogeneous structure with the region of high vacancy concentration surrounded by the envelope of interstitials. When the dose of implantation is quite moderate and evry track is individual and inclosed by undisordered lattice, so the pressure may appear due to lack of coincidence of lattice parameters of the ideal matrix and the track region. Higher implantation doses lead to mixing the mentioned structure and as a result to disappearance of the pressure.

The annealing of the irradiated cBN at temperatures ~400°C and higher results in considerable reduction of the pressure what is the sequel of effective defect structure transformation. The decrease of the irradiated layer heterogeneity at these temperatures is confirmed by narrowing of zero-phonon lines due to removal of nonhomogeneous broadening and weakening of RC-centers electron-phonon coupling.

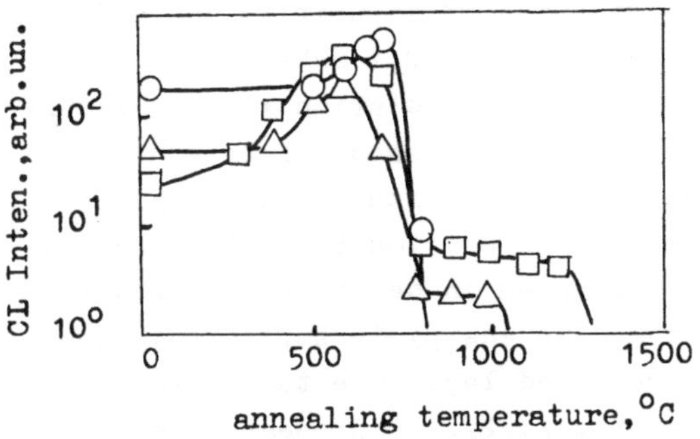

Fig. 2. RC-2 center luminescence intensity versus annealing temperature in cBN implanted with: O - B-ions, 92 MeV, $3x10^{15}$ cm^{-2}; Δ - Co-ions, 64 MeV, $6x10^{14}$ cm^{-2}; □ - Xe-ions, 130 MeV, $1.2x10^{14}$ cm^{-2}

The main stage of RC-centers annealing is 700÷800°C what coordinates with the data obtained on the samples implanted with conventional energy ions. But in contradiction to that case RC-centers in cBN implanted with high energy ions do not anneal completely up to temperature 1200°C. This effect is more (see fig. 2) prominent in the samples bombarded with hevier ions. As well as the efficiency of formation of the center with 2.09 eV zero-phonon line is much higher after high energy ion implantation. These facts testify to the presence of probably two types of regions in the layer bombarded with energetic ions. One of them is characterized by high concentration of defects, strong nonhomogeneous distortion, higher thermal stability and strong electron-phonon coupling with long-wave acoustic modes. The second one has the common defect structure of implanted layer when the usual electron-phonon coupling with undistorted quasilocal modes is dominant.

(1) Zaitsev A.M., A.A.Melnikov, and V.F.Stelmach: Working Meeting on Ion Implantation in Semiconductors and Other Materials and Ion Beam Devices. Ext.Abstr., Balatonaliga (Hungary), 1985, 97-98
(2) Northcliffe L.C., R.F.Shilling: Nucl.Data Tables, Acad.Press. A70 1970
(3) Fleisher R.L., P.B.Price, and R.M.Walker: J.Appl.Phys. 36 (1965) 3645-3652
(4) Tombrello T.A., G.R.Wie, N.Iton, and T.Nakayama T.: Phys.Lett. A100 (1984) 42-44
(5) Seiberling L.E., J.E.Griffith, and T.A.Tombrello: Rad.Eff.52 (1980) 201-210
(6) Dartyge E., L.P.Durand, J.Langevin, and M.Maurette: Phys.Rev. B23 (1981) 5213-5229

LUMINESCENCE OF DIAMOND IMPLANTED WITH HIGH ENERGY CARBON IONS

V.S.Varichenko, A.M.Zaitsev, A.A.Melnikov, V.F.Stelmach[1)]

Defect production and depth distribution of impurities and defects in crystals implanted with ions of energy above 0,1 MeV/a.e.m. differ considerably from those observed after ion implantation with conventional (tens and hundreds of keV) energies. The reasons of this distinction are deeper penetration of ions when the expression $R_p \gg \Delta R_p$ is correct (R_p-projective range, ΔR_p -struggling of the projective range), dominant role of electron stopping processes in defect production, proceeding of nuclear reactions when the ions energy is sufficient, appearance of high pressure in the irradiated layer due to restraining action of surrounded undisordered matrix /1,2/. At present time little information on specific properties of impurity-defect structure of high energy implanted semiconductor crystals is available /3,4/, so any investigations in this direction seem to be urgent.

Diamond is the promissing material when one needs to study defect production effects in irradiated and especially in implanted solid state. Indeed possessing the strongest interatomic covalent bonds and the most stable impurity-defect structure diamond is the unique substance to study the formation processes of such very nonequilibrium and nonhomogeneous structures as the semiconductor implanted layers are.

The present paper is devoted to investigation of luminescent properties of diamond crystals implanted with carbon ions of energy 80MeV and doses $10^{14} \div 1.2\times10^{16}$ cm^{-2}. The samples were IIa natural diamond (110) plates with nitrogen contentsabout 10^{15} cm^{-3}. Defect structure of the irradiated layers was investigated with cathodoluminescence (CL, the exciting electron energy was 15 keV) and photoluminescence (PL, wavelength of the exciting ligth was 530 nm) techniques at temperature 80K. The postimplantation one-hour isochronal annealing was performed in vacuum $\sim10^{-4}$ Pa at temperatures up to 1400°C. The depth distribution profiles of the luminescent centers were obtained with ion sputtering techique.

In CL spectra of the initial samples only the broad band with maximum 440 nm (A-band) is observed. The origin of this band is perhaps dislocations /5/. CL spectra of as-implanted samples reveal strong zero-phonon lines (ZPL) of well known radiation diamond centers 741 nm (GR1 center, neutral vacancy /6/), 470 nm (TR12 center, probably double Frenkel pair /7/) and 503.5 nm (3H center, unknown nature /8/). Besides these lines one also can see strong structures with ZPL 575 nm and 389 nm

---

1)Physics Department of the Byelorussian State University Leninskii prospekt 4, 220080 Minsk, USSR

what contradicts the data obtained on the samples implanted with carbon ions of conventional energies. In PL spectra of as-implanted samples only GR1 center is seen.

The subsequent annealing of the implanted samples causes the drastic changes in the spectra. The nitrogen containing centers with ZPL 575 nm and 389 nm are present after all annealing temperatures. At the temperatures 500÷600°C new ZPL 412; 484.5; 510.8 nm appear which anneal out together with GR1, TR12 and 3H cenrets at 800÷1000°C. At temperatures above 700°C ZPL 509; 511.5; 563; 590 nm and typical for implanted IIa diamonds B-banb /7/ apper. Further annealing at the temperature 1300°C leads to the origin of many narrow ZPL among which lines 553.5; 573.5 and 602 nm are the strongest. Then at 1400°C ZPL 602 nm increases very much, new ZPL 418; 488.3; 540.6 and 614.7 nm apper and all above mentioned ZPL weaken. Such variety of luminescent centers is unusual for diamonds implanted with ions of conventional energies and is characteristic feature just of high energy ion implantation.

At the temperature 1400°C B-band changes its form and well resolved phonon structure with characteristic phonon 65 meV may be seen. Such increase of electron-phonon coupling with 65 meV mode is probably due to weakening of coupling with long-wave acoustic modes.

The "ion-doped" layer can be determined taking into account the fact, that high energy ion implantation introduces 3H center into the irradiated but not doped layer while the centers GR1 and TR12 are introduced through the whole irradiated layer /2/. Estimated in that way value of R_p is equal to 72μm and coincides with theoretically prodicted one well /9/.

Specific feature of the defect structure of the implanted samples is the narrow layer with much higher defect concentration which takes place at the depth about (marked with arrow, fig.a) 68μm. The origin of this layer may be probably due to the ion-explosion defect production mechanism which is the most effective when the ions velocity becomes equal with electrons velocity in deep shells of the matrix atoms. The strong disorder of the lattice in the layer results in stimulation of nonradiative relaxation processes of excited optical centers and therefore essential weakening of luminescence. This explanation is supported by the strong broadening of ZPL of GR1 center (fig.c).

The spatial distribution of the centers with ZPL 509; 511.5; 563; 590 nm and B-band has the prominent maximum at the depth 56÷62μm(fig.b). Such a maximum is not typical for GR1, TR12 and 3H centers. One can explain this fact supposing the interstitial nature of the corresponding centers concentration of which increases due to introduction of additional carbon atoms locating predominantly in interstitial positions.

The present data reveal that the high energy ion implantation results in not only to deeper impurity atoms penetration, but is origin of formation the complicated spatial defect structure of irradiated layer the electron stopping mechanism plaing the most important role.

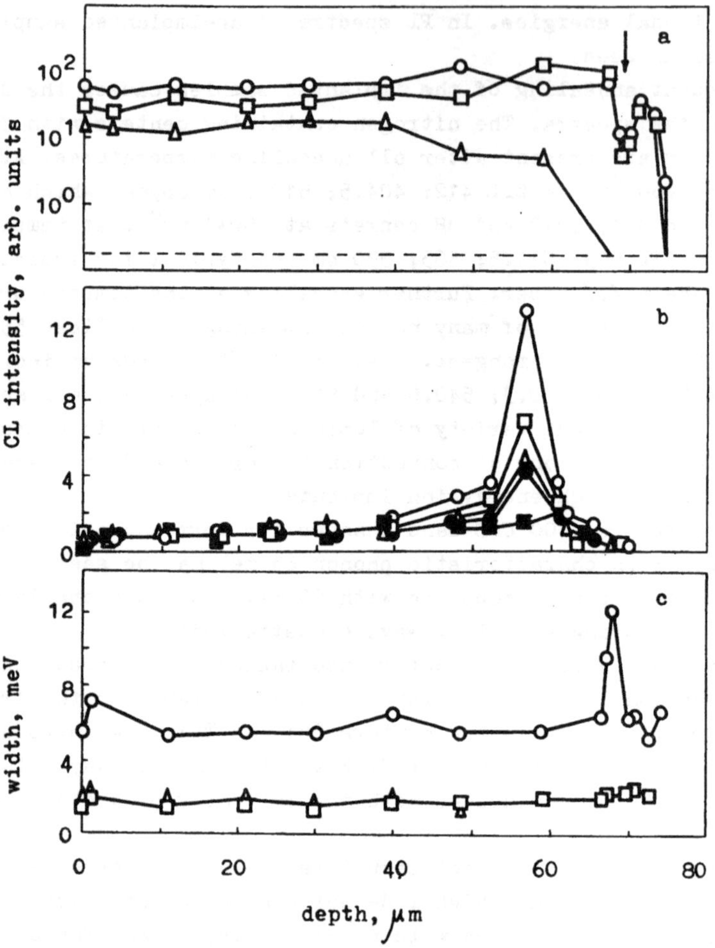

Fig. The depth distribution of intensity (a,b) and width (c) of luminescence centers zero-phonon lines in diamond implanted with carbon ions; 80 MeV, $10^{14} \div 1.2 \times 10^{16}$ cm^{-2}, a,c) ○ GR1 (741 nm), □ TR12 (470 nm), △ 3H (503.5 nm); b) △ 509 nm, ○ 511.5 nm, ● 563 nm, ■ 590 nm, □ B-band. The broken line shows the detection limit.

(1) Fahrner W.R., K.Heidemann and P.Schötle: Phys.Stat.Sol. A70(1982) 463-472
(2) Varichenko V.S., A.M.Zaitsev and V.F.Stelmach: Phys.Stat.Sol.A95 (1986) K123-K126
(3) Ziegler I.F.: Nucl.Instrum. and Meth. B6 (1985) 270-282
(4) Byrne P.F.: J.Appl.Phys. 54 (1983) 1146-1147
(5) Yamamoto N., I.Spence and D.Fathy: Phyl.Mag. B49 (1984) 609-629
(6) Field I.E.: The Properties of Diamond, Academic Press, London 1979
(7) Gippius A.A., A.M.Zaitsev and V.S.Vavilov: Fiz.Tekh.Poluprov. 16 (1982) 404-411
(8) Zaitsev A.M., A.A.Gippius and V.S.Vavilov: Fiz.Tekh.Poluprov. 16 (1982) 397-403
(9) Hubert F., A.Fleury, R.Bimbot and D.Gardes: Ann.Phys.Suppl.,Fr.1980

MODIFICATION OF SOLIDS DUE TO THE EXPOSURE TO HIGH TEMPERATURE PLASMAS

Rainer Behrisch+)

Abstract

In high temperature plasma experiments the solid walls surrounding the plasma are subjected to an intense load of plasma particles and energy. The surface-layers are modified by ion implantation, by deposition of material from other wall areas, by sputtering, evaporation, sublimation and eventually also by surface cracking and surface melting. These modifications present a concern for the vessels of all plasma discharges but they will be a very critical problem in a future DT-fusion reactor. The released wall atoms, which may enter the hydrogen plasma are impurities, which cool the plasma by radiation; the wall modifications will limit the lifetime of the vessel wall and implanted tritium presents an inventory problem. One major problem in fusion research is to find a wall material, to design the wall structure and to operate the plasma so that the wall erosion and the wall modifications are small and tolerable.

Introduction

Since about 1958 /1/, there is a worldwide increasing effort /2,3/ to develop thermonuclear fusion as a useful energy source on earth. Major problems in building such a fusion reactor are to confine a D,T-plasma at sufficient densities n for a time τ , with $n\tau > 10^{14}$ sec cm^{-3}, and to heat it to ion temperatures T_i of about 12 keV ($\gtrsim 100$ Mil ^{o}K), so that more energy is released in D-T fusion reactions than was needed to heat the plasma (For a ^{3}He,D fusion plasma temperatures of 55 keV are needed while the product $n\tau$ is about the same). The most promising scheme to confine the plasma particles is a strong magnetic field, either toroidally closed and nested magnetic surfaces (like in a Tokamak or a Stellarator) or in a linear geometry with increasing magnetic fields toward both ends (like in a Mirror). The plasma moves freely along magnetic field lines of constant field strength whilst the diffusion perpendicular to the magnetic surfaces is reduced by about three orders of magnitude and the plasma is partly reflected from areas of increasing field strength. The magnetically confined plasma has to be contained in a solid vessel, mainly to isolate it from the surrounding atmospheric pressure. In today's plasma experiments temperatures of about 10 keV and values for $n\tau$ close to 10^{14} sec cm^{-3} have been reached independently, however the product $n\tau T_i$ is still about a factor 50 smaller than is needed to ignite a D,T fusion plasma /4/.

+)Max-Planck-Instiut für Plasmaphysik, EURATOM Association
8046 Garching bei München, FRG

Due to the diffusion perpendicular to the magnetic surfaces the magnetic confinement is limited and the plasma particles finally impinge on the surrounding vessel walls. The plasma load on the vessel walls is not uniform. There are generally one or several limiters which are solid material blocks mounted at the vessel walls (see Fig. 1 /5/).

Fig. 1 View into the vessel of the Joint European Torus (JET) /5/. At the left side two limiters are seen and the ICRH Antenna protected with carbon tiles. The inner wall is also covered with carbon tiles, while the vessel is made of inconel.

They protrude furthest into the plasma and thus determine a last closed magnetic surface also named separatrix. The major plasma load occurs near the tip and at the sides of these limiters where magnetic field lines intersect. The plasma load decreases expontentially with the distance from the separatrix and the e-folding length is typically a few cms. The vessel walls receive only a reduced plasma flux, the major particle load is a flux of energetic neutral atoms which are created by charge exchange processes between slow neutral atoms and hot ions in the plasma and a flux of photons and soft x-rays emitted from the plasma.

The last closed magnetic surface may also be produced by currents in external coils as applied in divertor tokamaks. Particles diffusing out of the separatrix are guided onto divertor plates which then receive a similar plasma load like the limiters.

Plasma solid interaction processes

When the plasma hits the vessel walls, matter in two extreme states interacts. There is no thermal equilibrium between the plasma and the vessel walls and the phenomena have to be understood and described by the different atomic processes which are initiated by high fluxes of energetic ions, electrons neutrals and photons and as a consequence a high heat deposition or a thermal shock. For a burning D,T-plasma with the reaction

$$D(12\ keV) + T(12\ keV) \rightarrow {}^4He\ (3.5\ MeV) + n\ (14\ MeV)$$

the walls are bombarded in addition by ^4He-particles with energies up to 3.5 MeV (which contain 20% of the fusion energy) and by neutrons with energies up to 14 MeV (containing 80% of the fusion energy). In today's high temperature plasma experiments the total hydrogen particle fluxes are in the range of 10^{14} to 10^{18} ions and neutrals per cm^2 sec with energies of typically 10 eV to a few 100 eV having a high energy tail up to the 100 keV range. The average power load from the plasma onto the surfaces of the vessel walls is in the range of 1 to 10 W/cm^2 and it will be a factor of about 10 higher in a D,T-fusion reactor. Due to the nonuniform load a maximum power of 10 to 100 kW/cm^2 can occur locally, preferably at limiters and divertor plates, mainly during unstable plasma discharges.

This particle bombardment from the plasma initiates several effects at the vessel walls as:

- erosion by sputtering. A fraction of the released wall atoms enter the central plasma where they represent impurity atoms which cool the plasma by radiation. The tolerable impurity concentrations in a DT-plasma are below 1 % for atoms with low atomic number Z (as C or O), they are below 10^{-3} for medium Z atoms (as Fe, Ni, Cr) and they are below 10^{-4} for high Z atoms (as Mo, W, Ta). These impurities will finally also come back and bombard the vessel walls. They may be implanted and will also cause sputtering. This erosion and redeposition of limiter and wall material causes a redistribution of all materials ever used at the vessel walls.
- Hydrogen implantation into the surface layers of the vessel walls. The implanted hydrogen will diffuse back to the surface or into the bulk and it may be trapped at damage sites in the wall material.
- Modification of the vessel walls due to the energy load deposited by ions, electrons and electromagnetic radiation.
- Finally an electrical field builds up between the plasma and a solid surface which may cause the initiation of electrical arcs.

In the following we will mostly report about wall modifications observed in fusion experiments of the tokamak type because they have been investigated in most detail. However the effects are in principle similar to those in stellerators or mirror fusion devices and other high temperature plasma experiments. The plasma wall interaction processes leading to erosion and deposition have further achieved large technical interest for controlled surface modifications /6/.

Wall modifications

In order to understand the plasma-wall interaction phenomena in respect to a control of the plasma density and the impurity concentration in a plasma as well as to control the limiter and vessel wall distructions, detailed investigations about vessel wall modifications have been performed at most of today's large plasma experiments /7-29/. Generally parts of the vessel walls and of limiters and sometimes special wall samples have been taken out and cut into small pieces for a detailed surface layer analysis which was performed mostly by ion beam methods (RBS, PIXE, nuclear reactions, SIMS) and scanning electron microscopy (SEM). At some wall parts ^{13}C depth markers had been implanted and their shift relative to the surface was measured with RBS and SIMS after plasma exposure /30,31/, and also the geometrical shape was precisely measured before and after the exposure to a series of plasma discharges /29/.

Results of limiter and wall modifications

Due to its resistance to high thermal loads without melting carbon is used for the limiters of all large tokamaks, whilst the vessel walls are made of stainless steel or inconel. Besides erosion at the carbon also a metal deposition from the vessel walls is found on these limiters after several plasma discharges. This is shown at an example of a JET limiter in Fig. 2 /29/. Wall erosion as measured with ^{13}C depth markers and from the change in the geometrical shape is up to 0.2 mm at central areas and deposition is about 0.02 mm at the sides. In the lower part of the figure the results of the PIXE analysis are ploted showing considerable metal deposition also at the eroded areas.

Such a distribution, showing erosion and minima in the deposition at the central area exposed directly to the main plasma and maxima in the deposition at the sides which are more distant to the central plasma has been observed at all limiters and areas acting as limiters. In addition on all limiters a deuterium deposition was measured, about twice the amount of metals and with a similar distribution /22/.

The modifications found at the vessel walls are similar to those at the limiters. Erosion and a simultaneous deposition of all materials ever exposed to the plasma was detected. Erosion was measured with carbon probes attached to the vessel walls which had been implanted with ^{13}C depth markers /31/. After 1800 tokamak discharges in JET an averaged local erosion of a carbon layer of up to 0.6 μm was observed, while at other areas a carbon deposition of up to 0.4 μm was found /31/. Deposition of metals

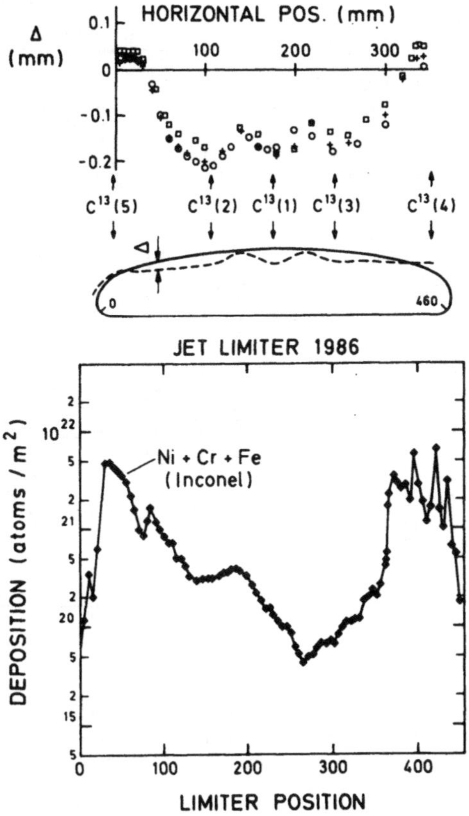

Fig. 2 Global erosion and deposition measurements at the central tile of a JET limiter after exposure to about $8 \cdot 10^4$ discharge seconds in 1986. Data were taken by the measurement of the change in the geometrical shape of the limiter tile. In the lower figure the metal deposit on this limiter tile is shown, demonstrating deposition also at erosion dominated areas /29/.

was also measured with carbon probes whilst for the detection of carbon deposition nickel and inconel probes were used. In addition the deposition of the discharge gas i.e. hydrogen and deuterium in the surface layers of several μm was measured in all wall probes with the $^3He(d,p)^4He$ nuclear reaction and with the elastic recoil detection method /10,14,16,20/. Some results of the metal depositon, the carbon deposition and the deuterium deposition as measured at JET are shown in Fig. 3 and Fig. 4.

The amount of deposited deuterium is always closely related to the carbon deposition and the ratio of both depositions is about 3. This corresponds to the maximum hydrogen concentration which can be trapped in carbon. This was confirmed also by depth profiling measurements /20/. The toal amount of deuterium trapped in the surface layers of the limiters is about 10^{22} D-atoms, whilst it as about 10^{23} D-atoms at the vessel walls. This corresponds to more than 100 times the D inventory in the

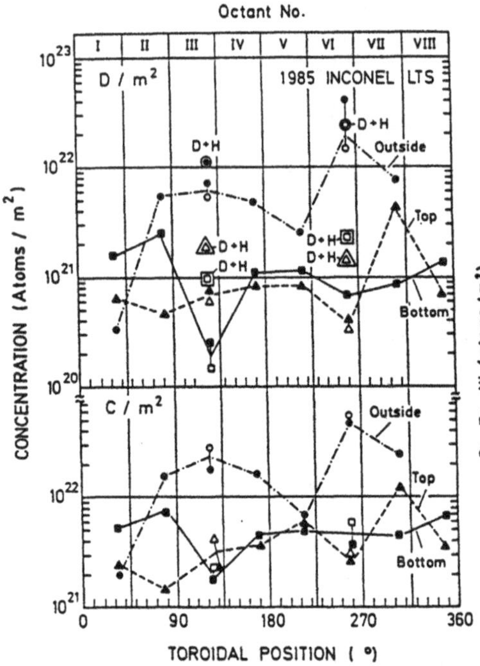

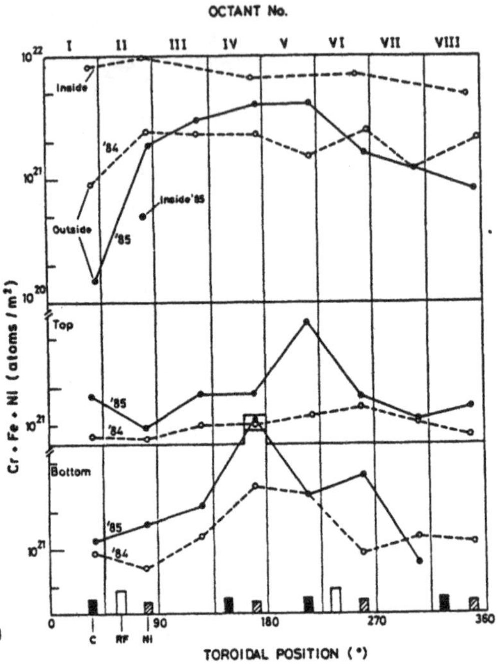

Fig. 3 (left) Toroidal distribution of the D,(D+H) and C deposition on inconel wall samples of JET from top, bottom and outside positions. Note the close correlation between the D and the C deposition.

Fig. 4 (right) Toridal distribution of the metal deposition on carbon wall samples of JET from top, inside, bottom and outside positions. The large deposition at the inside during the 1984 experimental compaign occured when this side was not yet protected by carbon tiles as shown in Fig. 1.

plasma during a discharge. Finally the deposited layers had been analysed with the scanning electron microscope. Fig. 5 to 7 show areas from the center of the limiter

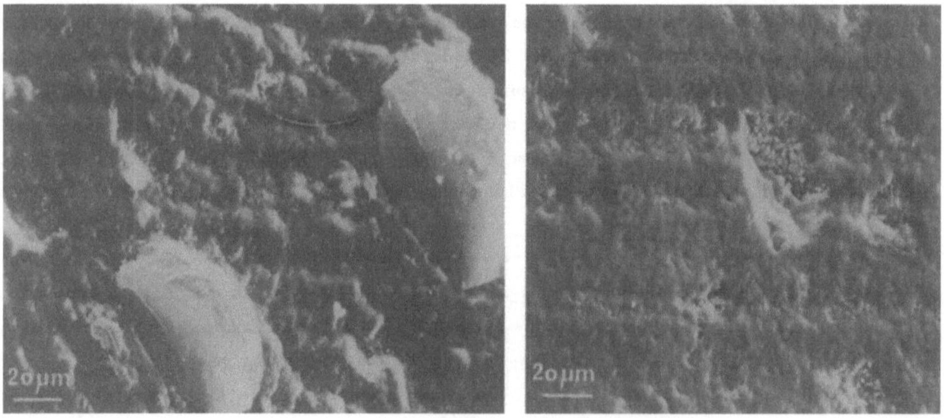

Fig. 5 Metal droplets at the central area of a JET limiter. The large metal droplets (left) are partly eroded whilst the very small droplets in a groove (right) are likely produced by agglemeration of atomic deposits.

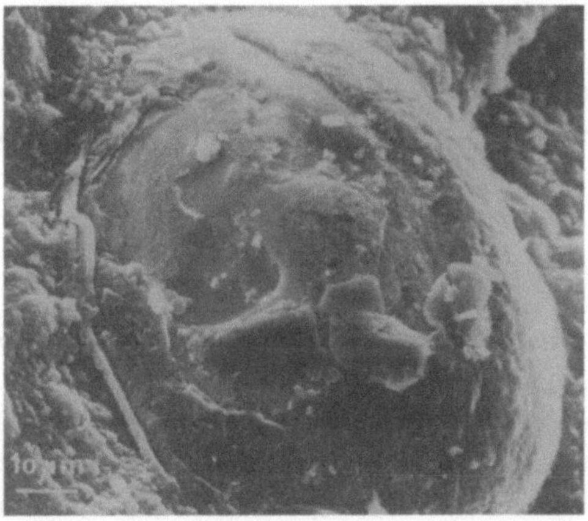

Fig. 6 Large metal droplet at the side of a JET limiter, covered with a carbon layer which is partly flaked.

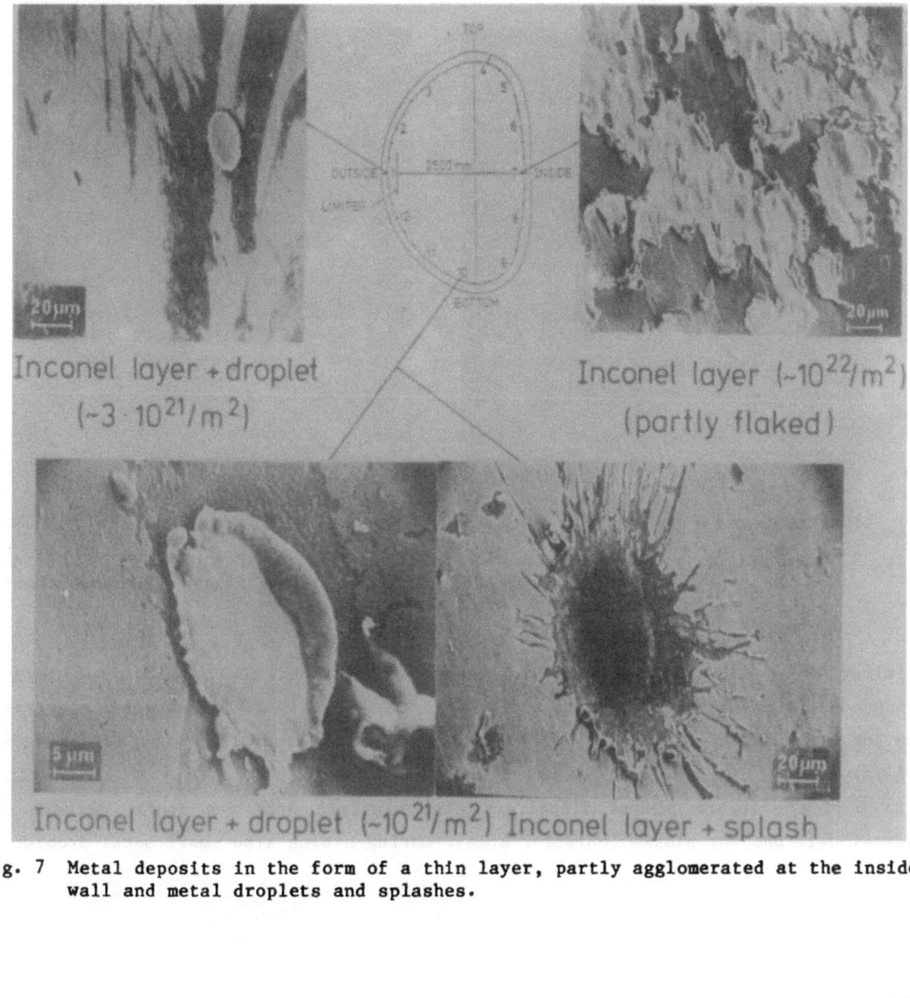

Fig. 7 Metal deposits in the form of a thin layer, partly agglomerated at the inside wall and metal droplets and splashes.

(Fig. 5), from the sides of the limiter (Fig. 6) and from the first wall (Fig. 7). At the central area of the limiter large metal droplets are found which are partly eroded. In addition mostly in grooves very small droplets can be detected. At the sides of the limiter again large droplets are seen, however they are mostly covered up with a carbon layer. At the vessel walls an atomically deposited layer is found. This is partly melted as on the inside or partly eroded by metal splashes impinging on the vessel walls.

Discussion

There are three major erosion processes which contribute to the redistribution of the wall material.

i) During unstable and disruptive plasma discharges some wall areas may recieve an energy load in the range of kJ during a time of ms to s. These thermal shocks lead to surface melting, emission of droplets and to surface cracking during the cooling period. Melted areas and surface cracks have been found in JET mostly at the inside metal protection plates (Fig. 8) when they were not yet covered by carbon tiles. At carbon tiles cracks have been found (Fig. 9). The large metal droplets found on some

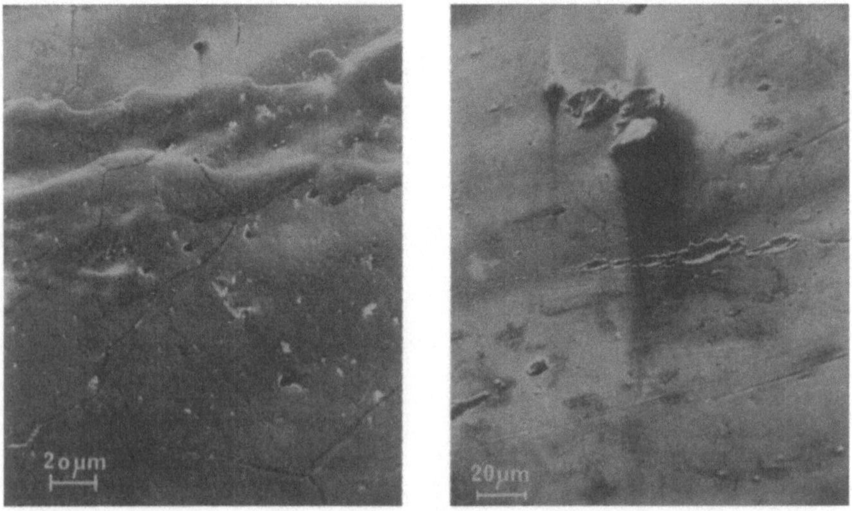

Fig. 8 Melted areas and surface cracks (left) and traces of a unipolar arc (right) on the inner wall metal protection plates of JET after the 1984 experimental period.

limiter areas likely originate from such wall melting, their number was largely reduced after the carbon inner wall tiles had been installed. The material evaporated during the thermal pulse /34/ will contribute to the atomic deposits at the vessel walls.

ii) During unstable plasma discharges also electric arcs can ignite between the vessel wall and the plasma (unipolar arcs). During arcing also small metal droplets

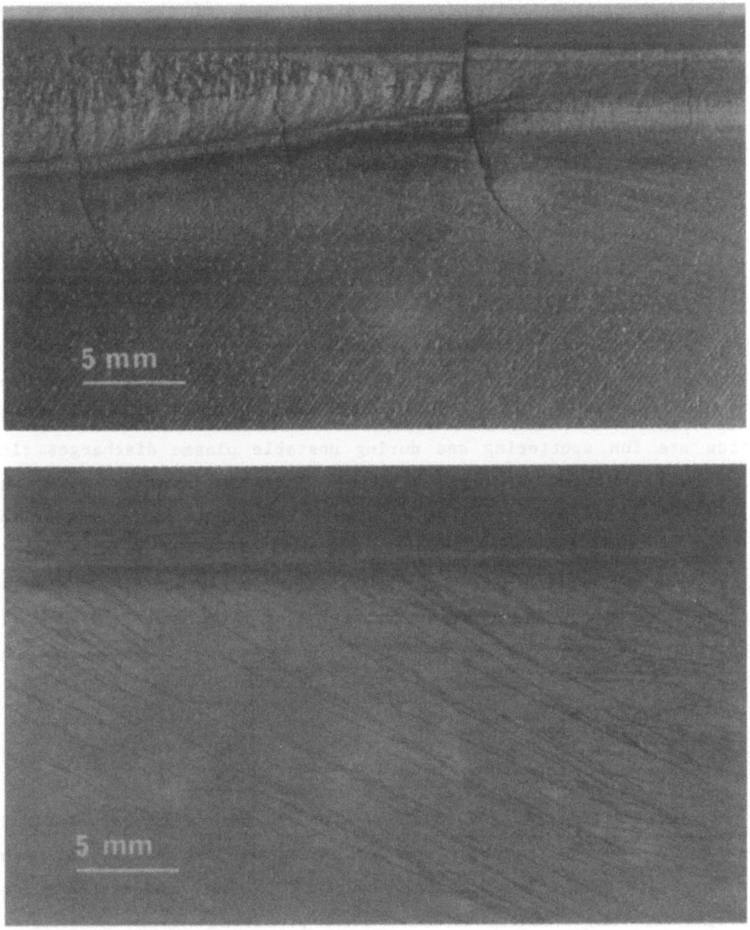

Fig. 9 Largely eroded area and cracks (top) as well as traces of electrical arcs (bottom) at a limiter tile of the JET vessel after the 1986 experimental period.

are released and the metal sphlashes seen on the vessel walls may originate from such electrical arcs. Arc traces have been found on nearly all areas of the vesel walls, but predominantly on limiters and (Fig. 9) protection plates.

iii) The bombardment of the walls by ions and energetic neutrals causes sputtering /32,33/. Generally the amount of material redeposited at the vessel walls is larger than can be explained by sputtering alone. However, sputtering is a major process for introducing wall atoms into the plasma and for redistributing the material at the limiters /33,34/. In the central part of the limiter directly exposed to the plasma the ion energies are sufficiently large to effectively erode the limiter by sputtering. This is confirmed by the erosion of the metal droplets at this area (Fig. 5). The sides of the limiter are more distant from the central plasma, the plasma temperature is lower, the energy of the bombarding particles is reduced and may ly below the threshold for sputtering. Thus atoms are deposited leading to the built up of the thick layers (Fig. 6). The material eroded from the central part of the limiter is only partly redeposited at the sides, some of the eroded material is deposited on the vessel walls. The carbon deposition found on the inconel samples from the vessel wall originates both from the limiters and the carbon tiles at the inner wall of JET.

Conclusions:

Investigations of the surface layers of the vessel walls in high temperature plasma experiments show a redistribution of all wall materials ever exposed to the plasma. At the areas more close to the central plasma a larger erosion than deposition was measured (erosion dominated) while at areas more distinct from the plasma deposition is larger than erosion (deposition dominated). The major processes causing this redistribution are ion sputtering and during unstable plasma discharges electrical arcs and thermal pulses, removing material from the vessel walls which is subsequently redeposited in the form of an atomic layer and as droplets or splashes. If unstable plasma operation can be avoided all wall areas get finally covered with the material from the erosion dominated areas. The redistribution of the material from the vessel walls can lead to the incorporation of large amounts of the discharge gas into the redeposited layers.

Acknowledgements

It is a great pleasure to thank the colleagues of the plasma boundary topic group at JET, especially P. Coad, J. Ehrenberg and G.M. McCracken and several colleagues from the Surface Physics Division in Garching, especially J. Roth and Mrs Wang Wenmin for the collaboration and may discussions about these problems.

References

/1/ Int. Conf. on peaceful uses of atomic energy, Geneve 1958

/2/ INTOR (International Tokamak Reactor) project, IAEA, Vienna, zero phase (1980)

/3/ Gorbatschov-Reagan Initiative to build an International tokamak experimental reactor (ITER), Geneve, 1986

/4/ JET-team presented by P.H. Rebut. Plasmaphysics and Controlled Nuclear Fusion Research Proc. 11th International Conference Kyoto (1986), IAEA, Vienna, Vol 1 (1987) 31

/5/ R. Bickerton et al., Plasmaphysics and Controlled Fusion 28 (1986) 55

/6/ Symposium on plasma treatment of surfaces, science and technology, Proc. 7th EPS conference "Trends in Physics", Helsinki, Aug. 1987, Physica Scripta (1987)

/7/ P. Staib, R. Behrisch, W. Heiland, G. Staudenmaier, Proc. 7th Europ. Conf. on Contr. Fusion and Plasmaphysics, Lausanne, Sept. (1975) 139

/8/ P. Staib, G. Staudenmaier, J. Nucl. Mat. 78/77 (1978) 78

/9/ R. Behrisch, R.S. Blewer, H. Kukral, B.M.U. Scherzer, H. Schmidl, P. Staib, G. Staudenmaier, and Group de TFR, J. Nucl. Mat. 78&77 (1978) 437

/10/ B.M.U. Scherzer, R. Behrisch, R.S.F. Blewer, H. Schmidl, and Group de TFR, Proc. 10th Symp. on Fusion Technology, Padua, Italy, Sept. 1978

/11/ Ph. Staib, H. Kukral, E. Glock, G. Staudenmaier, J. Nucl. Mat. 111/112 (1979) 173

/12/ G.M. McCracken, P.E. Stott, Nucl. Fus. 19 (1979) 889

/13/ D. Hildebrandt, M. Laux, J. Lingertat, P. Pech, H.-D. Reiner, H. Strusny, H. Wolff, Beitr. zur Plasmaphysik 23 (1980) 313

/14/ J.H. Shea, C.J. Sofield, G.M. McCracken, J.H. Evans, L.B. Bridwell, M. Malinowski, J. Nucl. Mat. 93&94 (1980) 299

/15/ S.A. Cohen, R. Budny, G.M. McCracken, M. Ulrickson, Nucl. Fusion 21 (1981) 233

/16/ B. Doyle, B. Wampler, Brice, J. Nucl. Mat. 103/104 (1981) 513

/17/ R. Behrisch, P. Børgesen, J. Ehrenberg, B.M.U. Scherzer, B.D. Sawicka, J.A. Sawicki, J. Nucl. Mat. 128/129 (1984) 470

/18/ G. Bobrovskii and the IAE group, the CIEP group, J. Nucl. Mat. 145-147 (1987) 172

/19/ J. Ehrenberg, R. Behrisch, J.P. Coad, L. de Kock, D.H.J. Goodall, G.M. McCracken, A.P. Martinelli, J. Nucl. Mat. 145-147 (1987) 741

/20/ R. Behrisch, J. Ehrenberg, M. Wielunski, A.P. Martinelli, H. Bergsaker, B. Emmoth, L.de Kock, J.P. Coad, J. Nucl. Mat. 145-147 (1987) 723

/21/ R. Behrisch, J. Ehrenberg, H. Bergsaker, J.P. Coad, L. de Kock, B. Emmoth, H. Kukral, A.P. Martinelli, G.M. McCracken, J.W. Partridge, J. Nucl. Mat. 145-148 (1987) 731

/22/ H. Bergsaker, R. Behrisch, J.P. Coad, J. Ehrenberg, B. Emmoth, S.K. Erents, G.M. McCracken, A.P. Martinelli, J.W. Partridge, J. Nucl. Mat. 145-147 (1987) 727

/23/ A.P. Martinelli, R. Behrisch, H. Hammer, J. Hackmann, J. Nucl. Mat. 145-147 (1987) 755

/24/ The TFR Group, A.P. Martinelli, R. Behrisch, W. Englert, J. Nucl. Mat. 145-147 (1987) 681

/25/ E. Franconi, F. Brossa, J. Nucl. Mat. 145-147 (1987) 691

/26/ R.T. McGrath, B.L. Doyle, J.N. Brooks, A.E. Pontau, G.J. Thomas, J. Nucl. Mat. 145-147 (1987) 660

/27/ W.R. Wampler, B.L. Doyle, A.E. Pontau, J. Nucl. Mat. 145-147 (1987) 353

/28/ H. Amemiya, S. Ishii, H. Oyama, Y. Sakamoto, Y. Hori, K. Akaishi, N. Noda, H. Minagawa, T. Yamashina, J. Winter, T. Banno, H.G. Esser, P. Wiehold, F. Waelbroek, J. Nucl. Mat. 145-147 (1987) 647

/29/ J.P. Coad, R. Behrisch, J. Roth, L. de Kock, J. Ehrenberg, G. Israel, D.H.J. Goodall, W. Wang, M. Wielunski, Proc. 14th. Europ. Conf. Contr. Fusion and Plasmaphys. Madrid, Spain, June (1987)

/30/ J.B.Roberto, J. Roth, E. Taglauer, O.W. Holland, J. Nucl. Mat. 128&129 (1984) 244

/31/ J. Roth, J. Ehrenberg, K. Wittmaack, J.P. Coad, J.B. Roberto, J. Nucl. Mat. 145-147 (1987) 383

/32/ G.M. McCracken, J. Nucl. Mat. 93&94 (1980) 3

/33/ R. Behrisch, J. Roth, G. Staudenmaier, H. Verbeek, Nucl. Instr. Meth. in Phys. Res. B19/20 (1987)

/34/ G.M. McCracken, J. Ehrenberg, P.E. Stott, R. Behrisch, L. de Kock, J. Nucl. Mat. 145-147 (1987) 621

STRUCTURE AND PROPERTIES OF NANOMETER-SIZED SOLIDS

H.-E. Schaefer [1], R. Würschum [1], R. Birringer [2], and H. Gleiter [2]

Abstract

Nanocrystalline materials are ultrafine-grained polycrystalline solids with a substantial fraction of atoms located in the disordered interfacial structure which gives rise to novel physical properties and which promises attractive technological applications. In the present review studies of x-ray diffraction, Mössbauer spectroscopy and positron lifetime measurements are summarized indicating a wide interatomic distance distribution in the interfacial structure. The thermal, elastic, and magnetic properties support this hypothesis. The internal friction behaviour as well as the high diffusivity and reactivity is outlined. In addition to the studies on nanocrystalline metals and semiconductors an overview is given of the properties of nanocrystalline metal oxides. Some future prospects are discussed briefly.

1. Introduction

Nanometer-sized crystalline materials (called here nanocrystalline materials) have been first studied by Gleiter and associates (Gleiter and Marquardt, 1984; Birringer, Gleiter, Klein, und Marquardt, 1984; Birringer, Herr, and Gleiter; 1986). They are polycrystalline materials with very small crystallites about 5 to 10 nm in diameter. This small size gives rise to a high number of interfaces (600 m^2/cm^3) with many different, randomly distributed orientation relationships ($6x10^{19}/cm^3$) and a substantial fraction of atoms (20-50%) located in these interfaces. Therefore, the nanocrystalline materials may be considered to comprise two structural components as sketched in Fig. 1: Crystallites which have long-range order and the disordered interfacial component which represents all the different interatomic spacings in the various types of interfaces and which, therefore, is expected to show on the average a wide interatomic distance distribution and appears to exhibit no short-range order characteristic for glassy materials. This disordered interfacial structure represents a considerable volume fraction of the nanocrystalline solid and therefore may be investigated by macroscopic volume measuring techniques in addition to microscopic methods. It promises novel physical properties as it is not confined to the restrictions of phase equilibria. The understanding of these properties may be favoured by a direct comparison to the well known data of the chemically identical crystalline materials.
For technological applications the high diffusivity and reactivity of nanocrystalline materials, the use for catalytic and absorption processes, the alloying of conventional-

[1] Institut für Theoretische und Angewandte Physik der Universität Stuttgart, Postfach 80 11 40, D-7000 Stuttgart 80, FRG

[2] Institut für Werkstoffwissenschaften, Universität des Saarlandes, FB 12.1, Bau 2, D-6600 Saarbrücken, FRG

ly immiscible components as well as the ductile behaviour of nanocrystalline metal oxides may be attractive.

After a short description of the preparation of nanocrystalline materials a survey of recent studies on the structure and the properties of nanocrystalline metals, semiconductors and metal oxides is given in Sects. 3 and 4. Finally some future prospects are outlined.

2. Preparation of nanometer-sized materials

For the preparation of nanocrystalline materials the noble gas condensation method is used (Kimoto, Kamiya, Nanoyama, and Ueda, 1963; Granqvist and Buhrman, 1976). In this technique ultrafine particles are produced in a vacuum chamber (preferably in a bakeable ultrahigh vacuum (UHV) chamber) by evaporation of a metal in a high-purity noble gas atmosphere and condensation to yield small crystallites (5-10 nm). The crystallite size can be controlled by the pressure and type of the inert gas as well as by the metal evaporation rate. The crystallites are collected by a convective flow at a liquid-nitrogen cold finger where they are stripped off after evacuation of the chamber in order to compact them uniaxially in a high-pressure device ($p < 5GPa$). In this way nanocrystalline metal specimens (n-metals) with a diameter of 5 to 8 mm, a thickness of about 0.05 to 1 mm, a mass density of about 85% of the bulk density, and impurity contents of less than 0.5 wt.% are produced (Zhu, Birringer, Herr, and Gleiter, 1987). Two-component nanocrystals or mixtures of different particles can be produced using two evaporation sources. Nanocrystalline metal oxides, e.g., n-TiO_2, are prepared by oxidation of metallic crystallites after condensation (Karch, Birringer, and Gleiter, 1987; Siegel, Hahn, Ramasamy, Li, Lu, and Gronski, 1987).

The preparation of n-Si is performed at present by chemical transport and deposition on monocrystalline Si substrates in a hydrogen plasma by DC glow discharge (Veprek, 1984). This procedure yields nanocrystalline Si (n-Si) layer thicknesses up to 100 μm, crystallite sizes of 3 to 10 nm, mass densities near to the bulk density, and hydrogen contents of about 4 at.%. Inert-gas condensation of n-Si for preventing the specimens from incorporation of H is desirable in the present studies.

3. Structural investigations

Investigations of the structure of nanocrystalline metals by x-ray diffraction experiments with accompanying simulation calculations have been performed on nanocrystalline iron by Zhu et al. (1987). In these experiments, discrete intensity maxima due to the scattering from the crystallites and an increased background intensity due to the atoms which form the interfaces have been observed. For modelling the experimental results the x-ray interference function was calculated by summing over all orientations of a cube-edged α-Fe crystallite with a size determined from transmission electron microscopy. In order to simulate the scattering from the interfacial structure the atoms of the two outer layers of the model crystallite were displaced in random directions. The best fit to the experimental interference function was obtained in the model calculations by dis-

placing the two outer layers in random directions by 50% and 25% of the nearest neighbour distance, respectively, and by using the experimentally determined size distribution of the crystallites (Hort, 1986). No good fit was achieved by larger atomic displacements, by a higher number of displaced layers, by using a smaller crystallite size, or by considering isolated randomly orientated crystallites. Furthermore, the details of the experimental interference function could not be reproduced by using a short-range ordered model of the interfacial structure in the simulation calculations.
In addition, the probability function for the interatomic distances (see Fig. 2) was deduced (Zhu et al., 1987) for the both models either assuming the interfacial structure to be short-range ordered, or to consist of randomly arranged atoms, the model which best fits the experimental results. The latter result (see Fig. 2) shows a wide interatomic distance distribution, which is obviously characteristic for nanocrystalline metals, whereas the other result shows the typical features of glassy metals.
X-ray diffraction measurements on nanocrystalline Si deposited by the glow-discharge technique (Veprek, 1984) also show the typical crystalline reflections and no interfacial diffraction characteristic for amorphous Si has been observed. The question of the interfacial structure in n-Si was additionally studied by Raman scattering experiments (Iqbal and Veprek, 1982). A signal (Raman shift $\Delta\nu$ = 480 cm^{-1}) which also appears in amorphous Si is ascribed to a Si-Si shear mode in the nanocrystalline interfaces. One could imagine that this mode also occurs in a disordered interfacial structure which is different from the amorphous one.
In EXAFS (extended x-ray absorption fine structure) measurements on nanocrystalline Cu and Pd the volume fraction of the fine structure intensity equivalent to the interfaces is missing and no signal typical for short-range order appears (Haubold, Lengeler, Birringer, and Gleiter, 1987). This again supports the hypothesis of an interfacial structure without interatomic correlation in the distance distribution.

A further technique for structural investigations is available by means of **Mössbauer spectroscopy**. The Mössbauer spectrum (Fig. 3) measured on nanocrystalline Fe (Herr, Jing, Birringer, Gonser, and Gleiter, 1987) may be decomposed into two components one of which represents the characteristic six-line spectrum of crystalline bulk iron. The second component is characterized by an increase of the hyperfine magnetic field H, of the line width, and of the isomer shift, compared to bulk iron, and is ascribed to the interfacial structure since it vanishes when the interfaces are removed by high-temperature annealing (Hort, 1986). All these features support the model of a wide interatomic distance distribution and a reduced mass density in the interfacial structure: The increased isomer shift is significant for a decreased electron density at the Mössbauer probe whereas the enhanced hyperfine field is, in the case of α-iron, well expected for an increase of the nearest neighbour distance, and the broadening of the Mössbauer lines again point to a broad variation of the interatomic distances.

Positron annihilation, and in particular **positron lifetime spectroscopy** is a most appropriate nuclear technique (see Hautojärvi, 1979) for investigating the size of free volumes in solids, e.g., vacancies, their agglomerates, and larger voids. This technique has been applied in order to study free volumes and the interfacial structure of nanocrystalline metals and semiconductors (Schaefer and Würschum, 1987; Würschum, Scheytt,

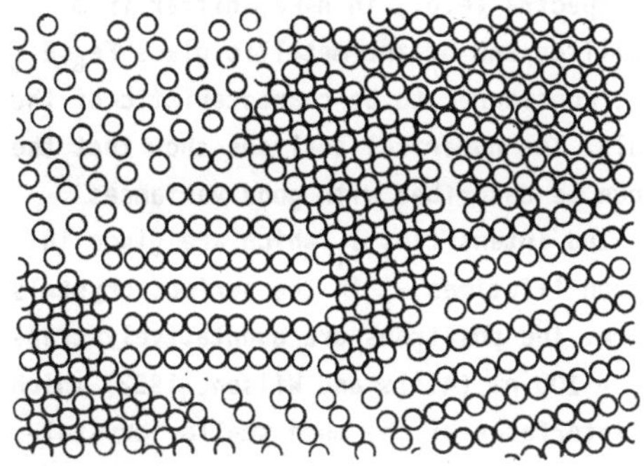

Fig. 1: Two dimensional sketch of a nanocrystalline material indicating the variety of interatomic distances in the different interfaces. In addition to that, atomic relaxation has to be expected within the interfaces which is not shown in the figure.

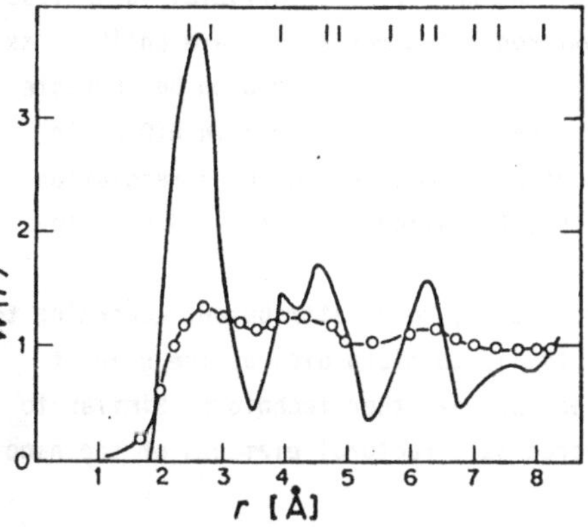

Fig. 2: Probability functions W(r) of the interatomic distances computed for a short-range ordered structure (——) and for the interfacial structure (ooo) of n-Fe (Zhu et al., 1987) which is simulated by two randomly displaced outer shells of a model crystal (see text) the calculated interference function of which fits the experimental x-ray diffraction results.

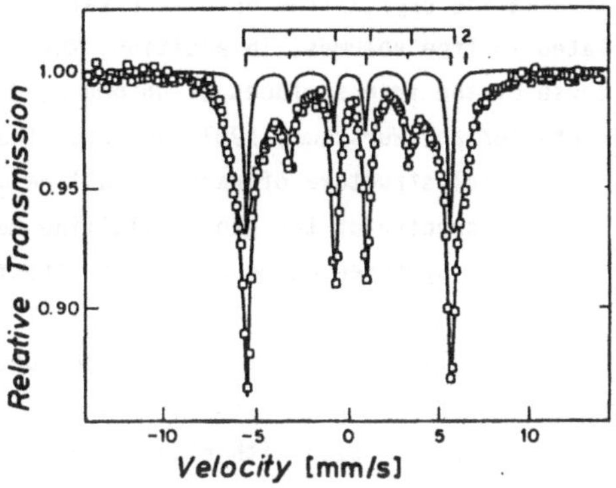

Fig. 3: Mössbauer spectrum of n-Fe (□□□) measured at 77K (Herr et al., 1987). The spectrum was decomposed into two subspectra (——; 1:sharp lines, 2:broad lines).

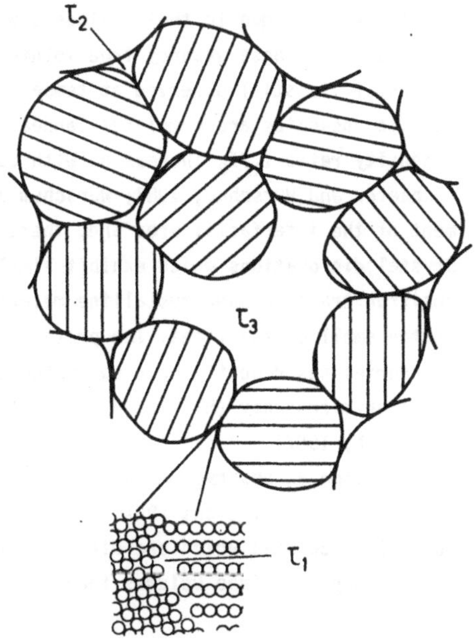

Fig. 4: Schematic two-dimensional arrangement of the crystallites in a nanocrystalline material with an interfacial model structure shown in the inset. The various sites of positron annihilation attributed to the positron lifetimes τ_1, τ_2, and τ_3 in nanocrystalline metals (see text) are indicated.

and Schaefer, 1987; Schaefer, Würschum, Scheytt, Birringer, and Gleiter, 1987a; Scheytt, Würschum, Schaefer, Veprek, and Gleiter, 1987), the results of which will be overviewed in the following paragraph.

The mean positron lifetimes $\bar{\tau}$ in nanocrystalline Fe, Cu, Pd and Si, which are longer than in the crystalline bulk materials, show that the positrons are trapped and annihilated in free volumes. In addition, the lifetime spectra (e.g., in n-Fe) differ from those of the free surfaces of uncompacted crystallites and of the metallic glass $Fe_{85}B_{15}$ (Schaefer and Würschum, 1987) and are, therefore, considered to be characteristic of the interfacial structure of nanocrystalline materials. The numerical analyses show that the lifetime spectra of the nanocrystalline metals can be described with good variances ($\chi^2 < 1.1$) by three components with lifetimes τ_i and intensities I_i which are given in Tab. 1 together with results for n-Si. The positron lifetimes τ_f in the perfect crystals and τ_{1V} in lattice vacancies are also quoted there. The positrons are quantatively trapped due to their crystalline diffusion length $L_+ \approx 100$ nm (Mills and Wilson, 1982) which enables them to reach with high probability the interfaces of the much smaller crystallites.

The positron lifetimes measured in nanocrystalline metals (see Tab. 1) can be interpreted as follows (see Fig.4):

- The lifetimes τ_1 are attributed to interfacial free volumes of the size of about a lattice vacancy due to their values similar to those observed in lattice vacancies (see Tab. 1). These vacancy free volumes are expected to be located in the interfaces because τ_1 is still observed at temperatures (600 K in n-Fe, 650 K in n-Cu) where lattice vacancies are completely annealed out (Schaefer, 1982), and because of the intensity ratio I_1/I_2 increasing with compacting pressure in n-Fe and in n-Cu (Schaefer and Würschum, 1987; Würschum et al. 1987) which is ascribed to an enhancement of the interfacial area and, therefore, of the number of interfacial vacancies. Crystal dislocations which exhibit similar positron lifetimes as τ_{1V} are unlikely as positron traps in nanocrystalline materials because they are assumed to be unstable in the small crystallites. The presence of vacancy-like free volumes in NiO grain boundaries has recently been suggested by a contrast comparison of high resolution electron microscopy (HREM) and image simulation calculations (Merkle, Reddy, Wiley, and Smith, 1987).
- The lifetime τ_2 observed with substantial intensity is to be attributed, according to theoretical calculations (Puska and Nieminen, 1983), to microvoids of the size of about 10 vacancies which are difficult to detect by any other technique. Similar to the vacancy size free volumes they are considered as structural elements of the nano-

Table 1: Positron lifetimes τ_i (ps) and intensities I_i (%) measured on nanocrystalline Cu, Pd, and Fe, compacted under the pressure p, and on n-Si deposited by DC glow discharge (DC-GD) on a Si monocrystal (Würschum et al., 1987). For comparison, positron lifetimes measured in the free bulk state (τ_f), in monovacancies (τ_{1V}), and after plastic deformation ($\tau_{plast.}$) in the corresponding bulk crystals are given (see references in the paper of Würschum et al., 1987).

	nanocrystalline materials						crystalline materials		
	p (GPa)	τ_1	τ_2	I_1/I_2	τ_3	I_3	τ_f	τ_{1V}	τ_{plast}
Cu	5.0	165±3	322±4	0.57	2600± 300	0.2±0.0	112	179	164
Pd	5.0	142±3	321±6	0.72	700± 100	2.6±0.7	96	Pt:168	171
Fe	4.5	185±5	337±5	0.63	4100± 400	8.5±0.5	106	175	167
Si	DC-GD	212±8	400±2		10900±1500		213	266	

crystalline materials located at the intersections of interfaces (see Fig. 4) because microvoids of this size are not stable in the bulk crystals under the annealing treatments mentioned above.

- The long-lived component (τ_3) observed in nanocrystalline metals and in n-Si is ascribed to ortho-positronium (o-Ps) formation at the surfaces of larger voids of the size of, e.g., a crystallite. A detailed discussion of τ_3 and I_3 would require information on the size, the surface impurity contamination and the concentration of these voids (Schaefer, Würschum, Birringer, and Gleiter, 1987b).

Some indication for free volumes smaller than vacancies may be deduced from the decrease of the mean lifetime $\bar{\tau}$ in n-Fe and n-Cu between 295 K and 77 K. Due to a small positron binding energy these smaller free volumes may be incapable of trapping at 295 K but contribute with lifetimes smaller than τ_{1V} at lower temperatures.

The dominant positron lifetime component of $\tau_2 \approx 400$ ps in nanocrystalline Si is attributed to microvoids containing about 10-15 vacancies in analogy to the detailed discussion of the results in hydrogenated amorphous Si (Schaefer, Würschum, Schwarz, Slobodin, and Wagner, 1986;1987c). According to the thermal stability of τ_2 (Würschum et al.,1987) these microvoids are assumed to be located in interfacial intersections similar to nanocrystalline metals. The short-lived component (τ_1) can be quantitatively ascribed to positron annihilation in the nanocrystalline Si-substrate. It should be pointed out that no positron lifetime characteristic for vacancy-size free volumes could be detected in n-Si. This could either originate (Würschum et al., 1987) from a lack of this type of free volumes due to strong atomic relaxation in the interfaces or from a positive charge of these free volumes inhibiting positron trapping.

Positron lifetime measurements also have been used to study the interfacial structure and the sintering behaviour of the nanocrystalline metal oxide TiO_2 (Siegel et al., 1987). However, for the interpretation of positron lifetime measurements in nanocrystalline metal oxides a detailed understanding of the positron states in the defect-free and in defected bulk oxides (Schaefer, Forster, and Mundy, 1987d) is required.

In concluding this section it should be emphasized that in addition to x-ray diffraction and Mössbauer spectroscopy positron lifetime measurements supply a valuable tool for investigating the structure of nanocrystalline materials. By the detection of free volumes in the interfaces they support the hypothesis of a wide interatomic distance distribution in the nanocrystalline interfacial structure.

A reduced mass density in the interfacial structure has also been observed by small angle neutron scattering (SANS) on n-Pd (Jorra, 1987).

Further studies with nuclear probes, e.g., angular correlation of the positron annihilation radiation, perturbed angular correlation, muon spin rotation, etc., should be encouraged.

4. Properties of nanometer-sized materials

Many physical properties of a solid, e.g., specific heat, thermal expansion, elastic constants, magnetic properties, etc., depend on its interatomic neighbourhood. Therefore, changes of the interatomic distances in nanocrystalline materials are expected to drastically change these properties in comparison to the chemically identical bulk crystalline materials.

The **specific heat** C_p in nanocrystalline Cu and Pd (Rupp and Birringer, 1987; see Tab. 2) was observed to be higher than in the corresponding crystalline metals. From the measurements of $C_p(T)$ on n-Pd compacted at 295 K an excess entropy per interfacial atom can be estimated to about 3.3 k_B at 300 K (Korn, Morsch, Birringer, Arnold, and Gleiter, 1987). This must be discussed in terms of anharmonicities and lower frequencies of atomic vibrations compared to crystalline materials as well as by multiple atomic equilibrium sites in the interfaces. In addition an enhanced **thermal expansion** (see Tab. 2) is observed in n-Cu (Rupp and Birringer, 1987). The increased C_p in nanocrystalline Cu and Pd is reduced by an annealing treatment at 750 K when the crystallites grow to sizes up to 20 nm. A considerable reduction of C_p and an enhancement of the shear modulus (Weller and Schaefer, 1987; see Fig. 5) in n-Pd after annealing at 380 K, where practically no crystallite growth can be observed, indicate a structural reordering in the interfaces at rather moderate temperatures. Low-temperature compaction of nanocrystalline materials could enable further investigation of these interfacial reordering phenomena.
Differences in the phonon spectra of the crystalline and the interfacial nanocrystalline structure are indicated, in addition, by the increased temperature variation of the integral intensity I of the broadened interfacial Mössbauer component (see Sect. 3) in n-Fe (Herr et al., 1987). In terms of a Debye model, this temperature variation, which may be described by a reduced Debye-Waller factor, yields a Debye temperature θ_D about 120 K lower than in crystalline iron.

The **elastic moduli** and the **internal friction** of nanocrystalline materials have been investigated at MHz (Korn et al., 1987) or at 1 Hz frequencies (Weller and Schaefer, 1987) in order to gain further information on the interatomic potentials and on the dynamic relaxation processes in the interfaces, respectively.
The Young's modulus E and the shear modulus G measured on nanocrystalline specimens is much smaller than the polycrystalline moduli in the cases of CaF_2 and Pd (see Tab. 2).

Table 2: Properties of nanocrystalline materials compared to those of the course-grained polycrystalline materials

Property	material	nanocrystalline	crystalline
specific heat C_p at 295K (J/gK)[a]	Pd	0.37	0.24
thermal expansion ($10^{-6}K^{-1}$)[a]	Cu	31	16
elastic moduli (GPa) (Young's modulus E ; shear modulus G)	Pd	88;32[b] 35[c]	123;43
	CaF_2[b]	38;19	111;42
specific saturation magnetization at 4 K (10^3 A m^2/g)[d]	Fe	130	222
diffusion activation enthalpy (eV)	Ag in Cu	0.51[e]	2.0[f]
	Cu in Cu	0.64[g]	2.04[h]

a) Rupp and Birringer (1987)
b) Korn et al. (1987)
c) Weller and Schaefer (1987)
d) Birringer et al. (1986)
e) Strauß (1986)
f) Barreau et al. (1970)
g) Horvath et al. (1987)
h) Maier (1977)

It should be pointed out that the reduced nanocrystalline moduli originate from the still lower moduli of the interfacial structure. In terms of a simple model of cube-shaped crystallites (6 nm) and an interfacial thickness of 1 nm the interfacial modulus E_i for n-Pd can be estimated from the values presented in Tab. 2 to give E_i = 40 GPa. The reduction of the elastic moduli in n-CaF_2 and n-Pd may be ascribed to the increased interatomic distances in the interfaces if the same interatomic potentials in the interfaces and in the lattice are assumed.

The temperature variation of the shear modulus G which is proportional to the frequency square f^2 of torsional vibrations is shown in curve a of Fig. 5 for n-Pd compacted at 295 K. The decrease of G between 6 K and 120 K is typical for the elastic constants of crystalline Pd (Walker, Ortelli, and Peter, 1970) and can be correlated to the anomalous behaviour of the paramagnetic susceptibility which may be due to a rearrangement of the 4d electrons in the Fermi surface (Belmahi, Merck, Perréard, Peter, Walker, and Schrieffer, 1966).

A narrow irreversible increase of G by about 8% is observed in n-Pd near 400 K together with the decrease of the specific heat C_p (Rupp and Birringer, 1987) and an increase in density (Korn et al.,1987). These observations are ascribed to a structural reordering within the crystallite interfaces showing that the interfacial structure may occur in different metastable states. The state of lower energy is reached when the interfacial free volume is reduced. This is consistent with the theoretical considerations of Seeger and Schottky (1959) on large angle-grain boundaries.

After annealing at 640 K (curve b in Fig. 5) where crystallite growth may have occured a general increase of G between 6 K and 640 K is observed.

The internal friction Q^{-1} measured in n-Pd below 300 K is small and decreases even more after 640 K annealing. The internal friction increasing steeply above 300 K may be tentatively described by a model of viscous sliding of interfacial boundaries in analogy to

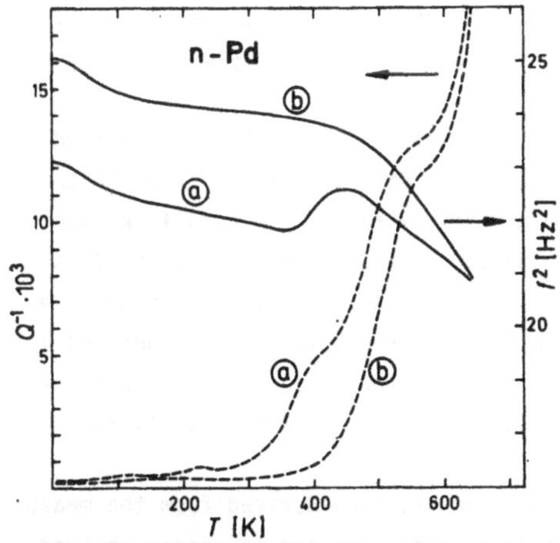

Fig. 5: Temperature variation of the internal friction Q^{-1} and the shear modulus $G \sim f^2$ measured on n-Pd with a heating rate of 100K/h (a) after 5GPa-compaction at 295K and (b) in a second run after heating up to 640K.

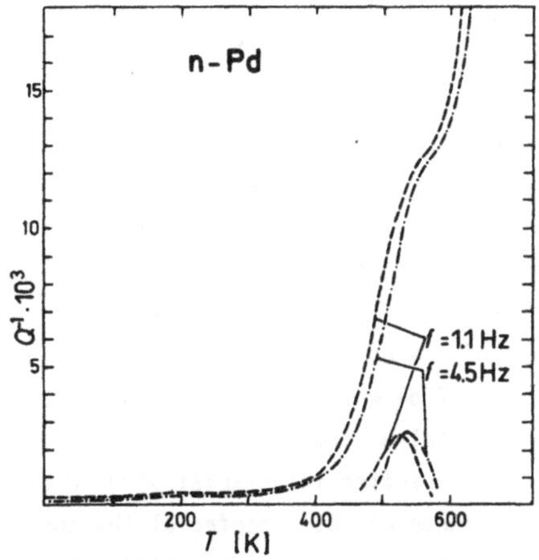

Fig. 6: Temperature variation of the internal friction Q^{-1} measured on n-Pd with two measuring frequencies f after annealing at 640K (Weller and Schaefer, 1987).

the grain boundary sliding suggested for polycrystalline Al (Ke, 1986) or Al bicrystals (Iwasaki, 1985). As observed for the Q^{-1}-peak in Al ascribed to grain boundary sliding (Ke, 1987) the internal friction in n-Pd above 300 K shifts to higher temperatures after annnealing (see Fig. 5). The general increase of Q^{-1} above 300 K as well as the shoulder at 520 K, which in Fig. 6 is additionally shown as a Q^{-1} maximum after background subtraction, are shifted to higher temperatures with increasing measuring frequency. This indicates thermal activation of the interfacial loss processes. From the present results an activation energy of 1.3 eV is estimated for the 520K-process assuming a pre-exponential factor $f_0=10^{13}/s$. However, the temperature shift of the 520K-process with frequency variation (see Fig. 6) is smaller than predicted by these figures and additional studies are required in order to elucidate the underlying processes.

The **ferromagnetic properties** of a solid characterized by the crystal anisotropy and the exchange interaction are also expected to be rather sensitive to alterations of the interatomic distances. In n-Fe the saturation magnetization is considerably reduced (see Tab. 2) with respect to bulk material. The fast temperature decrease of the interfacial magnetic hyperfine field in n-Fe (see Sect. 3) observed by Mössbauer spectroscopy (Herr et al., 1987), indicates a lower ferromagnetic order-disorder temperature T_C than in crystalline iron. In n-Er the magnetic phase transitions observed in coarse-grained Er are shifted or vanish, depending on the crystallite size, and a low-temperature superparamagnetic or spinglass-like phase appears (Cowen, Stolzman, Averback, and Hahn, 1987).

Unconventionally **high diffusivities** of Ag (Strauss, 1986), Cu (Horvath, Birringer and Gleiter, 1987) or Bi (Höfler, Hahn, and Averback, 1987) in nanocrystalline Cu or of Cu and Ag in n-Pd (Höfler et al., 1987) have been observed recently by various techniques. For n-Cu self-diffusivities (Horvath et al., 1987) about 10^{14} times higher (at 393 K) and an estimate of the self-diffusion enthalpy much lower than for bulk diffusion (see Tab. 2) has been reported. Obviously the diffusion of substitutional atoms is strongly enhanced in the interfaces due to the high density of free volumes and may be compared to the behaviour of fast diffusors (e.g., Cu in Pb) which predominantly migrate via interstitial sites (LeClaire, 1978).
A diffusion behaviour totally different from that of metal atoms is observed in n-Pd for hydrogen (Mütschele and Kirchheim, 1987a) which diffuses on interstitial sites in the bulk material. At low H concentrations, C_H, the H diffusivity, D_H, in n-Pd is reduced in comparison to the bulk crystal due to the trapping at interfacial free volumes. These volumes are filled up at higher concentrations giving rise to a diffusivity exceeding that of the bulk crystal. From $D_H(C_H)$ a distribution of interfacial H-site energies is derived from which it is concluded that the distortion of the Pd-polyhedra around the interfacial H atoms is larger than around the octahedral sites in the crystal. This indicates a softer interfacial structure.
An estimate of the interfacial thickness of about 1 nm has been derived from the measurement of the pressure-composition isotherms in nanocrystalline and in coarse-grained Pd (Mütschele and Kirchheim, 1987b). This is in good agreement with the results of x-ray diffraction studies (see Sect. 3).

The **reactivity** of nanocrystalline materials has been investigated by the formation of the intermetallic compound Pd_3Bi from a Bi film on a nanocrystalline Pd specimen (Höfler et al., 1987). This reaction starts at a rather low temperature of about 395 K. A mixture of nanocrystalline Cu und Er particles has been shown to react to the equilibrium compound CuEr with a CsCl structure by room temperature compaction alone (Hahn and Averback, 1987; Siegel and Hahn, 1987). These first experiments indicate a new method for producing bulk compounds at relatively low temperatures by making use of the potential of nanocrystalline materials.

Investigations of **nanocrystalline metal oxides** offer new aspects for the improvement of ceramic materials. In nanocrystalline TiO_2 (rutile) of 12 nm mean grain diameter the sintering process already starts at 800 K (Siegel et al., 1987). After a thermal treatment at 1050 K the Vickers hardness of monocrystalline TiO_2 is exceeded, a hardness which requires in the case of coarse-grained (1.3 μm) TiO_2 a sintering treatment at about 1600 K. This shows clearly that the material processing temperatures can be considerably lowered by the use of nanocrystalline starting materials.
The application of ceramics with their technologically attractive properties is often limited by their low-temperature brittleness. Karch et al. (1987) demonstrated on nanocrystalline CaF_2 and TiO_2 (8 nm) that this limitation can be overcome. These materials appear to be ductile at ambient temperature due to a strongly enhanced grain boundary creep rate which originates from the small diffusion lengths required in nanocrystalline materials and from the increased interfacial diffusivities. The interpretation of this plastic behaviour by a thermally activated process is supported by the ductile-brittle transformation observed at higher deformation rates.

5. Future prospects

Although a number of structural and physical properties have been studied recently the investigation of nanocrystalline materials is still in an early stage. The development of processes for the production of nanocrystalline materials in larger quantities and a wide variety of compositions is required. Some areas of future interest in basic research are pointed out in the following.
Investigation of various metastable configurations of the interfaces and their reordering or relaxation processes would be aided by the compaction of nanometer-sized particles at low temperatures or the preparation of nanocrystalline specimens from high-melting materials. Correlated theoretical studies of the compaction process and of the thermodynamical properties of interfaces would be helpful.
The properties of nanocrystalline materials should be studied as a function of the carefully characterized crystallite size. In particular, it would be interesting to diminish crystallite size even further in order to increase the fraction of the disordered interfacial atoms and to study the stability of nanocrystalline materials.
Further structural and chemical characterization of the interfaces is indispensible for understanding the reactivity, the alloying behaviour, and the absorption processes.
A wide and promising field is open for the synthesis and investigation of nanocrystalline composites.

Acknowlegdgements

We thank H. Hahn, R.W. Siegel, E. Jorra, and G. Wallner for making available their data prior to publication. The authors are indebted to C.A. Wert and M. Weller for critically reading the manuscript. Financial support by the Deutsche Forschungsgemeinschaft and the German Bundesminister für Forschung und Technologie is acknowledged.

References

Barreau, G., G. Brunnel, G. Ciceron, and P. Lacombe, Comptes Rendues Acad. Sci. Paris 270 (1970) 516

Belmahi, O., M. Merck, E. Perréard, M. Peter, E. Walker, and J.R. Schrieffer, Helv. Phys. Acta 39 (1966) 338

Birringer,R., H. Gleiter, H.-P. Klein, and P. Marquardt, Phys. Letters 102A (1984) 365

Birringer, R., U. Herr, and H. Gleiter, Suppl. to Trans. Jap. Inst. Metals 27 (1986) 43

Cowen, J.A., B. Stolzman, R.S. Averback, and H. Hahn, J. Appl. Phys. 61 (1987) 3317

Gleiter, H. and P. Marquardt, Z. Metallkunde 75 (1984) 263

Granqvist, C.G. and R.A. Buhrman, J. Appl. Phys. 47 (1976) 2200

Hahn, H. and R.S. Averback (1987), to be published

Haubold, T., B. Lengeler, R. Birringer, H. Gleiter, (1987), to be published

Hautojärvi, P., Ed., Positrons in Solids, Springer-Verlag, Berlin, Heidelberg, New York (1979)

Herr, U., J. Jing, R. Birringer, U. Gonser, and H. Gleiter, Appl. Phys. Lett. 50 (1987) 472

Höfler, H.-J., H. Hahn, and R. Averback (1987), to be published

Hort, E., Diplomarbeit, Universität des Saarlandes, Saarbrücken (1986)

Horvath, J., R. Birringer, and H. Gleiter, Solid State Commun. 62 (1987) 319

Iqbal, Z., and S. Veprek, J. Phys. C 15 (1983) 377

Iwasaki, K., phys. stat. sol. (a) 90 (1985) K35

Jorra, E., Diplomarbeit, Universität München (1987)

Karch, J., R. Birringer, and H. Gleiter, Nature, (1987), to be published

Ke, T.S., Suppl. to Trans. Jap. Inst. Metals 27 (1986) 679

Kimoto, K., Y. Kamiya, M. Nanoyama, and R. Ueda, Jap. J. Appl. Phys. 2 (1963) 702

Korn, D., A. Morsch, R. Birringer, W. Arnold, and H. Gleiter, J. Physique C, (1987), to be published

LeClaire, A. D., J. Nucl. Mater. 69/70 (1978) 70

Maier, K., phys. stat. sol. (a) 44 (1977) 567

Merkle, K.L., J.F. Reddy, C.L. Wiley, and D.J. Smith, J. Physique C (1987), to be published

Mills, A.P. and R.J. Wilson, Phys. Rev. A26 (1982) 490

Mütschele, T. and R. Kirchheim, Scripta metall. 21 (1987a) 135

Mütschele, T. and R. Kirchheim (1987b) to be published

Puska, M.J. and R.M. Nieminen, J. Phys. F13 (1983) 333

Rupp, J. and R. Birringer, Phys. Rev. Letters (1987), to be published

Schaefer, H.-E., in: Positron Annihilation, Ed. P.G. Coleman, S.C. Sharma, and L.M. Diana, North Holland Publ. Co., Amsterdam (1982) p. 483

Schaefer, H.-E. and R. Würschum, Phys. Letters A119 (1987) 370

Schaefer, H.-E., R. Würschum, R. Schwarz, D. Slobodin, and S. Wagner, Appl. Phys. A40 (1986) 145; ibid. A43 (1987c) 295

Schaefer, H.-E., R. Würschum, M. Scheytt, R. Birringer, and H. Gleiter, Mater. Sci. Forum 15/18 (1987a) 955

Schaefer, H.-E., R. Würschum, R. Birringer, and H. Gleiter (1987b), to be published

Schaefer, H.-E., M. Forster, and J.N. Mundy, presented at the International Conference on Electronic Structure and Phase Stability in Advanced Ceramics, Argonne, U.S.A., (1987d), and to be published

Scheytt, M., R. Würschum, H.-E. Schaefer, S. Veprek, and H. Gleiter, Verhandl. DPG (VI) 22 (1987)I, M-24.2

Seeger, A. and G. Schottky, Acta Met. 7 (1959) 495

Siegel, R.W. and H. Hahn, in: Current Trends in the Physics of Materials, Ed. M. Yussouff, World Scientific Publ. Co., Singapore (1987), to be published

Siegel, R.W., H. Hahn, S. Ramasamy, Li Zongquan, Lu Ting, and R. Gronsky, J. Physique C, (1987), to be published

Strauss, R., Diplomarbeit, Universität des Saarlandes, Saarbrücken (1986)

Veprek, S., in: Poly-Micro-Crystalline and Amorphous Semiconductors, Eds. P. Pinard and S. Kalbitzer, Les éditions de Physique, Les Ullis (1984) p. 425

Walker, E., J. Ortelli, and M. Peter, Phys. Letter 31A (1970) 240

Weller, M. and H.-E. Schaefer, (1987), to be published

Würschum, R., M. Scheytt, and H.-E. Schaefer, phys. stat. sol. (a) 102 (1987) 119

Zhu, X., R. Birringer, U. Herr, and H. Gleiter, Phys. Rev. B35 (1987) in the press

L A S E R S   F O R   U L T R A S H O R T
L I G H T   P U L S E S

by

Joachim Herrmann / Bernd Wilhelmi

1987 - 302 pages - 143 figures - 10 tables

price: GDR 38,-- M

order number: 763 621 7

ISBN 3-05-500 060 - 9

The book

- presents basic knowledge of the interaction of light with atomic systems, photoinduced physical and chemical elementary processes and of lasers
- discusses the fundamental principles of ultrafast measuring techniques
- explains the generation of picosecond and femtosecond pulses in actively and passively modelocked and synchronously pumped lasers with reference to current theoretical and experimental results
- describes frequency conversion and pulse compression methods for generating ultrashort light pulses
- deals with ultrafast spectroscopy and ultrafast optoelectronics
- is designed for students of physics, chemistry, biology, and electronics as well as for teachers, engineers and scientists of these disciplines. It will also be useful for anyone requiring an introduction to the fundamentals of lasers for ultrashort light pulses and their application.

Akademie-Verlag Berlin

Bei Fragen zur Produktsicherheit wenden Sie sich bitte an:
If you have any questions regarding product safety,
please contact:

Walter de Gruyter GmbH
Genthiner Straße 13
10785 Berlin
productsafety@degruyterbrill.com